AF618509

THE INTERACTIONS BETWEEN SEDIMENTS AND WATER

THE INTERACTIONS BETWEEN SEDIMENTS AND WATER

Proceedings of the 7th International Symposium,
Baveno, Italy
22-25 September 1996

Edited by

R. DOUGLAS EVANS
JOE WISNIEWSKI
and
JAN R. WISNIEWSKI

Reprinted from *Water, Air, and Soil Pollution*
Volume 99, Nos. 1–4, October 1997

SPRINGER-SCIENCE+BUSINESS MEDIA, B.V.

A C.I.P. Catalogue record for this book is available from the Library of Congress.

ISBN 978-94-010-6339-5 ISBN 978-94-011-5552-6 (eBook)
DOI 10.1007/978-94-011-5552-6

Printed on acid-free paper

WATER, AIR AND SOIL POLLUTION / *Volume 99 Nos. 1–4 October 1997*

PREFACE

The interactions between sediments and water have received increased emphasis over the last two decades, leading to a greater understanding of physical, chemical and biological processes in a wide range of aquatic systems including river, lake, estuarine and coastal systems. This new understanding has fostered research and assessment projects, and focused awareness on the implications of sediment/water interactions for policy and regulation.

The international symposium on "The Interactions Between Sediments and Water" held in Baveno, Italy from 22-26 September 1996 was the seventh in a series of symposia held since 1976. All focused on issues relating to processes which occur near the sediment/water interface as well as responses to interactions between sediments and water. Previous symposia were held in Amsterdam in 1976, Kingston (Canada) in 1982, Geneva in 1984, Melbourne in 1986, Uppsala in 1990 and Santa Barbara in 1993.

Almost 200 oral and poster presentations were made by attendees representing 27 nations, including dozens of participants from developing countries. The scientific sessions were organized around the following themes: sediment/water dynamics, contaminant interactions, sediment/nutrient relationships, sediments as historical records of deposition, sediment/water science in remote areas, sediment/organism interactions, and novel methods and analytical quality control. In addition, an interactive workshop was held which involved all participants and included discussions of the key issues concerning the flux of dissolved compounds into and out of the sediment layer. Methodological approaches and problems which require further attention were highlighted.

Sediments are a driving force for many elemental cycles in aquatic systems and are recognized as one of the largest sources of in-place pollutants. It is hoped that the results of this symposium will offer new insights into the behaviour and role of sediments in aquatic systems and will also guide those with an interest in reducing sediment toxicity levels on local, regional or global scales and assist those concerned with the development of management strategies, policy and legislation. Until these goals are reached, impacts from sediment contamination will continue to be a growing concern to the international environmental community, experts in science and industry, and government and private organizations. The next symposium, to be held in Beijing, China in the late summer of 1999, will demonstrate how successful the Baveno conference has been in defining the work needed to advance our understanding of the water/sediment interface and its importance to the understanding of how streams, lakes estuaries and oceans interact.

ACKNOWLEDGMENTS

We acknowledge both the individual contributions of symposium participants who delivered poster and oral presentations and who completed papers for the symposium proceedings, and the efforts of all of the session chairs who kept discussions active and interesting. We acknowledge the assistance of the Scientific Committee in the task of selecting themes and presentations for the conference. Finally, we acknowledge past and present symposium chairs, and presidents and directors of IASWS for their leadership and guidance in maintaining a successful symposium series.

We acknowledge and appreciate the support of the sponsors: the University of Milan, the National Research Council of Italy, the International Association of Sediment Water Science and, particularly, the Electric Power Research Institute, whose generous support made the publication of this special issue possible.

INTERACTIONS BETWEEN SEDIMENTS AND WATER SUMMARY OF THE 7TH INTERNATIONAL SYMPOSIUM

R. D. EVANS[1], A. PROVINI[2], J. MATTICE[3], B. HART[4] AND J. WISNIEWSKI[5]

[1]Trent University, Environmental Sciences Centre, Peterborough, ON K9J 7B8, Canada, [2]Universita di Milano, Dipartimento di Biologia sez. Ecologia, Via Celoria, 26, 20133 Milano, Italy,[3]Electric Power Research Institute, 3412 Hillview Avenue, Palo Alto, California 94304-1395, USA, [4]Monash University, Water Studies Centre, PO Box 197, Caulfield East, 3245, Melbourne, Australia, [5]Wisniewski & Associates, Inc., 6862 McLean Province Circle, Falls Church, Virginia 22043, USA

Abstract. The interactions between sediments and water have received increased emphasis over the last two decades, leading to a greater understanding of physical, chemical and biological processes in a wide range of aquatic systems including river, lake, estuarine and coastal systems. This new understanding has fostered research and assessment projects, and focused awareness on the implications of sediment/water interactions for policy and regulation. The international symposium on "The Interactions Between Sediments and Water" held in Baveno, Italy from 22-26 September 1996 was the seventh in a series of symposia held since 1976. All focused on issues relating to processes which occur near the sediment/water interface as well as responses to interactions between sediments and water. Previous symposia were held in Amsterdam in 1976, Kingston (Canada) in 1982, Geneva in 1984, Melbourne in 1986, Uppsala in 1990 and Santa Barbara in 1993. Almost 200 oral and poster presentations were made by attendees representing 27 nations, including dozens of participants from developing countries. The scientific sessions were organized around the following themes: sediment/water dynamics, contaminant interactions, sediment/nutrient relationships, sediments as historical records of deposition, sediment/water science in remote areas, sediment/organism interactions, and novel methods and analytical quality control. This paper attempts to summarize the key highlights of the various sessions.

Key words. Sediments, water, water dynamics, contaminant interactions, nutrient cycles, historical indicators, organism interactions, remote areas, quality control

1. Introduction

The study of sediment/water interactions crosses disciplinary and ecosystem boundaries, and is essentially the study of linkages. Marine, estuarine and freshwater studies were all well represented. It was apparent from many of the presentations that the observations, models and conclusions from one type of system often inform the work of others in different media. Not only did the papers presented span the range of types of aquatic systems, they also covered environments from the most remote and presumably pristine, such as Antarctica, to some of the most highly urbanized, such as the Venice lagoon. Sediments themselves are a link between the present and the past. Much of our understanding about past conditions in aquatic systems comes from the study of sediment cores.

Water, Air and Soil Pollution **99**: 1-7, 1997.

The presentations were organized into several theme areas including sediment/water dynamics, sediment/contaminant interactions, the role of sediments in nutrient cycles, sediments as historical indicators, sediment/organism interactions, remote areas and novel methods. For many of the papers it is difficult to assign them to a specific section because of the strong linkages between areas. For example, several papers present evidence for elemental speciation changes in sediments affected by redox reactions (e.g., contaminant interactions, section 2). Redox reactions are affected by physical processes (section 1) and affect the microbial activity in sediments (section 6). Thus the work presented during the symposium should be viewed as a continuum, rather than several discrete areas of focus.

The Sediment/Water Interactions Symposia bring together researchers with widely differing backgrounds. These include chemists, physicists, earth scientists and biologists. The collaboration of all these groups is often necessary to understand the complex functioning of sediment/water interactions. We are constantly seeking people from other research backgrounds to further enhance our understanding of the role of particles in aquatic systems; Mattice et al. (1997) referred to one such area. Using examples of the metal toxicants mercury and selenium, he argued for enhanced interactions between physical and chemical sedimentologists and aquatic toxicologists. Hopefully future meetings will have additional participants with new and different research interests to further compliment our understanding of aquatic ecosystem functioning.

Of course, the most important linkages of this symposium are those among research groups and between individual researchers. Out of the relationships built during this symposium come the ideas and the collaborations which undoubtedly will be the highlights of future meetings. A few of the highlights and significant developments presented in Baveno are discussed below.

2. Sediment / Water Dynamics

There were many platform talks and posters presented as part of five separate sessions that dealt primarily with physical aspects of sediment/water dynamics. The conference began with a paper by Meade (1996) in which the importance of time and space scales to sediment dynamics was demonstrated. Using data from many of the largest rivers in the world, Meade demonstrated that water flow and sediment generation are not well correlated, with small tributaries often contributing a disproportionate sediment load. The temporal and spatial significance of storage in the floodplain and banks was emphasized which implies that the impacts of contaminant storage can be problematic in these large systems.

Several presentations of new or improved models for prediction of particle dynamics indicate that our understanding of factors affecting particle movement is improving, albeit slowly. Of particular importance is the role of particle aggregations or flocs. Several papers demonstrated the importance of understanding the behaviour of flocs, both for modelling particle dynamics and for predicting the fate of particulate-bound contaminants. Previous models to predict sediment transport have failed to incorporate the movement of less dense but large flocs or aggregates in some energy regimes. As a result, models often under-predict sediment mass-transfer, especially in the fine-grained end of the spectrum which is also significant for contaminant transport. Particularly with the help of new imaging techniques, formation rates and transport models are now including flocs and are becoming much better predictively.

Several papers explored the use of natural or anthropogenic radionuclides as tracers of sediment particle movement. Although much work remains before good predictive models of sediment resuspension are developed, several studies indicated the importance of local or episodic events in causing resuspension, particularly those that increase water circulation and velocity. Current estimates suggest that much of the flux of particles collected by bottom sediment traps originate as resuspended material. The importance of short interval measurements to valid interpretation of sources of particles in the water column was demonstrated. Of particular interest were data presented that suggest that density currents caused by geochemical changes such as redox cycles could also be responsible for sediment resuspension.

The implications of sediment resuspension for contaminant movement back into the water column was the focus of several papers. A significant issue in this respect is the depth of the physically and biologically active layer. More information is required for us to adequately define this layer which undoubtedly plays a determining role in fate of contaminants.

3. Sediment / Contaminant Interactions

Four sessions and two plenary presentations were focused on the issue of contaminants associated with particles and the role of sediments in the fate of anthropogenic substances. Eisenreich (1996) used a mass balance approach to examine the role of sediments in the fate of PCB's in both small and large lakes in North America. The role of settling particles both as a sink for contaminants and ultimately as a record of deposition was highlighted. Of particular interest to many was the observation that efflux of PCB's to the atmosphere from the water column is a significant process probably controlled by organic carbon concentrations in the aquatic system.

The kinetics of pollutant sorption and desorption were the subject of several presentations. Most models assume equilibrium and steady-state conditions. However, some data presented suggest that both sorption and desorption kinetics can be quite slow. There are many implications of this phenomenon for modellers of particle - contaminant interactions. Coupled with this is the issue of data quality. Several presentations discussed the problems for model development with poor quality data. The quality of data is, in part, a function of study design and in part a result of analytical limitations.

Sediment are a significant source of in-place pollutants and potential remobilization of those pollutants was the theme of several studies. One approach to determining this potential is a laboratory or mesocosm experiment. From results presented during the symposium, remobilization is not always a large fraction of the total sedimentary concentration for trace metals. The situation for organic pollutants remains less clear. Methodological development remains a key issue for remobilization studies, as highlighted by Forstner (1996).

Several papers dealt with metal speciation changes or changes in redox chemistry. The extent to which these changes control contaminant cycling is unclear at present, but several papers highlighted the significant changes which can occur in the speciation of sediment-bound metals under changing redox conditions. The role of bacteria in pollutant dynamics is now receiving much-needed attention (see section 6). As with other aspects of sediment/contaminant interactions, methodological issues concerning metal speciation in sediments remain unresolved. Our understanding of contaminant - particle interactions is limited by suitable analytical and experimental procedures. It is to be hoped that some of the new approaches presented during this symposium will help to resolve outstanding issues.

In addition to sessions concerning the interactions between contaminants and sediment, a session was held on the evaluation of sediment quality and remediation potential. As noted earlier, sediments are a significant source of in-place pollutants and in many areas remediation is required. A paper by Huet (1996)of the Organization for Economic Cooperation and Development (OECD) outlined the development of test guidelines for the OECD and the future regulatory climate. While remediation techniques were not discussed widely during this symposium, it is proposed that future meetings address this area more thoroughly.

4. Role of Sediments in Nutrient Cycles

Four sessions dealt with aspects of nutrient cycles in lakes and the role of sediments as both a source and a sink for nutrients. The role of sediments in nutrient cycles has been recognized since the first Sediment/Water Interactions Symposium. However, our predictive capabilities with respect to nutrient fluxes into and out of sediments remain poor.

It is well known that sediments are a significant pool of nutrients in all aquatic systems and under certain circumstances can be a net source of nutrients. The role of sulphate reduction and release of soluble iron in releasing sediment bound phosphorus was illustrated in several presentations. Perhaps surprising, however, is the magnitude of phosphorus release, as much as a doubling over pre-stratification concentrations, seen in some lakes. Sediment resuspension can also be a significant cause of P release. Newer evidence suggests that bacterial degradation of organic matter under anaerobic conditions may also be a significant source of phosphorus release from sediments.

It was illustrated by one presenter that there have been no significant developments in formulating a modelling framework for nutrient behaviour in lakes in many years. He proposed a new category of model but at this stage it remains largely untested. For all questions concerning the role of sediments in nutrient cycles, the issue of scale, both spatial and temporal, remains problematic. Better ways of integrating ecosystem variability into our models are required.

Several presentations dealt with the role of sediments in cycling of elements other than phosphorus, such as carbon and sulfur. One paper demonstrated the role of sediments in internal alkalinity generation and the seasonality of that process. At this symposium, there was a greater emphasis on marine and estuarine studies than at past symposia. It was clear from the questions and the presentations that more cross-fertilization between the marine and freshwater groups is beginning to occur.

5. Use of Sediments as Historical Indicators

Two sessions were held in which the use of sediments as repositories of physical, chemical and biological information was examined. Sediment deposits build through time incorporating materials from the water column and the atmosphere. At every symposium several studies have examined the use of sediment core profiles of contaminants and other elements as recorders of change in atmospheric or catchment processes. Several presentations at this symposium examined this potential use for sediment deposits. Appleby (1997) presented a paper on techniques for dating various sedimentary horizons. He illustrated the linkage between atmospheric fluxes of elements and the sedimentary record.

The implications of mixing at the sediment/water interface for the use of sedimentary profiles was highlighted by Appleby and others. Clearly there is a need for the types of information being generated by those studying particle dynamics for inclusion in models for reconstructing historical changes in water chemistry from sediment cores. In the past, much of the sediment reconstruction has focused on elements or contaminants. At this symposium, there were several presentations using biomarkers of paleoproductivity and human change. Some of these included diatoms, algal and bacterial pigments, signature lipids and ostracods. It is encouraging to see the many new approaches to paleo-reconstruction and we expect that this trend will continue as our understanding of these biomarkers improves. The range of biomarkers will expand and allow us to choose the appropriate ones for a specific purpose.

6. Sediment / Organism Interactions

Two sessions were devoted to the role of bentic organisms in defining the sedimentary environment. The cycling of many substances in sediments is largely determined by the activity of organisms living in sediments. Our appreciation for the role of benthic organisms was greatly improved by some remarkable video footage presented by Reynoldson (1996) as part of his paper. Many attendees commented on their greater understanding of the dynamic nature of benthic activity following the presentation. Of interest, however, was his conclusion that, despite marked differences in the behaviour of different types of benthos, the result on chemical profiles in the sediment column was minimal.

The role of organisms in sediment/water interactions can be studied in many ways. Often organisms are removed from their natural situation and brought into the lab for study. While this is perhaps necessary in some cases, several speakers argued for a more realistic approach using *in situ* studies or using mesocosm experiments in the aquatic systems. The benefits of added realism with respect to the real world appeared to be worth the difficulties encountered. The use of stable C and N isotopes to measure benthic structure is an emerging field (see section 7).

The role of the microbial portion of the sediment dwelling biota was the focus of two sessions. This is a relatively new area and, judging by the observations presented, one which will surely get more attention at future symposia. Bacteria were shown to be the driving force in many chemical processes, including nutrient cycles, organic matter mineralization and sulfur cycles.

7. Studies of Remote Areas

There are still many parts of the globe for which we lack any basic information about aquatic systems. Sediments can be used as a means of obtaining information from remote regions quickly. A session on remote areas included five presentations, one examining organic contaminants in the Himalayas and four concerning sedimentary processes in the marine environment adjacent to Antarctica. Remote regions, apart from their intrinsic interest as unknowns, are often assumed to have a value as end-members of geographical and concentration gradients. Several of the presentations suggest that this hypothesis needs evaluation on a case-by-case basis; long range transport of contanimants, for example, may mean that remote environments are not substantially different than others for particular chemicals. These papers indicate clearly the importance of and need for further work in remote regions and we would encourage more presentations on this topic at the next meeting.

8. Novel Methods and Quality Control

With every symposium comes several new approaches to the study of sediment/water interactions; this meeting being no exception. Among the many new developments, a technique for measuring pore water profiles at sub-centimeter intervals using thin film technology was described. Camera techniques for studying floc formation in turbulent streams were presented and a new acoustic procedure for monitoring the stability of sediments was detailed. At the analytical level, the use of stable isotopes, particularly C and N, for a variety of purposes is a significant change in the past few years. Several papers attested to the valuable information which can be obtained using stable isotopes and it is clear that the methods have a significant future in the study of sediment/water interactions.

Along with new methods comes the need to make sure that all methods are providing the highest quality data. A paper by Muntau (1996) highlighted the importance of knowing the uncertainty associated with our measurements and the need for proper quality assurance programs. This is a recurrent theme at all symposia, however, it was clear from several presentations that the need to remind ourselves of the importance of accurate and precise data is ongoing.

9. Sediment Porewater Measurements Workshop

A workshop which examined techniques for measuring porewater fluxes of compounds across the sediment water interface was held during the symposium. Techniques discussed for determining fluxes across the interface included porewater peepers, thin film devices, benthic chambers, porewater separation techniques, like centrifugation or dialysis, and modelling. All present agreed that there is no one preferred approach to the issue. The choice of technique is dependent on the substance being measured, the concentration gradients present, the resolution required and the ultimate objective of the study. The need for inter-lab and inter-technique comparisons was evident. However, such comparisons will depend, in part, on these same variables, reinforcing the conclusion that there is probably no single "best" approach.

10. Summary

In summary, there were many presentations, both poster and platform, which could be identified as highlights of the 7th Sediment/Water Interactions Symposium. It is, of course, not possible to review them all. What is apparent, however, from this brief summary of the proceedings of the symposium, is that the research area of sediment/water interactions is flourishing and has many important insights to offer society for the management of aquatic resources. We look forward with anticipation to the next symposium in China in 1999.

Acknowledgements

We acknowledge both the individual contributions of symposium participants who delivered poster and oral presentations and who completed papers for the symposium proceedings, and the efforts of all of the session chairs who kept discussions active and interesting. We acknowledge the assistance of the Scientific Committee in the task of selecting themes and presentations for the conference. Finally, we acknowledge past and present symposium chairs, and presidents and directors of IASWS for their leadership and guidance in maintaining a successful symposium series.

We acknowledge and appreciate the support of the sponsors: the University of Milan, the National Research Council of Italy, the International Association of Sediment Water Science and, particularly, the Electric Power Research Institute, whose generous support made the publication of this special issue possible.

References

Appleby, P.G.: 1997, *Water, Air and Soil Pollut.* This volume.

Eisenreich, S.J.: 1996, Paper presented at the *Interactions Between Sediments and Water: 7th International Symposium*, 22-25 September 1996.

Forstner, U: 1996, Paper presented at the *Interactions Between Sediments and Water: 7th International Symposium*, 22-25 September 1996.

Huet, M.C.: 1996, Paper presented at the *Interactions Between Sediments and Water: 7th International Symposium*, 22-25 September 1996.

Mattice, J. et al.: 1997, *Water, Air and Soil Polut.* This volume.

Meade, R.H.: 1996, Paper presented at the *Interactions Between Sediments and Water: 7th International Symposium*, 22-25 September 1996.

Muntau, H.: 1996, Paper presented at the *Interactions Between Sediments and Water: 7th International Symposium*, 22-25 September 1996.

Reynoldson, T.B.: 1996, Paper presented at the *Interactions Between Sediments and Water: 7th International Symposium*, 22-25 September 1996.

INVESTIGATING SPATIAL PATTERNS OF OVERBANK SEDIMENTATION ON RIVER FLOODPLAINS

D. E. WALLING and Q. HE

Department of Geography, University of Exeter, Exeter, EX4 4RJ, UK.

Abstract. Attempts to study spatial patterns of overbank sedimentation on river floodplains commonly face important operational and sampling problems in documenting deposition rates. Recent advances in the application of fallout radionuclides (^{137}Cs and unsupported ^{210}Pb) to the estimation of medium-term rates of overbank sedimentation offer an essentially unique opportunity to assemble detailed distributed data sets for medium-term deposition rates. Such data can afford a valuable basis for investigating the complex relationship between sedimentation rates and floodplain microtopography and morphology and flow hydraulics. This paper reports the results of an investigation of the spatial pattern of overbank sedimentation rates on the floodplain of the River Culm, Devon, UK. Caesium-137 and unsupported ^{210}Pb measurements have been used to establish the general pattern of deposition rates along an 11 km reach and more intensive measurements have been employed to document the detailed pattern existing within a small area subject to frequent inundation. The resultant data have been used as a basis for interpreting the major controls on the observed patterns.
Key words. Floodplains, sedimentation, caesium-137, unsupported lead-210, overbank deposition, deposition rates, fallout radionuclides.

1. Introduction

River floodplains have attracted increasing attention by geomorphologists, hydrologists and sedimentologists in recent years. This interest reflects, at least in part, the ecological significance of floodplains within the landscape, their role as a buffer between the river and the surrounding land, and also a growing awareness of their potential significance as sinks for river-borne sediment and associated nutrients and contaminants (cf. Anderson *et al.*, 1996). Because of their dynamic nature, sediment deposited in river floodplains may be reworked in the future and may thus also constitute a problem for future river management (cf. Leenaers and Schouten, 1989).

In considering the geomorphological evolution of floodplains and their role as sediment sources and sinks, attention is commonly directed either to the coarse channel deposits and the interaction between channel migration and floodplain construction and destruction (cf. Wolman and Leopold, 1957) or to the fine overbank deposits which mantle large areas of most floodplains and result in vertical accretion of the floodplain surface. For many lowland river floodplains, particularly those where channelisation and river training works limit or prevent channel migration, the overbank deposition of fine sediment will represent the dominant component of floodplain evolution. In order to develop an improved understanding of this component of floodplain development, more information on rates and patterns of overbank sediment deposition is required.

Existing approaches to documenting rates of overbank sedimentation on river floodplains have included the use of sedimentation traps (e.g. Lambert and Walling, 1987; Asselman and Middelkoop, 1995), post event surveys of the deposits resulting from individual floods (e.g. Brown, 1983; Marriott, 1992) and the identification of

Water, Air and Soil Pollution **99**: 9-20, 1997.

datable levels within the overbank deposits (e.g. Hupp, 1988; Lewin and Macklin, 1987). Each of these approaches involves a number of practical and logistical difficulties. Recent advances in the application of the environmental radionuclides, caesium-137 (^{137}Cs) and unsupported lead-210 (^{210}Pb), to documenting rates and patterns of floodplain sedimentation can be viewed as an extension of the use of datable levels, wherein down-profile variations in the concentrations of these radionuclides can provide a basis for establishing the recent chronology of overbank sediment deposits and measurements of total radionuclide inventories can be used to estimate average sedimentation rates (cf. He and Walling, 1996; Walling and He, 1993, 1994, 1997). Use of fallout radionuclides in this context offers many advantages over other methods for documenting overbank floodplain sedimentation, including the general applicability of the approach to a wide range of environments, the medium-term timescales involved (i.e. ca. 35 years for ^{137}Cs and 100 years for unsupported ^{210}Pb) and the potential for assembling data for a large number of points on a floodplain. This paper reports the results of a study of the spatial pattern of recent overbank sedimentation on the floodplain of the River Culm, Devon, UK, based on fallout radionuclide measurements.

2. Use of fallout radionuclides to documents rates and patterns of overbank sedimentation on river floodplains

Detailed discussion of the basis for using ^{137}Cs and unsupported ^{210}Pb measurements to estimates rates of overbank sedimentation on river floodplains lies beyond the scope of this paper and the reader is referred to He and Walling (1996), Walling and He (1993, 1994, 1997) and Walling *et al.* (1996) for further details of the approach. In essence, it exploits the fact that ^{137}Cs and unsupported ^{210}Pb accumulate within accreting overbank sediment deposits as a result of both direct atmospheric fallout to the floodplain surface and deposition of suspended sediment representing material mobilised from the upstream catchment by erosion which will also contain ^{137}Cs and unsupported ^{210}Pb deposited as fallout on the catchment surface. Both the total inventory or amount of the radionuclide contained in a floodplain sediment profile and its vertical distribution will therefore differ from that of a natural undisturbed soil above the level of flood inundation which will only receive inputs associated with direct atmospheric fallout and which will be characterized by a stable non-accreting surface. Figure 1 presents a typical example of the vertical distribution of ^{137}Cs and unsupported ^{210}Pb in both undisturbed soils and overbank floodplain sediments. In the case of ^{137}Cs, which is an artificial fallout radionuclide with a half-life of 30.17 years which was produced by the atmospheric testing of thermonuclear weapons primarily during the late 1950s and the 1960s, the depth distribution of the radionuclide in the floodplain sediment reflects the temporal pattern of fallout. The peak ^{137}Cs activity found at ca. 10 cm depth can thus be ascribed to 1963, the time of maximum fallout, and the dating of this level provides a means of estimating the average sedimentation rate over the past ca. 30 years. Comparison of the total ^{137}Cs inventory for the floodplain core with those of adjacent natural undisturbed soils, in order to establish the excess inventory associated with sediment deposition, provides an alternative means of estimating the average sedimentation rate over the period since the onset of significant radiocaesium fallout (i.e. since ca. 1954). In the latter case, only a single measurement of the total inventory of the bulk core is required, although the effects of variation in the grain size of the sediment on the ^{137}Cs inventory must also be considered. Because of the time consuming nature of ^{137}Cs measurements, use of whole core inventory values to estimate sedimentation rates provides greater scope for assembling information for a large number of points on the floodplain, as would be needed to investigate spatial patterns of sedimentation.

The depth distributions of unsupported ^{210}Pb presented in Figure 1 are significantly different from those of ^{137}Cs, since, although unsupported ^{210}Pb is also a fallout radionuclide, it differs from ^{137}Cs in two important respects. First, it is of natural origin, representing a product of the ^{238}U decay series with a half-life of 22.26 years. Secondly, because of its natural origin, the annual fallout may be treated as essentially constant through time. It therefore affords a means of estimating deposition rates over somewhat longer periods (i.e. 50-150 years). Information on the vertical distribution of unsupported ^{210}Pb concentrations in overbank floodplain sediments can again be used to date specific levels and thereby estimate deposition rates, but He and Walling (1996) have also demonstrated how, as in the case of ^{137}Cs, a single measurement of the total unsupported ^{210}Pb inventory for a bulk sediment core can be used to estimate the average rate of accretion at the point where the core was collected. Again use of single bulk core measurements affords potential for estimating deposition rates for a large number of points on a floodplain which can be used to document spatial patterns.

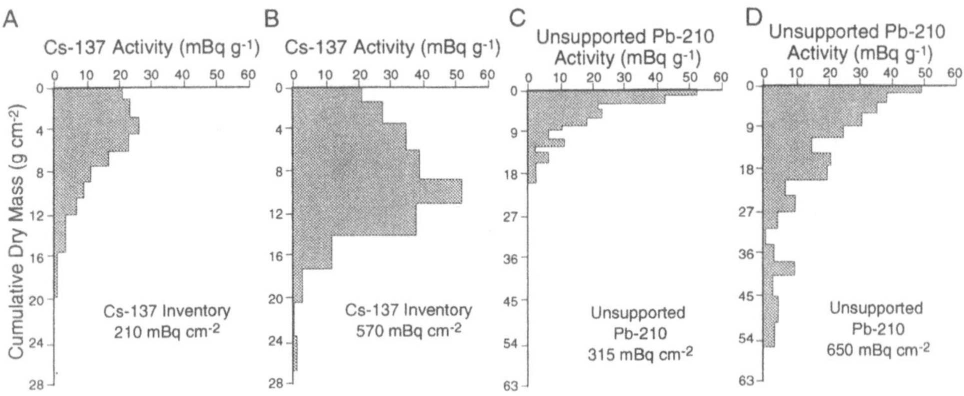

Fig. 1. Representative examples of the vertical distribution and total inventory of ^{137}Cs and unsupported ^{210}Pb in undisturbed permanent pasture above the level of inundating floodwater (A, C) and in an adjacent area of the floodplain of the River Culm, Devon, UK (B, D).

3. The study area

The study reported focussed on a 11 km stretch of the lower reaches of the River Culm, in Devon, UK, illustrated in Figure 2. At its confluence with the River Exe at Stoke Canon, the river drains a catchment area of 276 km^2. The mean annual precipitation and runoff for the catchment are estimated to be ca. 925mm and 510mm respectively. The lower reaches of the river between Cullompton and Stoke Canon meander across a well-developed floodplain which averages about 450 m in width. Overbank flooding is relatively frequent during the winter months and substantial inundation of this floodplain area typically occurs on about seven occasions each year. Depths of inundation vary according to the local topography of the floodplain, but in the middle reaches floodwater depths are typically about 40 cm for the mean annual flood and ca. 70 cm for a 50 year flood. Land use on the floodplain is almost exclusively permanent pasture. Previous work undertaken on this river has indicated that the area of floodplain investigated is an important sediment sink and that approximately 28% of the annual suspended sediment load entering the reach may be deposited on the floodplain during overbank events (Walling *et al.*, 1986). Such conveyance losses are equivalent to an average accretion rate for overbank sediment of ca. 500 g m^{-2} $year^{-1}$ (0.5 mm $year^{-1}$).

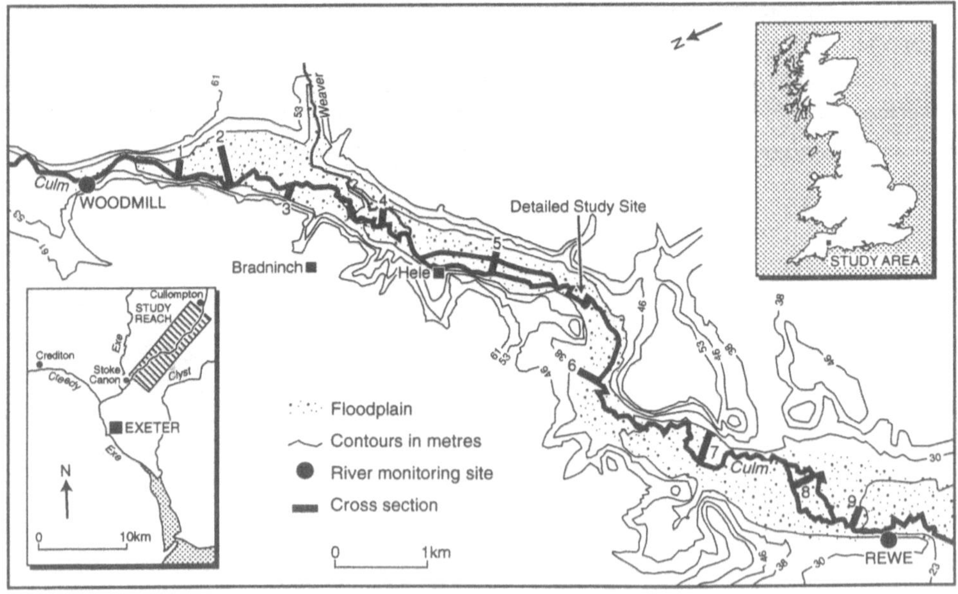

Fig. 2. The study reach of the River Culm, Devon, UK, showing the nine cross sections and the detailed study site.

4. Methods

In order to consider the spatial variability of sedimentation rates at different scales, the field sampling programme aimed at collecting sediment cores for subsequent ^{137}Cs and unsupported ^{210}Pb assay was structured to include two nested components. Firstly, sediment cores were collected from 9 cross-sections of the River Culm floodplain, located as shown on Figure 2, in order to provide a general assessment of the longitudinal and lateral variability of overbank sedimentation along the study reach. In each case, the cross-section extended from the river channel to the outer margin of the area commonly inundated during flood events and cores were collected at regular intervals along these transects with spacings of 14m-28m according to the total distance involved. A total of 87 cores was collected from these profiles. Secondly, a small area of floodplain in the middle of the study reach near Silverton Mill subject to frequent inundation (cf. Figure 2) was selected for more detailed investigation. In this case, 274 sediment cores were collected at the intersections of a 12m grid. In most instances the sediment cores were collected from the floodplain using a motorised percussion corer equipped with a 6.9cm diameter core tube. Cores were commonly ca. 75 cm in length and a small sample was collected from the base of each core for subsequent radionuclide assay to ensure that the core had penetrated to the full depth of the ^{137}Cs and unsupported ^{210}Pb profiles. Where cores were collected for subsequent sectioning, a larger diameter 12 cm core tube was used. In addition, samples of surface sediment were collected from a point immediately adjacent to each core. Since the procedures used to derive estimates of sedimentation rate from the values of excess ^{137}Cs or unsupported ^{210}Pb for the individual cores require information on the fallout radionuclide content of freshly deposited sediment, a number of sediment traps were installed at various locations on the floodplain prior to flood events and retrieved immediately after the flood receded. All bulk cores and other samples were air dried, ground and homogenised prior to measurement of their ^{137}Cs and unsupported ^{210}Pb content by gamma spectrometry. Values of unsupported ^{210}Pb activity were derived from the measurements of total ^{210}Pb activity by subtracting the ^{226}Ra supported ^{210}Pb activity. Count times were typically 10 hours and produced values of ^{137}Cs and unsupported ^{210}Pb activity with a precision of ca. ±5% and ±10% respectively, at the 95% level of confidence. Where required, measurements of the grain size composition of sediment samples were undertaken by laser diffraction (Malvern Mastersizer), after appropriate pre-treatment.

5. Results

The estimates of average annual sedimentation rates obtained from the radionuclide measurements undertaken on the individual cores were checked for consistency against independent evidence of short-term rates of overbank sedimentation on the floodplain of the lower River Culm provided by sediment traps reported by Simm (1993). Although the latter data relate to a short period (ca. 1 year) and may not be representative of the longer-term, the values were in close agreement with the longer-term estimates provided by the radionuclide measurements and inspired confidence in the resulting data. Further discussion and interpretation of these data is usefully approached by considering, firstly, the results obtained from the nine transects located along the study reach, and, secondly, the more detailed information on the spatial variation of sedimentation rates obtained from the study site near Silverton Mill.

5.1. REACH-WIDE PATTERNS

The estimates of average annual sedimentation rate derived from the ^{137}Cs and unsupported ^{210}Pb measurements for the nine transects are presented in Figure 3 along with information on the topography of the individual transects. In all cases, there is very close agreement between the values based on the ^{137}Cs and the unsupported ^{210}Pb measurements. Because the sedimentation rate estimates derived from the ^{137}Cs measurements represent average values for about the past 40 years (i.e. since the onset of significant ^{137}Cs fallout in the mid 1950s), whereas those derived from the unsupported ^{210}Pb measurements represent average values for the past ca. 100 years, the close correspondence between the two sets of estimates suggests that there has been little change in sedimentation rates over the past 100 years. However, there is some evidence that recent sedimentation rates (i.e. based on the ^{137}Cs measurements) are generally slightly lower than the longer-term values (based on ^{210}Pb measurements). This could be an artifact of the procedures used to estimate deposition rates from the two sets of fallout radionuclide measurements, but it may also reflect subtle changes in the

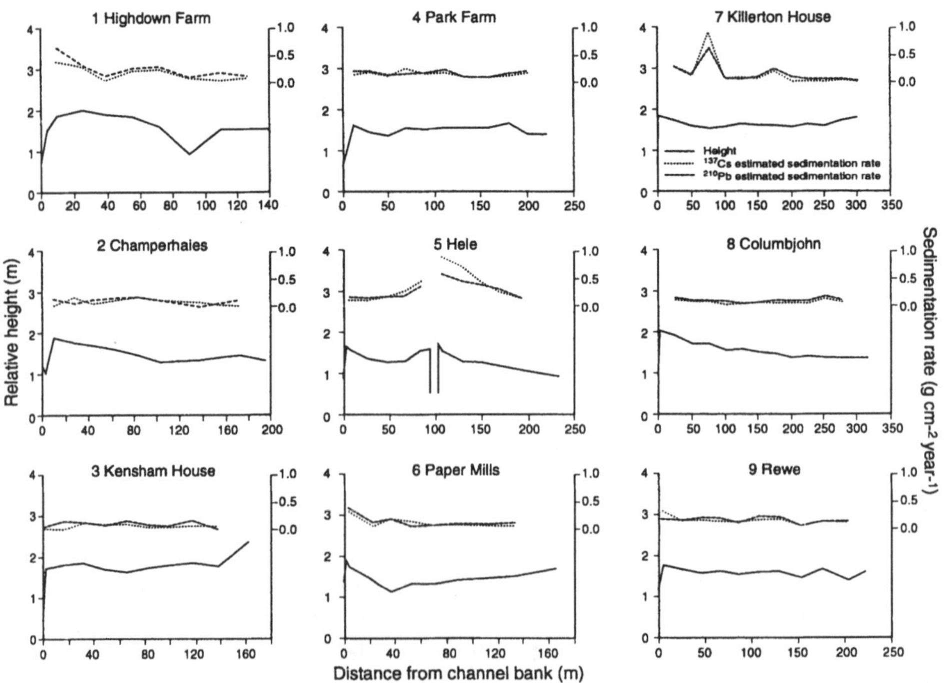

Fig. 3. Sedimentation rates estimated for the nine cross-sections of the floodplain of River Culm shown on Figure 2 using ^{137}Cs and unsupported ^{210}Pb measurements. The topography of the cross sections is also shown.

magnitude and frequency of overbank flood events during the period involved or the effects of slightly reduced water depths associated with the progressive aggradation of the floodplain surface.

TABLE I

Mean rates of sedimentation estimated for the nine floodplain cross sections used in the study

Cross-section No.	^{137}Cs estimates (g cm^{-2} year^{-1})	Unsupported ^{210}Pb estimates (g cm^{-2} year^{-1})
1	0.12	0.25
2	0.08	0.09
3	0.06	0.09
4	0.10	0.12
5	0.33	0.26
6	0.10	0.12
7	0.15	0.14
8	0.05	0.08
9	0.13	0.13
Overall Mean	0.13	0.15

For the 87 cores collected from the nine cross sections established along the reach, the average deposition rate was 0.13 g cm^{-2} year^{-1} based on the ^{137}Cs measurements and 0.15 g cm^{-2} year^{-1} based on the unsupported ^{210}Pb measurements. The average deposition rates associated with the individual cross-sections are listed in Table 1. Both the values presented in Table 1 and the more detailed information on the range of sedimentation rates associated with the individual cross sections presented in Figure 3 show no clear longitudinal trend down the reach, such as might be expected in response to the influence of a progressive reduction in the suspended sediment load through the reach due to deposition losses or of the general downstream increase in floodplain width. Rather, the rates of deposition documented for the individual cross sections would appear to reflect control by local conditions including channel geometry, overbank flow patterns and floodplain morphology and microtopography. A clearer trend is, however, evident for lateral variation across the floodplain. All sites evidence a general decline in deposition rate with increasing distance from the channel, which is consistent with the diffusion effect documented by other workers (e.g. Marriott, 1992) and the general reduction in frequency and duration of inundation in the outer zones of the floodplain. On average, sedimentation rates on the floodplain close to the channel are about 4 times greater than those found at the outer margin. The relationship between sedimentation rate (R (g cm^{-2} year^{-1})) and distance from the riverbank (Y (m)) is plotted for all 87 cores for both the ^{137}Cs and unsupported ^{210}Pb derived estimates of sedimentation rate in Figure 4. In both cases a simple exponential function has been fitted to the data to describe the overall trend of the decrease in sedimentation rate with increasing distance from the channel. The r^2 values for the fitted function are 0.19 for the ^{137}Cs-derived estimates and 0.16 for the ^{210}Pb-derived estimates.

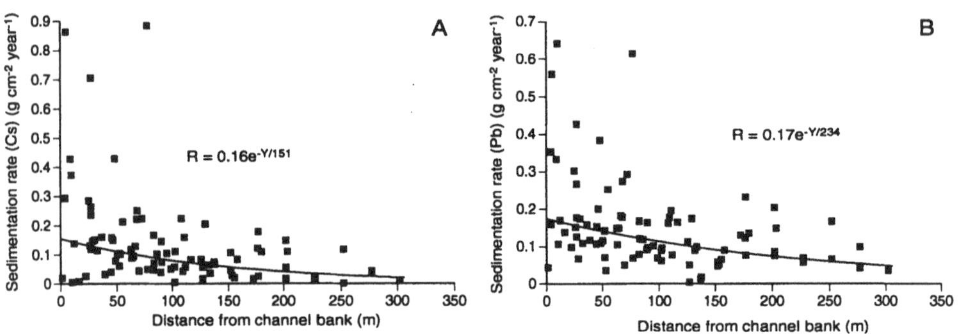

Fig. 4. Relationships between sedimentation rate and distance from the channel established for the floodplain cross-sections shown on Figure 2.

5.2. DETAILED PATTERNS

Figure 5 provides further information concerning the location and topography of the study area at Silverton Mill for which a more detailed investigation of the spatial pattern of overbank deposition was undertaken using a network of 274 sediment cores. In general terms it comprises a depression within a meander bend, bordered by an elevated bank margin or natural levee, and a linear depression along the outer margin of the floodplain. The mean annual sedimentation rates within the area estimated from the ^{137}Cs and ^{210}Pb measurements are shown in Figures 6A and 6B. The two sets of estimates are similar in terms of both magnitude and overall pattern and this again suggests that rates of sedimentation have been relatively constant during the past 100 years. There is, however, some evidence that sedimentation rates at this site may have increased slightly since the 1950s, because the estimates derived from the ^{137}Cs measurements are in places slightly greater than those derived from the unsupported ^{210}Pb measurements. Overall, the rates of sedimentation documented at this site are somewhat greater than those reported in Table 1 for the overall study reach. This tendency reflects the local site conditions and more particularly the greater frequency of inundation.

The sedimentation rates shown on Figure 6 show clear evidence of enhanced rates of deposition close to the channel and a general decrease with increasing distance from the channel. This pattern reflects the transfer of suspended sediment from the channel onto the adjacent floodplain both by diffusion and by convection associated with the changing orientation of the channel. In the latter case, the coarser sediment with higher settling velocities will be deposited close to the channel, causing a progressive reduction in the concentration of coarser particles in flows across the floodplain in response to deposition within the intervening zone. The influence of the local microtopography is also apparent in the increased sedimentation rates observed in some of the depression areas lying along the southeast margin of the study area. These reflect the increased depth of floodwater in these areas during inundation, since the amount of sediment deposited per unit area will be proportional to the total mass of sediment in the overlying water column, which is, in turn, proportional to the water depth. The pattern of sedimentation in this area is therefore generally consistent with the models proposed by other workers such as James (1985), Pizzuto (1987) and Marriott (1992).

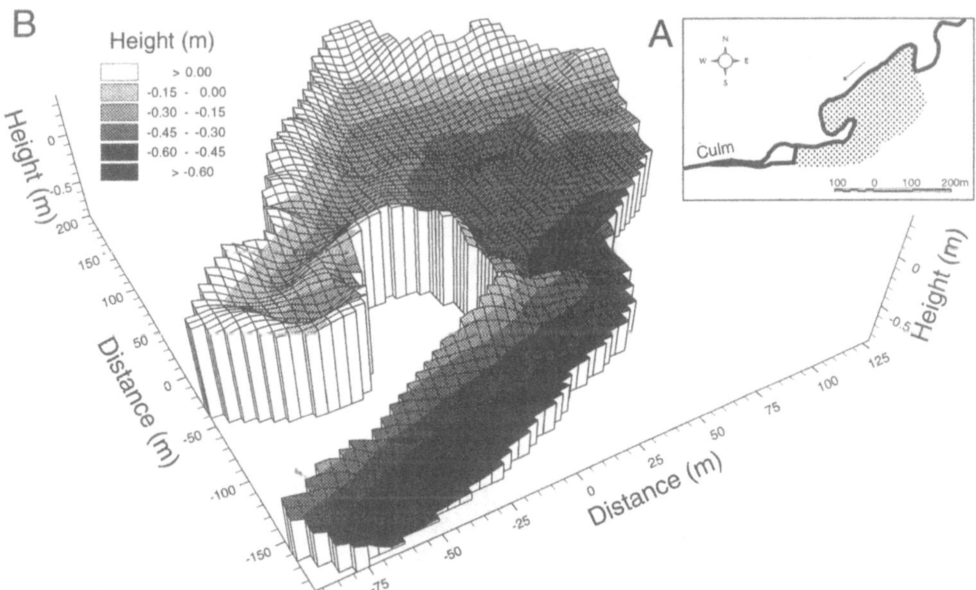

Fig. 5. The location of the study site at Silverton Mill (A) and the local microtopography of the site (B).

An attempt has been made to represent the combined effects of distance from the channel and floodwater depth on the observed rates of deposition (R (g cm^{-2} year^{-1})) by developing a simple deposition function. Following Howard (1992), overbank deposition on a floodplain can be seen as comprising two components. The first represents the deposition of the coarser fraction across the floodplain at a rate proportional to the floodwater depth and its concentration, which is assumed to decrease exponentially with increasing distance (Y(m)) from the nearest channel. The second reflects the deposition of fine particles, for which the sedimentation rate can be considered to be directly proportional to the floodwater depth (Z (m)) (defined as the difference between the maximum floodplain height and the local floodplain height), since, in the case of fine sediment, concentrations can be assumed to be relatively constant across the floodplain. For a specific point on the floodplain it is reasonable to assume that the concentration of fine sediment particles in the overlying water will be relatively uniform in the vertical, but the vertical profile of coarse particles is likely to be characterised by higher

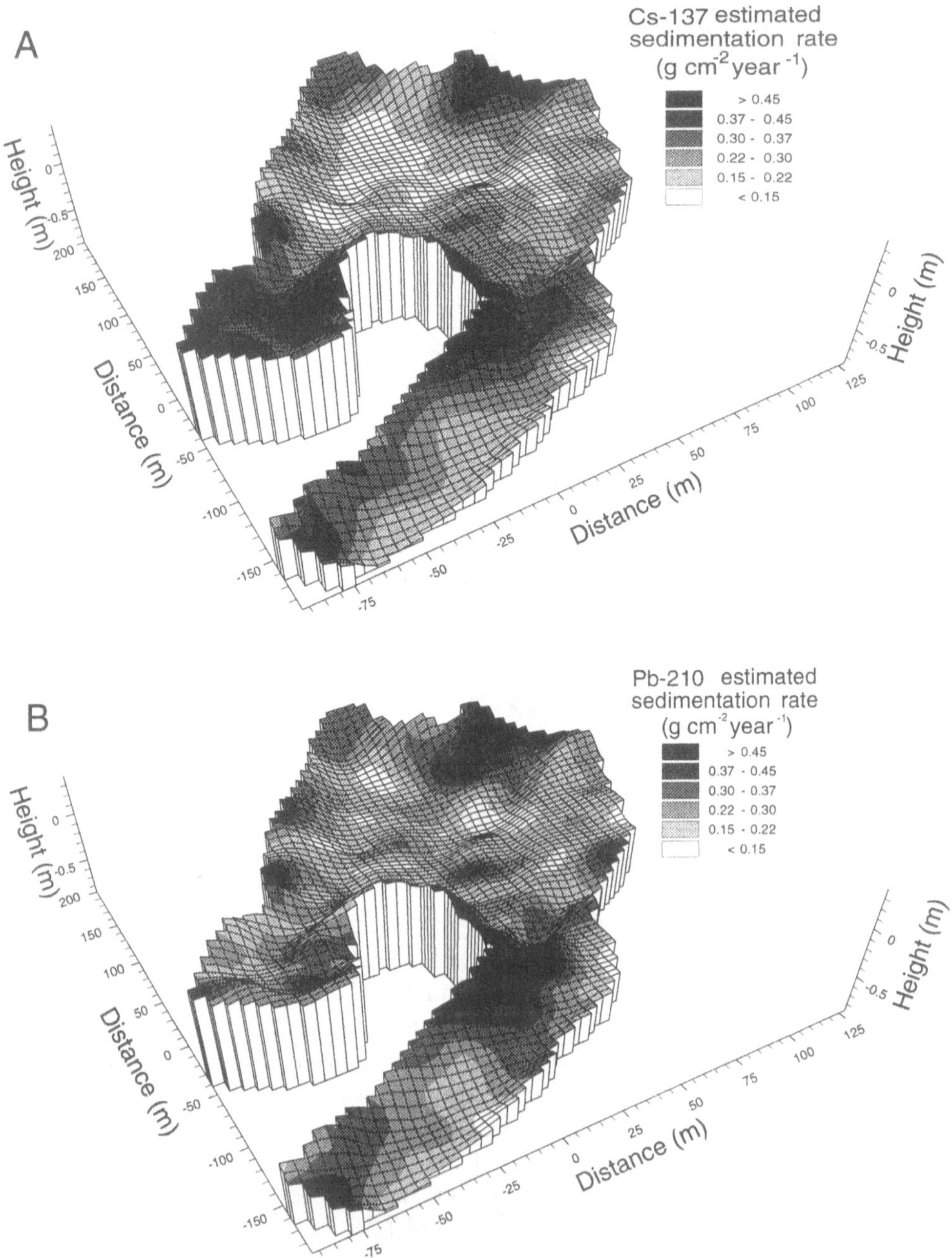

Fig. 6. Deposition rates across the floodplain study site at Silverton Mill estimated from ^{137}Cs (A) and unsupported ^{210}Pb (B) measurements.

concentrations close to the floodplain surface. These mechanisms may be represented by the following function:

$$R = \alpha Z^{\beta} e^{-Y/\gamma} + \delta Z \tag{1}$$

The first term on the right hand side of Equation 1 represents the deposition of the coarser particles and the second term the deposition of fine-grained sediment. α and δ are the average concentrations of the coarse and fine fractions of suspended sediment respectively. The constant γ represents the general decline in the concentration of the coarse fraction with increasing distance from the channel. The parameter β serves as a correction factor to take account of the non-uniform vertical distribution of the coarse fraction in the floodwater. Fitting Equation 1 to the sedimentation rates estimated from the ^{137}Cs measurements gives:

$$R = 0.68Z^{0.75} e^{-Y/15} + 0.08Z + 0.12 \tag{2}$$

with R^2=0.43. The sedimentation rates estimated from the unsupported ^{210}Pb measurements give a similar result viz:

$$R = 0.49Z^{0.80} e^{-Y/15} + 0.25Z \tag{3}$$

with R^2=0.41. These results presented graphically in Figure 7 confirm the general influence of both distance from the channel and floodwater depth in accounting for the spatial pattern of sedimentation rates in the study area, but it is clear that a fuller explanation of the detailed pattern would need to take account of the additional influence of the local microtopography and channel geometry and their interaction with the flow and sediment transport processes.

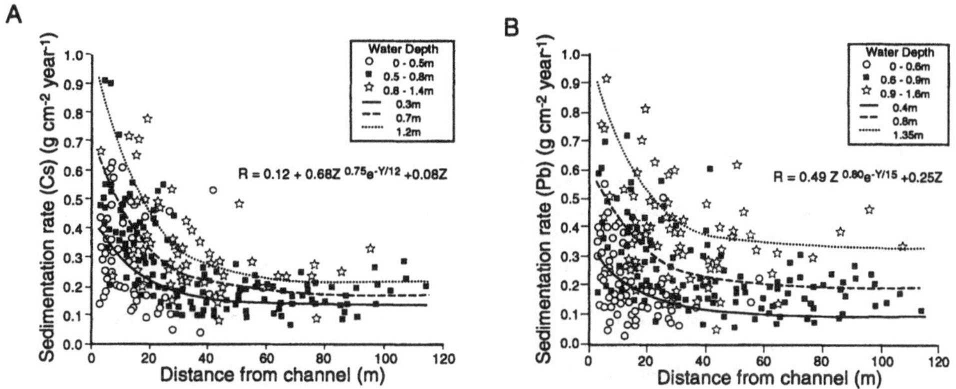

Fig. 7. The interrelationships between sedimentation rate and distance from the channel and water depth established for the floodplain study site at Silverton Mill based on ^{137}Cs (A) and unsupported ^{210}Pb (B) measurements.

6. Perspective

This paper has attempted to demonstrate the potential for using fallout radionuclide measurements as a basis for investigating rates of overbank sedimentation on river floodplains. The approach offers an essentially unique means of assembling both general information on the rates of sedimentation involved and detailed distributed data sets. The latter can in turn be used to explore the complex relationships between sedimentation rates and floodplain microtopography and morphology and flow hydraulics, and to provide a valuable basis for model development and validation.

Acknowledgements

The authors gratefully acknowledge the financial support of the UK Natural Environment Research Council for work on floodplain sedimentation (Research Grant GR3/8633), the co-operation of landowners in permitting access to field sites and sample collection, and the assistance of Mr. Jim Grapes with sample analysis.

References

Anderson, M. G., Walling, D. E., and Bates, P.: 1996, *Floodplain Processes*. Wiley, Chichester.

Asselmann, N. E. M. and Middelkoop, H.: 1993, *Earth Surface Processes and Landforms* **20**, 481-499.

Brown, A. G.: 1983, *Revue de Geomorphologie Dynamique* **32**, 95-99.

He, Q. and Walling, D. E.: 1996, *Earth Surface Processes and Landforms* **21**, 141-154.

Howard, A. D.: 1992, *Lowland Floodplain Rivers: Geomorphological Perspectives* (ed. by P. A. Carling and G. E. Petts), 1-41. John Wiley & Sons, Chichester.

Hupp, C. R: 1988, *Flood Geomorphology* (ed. by V. R. Baker, R. C. Kochel and P. C. Patton), 335-356. Wiley-Interscience, New York.

James, C. S.: 1985, *J. Hydraulic Res.* **23**, 435-452.

Lambert, C. P. and Walling, D. E.: 1987, *Geografiska Annaler* **69A**, 47-59.

Leenaers, H. and Schouten, C. J.: 1989, *Sediment and the Environment* (Proceedings of the Baltimore Symposium). *IASH Publ. no.* **184**, 75-83.

Lewin, J. and Macklin, M. G.: 1987, *International Geomorphology, Part I* (ed. by V. Gardiner), 1009-1027. Wiley.

Marriott, S.: 1992, *Earth Surface Processes and Landforms* **17**, 687-697.

Pizzuto, J. E.: 1987, *Sedimentology* **34**, 301-317.

Simm, D. J.: 1993, The deposition and storage of suspended sediment in contemporary floodplain systems: A case study of the River Culm, Devon. *Unpublished PhD Thesis*. University of Exeter.

Walling, D. E., Bradley, S. B. and Lambert, C. P.: 1986, *Drainage Basin Sediment Delivery* (Proceedings of the Albuquerque Symposium, August 1986). *IAHS Publ. no.* **159**, 119-131.

Walling, D. E. and He, H.: 1993, *Tracers in Hydrology* (Proceedings of the Yokohama Symposium). *IAHS Publ. no.* **215**, 319-328.

Walling, D. E. and He, Q.: 1994, *Variability in Stream Erosion and Sediment Transport* (Proceedings of the Canberra Symposium). *IAHS Publ. no.* **224**, 203-210.

Walling, D. E. and He, Q.: 1997, *Catena* (in press).

Walling, D. E., He, Q. and Nicholas, A. P.: 1996, *Floodplain Processes* (ed. By M. G. Anderson, D. E. Walling and P. Bates), 399-440. Wiley, Chichester.

Wolman, M. G. and Leopold, L. B.: 1957, *U.S. Geological Survey Professional Paper* **282-C**, 109pp.

EFFECTS OF BULK DENSITY ON SEDIMENT EROSION RATES

RICH JEPSEN, JESSE ROBERTS, and WILBERT LICK

Department of Mechanical and Environmental Engineering, University of California, Santa Barbara, CA 93106

Abstract. By means of a recently developed flume, sediment erosion rates as a function of shear stress and with depth in the sediments have previously been determined for relatively undisturbed sediments from several rivers and lakes. These experiments demonstrated that erosion rates depended on at least the following parameters: bulk density (or water content) of the sediments, particle size distribution as well as mean particle size, mineralogy, organic content, and amounts and sizes of gas bubbles. In order to isolate and quantify the effects of one of these parameters, the bulk density, additional experiments have been done with reconstructed sediments and are reported here. These experiments first determined the bulk density as a function of depth in the sediments for three different types of sediments, for three different sediment core lengths, and for compaction times varying from 1 to 60 days. For each of these sediment cores and compaction times, the erosion rate as a function of shear stress and with depth was then measured and related to the local bulk density of the sediment. The results demonstrate that, for a particular sediment and shear stress, the erosion rate is a unique function of the bulk density and can be expressed as a product of powers of the shear stress and bulk density.

Key words: sediment erosion, bulk density, depth dependence, shear stress, compaction time.

Introduction

Many contaminants are sorbed to sedimentary particles and are buried at depths of up to several meters in the bottom sediments of rivers, lakes, estuaries, and near-shore areas of the oceans. A major question is whether these buried sediments and associated contaminants can be exposed and transported during large floods and storms. In order to answer this question, a knowledge of the erosion rates of sediments at high shear stresses (up to stresses on the order of 100 dynes/cm^2) and with depth in the sediments (down to a meter or more) are needed.

A major limitation of most existing laboratory and field devices for measuring erosion rates is that they can only be used to measure sediment erosion at shear stresses below about 10 dynes/cm^2; because of this, these devices are only capable of eroding the surficial layers of the sediments, usually only the top few millimeters. In order to measure the erosion of sediments at high shear stresses and with depth, a unique flume (called Sedflume) has been designed, constructed, and tested (McNeil et al, 1996; Taylor and Lick, 1996). By means of this device, erosion rates of relatively undisturbed sediments from the Trenton Channel of the Detroit River in Michigan, the Lower Fox River in Wisconsin, and Lake Michigan have been determined. These field tests have

Water, Air and Soil Pollution **99**: 21-31, 1997.

illustrated the large differences in sediment erosion rates (by as much as several orders of magnitude) at different sites, with depth in the sediments, and as a function of shear stress.

In addition to this quantitative data on erosion rates, these tests qualitatively determined that erosion rates depended on at least the following parameters: bulk density (wet) or water content of the sediments, particle size distribution as well as mean particle size, mineralogy, organic content, and amounts and sizes of gas bubbles. Since the measurements of these bulk properties of sediments are standard and/or can be done relatively easily, it would be useful if sediment erosion rates could be predicted from a knowledge of these bulk properties. However, at present, this can not be done because of our lack of quantitative data on erosion rates as a function of these parameters. For this reason, a series of tests has been initiated by us to investigate the effects of various parameters on the erosion rates of sediments. The experiments described here were designed to isolate and measure the effects of the bulk density on erosion rates.

The bulk density is primarily related to the water content of a sediment and can be highly variable as the water content of the sediment changes due to deposition and subsequent compaction. As a column of sediment consolidates after deposition, the bulk density of the sediment generally increases with depth and time as the pore waters are expelled from the sediment and are transported to the surface. However, this general increase of the bulk density with depth and time may be non-monotonic due to vertical variations in the pore water transport, as will be seen below. This dependence of the bulk density of the sediments at a certain depth on the transport of water to the sediments above and from the sediments below indicates that the bulk density of a particular layer of sediment depends on the depth of the compacting sediment (or, in the case of laboratory experiments, on the length of the sediment core) as well as on the compaction time. As will be seen, erosion rates for a particular sediment depend critically on the bulk density and decrease rapidly as the bulk density increases.

In order to first understand and quantify the variations of bulk density, the bulk density was experimentally determined as a function of depth for three different types of sediment, for three different sediment core lengths, and for compaction times varying from 1 to 60 days. For each of these sediment cores, the erosion rates as a function of applied shear stress and with depth were then measured by means of Sedflume and subsequently related to the local bulk density of each sediment.

In the following section, the experimental procedures are briefly described. The results of the experiments and a discussion of the results are given in the third section. A summary and concluding remarks are given in the final section.

Experimental Procedures

Measurements of erosion rates of different types of sediments with varying bulk densities were made by means of Sedflume (see Fig. 1). This flume is essentially a straight flume which has a test section with an open bottom through which a rectangular cross section coring tube containing sediment can be inserted. This coring tube is 1 m long and has a cross-section which is 10 cm by 15 cm. Water is pumped through the flume at varying rates and produces a turbulent shear stress at the sediment-water interface in the test

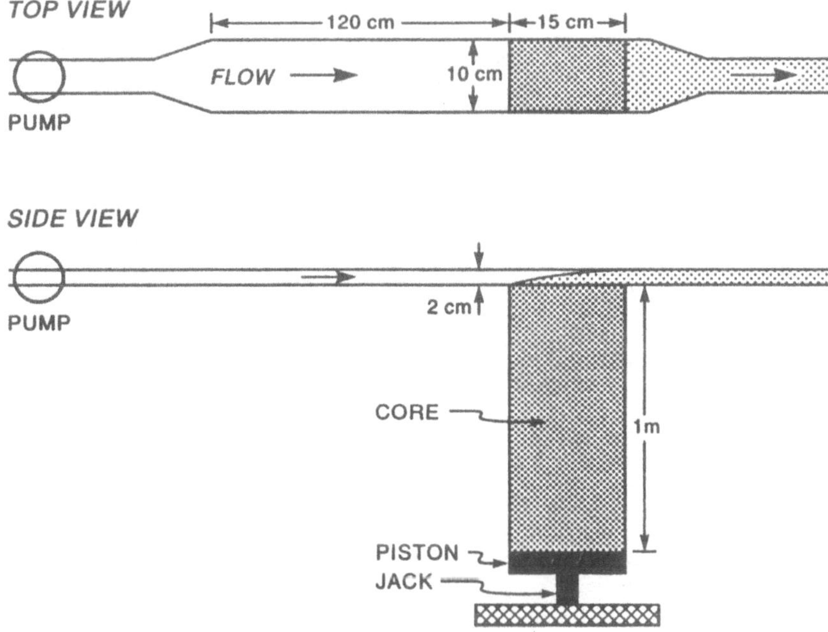

Fig. 1. Schematic diagram of Sedflume.

section. This shear stress is known as a function of flow rate from standard pipe flow theory. As the shear produced by the flow causes the sediments in the core to erode, the sediments are continually moved upwards by the operator so that the sediment-water interface remains level with the bottom of the test and inlet sections. The erosion rate is then recorded as the upward movement of sediments in the coring tube. The results are reproducible within a $\pm$ 25% error and are independent of the operator. For more details of the procedure and apparatus, see McNeil et al (1996).

Sediments to be used in Sedflume can either be obtained relatively undisturbed from the bottoms of surface bodies of water by coring or by reconstructing sediments in the laboratory. In the present experiments, reconstructed sediments were used so that the bulk parameters of the sediments could be carefully controlled. Three types of sediments were used in the tests. These were from the Detroit River in Michigan, the Fox River in Wisconsin, and a slough near Santa Barbara, California. Their disaggregated particle size distributions are shown in Fig. 2. Their average disaggregated particle sizes were 12, 20, and 35 μm, while their organic carbon contents were 3.3, 4.1, and 1.8% respectively. The mineralogies of the three sediments were approximately determined by means of X-ray powder diffraction using a Philips X'Pert Model PW 3040. The relative composition of each of the sediments was as follows: quartz, muscovite, dolomite, and calcite for the Detroit River; quartz, albite, dolomite, illite, and kaolinite for the Fox River; and albite, quartz, and microcline for the Santa Barbara slough.

In order to obtain different bulk densities for each sediment for the erosion tests, sediment cores were prepared as follows. Fifty to a hundred gallons of each wet sediment were placed in a 150 gallon cylindrical tank and mixed with water for 15 to 30 minutes until the sediment-water mixture was homogeneous. The amount of water added was

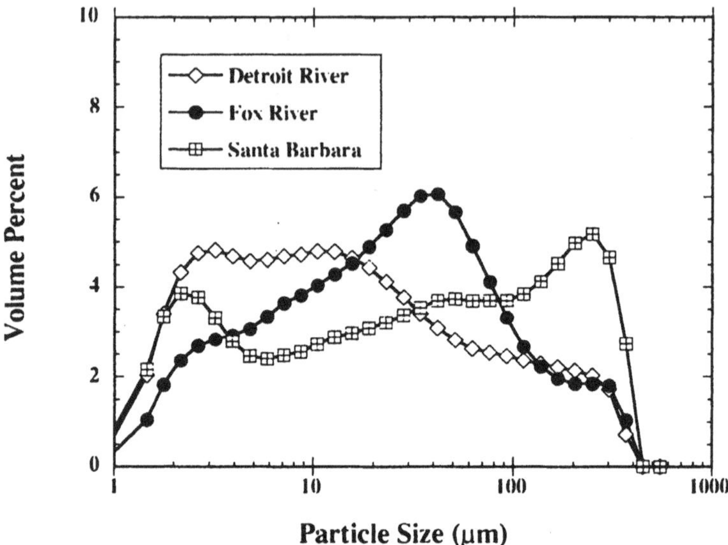

Fig. 2. Disaggregated particle size distributions for sediments from the Detroit River, Fox River, and Santa Barbara slough.

enough to allow the mixture to be fluid, but care was taken to also keep the mixture thick so that stratification of the sediment due to differential settling of the particles did not occur. The sediment mixtures were then poured into the coring tubes to depths of 20, 40, and 80 cm. These cores were then allowed to compact for 1, 2, 5, 12, 21, 32, 60, or 100 days. Duplicate analysis cores were checked for repeatability and to ensure uniformity in particle size and organic content for all depths and compaction times.

In order to determine the bulk density of the sediments at a particular depth and compaction time, the sediment analysis cores were frozen, sliced into 3 to 4 cm sections, and then weighed (wet weight). They were then dried in the oven at approximately 75°C for 2 days and weighed again (dry weight). The water content W is then given by

$$W = \left(\frac{m_w - m_d}{m_w} \right) \tag{1}$$

where m_w and m_d are the wet and dry weights respectively. The bulk density ρ(gm/cm^3) is related to the water content by

$$\rho = \frac{\rho_s \rho_w}{\rho_w + (\rho_s - \rho_w) W} \tag{2}$$

where ρ_s = 2.6 gm/cm^3 and ρ_w = 1.0 gm/cm^3. The total error in bulk density measurements was within $\pm$0.001 gm/cm^3.

Average particle sizes and particle size distributions were determined by use of a Malvern Particle Sizer for particle diameters between 0.5 and 600 μm. To do this, a small amount of sediment was mixed with water and disaggregated in a Waring blender.

Approximately 1 mL of this solution was then used for analysis by the particle sizer. From these measurements, the distribution of grain sizes and the average grain size as a function of depth were obtained.

The procedure for measuring the erosion rates of the sediments as a function of shear stress and depth was as follows. The sediment cores were prepared as described above and then moved upward into the test section until the sediment surface was even with the bottom of the test section. A measurement was made of the depth to the bottom of the sediment in the core. The flume was then run at a specific flow rate corresponding to a particular shear stress. Erosion rates were obtained by measuring the remaining core length at different time intervals, taking the difference between each successive measurement, and dividing by the time interval.

In order to measure erosion rates at several different shear stresses using only one core, the following procedure was generally used. Starting at a low shear stress, the flume was run sequentially at higher shear stresses with each succeeding shear stress being approximately twice the previous one. Generally about three shear stresses were run sequentially. Each shear stress was run until at least 3 mm but not more than 2 cm were eroded. The time interval was recorded for each run with a stop watch. The flow was then increased to the next shear stress, and so on until the highest shear stress was run. This cycle was repeated until all of the sediment had eroded from the core. If after three cycles a particular shear stress showed a rate of erosion less than 10^{-4} cm/s, it was dropped from the cycle; if after many cycles the erosion rates decreased significantly, a higher shear stress was included in the cycle.

Results

BULK DENSITY

Bulk densities were determined as a function of depth for each of the three sediments described above; for 20, 40, and 80 cm sediment core lengths; and for compaction times of 1, 2, 5, 12, 21, 32, and 60 days. The densities of sediments compacted for 100 days could not be accurately determined because gas pockets had formed and the present method for measuring bulk density can not account for the presence of gas bubbles.

Bulk densities for the 20, 40, and 80 cm cores from the Detroit River are shown in Figs. 3a, b, and c. It can be seen that the densities increase with time and generally increase with depth. This is to be expected as the pore waters are forced upwards and out of the bottom sediments due to the weight of the overlying sediments. The increase with time is most rapid initially and then decreases as time increases and compaction slows. It can also be seen that there are deviations from a monotonic increase of density with depth. These deviations tend to appear between 5 and 21 days and are due to increases in water content at a particular depth as the waters above this layer are transported upwards slower than the waters below this layer. This trapping effect disappears by 60 days in all cores. The depth at which the trapping effect is maximum increases as the core length increases.

The bulk densities for the 20 cm cores from the Fox River and Santa Barbara slough are shown in Figs. 4 and 5. It can be seen that the general character of the dependence of bulk density on depth for these two sediments is the same as that for the Detroit River.

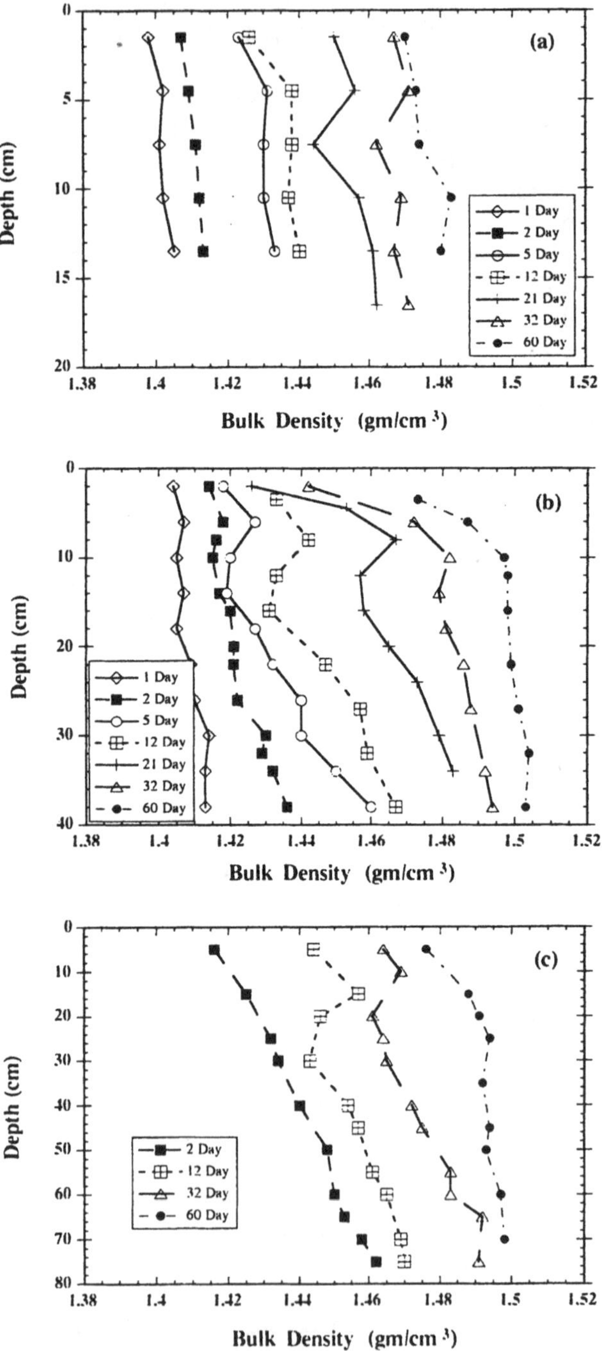

Fig. 3. Bulk density as a function of depth for the Detroit River cores at compaction times of 1, 2, 5, 12, 21, 32, and 60 days. (a) 20 cm cores. (b) 40 cm cores. (c) 80 cm cores.

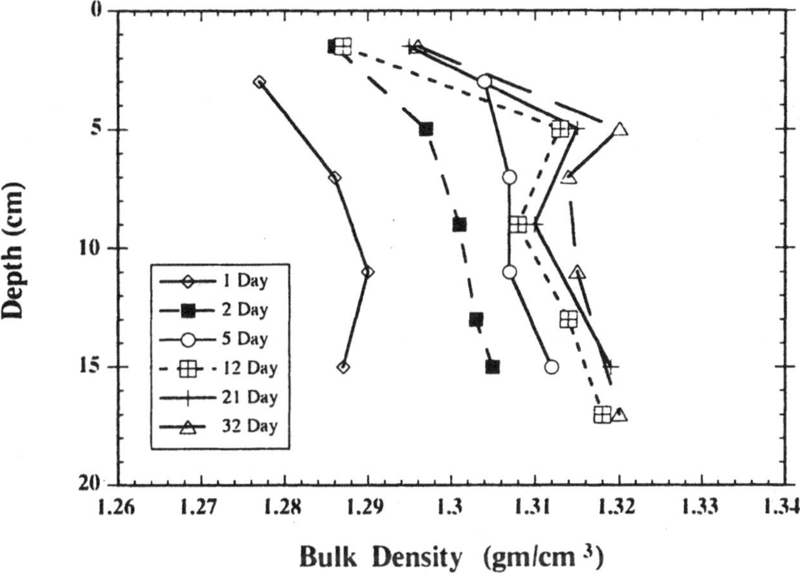

Fig. 4. Bulk density as a function of depth for the 20 cm cores of Fox River sediments.

This is also true for the 40 and 80 cm cores. It can also be seen that the ranges of densities for each type of sediment are quite different, i.e., the Detroit River density varies from 1.405 to 1.505, the Fox River density varies from 1.275 to 1.335, while the Santa Barbara density varies from 1.67 to 1.9. Trapping of water is evident in the Fox River cores and is present, but to a lesser extent, in the Santa Barbara cores.

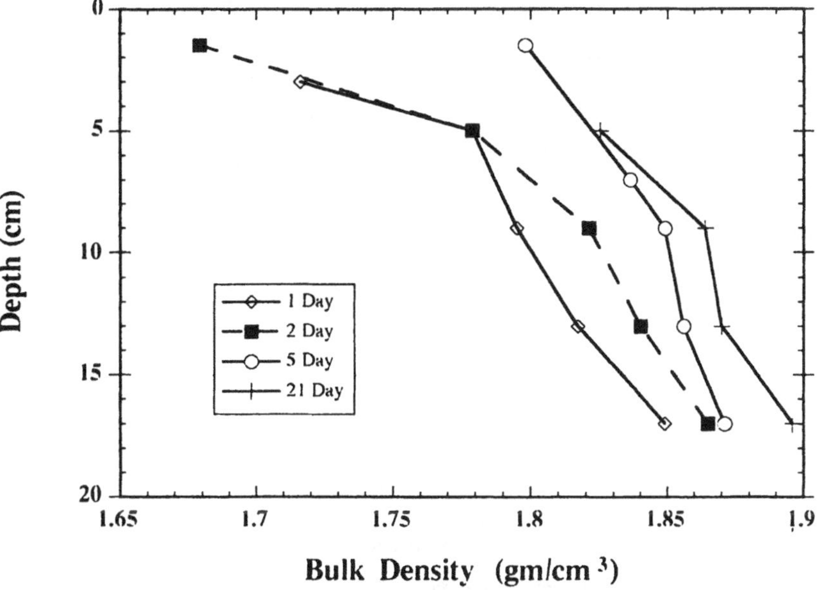

Fig. 5. Bulk density as a function of depth for the 20 cm cores of Santa Barbara slough sediments.

EROSION RATES

Sediment cores were prepared, inserted into Sedflume, and were then eroded by the procedures described above. By this means, erosion rates as a function of shear stress and with depth were obtained for all three types of sediments, for all three core lengths, and for compaction times from 1 to 60 days. Results for the 40 cm Detroit River core at a 21-day compaction time are shown in Fig. 6 for shear stresses of 16, 32, 64, and 96 dynes/cm^2. It can be seen that erosion rates increase rapidly with shear stress and generally decrease with depth throughout most of the core, although not monotonically. These results are typical of the data obtained from the other cores. From a comparison of the erosion rates in Fig. 6 with the bulk densities for the same core shown in Fig. 3b, one can see that a direct relation between the erosion rate and bulk density exists, i.e., as the bulk density increases, the erosion rate decreases. The effect of the trapping of water at the 15 cm level on both the bulk densities and erosion rates can be clearly seen. Figure 6 also shows a thin surficial layer (1 to 3 mm thick) which was very difficult to erode. This layer was discolored (usually orange or light brown) and very cohesive but had the same particle size distribution as the rest of the sediment. This layer tended to peel off in slabs, up to 5 cm in width, rather than eroding particle by particle, or in small chunks of sediment.

All of the data for erosion rates as a function of bulk density for the Detroit River sediments at shear stresses of 2, 4, 8, 16, 32, and 64 dynes/cm^2 are shown in Fig. 7a. The rapid decrease of the erosion rate as the bulk density increases can be clearly seen. The data can be approximated by an equation of the form

$$E = A \tau^n \rho^m \tag{3}$$

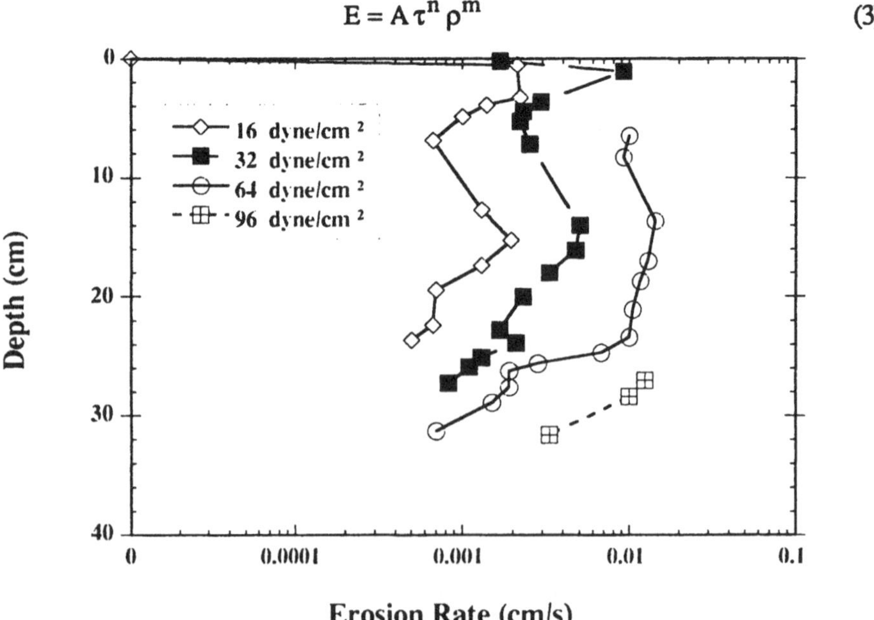

Fig. 6. Erosion rates as a function of depth for the 40 cm cores of Detroit River sediments which have compacted for 21 days. Erosion rates for shear stresses of 16, 32, 64, and 96 dynes/cm^2 are shown.

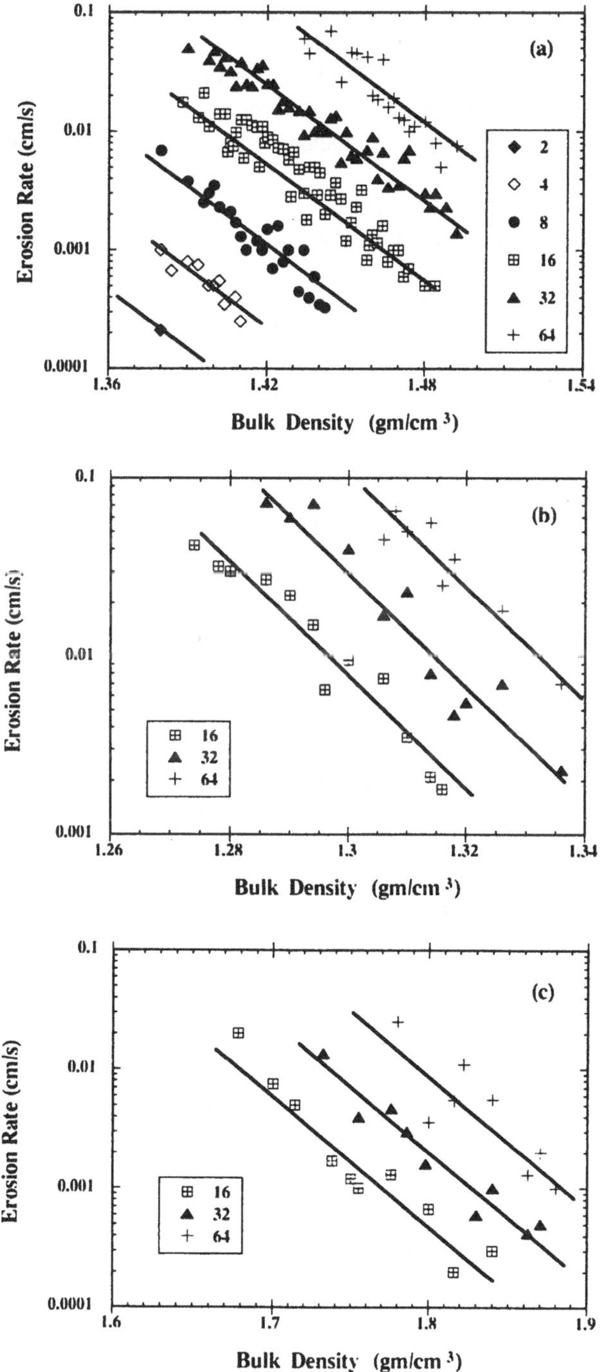

Fig. 7. Erosion rates as a function of bulk density and for different shear stresses (dyne/cm2). Sediments from (a) the Detroit River, (b) the Fox River, and (c) the Santa Barbara slough.

where E is the erosion rate (cm/s), τ is the shear stress (dynes/cm^2), ρ is the bulk density, and n, m, and A are constants. For each shear stress, the erosion rate as a function of bulk density is shown as an essentially straight line in Fig. 7a. It can be seen that the above equation summarizes the data quite well. This figure also demonstrates that for this sediment the erosion rate for a particular shear stress is a unique function of the bulk density.

Results for the erosion rate as a function of bulk density for sediments from the Fox River and the Santa Barbara slough are shown in Figs. 7b and c. It can be seen that the general dependence of the erosion rate on shear stress and bulk density is the same as that for the Detroit River sediments. Because of this, the data for each sediment can again be approximated by Eq. (3). However, since the dependence of the erosion rate on shear stress and bulk density is different for each sediment (compare Figs. 7a, b, and c), the constants A, n, and m are also different for each sediment. These constants are shown in Table I.

Summary and Concluding Remarks

By means of the experiments described here, the effects of sediment bulk density on erosion rates were measured. The bulk density is primarily a function of the water content of the sediment and varies as the water content changes due to sediment deposition and subsequent compaction. In order to first understand and quantify the variations of bulk density, the bulk density was experimentally determined as a function of depth for three different types of sediments, for three different sediment core lengths, and for compaction times varying from 1 to 60 days. For each of these sediment cores and compaction times, the erosion rate as a function of shear stress and with depth was then measured.

From these experiments, the following was determined. (1) The bulk density of the sediments generally increases with depth and time, but trapping of water can cause a local decrease in the density. (2) The location and magnitude of this density decrease due to

TABLE I

Erosion parameters for sediments from the Detroit River, Fox River, and Santa Barbara slough

	n	m	A
Detroit River	2.23	–56	3.65×10^3
Fox River	1.89	–95	2.69×10^6
Santa Barbara	2.10	–45	4.15×10^5

trapping of water depend on the sediment core length and type of sediment. (3) For each sediment and shear stress, the erosion rate is a unique function of bulk density and decreases as the bulk density increases. (4) For each sediment, the erosion rate can be approximated as $E = A\tau^n \rho^m$ where A, n, and m are constants which depend on the type of sediment.

Acknowledgements

This research was supported by the U.S. Environmental Protection Agency and by the University of California Campus Laboratory Collaboration Program. We would like to thank James Boles and Dave Pierce of the Geological Sciences Department at UCSB for their help with the mineralogical properties, and Blair Erbstoeszer and Artur Hanski for their help in measuring sediment bulk properties.

References

McNeil, J., Taylor, C., and Lick, W.: 1996, *J. Hydraulic Engineering*, **122**, 316-324.

Taylor, C. and Lick, W.: 1996, *Erosion Properties of Great Lakes Sediments*, UCSB Report.

THE EFFECT OF FLOCCULATION ON THE SIZE DISTRIBUTIONS OF BOTTOM SEDIMENT IN COASTAL INLETS: IMPLICATIONS FOR CONTAMINANT TRANSPORT.

T.G. MILLIGAN AND D.H. LORING
Marine Environmental Sciences Division, Fisheries and Oceans Canada, Bedford Institute of Oceanography, P.O. Box 1006, Dartmouth, N.S. B2Y 4A2

Abstract. Grain size is the most basic of classification criteria for sediments. The size distribution of a given sediment records the physical transport processes involved in its formation. By using precise grain size analysis and the model of Kranck et al. (1996a,b), it is possible to break down a sediment into the three major components from which it was formed: material deposited as flocs, material deposited as single grains from suspension, and material carried under higher energy conditions. With this method, both the amount of material deposited in a flocculated state and the maximum size, or floc limit, of the particles composing the floc can be determined. Changes in floc limit indicate changes in the aggregation dynamics of the system. As most trace metals and many other contaminants associate closely with the fine particle fraction of sediments, it is important to determine both the areal distribution and reworking history of the floc settled portion of a sediment. This paper discusses the application of the method to coastal inlets in Atlantic Canada and examines the relationship between proportion of floc-settled material and trace metal concentrations. Disaggregated inorganic grain size distributions are also used to illustrate changes in the aggregation dynamics in areas of intense aquaculture.

Keywords: Aquaculture, aggregation, flocculation, grain size, sediment, trace metal.

1. Introduction

The disaggregated inorganic grain size of sediment from marine and freshwater environments can be used to describe depositional conditions in terms of aggregation dynamics, energy and frequency of reworking. Using this information it is possible to determine the fate of contaminants reactive with both colloidal and particulate phases and to identify changes in the aggregation dynamics of a body of water. As part of a study of metallic contamination (Loring et al., 1996), surface sediment samples collected from 11 Nova Scotian coastal embayments were described in terms of method of formation, reworking history, and amount of material deposited in a flocculated form. In a separate study to assess the impacts of organic enrichment from marine aquaculture in Letang Inlet, N.B., core samples collected under cages and at control sites were examined for changes in grain size distribution resulting from salmon farming (Hargrave et al., 1995). By applying the model presented by Kranck (1993) and Kranck et al. (1996a,b) for the development of disaggregated inorganic size spectra of bottom sediments the processes involved in the formation of the samples collected in these studies can be determined.

Understanding the behaviour of the fine particle fraction of sediments deposited from suspension is essential to the study of contaminants in the aquatic environment. Trace metals and other contaminants such as organic pesticides are closely associated with colloidal and particulate material in suspension in both marine and freshwater environments (Muller, 1996). Through flocculation with the naturally occurring suspended particulate matter, the fate of surface active contaminants will ultimately be controlled by the deposition of the aggregates formed. This association with fine particulates is most often accounted for by normalising metallic concentration to the fraction of sediment below some size limit e.g. <63 μm (Loring and Nota, 1968), <53 μm

(Loring, 1988), <16μm (Zwolsman et al., 1996). The arbitrary selection of a particle size does not take into account aggregation dynamics and the effect of flocculation on the final size distribution of a sediment. Required for both transport and depositional studies is a method to determine the amount of material in a sediment which has been deposited in the form of aggregates.

Models of particle aggregation (Jackson, 1995, Hill and Nowell, 1995) and laboratory studies (Kranck, 1980) indicate that particle concentration has a strong effect on aggregation rate. Aggregation efficiency also controls floc formation rate with the 'stickiness' of particles in suspension being an important factor in both the size and stability of the flocs formed (Kranck and Milligan, 1980; Muhle, 1993). In regions where particle concentration or particle stickiness is increased, an increase in aggregation rate and maximal floc size would be expected to occur. The development of fluff layers in areas of extensive dredging is one example of particle flux increasing as a result of increased sediment concentration (Kranck and Milligan, 1989). The effect of particle "stickiness" has been well documented in laboratory studies of clay mineral flocculation using different flocculants (e.g. Tambo and Hozumi, 1979; Spicer and Pratsinis,1996) and observed in the aggregation of diatoms (Kepkay et al. 1993; Niven et al., 1996). Aquaculture operations are potential sources of fine particles which could be expected to have increased stickiness due to bacterial degradation. Changes in the aggregation dynamics in coastal embayments would result in changes in the grain size distributions over time that should be detected in core samples.

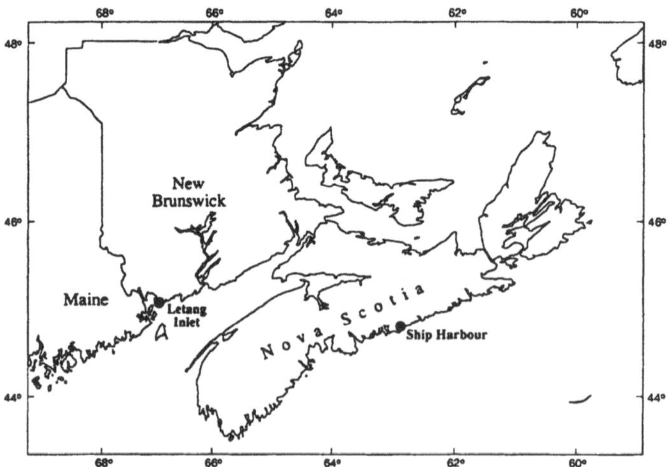

Fig. 1. Map of Eastern Canada showing location of sample sites.

This paper describes the sedimentary environment in two coastal inlets on the Atlantic coast of Canada in terms of formation processes determined from the disaggregated inorganic grain size analysis of bottom samples (fig. 1). In Ship Harbour, Nova Scotia, the size structure of the bottom sediments is used to show the transition from unsorted riverine sediments through highly flocculated sediments, to wave sorted sediments at the mouth of the inlet. The relationship between metal concentration, aggregation and degree of sorting is demonstrated using the size distributions of the sediments collected. In

Letang Inlet, New Brunswick, the descibed method of detailed size analysis of bottom sediments is used to show changes in the aggregation dynamics in a macro-tidal environment undergoing intense aquaculture development.

2 Methods

Surficial sediments were collected from 28 stations in Ship Harbour, using a 0.1 m^2 Eckman grab. Detailed methods and results for Ship Harbour and 10 other coastal embayments in Nova Scotia can be found in Loring et al. (1996). Subsamples for trace metal and inorganic grain size were taken from the top 1 cm. Disaggregated inorganic grain size spectra were determined by electro-resistance particle sizing using the techniques described by Milligan and Kranck (1991). Sample analyses were carried out using a Coulter Multisizer IIE over a size range from 0.7 - 400 μm. Subsamples were digested in an excess of 35% H_2O_2 and suspended in 1% NaCl before disaggregation with a sapphire tipped sonic probe. Results were plotted as frequency distributions, or frequency spectra, of $\log_{10}$ equivalent weight % of sediment, determined from the volume in logarithmically equal size classes using a specific gravity of 2.65 kg m^{-3} vs. $\log_{10}$ diameter in 1/5 ϕ mid point size classes i.e. diameter doubling every five channels. The size distribution data were normalised to the total weight in the size range analysed. It is important to note that instruments which use iterative or interpretative techniques to derive best fits to assumed size distributions will not produce the same results. For example, Laser diffraction instruments fail to resolve size distributions in which some portion of the curve has a straight line distribution with a slope close to 0, such as are found in unsorted sediments and in the floc settled portion of most sorted sediments.

Samples for chemical analysis were initially stored in plastic vials, homogenised and oven-dried at 60° C. Subsamples were removed for Hg analysis and the remainder was re-dried at 110° C. Total Al, Cd, Cr, Cu, Fe, Li, Ni, Pb, V and Zn concentrations were determined using a 0.2 g air dried sample by atomic adsorption spectrophotometry (AAS) using techniques described by Loring and Rantala (1992) after digestion with a combination of hydrofluoric and aqua regia in a microwave oven (Rantala and Loring, 1989).

Sediment cores from the Letang Inlet, were collected at 11 salmon aquaculture sites and 11 control sites. Sediment profiles of disaggregated inorganic grain size, water content, redox potential, ammonium, sulphides, sulphate, organic carbon and nitrogen, benthic dissolved O_2 and CO_2 exchange and biomass were determined. A detailed description of sampling methods and results are provided by Hargrave et al. (1995). Samples for grain size analysis were taken at 2 cm intervals through holes drilled in the side of the core liner and were analysed using the techniques described above. A Benthic Enrichment Index (BEI) was determined for each site from the oxidation reduction potential (Eh) and organic carbon concentration (C_{org}) in the sediment (Hargrave, 1994).

The disaggregated inorganic grain size spectra were analysed to determine the fraction of material deposited as flocs, single grains from suspension, and through higher energy bottom transport using the model described by Kranck et al. (1996a,b). In their model, Kranck et al. (1996a,b) presented a method to describe the generation of disaggregated inorganic grain size spectra from sediment deposited from suspension. This model, based on grain size analysis carried out with Coulter Counters (Milligan and Kranck, 1991), separates the inorganic fraction of a sediment into three component parts; material settled in a flocculated form, material deposited as single grains from suspension and material which has undergone further suspension sorting, usually as a result of high energy events. The simplified form of the equation describing the size distribution is

$$C = Q_o w^{m+n} e^{-Kw} \qquad (1)$$

where C is the concentration in any one size class, Q_o is the concentration at a reference settling velocity, w is the settling velocity calculated using Gibbs et al. (1971). The variable m is the slope of the source suspension as defined by the straight line portion of the curve starting at 0.7 µm and has a value close to 0 (Kranck and Milligan, 1983, 1985, Kranck et al. 1996a). Material in this region of the curve, termed the floc tail, has been deposited in a flocculated form from the suspension. The exponent n has integer values of 0 for the floc settled portion of the curve, 1 for the portion of the curve derived from single grain deposition from suspension, termed a one round distribution, and values > 1 for the coarsest fraction of the sediment which has been subjected to further sorting (Kranck, 1993, Kranck et al. 1996a,b). K describes the maturity of the deposition process, i.e. what fraction of the coarse grains have settled. A linearized form of the equation can be fitted to the inorganic grain size spectra using a least square method and the values of Q, m, n and K can be evaluated. Using these values, the component parts of the size distribution and their overall contribution to the total sediment can be calculated and the percentage of material deposited in a flocculated form determined (fig. 2). A floc limit describing the degree of flocculation in terms of the size of the particles incorporated into and deposited as flocs was determined from the intercept between the single grain and floc settled portions of the sediment (fig. 2) (Schell, 1996).

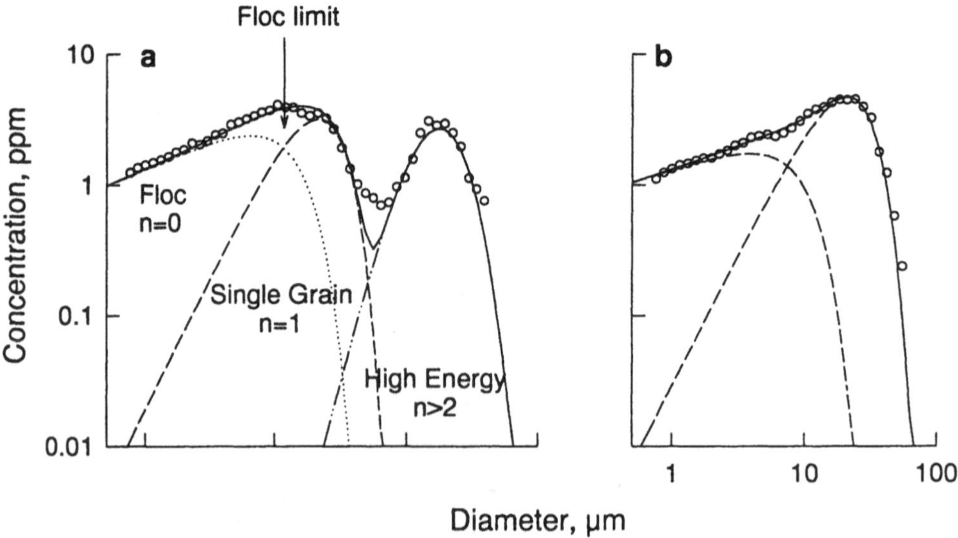

Fig. 2: Disaggregated inorganic grain spectra showing floc, single grain (1 round) and high energy (multi round) components of a sediment. Fitted curves derived from least squares fit of equation 1 to data points (circles). Floc limit is defined by the diameter of the intercept between the floc curve (dotted) and the one round curve (dashed). Curve a is from Letang Inlet, N.B., curve b is from Ship Harbour N.S.

3 Results And Discussion

Regional Description: Four distinct sedimentation regions were identified using the size analysis of sediments from 28 stations in Ship Harbour. Ship Harbour is a fjord-like inlet 8.3 km long and 0.3 km wide, with water depths >25 m in the inner portion of the harbour and a shoal <4 m at a sill located inshore of Lower Ship Harbour (fig. 3). It is mostly surrounded by small farms and woodlots and has a low population with no industrial development. Mussel farming is carried out in the inner reaches.

The four depositional zones identified in Ship Harbour from the inorganic grain size spectra show the transition from unsorted riverine sediment at the head of the embayment through a region of floc-dominated sedimentation in the central region to well sorted fine sands and sands with only minor amounts of floc deposited sediment at the seaward end (fig. 3). Unsorted erosional products such as those derived from tills have straight line distributions when plotted in log-log format (Kranck and Milligan, 1983, 1985). At the head of Ship Harbour, the size distribution of the river bottom sediment, (fig. 3, 2), is similar to that of the eroding till cliff, (fig. 3, 1), except that the distribution has been truncated at the coarse end in response to the lack of competence of the riverine flow for transporting large grains. This straight-line source distribution, modified by input of suspended material from the offshore, is preserved in the floc tail in sediments formed within the bay (Kranck and Milligan, 1983).

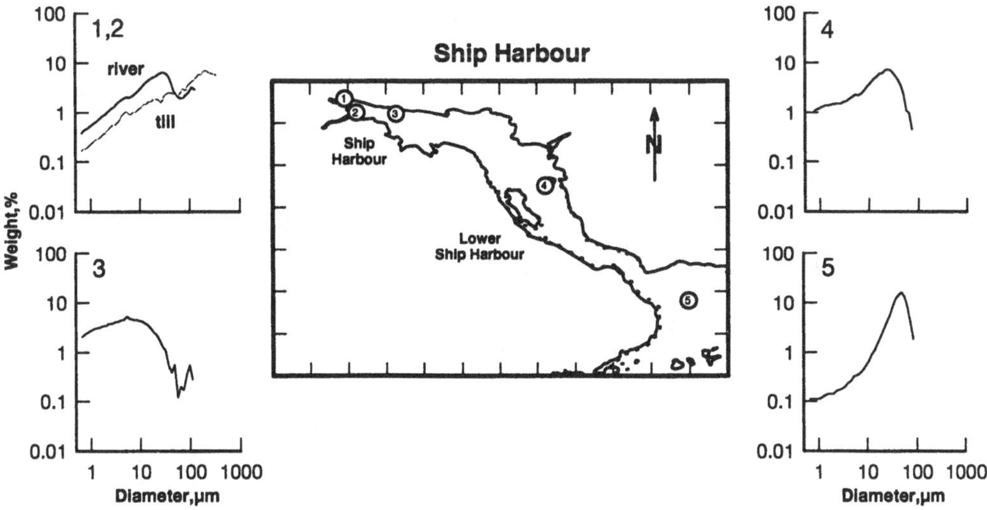

Fig. 3: Map of Ship Harbour N.S. showing typical inorganic size distribution for 4 sedimentary regions

In the central region of Ship Harbour, the constituent grains forming the sediments have been predominantly deposited in a flocculated form, with only very small amounts of material being deposited as single grains (fig 3, 3). The lack of a sorting and the absence of higher round sediment at the coarse end of the distribution indicates that little or no resuspension occurs in this region, likely as a result of the sill. Based on the grain size distribution data, this deep region of Ship Harbour represents a depositional area for fine-grained aggregates. Turbulent energy from currents and wave action will remove part of the floc settled material from a sediment and increase both the size and degree of sorting at the coarse end of the distribution. In Ship Harbour, samples seaward of the sill show coarser grain sizes and are sorted to higher round distributions (fig. 3, 4). At the offshore stations only very small amounts of floc settled material (<5%) are present in well sorted fine to medium sands (fig 3, 5).

3.1 METAL CONTAMINATION

The sediments of inlets such as Ship Harbour can also be described in terms of their metallic concentration. Trace metal analysis of the sediments (Loring et al., 1996) showed a high degree of metallic contamination in the inner bay. With the exception of Cadmium (Cd), 70-90% of the detrital mineral fraction is carried in the fine grained host minerals and occurs with other fine grained material (Loring, 1988). Cd is associated with the colloidal fraction in marine environments (Muller, 1995) and hence is deposited with the non-detrital material as part of flocs. Total surface area in both cases is the dominant factor controlling the deposition of these materials from suspension, and hence deposition will be governed by the aggregation dynamics of the system (Honeyman and Santschi, 1986). Analysis of the correlation matrices for the metals in Ship Harbour showed anomalously high concentrations of Zinc (Zn) and Cd were associated with iron monosulphides and/or pyrite formed *in situ* in the organic rich, fine-grained sediments. Elevated concentrations in the upper region of the harbour are the result of the predominance of floc settled material and the lack of resuspension of the sediments in this zone. Because the grain size of most sediments in Ship Harbour is <63 μm, the metal concentrations associated with the fine particulate fraction show poor correlation with this fraction. For example, Zn and Cd plotted against the log of the <63 μm fraction return r^2 values of 0.34, 0.14 respectively for the fitted regression lines (fig. 4). The expected relationship to the fine particle fraction can be seen by plotting concentration with the log of the <16 μm fraction (fig. 4). Although the <16 μm fraction shows a relationship to metal concentration (r^2= 0.89, 0.44 for Zn and Cd respectively) the fine particulate fraction deposited in the form of aggregates is not completely disassociated from the larger, less active individually deposited grains. For the sample from the inner harbour depicted in figure 2, 40% of the material <16 μm was deposited as single grains. Plotting metal concentration against % deposited as flocs changes the slope of the regression line (r^2= 0.88, 0.41 for Zn and Cd respectively) and increases the resolution of the curve for samples similar to that depicted in figure 3,5 where only a very small portion of the curve is floc settled. In this sample, the fine grained nature of this well sorted sediment results in a substantial amount of the single grain portion of the curve being included in the <16 μm fraction.

Plots of metal concentration vs. % floc from the ten other inlets studied showed similar patterns for most metals (Loring et al., 1996). In inlets where no sill was present, the relationship was complicated by the mixing of offshore water with that introduced from runoff and river input (Muller, 1996). Similar to Ship Harbour, however, the relationship with % floc removed the effect of single grain settling in the fine grained sediments.

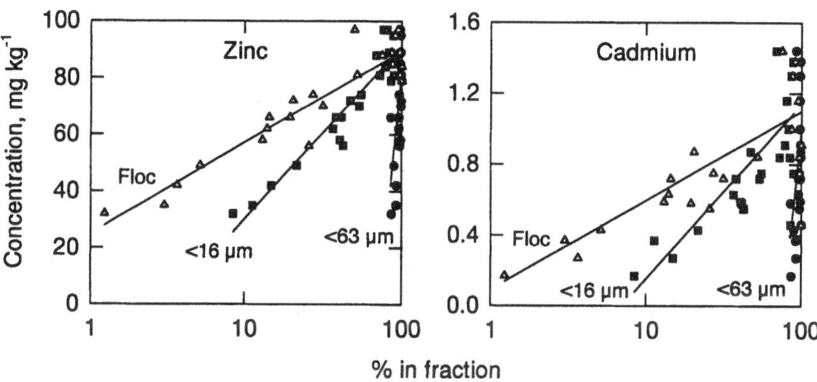

Fig. 4: Relationship between concentration of zinc and cadmium in sediments from Ship Harbour, N.S. and different methods of characterizing the fine particle fraction. X-axis represents the % of total sediment <63 µm (circles), <16 µm (squares) and deposited as flocs (triangles).

3.2 AGGREGATION DYNAMICS

Back Bay is part of the Letang Inlet system located in an area of intense salmon aquaculture in South-western New Brunswick. Connected to the Bay of Fundy, it is subject to tides on the order of 6-8 m. Sedimentation rates in the area are high with carbon fluxes ranging from <1 g C m^{-2} d^{-1} in exposed areas to >15 g C m^{-2} d^{-1} near salmon cages (Hargrave, 1994). Inorganic size spectra for the Letang Inlet sediments have a surprisingly constant shape, with a flat floc tail, slope 0.22 ± 0.05, and a prominent one round single grain peak. Samples from locations with high currents and those exposed to storm waves have additional higher round peaks indicating periodic reworking of the bottom sediment. Most cores collected showed a uniform size distribution down core. In Back Bay samples from a core collected at a control site remote from the salmon cages showed a distinct change in the relative amounts of floc and single grain settled material toward the top of the core (fig. 5a). A floc tail and one round single grain peak, typical of the Letang Inlet, can be seen in the samples below 9 cm depth. With decreasing depth in the core, the amount of material deposited as flocs increases and eventually overlaps the single grain settled portion of the curve. The intercept between the floc and one-round portions of the curve which defines the floc limit shows, a steady increase from 2-3 µm below 8 cm in the core to greater than 7 µm at the surface (fig. 5b). Floc limit at the bottom of the core conforms with other sediments from similar environments within the Letang Inlet.

The physical model for the Letang region described by Trites and Petrie (1995) indicated that material released within the Letang Inlet would accumulate in Back Bay. Hargrave et al. (1995) assigned the sediments in Back Bay a negative BEI based on the

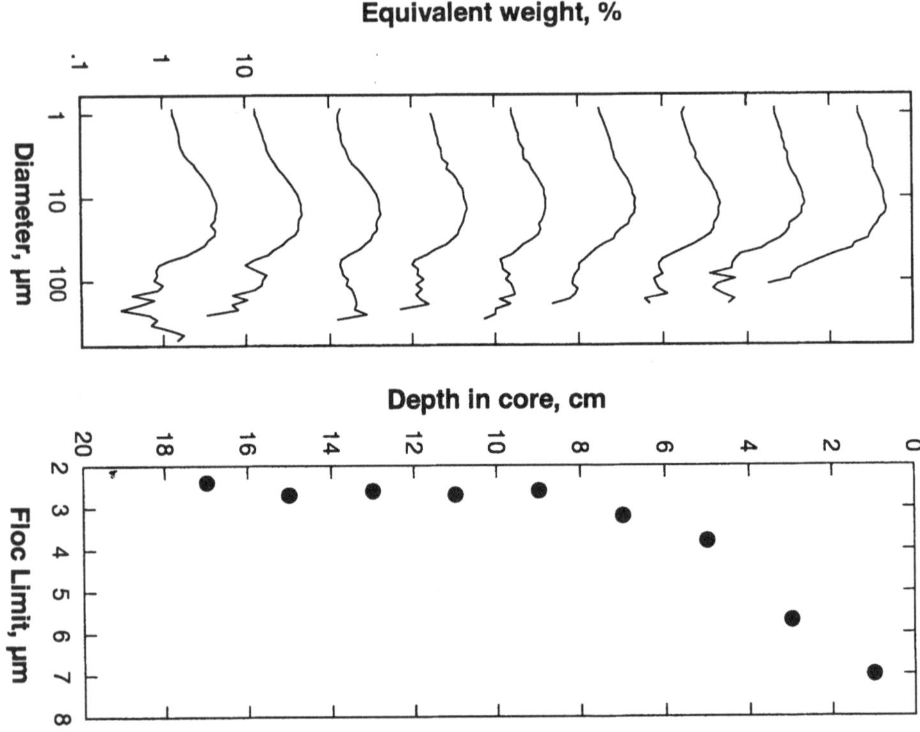

Fig. 5: Plots of disaggregated inorganic grain size distributions showing change in relative amount of floc settled material towards the top of a core collected in Back Bay N.B. (a) and the floc limit value determined from the size spectra (b). Log-log size spectra are offset by one decade on the Y axis to correspond with depth.

water content, Eh and organic carbon content in the core described above. With the exception of Back Bay, negative values were only associated with the sediments directly under salmon cages where high organic loading as a result of accumulation of faeces and excess food was found. New models of aggregation suggest that an increased concentration of particulates or an increase in collision efficiency, the likelihood that colliding particles adhere, will increase the maximum size of the particles included in the aggregates formed (P.S. Hill, pers com.). Particle adhesion and floc stability depend on the effect of polymer concentration on particle surfaces (Muhle, 1993) and maximum floc size increases with increased concentrations of flocculant (e.g. Tambo and Hozumi, 1979, Spicer and Pratsinis, 1996). An increased aggregation rate would result in more rapid settling of the material in suspension, and, when coupled with increased adhesion with the bottom sediment, would increase the accumulation of inorganic material in the bay. Muschenheim et al., (1989) showed that dissolved organic carbon is crucial in the formation of flocculated suspensions. The high carbon flux found in Letang Inlet suggests that both concentration and 'stickiness' have increased due to particulate loading and bacterial degradation associated with the intense aquaculture in the area. Although it is obvious that a change in the floc limit has occurred within the top 8 cm of the core collected in Back Bay, the exact reason for this change has not yet been determined.

4 Conclusion

The disaggregated inorganic grain size of sediments collected in both fresh and salt water environments contains information about deposition and aggregation dynamics of the system from which they were derived. From the slope of the source materials preserved in the floc tail to the energy represented by the maximum particle size, the disaggregated grain size distribution records the physical processes which formed the sediment. By examining the relative portion of the three components comprising a sediment: floc, one-round, and multi-round, the behaviour and fate of material transported and deposited in association with or as fine particles can be determined. Particle size analysis of bottom sediments in coastal embayments can provide a relatively inexpensive way of describing the sedimentary environment and to determine regions which could be impacted as a result of transport of contaminants associated with flocculated material in suspension. For surface active contaminants, determining the portion of the sediment deposited as aggregates provides higher precision than arbitrarily assigned size classes. Subtle changes in the aggregation dynamics in regions of fine particle deposition can be identified using the model of Kranck (1996 a,b). The method described above has application to any study concerning alteration of turbulence levels or changes in the concentration or composition of the sediment in suspension.

Acknowledgements

Our thanks to K. Saunders and A. Prior who carried out the particle size distribution analysis and R.T.T. Rantala who did the trace metal analysis. We appreciate the effort of our reviewers and thank them for their assistance. This work was supported by the toxic chemicals component of the Green Plan.

References:

Gibbs, R.J., Matthews, M.D. and Link, D.A.: 1971, *J. Sed. Petrol.* **41**, 7-18.
Hargrave, B.T.: 1994, *Can. Tech. Rep. Fish. Aquat. Sci.*, **1949**, 79-91.
Hargrave, B.T., Phillips, G.A., Doucette, L.I., White, M.J., Milligan, T.G., Wildish D.J. and Cranston, R.E.: 1995, *Can. Tech. Rep. Fish. Aquat. Sci.* **2062**, v + 159p.
Hill, P. S. and Nowell, A. R. M.: 1995, *J. Geophys. Res.*. **100(C11):** 22,749-22,763.
Honeyman, B.D. and Santschi, P.H.: 1989, *J. Mar. Res.* **47**, 951-992.
Jackson, G.A., 1995, *Deep Sea Res. II.* **42**, 1, 159-184.
Kepkay, P.E., Niven, S.E.H. and Milligan, T.G.: 1993, *Mar. Ecol. Prog. Ser.* **100**, 233-244.
Kranck, K., 1980, *Can. J. Earth Sci.* **17**, 1517-1526.
Kranck, K., 1993, *Arch. Hydrobiol./Suppl.* **75**, 299-309.
Kranck, K. and Milligan, T.: 1980, *Mar. Ecol. Prog. Ser.* **3**, 19-24.

Kranck, K. and Milligan, T.: 1983. *Mitteilungen aus dem Geologisch-Paleontologisch Institut der Universitat Hamburg*, **52**, Degens E.T. (ed).

Kranck, K. and Milligan, T.G.: 1985, *Geomarine Letters*, **5**, 61-66.

Kranck, K. and Milligan, T.G.: 1989. *Can. Tech. Rep. Hydrogr. Ocean Sci.*, **112**: iv, 61 pp.

Kranck, K., Smith, P.C. and Milligan, T.G.: 1996a, *Sedimentology*, **43**, 589-596.

Kranck, K., Smith, P.C. and Milligan, T.G.: 1996b, *Sedimentology*, **43**, 589-596.

Loring, D.H.: 1988 *Can. Bull. Fish. Aquat. Sci.*, **220**, 99-122.

Loring, D.H. and Nota, D.J.G.: 1968, *J. Fish. Res. Board Can.*, **182**: 147 pp

Loring, D.H. and Rantala, R.T.T.: 1992, *Earth Sci. Rev.*, **32**, 235-283.

Loring, D.H., Rantala, R.T.T. and Milligan, T.G.: 1996, *Can. Tech. Rep. Fish. Aquat. Sci.* **2111**: vii+268pp.

Milligan, T.G. and Kranck, K.: 1991, *Theory, Methods and Applications of Particle Size Analysis*. Cambridge University Press, New York, 109-118.

Muhle, K.: 1993, *Coagulation and Flocculation: theory and application*, Lib. Congress Surfactant Sci. Ser., **47**, 355-390.

Muller, F.L.L.: 1996, Mar. Chem., **52**, 245-268.

Muschenheim, D.K., Kepkay, P.E. and Kranck, K.: 1989, *Neth. J. of Sea Res.*, **23(3)**, 283-292.

Niven, S.E.H., Kepkay, P.E. and Boraie, A.: 1995, *Deep Sea Res. II*, **42**, 1, 257-273.

Schell, T. M: 1996, *Proximal to distal trends of the floccculation limit of fine-grained turbidites*. Dalhousie University MSc. thesis, 178pp.

Spicer, P.T. and Pratsinis, S.E.: 1996, *Water Res.*, **50**, 5, 1049-1056.

Rantala, R.T.T. and Loring, D.H.: 1989, *Anal. Chem.*, **220**, 263-267.

Tambo, N. and Hozumi, H.: 1979, *Water Res.*, **13**, 421-427.

Trites, R.W. and Petrie, L.: 1995, *Can Tech. Rep. Hydrogr. Ocean Sci.*, **163**: 53pp.

van Leussen, W.: 1988, *Physical Processes in Estuaries*, Springer Verlag, Berlin, New York, 404-426.

Zwolsman, J.J.G., van Eck, G.T.M. and Burger, G.: 1996, *Estuaine, Coastal Shelf Sci.*, **43**, 55-79.

THE FRESHWATER FLOC: A FUNCTIONAL RELATIONSHIP OF WATER AND ORGANIC AND INORGANIC FLOC CONSTITUENTS AFFECTING SUSPENDED SEDIMENT PROPERTIES

I.G. DROPPO[1,2], G.G. LEPPARD[1,3], D.T. FLANNIGAN[3] AND S.N. LISS[4]

[1]National Water Research Institute, PO Box 5050, Burlington, Ontario, Canada, L7R 4A6, [2]Department of Geography, University of Exeter, Exeter EX4 4RJ, UK, [3]Department of Biology, McMaster University, Hamilton, Ontario, Canada, L8S 4K1,[4]Department of Applied Chemical and Biological Sciences, Ryerson Polytechnic University, Toronto, Ontario, Canada, M5B 2K3

Abstract. Flocculated fine-grained sediment is a complex matrix of microbial communities and organic (detritus, cellular debris and extracellular polymers) and inorganic material. Suspended flocs within any aquatic system play a significant ecological role as they can regulate the overall water quality through their physical, chemical and/or biological activity. This paper investigates the complex structural matrix of riverine flocs over a large range of magnifications using correlative microscopic techniques. The significance of floc structural characteristics [(size, shape, porosity, density, inorganic composition, organic composition (bacteria and fibrils)] on the physical (eg. transport and settling), chemical (eg. adsorbing/transforming contaminants and nutrients), and biological (eg. biotransformation and habitat development) behaviour of a floc is investigated. Results suggest that it is the floc's internal structure that has a significant impact on controlling the above floc behaviours. This internal structure is complex and is often dominated by the existence of a three-dimensional matrix of fibrillar material secreted by the active microbial community within the floc. This matrix, in conjunction with the inorganic and bioorganic (active and inactive) constituents of a floc, provides an intricate pore structure that may result in water being an important bound component of a floc. These complex interactive structural and functional properties of a floc are considered to influence a floc's behaviour both physically in how it is transported or settled, chemically in how it adsorbs/transforms contaminants and nutrients, and biologically in how it develops a diverse microhabitat capable of modifying the structural, chemical and biological makeup of the floc.

Key Words. freshwater floc, flocculation, bacteria, fibrils, pores, inorganic particles, settling, size and structure.

1. Introduction

Flocculation of fine-grained sediment significantly alters the hydrodynamics of the constituent particles and will affect the transport of particle-bound contaminants. Discrete primary particles that typically would not settle in a given flow may settle when incorporated within a floc. Flocs can be viewed as individual microecosystems with autonomous and interactive physical, chemical and biological functions or behaviours operating within the floc matrix. These microecosystems also continuously interact with their surroundings, as the medium in which they are transported provides the flocs with building materials, energy, nutrients and chemicals for biological growth, chemical reactions and morphological development. While flocs can regulate their own environment they also are known to have the ability to regulate surrounding water quality by their physical, chemical and/or biological activity in both natural and engineered systems (Liss *et al.*, 1996; Decho, 1990; Leppard, 1985).

While the existence of flocculated particles within aquatic systems has been recognized for some time, the significance of these on the hydrodynamics of the sediment has only more recently been explored. The change in the sediment transport behaviour has been attributed to the changes in particle size, density and porosity brought about by flocculation (Li and Ganczarczyk 1986, 1987; Liss *et al.*, 1996; Droppo and Ongley, 1994). While these gross scale floc properties are known to affect the physical transport of sediment, there is still a fundamental lack of knowledge on the controlling factors of

Water, Air and Soil Pollution **99**: 43–54, 1997.

flocculation and how floc architecture influences its physical, chemical and/or biological behaviour. This lack of knowledge is attributed in part to the inadequacy of single observational techniques to observe the whole realm of the floc structure and processes from the fine (submicron) to the gross scale (micron).

Studies of floc structure generally require techniques such as conventional optical microscopy (COM) and scanning electron microscopy (SEM). For the observation, measurement and modelling of sediment structure and behaviour (ie. transport and settling), COM is often the technique of choice (Droppo and Ongley, 1992, 1994; Li and Ganczarczyk, 1986). This scale (>1 μm) is used because the outward behaviour of flocs can be observed at this resolution. Furthermore, from an engineering stand point, flocs greater than 1.0 μm will represent the majority of the mass for sediment transport and sedimentation studies.

While the outward floc structure and behaviour may be derived from COM, the internal architecture of the floc cannot be resolved at this scale. This internal structure is believed to influence the outward behaviour of the floc (Liss *et al.*, 1996; Heissenberger *et al.*, 1996). High resolution techniques such as transmission electron microscopy (TEM) and scanning confocal laser microscopy (SCLM) have revealed a consistent presence of a fibrillar extracellular polymeric material that bridges and binds primary particles (organic and inorganic) together within the floc matrix (Liss *et al.*, 1996). These fibrils, in conjunction with the active biological component of the floc matrix, may influence the chemical, physicochemical and biological processes within the floc itself and in the natural aquatic system as a whole (Leppard, 1995, 1993, 1985; Decho, 1990). The significance of the internal floc structure on the outward gross behaviour of the floc is, however, still poorly understood.

To evaluate the internal structure of freshwater flocs and its potential impact on floc behaviour, we have examined numerous flocs with correlative microscopy (Liss *et al.*, 1996; Leppard, 1992). Correlative microscopy is a means of compensating for the artifacts and resolution constraints imposed on results when only one microscopy and related techniques are used. Correlative microscopy is a strategy that uses different microscope types and their ancillary techniques to derive multiple levels of information from a given specimen. By observing flocculated material over a full range of magnifications (>1.0 mm to <1.0 μm), this paper provides a better understanding of how floc form in both its gross and fine scales can affect the function or behaviour of a freshwater floc.

2. Experimental

2.1. Study Sites

Sample sites were chosen to provide information on freshwater floc structure and not for the characterization or spatial comparison of specific sites and their floc structure. Suspended and bed sediments were collected from a variety of freshwater systems including both riverine (Nith River and 14-Mile Creek, Ontario, Canada) and lacustrine environments (Hamilton Harbour and Port Stanley Harbour, Lake Ontario, Ontario, Canada). These sites have been previously studied with their hydrological and geomorphological characteristics documented by Droppo and Jaskot, 1995 and Liss *et al.*, 1996 (Nith River), Ongley, 1974 (14-Mile Creek), Amos and Droppo, 1996 (Hamilton

Harbour), and Skafel and Krishnappan, 1995 (Port Stanley Harbour).

2.2. CORRELATIVE MICROSCOPY

2.2.1. *Conventional Optical Microscopy for Gross Structural Determination*

Samples were collected following the method of Droppo and Ongley (1992). This method allows for the non-destructive direct sampling, observation and measurement of flocculated material within a settling column (plankton chamber). The flocs that settle onto the plankton chamber's microscope slide were imaged using a Zeiss Axiovert 100 microscope interfaced with an image analysis system (Northern Exposure™ - Empix Imaging, Inc.). Northern Exposure™ allows for the determination of floc morphological characteristics and grain-size distributions down to a lower resolution of approximately 2 μm at 100 X magnification (10 X objective). Samples from the 14-Mile Creek were stabilized in agarose following the methods of Droppo *et al.* (1996) for subsequent use with the higher resolution microscopes.

2.2.2. *Scanning Confocal Laser Microscopy for Internal Floc Matrix Determination*

Samples stabilized in agarose were stained with a solution of wheat germ agglutinin conjugated to Texas Red™ (WGA-TR, Molecular Probes, Inc., USA). This molecular probe is specific to many of the polysaccharides present in the fibrillar material (those containing N-acetyl-D-glucosamine). The flocs were then washed three times with 0.1 M phosphate buffer (pH 7.0). The agarose disk was then divided into multiple sections (sections also used for TEM). One section was placed on a standard microscope slide and covered by a cover slip which was sealed around the edges with nail polish (prior to sealing, the area under the cover slip was flooded with buffer solution). The sample was imaged using a Zeiss Micro System LSM (Model LSM 10 BioMed). The SCLM was equipped with an argon laser (emission lines at 418 and 514 nm) and a 63x (1.4 na) objective with an electronic zoom of 30X. Image slices were collected at 1 μm vertical intervals and reassembled to view the whole floc volume specific to the stained component.

2.2.3. *Transmission Electron Microscopy for Ultrastructure Determination*

More detailed ultrastructural observations of sediment samples (discrete or in agarose) were made by preparing the samples for TEM following the four-fold multi-preparatory technique (Liss *et al.*, 1996). This technique allows for the enhanced observation of specific components of the floc such as cells and polymeric material. Ultrathin sections were imaged in transmission mode (TEM) at an accelerated voltage of 80 kV using a JEOL 1200 Ex II TEMSCAN scanning transmission electron microscope.

2.2.4. *Stereoscopic Microscopy for Settling Velocity, Density and Porosity Determination*

Settling experiments were performed similar to that of Li and Ganczarczyk (1987), however with the use of a Nikon SMZ-2T stereoscopic microscope and image analysis system. A drop of sediment collected with a wide mouth pipette (3.74 mm) from a gently homogenized sample bottle was introduced into an insulated 2.5 L capacity settling column. The settling column itself is 5 x 10 x 50 cm with the stereoscopic microscope focusing on a plane inside the column at a distance of 35 cm from the top. The long settling distance relative to the size of the particles is required to damp out any turbulence or settling irregularities resulting from the initial introduction of the sediment and to allow the sediment to reach terminal velocity prior to detection.

As the flocs pass through the field of view of the microscope they are video taped on a SVHS VCR through a CCD camera interface. Using Northern Exposure™ (Empix Imaging Inc.) image analysis software, the settling velocity was derived by digitally overlaying two video frames of a known time interval apart. In this way the same particle appears on the newly combined image twice and the distance of settling and the particle size can be digitized. The data are automatically sent to a spreadsheet where density and porosity can be mathematically estimated.

The density of a floc was estimated using Stokes' Law (Equation 1). As Stokes' Law is based on the settling of single impermeable spherical particles in a laminar region (Reynolds Number < 0.5), it is not ideal for the determination of floc density because of the heterogeneous structure and irregular shape of the floc (Hawley, 1982; Burban *et al.*, 1990). Nevertheless Stokes' Law or a modification thereof has often been used to determine the wet density of singular flocs (Tambo and Watanabe, 1979; Gibbs, 1985; Li and Ganczarczyk, 1987; Fennessy *et al.*, 1994), and does provide an indication of how aggregate settling velocity, density, and porosity are related to aggregate size.

Equation 1

$$\varpi = \frac{1}{18} D^2 (\rho_f - \rho_w) \frac{g}{\mu}$$

Where: ϖ = settling velocity, D = diameter, ρ_f = wet density of the floc, ρ_w = density of the water, and μ = dynamic viscosity (kinemetic viscosity x ρ_w).

As ϖ and D are derived from the image analysis system and ρ_w and μ are constants for a given water temperature, the wet density of the floc (ρ_f) can be calculated. Densities are expressed as excess density ($1-\rho_f$)).

Following the method of Li and Ganczarczyk (1987), the floc porosity can be expressed by a mass balance equation (Equation 2) assuming a typical density of dried silt and clay of 1.65 gm/cm^3.

Equation 2

$$\varepsilon = \frac{\rho_s - \rho_f}{\rho_s - \rho_w}$$

where: ε = floc porosity, and ρ_s = density of the dried solid material.

3. Results and Discussion

Lacustrine and riverine flocs have a complex composition and may be structurally and functionally dynamic due to the multiple interactive processes operating between the various physical, chemical and biological factors (Figure 1). The common thread for most flocculated natural aquatic sediments, however, is that they all have a living and active biological component in conjunction with inorganic and non-living biological particles.

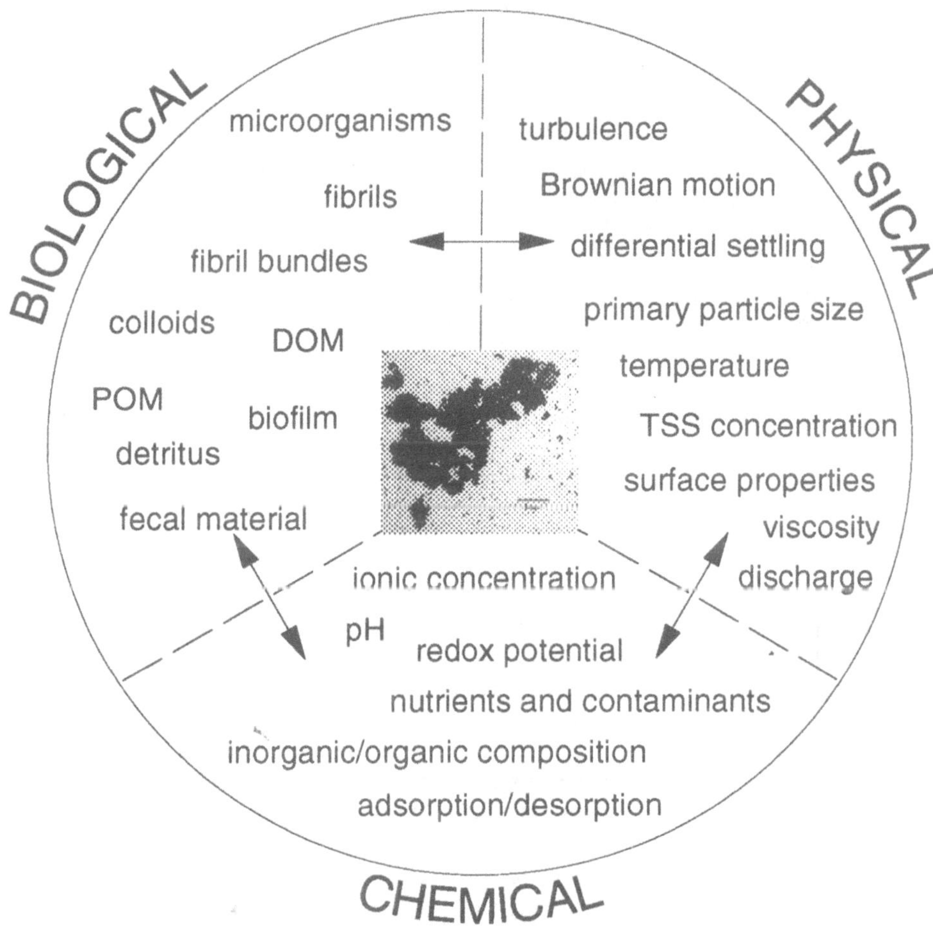

Fig. 1. Interactive physical, chemical and biological factors influencing floc development.

The gross structural characteristics of two riverine flocs (Nith River) can be observed in Figure 2a and b. Each appears heterogeneous in terms of spatial density across the flocs with the presence of macro-pores. All flocs observed were composed of organic and inorganic particles mixed together. Inorganic grains were silts and clays with generally no sand particles present. The majority of organic material can not be differentiated or speciated at this gross scale (only a difference in shading), however diatoms and other large living organisms were occasionally observed within the floc matrix.

Using COM in conjunction with image analysis, floc size distributions were determined (an example from the Nith River is given in Figure 3a). While the descriptive statistics may vary between samples and sites, the same trends were observed for all. Percent by number distributions were positively skewed while percent by volume distributions were negatively skewed. This suggests that while the majority of the mass is in the larger size classes the majority of the particles is in the smaller size classes. This has significant implications for the modelling of contaminant movement as the larger flocs will likely settle out much faster than the finer flocs that will travel with their associated contaminant much further within the system. An understanding of the floc distributions (by number and volume) is thus important for the development of sediment and contaminant transport models.

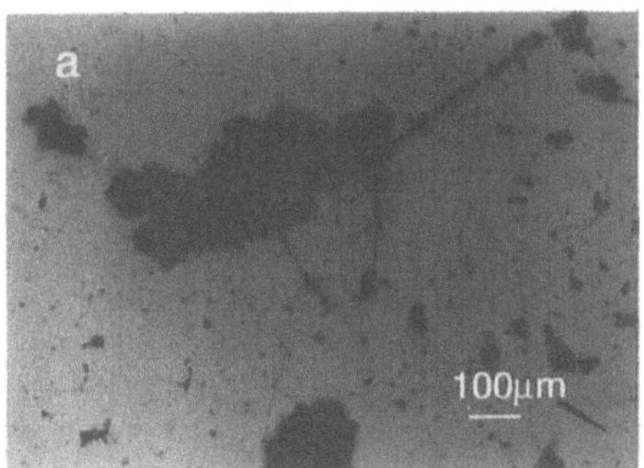

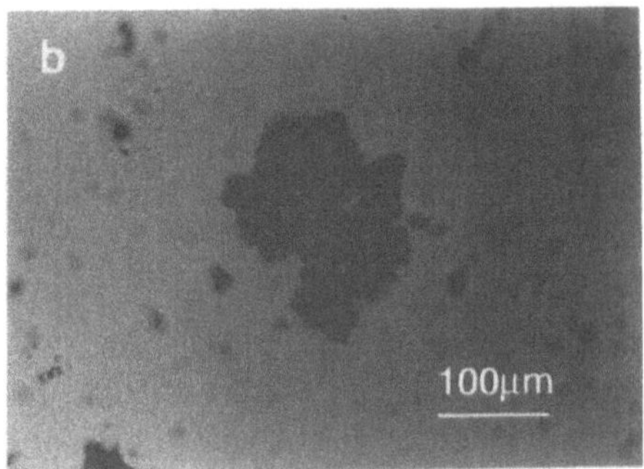

Fig. 2a and b. COM observations of flocculated sediment from the Nith River.

Many other floc morphological characteristics also are important when studying or modelling floc behaviour. Figures 3b to 3d illustrate three such additional parameters of longest dimension, width, and shape factor which are known to affect settling (Li and Ganczarczyk, 1987). The flocs observed generally were irregular in shape but tended towards a circular shape with median values ranging from 0.6 to 0.7 (1.0 = a perfect circle). Longest dimensions and widths yielded similar distributions to the equivalent spherical diameters with slight variations reflective of the non-spherical shape of flocs

Knowledge of floc settling velocity, density and porosity also is useful for the characterization of sediment environments and for the study and modelling of sediment and contaminant transport. Figure 4a illustrates the settling velocities of eroded aggregated particles from the bed sediments of Hamilton Harbour, Lake Ontario (Amos and Droppo, 1996). In the observed positive relation between floc size and settling velocity, a significant amount of variability ($R^2 = 0.47$) is evident but not surprising since the majority of the sediment is present in a flocculated form. Factors that can affect the settling rate of a floc and impose variability in the results are floc composition (proportion organic and inorganic), shape, porosity and water content. While settling velocity increases with floc size, Figure 3b demonstrates that the density decreases and the porosity increases with floc size. This relationship also has been observed by Li and Ganczarczyk (1987), Logan and Hunt (1988) and Tambo and Watanabe (1979). As a floc grows, its number of linkages to additional particles or flocs increases creating additional pores. This increased porosity leads to an increase in water content that forces the density of the floc towards the density of water and tends to reduce the settling velocity of the floc.

Logan and Hunt (1987, 1988) found that the development of porous flocs was advantageous to the microbial community as substrate removal was up to 60% greater than for dispersed bacteria. This was attributed to advective flow (calculated to be up to 100 $\mu m\ s^{-1}$) through the floc pores rather than to traditional molecular diffusion theory. Li and Ganczarczyk (1988) proposed that, if there is significant flow through the floc's pores, then the floc's settling velocity may be increased rather than reduced. Sherman (1953) on the other hand calculated that flow through flocs studied from Lake Mead was likely

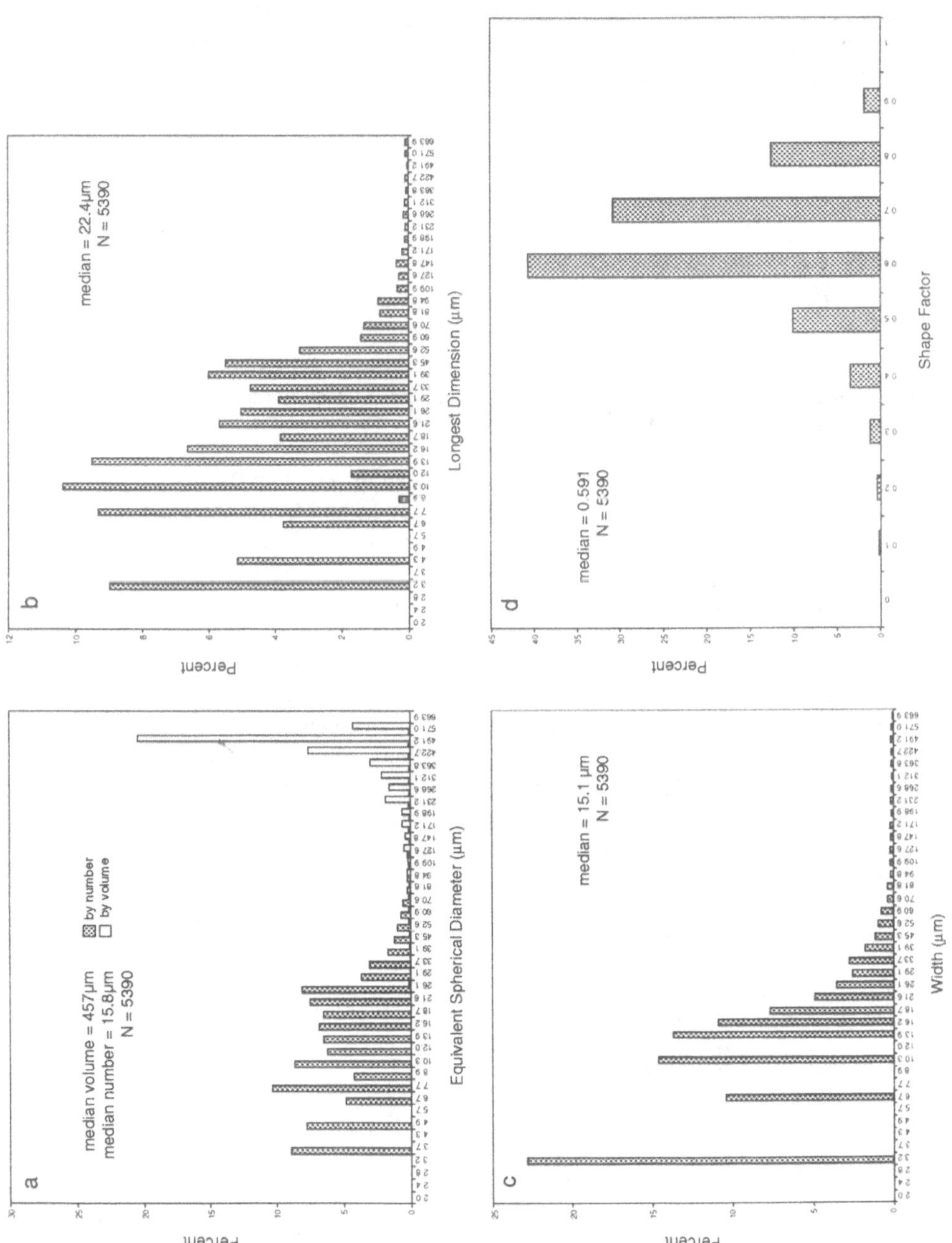

Fig. 3. Example of COM and image analysis floc morphological characterization for the 14-Mile Creek. (a) Equivalent spherical diameter distributions by volume and number, (b) particle longest dimension distribution, (c) particle width distribution, and (d) particle shape factor distribution.

negligible and that the bound water was thus an integral part of the floc structure. If water does move through a floc, it may act as a food stream or supply for the biological component and also will bring additional floc building material of organic and inorganic composition into the floc matrix. Otherwise a floc may assimilate nutrients and contaminants primarily through diffusional gradients.

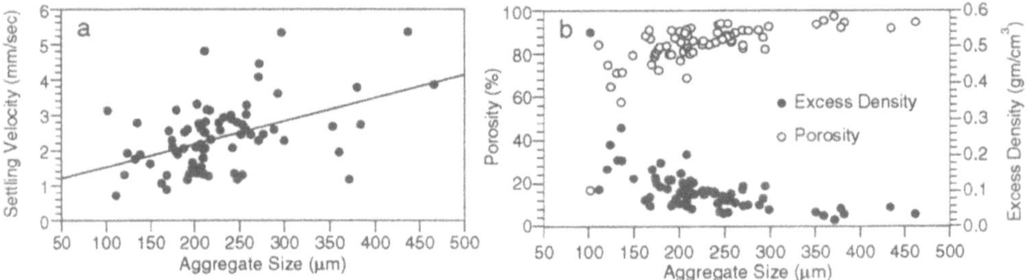

Fig. 4. The relationship of floc size to (a) settling velocity, and (b) porosity and excess density for eroded flocculated particles from Hamilton Harbour bed sediment (from Amos and Droppo, 1996).

The porosity of flocculated sediment usually is calculated from the density as inferred from measured floc settling velocity (as above and by Li and Ganczarczyk, 1987; Tambo and Watanabe, 1979). Porosity is expressed as a percentage which is used mathematically for the prediction or inference of floc hydrodynamics (Li and Ganczarczyk, 1987) or substrate uptake by natural or wastewater flocs through advective flow and diffusion gradients (Logan and Hunt, 1987 and 1988). Porosity is a complicated issue as the definition or identification of pores is highly dependent on the resolving power used for their observation. Pores observed with COM such as those in Figure 2a and b appear to be devoid of materials and appear to be significant open channels for the movement of water through the floc as it is transported or settled. While we cannot observe the same pores as those viewed with COM, TEM observations often reveal that many of these macro-pores are in fact filled with a complex network of cells and fibrils (Figure 5). As such, calculations of pore water movement and implications to floc settling and contaminant transport that do not consider the internal structure of the pores may be erroneous.

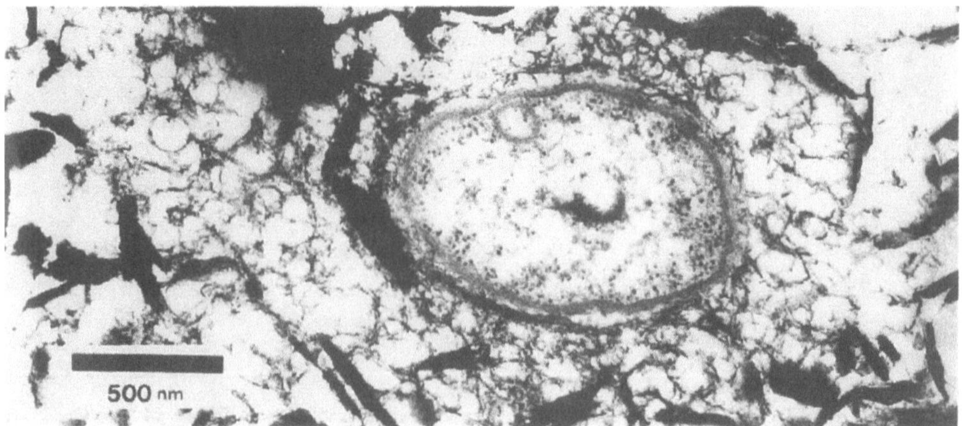

Fig. 5. TEM micrograph of eroded floc from Port Stanley showing extensive fibril network among clay particles (floc stabilized in glutaraldehyde/ruthenium red fixative).

The fibrils illustrated in Figure 5 are the dominant physical bridging mechanism between the organic and inorganic constituents of the natural flocs observed. These polymeric fibrils produced by the bacteria serve as a means of attachment and nutrient assimilation for the many cells that colonize the flocculated material (Costerton *et al.*, 1987). The flocs exhibit a pseudoplastic nature that is likely to be related to the flexible nature of the fibrillar material (Droppo and Ongley, 1992). We believe this polymeric material to be the dominant material for the development and stabilization of flocculated material (Droppo and Ongley, 1994). This does not exclude electrochemical flocculation completely (inorganic particle-particle contact is observed in Figure 5 - inorganic particles were composed of Fe, Al/Si or Si elements as derived from energy dispersive spectroscopy), but rather that it appears to be less significant than the biological flocculation within the natural systems we have examined. In no case was a floc observed to be aggregated solely by electrochemical flocculation. A significant quantity of fibrils was always present.

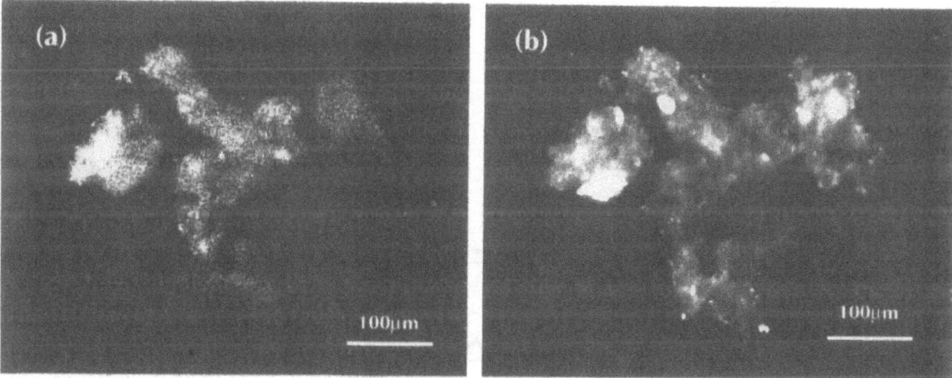

Fig. 6. SCLM images of polymeric material of a floc from 14-Mile Creek. (a) single optical section, (b) computerized reconstruction (stacking) of all optical sections.

SCLM helps to visualize in three-dimensions the significance of fibrillar material to floc building and behavioural (transport and settling) characteristics. The use of the molecular probe wheat germ agglutinin in conjunction with SCLM allows for the observation and estimation of the polymeric fibrillar material. Figure 6a illustrates a single optical section from a 14-Mile Creek floc, while Figure 6b represents the total floc volume as derived from the computerized stacking of optical sections obtained at 1 μm intervals. It is evident that the fibrillar material of natural flocs is a significant constituent component of the floc. In association with all other components this extracellular polymeric substances will influence floc size, floc density (increases water content), contaminant and nutrient adsorption and/or modification (large surface areas) and assist in maintaining diffusional and electrochemical gradients.

4. Conclusion

Flocculation, because it alters the hydrodynamic properties of particles in transport, significantly influences the fate and effect of sediment and associated contaminants. It was found that the complex structure and composition of a floc will have a significant influence over its physical, chemical and biological behaviour. An important observation was the apparent structural dominance of the fibrillar extracellular polymeric material within freshwater flocs. These fibrils are believed to be the dominant material for the development and stabilization of flocculated material. Each general component of a floc (organic and inorganic particles plus water and pores) is diverse in itself and can possess a specific function within a floc. The interactions of these constituents and their functional processes can result in the modification of a floc's behaviour, physically in how it is transported and settled, chemically in how it adsorbs and transforms contaminants and nutrients, and biologically in how it develops a diverse microhabitat capable of modifying the structural, chemical and biological makeup of a floc. These interactions and functions are summarized in Figure 7.

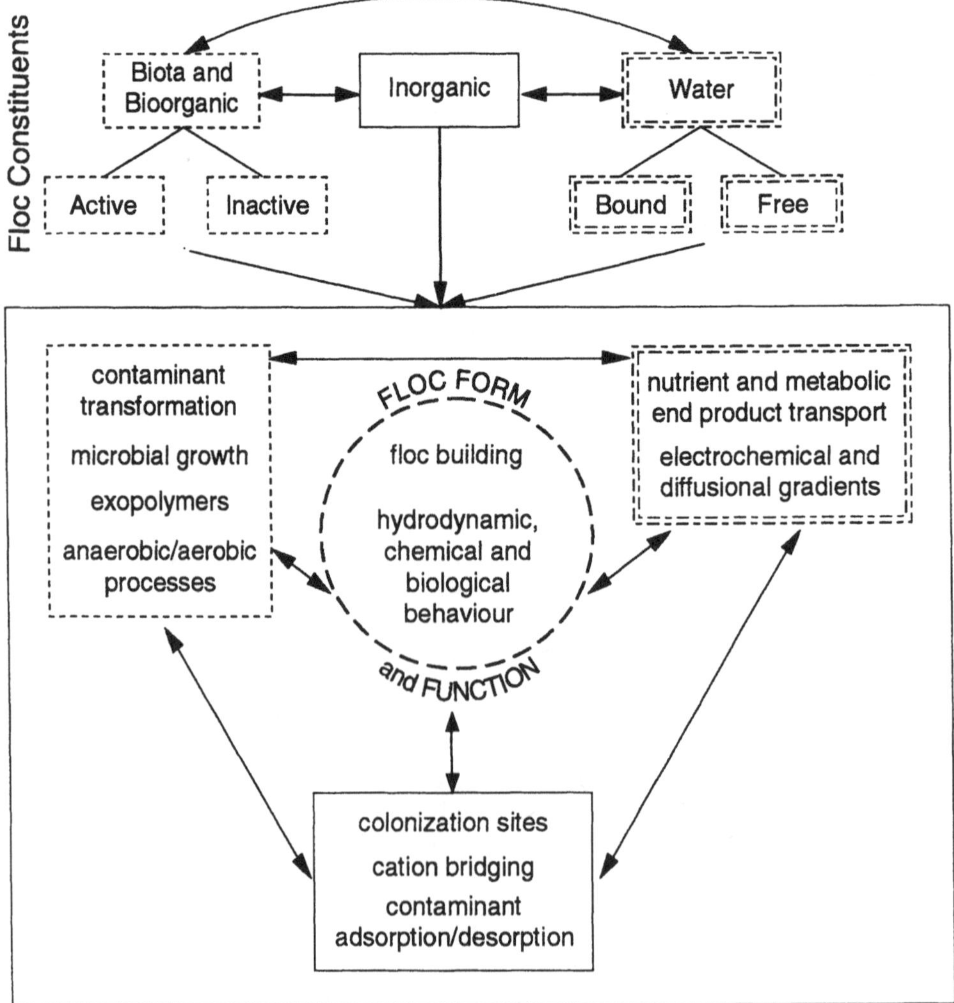

Fig. 7. Conceptual model of floc form and function.

Acknowledgements

This work was supported by Environment Canada and by a National Science and Engineering Research Council of Canada Strategic Grant, STR0167324, awarded to S.N.L., G.G.L. and I.G.D. We thank M.M. West and C. Jaskot for their analytical assistance, Dr. M. Skafel for the provision of Port Stanley sediment and Dr. K.N. Irvine and three anonymous reviewers for their critical review of the manuscript.

References

Amos C.L. and Droppo, I.G.: 1996, 'The stability of remediated lake bed sediment, Hamilton Harbour, Lake Ontario, Canada'. Geological Survey of Canada Open File Report # 2276.

Burban, P.-Y, Xu, Y,-J, McNeil, J. and Lick, W.: 1990, J. Geophys. Res.**95**(C10), 18213-18220.

Costerton, J.W., Cheng, K.-L., Geesey, G.G.: Ladd, T., Nickel,J.C., Dasgupta, M., and Marrie, T.J.: 1987, Annu. Rev. Microbiol. **41**, 435-464.

Decho, A.W.: 1990, Oceanogr. Mar. Biol. Annu. Rev. **28**, 73-153.

Droppo, I.G. and Ongley, E.D.: 1992, Wat. Res.**26**, 65-72.

Droppo, I.G. and Ongley, E.D.: 1994.Wat. Res. 28(8): 1799-1809.

Droppo, I.G. and Jaskot, C.: 1995, Environ. Sci. Technol**29**, 161-170.

Droppo, I.G., Flannigan, D.T., Leppard, G.G., Jaskot, C. and Liss, S.N.: 1996, Appl. Environ. Microbiol. **62**, 3508-3515.

Fennessy, M.J., Dyer, K.R. and Huntley, D.A.: 1994, Mar. Geol.**117**, 107-117.

Gibbs, R.J.: 1985, J. Geophys. Res.**90**(C2), 3249-3251.

Hawley, N.: 1982, J. Geophys. Res.**87**(C12), 9489-9498.

Heissenberger, A., Leppard, G.G. andHerndl, G.J.: 1996, Mar. Ecol. Prog. Ser.**135**, 299-308.

Leppard, G.G.: 1985, Water Pollut. Res. J. Can.**20**(2), 100-110.

Leppard, G.G.: 1992, Analyst**117**, 595-603.

Leppard, G.G.: 1993, 'Organic flocs in surface waters: their native state and aggregation behavior in relation to contaminant dispersion', in S.S. Rao (ed.), *Particulate Matter and Aquatic Contaminants*. Lewis Publ., Chelsea, MI, pp. 169-195.

Leppard, G.G.: 1995, Sci. Total Environ.**165**, 103-131.

Li, D.-H and Ganczarczyk, J.: 1986, Water Poll. Res. J. Can**21**, 130-149.

Li, D.-H and Ganczarczyk, J.: 1987, Wat. Res.**21**, 257-262.

Li, D.-H and Ganczarczyk, J.: 1988, Wat. Res.**22**, 789-792.

Liss, S.N., Droppo. I.D., Flannigan, D.T. and Leppard, G.G.: 1996, Environ. Sci. Technol**30**, 680-686.

Logan, B.E. and Hunt, J.R.: 1987, Limnol. Oceanogr**32**, 1034-1048.

Logan, B.E. and Hunt, J.R.: 1988, Biotechnol. Bioeng.**31**, 91-101.

Ongley, E.D.: 1974, 'Hydrophysical characteristics of Great Lakes tributary drainage, Canada'. Pollution and Land Use Activities Reference Group (PLUARG), International Joint Commission, Windsor, Ontario, Canada.

Sherman, I.: 1953, Trans. AGU.,**34**, 394-406.

Skafel, M.G. and Krishnappan, B.G. 1995. 'Deposition of fine-grained sediment under wave action', in *Proceedings of the 1995 Canadian Coastal Conference*Dartmouth, Nova Scotia, Canada, pp. 767.

Tambo, N. and Watanabe, Y.: 1979, Wat. Res.**13**, 409-419.

FLOCCULATION OF PARTICLES BY FLUID SHEAR IN BUFFERED SUSPENSIONS

C. H. TSAI and J. Q. HU
Department of Oceanography, National Taiwan Ocean University, Keelung, Taiwan
(e-mail: chtsai@ntou66.ntou.edu.tw)

Abstract. Flocculation experiments were carried out by using a Couette type of viscometer for applying uniform fluid shears to freshwater suspensions of estuarine fine-grained particles. The suspensions were buffered by a maleic acid/ammonium hydroxide system. The objectives of the study are to investigate the relationship between the steady-state sizes of flocculated particles and applied shear at various pH values. The results showed that the rate of particle flocculation increased with the addition of the buffer. With a particle concentration of 100 mg/l and pH of 6.0, the floc size decreased with the increasing shear. While with a pH of 7.5, the floc size increased with the shear until it reached a maximum value at the shear of 400 s^{-1}, and then the size decreased as the shear increased further to 500 s^{-1}. The maximum size shear was found at 200 s^{-1} for pH 7.9. By increasing the pH value to 9.0, there was no particle growth with any applied shear. The electrophoretic mobility of the particles was measured. It was found that for any constant shear the floc size is inversely proportional to the absolute value of the mobility. That is, the more repulsive the force between particles, the smaller the floc size. The effect of pH on floc size was not uniform within the pH range tested. The results demonstrated that the higher the shear the more significant the changes of floc size due to the pH effect.

Key words: Flocs, median size, fine-sediments, pH, shear, size spreading, probability of cohesion, electrophoretic mobility.

1. Introduction

Flocculation of particles is a dynamic process in which two competing actions occur simultaneously. These two are particle aggregation and disaggregation; the former increases the size of the floc while the latter decreases the size. These two actions are mostly direct results of particle collision. Fluid shear is an important factor in causing particle collision, especially for the water environment in the river, coastal water and water near the seabed. In a shear field, individual suspended particles are carried by the water at different speed and thus particles can collide with each other. The most apparent consequence of particle flocculation is the change of floc sizes, which is also the topic of most related studies. Many flocculation experiments have shown that, and it has also been widely accepted, that the steady-state floc size is inversely related to the fluid shear (Boadway, 1978; Tsai et al., 1987; Burban, 1989 and Williams et al., 1992). That is, the larger the shear the smaller the flocs, and its rationale is simple. Flocs are fragile and easy to break; large fluid shear tends to break flocs. However, Dyer (1989) argued that at very low shear, where particle collision does not occur very often, increasing the shear should increase the collision frequency, since which is proportional to the fluid shear (Ives, 1978). Hence, he proposed, without experimental evidence, that under low shear conditions an increase in fluid shear should increase the floc size. He also suggested that the sizes of flocs increase with the shear up to a point. Larger than this shear, further

Water, Air and Soil Pollution **99**: 55-62, 1997.

increase of shear reduces the floc size, since the effect of floc breakage becomes more important.

Tsai and Hwang (1995) carried out flocculation experiments using sediments from Tanshui river estuary, Taiwan. They showed that the steady-state floc size was proportional to the fluid shear, similar to the low shear case suggested by Dyer (1989). In that study, they also obtained the initial probability of cohesion by measuring the initial time rate of particle number reduction (Smoluchowski, 1917). This number can be seen as an indication of particle stickiness for a given sediment. It was found that the initial probability of this sediment was 0.005. Much smaller than the probability of 0.15 calculated by Lick and Lick (1988) for Lake Erie sediment used in Tsai et al. (1987), in which the size was proportional to the shear. Tsai and Hwang (1995) postulated that since the Tanshui sediment was much more difficult to flocculate, increasing fluid shear should promote the floc growth. Hence, the floc size increased as the shear increased.

In this study, flocculation experiments similar to that of Tsai and Hwang (1995) were carried out with the same sediment from Tanshui estuary, except that sediment particles were in buffered suspensions. In this test, the particle suspension was added with maleic acid/Ammonium hydroxide buffer system. The reasons for choosing this buffer system were twofold. It was found, in the preliminary testing, that this system increased the rate of flocculation. By increasing the flocculation rate, one can investigate the floc size and fluid shear relationship with the raised flocculation rate, and this is the first objective of this study.

Secondly, with this buffer system, it is possible to study the flocculation behaviors (or size and shear relationship) of the sediment at different pH. Among all factors that can influence flocculation behaviors, pH is one of the most important factors. It has been generally recognized that the negative charge of particles will increase when the pH is raised, making the particles more repulsive and thus more difficult to flocculate (Gerritsen and Bradley, 1987). The second objective of this study is to investigate the floc size and shear relationship for particles in the present buffer system at various pH values. In the following, the experimental method is first described, followed by the results of the experiments. A summary and concluding remarks is given lastly.

2. Experimental

Pai et al. (1990) prepared a maleic acid/ammonium hydroxide buffer system, which was used for preconcentrating trace metals from seawater. The buffer system used in this study was essentially prepared the same way as was suggested in that article. A stock buffer solution was made by first adding 58.01 gm of maleic acid into ca. 600 ml of DI water. The pH of the solution was adjusted to about 6.5 by adding concentrated ammonium hydroxide and the solution was then diluted to ca. 1000 ml. Four hundred ml of this stock buffer solution was added with various amounts of concentrated/or 2N hydrochloric acid or concentrated/or 2N ammonium hydroxide (as shown in the Table 1. in Pai et al., 1990) for adjusting the pH to pre-designated value. The working buffer was then prepared by diluting this solution to 500 ml with filtered freshwater, together with minor pH adjustments. In this study, the pre-designated pH values for working buffers were: 6.0, 7.5, 7.9 and 9.0.

The sediments, the same as that used in Tsai and Hwang (1995) consisted of fine silt and clay, and the predominant clay minerals were chlorite, illite, montmorillonite and kaolinite. The procedures for preparing sediment suspension were identical to that reported therein. The sediment suspensions were prepared by mixing the particles less than 20 μm with 0.4-μm-filtered freshwater, and its concentration was adjusted to the predetermined values. The suspension had a log-normal size distribution with a median diameter of 5 to 7 μm. This suspension was then mixed with the working buffer with predetermined pH at 5% (v/v) ratio. Since the amount of the working buffer was small, the particle concentration was not re-adjusted. The size distribution did not vary with the pH and particle concentration appreciably.

The procedures of flocculation experiments were reported in Tsai and Hwang (1995). Without going into details, it is sufficient to mention the following. The flocculator is a Couette type flocculator (see Fig. 1 in Tsai and Hwang, 1995) with a length-to-gap ratio of 125. The inner radius of the annular chamber is 2.3 cm and the outer radius 2.5 cm. By rotating the outer cylinder, a near uniform shear is created in the annular chamber. The fluid in the chamber becomes unstable at the shear of 1034 s^{-1} (van Duuren, 1968). Before each test, the viscometer was thoroughly cleaned and immersed in DI water and ultrasonic bath for one day. Then the viscometer was filled with the particle suspension, and a fluid shear was applied. During the course of the experiment, the suspension in the chamber was withdrawn at times for floc size measurement. Although the initial particle size distribution had a log-normal shape, the size distribution deviated from that shape as the particle flocculated and larger flocs appeared. Then the floc size distribution became near log-normal again at larger size range than the initial one and the median diameter kept steady. The tests were kept running for at least 8 more hours, so that one can be certain that the steady-state was reached. In this test, the fluid shear varied between 100 s^{-1} and 500 s^{-1}, and the particle concentrations were 100 mg/l and 400 mg/l.

The floc size distribution was measured by a Galai CIS-1 laser particle sizer, in which a fine, rotating laser beam scans a sample chamber(Aharonson et al., 1986). Time length of beam blockage by the scanned particles is used to determine the size of the particles. This instrument is well suited for measuring the floc size, since the laser beam does not break up flocs. Particularly, it has also been found that this instrument performed well in measuring natural flocs sampled in the ocean (Tsai, 1996). The electrophoretic mobility (EM), a measure of repulsive force (Gerritsen and Bradley, 1987), of the particles was measured by a Zeta meter 3.0+ (Zeta Meter, Inc.). The procedures for measuring EM is similar to that reported in Tsai (1995). Table I shows the mean and standard deviation of EM, equivalent mean Zeta potential (ZP) and conductivity of various buffered and

TABLE I

Electrophoretic mobility (EM), Zeta potential and conductivity of the particle suspension. (EM: the mean of over 100 tracked particles with unit of μm s^{-1} (V $cm^{-1})^{-1}$, V: voltage, SD: the standard deviation, ZP: the Zeta potential corresponding to the mean EM value with the unit mV, and Conductivity: with unit of μS cm^{-1}).

	pH				
	6.0	7.5*	7.5	7.9	9.0
EM	-1.54	-1.57	-1.21	-1.27	-1.81
SD of EM	0.27	0.28	0.22	0.22	0.28
ZP	-19.5	-19.9	-15.3	-16.1	-22.9
Conductivity	6740	190	6550	6710	7870

unbuffered particle suspension. The negative values in EM and ZP indicate particles carry negative charges, and the lower the absolute value the lower the charges. Results showed that the EM values varied in a narrow range between -1.21 and -1.81 $\mu m\ s^{-1}$ (V $cm^{-1})^{-1}$, where V is voltage. Particles were least charged at pH 7.5, and highest charged at pH 9.0, while at pH 6.0 the EM value was about the same as the value measured in freshwater without the buffer. The table also shows that the conductivity of all buffered suspensions was kept at around 7000 $\mu S\ cm^{-1}$.

3. Results

3.1 Temporal history of floc growth

The temporal variation of the median diameter and the size distribution at steady-state for all tests all exhibited similar trend to that reported in earlier similar studies. The flocs grew rapidly at first (in the first hour), then more slowly. At the final stage of the growth, the floc size settled to a steady-state value with slight up and down fluctuation. The floc size distributions at steady state were close to log-normal shapes.

The times to reach steady state, listed in Table II, are between 3 and 12 hours, depending on the pH and particle concentration. For the same shear and concentration, the test with a pH of 6.0 took the least time, and the time increased with pH value. All tests with pH 9.0 produced no floc growth. From the table, one can see that for the same shear and pH tests with higher particle concentration reach steady states faster. Also included in the table are the steady-state times for tests without the addition of buffer system (Tsai and Hwang, 1995). As can be seen, by adding the buffer system, flocculation completed 3 to 5 times faster than those without the buffer.

3.2 Steady-state floc size and fluid shear

The steady-state floc sizes obtained for tests with various pH values and concentration as a function of the applied shear are shown in Fig. 1. Notably, for most cases, with the same pH and particle concentration, the floc size increased with the fluid shear up to a certain shear (called maximum size shear hereafter). For the tests with particle concentration of 100 mg/l, there was no maximum size shear for buffered suspension with pH 6.0. In this series, the floc size decreased with shear slightly between 100 s^{-1} and

TABLE II

Times (hr) to reach steady state for tests with particle concentration of 100 mg/l and 400 mg/l(in parenthesis). (*: for tests without buffer; +: data from Tsai and Hwang (1995); **: no significant growth after 20 hours.)

	Shear (s^{-1})			
pH	100	200	400	500
6.0	5	7	7	5
7.5*	$19^{+}(15^{+})$	(14^{+})	$25(16^{+})$	
7.5	7(3)	6(6)	8(5)	7(7)
7.9	12	11	10	9
9.0	(**)		(**)	**

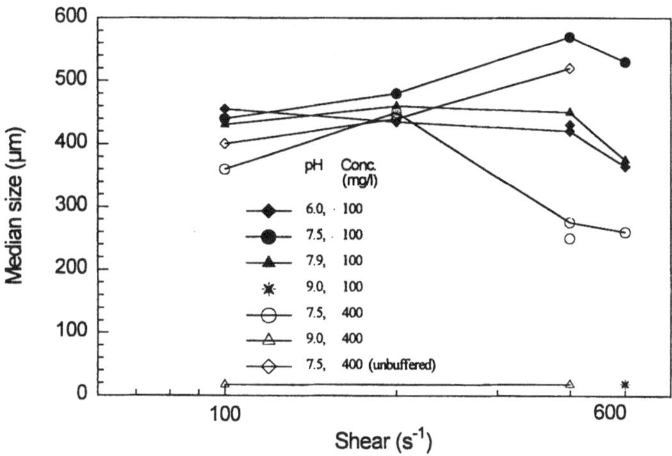

Fig. 1 Steady-state floc size as a function of fluid shear at various pHs and particle concentrations. Sizes for unbuffered suspension are from Tsai and Hwang (1995).

400 s^{-1}, from 455 μm to 420 μm, and then dropped to 365 μm as the shear increases to 500 s^{-1}. With the same particle concentration, the size for pH 7.5 increased from 440 μm to 570 μm as the shear varied from 100 s^{-1} to 400 s^{-1}. The maximum size shear occurred at 400 s^{-1} for this case, and the floc decreased to 530 μm for the shear of 500 s^{-1}. With the pH of 7.9 and concentration 100 mg/l, the size for the shear at 100 s^{-1} was 430 μm, not much different from that of pH 7.5. The size increased to 460 μm at the maximum size shear of 200 s^{-1}, then decreased a little to 450 μm at the shear of 300 s^{-1} and to 375 μm at the shear of 400 s^{-1}. The tests with pH 9.0, concentration 100 mg/l and a shear of 500 s^{-1} did not produce significant floc growth. The floc size was only 18 μm.

As the particle concentration was increased to 400 mg/l, for the tests with pH 7.5, the floc size variation with the applied shear was more significant. With the shear of 100 s^{-1},

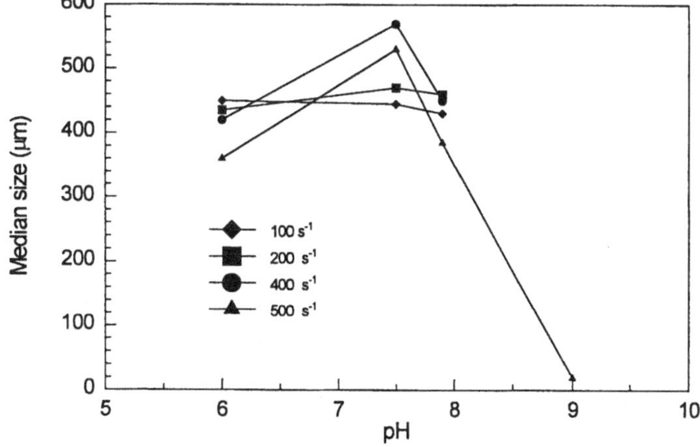

Fig. 2 Steady-state median floc size as a function of pH for the tests with particle concentration of 100 mg/l.

the floc size was only 360 μm. The maximum size shear was 200 s^{-1}, which produced a median size of 450 μm. The floc size reduced to 275 μm and 265 μm as the shear increased to 400 s^{-1} and 500 s^{-1}, respectively. Again, tests with the concentration of 400 mg/l, pH 9.0 and shears at 100 s^{-1} and 400 s^{-1} did not have any floc growth.

Fig. 1 also includes the floc size obtained without the buffer system (Tsai and Hwang, 1995), and with a pH 7.5. One can see that without the buffer the floc size with a shear of 100 s^{-1} was 400 μm, and that it increased to 440 μm and 520 μm as the shear increased to 200 s^{-1} and 400 s^{-1}, respectively. Comparison of this results with that obtained by tests having the buffer system, showed that for the shears of 100 s^{-1} and 200 s^{-1}, the floc size did not change greatly because of the addition of the buffer. However, at the shear of 500 s^{-1}, by adding the buffer system, the floc size, instead of keeping on increasing, reduced tremendously.

3.3 Floc size and pH

From Fig. 2, one can see that the effect of pH on steady-state floc size was shear dependent. For small shears, 100 s^{-1} and 200 s^{-1} the floc size did not change appreciably with pH. However, as the shear was raised to 400 s^{-1} and 500 s^{-1}, the size increased as the pH changes from 6 to 7.5. At this pH, the floc size reached a maximum. Then, the size decreased as the pH increased to 7.9 and 9.0. It is to note that for the same applied shear the sizes obtained for pH 6 were only slightly smaller than that for pH 7.9. Moreover, the sizes obtained with a shear of 400 s^{-1} were larger than that with the shear of 500 s^{-1} within the pH range studied. This is because 500 s^{-1} was larger than the maximum size shear for most cases.

3.4 Spreading of the floc size distribution

In order to determine the spreading of the floc size distribution relative to its median size, a coefficient of spreading (CS) is defined as:

$$CS=SD/Dm \tag{1}$$

where SD and Dm are the standard deviation and median diameter of the floc size distribution, respective. Fig. 3 plots the spreading coefficient versus the applied shear. It is to emphasize that if the coefficient of deviation was used instead of CS, the results do not differ appreciably from that shown in Fig. 3. As can be seen, for each pH and particle concentration, CS varied with the shear. For the tests with concentration of 100 mg/l and pH 6.0, CS increased slightly as shear increased from 100 s^{-1} to 400 s^{-1}, hovering around 50%. Then the CS jumped up to 60 % for the shear of 500 s^{-1}. The tests with pHs 7.5 and 7.9 showed that CS decreased somewhat as shear increased from 100 s^{-1} to 400 s^{-1}. Then it also jumped up for the 500 s^{-1} shear. For the tests with pH 7.5 and concentration 400 mg/l, the curve clearly was "v" shaped, with the minimum CS value of 45 % occurred at the shear of 200 s^{-1}. Without the buffer, the coefficient of spreading decreased with the increase of the fluid shear.

By comparing Figs. 3 and 1, one can see that the CS and shear relationship is exactly in reverse of that between the size and shear. It clearly indicated that when the flocs were bigger, their spreading coefficient was smaller. This means that as the shear was

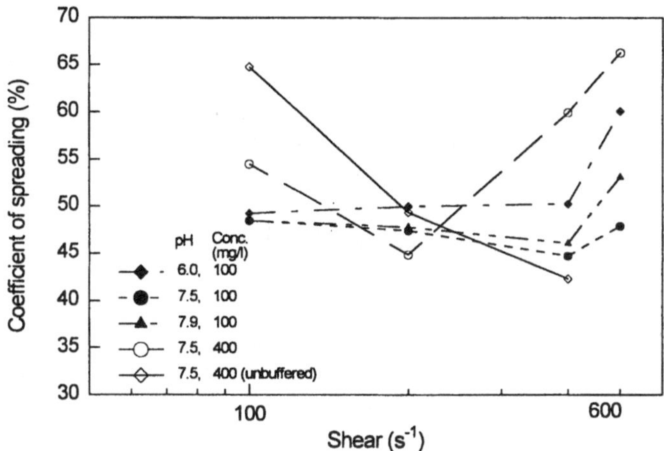

Fig. 3 Spreading of steady-state floc size distribution as a function of the fluid shear at various pHs and particle concentrations.

increased from a low value, the flocs grew bigger with the shear, and the floc size distribution became narrower, and more centralized, showing the effect of increasing degree of flocculation. At the maximum size shear, the flocs were at their largest size and the size distribution was most centralized. Then when the shear was higher than the maximum size shear, more flocs were broken up than been formed. Because of this the floc size distribution spread over a wider size range, and the spreading increased.

4. Summary and Concluding Remarks

One effect of adding this buffer system was the increase of the rate of flocculation. However, the most significant change in flocculation behaviors due to the buffer system was its changes in the relationship between the steady-state floc size and the applied shear. Without the buffer system, Tsai and Hwang (1995) found, using the same sediments and within their ranges of variables tested, that the steady-state size increased with the fluid shear. With the added buffer system, the particles were made easier to stick together, and a monotonous increasing size and shear relationship no longer existed. In most cases the floc size increased as the shear increased, but there also existed a maximum size shear. The floc size decreased as the shear was further increased. This result seems to suggest that, for the same pH, with the buffer, the size-shear curve was shifted left, toward the smaller shear direction. The results also showed that the shape of these size-shear curve and the location of the maximum size shear varied with the particle concentration and pH. Since pH affects the electrochemical properties of the particles, any other factors that change these properties, such as organic matter and pollutants, may change the shape of this curve and the value of maximum size shear as well.

This study found that particles in suspensions with different pH have different electrophoretic mobility (or repulsive forces), albeit the variation of EM in the range of pH tested was not large. Overall, the floc size was inversely proportional to the magnitude of the repulsive force between particles, when the shear was held constant.

That is, the smaller the absolute value of electrophoretic mobility value, the larger the floc size. Moreover, size change due to pH was not uniform over the range of shear tested. The pH had much less influence on the floc size when the fluid shear was below 200 s^{-1} than when the shear was larger. With a pH 9.0, the particles were unable to flocculate, even when a shear of 500 s^{-1} was applied.

It is to note that the EM value for the unbuffered suspension was very close to that of suspension buffered at pH 6.0, although the former had a much less conductivity (Table 1). However, the flocculation behaviors between these suspensions were significantly different. The effect of repulsive force between particles can not fully explain this difference. It seems that the buffer system itself has the effect of speeding up the flocculation rate, possibly through the bridging effect.

Although, the floc size variation with pH for any constant shear can be explained by the changes of EM value, the changes of maximum size shear due to pH and particle concentration can not be easily explained. However, there seems to be a correlation between pH and maximum size shear. Which is the less repulsive between particles the higher the maximum size shear. However, the reasons for this are still unclear. To understand them, one may need to carefully study the probabilities of cohesion and disaggregation under various pHs and concentrations, since the balancing effect of these two probabilities determines the floc size at steady state.

Investigation on the floc size distribution at steady states found that the spreading of the distribution was also a function of the pH, fluid shear and particle concentration. Most importantly, the curve describing the relationship between the size spreading and the fluid shear was exactly opposite to that between the size and shear. That is, the larger the flocs the smaller the size spreading, indicating a higher degree of aggregation.

Acknowledgments

This study was partially funded by the National Science Council (NSC81-0421-E-019-02-Z). Their support is much appreciated. We also like to thank Professor Su-Cheng Pai for his advises on the buffer system.

References

Aharonson, E.F., Karasikov, N., Roitberg, M. and Shamir, J.: 1986, J. Aerosol Sci., **17**, 530-536.
Boadway, J.D.: 1978, J. Environ. Engr. Div., **104**, 901-915.
Burban, P.Y. Lick, W. and Lick, J.: 1989, J. Geophy. Res., **94**(C6), 8323-8330.
Dyer, K.R.: 1989, J. Geophy. Res., **94**(C10), 14327-14339.
Gerritsen, J. and Bradley, S.W.: 1987, Limnol. Oceanogr., **32**(5), 1049-1058.
Ives, K. J.: 1978, In K. J. Ives (ed), The Scientific Basis of flocculation, Sijthoff and Noordhoff International Publishers, B.V., Alphen aan den Rijn, The Netherlands, 37- 61.
Lick, W. and Lick, J.: 1988, J. Great Lakes Res., **14**, 514-523.
Pai, S.C., Chen, T.C., Wong, G.T.F. and Hung, C.C.: 1990, Analyt. Chem., **62**(7), 774-777.
Smoluchowski, M.: 1917, Zeitschrift fur Physikalische Chemi, **92**, 129-168.
Tsai, C. H. Iacobellis, S. and Lick W.: 1987, J. Great Lakes Res., **13**, 135-146.
Tsai, C. H. and Hwang, S. C.: 1995, Mar. Freshwater Res., **46**, 383-392.
Tsai, C.H.: 1996, Marine Geology, **134**, 95-112 .
van Duuren, F.A.: 1968, J. Sanit. Engr. Div., **94**(SA4), 671-682.
Williams, R.A. Peng, S.J. and Naylor, A.: 1992, Powder Technol., **73**, 75-83.

PARTICLE SIZE SELECTIVITY CONSIDERATIONS IN SUSPENDED SEDIMENT BUDGET INVESTIGATIONS

P.M. STONE and D.E. WALLING

Department of Geography, University of Exeter, Exeter EX4 6QX, England.

Abstract. The delivery of suspended sediment from drainage basins has frequently been quantified in mass terms by use of the sediment budget approach, which identifies sources, storage and output of mobilised sediment. An attempt is presented here to define the main components of a generalised suspended sediment budget for a drainage basin in Devon, U.K. in terms of particle size characteristics and grain size selectivity, rather than total amounts of sediment. Samples of sediment mobilised from the hillslopes, fluvial suspended sediment and suspended sediment deposited on the river bed were all collected for particle size characterisation. These samples were then treated to remove organic matter and their chemically dispersed (absolute) particle size composition was measured using a Coulter LS130 laser granulometer. Where possible, measurements of the natural *in-situ* particle size distribution (effective particle size) were also undertaken. Samples were collected at different times of the year so that temporal variation of hydrometeorolgical and ground conditions was represented. Comparison of the results for the different components of the delivery process shows that significant particle size selectivity occurs in the mobilisation and transfer of sediment from the hillslopes to the basin outlet. This reflects the particle size selectivity of detachment, transport and deposition processes, which is in turn influenced by the aggregation or flocculation (effective particle size) of the sediment.

KEYWORDS: suspended sediment delivery, particle size selectivity, mobilised sediment, fine-grained bed sediment

1. Introduction

Suspended sediment budget studies have quantified sources and sinks in the suspended sediment delivery system and the fluxes involved (e.g. Walling, 1988; Sutherland and Bryan, 1991) but there have been very few attempts to express the suspended sediment budget in terms of the particle size composition or particle size selectivity involved. Particle size selectivity occurs if a sediment sample is enriched or depleted in certain particle sizes when compared to its source, due to the erosion, transportation or deposition processes operating. Existing investigations of particle size selectivity have considered sediment mobilisation from hillsplopes (e.g. Parsons *et al.* 1991), transport of suspended sediment in rivers (e.g. Walling and Moorehead, 1987), deposition of suspended sediment on the river bed (e.g. Tipping *et al.*, 1993; Droppo and Stone, 1994) or overbank suspended sediment depositions (e.g. Lambert and Walling, 1987). Most concentrate on individual components of the sediment budget and few consider the particle size selectivity of sediment delivery (e.g. Sutherland and Bryan, 1989). The aim of this paper is quantify the sediment budget of a river in terms of its particle size characteristics and particle size selectivity occurring during sediment delivery.

2. Study Area and Methods

The 46km^2 drainage basin of the River Dart, a tributary of the River Exe, Devon, U.K. was chosen for the study. The underlying geology is Upper Carboniferous Culm Measures (sandstones interbedded with shales, mudstones and siltstones) and the land

Water, Air and Soil Pollution **99**: 63-70, 1997.

use as classified by the Ministry of Agriculture, Fisheries and Food (MAFF) 1994 statistics comprises 10.8% cultivated, 82.2% grassland and 7% woodland. The topography is characterised by well developed interfluves and incised valleys (average slope angle 11°). Mean annual precipitation is almost 900mm, with 60% falling during the winter season. Existing suspended sediment tracing studies suggest that over the longer-term ca. 79% of the sediment yield is derived from surface sources and ca. 21% from channel banks (Walling *et. al.* 1993). Pasture areas, frequently poached by livestock in winter, represent the dominant sediment source. The specific suspended sediment yield from the basin has been estimated to be ca. 58 t $km^{-2}year^{-1}$ (Walling and Webb, 1987). The main components of the suspended sediment budget were identified as sediment mobilisation from the basin hillslopes, transport of suspended sediment by the river and the deposition of suspended sediment on the bed of the river. The particle size characteristics and selectivity of these three main components were investigated.

A field portable rainfall simulator based on the design of Bowyer-Bower and Burt (1989) was used to generate rainfall of 7 mm/hour intensity for 3 hours. This design storm was representative of natural conditions that caused significant suspended sediment transport in the river throughout the year. Sediment samples from pasture land (the dominant source of suspended sediment) were collected at the base of the simulator plots. Seasonal variation in sediment mobilisation was investigated on a monthly basis using plots with similar slope and land use at Well Farm. Single simulation runs on similar plots at other locations in the basin indicated that the results from Well Farm were representative of sediment mobilisation from pasture throughout the basin. Samples of the top 1 cm of soil adjacent to the plots were compared with the samples of mobilised sediment to determine whether particle size selectivity was occurring.

Suspended sediment samples were collected from the river channel during natural flood events when the sediment concentration exceeded $100mgl^{-1}$. Samples were either collected manually using a USDH 48 depth integrating sampler or using automatic rising stage siphon samplers. The latter were deployed at 5 sites within the basin to determine whether spatial variation in the particle size composition of suspended sediment existed.

A sealed, removable bed trap based on the design of Phillips (1996) and containing natural bed sediment was used to collect samples of suspended sediment deposited on the river bed. The trap was initially filled with sediment < 1 mm. During flood events the seal was removed and any sediment in the trap smaller than 1 mm could be assumed to have been deposited during the event studied. After the event, the trap was removed from the river, the contents resuspended in clear water and samples collected at three stages of resuspension (stage 1 to 3) with increasing depth of agitation of the trap contents. The trap was located at the basin outlet and the spatial representativeness of the results obtained was tested by mobilising deposited sediment at five other locations throughout the basin at low flow following a flood event. An area of the bed was isolated within a steel cylinder and the surface agitated with a stick, allowing samples of resuspended sediment to be collected from the water trapped in the cylinder. These samples did not relate to a specific flood event but were not significantly different

(Mann-Whitney U test 95% confidence) from samples collected from the bed trap during the preceding flood event. No significant spatial variation in deposition was identified.

A Coulter LS 130 laser granulometer with a measurement range of 0.1 to 800μm was used to measure the particle size distribution of samples. Hydrogen peroxide was added to each sample to remove the organic fraction, and each was resuspended in sodium hexametaphosphate and ultrasonically dispersed so that a measure of the absolute particle size distribution (Ongley *et al.* 1981) could be made using the laser granulometer. Assessment of particle size selectivity associated with the various components of the suspended sediment budget was based on the absolute particle size distributions. Where possible, some suspended sediment samples and samples of mobilised sediment were not allowed to settle, but were quickly returned to the laboratory and their particle size composition measured without pre-treatment, using the laser granulometer. Such data were designated the effective particle size distribution by Ongley *et al.* (1981), and reflect the presence of composite particles (aggregates/flocs), and primary particles. Whilst not providing a true measure of the *in situ* particle size distribution, Phillips and Walling (1995) suggest that this measurement technique produces adequate results. The absolute particle size distribution of the same samples were subsequently measured after treatment for comparison. Composite particles associated with samples collected from the hillslopes are referred to as "aggregates" whilst those associated with the suspended sediment samples collected from the river are termed "flocs". Comparison of particle size distributions was based on clay (<2μm), four silt (2-5μm, 5-10μm, 10-20μm and 20-63μm) and sand (>63μm) size classes. Any enrichment or depletion in each size classes was ascribed to particle size selectivity.

3. Results and Discussion

3.1. PARTICLE SIZE SELECTIVITY OF SEDIMENT MOBILISATION FROM THE HILLSLOPES

A comparison of the average absolute particle size distribution of samples of soil and mobilised sediment collected on a monthly basis from the Well Farm site (Figure 1) indicates that there is a particle size selective mobilisation of sediment finer than 20-60μm, whilst the sand-sized material remains *in-situ.* This result is based on average values and although the soil sample did not display a significant seasonal variation (maximum standard deviation 2.24 for the 10-20μm silt class), the mobilised sediment samples showed significant seasonal variation (Figure 2). A statistically significant difference (Mann-Whitney U test 95% level of confidence) was identified between samples collected during the October-March period (autumn/winter season) and April-August period (spring/summer season). The particle size distributions of autumn/winter samples were generally finer than the spring/summer samples. During the autumn/winter, the degree of grass cover decreased and the soils became saturated. Runoff volumes were greater in the autumn/winter than spring/summer but there would appear to be an inverse relationship between flow and the size of sediment transported.

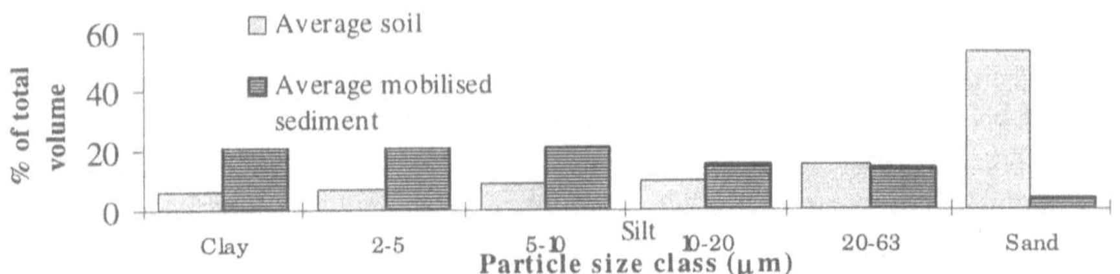

Fig. 1 The averaged absolute particle size distributions for the soil samples and for mobilised sediment in runoff collected at Well Farm on a monthly basis.

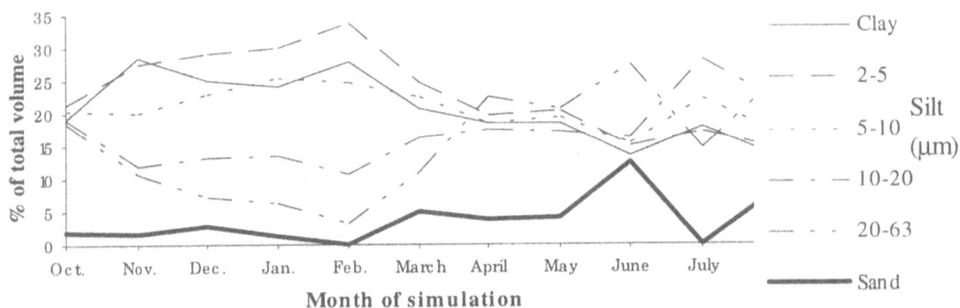

Fig. 2 The relative contribution of each of the absolute particle size classes for the samples of mobilised sediment collected during the monthly simulation runs undertaken at Well Farm.

The effective particle size distribution of samples of mobilised sediment collected from monthly simulation runs also evidenced seasonal variation. The effective and absolute particle size distributions for a sample from each season are compared in Figure 3.

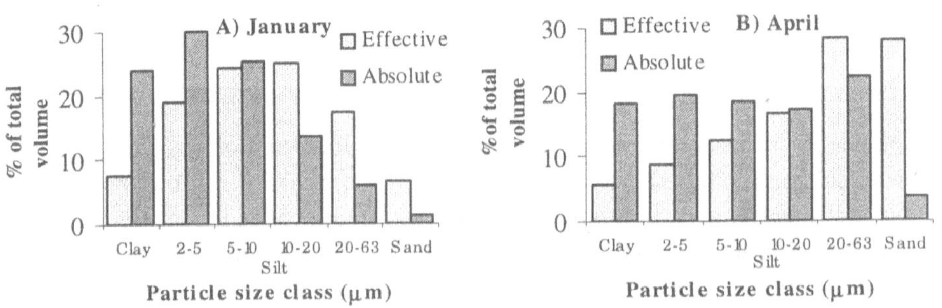

Fig. 3 A comparison of the effective and absolute particle size distributions of samples collected using the rainfall simulator in January (A) and April (B)

A comparison of the effective particle size distributions from each season shows that a greater amount of primary and aggregate particles in the 20-63μm silt and sand size classes are mobilised during the spring/summer whilst greater amounts of 2-5μm, 5-10μm and 10-20μm silt are mobilised during the autumn/winter. The aggregates associated with mobilised sediment are therefore larger in the spring/summer than in the autumn/winter. When these aggregate particles are chemically dispersed to produce the primary particle composition (absolute particle size distribution) there will be an enrichment of particles finer than the aggregates. Since the effective particle size of autumn/winter samples was finer than spring/summer the resultant absolute particle size is also finer. The seasonal difference in absolute particle size previously noted can therefore be regarded as supply-controlled.

3.2. THE PARTICLE SIZE SELECTIVITY OF SUSPENDED SEDIMENT TRANSPORT

A comparison of the absolute particle size distribution of mobilised sediment samples obtained from the monthly simulations with the average absolute particle size distribution of suspended sediment samples taken from floods at a similar time (Figure 4) suggested that in general the suspended sediment is coarser than the mobilised sediment in the autumn/winter but finer in the spring/summer. This could either be caused by seasonal variations in river flow such that increasing discharge transports increasingly larger particles or could be related to sediment supply factors. The relationship between discharge and sand sized sediment for both effective and absolute particle size distributions (Figure 5) of all suspended sediment samples

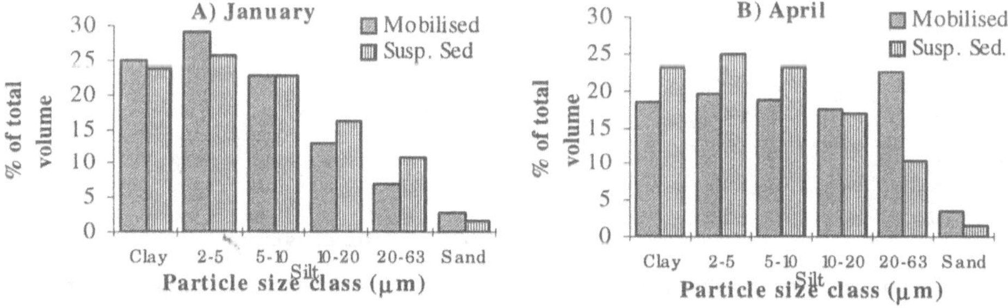

Fig. 4 A comparison of the absolute particle size composition of mobilised sediment and suspended sediment. The samples in (A) were collected during autumn/winter conditions and those for (B) during spring/summer

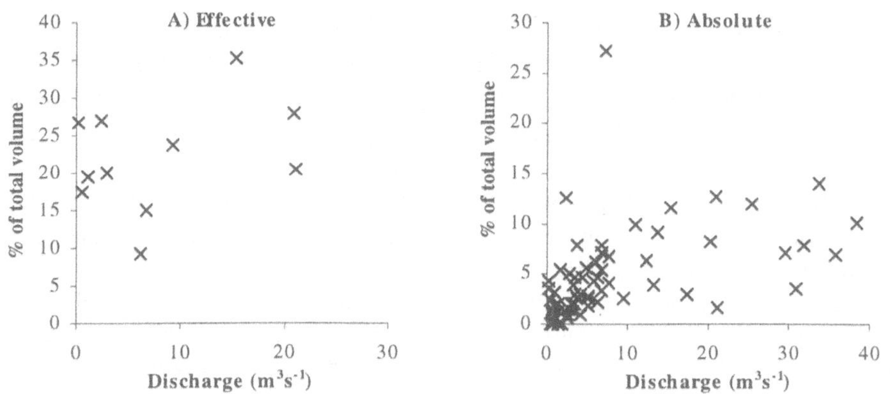

Fig. 5 The relationship between the % sand content of suspended sediment and discharge for both the effective (A) and absolute (B) particle size distributions of suspended sediment

collected evidenced little relationship. The seasonal variation noted above would therefore appear to be supply-controlled..

It is recognised that only the particle size characteristics of sediment mobilised from pasture land are considered in this study. This source was suggested by Walling *et al.* (1993) to be the main surficial source and it represented the dominant land use in the basin. Other sources of suspended sediment such as channel banks are noted but Walling *et al.* (1993) suggested that their contribution was relatively small and did not vary seasonally. Comparison of suspended sediment with sediment mobilised from intensively grazed pasture land was there felt to be valid. Figure 6 shows that, as with mobilised sediment (Figure 2), the average particle size composition of suspended sediment evidenced seasonal variation, but in this case the spring/summer samples contain greater proportions of finer sediment than those for the autumn/winter. This would be expected due to the reduced competence of runoff to transport coarser sediment

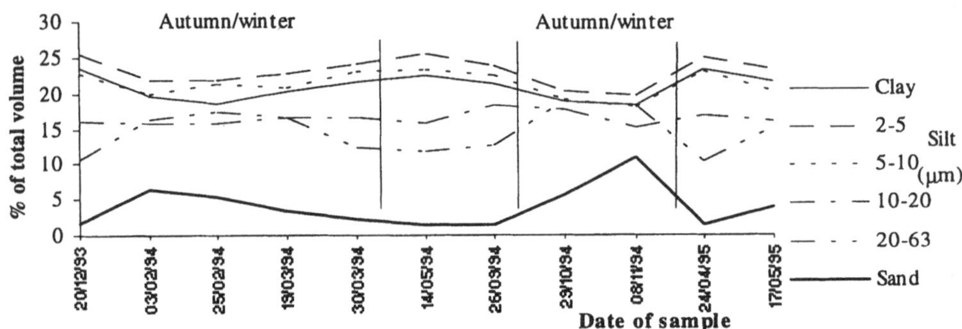

Fig. 6 Inter-storm variation in the absolute particle size composition of suspended sediment samples collected during flood events.

from the hillslopes to the channel during the spring/summer. However, when the suspended sediment and mobilised sediment are compared (Figure 3) the suspended sediment is seen to be finer than the mobilised sediment in spring/summer but coarser in autumn/winter, the latter suggesting deposition of fine sediment on the hillslopes. This may result from the deposition of large aggregate particles from the sediment transported from the hillslopes to the river channel. Figure 3 implied that autumn/winter aggregate particles were composed of finer primary particles than spring/summer aggregates Therefore, deposition of aggregates would cause a net reduction in the importance of finer sediment size classes in the absolute particle size distribution of the suspended sediment samples collected during the autumn/winter.

A comparison of representative effective particle size distributions and their equivalent absolute particle size distributions for suspended sediment collected during the autumn/winter and spring/summer seasons is shown in Figure 7. Both the autumn/winter effective and absolute particle size distributions are coarser than those for the spring/summer sample, the opposite trend to that noted for the mobilised sediment samples. This confirms that the larger aggregate particles mobilised from the hillslopes in the spring/summer did not reach the channel.

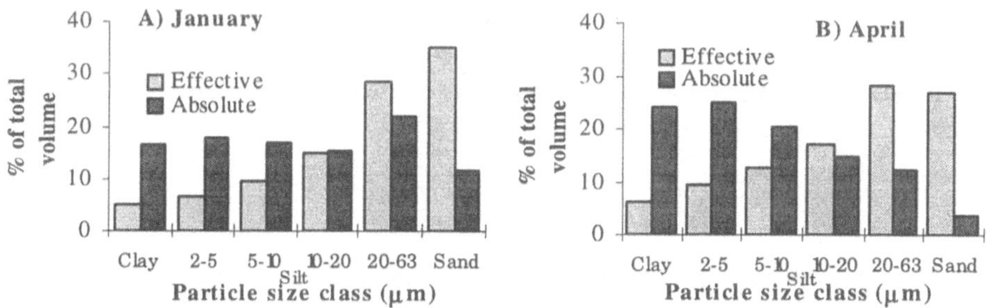

Fig. 7 A comparison of the effective and absolute particle size composition of suspended sediment. The results presented are for a sample collected in autumn/winter (A) and spring/summer (B)

3.3. THE DEPOSITION OF SUSPENDED SEDIMENT ON THE BED OF THE RIVER

A comparison of the samples of deposited suspended sediment recovered from the bed trap suggested that the degree of disturbance involved in resuspending the sediment did not influence the absolute particle size composition (Figure 8). A comparison with the absolute particle size distribution of suspended sediment also showed that coarser suspended sediment particles are selectively deposited. The finer size classes are, however, also deposited, reflecting the role of the effective particle size of the suspended sediment in the selective deposition of particles. The settling velocities of primary particles in the finest size classes are probably too low for deposition to occur, but they may be deposited as a components of flocs which have higher settling velocities than the particles of which they are composed.

Rising limb suspended sediment samples collected along the course of the river at similar times during the flood wave and samples of sediment deposited on the river bed, collected from similar locations using the cylinder showed no significant spatial variation

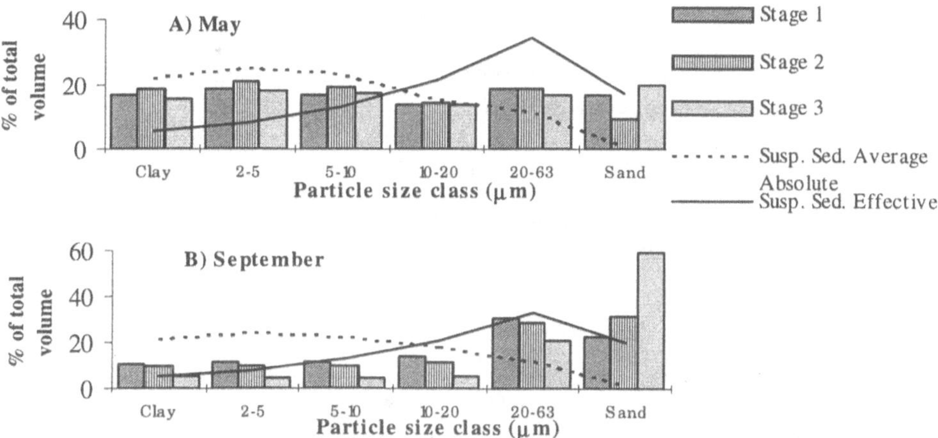

Fig. 8 The absolute particle size composition of samples collected during the three stages of resuspension of suspended sediment deposited in the bed trap during the events of 14/5/94 (A) and 26/9/94 (B). Representative effective and absolute particle size distributions for suspended sediment collected from the same event are shown.

(Mann-Whitney U test 95%). Whilst the deposition of suspended sediment onto the river bed is particle size selective, the total amounts deposited are relatively small and this appears to have no effect on the particle size composition of suspended sediment, which did not change significantly downstream.

4. Conclusion

There is clear evidence of particle size selectivity in the delivery of suspended sediment from the hillslopes to the basin outlet of the River Dart. Sediment finer than sand was preferentially mobilised from the hillslopes and most of the sand-sized sediment remained *in situ*. Both the effective and absolute particle size composition of this mobilised sediment varied seasonally. There was no clear relationship between suspended sediment particle size characteristics and river flow, suggesting that any variation observed was a result of supply factors rather than transport factors. Seasonal variation was also identified in the particle size composition of suspended sediment. A comparison of mobilised and suspended sediment suggested that particle size selectivity was occurring in the transport of sediment to the river channel, such that not all of the mobilised sediment reached the river. Some particle size classes of suspended sediment are selectively deposited in the river bed. This deposition is influenced by the effective particle size characteristics of the suspended sediment, but it appears not to have any effect on the particle size composition of the suspended sediment due to the relatively small amounts of deposition involved.

Acknowledgements

The authors gratefully acknowledge the U.K. Natural Environment Research Council in providing a Research Studentship to P.M.S. and the co-operation of landowners.

References

Bowyer-Bower, T.A.S. and Burt, T.P.: 1989, *Soil Technology*. **2** 1-16.
Droppo, I.G. and Stone, M.: 1994, *Hydrological Processes*. **8**, 101-111
Lambert, C.P. and Walling, D.E.: 1987, *Geografiska Annaler*. **69A 3-4**, 393-404.
Ongley, E.D., Bynoe, M.C. and Percival, J.B.: 1981, *Can J.Earth Sci.* **18**, 1365-1379.
Parsons, A.J., Abrahams, A.D. and Shiu-Hungluk Luk.: 1991, *Earth Surf. Proc. Landforms* **16**, 143-152.
Phillips, J.M. and Walling, D.E.: (1995), *Water Res.* **29(11)** 2498-2508.
Phillips, J.M.: 1996, *The Effective Particle Size Characteristics of Fluvial Suspended Sediment.* Unpublished Thesis, University of Exeter.
Sutherland, R.A. and Bryan, R.B.: 1989, *Catena* **16**, 189-204.
Sutherland, R.A. and Bryan, R.B.: 1991, *Earth Surf. Proc. Landforms* **16**, 383-398.
Tipping, E., Wolf, C. and Clarke, K.: 1993, *Hydrological Processes*. **7**, 263-277.
Walling, D.E.: 1988, *J. Hydrol.* **100**, 113-141.
Walling, D.E. and Moorehead, P.W.: 1989, *Hydrobiol.* **176/177**, 125-149.
Walling, D.E., Woodward, J.C. and Nicholas, A.P.: 1993, *Tracers in Hydrology (Proceedings of the Yokohama Symposium, July 1993)* IAHS Pub. No. 215, p329-338.
Walling, D.E. and Webb, B.W.: 1987, *Sediment Transport in Gravel-bed Rivers.* Ed. Thorne, C.R., Bathurst, J.C. and Hey, R.D. John Wiley and Sons. p691-723.

SPATIAL VARIABILITY OF THE PARTICLE SIZE COMPOSITION OF OVERBANK FLOODPLAIN DEPOSITS

Q. HE and D. E. WALLING

Department of Geography, University of Exeter, Exeter, EX4 4RJ, UK

Abstract. An important feature of overbank floodplain deposits is the spatial variability of their particle size composition. Analysis of such spatial variability can assist in developing an improved understanding of the transport and deposition of suspended sediment on river floodplains during overbank flood events, in investigating the fate of sediment-associated contaminants and in calibrating existing floodplain sediment deposition models. The study reported in this paper investigates the spatial variability of the grain size composition of overbank floodplain deposits at different spatial scales, through analysis of surface sediment samples collected from frequently inundated floodplain sites on the Rivers Culm, Stour and Severn in the UK. Significant lateral and downstream variations in the grain size composition of the sediment deposits have been documented at the study sites, and the results obtained have been interpreted in terms of the processes governing overbank floodplain flow and sediment transport and deposition, which are influenced by a number of factors including floodplain geometry and topography.

Key words. River floodplains, overbank sedimentation, floodplain deposits, grain size composition, spatial variability.

1. Introduction

River floodplains have been widely recognised as important sinks for storing suspended sediment and associated contaminants mobilised from upstream catchments (cf. Rang and Schouten, 1989; Marron, 1992; Walling *et al.*, 1996). Most of the deposition of suspended sediment during periods of floodplain inundation is associated with overbank flood deposits, which commonly represent an important component of floodplain development and evolution (cf. Wolman and Leopold, 1957; Lewin, 1978). Both the suspended sediment transported by a river and the resultant overbank floodplain deposits can exhibit significant spatial variation in grain size composition because of the dynamic nature of sediment mobilisation, transport and deposition (cf. Marriott, 1992; Walling, 1996; Walling *et al.*, 1996). Since many contaminants are associated with fine-grained sediment, investigation of the spatial variability of the particle size composition of floodplain sediment deposits is an important requirement for studying the cycling and fate of these contaminants in fluvial systems. Furthermore, existing physically-based overbank floodplain flow and sediment transport models and empirical sediment deposition models provide a useful basis for representing overbank processes. Information concerning the grain size composition of floodplain sediments is crucial for calibrating and validating such models (cf. James, 1985; Pizzuto, 1987; Howard, 1992).

Against this background, there is a need for information concerning the spatial variability of the grain size composition of overbank floodplain deposits, and there have been numerous studies aimed at documenting this variability. In general, the sediment deposited in areas near the river channel has been found to be coarser than that deposited in areas further away from the channel (e.g. Brown, 1983; Pizzuto, 1987; Marriott;

Water, Air and Soil Pollution **99**: 71-80, 1997.

1992; Asselman and Middelkoop, 1995; Walling *et al.*, 1996). Further investigation of the cross-valley and downstream variability in the particle size composition of overbank floodplain deposits at both larger and smaller reach scales is, however, needed in order to examine the role of river plan-form and floodplain topography in influencing overbank sediment transport and deposition processes.

The present study attempts to investigate the spatial variability of the grain size composition of contemporary overbank floodplain deposits through analysis of surface sediment samples collected from floodplain sites on three UK rivers, namely, the River Culm in Devon, the River Stour in Dorset and the River Severn in Shropshire. Results obtained from the study demonstrate that significant spatial variability in the particle size composition of the sediment deposits exists at these sites, and an attempt has been made to interpret this spatial variability in terms of local floodplain geometry and topography.

2. The study sites

The study rivers and their drainage basins vary in catchment size and relief, soil type, geology, and land use. The River Culm has a total drainage area of ca. 276 km^2, while the River Severn at Buildwas drains an area of ca. 3717 km^2 (Figure 1). The River Stour has a drainage area of ca. 523 km^2 at Hammoon (Figure 1). Average values of the median grain size or d_{50} for the absolute particle size distribution (chemically-dispersed mineral fraction) of the suspended sediment transported by the rivers are ca. 2.9 µm for

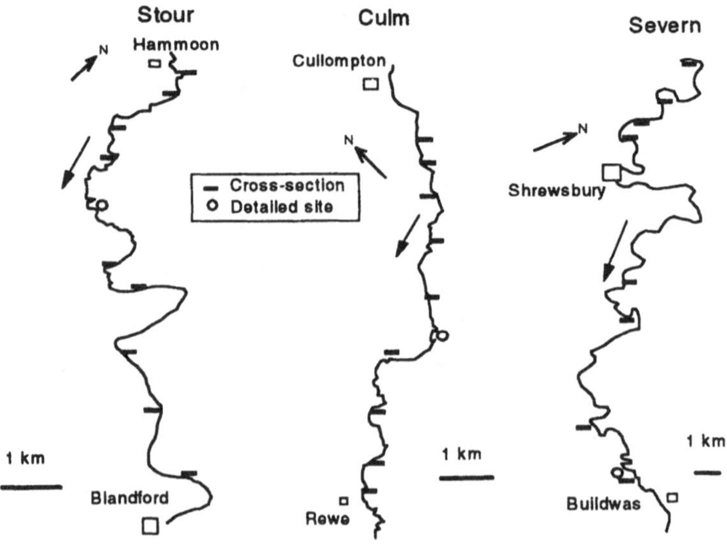

Fig. 1. The study reaches of the floodplains of the Rivers Stour, Culm and Severn and the locations of the sediment sampling sites.

the River Stour, 6.5 μm for the River Culm and 7.6 μm for the River Severn. This variation reflects both catchment geology and morphology. The study sites are situated on the middle or lower reaches of the rivers and Figure 1 locates the floodplain sites studied. These study sites have supported permanent pasture for several decades. They are inundated frequently during large storm events and are characterised by rates of overbank sediment deposition of the order of 0.20 g cm^{-2} yr^{-1}.

3. Field sampling and laboratory analysis

To provide information on cross-valley and downstream variation of the grain size composition of overbank floodplain deposits, representative cross-sections 1-6 km apart were selected within the study reaches (9 on the Rivers Stour and Culm and 8 on the River Severn) (cf. Figure 1) and 8-13 surface sediment samples (12-30 m apart and ca. 1 cm thick) were collected from each cross-section for particle size measurement using a trowel. These surface sediment samples represent recent sediment deposited on the floodplains during the past few years. In addition, suspended sediment samples were collected during flood periods and analysed for their particle size composition. To document spatial variation of the grain size composition of the deposited sediment at a smaller scale, detailed sampling at one particular site within each of the study reaches was also undertaken (cf. Figure 1). In the case of the River Stour, 97 surface sediment samples were collected from the floodplain near Chisel Farm based on a 25m×25m grid. At the detailed study site near Buildwas on the River Severn, 133 surface sediment samples were collected based on a 25m×25m grid. For the River Culm, 274 surface sediment samples were collected from the detailed site near Silverton Mill at the intersection of a 12m×12m grid.

All sediment samples collected were air-dried, gently ground and passed through a 2 mm mesh sieve. Sub-samples (2-10g) were then separated into sand (>63 μm) and clay and silt (< 63 μm) fractions by sieving following standard procedures. The absolute grain size distributions of the <63μm fractions of the samples were measured by laser-diffraction (Malvern MasterSizer) after pretreatment to remove the organic fraction and chemical dispersion, and the proportions of clay (<2.0μm) and silt (2.0-63μm) were calculated.

4 Results and discussion

4.1. CROSS-VALLEY VARIATION OF THE GRAIN SIZE COMPOSITION OF FLOODPLAIN DEPOSITS

Figure 2 plots the percentage clay P_c (%) and sand P_s (%) content of overbank sediment against distance Y (m) from river channel for the surface sediment samples collected from the cross-sections along the three rivers. The sediment from the River Stour floodplain is characterised by a higher clay content than the sediment from the floodplains of the River Culm and River Severn, reflecting the fine-grained nature of the suspended sediment transported by the River Stour. Distance from the channel has frequently been identified as an important control on the grain size composition of deposited sediment, and Figure 2 shows that the clay content of the overbank sediment deposits generally increases with increasing distance from the river channel, while the sand content decreases towards the outer margins of the floodplain. These trends are consistent with findings of other researchers such as Pizzuto (1987) and Marriott (1992). The rapid reduction in sand content in the sediment with increasing distance from the river channel in the area close to the channel suggests that the coarse fractions of the suspended sediment are transported onto the floodplain from the channel primarily through diffusive processes, because the energy associated with overbank floodplain flow is generally relatively low (cf. James, 1985; Pizzuto, 1987). Taking account of the

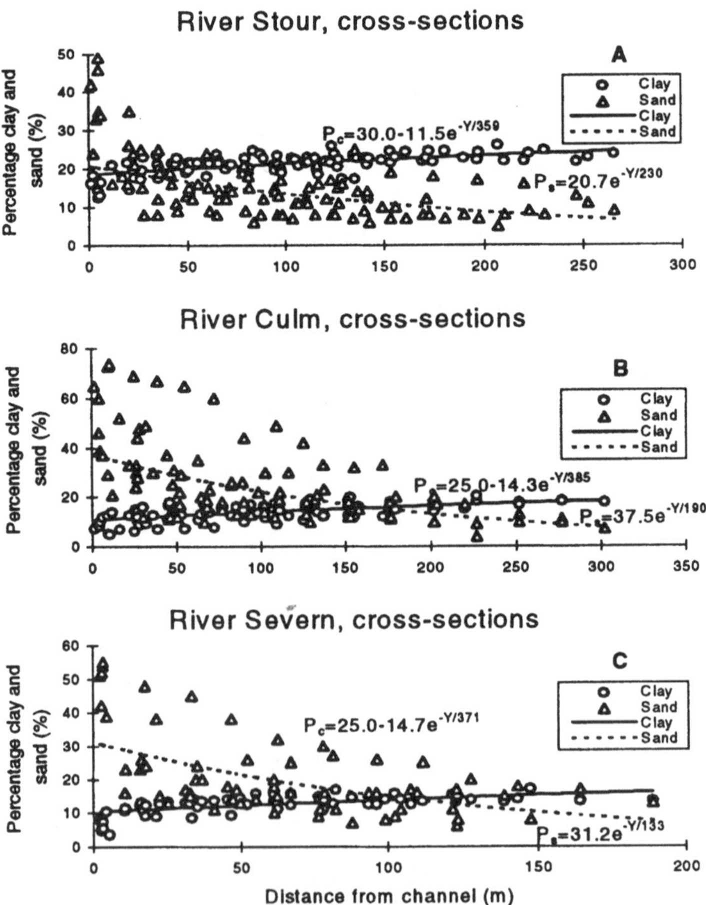

Fig. 2. Cross-valley variation in the grain size composition of overbank sediment deposits collected from cross-sections of the floodplains of the River Stour (A), River Culm (B) and River Severn (C).

trends shown in Figure 2, an exponential function has been employed to describe the general relationship between the particle size composition of the sediment (P_c and P_s) and distance from the river channel (Y) for the individual rivers. For clay content,

$$P_c = \begin{cases} 30.0 - 11.5e^{-Y/359} & (r^2 = 0.32) \quad \text{River Stour} \\ 25.0 - 14.3e^{-Y/385} & (r^2 = 0.39) \quad \text{River Culm} \\ 25.0 - 14.7e^{-Y/371} & (r^2 = 0.38) \quad \text{River Severn} \end{cases} \tag{1}$$

For sand content,

$$P_s = \begin{cases} 20.7e^{-Y/230} & (r^2 = 0.39) \quad \text{River Stour} \\ 37.5e^{-Y/133} & (r^2 = 0.41) \quad \text{River Culm} \\ 31.2e^{-Y/133} & (r^2 = 0.36) \quad \text{River Severn} \end{cases} \tag{2}$$

The rates of increase in the percentage clay content with increasing distance from the channel for the three rivers are similar. However, the rates of decrease in the percentage sand content for the Culm and Severn sediment samples are higher than those for the Stour floodplain samples. This feature may reflect contrasts in the particle size composition of the suspended sediment transported by these rivers, the magnitude of the river discharge, the river geometry, and the width and topography of the floodplains between the study reaches. Overall, however, the degree of explanation provided by the exponential functions is relatively low, suggesting that on these floodplains, factors other than distance from the channel are important in controlling the grain size composition of deposited sediment.

4.2. DOWNSTREAM VARIATION OF THE GRAIN SIZE COMPOSITION OF FLOODPLAIN DEPOSITS

To investigate the extent of downstream variation in the grain size composition of the overbank flood deposits through the study reaches, mean values of the percentage clay and percentage sand content and of the d_{50} for each cross-section were assumed to represent the particle size composition of that section and Figure 3 plots their downstream variation for the three floodplain reaches along with information on the variability of these parameters within each cross-section ($\pm 1\sigma$). In the case of the Rivers Stour and Severn, both the mean clay and sand contents exhibit appreciable variation between the cross-sections, but no statistically significant relationships between the three grain size parameters and distance downstream exists. Although downstream variation in the grain size composition of suspended sediment will frequently exist and will influence the grain size composition of the floodplain deposits (cf. Walling and Moorehead, 1989), local flow patterns on the floodplains can also be expected to exert an important influence on the mean grain size composition for each cross-section. For example, in the case of the River Stour, cross-section 4 is located upstream of a bridge, and the mean sand content in the sediment is significantly lower ($\alpha<0.05$) than that associated with the sediment from the other sections. On the other hand, the relatively high mean clay content in the sediment from cross-section 9 may reflect the relatively flat and broad nature of this section. In the case of the River Culm, however, there are statistically significant relationships between the clay content (r^2=0.74), sand content (r^2=0.55) and d_{50} (r^2=0.58) of the sediment deposits and distance downstream (P>95%). The mean clay content of the cross-sections increases downstream, whereas the mean sand content decreases, with cross-section 1 evidencing the lowest clay content and highest sand content. To a certain degree, this trend reflects a downstream fining of the suspended sediment transported through this reach as observed by Walling and Moorehead (1989). This, in turn, reflects the importance of sediment accumulation on the floodplain of this river associated with overbank flooding events. In this case, floodplain sedimentation within the study reach can account for reductions in the total suspended sediment load transported by the river by as much as 28% (cf. Walling and Bradley, 1989).

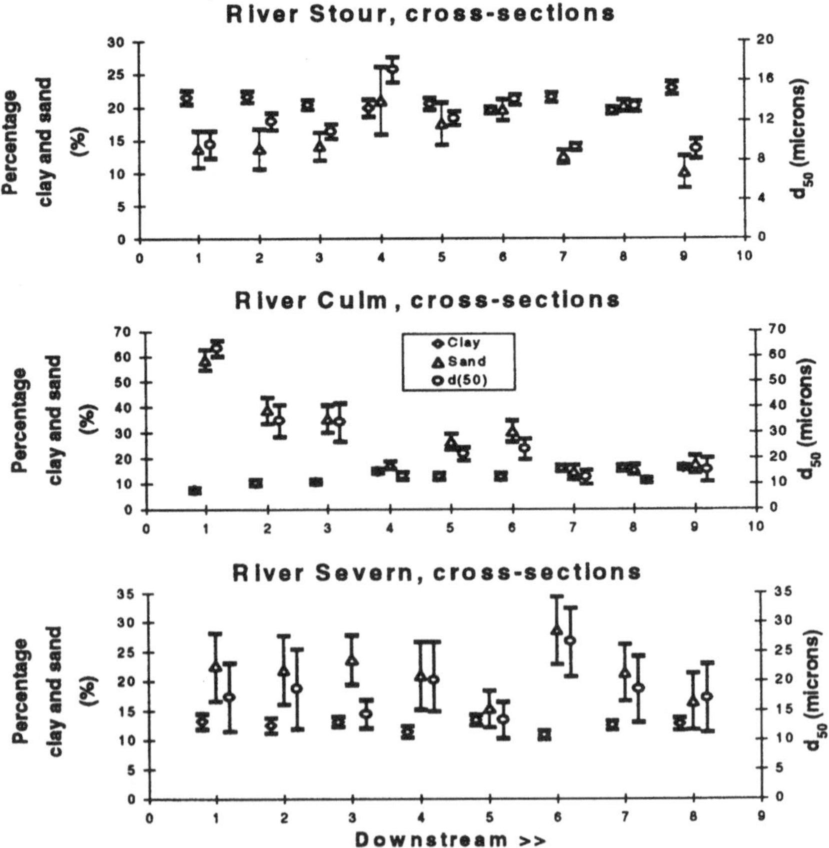

Fig. 3. Downstream variation in the grain size composition of overbank floodplain deposits for the study reaches of the River Stour (A), River Culm (B) and River Severn (C) (the error bars represent $\pm 1\sigma$ of the mean).

4.3. SMALL SCALE VARIATION OF THE GRAIN SIZE COMPOSITION OF FLOODPLAIN DEPOSITS

For small areas on a floodplain such as those represented by the detailed study sites, local variability in the particle size composition of the sediment deposits will be controlled by the suspended sediment properties and the hydraulics of the local floodplain flow, which will in turn be influenced by river discharge as well as the local floodplain geometry and topography. Figure 4 depicts the spatial distribution of the percentage clay and silt content of the floodplain sediment for the detailed study sites at Chisel Farm on the River Stour, Silverton Mill on the River Culm and Buildwas on the

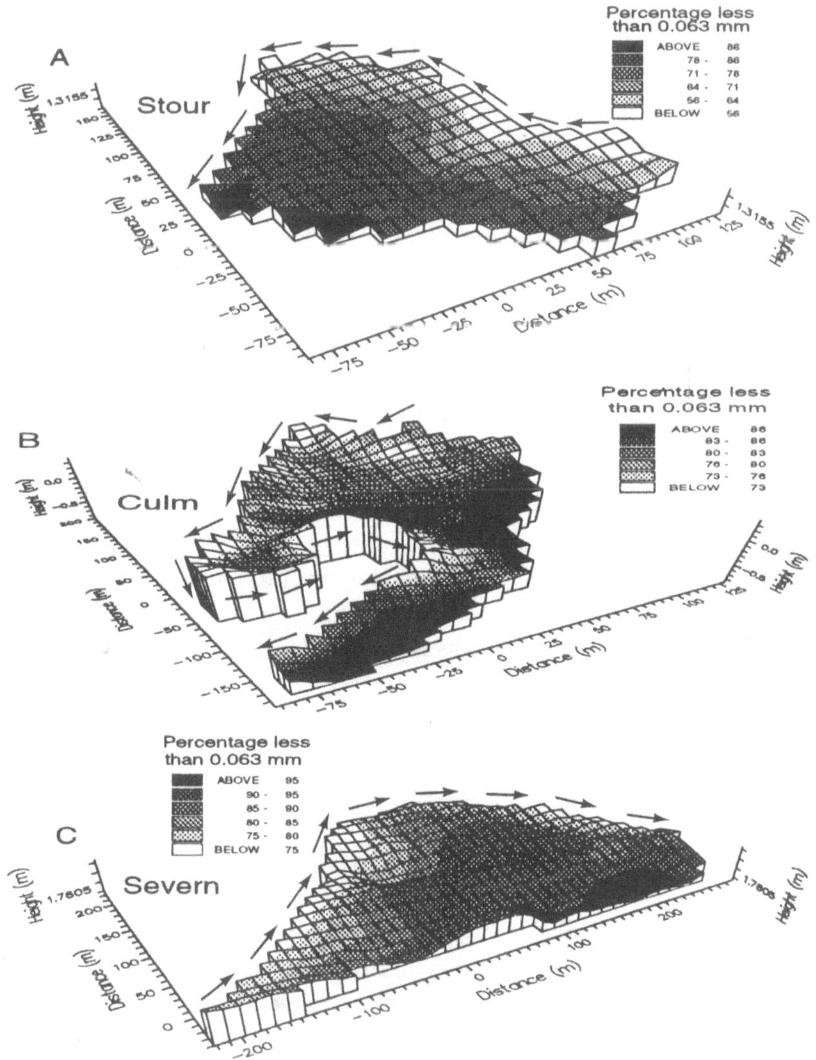

Fig. 4. The spatial variation of the clay and silt content of surface sediment for the detailed study sites at Chisel Farm on the River Stour (A), Silverton Mill on the River Culm (B) and Buildwas on the River Severn (C).

River Severn. Significant spatial variation of the grain size composition of the sediment exists at all three sites. Distance from the channel clearly exerts an important influence, since the sediment is generally coarser in areas near the river channel and finer in areas near the floodplain outer margins. The influence of the local river plan-form on the particle size distribution of the overbank sediment deposits is, however, also apparent. For example, the spatial distribution of the percentage clay and silt content of the overbank sediment deposits at the Stour study site near Chisel Farm (Figure 4A) clearly shows that the upstream areas near the river channel, which are depleted in clay- and silt-sized particles, extend further into the middle of the floodplain than the equivalent areas downstream. This suggests that the transport of coarser fractions onto the floodplain from the river channel in the former situation is principally associated with convective overbank flow, while in the latter case it is associated with diffusive processes. The spatial distribution of the particle size composition of the sediment at the Buildwas site (Figure 4C) on the Severn is similar to that of the sediment at the Chisel Farm site. In contrast to the transport of coarse sediment particles, the transport of finer suspended material over the floodplain is primarily associated with convective overbank flow, and this implies that the concentrations of fine sediment particles in the overlying floodwater will be relatively uniform across the floodplain surface. Therefore, rates of deposition of the fine fractions will be primarily governed by floodwater depth and thus the local floodplain topography, with higher rates of deposition occurring in depression areas. The overall grain size distribution of the sediment on a floodplain will therefore also be influenced by the local floodplain topography. Although detailed interpretation of the spatial distribution of the grain size composition of deposited sediment requires consideration of floodplain flow hydraulics and the dynamics of sediment-water interactions, here an attempt has been made to take account of the influence of floodplain geometry and topography in interpreting the results obtained from the detailed study sites. The following empirical functions were derived to represent the relationship between the percentage clay and silt content P_{cs} (%) and both distance from the river channel Y and relative floodwater depth Z (m) in these areas:

$$P_{cs} = \begin{cases} 70.4 + 13.3Z - 29.9e^{-Y/12.0} & (R^2 = 0.40) \quad \text{River Stour} \\ 75.4 + 14.4Z - 12.8e^{-Y/15.0} & (R^2 = 0.47) \quad \text{River Culm} \\ 90.8 + 2.7Z - 30.7e^{-Y/25.0} & (R^2 = 0.77) \quad \text{River Severn} \end{cases} \tag{3}$$

The constants on the right-hand side of Equation 3 reflect the base level employed in calculating the relative floodwater depths on the floodplains. Figure 5 plots the relationship between the measured grain size composition and distance from the channel and relative floodwater depth for sediment from the floodplain surface, and the results of fitting Equation 3 to these data have been superimposed on Figure 5. There is generally a trend for the clay and silt content of the sediment to increase with increasing floodwater depth. However, it is stressed that in the present study attention has focussed on the spatial variability of the absolute particle size distribution (chemically dispersed mineral fraction) of the overbank deposits. It is recognised that a substantial proportion of the suspended sediment is transported and deposited as aggregates rather than primary particles and that the in-situ (effective) particle size distribution of the deposited sediment will be significantly different from the absolute particle size distribution (cf. Walling and Moorehead, 1989; Droppo and Ongley, 1994; Phillips and Walling 1995; Walling, 1996). The influence of controls such as relative floodwater depth may therefore be less clearly evident when considering the absolute, as distinct from the effective, grain size distribution of the deposited sediment.

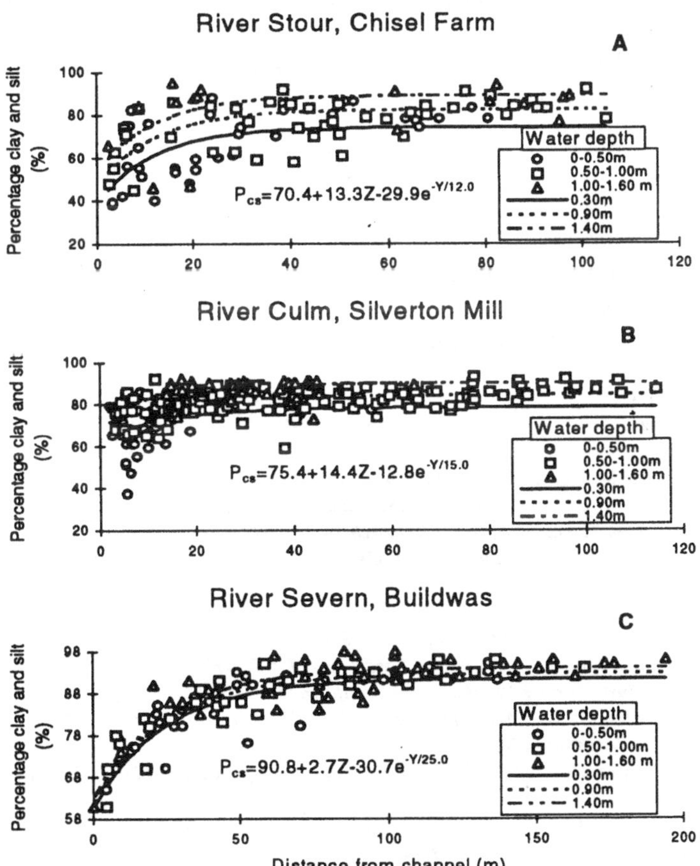

Fig. 5. The relationship between the grain size composition of surface sediment and distance from the channel and relative water depth for the detailed study sites at Chisel Farm on the River Stour (A), Silverton Mill on the River Culm (B) and Buildwas on the River Severn (C).

5. Conclusions

The results presented above demonstrate that the grain size composition of overbank floodplain deposits exhibits significant spatial variability. Cross-valley and downstream variations in the grain size composition of surface sediments representative of contemporary floodplain deposits have been documented within the study reaches. This spatial variability reflects the interaction between overbank floodplain flows and suspended sediment transport and deposition at these sites, which is in turn influenced by the local river plan-form and floodplain topography. The results from the three detailed study sites have further demonstrated the importance of such influence on local variability in the grain size composition of overbank floodplain deposits. The information obtained from this study has significant implications for understanding the development and evolution of river floodplains which are an important component of the fluvial system and for establishing the fate of sediment-associated contaminants. The results could also afford a useful basis for validating existing models of overbank sediment transport and deposition.

Acknowledgement

The financial support of the UK Natural Environmental Research Council (Research Grant GR3/8633) for the work reported in this contribution and the cooperation of landowners in allowing access to field sites and the collection of floodplain sediment samples are gratefully acknowledged.

References

Asselman, N. E. M. and Middelkoop, H.: 1995, *Earth Surface Processes and Landforms* **20**, 481-499.

Brown, A. G.: 1983, *Revue de Geomorphologie Dynamique* **32**, 95-99.

Droppo, I. G. and Ongley, E. D.: 1994, *Water Research* **28**, 1799-1890.

Howard, A. D.: 1992, *Lowland Floodplain Rivers: Geomorphological Perspectives* (ed. by P. A. Carling and G. E. Petts), 1-41.Wiley.

James, C. S.: 1985, *J. Hydraulic Res.* **23**, 435-452.

Lewin, J.: 1978, *Progress in Physical Geography* **2**, 408-437.

Marriott, S.: 1992, *Earth Surface Processes and Landforms* **17**, 687-697.

Marron, D. C.: 1992, *Earth Surface Processes and Landforms* **17**, 675-685.

Phillips, J. M. and Walling, D. E.: 1995, *J. Marine and Fresh Water Research* **46**, 349-457.

Pizzuto, J. E.: 1987, *Sedimentology* **34**, 301-317.

Rang, M. C. and Schouten, C. J.: 1989, *Historical Changes of Large Alluvial Rivers: Western Europe* (ed. by G. E. Petts), 127-142. Wiley.

Walling, D. E.: 1996, *Arch. Hydrobil. Spec. Issues Advance. Limnol.* **47**, 1-27.

Walling, D. E. and Bradley, S. B.: 1989, *GeoJournal* **19**, 53-62.

Walling, D. E. and Moorehead, P. W.: 1989, *Hydrobiologia* **176/177**, 125-149.

Walling, D. E., He, Q. and Nicholas, A.: 1996, *Floodplain Processes* (ed. by M. G. Anderson, D. E. Walling and P. D. Bates), 399-440. Wiley.

Wolman, M. G. and Leopold, L. B.: 1957, *U.S. Geological Survey Professional Paper* **282-C**, 109pp.

VARIATIONS OF BED SURFACE SEDIMENT SIZE IN A CHANNEL BEND

CHIN-LIEN YEN[1] and YAN-LANG LIN[2]

[1]*Dept. of Civil Engineering & Hydraulic Research Lab., National Taiwan University, Taipei, Taiwan, R.O.C.*
[2]*Agricultural Engineering Research Center, Chungli, Taiwan, R.O.C.*

Abstract. Variations of sediment size and its gradation of the bed surface layer in a channel bend with nonuniform sediment are investigated experimentally. Four groups of sediment with the same initial median diameter (D_o) but different initial size gradation (σ_o) have been used for experiments which were run until the equilibrium bed topography was achieved. Analyses of experimental data have yielded the following results: (1) The time of equilibrium for bed evolution decreases as σ_o increases; (2) the median size of sediment (D) for a given section in the bend increases with increasing distance from the inner bank towards the outer bank, and it also increases with increasing σ_o ; (3) the value of D/D_o along the inner bank decreases with increasing σ_o , and it also shows a gradual decrease in the upper half of the bend and a slight recovery in the lower half; and (5) the transverse variation of σ value exhibits a general trend increasing from the inner bank towards the outer bank.

Key words: channel, bend, sediment, size, gradation, sorting, bed, topography

1. Introduction

The hydraulics of sediment transport in channel bends is much more complex than that in straight channels. The complexity arises from the fact that the nonuniform sediment in a bend is subject not only to the longitudinal transport but also to the transverse transport and sorting by the secondary flow inherently associated with the bend. Traditionally, hydraulic engineers pay a great deal of attention to the characteristics of cross-sectional geometry in alluvial bends because of their concerns about river bank protection and navigation safety. However, the variation of sediment size in channel bend may have important ecological implications, in addition to its engineering significance, as it can affect the habitat of aquatic life.

In the past few decades, much progress has been made in understanding the hydraulics of flow and sediment transport in channel bends. Most of these past investigations emphasized the rate of sediment transport and channel morphology as related to flow conditions. Relatively few studies dealt with variation of sediment size within the bend. This fact renders accurate description of size variation in natural bends impossible.

Past investigations of flow in channel bends may be categorized into four groups: laboratory experiments, field surveys, theoretical analysis and numerical modeling. In the first category, many experiments have been carried out on bends with rigid (Yen, 1965) as well as movable boundaries (Yen, 1967; Yen & Ho, 1990). In the latter case only a very limited number of experimental studies dealing with nonuniform sediment which were carried out for the purpose of verifying some theoretical analyses can be found in the literature (Lee, 1984). In the case of field investigations, although these normally deal with nonuniform sediment, it is difficult to derive any general conclusions from field data because of the problems of conducting systematic research under field conditions (Bridge, 1976). Although theoretical analyses of flow in channel bends have advanced in many aspects, they have almost all concentrated on bends with uniform sediment. In the last decade, there have been many numerical models developed for simulating bend

Water, Air and Soil Pollution **99**: 81-88, 1997.

flow. However, none of these investigations has specifically taken the problem of nonuniform sediment into consideration.

Due to the action of secondary flow, there are two distinct features concerning the transport of nonuniform sediment in channel bends. The first is the sorting effect resulting in the accumulation of finer particles along the inner bank of the bend and the appearance of coarser particles along the outer bank. The second is the armoring effect that limits the scouring along the outer bank to a depth less than that for uniform sediment. The sorting phenomenon is brought about by the secondary flow which is responsible for the development of the bed topography. The development of the bed topography results in an increasingly steeper transverse bed slope. This makes it more difficult to move coarser particles up the slope laterally, and eventually only the finer ones can reach the inner side of the bend. This process continues until an armoring layer consisting of coarser particles is formed along the outer bank. At this time, the bed topography is stabilized.

The purpose of the present paper is to investigate longintudinal, as well as transverse, variations in sediment size and its gradation, under the action of the secondary flow inherited in a channel bend. The focus of this study is on the effect of the initial sediment size gradation on these variations in a bend with stabilized bed topography.

2. Problem Analysis

The size gradation and mean size of sediment on the surface of the stabilized topography is a function of a large number of variables which represent flow condition, bend geometry, initial sediment size gradation and its representative size. Since the present study focuses only on the variations of sediment size gradation, the dependent variables to be considered are the mean diameter of sediment D and the geometrical standard deviation of size gradation σ. The independent variables that may have an effect on the dependent variables include:

Channel geometry: $B, R_c, \theta_c, S_o, R, \theta$

Flow condition: H_o, V_o

Fluid properties: ρ, μ, γ

Sediment properties: $\rho_s, D_o, \sigma_o, \omega$

where B = channel width (m); R_c = radius of curvature of the bend (m); θ_c = central angle of the bend (°); R = radial coordinate (m), θ = streamwise angular coordinate (°); H_o = depth of incoming flow (m); V_o = mean velocity of incoming flow (m/s); ρ = density of water (kg/m^3); γ = specific weight of water (N/m^3); μ = dynamic viscosity of water (N·s/m^2); ρ_s = density of sediment particle (kg/m^3); D_o = mean diameter of initial bed sediment (mm); σ_o = geometrical standard deviation of initial bed sediment (0); and ω = terminal fall velocity of the mean sediment particle (m/s).

Organizing the variables listed above into dimensionless parameters and expressing the dependent variables as functions of the independent variables, one can obtain the following relationship:

$$\frac{D}{D_o}, \sigma = f_{1,2}\left(\frac{R}{R_c}, \theta, \frac{B}{H_c}, \theta_c, S_o, \boldsymbol{R}, \boldsymbol{F}, \frac{\rho_s}{\rho}, \frac{\omega}{V_c}, \frac{\rho_s}{\rho}, \frac{\omega D_o \rho}{\mu}, \sigma_o\right) \qquad \text{.............(1)}$$

where $\boldsymbol{R} = V_o H_o \rho / \mu$ and $\boldsymbol{F} = V_o / \sqrt{\gamma H_o / \rho}$.

For the present study, the bend geometry, fluid properties and sediment properties, except σ_o, are kept constant. Therefore, Eq.(1) can be reduced to

$$\frac{D}{D_o}, \sigma = f_{3,4}\left(\frac{R}{R_c}, \theta, \sigma_o\right) \qquad \text{...............................(2)}$$

Equations (1) and (2) are used for analyzing the data obtained in the experiments carried out for the present study as described below.

3. Experiments

A laboratory channel bend with θ_c = 180°, R_c = 4 m and B = 1 m was employed in the experiments for the present study. The bend was connected with a straight upstream and downstream reach, each having a length of 12.5 m. A water discharge of 0.0315 m^3/s was supplied from a constant head tank. This particular discharge was chosen on the basis that an adequate quantity of sediment could be transported in the bend as bed load and yet there was no significant amount of suspended load. Before the beginning of each experimental run, the channel bed was filled with sediment in such a way as to obtain a longitudinal surface slope of 0.002. Under these conditions, the mean velocity of the flow was 0.427 m/s.

Four groups of sediment were used for the experiments. They all have the same D_o of 1 mm, but different σ_o values of 1.5, 2.0, 2.5 and 3.0. The source material used to produce the sediment for the experiments containing the size range 0.125 mm to 8.0 mm was first segregated into several size classes by sieving. The size gradation curve for each specific σ_o value was then subdivided into intervals in accordance with those of the segregated source material. The amount of sediment in each class used to produce the required bed material was determined from the size gradation curve. The sediment was then thoroughly mixed. About 8,000 kg of graded sediment was required for each experimental run.

The sediment was then placed in the channel bend to a thickness of 15-20 cm, being thinner in the downstream portion. The bed surface was carefully smoothed to achieve the specified longitudinal slope and level cross-sections. Then the channel was filled with water to a prescribed depth. The filling process was undertaken slowly to avoid disturbance to the bed surface. After completion of the preparatory work, water was released to run through the channel bend and allow development of the bed topography. At time intervals of every 2 hours, the flow was carefully stopped and the water drained to permit measurement of bed elevations and sampling of bed sediment at prescribed locations. Flow was restarted after the completion of these measurements and the sampling. Sediment sampling was undertaken by pouring hot paraffin wax onto a small surface area at specified sites. The solidified wax with sediment adhered to it was taken out and melted in hot water to separate the sediment particles. Samples were taken from seven sections in the bend and several from each of the straight reaches. In this paper only the mean sediment size and the size gradation for the bend with stabilized bed topography are presented. Additional information obtained during the development of the bed topography is available elsewhere (Yen & Lin, 1990; Yen & Lin, 1991; Lin, 1991).

4. Results and Discussions

4.1 EQUILIBRIUM TIME

The equilibrium time, defined as the time required to develop a stabilized bed topography, is established by examining carefully the variation in bed elevation with respect to time at each location. When all the bed elevations achieve stable values (except for minor fluctuations), the time is recorded as the equilibrium time. The equilibrium times so obtained for all runs are listed in Table I, showing that as the initial sediment gradation σ_o increases the equilibrium time decreases.

TABLE I

Equilibrium Time as a Function of σ_o

σ_o	$t_e(u_* - u_{*c})/B$	
	at $\theta = 165°$	at $\theta = 180°$
1.5	0.060	0.063
2.0	0.073	0.077
2.5	0.086	0.094
3.0	0.092	0.102

Note: t_e = equilibrium time (hr); u_* = shear velocity (m/s);
u_{*c} = critical shear velocity (m/s); B = channel width (m).

4.2 SEDIMENT SIZE

Figs. 1(a), 1(b), 1(c) and 1(d) show the variations of mean sediment size for the four different initial values of sediment gradation (σ_o) of 1.5, 2.0, 2.5 and 3.0, respectively. From these figures, one can see clearly the influence of σ_o on mean sediment size. Generally speaking, the value of D/D_o increases with increasing σ_o and also with R/R_c. The range of variation across a cross-section is greater for larger values of σ_o. For a given value of σ_o, the range of variation in D/D_o increases steadily in the streamwise direction until about $\theta = 90°$ where the largest range of D/D_o appears. Thereafter the range of variation drops slightly and then remains practically unchanged. The ranges of D/D_o at $\theta = 90°$ are listed in Table II .

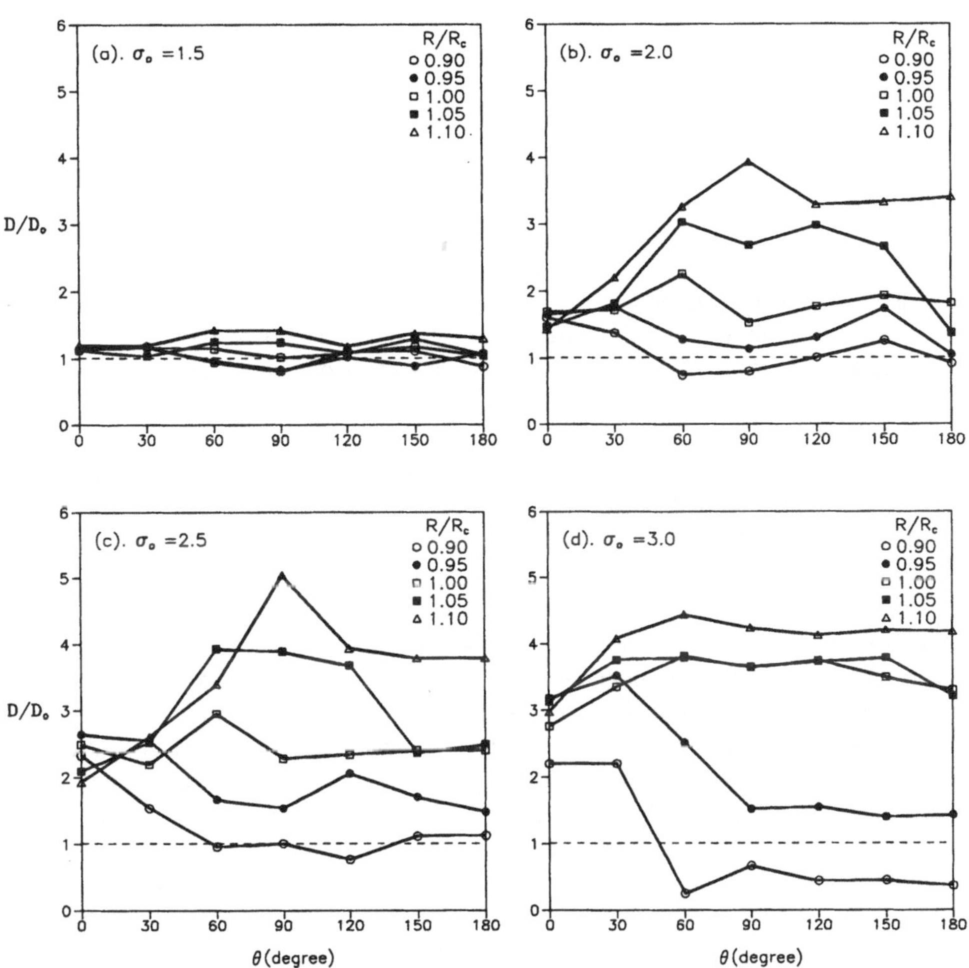

Fig.1 Variations of D/D_o on the Bed Surface

From the data presented in Table Ⅱ, one can see that the mean sediment size along the outer bank (D_2) increases more rapidly with σ_o than that along the inner bank (D_1). However, both decrease at $\sigma_o = 3.0$. Nevertheless, the ratio of D_2/D_1 increases steadily as σ_o increases. This phenomenon may be explained by the fact that in the case of $\sigma_o =$ 3.0 there is likely to be a sufficiently large amount of small particles which are easily scoured away from the outer bank area and transported to the inner bank area to form a steeper lateral slope such that a larger proportion of medium particles are not able to move up the slope. This then results in the mean diameters for both the inner and outer bank areas being smaller than those in the case with $\sigma_o = 2.5$.

TABLE II

Ranges of Sediment Size as a Function of σ_o

σ_o	D_1/D_o (inner bank)	D_2/D_o (outer bank)	D_2/D_1
1.5	0.81	1.42	1.75
2.0	1.01	3.93	3.89
2.5	1.01	5.04	4.99
3.0	0.67	4.24	6.97

From Fig. 1, it can also be observed that D/D_o decreases along the inner bank in the upper half of the bend and then increases slightly in the lower half. On the contrary, it increases fairly rapidly along the outer bank in the upper half of the bend and then reduces at around $\theta = 90°\sim100°$. Thereafter, no significant change can be observed.

4.3 SEDIMENT GRADATION

Figs. 2(a), 2(b), 2(c) and 2(d) show the variations of geometrical standard deviation of sediment size gradation in the bend for σ_o values of 1.5, 2.0, 2.5 and 3.0, respectively. In general, the sediment size gradation σ on the surface of the stabilized bed topography decreases with increasing σ_o. Its range of variation also increases with increasing σ_o. The σ value decreases in the streamwise direction in the upper half of the bend and recovers to some extent in the lower half. From Fig. 2(a) for $\sigma_o = 1.50$, one can observe that σ is greater than σ_o throughout the entire bend except for the point at $R/R_c = 1.10$ and $\theta = 90°$. In this case of low σ_o value, the sediment size is rather uniform and therefore the proportion of finer particles is small. These finer particles can readily find ample space to hide behind the coarser ones. Consequently, the particles that are transported away are most probably those of medium size, and the material on bed surface becomes less uniform, i.e. σ is greater than σ_o.

For the case of $\sigma_o = 2.0$, the number of locations where σ is less than σ_o increases to about one third of the total. These occur in the area near the inner bank, covering the region $R/R_c = 0.9 \sim 0.95$ and $\theta = 60° \sim 90°$. As for the case with $\sigma_o = 2.5$, most σ values are less than σ_o and there are only four locations where σ becomes greater then σ_o. Three of them are located in the upstream area of the bend, and the other where $R/R_c = 1.05$ and $\theta = 90°$. In the case of $\sigma_o = 3.0$, there are only three locations where σ is greater than σ_o. All of these occur at the entrance section of the bend. It is also noteworthy that the difference between σ_o and the minimum value of σ in the bend increases with increasing σ_o.

The decrease in σ with increasing σ_o described above can be explained by the fact that the proportion of finer particles is greater when σ_o is larger. A larger portion of these finer particles is easily transported away as they cannot find sufficient space behind the coarser ones to hide. Therefore, as the bed topography develops, the surface material becomes more uniform and the σ value decreases.

From Fig. 2, one can also observe that there is a general trend of increase in σ value from the inner bank area towards the outer bank area although a significant degree of fluctuation in σ values exists across a given cross-section.

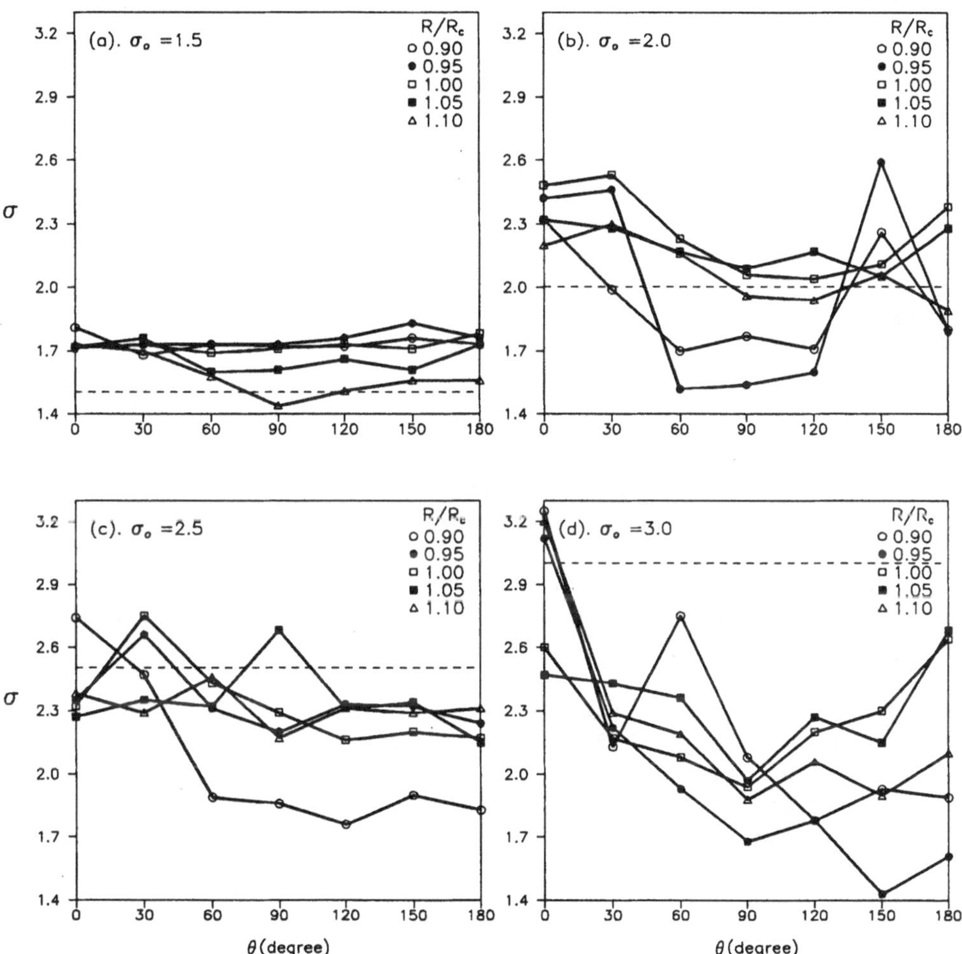

Fig. 2 Variations of σ on the Bed Surface

5. Conclusions

From the results presented above, the following conclusions can be drawn:

(1) The equilibrium time for bed evolution in the channel bend decreases as the initial sediment size gradation σ_o increases.

(2) In general, the median size of sediment D at equilibrium increases with increasing σ_o. For a given section in the bend, it also increases with increasing distance from the inner bank towards the outer bank.

(3) The ratio of D to its initial value D_o along the inner bank decreases in the upper half of the bend and then increases slightly in the lower half. The opposite trend is found along the outer bank.

(4) In general, the size gradation σ at equilibrium decreases with increasing σ_o. It also shows a gradual streamwise decrease in the upper half of the bend, and recovers somewhat in the lower half.

(5) The transverse variation of σ values exhibits a general trend of increasing from the inner bank towards the outer bank.

Acknowlegements

The results reported herein are based on research work conducted at the Hydraulic Research Laboratory, National Taiwan University (HRL-NTU), sponsored by National Science Council (NSC), Republic of China under Grant No. NSC-77-0410-E-002-39. The authors would like to express their sincere gratitudes to both HRL-NTU and NSC.

References

Bridge, J. S.: 1976, *Sedimentology*, **23**, 407-414.

Lee, H. Y.: 1984, Flow and Bed Characteristics in Alluvial Channel Bends, Ph. D. Dissertation, Univ. of Iowa.

Lin, Y. L.: 1991, Effects of Sediment Gradation on Bed Topography and Bed Surface Sediment Distribution in a Channel Bend, Ph. D. Dissertation, National Taiwan Univ.

Yen, B. C.: 1965, Characteristics of Subcritical Flow in a Meandering Channel, Ph. D. Dissertation, Univ. of Iowa.

Yen, C. L.: 1967, Bed Configuration and Characteristics of Subrictical Flow in a Meandering Channel, Ph. D. Dissertation, Univ. of Iowa.

Yen, C. L. and Ho, S.Y.: 1990, *Jour. of Hydraulic Engineering*, ASCE, **116(4)**, 544-562.

Yen, C. L. and Lin, Y. L.: 1990, *Proc. 7th Congress APD-IAHR*, Beijing, 213-217.

Yen, C. L. and Lin, Y. L.: 1991, Res. Report No. **110**, Hydraulic Res. Lab., National Taiwan Univ. (in Chinese).

TRANSPORT CHARACTERISTICS OF TILE-DRAIN SEDIMENTS FROM AN AGRICULTURAL WATERSHED

M. Stone
School of Urban and Regional Planning and Department of Geography, University of Waterloo, Waterloo, Ontario, N2L 3C5

B.G. Krishnappan
Aquatic Ecosystems Protection Branch, National Water Research Institute, Burlington,Ontario, L7R 4A6

Abstract: The use of tile drains for subsurface drainage in agricultural watersheds has created concern for the delivery of sediment to receiving waters and potential undesirable effects on surface and subsurface water quality. In this study, transport characteristics of sediment from tile drains in an agricultural watershed of the Thames River, near Kintore, Ontario, Canada were tested in a 5 m diameter, rotating circular flume located at the National Water Research Institute in Burlington, Ontario. Tile drain sediments were collected and mixed with river water at different speeds in the flume to study transport processes such as deposition, erosion and flocculation as a function of bed shear stress. During deposition and erosion experiments, water samples were collected to determine changes in the concentrations of cations, anions and dissolved organic carbon. The results show that tile drain sediments have a tendency to flocculate when subjected to a range of shear stresses. The median diameter (D_{50}) of the floc size distribution reached a maximum value at a shear stress of 0.169 Nm^{2} which can be considered an optimum shear stress for flocculation for this sediment. The critical shear stress at which all of the initially suspended sediment deposited to the flume bed was measured as 0.056 Nm^{2}. The pH and cation concentrations remained relatively constant during erosion and deposition experiments. Anion concentrations were more variable, most likely due to the presence of bacteria which could have also played a role in the flocculation mechanism of tile drain sediment.

Key Words: Tile drains, cohesive sediment, flocculation, rotating flume

1. Introduction

Subsurface drainage is a common agricultural water management practice in areas with seasonally perched water tables or shallow groundwater. In southern Ontario, agricultural drainage activity began during the middle of the last century, but since 1960, it has increased both in extent and intensity (Kelly, 1975). The extent of agricultural drainage has created concern for its potential undesirable effects on surface and subsurface water quality. Tile drainage and agricultural runoff are sources of bacteria (Panti *et al.* 1984; Palmateer *et al.* 1993), contaminants (Kladviko *et al.* 1991; Mostaghimi *et al.* 1992; Richards and Baker, 1993) and suspended solids (Culley *et al.* 1983).

Tile drains can selectively transport fine-grained sediment from soils to receiving freshwater fluvial systems (Grass *et al.* 1979). However, there is a lack of information regarding both the delivery and transport characteristics of this material in agricultural watersheds (Stone and Droppo, 1994). Such information is required to model the source, transport and fate of cohesive sediment and associated contaminants for management purposes. In this paper, the transport characteristics of tile drain sediment collected from a southern Ontario agricultural watershed are studied in the laboratory using a rotating annular flume. Transport characteristics (erosion, deposition,

Water, Air and Soil Pollution **99**: 89-103, 1997.

flocculation) of the sediment are measured and related to water chemistry for a range of turbulent flow conditions.

2. Methods

2.1 STUDY AREA:

The study area is located in a headwater catchment of the Thames River, near Kintore Ontario. Silt-loam soils at this site have developed in silty alluvial deposits that overlie calcareous loam till. The study site has been drained by tile drains for over thirty years by a series of twelve tile drains that run perpendicular to the stream. Twelve soil pits were dug at the downslope edge of an experimental plot to sample sediment in the exposed tiles. The drains connect at a header before discharging into the stream. Approximately 5 kg of sediment was collected from tile drains for the flume experiments and additional samples of soils (directly above the tile drains), as well as river bottom sediment (above and below the tile drain outlet), were collected to determine sediment geochemistry and mineralogy. Approximately 100 m upstream from the tile drain outlet, 500 L of river water was collected for the flume experiments.

2.2 CHEMICAL ANALYSIS OF SEDIMENT AND WATER:

Concentrations of major elements (SiO_2, Al_2O_3, Fe_2O_3, MgO, CaO, K_2O, TiO_2, MnO, P_2O_5) and mineralogy of the sediment were determined using X-ray fluorescence (XRF) and X-ray Powder Diffraction (XRPD), respectively, at McMaster University, Hamilton, Ontario. Results are reported as percent dry weight. Accuracy of the samples was checked by running Canadian Reference Standards and comparing the stated reference values for major elements.

Water samples collected from the river, tile outlet and flume over a range of turbulent flow conditions for both the erosion and deposition experiments were analyzed for cations, anions and dissolved organic carbon (DOC). The cations including calcium (Ca), magnesium (Mg), sodium (Na), potassium (K) and silica (Si) were analyzed using a Thermo JarralAsh Inductively Coupled Spectrometer. The instrument was calibrated with commercially available EPA certified standards following EPA Contract Laboratory Protocol. The samples are filtered (0.45 μm) and acidified to a pH <2 with trace HCl. The anions including nitrate (NO_3^-), sulphate (SO_4^{2-}) and chloride (Cl^-) were analysed using a Dionex System 2000 ion chromatograph. Alkalinity was measured by titrating the sample to a pH 4.3 end point. The total alkalinity is reported as ppm HCO_3. Dissolved organic carbon (DOC) was analyzed using a Dohrmann DC 190 Carbon Analyzer. Cation and anion data were used to calculate ionic strength.

2.3 ROTATING FLUME EXPERIMENTS:

2.3.1 Description of the Flume:

The rotating, circular flume is 5.0 m in diameter, 0.30 m wide, 0.30 m deep and rests on a rotating platform 7.0 m in diameter. The flume assembly consists of a top cover, which fits inside the flume with close tolerance (~1.5 mm on either side). The height of the top cover can be adjusted so that it makes contact with the water surface inside the flume. By rotating the flume and the top cover in the opposite direction at different speeds, it is possible to generate nearly two-dimensional flow fields with different turbulent shear stresses and turbulence intensities. Complete details of the flume and its flow characteristics are found in Krishnappan (1993).

2.3.2 Instrumentation:

The flume is equipped with a Laser Doppler Anemometer (LDA) that operates in the back-scatter mode to measure the flow field. The Laser and optics are mounted on an optical bench, positioned on the rotating platform beside a glass window of the flume. The optical bench can be traversed both in the vertical and horizontal directions. In this configuration, the instrument measures the tangential and vertical velocity components from which the time average velocities and turbulent shear stress acting on the horizontal plane in the tangential direction can be extracted. Measurements made using this instrument were used to verify a 3-D numerical model of turbulent flows in rotating flume assemblies (Petersen and Krishnappan, 1994; Krishnappan *et al.* 1994).

The size distribution of suspended sediment in the flume is measured with a Malvern Particle Size Analyzer that operates on the Fraunhoffer diffraction principle (Weiner, 1984). The instrument is mounted in a cradle beneath the flume and operates in the continuous flow-through mode. The flow-through cell of the Malvern is connected to a sampling tube fitted through the bottom of the flume. The tube faces the flow and extends up to mid-depth at the centre of the flume. The sediment suspension is drawn continuously from the flume by gravity through the sample cell then returned to the flume. The sampling tube length to the cell is kept at a minimum to reduce changes in floc structure due to the flow field that exist within the tube. With this arrangement, the instrument is capable of measuring the in-situ size distribution of the sediment flocs in the flume. The flume is also fitted with sampling ports for drawing sediment water mixture for the purpose of determining the concentration of suspended sediment and for chemical analysis. The sediment concentration was determined by the filtration method (Environment Canada, 1988).

2.3.3 Experimental Procedure:

Two types of experiments were conducted in the flume. The first type dealt with deposition behavior while the second type examined erosion characteristics of the sediment. For the deposition tests, the sediment-water suspension was thoroughly mixed by first re-suspending any deposited sediment using a brush and then further breaking-up flocs with an electric blender. The depth of water inside the flume was adjusted to 12.0 cm. The top cover was then lowered until it penetrated the water surface by about 3 mm to ensure proper contact with the water in the flume. Trapped air was removed by rotating the top cover while the flume was kept stationary. To begin the deposition test, the flume and the top cover were rotated in opposite directions at their maximum speeds for twenty minutes to thoroughly mix the sediment and water and to further break up the flocs. During this high speed operation, sediment samples were collected for sediment concentration measurements every five minutes and the size distribution was determined every two minutes with the Malvern Particle Size Analyser. Water samples for chemical analysis were collected at the 10 minute mark. After 20 minutes, flume speeds were reduced to a particular bed shear stress and maintained for a period of about five hours.

Additional water samples were collected for chemical analysis at the 60 minute and 240 minute mark for each shear stress condition.

For the erosion experiment, the sediment-water mixture was left undisturbed for a period of 114 hours to allow the sediment to settle and age on the flume bed. The speeds of the flume and top cover were increased in steps so that the shear stress is applied as a stair-case function. At each step, the sediment concentration in the flume and size distribution of the eroded sediment were measured as a function of time. Water samples were also collected at each shear stress level for chemical analysis.

3 Results

3.1 DEPOSITION RESULTS:

3.1.1 Suspended sediment concentration:

The time variation of suspended sediment concentration during the deposition experiments are shown in Fig.1. For a particular shear stress, the sediment concentration decreases at a faster rate at the start of the experiment then gradually reaches a steady state value. The time to reach steady state value and the magnitude of the steady state concentration depends on the bed shear stress. The time to reach steady state decreases as shear stress increases and the magnitude of steady state concentration increases as bed shear stress increases. The bed shear stress values in these experiments range from 0.056 Nm^{-2} to 0.324 Nm^{-2} and sediment deposition occurred under all shear stress conditions examined. At the lowest shear stress, the sediment concentration drops from the initial 250 mgl^{-1} to about 30 mgl^{-1}. If the bed shear stress were slightly lower than this value, all of the initially suspended material would have deposited. Such a shear stress condition is termed the critical shear stress for deposition. At the maximum shear stress tested, the steady state concentration is about 150 mgl^{-1} which represents about 60% of the initially suspended material. Therefore, we see that deposition of sediment has occurred in all shear stress conditions tested.

3.1.2 Size distribution of suspended sediment flocs:

Changes in the median diameter (D_{50}) of the suspended sediment distributions plotted as a function of time are shown in Fig.2. From this figure, we could infer the flocculation behaviour of the sediment and its dependency on bed shear stress. When the bed shear stress is low (low turbulence level), the median diameter of the sediment distribution decreases as a function of time. For example, in the deposition experiment with the shear stress of 0.056 N/m^2, the median size drops monotonically from 20 μm to about 10 μm which implies that larger particles settle out with minimum flocculation leading to a decrease in median size. But as shear stress increases, flocculation of the sediment in suspension becomes predominant. For example, when the shear stress is 0.121 Nm^{-2}, the D_{50} increases from 20 μm to about 25 μm. With a further increase in shear stress (0.169 Nm^{-2}), the D_{50} increases to a high of 47 μm. However, the flocs were not able to withstand the high turbulence level with subsequent increases in shear stress. Hence, smaller increases in the D_{50} can be seen for the shear stress of 0.213 Nm^{-2}. For the highest shear stress (0.324 Nm^{-2}), the D_{50} remains around the initial value of 20 μm.

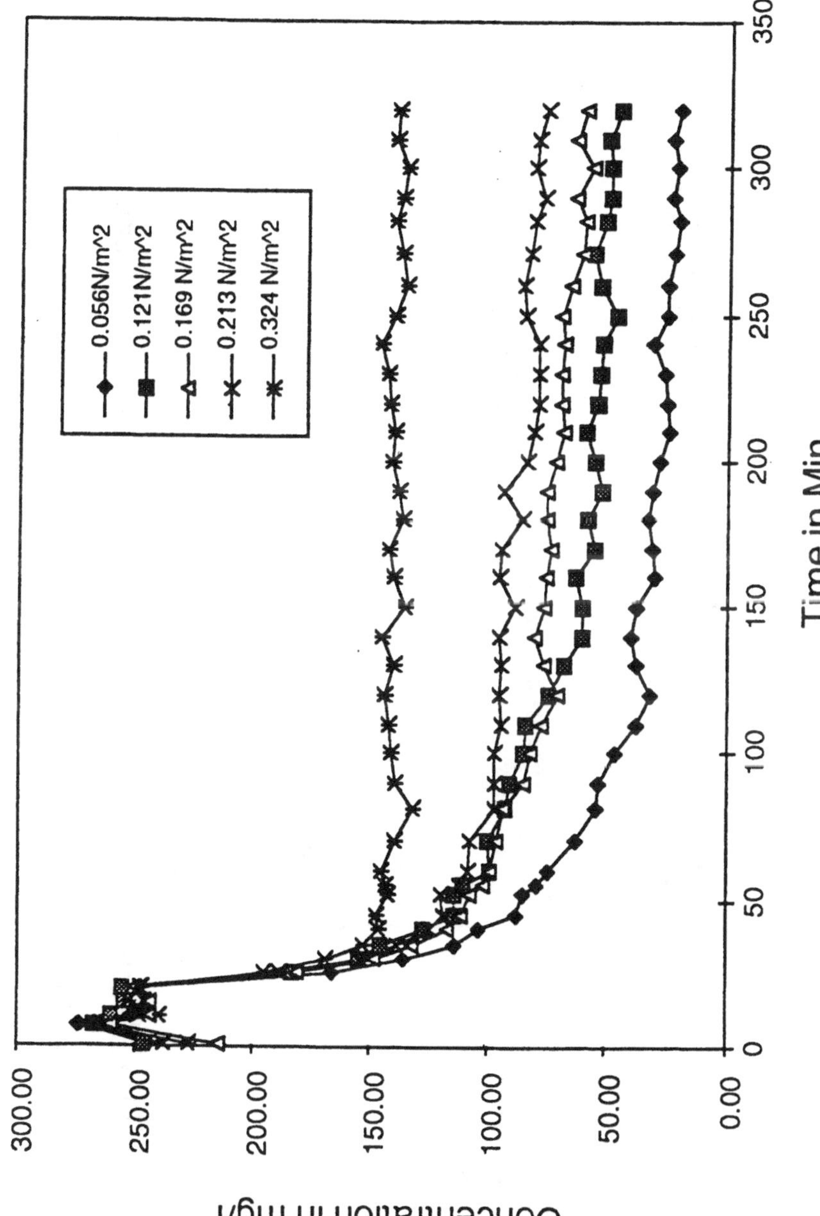

Fig. 1. Sediment concentration vs Time as a function of shear stress in deposition experiments.

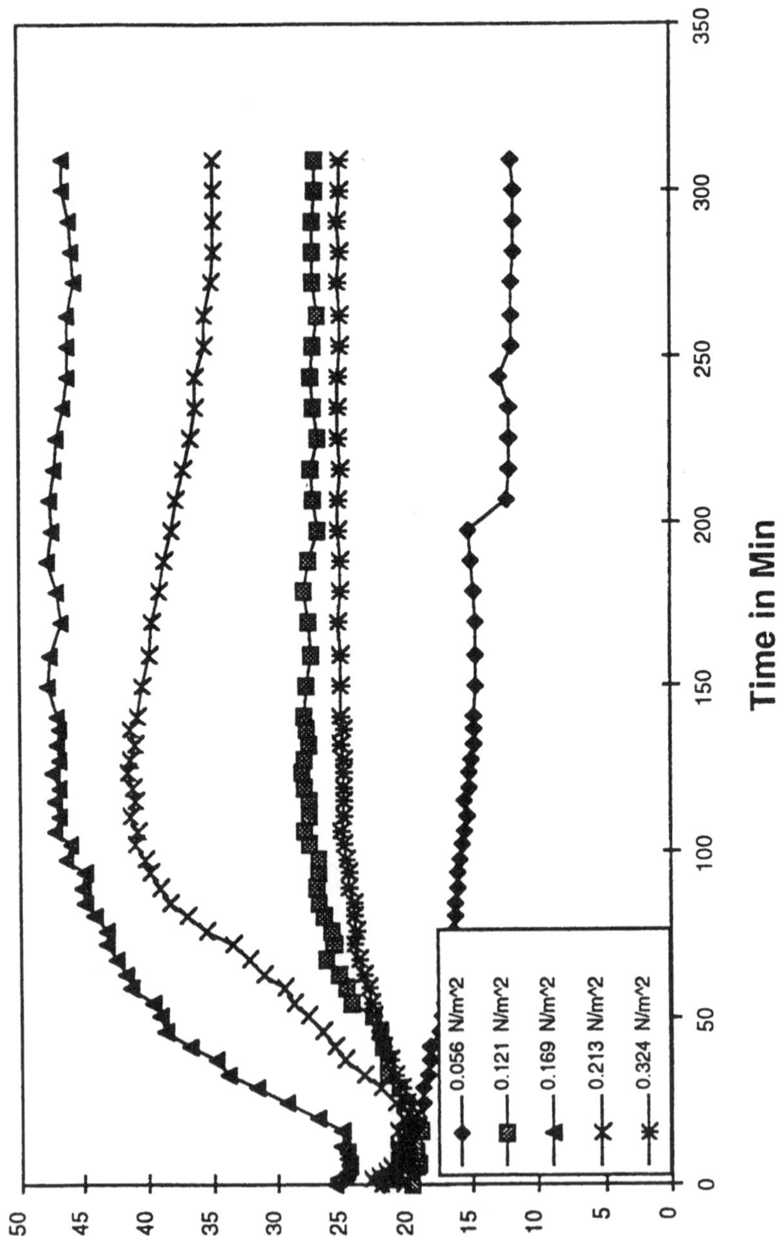

Fig. 2. The effect of shear stress on median diameter of sediment flocs in deposition experiments.

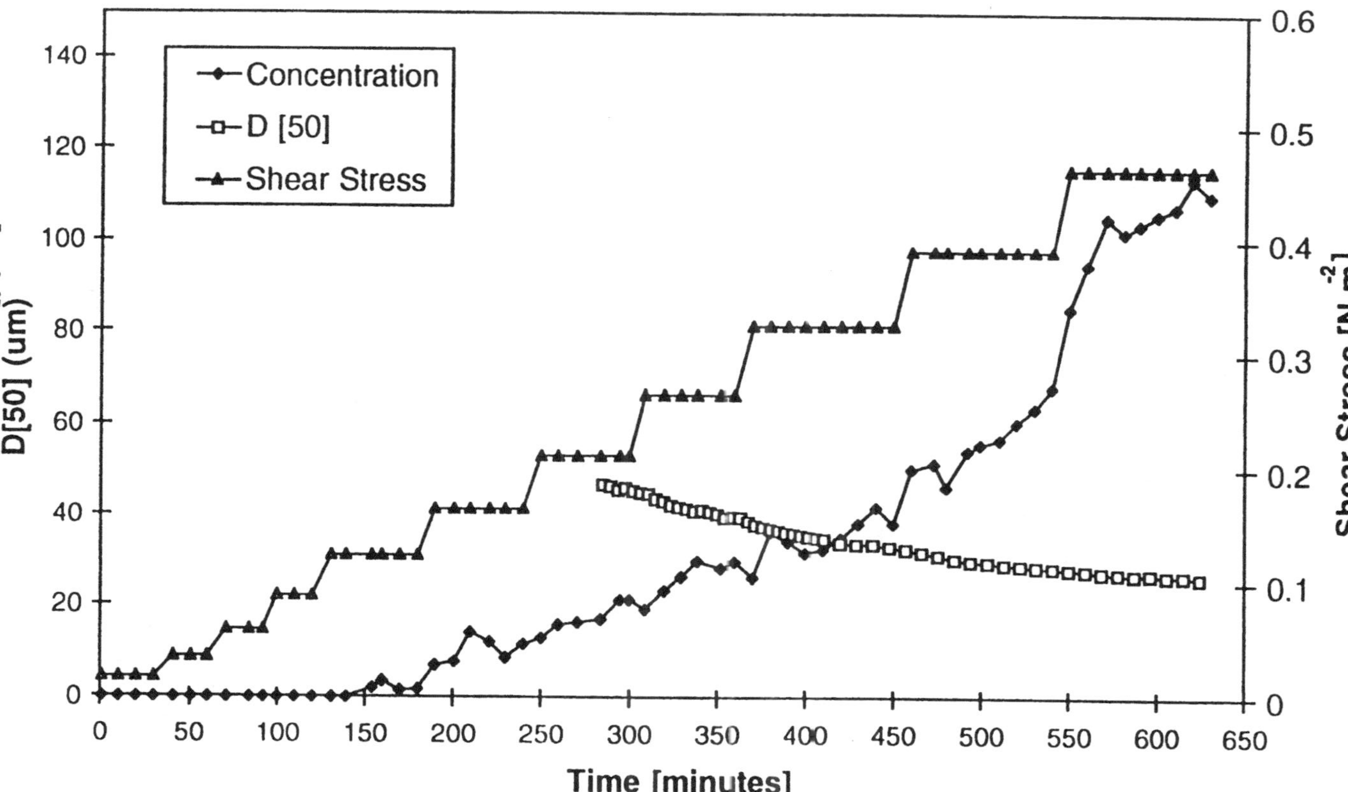

Fig. 3. Changes in sediment concentration, median diameter and shear stress as a function of time in the erosion experiment.

Table 1: Sediment Chemistry and Mineralogy (Mean and standard deviation as percent dry weight)

Sample Location	SiO_2	Al_2O_3	Fe_2O_3	MgO	CaO	Na_2O	K_2O	TiO_2	MnO	P_2O_5	LOI
1. Stream Sediment	63.4	8.2	3.3	2.4	4.2	1.4	2.3	0.34	0.078	0.35	14.03
Above Tile Drain	0.2	0.1	0.09	0.07	0.1	0.05	0.07	0.02	0.006	0.02	
2. Stream Sediment	49.5	5.3	2.3	4.0	17.0	1.2	1.6	0.19	0.051	0.15	18.71
Below Tile Drain	0.2	0.1	0.07	0.09	0.2	0.05	0.06	0.01	0.004	0.01	
3. Tile Drain Sediment	47.6	6.2	2.0	5.9	17.4	1.3	1.7	0.26	0.087	0.22	17.33
	0.2	0.1	0.07	0.1	0.2	0.05	0.06	0.02	0.007	0.02	
4. Surface Soils	69.7	11.0	5.8	2.8	2.6	0.99	3.0	0.52	0.180	0.66	2.75
	0.2	0.2	0.1	0.08	0.08	0.04	0.08	0.03	0.010	0.03	

Mineral Type	Stream Sediment Above Tile Drain	Stream Sediment Below Tile Drain	Tile Drain Sediment	Surface Soils
Quartz	~80	~60	~60	~80
Anorthite/Albite	5	17	15	15
Dolomite	4	10	16	—
Calcite	3	10	6	—
Other	5	3	3	3
Amorphous Organic	~3	—	—	~2

This result clearly shows that the tile-sediment has the tendency to flocculate and the flocculation process is dependent on bed shear stress.

3.2 EROSION RESULTS:

3.2.1 Concentration of the eroded sediment:

The concentrations of eroded sediment for a range of applied shear stress conditions as a function of time are shown in Fig. 3. Re-suspension of the sediment does not occur until the shear stress reaches a value of 0.121 Nm^{-2}, which is higher than the critical shear stress for deposition. Throughout the erosion experiment, higher values of shear stress levels were needed to bring concentrations comparable to the steady state concentrations that were established during the deposition experiment (Fig 2). In fact, at a shear stress of 0.462 Nm^{-2}, the re-suspended sediment concentration is about the same as the steady state concentration during deposition under a shear stress of 0.213 Nm^{-2}.

3.2.2 Size distribution of eroded sediment:

The largest D_{50} of sediment was observed at lower shear stresses and as the applied shear stress increased, the D_{50} of the eroded sediment decreased (Fig. 3). This seems to imply that the deposited sediment is flocculated and as the sediment is eroded, the flocs are dislodged from the bed and with increasing shear stress, the eroded flocs break up causing the size distribution to become finer.

3.3 SEDIMENT AND WATER CHEMISTRY:

The mineralogy and major element composition of tile drain and river sediment collected down stream of the tile outlet varies from that of surface soils and river sediment collected upstream of the tile outlet (Table 1). Due to weathering processes, tile drain sediment is depleted in Si, Al, K, Fe and P but enriched in Ca and Mg. The tile drains penetrate into the calcareous loam till which may account for the higher dolomite and calcite content observed in the drains and in the stream below the tile outlet. Compared to river sediments collected above the tile drain, increased levels of anorthite ($CaAl_2Si_2O_8$) and albite ($NaAlSi_3O_8$) are present in surface soils, tile sediments and stream sediment collected below the tile drain. This suggests that fine-grained surface materials are selectively transported through macropores into the tile drains. During subsequent irrigation or rainfall events, these fine-grained feldspar minerals will be resuspended and transported through the tiles to the stream.

The chemical characteristics of river water, tile drain water and stored river water are shown in Table 2. The higher Ca concentration in tile water is related to higher partial pressures of carbon dioxide (P_{CO2}) measured in soils which effectively lowers the solubility of calcite and promotes weathering of Ca from soils (Morel and Herring, 1993). The threefold increase of nitrate in tile water is attributed to leaching of chemical fertilizers and liquid manure applied to soils at the study site. River water collected for the flume experiments was stored at low temperature (4 °C) for 24 days. The chemistry of the river water changed during storage and lower concentrations of nitrate, sulphate and chloride were observed (Table 2). Analysis of the chemical data with the chemical equilibrium model $MINEQL^+$ showed that precipitation is not a removal mechanism for nitrate and sulphate from the water for the measured conditions

Table 2: Chemistry of River and Tile Drain Water

Sample					Concentration (mg/L)				
	Ca^{2+}	Mg^{2+}	Na^{+}	K^{+}	Si	Cl^{-}	NO_3^{-}	SO_4^{2-}	HCO_3^{-}
River Water	89.20	22.30	9.81	1.16	4.24	22.30	11.30	33.20	306
Tile Drain Water	106.00	21.40	7.40	0.73	4.98	12.30	44.20	20.90	305
Stored River Water	81.90	20.00	7.12	1.21	2.87	5.55	0.00	3.45	202

of pH and temperature. A second possible explanation for the observed differences is related to the reduction of nitrate and sulphate by bacteria. In waters initially containing oxygen but lacking a mechanism for reaeration, organic matter can be oxidized by, successively, dissolved O_2, NO_3^-, and SO_4^{2-} (Bender *et al*, 1977).

The chemical composition of flume water was measured at intervals for conditions of varying shear stress to examine potential effects of variable solution chemistry on depositional characteristics of cohesive sediments (Table 3). For all deposition experiments, the pH and concentrations of Ca, Mg, Na, K, Si, HCO_3^- did not vary (coefficient of variation, CV) by more than 5%. However, concentrations of NO_3^-, SO_4^{2-} and Cl^- were more variable (CV > 70%). This variability results from the fact that initial concentrations of NO_3^-, SO_4^{2-} and Cl^- measured in river water decreased by 100%, 90% and 75%, respectively during storage. Variation in water chemistry is also related to the order in which the depositional experiments were conducted. The first two deposition experiments were conducted on June 24 and 25, 1996 at a shear stress of 0.213 and 0.324 Nm^{-2}, respectively. During these two experimental runs, the concentration of the three anions were low (Table 3) but with successive flume runs on June 26, 27 and July 3 at a shear stress of 0.056, 0.121 and 0.169 Nm^{-2}, respectively, concentrations of NO_3^-, SO_4^{2-} and Cl^- returned to levels approximating the original chemical composition of the water (Table 2). This is likely due to reaeration of flume water where oxidation of ammonia and hydrogen sulphide would increase levels of nitrate and sulphate.

Changes in the chemical composition of flume water during erosion experiments are shown in Table 4. During this experiment, concentrations of Ca, Mg, Na, K, Si, HCO_3^- were relatively stable and generally did not vary by more than 5%. However, the concentrations of NO_3^-, SO_4^{2-} and Cl^- were more variable (41 to 49%).

4 Discussion and Conclusions

Deposition and erosion experiments in the rotating flume give quantitative information on the mode of transport of tile drain sediment when it enters the stream. If bed shear stress in the stream is less than the critical shear stress for deposition (0.056 Nm^{-2}), then all of the tile sediment entering the stream will be deposited on the stream bed. But, if the bed shear stress in the stream is greater than the critical shear stress for deposition, then only a portion of the sediment entering the stream is transported in suspension.

Table 3: Water Chemistry During Deposition Experiments. •

Date	Flume Run Shear Stress (Nm^2)	Time (min.)	pH	Ca^{2+}	Mg^{2+}	Na^+	K^+	Si	Cl^-	NO_3^-	SO_4^{2-}	HCO_3^-	DOC
				Concentration (mg/L)									
June-26	0.056	10	8.48	79.00	20.00	7.05	1.19	2.69	21.90	9.59	24.70	302	3.80
		60	8.48	83.00	20.90	7.50	1.35	2.89	21.70	10.80	29.50	295	3.85
		240	8.49	80.50	20.40	7.43	1.41	2.77	20.00	10.40	28.70	299	3.77
June-27	0.121	10	8.49	81.90	21.00	7.42	1.22	2.86	20.00	9.43	27.00	286	3.76
		60	8.47	79.80	20.20	7.32	1.14	2.81	20.80	9.97	28.20	292	3.47
		240	8.49	84.50	21.60	7.73	1.30	3.00	15.10	10.70	28.30	294	3.94
July-3	0.169	10	7.69	76.30	20.50	7.30	1.18	2.81	15.20	10.90	30.00	281	3.78
		60	8.45	74.20	19.90	7.06	1.14	2.89	14.60	9.91	26.80	281	3.51
		240	8.47	79.00	21.10	7.82	1.36	2.82	12.00	5.60	16.60	280	3.90
June-24	0.213	10	8.45	81.90	20.00	7.12	1.21	2.87	5.55	0.00	3.45	302	4.03
		60	8.45	88.10	22.00	7.38	1.12	2.97	4.42	0.00	0.00	307	4.10
		240	8.45	85.60	21.50	7.05	1.13	2.87	2.36	0.00	0.00	303	4.96
June-25	0.324	10	8.45	87.20	21.50	7.22	<1.00	2.75	1.52	0.00	0.00	292	4.44
		60	8.49	80.20	20.20	6.85	<1.00	2.74	1.35	0.00	0.00	302	4.41
		240	8.45	83.20	20.70	7.08	<1.00	2.86	0.00	0.00	0.00	296	4.25

Table 4: Water Chemistry During Erosion Experiment

Flume Run	Concentration (mg/L)										
Shear Stress (Nm^2)	pH	Ca^{2+}	Mg^{2+}	Na^+	K^+	Si	Cl^-	NO_3^-	SO_4^{2-}	HCO_3^-	DOC
0.090	8.40	80.50	21.30	7.06	1.02	2.94	12.30	8.11	24.50	280	3.94
0.169	8.40	79.30	20.70	6.88	<1.00	2.87	44.00	22.70	63.30	290	3.60
0.259	8.41	76.10	20.10	6.68	<1.00	2.66	20.70	11.60	31.20	281	3.88
0.390	7.39	78.60	20.30	6.92	<1.00	2.94	22.00	11.20	31.90	282	4.15
0.460	8.44	77.50	20.40	7.04	1.03	2.82	22.00	12.00	32.80	240	4.14

Thus suspended tile sediments will be carried long distances within the stream depending on variations in shear stress along the length of the stream and the textural composition of the stream bed. The erosion experiments give quantitative information on the erodibility of the deposited sediment which is useful for the development of transport models of sediment in receiving streams. A modelling framework that employs such information is developed by Krishnappan (1996).

The shear stresses measured from the flume experiments can be translated into easily measurable hydraulic parameters such as the average flow velocity or the flow rate if some simplifying assumptions are made regarding the flow in the natural system. If we assume that the flow is rough turbulent and uniform, then the relationship between the average velocity and the bed shear stress becomes:

$$\frac{U}{V_*} = 2.50 \ell n(11.0 \frac{R}{D_{65}}) \quad (1)$$

where U is the average flow velocity (ms^{-1}), *ln* is the natural logarithm, R is the hydraulic radius, D_{65} (m) is a representative size of the bed sediment and V_* is the shear velocity (ms^{-1}), which is related to the bed shear stress as follows:

$$V_* = \sqrt{\tau_0 / \rho} \quad (2)$$

Here, τ_0 is the bed shear stress (Nm^{-2}) and ρ is the density of water. Therefore, given the bed shear stress, hydraulic radius and a representative size of the bed material, the average velocity that corresponds to the bed shear stress can be calculated using equations 1 and 2.

The rotating flume experiments show that tile drain sediment has a tendency to flocculate when subjected to turbulent shear stress. This can be seen from the size distribution data measured during deposition and erosion experiments. In addition, the dual role of the turbulent shear stress in the flocculation process can also be inferred. For example, as sediment deposits in flows with low shear stress, the size distribution of the suspended sediment becomes finer and finer. This suggests that the sediment settles as individual particles resulting primarily in the settling of coarser fractions, thus leaving finer particles in suspension. As shear stress is increased, the size distribution of suspended sediment becomes coarser, which suggests that sediment particles in suspension collide and form flocs. The degree of flocculation reaches a maximum at a shear stress of 0.169 Nm^{-2} . With a further increase in shear stress, the floc sizes become smaller suggesting that the flocs are breaking up beyond the optimal shear stress for flocculation. Size distributions measured during the erosion experiments suggest that sediment eroded at lower shear stress values are coarser and as the shear stress is increased, flocs will break up causing the distribution to become finer. At high shear stress during the erosion experiment, the median size of the sediment flocs approach the size distribution established during the start of the deposition tests. The latter was produced by blending the sediment water mixture by an electric blender. This suggests that at the highest shear stress applied during the erosion test nearly all the flocs break up and the resulting size distribution can be regarded as that of primary particles.

The shear stress and the turbulence level created in the flume provide the collision mechanism, which is one of the two building blocks in the flocculation process. The second building block, namely, the coagulation mechanism is provided by either chemical coagulation or biological coagulation or both. Chemical coagulation is due to the balance between Van der Waals force of attraction and the electrostatic force of repulsion (Stumm and Morgan, 1981). The latter force is a function of the ionic strength of the suspending medium. With increasing ionic strength, the electric double layer is compressed and the electrostatic repulsion between particles is decreased (Stumm, 1992). Biological coagulation is due to polymeric fibrils secreted by bacteria which represent the dominant bridging mechanism of organic and inorganic components of flocs (Leppard, 1995; Liss *et al.* 1996). Chemical data from both deposition and erosion experiments show that pH as well as concentrations of DOC and cations remained relatively constant while the anions were more variable at various shear stress levels (Fig. 4). Ionic strength of the flume varied during erosion and deposition experiments. However, it is difficult to determine to what extent the observed changes in ionic strength can be attributed to changes in shear stress and the presence of bacteria. Still important and yet to be determined is the relative importance of bacteria and changes in solution chemistry on flocculation of the tile sediment for a range of turbulent shear stresses.

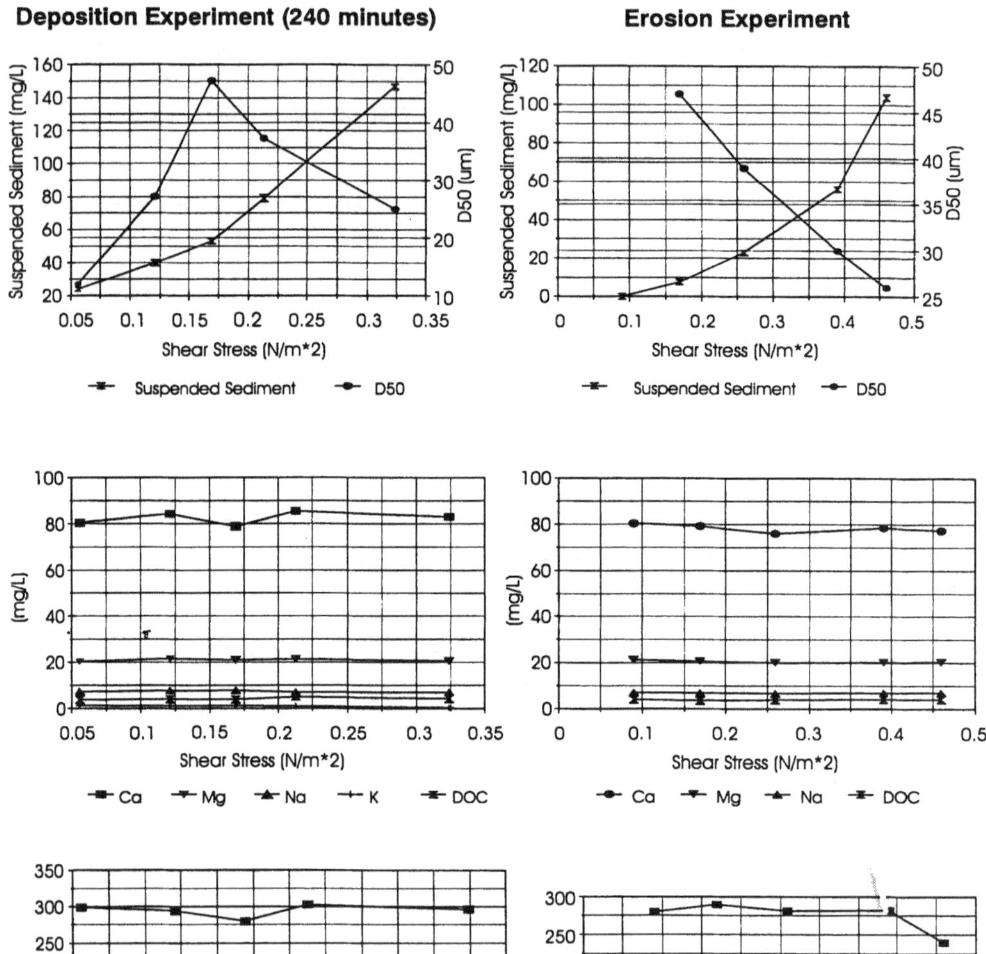
Deposition Experiment (240 minutes)
Erosion Experiment
Suspended Sediment (mg/L)
D50 (um)
Shear Stress (N/m*2)
Suspended Sediment
D50
(mg/L)
Ca
Mg
Na
K
DOC
Cl
NO3
SO4
HCO3

Acknowledgements

The authors wish to thank Robert Stephens of Aquatic Ecosystem Protection Branch of National Water Research Institute for assistance with sample collection and conducting the flume experiments. The comments of Dr. H. Saunderson, Dr. G. Wall and two reviewers were greatly appreciated. Funding was provided by a Natural Sciences and Engineering Research Council (NSERC) of Canada Operating Grant held by M. Stone.

References

Bender, M.L., Fanning, K. A., Froelich, P.N. and Maynard, V.: 1977, *Science*, **198**, 605-608.

Culley, J.B., Bolton, E.F. and Bernyk, V.: 1983. *J. Environ. Qual.* **12(4),** 493-503.

Environment Canada.: 1988. *Laboratory Procedures for Sediment Analysis*, Inland Waters Directorate, Ottawa.

Grass, L., Willardson, L., Lemert, R.: 1979. *Transactions of ASAE.* 1054-1057.

Morel, F.M. and Hering, J.G.: 1993. *Principles and Applications of Aquatic Chemistry.* John Wiley and Sons, N.Y.

Kelly, K.: 1975. *Canadian Geographer.* **4**, 279-298.

Klavidko, E., Van Scoyog, G., Monke, E., Oates, K. and Pask, E.: 1991, *J. Environ. Qual.* **20**, 264-270.

Krishnappan, B.G.: 1993. *J. Hydraulic Engineering*, ASCE. **119 (6)**, 758-767.

Krishnappan, B. G., Engel, P., and Stephens, R.: 1994. *National Water Research Institute*, Environment Canada, Report: 94-102

Krishnappan, B.G.: 1996. *National Water Research Institute*, Environment Canada, Report: 96-160.

Leppard, G.: 1995. Sci. Tot. Environ. **165,**103-121.

Liss, S., Droppo, I.G., Flannigan, D. and Leppard, G.: Environ. Sci. Technol. 30(2), 680-686.

Mostaghimi, S., Younos, T.M. and Tim, U.: 1992. *Water Resources Bull.* **28(3),** 545-552.

Palmateer, G., McLean, D., Kutas, W. and Meissner, M.: 1993, *In: Particulate Matter and Aquatic Systems.* Ed. S.Rao, Lewis Publishers.

Panti, N.K., Toxopeus, K., Tennant, A.D., and Hore, F.R.: 1984. *Water Res.* **18(2),** 127-132.

Petersen, O. and Krishnappan, B.G.: 1994. *J. Hydraulic Research*, **32 (4)** 483-494

Richards, R. and Baker, D.B.: 1993. *Environ. Tox. and Chem.* **12**, 13-26.

Stumm, W.: 1992. *Chemistry of the Solid-Water Interface*, Wiley, N.Y. p. 428.

Stumm, W. and Morgan, J.: 1981. *Aquatic Chemistry*, Wiley, N.Y. p.656

Weiner, B.B. 1984.: In *Modern Methods of Particle Size Analysis*, Ed. H.G.Barth. John Wiley and Sons.

SEDIMENT BUDGET ANALYSIS FOR RIVER RESERVOIRS

U. KERN and B. WESTRICH
Institute of Hydraulic Engineering, University of Stuttgart,
D-70550 Stuttgart, Germany, E-mail: kern@iws.uni-stuttgart.de

Abstract. The sediment budget of a reservoir on the Neckar River, Germany, was investigated by means of experimental and numerical methods. Field measurements of channel bathymetry show that sedimentation and erosion occur primarily in the lower backwater-influenced section of the reservoir, which stores approximately 350,000 m^3 of fine-grained deposits. Sediment load balances for two major storm events in Dec. 1993 and Apr. 1994 showed net erosion of 32,000±10,000 tonnes and 24,000±5,000 tonnes of sediment, respectively. A balanced sediment budget was found for a minor flood in Jan. 1995. In agreement with the field data, numerical simulation of sediment transport over a period of 45 years demonstrates that the river reservoir served initially as a sediment trap from 1950 to 1978, and since then as a temporary storage basin for sediment.

Key words: Suspended Particulate Matter, Sediment Budget, River Reservoir, Numerical Model, Sedimentation, Erosion

1. Introduction

Lock-regulation of rivers has a strong influence on the transport of suspended particulate matter (SPM). At low discharges, SPM settles in backwater sections of river reservoirs. During storm events, erosion of channel sediment may take place. Beside quantitative aspects, the need to evaluate the sediment budgets of river reservoirs arises from the fact that various pollutants are associated with fine-grained sediment.

In this field study, the sediment budget of a river reservoir is analysed (i) by echo-sounding surveys of channel morphology, (ii) by measuring loads of SPM at the boundaries of the reservoir during storm events and (iii) by numerical transport simulation over a period of 45 years.

2. Field Site

The 11 km long lock-regulated section investigated is located in the middle of a chain of 27 reservoirs on the Neckar River, Germany. It is bounded by hydraulic structures at its upstream and downstream end (Fig. 1). At station N3, the discharge of the Neckar River varies between 14 and 1650 m^3/s with an average (MQ) of 88 m^3/s. During the period 1988 to 1994, the annual sediment yield from the catchment area (A_D) of 7916 km^2 has ranged from 7.4 to 70.9 t/km^2. At its mouth, the MQ and A_D of the Enz River are 21 m^3/s and 2230 km^2, respectively.

Water, Air and Soil Pollution **99**: 105-112, 1997.

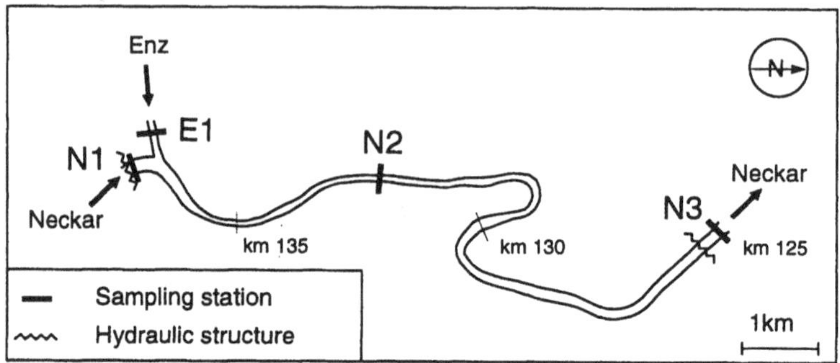

Figure 1. The Lauffen Reservoir on the Neckar River showing the location of the sampling stations E1, N1, N2 and N3.

3. Experimental and Numerical Methods

3.1. Channel sediment budgets

Several times between 1950 and 1994, the bathymetry of channel cross-sections was surveyed at a spacing of 100 m along the river reservoir. Until 1974, the sounding was performed manually using rods. Since then, a survey ship equipped with a laser-rangefinder (Lara 10, Fa. Krupp Atlas Elektronik) and an echo-sounder (Deso 20, Fa. Krupp Atlas Elektronik, frequencies: 33 kHz, 210 kHz) has been used. By integrating the bathymetric data over the channel width (w), an average geodetic height (z) of the sediment surface was calculated for each cross-section. Sediment volumes were obtained by multiplying w by z and integrating the resulting bed area over the length of the river section.

3.2. SPM load budgets

During flood events in Dec. 1993, Apr. 1994 and Jan. 1995, water samples were collected from the Neckar River at stations N1, N2 and N3 and from the Enz River at E1 (Fig. 1). Samples of 1 L were taken from the centroid of the flow near the water surface. Cross-sectional representativity of the sampling procedure was measured to be $< 2\%$ (Kern, 1995). The concentration of SPM was determined gravimetrically by filtration of the samples using 0.45 μm cellulose acetate membrane filters (Sartorius, Göttingen).

Discharges were obtained from gauging stations at N3 and E1, and calculated at N1 and N2. Event loads were calculated as follows: SPM values were interpolated and multiplied by the corresponding discharge values and the resulting transport rate was integrated over the entire flood period.

3.3. Numerical simulation

The numerical compartment model employed consists of one-dimensional models for flow and sediment transport, coupled with a model for a layered sediment bed. It accounts for the following governing physical processes: convective and dispersive transport, sediment mixing at the river bed, as well as deposition, consolidation and resuspension of sediment. The concepts and formulation of the numerical model as well as the calibration procedure employed are explained in more detail by Westrich *et al.* (1996) and Dreher and Westrich (1995).

4. Results and Discussion

4.1. Channel sediment budgets

With regard to sediment dynamics, the river reservoir may be divided into a morphologically almost inactive section upstream of km 130 and a section further downstream (see Fig. 2). The budgets of channel sediment volume shown in Fig. 3 have been calculated for the downstream part of the reservoir only. As the water surface area in this section is approximately 500,000 m^2, a sounding error of 0.01 m as measured over a concrete surface corresponds to an error of 5,000 m^3 in sediment volume or of 7 t in sediment mass (bulk density: 1,40±0,01 t/m^3, Mayer, 1988). In case of sediment mud, the sounding error is assumed to be larger (0.03 to 0.05 m) due to interference or ambiguity in echo-sounding signals at the water-sediment interface. For better interpretation of the results presented in Fig. 3, cumulative discharge plots for major flood events are provided in Fig. 4.

Between 1950, when the downstream hydraulic structure was completed, and 1973, more than 500,000 m^3 of sediment was deposited in the lower section of the reservoir. As shown in Fig. 2 (a), the deposits reached a maximum thickness of more than 3 m upstream of the hydraulic structure near Lauffen. The settling of suspended particles in the downstream section is due to the fact that the transport capacity of the flow decreases along the backwater curve at low and mean discharge in this part of the reservoir.

From 1973, periods of sediment accumulation and erosion alternate. Without exception, intervals that are marked by sediment erosion are related to short flood events with a peak discharge above 1200 m^3/s (Fig. 4). The greatest loss, involving 125,000 m^3 of sediment, was recorded between Aug. 1989 and Jun. 1991. Within this period, a peak discharge of 1640 m^3/s was measured in Feb. 1990, when a flood lasting

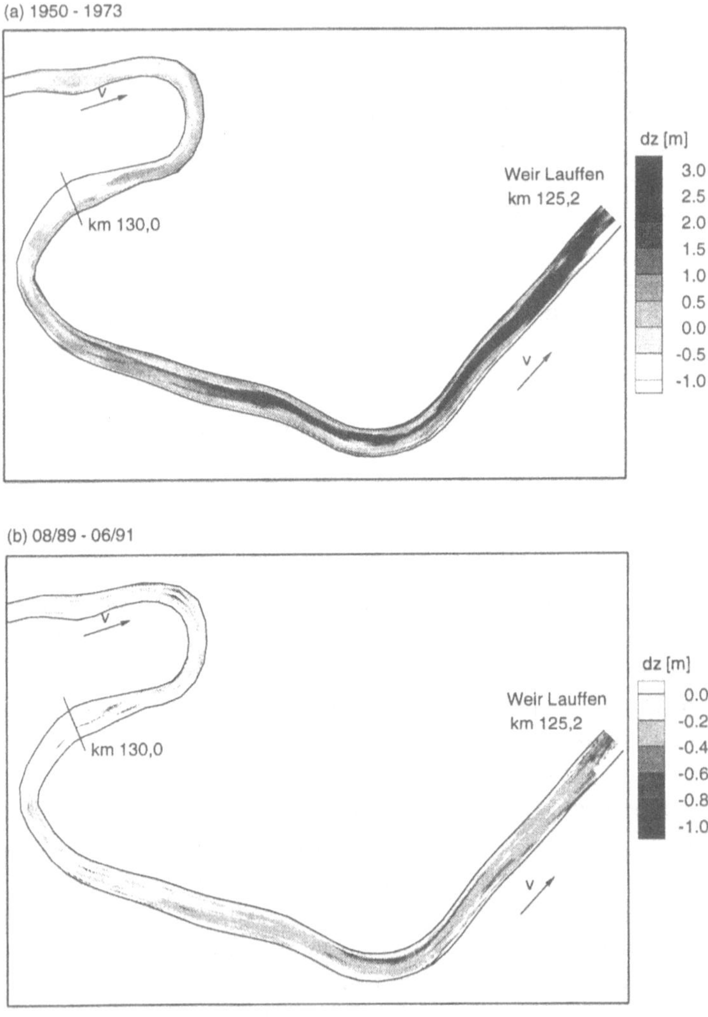

Figure 2. Change of channel bed elevation in the lower part of the reservoir between (a) 1950 and 1973, (b) Aug. 1989 and Jun. 1991

72 hours spilled a total of about $250 \cdot 10^6$ m^3/s of water. The corresponding erosion depth in the backwater section, shown in Fig. 2 (b), was 0.25 m on average and reached a maximum of about 1 m. Preferential erosion at the shallow inner side of the river bend at km 127 indicates that erosion depends not only on the bed shear stress, but also on the supply of erodible channel sediment.

During the period Mar. 1984 to Aug. 1989, the channel bed volume increased by 153,000 m^3. Within this period, the storm event in Mar. 1988 exceeded the arbitrarily chosen discharge of 300 m^3/s for

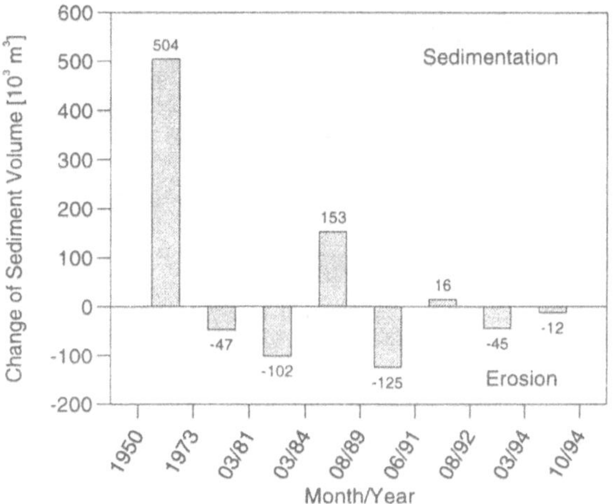

Figure 3. Changes of channel bed volume within the backwater section of the reservoir (Neckar km 125.2 to km 130.0)

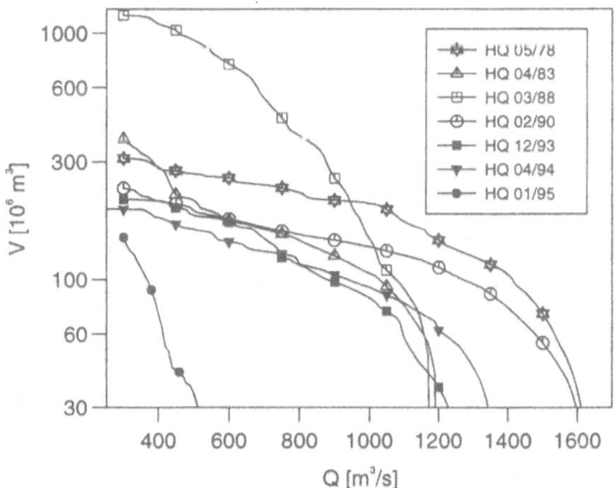

Figure 4. Cumulative discharge diagram of major flood events at station N3.

24 days resulting in a cumulative discharge of $1100 \cdot 10^6$ m^3. Obviously, sediment deposition is enhanced at moderate discharge, when the backwater effect is still apparent in the lower part of the reservoir and the influx of SPM from the catchment is increased.

4.2. SPM LOAD BUDGETS FOR FLOOD EVENTS

SPM loads have been determined for two major, three day flood events in Dec. 1993 and Apr. 1994 and for a minor flood in Jan. 1995 extending over a period of one week (see Fig. 4, Tab. I).

The mass balance for the Dec. 1993 event shows that the amount of SPM flowing into the river reservoir was 160,000 t, with the contribution of the Neckar River being twice that of the Enz River. The load of

Table I. SPM load budgets of flood events in Dec. 1993, Apr. 1994 and Jan. 1995. The table shows the event loads at the sampling stations shown in Fig. 1, the SPM mass balances Δm_1 and Δm_2, an estimate ε of the error affecting the sediment budgets, the flood duration t_s and the average sampling interval Δt_s.

Station	SPM load [10^3 tonnes]		
	HQ 12/93	HQ 04/94	HQ 01/95
E1	53	8	15
N1	107	192	52
N2	n.d.	204	69
N3	196	228	67
Δm_1=N3-(E1+N1)	36	28	0
Δm_2=N3-N2	n.d.	24	-2
ε	10	5	3
t_s [h]	81	72	168
Δt_s [h/sample]	6.8	1.8	2.6

n.d.: not determined

SPM leaving the study section amounted to 196,000 t. The difference between efflux and influx over the entire flood period of 81 hours was 36,000 t of SPM. For the Apr. 1994 event, the inflowing SPM load of 200,000 t from E1 and N1 corresponds closely to the load of 204,000 t at N2. At the downstream station E3, the event load of SPM was measured as 228,000 t. For the lower section of the reservoir between N2 and N3, a net erosion of 24,000 t was therefore calculated. The Jan. 1995 flood shows a balanced budget of 67,000 t of SPM entering and leaving the study revervoir.

In Tab. I estimates of the uncertainty of the event load budgets are given, accounting for the following factors (for details, see Westrich and Kern, 1996): Non-uniform distribution of SPM concentration in the cross-sections, uncertainty of the discharge values of the Enz River which was assumed to be 10 %, uncertainty of SPM concentration due to the analytical procedure and the discontinuous sampling strategy involving an interpolation of the SPM data. In Dec. 1993 the sparse sampling at a rate of one sample per 7 hours resulted in higher inaccuracy (10,000 t) for the load balance than that calculated for the subsequent events.

4.3. Numerical simulation

Fig. 5 shows the evolution of the sediment mass trapped in the reservoir during the period from 1950 to 1994. A period of sediment accumulation

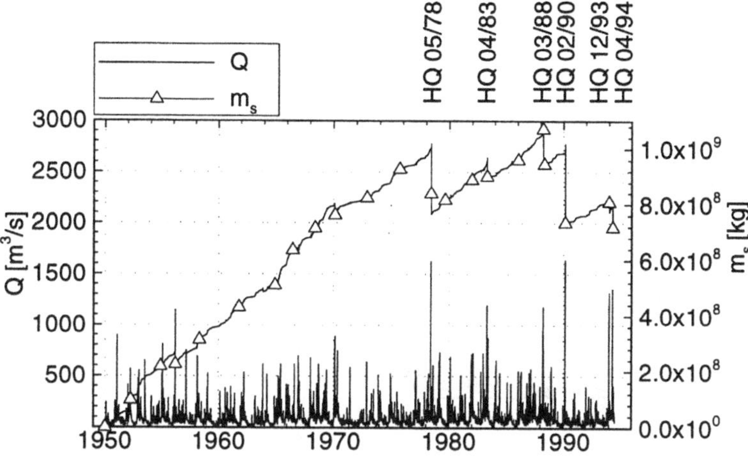

Figure 5. Graph of discharge and sediment deposited in the reservoir (1950 to 1994)

extending until 1978 can be distinguished from alternating deposition and scour in the subsequent period. Several factors account for this change in sediment dynamics: First, improvement of sewage treatment and installation of retarding basins for urban stormwater runoff have reduced the sediment load of the Neckar River. Second, between 1973 and 1968 13 hydraulic structures were constructed upstream of the study reservoir reducing SPM transport. Third, there was a lack of erosive floods in the initial period, followed by a sequence of significant flood events between 1978 and 1994 (Fig. 4). Finally, 500,000 m^3 of sediment corresponds at a horizontal water level to a reduction of the cross-sectional area of the backwater section by approximately 20 %. Hence, sediment transport capacity augmented especially at low and moderate discharges, when the constant headwater level controls the river stage in the downstream part of the reservoir. Since 1978, the flood events have not only prevented further accumulation of sediment, but they have uncovered deposits that were formed before 1970 (Fig. 4). Due to the higher erosion shear strength of these older sediments, the measured (Fig. 3) and calculated (Fig. 4) scour associated with the 1993 and 1994 flood events is small compared to events where a greater supply of erodible sediment was avaibable.

Fig. 6, where each hysteresis curve corresponds to an individual storm event, illustrates the meaning of sediment consolidation with respect to sediment erosion. For any arbitrarily chosen discharge at two stages of a storm event, the erosion rate is higher during the rising flood, when the uppermost, weak sediment layers are exposed to the flow, than during the falling flood, when older, more consolidated layers are subject to erosion. Most importantly, Fig. 6 indicates that erosion

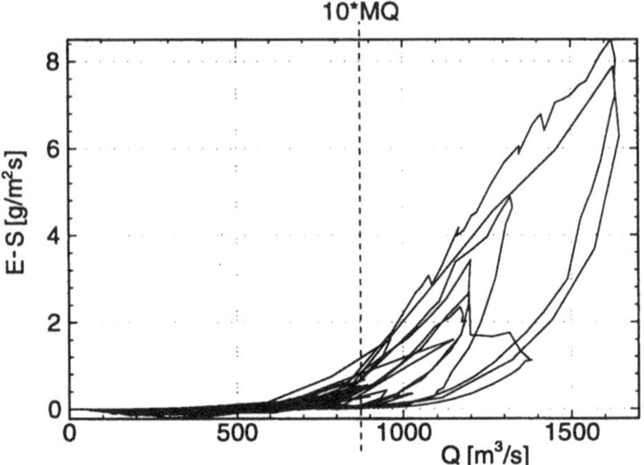

Figure 6. Mean rate of net erosion in the lower part of the reservoir versus discharge

rates increase rapidly as discharge exceeds the mean discharge by a factor of 10. This finding is in agreement with the experimental results presented in Fig. 3 and Tab. I.

5. Conclusions

River reservoirs provide permanent or temporary storages for fine-grained sediment depending on the amount and properties of the trapped sediment deposits, the hydrological regime, and the supply of SPM from the catchment. Accumulation of sediment takes place in backwater sections at low to moderate discharge. The transition from net sedimentation to net erosion can be characterized by a critical discharge, which was found to be approximately 10 times the mean discharge for the river reservoir investigated.

References

. Dreher, Th., Kern, U. and Westrich, B.: 1995, In J. Gardiner (ed.), *Hydra 2000 XXVIth IAHR Congress, Vol. 4*, Th. Telford, London, 199-204.
. Kern, U.: 1995, In J. Gardiner (ed.), *Hydra 2000 XXVIth IAHR Congress, Vol. 5*, Th. Telford, London, 73-78.
. Mayer, H.-J.: 1988, *DGM*, **32**, 27-34 (in German).
. Westrich, B. and Dreher, Th.: 1996, *Arch. Hydrobiol. Spec. Issues Advanc. Limnol.*, **47**, 373-380.
. Westrich, B. and Kern, U.: 1996, WB 96/23 (HG 237), Institut f. Wasserbau, Univ. Stuttgart, 186 pages (in German).

Address for correspondence: Ulrich Kern, Institute of Hydraulic Engineering, University of Stuttgart, D-70550 Stuttgart, Germany

EXPERIMENTAL TECHNIQUES FOR LABORATORY INVESTIGATION OF CLAY COLLOID TRANSPORT AND FILTRATION IN A STREAM WITH A SAND BED

AARON I. PACKMAN, NORMAN H. BROOKS, and JAMES J. MORGAN

Keck Laboratories of Hydraulics and Water Resources, Department of Environmental Engineering Science, California Institute of Technology, Pasadena, California 91125, USA

Abstract. The exchange of kaolinite clay between a stream and its associated sand bed was investigated in a recirculating flume. Bedforms at the sediment-water interface result in two bed-exchange mechanisms: the bedform shape induces an advective flow through the bed ('pumping'), and dune propagation causes the trapping and release of pore water ('turnover'). Chemical and electrostatic interactions then result in filtration of clay by the bed. In order to allow modeling of chemical effects, all flume materials must have defined chemical parameters. This required improvements in the flume water supply, construction of sand-washing equipment, and the use of defined clay and sand preparation procedures. Flume experimental results indicate that clay is extensively trapped by the bed. Advective pumping tends to carry clay to the deep bed where it can be permanently trapped, while turnover tends to continuously mix the upper layers of the bed, hindering penetration to the deep bed.

Key Words: kaolinite, colloid, stream, bedform, bed exchange, filtration, flume, hyporheic, pumping, turnover

1. Introduction

The continuous exchange of stream water with pore water in the stream bed and banks is an important process in determining the fate of pollutants in river systems. Where pollutants are strongly sorbed to fine sediments such as clay and silt, the transport of the pollutant is strongly dependent on the mobility of the sediment. To address this case, we have investigated clay-colloid transport as part of a research program on stream/bed exchange. Our approach is to conduct simplified, controllable laboratory experiments to obtain understandable results, which elucidate particular transport processes of natural streams. This approach was successfully used previously to develop fundamental, analytical models for bed-exchange of both conservative and sorptive solutes (Elliott, 1990; Elliott and Brooks, 1997ab; Eylers, 1994).

The purpose of this paper is to describe laboratory experiments that were used to investigate chemically-influenced particle exchange between a stream and its sand bed. The results of these experiments can be used to validate models that predict net colloid exchange based on measured physical and chemical variables. Previous flume studies of colloid exchange with sand beds did not include sufficient information on the chemistry of the system to reliably evaluate colloid-sand interactions (Packman and Brooks, 1995; Huettel et al., 1996).

2. Background

Recirculating flumes have been used for over 50 years to study sediment transport and other stream hydraulics problems. The flume is composed of a long rectangular channel with a pump at the downstream end and a return pipe to provide recirculation. To mimic conditions in a real stream, appropriate stream bed material is placed in the channel and water is continuously recirculated over the bed. This results in sediment transport and the development of bedforms, such as dunes and ripples. After running for a sufficient time, the flow will become uniform along the flume and all hydraulic parameters will approach a steady state. That is, the sediment transport rate and flow depth will become essentially constant over the entire flume and the average bedform shape will no longer change. Typical basic hydraulic parameters controllable in this system are: stream depth and velocity, and the bed sediment composition. The channel slope is set to match the energy gradient. Bedform geometry and the rate of sediment transport can be controlled by selecting the flow depth and velocity, and sand size.

The recirculating flume is a fairly good laboratory model for a reach of a real stream. For steady uniform flow, the rate of sediment transport is constant along the length of the flume, with the sediment transported to the downstream end being returned to provide an equal input at the upstream end. This constancy of sediment transport is an essential feature of river channels in equilibrium. However, the flume walls obviously restrict the flow to a straight channel, so that bank or meander processes are not included in this system. At the bedform scale, sediment transport is determined by the local velocity profile, which will be similar to that in a natural stream with the same average depth and velocity. Extensive research over many years has demonstrated that results from flume studies are transferable to natural rivers (Brooks, 1958; Vanoni, 1974 and 1984).

Smaller-scale experiments are necessary to evaluate relevant chemical processes. In particular, standard column experiments are used to investigate colloid filtration in flow through a porous medium. These experiments involve pumping a colloidal suspension through a column packed with larger sediment grains. Monitoring the colloid concentration of the effluent allows determination of the extent of filtration (Elimelech et al., 1995).

3. Description of Bed Exchange and Filtration

Before discussing experimental details, it is useful to consider the important processes that control clay transport in the flume. Both hydrodynamic exchange and chemical interactions will be discussed in general terms; for further details please see the indicated references.

There are two main mechanisms for water exchange between the stream and stream bed at the bedform scale (Figure 1). The first is an advective flow into and out of the bed due to the bedform-induced pressure variations along the sediment-water interface; we term this mechanism "pumping." The second is the trapping and release of pore water due to the deposition and scour of bed sediment as dunes move downstream; we call this "turnover." Pumping exchange is calculated from the pore-water velocity distribution, which results from the pressure variation over bedforms caused by the stream flow. Turnover exchange is calculated geometrically from the rate of erosion and deposition associated with the propagation of bedforms (Elliott, 1990; Elliott and Brooks, 1997a). Transport of reactive tracers can be modeled by including chemical effects in the prediction of exchange rates (Eylers, 1994).

Colloidal particles transported through porous media are primarily carried along by the fluid flow, but are also subjected to chemical and electrostatic forces which may cause them to become attached to stationary bed grains (Stumm, 1992; Elimelech et al., 1995). For the present study, these processes are modeled simply with an overall filtration

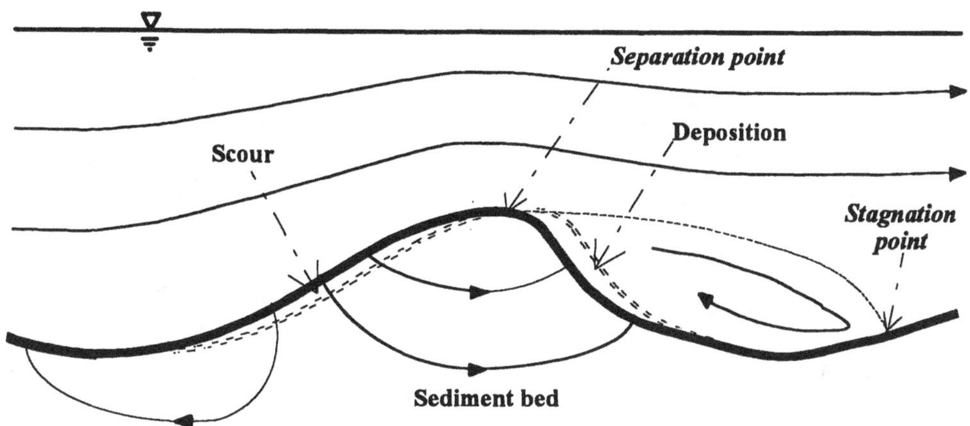

Figure 1. Drawing of a typical dune showing the flow pattern over and through the bedform. Exchange due to flow through the dune is called *pumping*, and exchange due to the scour and deposition associated with dune propagation is called *turnover*.

coefficient. In steady, uniform flow, filtration is empirically described by:

$$\delta C/\delta x = -\lambda C$$

where C is particle concentration, and λ is the filtration coefficient. This equation can be applied in the flume along any particle flow path. Changes in flow rate, colloid and collector properties, and water chemistry will all modify the filtration coefficient (Tien, 1989; Elimelech et al., 1995).

Column experiments were used to determine the filtration coefficient for clay transport through the bed sand under various chemical conditions. The background pH, ionic strength, and ionic composition of the clay suspension affects both the particle surface charge and the range of electrostatic forces in the suspension (Stumm and Morgan, 1996). Thus, the degree of filtration can be controlled by choosing appropriate chemical conditions. Column experiments will not be discussed in detail here, though filtration coefficients derived from two such experiments are presented in Table 1.

4. Experimental Equipment and Materials

4.1 RECIRCULATING FLUME

The major piece of experimental equipment used in this study is a tilting recirculating flume, shown in Figure 2, with a channel of length 12 m, width 26.5 cm, and depth 25.4 cm. It features a variable-speed pump, chemically resistant channel, and screw-jacks to modify the channel slope. A 1.5 m long glass window is installed in the flume wall over the entire depth in order to allow visual observation of the bed. A vertical series of sampling ports (1 cm spacing) allows collection of pore water via a long hypodermic needle and syringe (Figure 3).

Figure 2. Photograph of the downstream half of the flume. The recirculation pump is visible at the left of the picture, while the return pipe can be seen underneath the main channel.

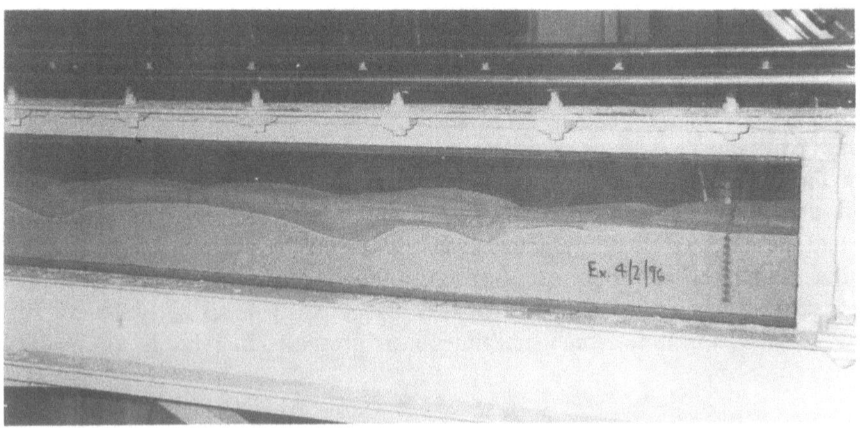

Figure 3. Photograph through the window, showing typical dunes formed at the bed surface and the septa-filled ports used to collect pore-water samples. Flow is from right to left.

Flume experiments were conducted in a defined chemical environment in order to properly characterize the interactions responsible for filtration of colloids in the stream bed. More than 500 l of water, 400 kg of sand, and 50-100 g of clay must be cleaned to a high standard in order to allow direct applicability of results from smaller-scale chemistry studies. As a result, the infrastructure of the Keck Hydraulics Lab was upgraded to provide large quantities of deionized water, and procedures were developed to ensure reproducible supplies of clay and sand.

4.2 WATER SUPPLY

Pasadena tap water was used in early experiments (Packman and Brooks, 1995). However, this water supply is both variable and high in dissolved solids (typically 700 mg/l TDS). ICP-MS analysis of a grab sample indicated calcium and magnesium concentrations in the mM range and an overall ionic strength of approximately 15 mM. Tap water pH is generally in the range 8.0 to 8.2. These background concentrations severely limited the range of interaction behaviors that could be investigated. Thus, deionized water was required in order to allow greater control of the chemical conditions governing clay trapping by the bed. A direct pipe from the Caltech deionized water supply was brought into the Hydraulics Laboratory, and connected to a 1000 liter polypropylene storage tank near the flume. A 5 mM sodium chloride (NaCl) solution was used as the baseline water composition. Additional experiments in the flume or column included ionic strength in the range 0.1 to 50 mM, pH from 4 to 9, and several different ionic compositions.

4.3 BED MEDIA

The bed material used in these experiments is an Ottawa silica sand with a mean diameter of 480 microns and a geometric standard deviation of 1.2. This material was selected because of its simple, homogeneous composition (99.8% silica).

Chemical cleaning was accomplished by fluidizing the sand in an upflow washing apparatus. The sand washer contains six lucite tubes, each 6 inches in diameter and 3 feet high, with mesh at the bottom to hold the sand in place. Tubes are removable for easier loading and unloading. This washing setup is used to clean more than 40 kg of sand per day to provide the 400 kg required for each flume experiment.

Deionized water was used for sand washing, with appropriate chemical agents to remove contaminants (Litton and Olson, 1993; Eylers, 1994). Due to the volumes of sand and water involved, the procedures recommended by Litton and Olson which involve dangerous acids or high temperatures could not be used. Thus a cleaning procedure was developed to remove loosely sorbed or attached contaminants, with the goal of restoring the sand surface to a similar condition before every experiment. The cleaning procedure, outlined below, features the use of strong acids and bases in moderate concentrations:

1. an initial rinse with deionized water (to remove free fines and scum)
2. acid wash for several hours with hydrochloric acid (HCl) at pH ~ 3.5 (to desorb metal ions and kill microorganisms)
3. an intermediate rinse with deionized water (for additional removal of acid, metal ions)
4. base wash overnight with sodium hydroxide (NaOH) at pH ~10.5 (to detach clay by increasing repulsive forces and subjecting particles to high fluid shear)
5. a final rinse with deionized water (for additional removal of base, clay)
6. storage in deionized water at appropriate pH for next experiment.

While this level of cleaning may not remove all contaminants, perhaps precluding direct comparison of filtration data with other researchers' results, it is sufficiently reproducible to allow comparison of results from different flume and column experiments.

4.4 CLAY

A clay preparation procedure was developed following the guidelines of Hunt's (1980) work on colloid coagulation. Kaolinite clay (from Georgia) was purchased from Ward's Natural Scientific Establishment. (See Van Olphen, 1977 for information on the properties of kaolinite and other clays.) Literature reports that Georgia kaolinite has only one major impurity, anatase at a concentration of a few percent (Newman, 1987). This composition was verified by X-ray diffraction and electron microprobe analysis.

As the clay is supplied in chunks, the first preparation step is to grind it using a mortar and pestle. The resulting coarse powder is placed in a high density alumina container with a number of balls of similar material, and tumbled in a ball mill of our own design for a period of at least two days. The fine powder produced by this grinding is then wet-sieved through a #325 steel mesh in order to remove particles larger than 45 microns. (Only a small fraction was removed in this step.) Sieved clay is stirred in five liters of 2 M NaCl solution for two days in order to entirely convert it to a homoionic sodium-kaolinite. The high concentration of sodium displaces all other cations on the kaolinite surface. After ion exchange, excess salt is washed from the kaolinite by multiple rinses with deionized water. Between rinses, the clay is allowed to settle in order to decant the old rinse water. The time required for settling is determined by the salt concentration, through the effect of ionic strength on kaolinite coagulation. The ionic strength has dropped to acceptable levels (on the order of 10 mM, all from NaCl) when the kaolinite takes days to settle as opposed to hours.

Prepared clay is stored in deionized water at concentrations up to 300 g/l. One preparation run yields approximately 500 g of clay, which is enough for about a half-dozen flume experiments. The kaolinite produced by this procedure has a mean spherical-equivalent diameter of 7 μm and a rather broad size distribution (very little mass in sizes less than 3 μm, but 20% greater than 10 μm), as determined by Coulter Counter analysis.

5. Procedure for Flume Experiments

5.1 SETUP

Prior to each flume bed-exchange experiment, the desired chemical and hydraulic conditions must be obtained. While it is easy to add the appropriate salts to modify ionic strength, pH adjustment is much more difficult due to slow buffering by particle surfaces and atmospheric bicarbonate. Both the clay and sand must be brought into equilibrium at the appropriate pH over a period of several days. Clay is stirred in a one- or two-liter flask, and sand is rinsed in large buckets or the flume itself. HCl and NaOH are used for pH adjustment. To start a flume experiment, sand is added to provide an 8-11 cm deep bed, and water to a yield a 10 cm deep stream. Then the flume is run at the desired flow conditions until a regular series of bedforms develops under uniform flow. This takes at least 12 hours, and sometimes requires several days. These physical and chemical conditions are maintained throughout the experiment.

5.2 DESCRIPTION

The transport experiment is composed of two separate phases. First, lithium is used as a conservative tracer in order to quantify exchange with the bed. Lithium ion is added in the form lithium chloride (LiCl), which is dissolved into 1-2 liters of flume water in a flask and then added back to the flume over one recirculation period. This results in the entire volume of recirculating water being rapidly brought to the same lithium concentration. Over time, the lithium concentration in the recirculating water drops as lithium-bearing stream water is exchanged with pore water. Eventually, both the pore water and the recirculating water have the same concentration, and net exchange ends. Typically, lithium is diluted to around 80% of its initial concentration. After the lithium-tracer portion of the experiment, kaolinite is added to the flume in slurry form, again over one recirculation, and the kaolinite transport to the bed is observed. Reduction in clay concentration in excess of dilution is the result of trapping by the bed.

5.3 SAMPLING AND ANALYSIS

Lithium and clay transport are measured in two different ways. The main data set is a time series of the stream-water concentrations, which decrease over time showing the net exchange of tracer between the stream and the bed. The stream-water concentration is obtained from a simple test-tube grab sample of the recirculating water. One sample from anywhere along the flume is sufficient to characterize the entire volume of recirculating water at any particular time. While the stream-water concentration decreases over time, it does so slowly and uniformly. Because the flume recirculation time (1-2 min.) is orders of magnitude faster than the characteristic bed-exchange time (hours), the stream is essentially well-mixed in the entire recirculating portion of the flume. Some downstream variation in concentration is typically observed early in the experiment due to nonuniform tracer addition, but these irregularities disappear after a short time (20-40 minutes).

A secondary data set is pore-water concentration profiles, derived from samples collected at 1 cm intervals in the bed through the sampling ports shown in Figure 3. These profiles show the penetration of tracer at a specific location in the bed, as opposed to the overall transport to the whole bed.

Since both stream- and pore-water samples are fundamentally similar, they may be analyzed in the same way. Clay concentrations are determined using a spectrophotometer, by calibration of the absorbance at a wavelength of 500 nm. An initial clay concentration of 200 mg/l was used for most flume experiments. Lithium was analyzed by ICP-MS, which allowed initial concentrations to be very small (30-50 μM).

Hydraulic parameters were measured for each experiment. The flow depth (and hence water volume) was selected prior to the experiment. Stream velocities were determined by measuring the recirculation flow rate with a Venturi meter and manometer, and then dividing by the flow cross-section in the channel. Stream and channel slopes were measured using a point gauge attached to the carriage which runs along the flume rails. (Slopes are calculated relative to still water levels.) Bedform parameters (height, wavelength) were initially obtained by visual and point gauge measurement, but a laser profiler was used in later experiments to gather bed surface data automatically and noninvasively. Manually collected data agree well with laser profiler results.

6. Results

While this paper is mostly focused on experimental materials and procedures, some results will be presented to demonstrate the effect of major variables. Figure 4 shows the lithium and clay transport to the bed for three different experiments; conditions for these experiments are given in Table 1.

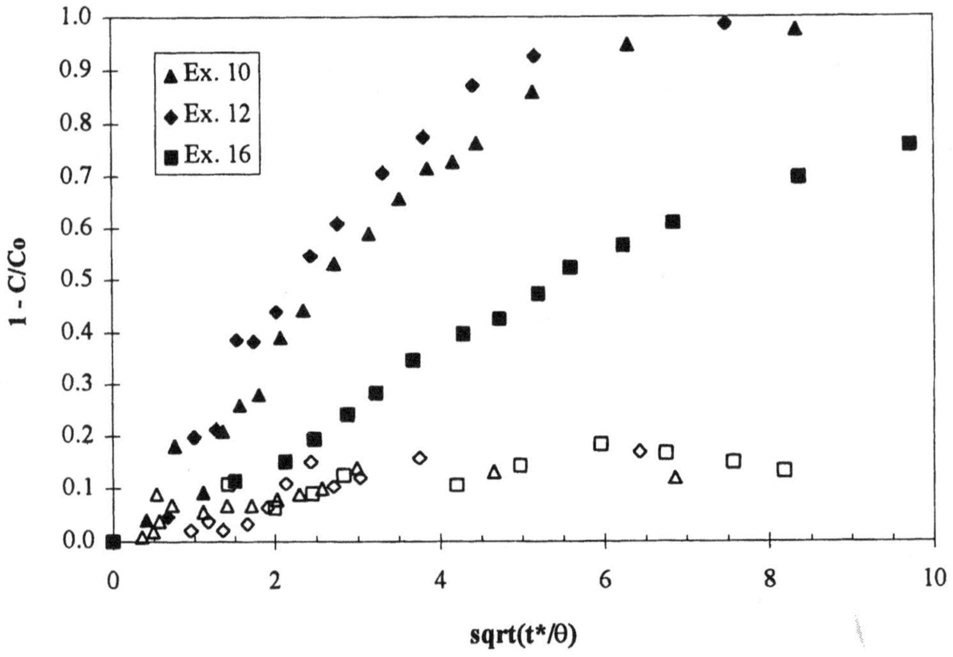

Figure 4. Results from several flume experiments. Closed symbols show clay exchange while open symbols are for lithium. C and Co are the current and initial concentrations of the tracer in the recirculating water. (t^*/θ) is time nondimensionalized to account for differences in the basic surface-bed exchange rate.

TABLE I

Conditions for the flume experiments depicted in Figure 4. Bedform parameters are averages.

Run No.	Stream Velocity (cm/s)	Bedform Height (cm)	Bedform Wavelength (cm)	Bedform Velocity (cm/min)	pH	Ionic Strength (mM)	Filtration Coefficient (1/cm)
10	15.7	1.5	28.6	0	6.9	5.0	0.65
12	15.8	1.5	30.5	0	8.0	5.3	0.20
16	34.3	1.7	33.0	1.1	7.9	5.0	0.20

Note that lithium exchange is very similar for each experiment, with 15-20% of the lithium initially added to the recirculating water ending up in the pore water. Final lithium concentrations can be predicted well merely by comparing the volume of recirculating water to the volume of pore water. That is, lithium is just diluted as a conservative tracer.

On the other hand, clay is not diluted conservatively, indicating that the mass of clay in suspension decreases over time. This is explained by the high filtration coefficients, which predict extensive trapping by the bed. Indeed, in the nonmoving bedform cases (runs 10 and 12) essentially all clay is trapped by the bed after a period of 24 hours. In spite of the different pH's (and thus different filtration coefficients) in these two runs, the filtration was sufficiently high in both so that essentially all clay that entered the bed was trapped there. In this case, advective pumping into the bed is the only exchange mechanism, and the stream bed acts like a horizontal filter. However, in the case of moving bedforms (run 16), more clay remains suspended in the recirculating water. We attribute this to the release of clay from pore water and sand grains due to turnover associated with dune movement. As formerly buried sand is scoured from the back of bedforms and subjected to the high-shear conditions of the stream flow, clay can be detached from the sand grains. Thus, while turnover causes rapid exchange with the upper bed layer, it hinders clay penetration to the deeper bed.

7. Discussion and Conclusions

A recirculating flume was used to study the exchange of kaolinite clay colloids with a sand bed. The hydraulics of transport in the flume is well understood. Colloid transport is the result of both hydraulic exchange and physiochemical interactions between the colloidal particles and bed sediment grains. These interactions are not observable directly in the flume, but can be investigated in separate column experiments. If column experiment results are to be used to interpret flume data, all materials must have the same specified chemical state in flume and column. In particular, the flume water composition must be controlled and all particles must have defined surface properties. For many hydraulics laboratories, extensive efforts would be required to provide the chemical control needed to do chemically-influenced transport studies in large experimental equipment such as flumes.

Flume experiment results indicate the important particle exchange processes in sand-bed stream systems. Interactions between clay and sand particles result in extensive clay trapping by the bed. Clay is subject to filtration in streams due to advective pumping through the sand bed. However, sediment transport causes bedform turnover which tends to release clay from the upper bed layers and hinders penetration into the deep bed.

A computer model is currently being developed to predict the net clay transport to the bed. The unique features of particle filtration and sedimentation will be added to previous computer models of pumping and turnover exchange (Elliott, 1990; Eylers, 1994).

Acknowledgments

This material is based upon the Ph.D. thesis research of the first author, supported by the U.S. National Science Foundation under award number BES-9421491. The authors thank two undergraduate assistants, Nathan Lee and Tad Fujioka, who helped collect and analyze data.

References

Brooks, N.H.: 1958, *Trans. ASCE*, **123**, 526-594 (reprinted in *Classic Papers in Hydraulics*, ASCE, 1992, 541-609).

Elimelech, M., Gregory, J., Jia, X., and Williams, R.: 1995, *Particle Deposition and Aggregation: measurement, modeling, and simulation*, Butterworth-Heinemann.

Elliott, A.H.: 1990, *Transfer of Solutes into and out of Streambeds*, Report No. KH-R-52, W.M. Keck Laboratory of Hydraulics and Water Resources, Caltech.

Elliott, A.H. and Brooks, N.H.: 1997a, *Water Resour. Res.* **33**, 123-136.

Elliott, A.H. and Brooks, N.H.: 1997b, *Water Resour. Res.* **33**, 137-151.

Eylers, H.: 1994, *Transport of Adsorbing Metal Ions between Stream Water and Sediment Bed in a Laboratory Flume*, Report No. KH-R-56, W.M. Keck Laboratory of Hydraulics and Water Resources, Caltech.

Huettel, M, Ziebis, W., and Forster, S.: 1996, *Limnol. Oceanogr.* **41**, 309-322.

Hunt, J.R.: 1980, *Coagulation in Continuous Particle Size Distributions, Theory and Experimental Verification*, Ph.D. Thesis, Caltech, 32-33..

Litton, G.M. and Olson, T.M.: 1993, *Environ. Sci. Technol.* **27**, 185-193.

Newman, A.C.D., ed.: 1987, *Chemistry of Clays and Clay Minerals*, Mineralogical Society, London, 22-26.

Packman , A.I. and Brooks, N.H.: 1995, *Mar. Freshwater Res.*, **46**, 233-6.

Stumm, W.: 1992, *Chemistry of the Solid-Water Interface*, Wiley-Interscience.

Stumm, W. and Morgan, J.J.: 1996, *Aquatic Chemistry*, 3rd ed., Wiley-Interscience.

Tien, C.: 1989, *Granular Filtration of Aerosols and Hydrosols*, Butterworths.

van Olphen, H.: 1977, *An Introduction to Clay Colloid Chemistry*, 2nd ed., Wiley-Interscience.

Vanoni, V.A.: 1974, *J. Hyd. Div., ASCE* **100**, HY3, 363-377.

Vanoni, V.A.: 1984, *J. Hyd. Eng.*,**110**, 1021-1057.

SEICHE-INDUCED RESUSPENSION IN LAKE KINNERET: A FLUORESCENT TRACER EXPERIMENT

B. SHTEINMAN, W. ECKERT, S. KAGANOWSKY and T. ZOHARY

Israel Oceanographic and Limnological Research, The Yigal Allon Kinneret Limnological Laboratory, P.O.Box 345, Tiberias 14102, Israel

Abstract. In warm-monomictic Lake Kinneret, wind-induced internal waves with amplitudes of up to 10 meters are common during April - October. This study was aimed to follow the horizontal and vertical transport of resuspended particles due to internal wave activity using fluorescently-dyed sediment particles (lake sediments and lyophilized algal cells) as tracers. Color-coded (5 colors) tracers were deployed along a transect perpendicular to the shore, and their dispersion was followed by recovery of labeled particles in sediment traps and in bottom sediment samples using epifluorescence microscopy. Wind-induced internal seiches as the driving force for resuspension were followed using a thermistor chain and a current velocity profiler. Examination of sediment trap and bottom sediment samples indicated particle transport from the hypolimnion to the epilimnion that could be linked to the seiche activity. Horizontal transport of particles was most distinct for littoral sediments whereby particles placed at 5 m depth were exposed to a strong long shore transport.

Keywords: fluorescent tracers, boundary mixing, sediment resuspension, sediment traps, Lake Kinneret, seiches

1. Introduction

Resuspension processes of bottom deposits in lakes are known to affect the cycling of sediments, nutrients, carbon and contaminants (Sanford, 1992; Evans, 1994). They are ubiquitous processes influenced by wind, currents and lake basin morphometry (Bloesch, 1995). Especially in shallow lakes where resuspension was shown to strongly enhance internal phosphorus loading, this process has major ecological implications for the eutrophication process (Istvanovics, 1988). In deep lakes that undergo thermal stratification the flux of resuspended phosphorus into the euphotic zone can be equal to or even higher than external P loading (Eadie et al., 1984; Dillon et al., 1990). One of the major driving forces for resuspension in stratified lakes are horizontal hypolimnetic bottom currents induced by internal seiches (Gloor et al., 1994).

Existing methods to assess resuspension in lakes include both direct (measurement of settling flux using sediment traps, turbidimetry in the water column or the determination of the radionuclide distribution (^{7}Be, ^{127}Cs) in bottom sediments) and indirect methods (mass balance calculations and modelling (Bloesch, 1994)). These methods do not provide information on the origin and the pathway of resuspended particles. The present study addressed these issues by introducing a fluorescent tracer technique into the research on resuspension.

Water, Air and Soil Pollution **99**: 123-131, 1997.

2. Methods

2.1 STUDY AREA

Lake Kinneret (12 x 22 km, 168 km^2) is a warm monomictic lake 209 m below mean sea level in the northern part of the Afro-Syrian rift valley at 32°N latitude (Serruya, 1978). Maximum and average depths are 42 m and 24 m, respectively. The lake is thermally stratified from April until December with surface water temperatures ranging 15°C - 30°C. Thermocline depth changes seasonally, from 13 m in spring to 15-17 m during the summer months (Hambright et al., 1994). During April-October daily westerly winds dominate the wind regime with wind speeds up to 12 m·s^{-1} (Serruya, 1975). These periodic winds create internal seiches with amplitudes up to 10 m. A characteristic feature of the lake is a winter-spring bloom of the large-celled (ca. 50 μm diameter) dinoflagellate *Peridinium gatunense* (Pollingher, 1986).

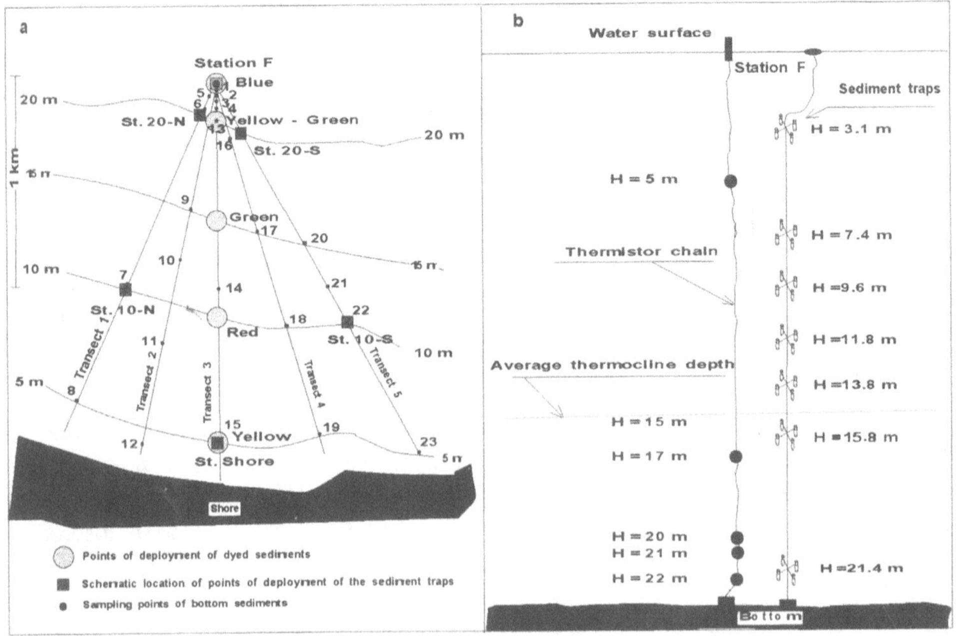

Fig. 1. Experimental setup. a) locations of tracer deployment, stations of sediment trap deployment and points of sediment samples (numbered). b) Array of sediment traps and thermistor chain at Station F.

2.2 EXPERIMENTAL

About 70 kg of Lake Kinneret bottom sediments (20 - 60 μm, silt fraction) were collected from each of the 5 locations where fluorescent tracer deployment was planned (Figure 1a). The air-dried sediments were fluorescently dyed with a custom-made glue as described by Shteinman and Imbar (1995). Lyophilized algal cells of *Peridinium gatunense*, collected with a 20 μm mesh net during the peak of the 1995 bloom, were marked in the same way. On August 9, 1995 dyed sediments were deployed by SCUBA divers on the lake bottom at three different water depths (10, 15, 20 m) along a transect perpendicular to the shore, between the Kinneret Limnological Laboratory and station F (Figure 1a). The respective tracer colors were red (10 m), green (15 m) and yellow-green (20 m). In addition, on August 11, yellow-dyed sediment particles were placed on the sediments at a shallow station (5 m) and blue-dyed *Peridinium* cells at station F (23 m). Sediment traps for the collection of resuspended material were moored on August 14 at 7 depths at station F (Figure 1b). Each set of traps designed after Koren (1995) consisted of four PVC tubes (inner diameter 50 mm) joined to a 250 ml collection bottle. Additional traps were deployed at stations 20 N (3 sets: 3.5, 9 and 18 m), 20S (3 sets; 5, 10 and 19 m), 10N (2 sets; 5 and 9 m), 10S (2 sets; 4 and 9 m) and at a shore station (1 set; 2.5 m). The traps remained installed for three days and were recovered on August 17. Samples from each set were mixed and filtered through pre-weighed GFC filters which were subsequently dried and re-weighed. The dried filters were analyzed for labeled particles by epifluorescence microscopy. In order to examine patterns of transport in the sediments, on August 17, 24 dredge samples were also taken within the experimental area along five transects from station F to the shore according to the sampling points indicated by their numbers in Figure 1a. Subsamples (10 g) were analyzed as described above.

During the three days of sediment trap exposure, wind-induced internal seiches were followed by means of a thermistor chain fixed at station F, providing temperature readings at 10 min time intervals from six depths (Figure 1b), and by repeated measurements of temperature (STD) and current velocity depth profiles within the experimental area as described by Shteinman and Gutman (1993).

2. Results

The isotherm diagram in Figure 2 as plotted from the thermistor chain data from station F reflects the changes in the thermal structure in between August 14 and August 17. During this time period, the thermocline oscillated in 24 hour intervals with an amplitude of about 4 meters between 13 and 17 m depth. The current velocity profiles (Figure 3) revealed inverse flow direction of horizontal surface and bottom currents with maximum velocities of 10 $cm \cdot s^{-1}$. The measured flow was directed either offshore or towards the shore.

The results from the microscopic enumeration of dyed particles that were collected in the

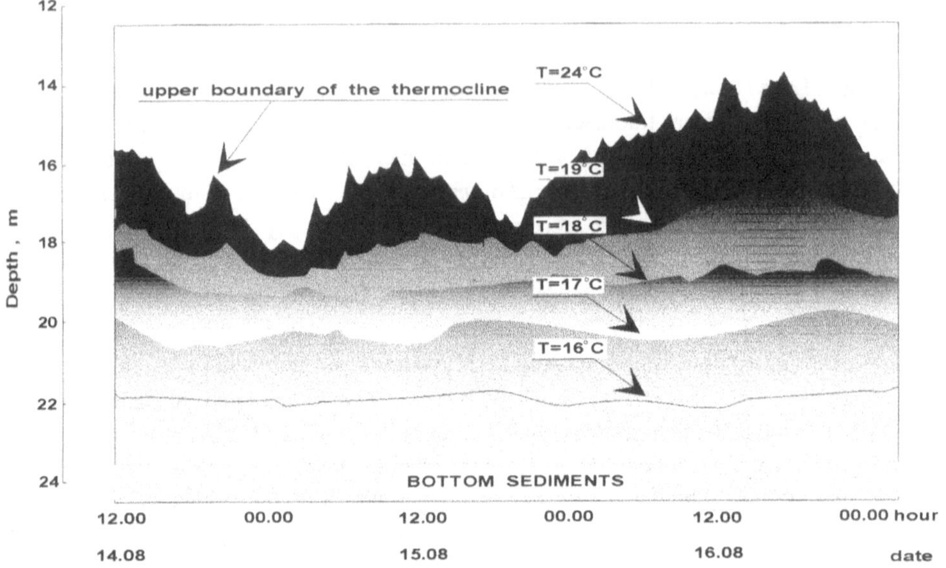

Fig. 2. Isotherm diagram from thermistor chain data measured at Station F, August 14 - 17, 1995.

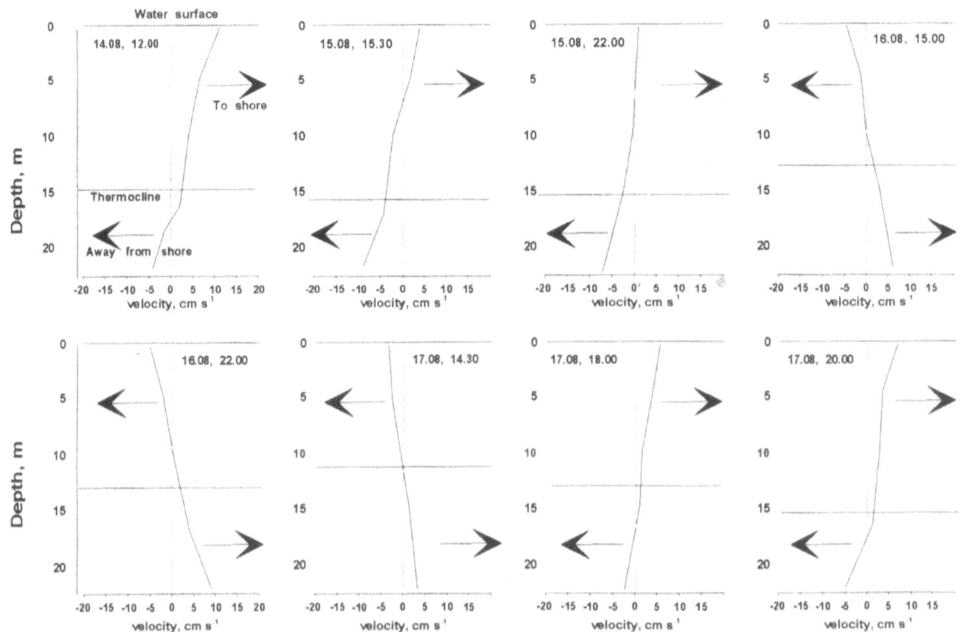

Fig. 3. Vertical profiles of current velocity measured at station F.

sediment traps are summarized in Figure 4 with numbers representing counts per 10 g of dry weight of material trapped in each set. Blue labeled *Peridinium* cells were recovered in the traps at stations F, 20N and 20S. The higher numbers of cells were found at station 20N at the depths of 9 m, 18 m and 3.5 m. A relatively smaller and vertically homogeneous amount was collected at station 20S. At station F, counts were highest in the deeper zones (15.8 m and 21.4 m) but some labeled algal material appeared in the upper water column in between 7.4 m and 13.8 m depth. Particles labeled by other colors were not recovered at station F. Yellow-green labeled particles that were deployed at a depth of 20 m were recovered in the material collected at stations 20S and 20N with counts being highest in the upper water column (3.5 m at 20N and 5 m at 20S) and relatively lower towards the sediment. At station 10S we collected material which was resuspended from the 5, 10 and 15 m sites. While counts increased with depth for the yellow and green labeled particles, representing resuspended material from 5 m and 15 m respectively, red dyed particles from 10 m appeared in the 4 m trap in relatively higher numbers than one meter above the sediment. The traps from station 10N were lost during the experiment. In the sediment trap deployed at the shore station at a depth of 2.5 m we counted 48 yellow and 18 red particles.

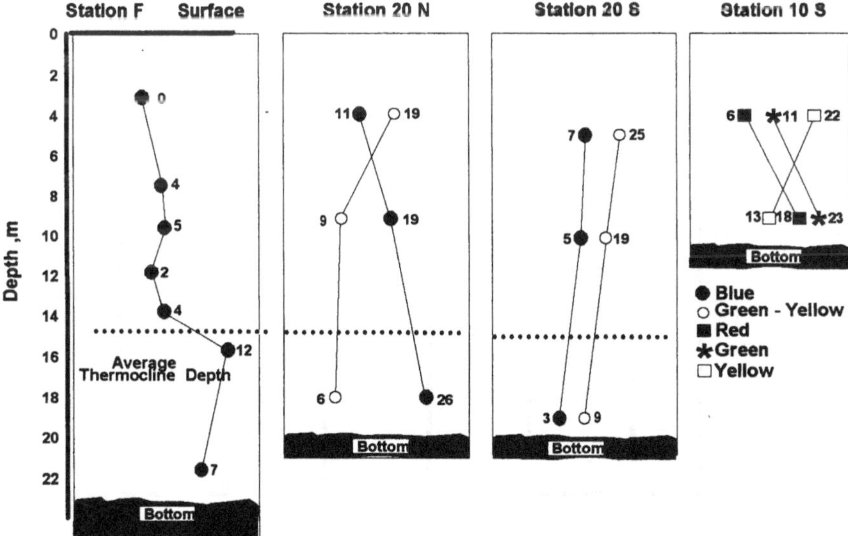

Fig. 4. Results of enumeration of fluorescent tracers recovered in sediment traps (in particle counts per 10 g dry weight)

The results from the enumeration of dyed particles in bottom sediments are summarized in Table 1. The sampling points refer to the pattern shown Figure 1a. Blue and yellow-green labeled particles were collected mainly in the vicinity of the site of their deployment while other markers were dispersed over a much bigger radius.

Table I

Numbers of dyed particles (per 10 g of sediment dry weight) collected in bottom sediments (samplin points are shown in Figure 1a).

Points	Colors				
	Blue	Yellow-Green	Green	Red	Yellow
St.F	23	0	0	0	0
1	19	0	0	0	0
2	10	0	0	0	0
3	5	5	0	0	0
4	2	37	0	0	0
5	0	35	0	0	0
6	0	0	28	0	0
7	0	0	0	14	5
8	0	0	0	24	9
9	0	37	21	0	0
10	0	0	0	65	2
11	0	0	0	71	5
12	0	0	0	66	51
13	0	97	19	0	0
14	0	0	8	15	0
15	0	0	0	19	23
16	5	23	0	0	0
17	0	0	8	0	0
18	0	0	3	0	9
19	0	0	0	0	16
20	0	0	5	0	0
21	0	0	0	0	7
22	0	0	0	0	2
23	0	0	0	28	0

3. Discussion

The analysis of the material collected in sediment traps indicates a net vertical transport of particles from all sites of tracer deployment. This vertical transport is of mayor importance considering hypolimnetic sediments. According to the thermocline fluctuations (Figure 2) in this study, the tracers deployed at 20 (yellow-green) and 23 m (blue) were subjected to hypolimnetic mixing . Particle resuspension below the thermocline was apparently driven by the relatively high current velocities near the lake bottom (Figure 3) indicating the existence of a turbulent benthic

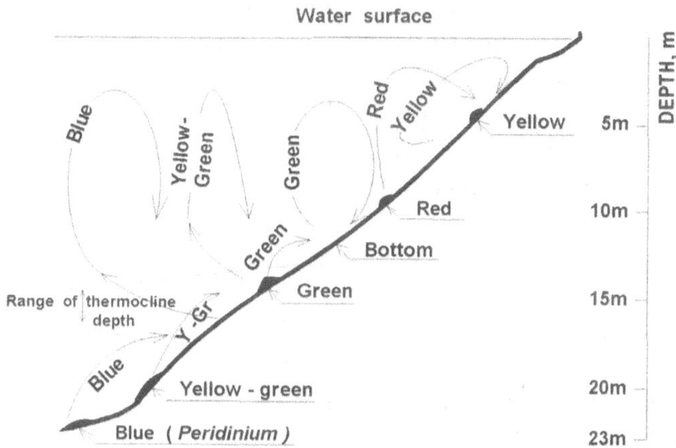

Fig. 5. Transport of labeled sediment particles based on trap catches of fluorescently dyed particles

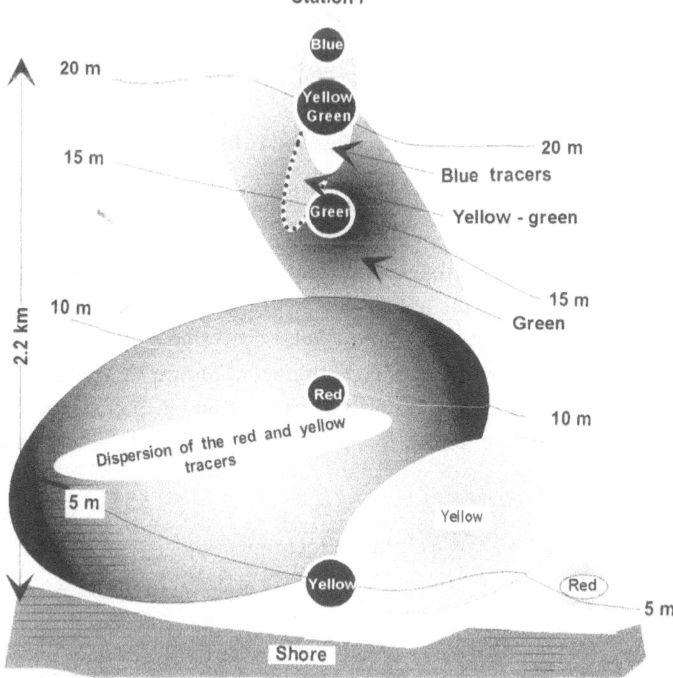

Fig. 6. Pattern of horrizontal sediment transport according to the recovery of labeled particles in dredge samples (data from Table I).

boundary layer (BBL). The dispersion of the BBL within the metalimnion can explain for the intrusion of hypolimnetic particles into the epilimnion (Imberger, 1994). Figure 5 summarizes these findings with arrows indicating the major vertical and horizontal transport directions according to the recovery of particles in sediment traps. The fact that no tracer other than blue was trapped at station F can have two explanations: One is a dominant horizontal transport towards the shore which corresponds to the recovery of blue tracers at stations 20N and 20S. Alternatively the greater dispersion towards deeper waters lowers the probability to entrap resuspended particles. According to the current velocity profiles (Figure 3) the direction of horizontal currents varied in the epilimnion as well as in the hypolimnion. Green labeled particles that were deployed at 15 m were in accordance to the thermocline fluctuation alternately exposed to epilimnetic and hypolimnetic mixing processes directed either offshore or towards the shore and therefore dispersed in a much wider radius as can be seen by their recovery at station 10S. The red and yellow tracers were exposed solely to eplimnetic mixing and therefore affected strongly by resuspension processes.

The dispersion of shallow sediments becomes evident from the redistribution of tracers in the bottom sediments (Figure 6) represented by the particle recovery in dredge samples (Table I). Here, red and yellow dyed particles were dispersed widely indicating intensive epilimnetic mixing and a long-shore transport pattern. The other tracers were dispersed perpendicular to the shore at a radius that was small for blue and yellow-green particles (hypolimnetic mixing) and relative bigger for green labeled particles (epi- & hypolimnetic mixing).

4. Conclusions

From our findings we can conclude that seiche induced currents and mixing processes in Lake Kinneret play an important role in the transport of resuspended bottom sediments and probably hypolimnetic water to the upper water column. Within the frame of this pilot study we restricted our sampling program to the experimental zone outlined in Figure 1a. Accordingly, the question of particle transport towards the center of the lake (sediment focussing) is not addressed. Nevertheless, fluorescent tracers showed to be a useful tool to qualitatively study resuspension processes.

Acknowledgements

This research was funded by a research grant (No 95-17-027) from The Israel Ministry for Energy and Infrastructure. We wish to thank the SCUBA divers Nir Koren and Alon Mindel for their help during the sampling program and Dr. K. D. Hambright his review of the manuscript..

References

Bloesch, J.: 1994, *Hydrobiologia* **284**, 13-18.

Bloesch, J.: 1995, *Mar. Freshwater Res.* **46**, 295-304.

Dillon, P. J., Evans, R. D. and Molot, L. A.: 1990, *Can. J. Fish. Aqu. Sci.* **47**, 1269-1274.

Eadie, B. J., Chambers, R. L., Gardner, W. S. and Bell, G. L.: 1984, *J. Great Lakes Res.* **10**, 307-321.

Evans, R. D.: 1994, *Hydrobiologia* **284**, 5-12.

Gloor, M., Wüest, A. and Münnich, M.: 1994, *Hydrobiologia* **284**, 59-68.

Hambright, K. D., Gophen, M., and Serruya, S.: 1994, *Limnol. Oceanogr.* **39**: 1234-1243.

Imberger, J.: 1994, in: R.Margalef (ed.) *Limnology Now. A Paradigm of Planetary Problems.* Elsevier.

Istvánovics, V.: 1988, *Wat. Res.* **22**, 1473-1481.

Koren, N.: 1993, MsC Thesis, The Haifa Technion.

Pollingher, U.: 1986, *Hydrobiologia* **138**, 127-138.

Sanford, L. P.: 1992, *Limnol. Oceanogr.* **37**, 1164-1178.

Serruya, C. (ed.): 1978, *Lake Kinneret*, Dr. W. Junk Publishers, The Hague.

Serruya, S.: 1975: *Verh. Internat. Verein. Limnol.* **19**, 73-87.

Shteinman, B. and Imbar, M.: 1995, *Advances in Hydroscience and Engineering*, Vol. II, Part B, Tsinghua Univ. Press, China. 2115-2122.

Shteinman, B. and Gutman, A.: 1993, in: S.S.Y. Wang (ed.) *Advances in Hydroscience and Engineering*, Vol. I, Part B, 1154-1160.

PROBING PARTICLE PROCESSES IN LAKE MICHIGAN USING SEDIMENT TRAPS

Brian J. Eadie
NOAA-Great Lakes Environmental Research Laboratory, Ann Arbor, MI. 48105

Abstract. Sediment trap sampling at an offshore site in southern Lake Michigan has continued for an 18 year period with a sampling frequency ranging from weekly to semi-annually. During the 6 month unstratified period sediment trap mass and tracer profiles are nearly constant and they have been used to describe the extent of sediment resuspension. After stratification, mass flux rapidly declines and particle tracers are removed from the epilimnion at the rate of 0.5-1 $m.d^{-1}$. Exponential profiles of mass flux clearly show the persistence of a benthic nepheloid layer. High frequency sampling with near-bottom sequencing traps show order of magnitude ranges in mass flux over a few day period.

1. INTRODUCTION

Lake Michigan, with a surface area of 57,800 km^2 and volume of 4,920 km^3, is the third largest of the North American Laurentian Great Lakes and the sixth-largest lake in the world. A population of approximately 14 million residing within the basin (EPA, 1987) coupled with a hydraulic residence time of 62 years (Quinn, 1992) has resulted in critical issues relating to anthropogenic contaminants. Rapid and efficient processes of sorption and settling through the average depth of 86m promotes internal removal of particle-reactive contaminants through sedimentation with the result that the large contaminant inventories presently reside in sediments. However, studies in the Great Lakes have shown that higher levels than expected, of these constituents, persist in the lakes if settling and burial were the sole transport process. Radiotracer studies with ^{239}Pu ($t_{1/2}$ = 25,000 years) and ^{137}Cs ($t_{1/2}$ = 30.2 years) indicated removal from the water and >95% transferred to sediments in a few years (Robbins and Edgington, 1975; Wahlgren et al, 1980; Eadie and Robbins. 1987). Although initial removal of particle-reactive tracers from the water is rapid, a small residual concentration in the water, either on particles, in biota or in solution, has diminished exponentially on time scales of decades. Studies of Lehman (1979), Eadie, et al. (1984), Robbins and Eadie (1991), have shown that the small amount remaining in the system is primarily the result of an annual cycle of sediment resuspension and redeposition releasing constituents from sediments back into the water. The long-term decline of ^{239}Pu and (decay-connected) ^{137}Cs in the lake has about a 20 year time constant (Wahlgren et al.,1980), which probably characterizes the net rate of incorporation of these tracers into permanent sediments (Robbins, 1982), a relatively long and inefficient process.

During the decades that these materials are part of the resuspendable pool, they constitute the major non-point source of nutrients and contaminants to the pelagic system. This material also serves as a food source for surface deposit feeders, suspension feeders, the microbial food web, and is probably the source for some of the material that makes up the benthic nepheloid layer (Chambers an Eadie, 1981; Eadie and Robbins, 1987) which plays a major role in coupling the inventory of constituents in surface sediments with overlying lake water throughout the year. New silty-clay materials erode from bluffs along the shore of Lake Michigan or from exposed glacio-lacustrine clays in relatively shallow waters and form temporary deposits of particles in

Water, Air and Soil Pollution **99**: 133-139, 1997.

patchy, transient reservoirs at the active sediment boundary layer. The materials in these transient reservoirs are biogeochemically transformed within the lake, then redistributed throughout the year by a spectrum of energetic events. Large episodic events resuspend and transport these materials from these temporary sinks to more permanent sinks with a small fraction becoming incorporated annually into the sediments of the depositional basins.

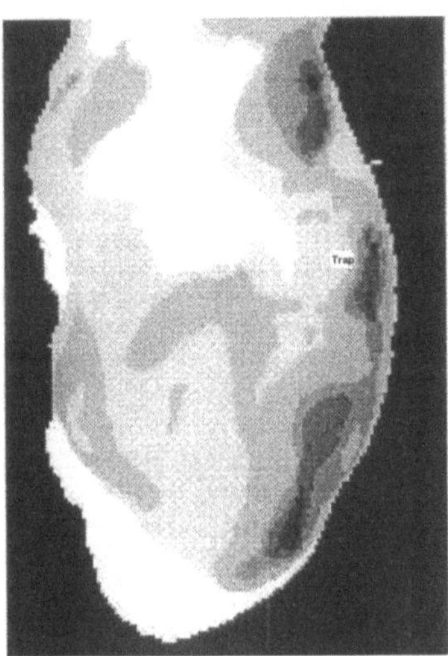

Figure 1. Sediment accumulation patterns in southern Lake Michigan as measured by the thickness of post-glacial sediment. The white region represents zero accumulation and the lightest gray represents accumulation of 0-1 m; the remainder of the contours are in 4m intervals with the darkest region having greater than 14m of accumulation. The sediment trap data discussed are primarily from a 100m deep station approximately 25 km off the eastern shore.

The resultant complex distribution of post-glacial sediment in the southern half of Lake Michigan is asymmetric (Lineback and Gross, 1972, Cahill, 1981), with the greatest accumulations found in a band about 20 km from the eastern shore and decreasing towards the deepest sounding in this basin (Figure 1). There is essentially no accumulation of sediment on the western side of the basin, with generally a thin 1-2 cm layer of 'floc' overlying exposed glacial till, glacio-lacustrine clay, or sand. It is mainly from these areas that portions of the inventory are resuspensed during the isothermal period and re-allocated to depositional sites.

The processes of particle flux and resuspension has been examined in Lake Michigan through the use of sediment traps since the mid 1970s (Wahlgren et al., 1980, Eadie et al, 1984). These cylindrical devices are moored at selected depths to intercept materials settling to the bottom. Traps provide an efficient tool for the collection of integrated samples of settling materials for detailed analysis. Measuring the mass collected allows us to calculate the gross downward flux of particulate matter and associated constituents and to calculate both mass and constituent settling velocities. Here we report on the results of some initial studies with sequencing traps from a station in southeastern Lake Michigan (figure 1) and place them in context with a long time series of earlier results.

2. METHODS

In this report we compare results from two types of traps. Our simple trap, based on reviews of various designs (Bloesch and Burns, 1980; Gardner, 1980a,b), is a cylinder 10 cm in diameter with an aspect ratio of 5:1 above the funnel opening to a 500 ml polyethylene bottle. In 1990 we developed a sequencing trap modified from the designs of Baker et al. (1988), and Jannasch et al., (1980). These are also cylindrical, but with an inner diameter of 20 cm and an aspect ratio of 8:1 above the funnel. A computer-controlled carrousel contains 23 sixty ml polyethylene bottles, which rotate under the funnel at preprogrammed intervals. An electric motor rotates the carousel and uses a single-pole detent switch to provide position feedback. A microprocessor-based controller, developed in house, runs the motor based on a schedule and records confirmation of each rotation using non-volatile memory. A battery pack allows up to two years of operation. Cylindrical traps have a high collection efficiency in low current lake environments and have proved satisfactory in many lake studies ((Bloesch and Burns, 1980; Eadie et al, 1984). The accuracy of calculated fluxes is poorly understood, but depends on the trap design, the types of particles in the fluid and the currents at the site (Gardner, 1980;, Hawley, 1988; Gardner, 1995).

In order to estimate trap collection precision and intercalibrate between the 10 cm diameter traps and the 20 cm diameter sequencing traps, a series of deployments were made in regions with a wide range of fluxes on specially constructed brackets to assure identical depth and exposure. The 20 cm traps used in these tests did not have sequencing capability, but were identical in other aspects. The traps were deployed as anchored arrays using subsurface buoyed 3/8" steel cable. The 500 ml bottles in the simple traps were poisoned with 25 ml of chloroform and filled with distilled water prior to deployment. The 60 ml polyethylene collection bottles in the sequencing trap were poisoned with 6 ml of chloroform and filled with distilled water immediately prior to deployment. This concentration of chloroform is an effective preservative (Lee et al., 1989) and results in a supersaturated solution, with beads of chloroform remaining after retrieval. The sequencing traps are deployed with the collection funnel feeding to an empty opening (no collection bottle). After a preprogrammed period of time the carousel will move the first collection bottle under the funnel. The remaining 22 bottles will follow in a preprogrammed sequence. After retrieval, the sample bottles are removed from the traps and to the laboratory in cold storage (4°C). The traps have on-board intelligence that records the time of each sequence and various system checks.

After arrival at the lab, the trap samples were allowed to settle in a refrigerator for a day, then overlying water was carefully siphoned off and the residual was freeze dried. After drying, samples were weighed and transferred into precleaned scintillation vials for storage in a freezer. All trap samples have been weighed on an analytical balance calibrated to within ± 1mg with known standard weights during each weighing session. Virtually all samples are greater that 100 mg, thus all mass weights have an accuracy and precision of less than 1 % (coefficient of variation).

3. RESULTS and DISCUSSION

Profiles of sediment traps were first deployed at the 100 m deep site (figure 1) in 1978 and sampling has continued with a frequency ranging from biweekly to semi-annually since then. The long-tern average mass fluxes measured from 1978-92 at the 100m deep station, 25 km offshore (figure 1), are presented in figure 2. Throughout the year, profiles of mass flux exhibit an exponential increase toward the bottom. From late December through early June, Lake Michigan is virtually isothermal and well mixed. Average fluxes during this period (figure 2) are high throughout the water column, but there is clear evidence of a benthic nepheloid layer (BNL). During the stratified period (June - December), the upper half of the water column becomes isolated from the large inventory of materials in the sediments, although episodic mixing does occur during upwellings.. A BNL is still clearly evident from the mass flux profile.

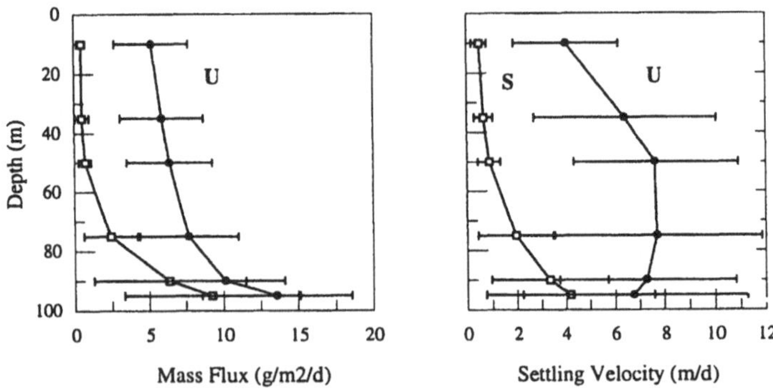

Figure 2. A 14 year synthesis of profiles of trap measured mass fluxes and settling velocities. Error bars represent 1 sd; replicates ranged from 14-54. The stratified and unstratified periods are designated with an S and U respectively.

Ensemble particle settling velocities, estimated from the ratio of mass flux to ambient suspended mater concentration (collected at deployment and retrieval) also show substantial differences between the two thermal periods (figure 2). During the stratified period, these calculated settling velocities in the epilimnion (0.5-1 $m.d^{-1}$) agree with those required to model the long-term behavior of fallout radiotracers (Robbins and Eadie, 1991). Settling velocities estimated for the BNL (several $m.d^{-1}$) shows clearly that frequent recharging of the BNL is required in order to maintain its observed persistence. The BNL is a regular feature in all of the

Great Lakes and appears to be composed primarily of resuspended sediments. In the shallow waters of the shelf and slope, surface and internal waves and occasional strong currents resuspend sediments sorting the particles and transporting them horizontally as well as vertically. The cycle of resuspension and redeposition has the effect of producing a resuspended pool composition which is relatively uniform throughout major basins of the lakes.

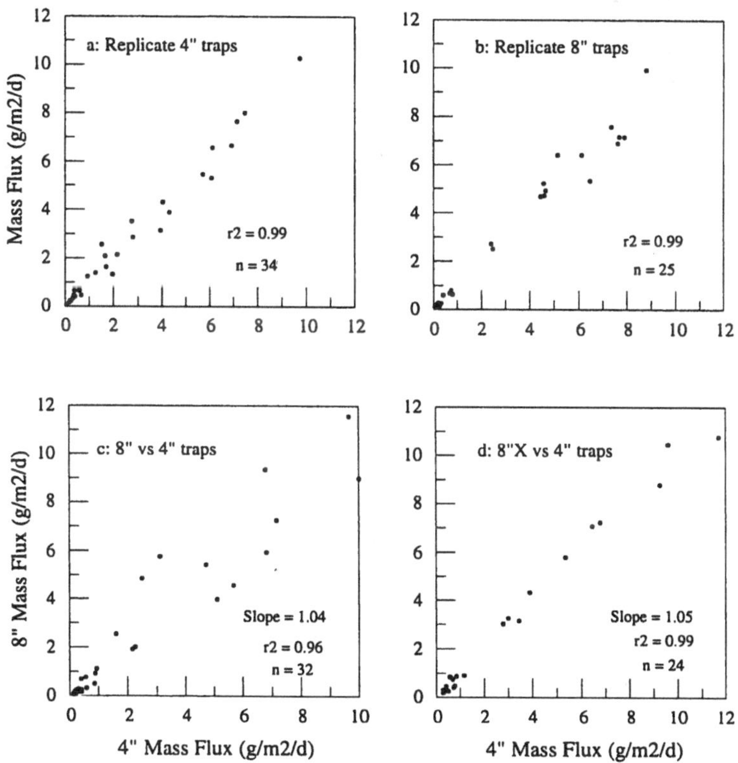

Figure 3. Comparisons of replicate trap flux measurements. a) Replicate 10 cm (4") diameter traps placed on brackets for simultaneous deployments. Correlation coefficient is high and the traps replicate with an average difference between pairs of ± 11%. b) A similar treatment of 20 cm (8") diameter traps with similar results. c) A comparison of 10 cm and 20 cm traps with 5:1 aspect ratios. The scatter was worse than for individual pairs of the same size. d) Comparison of 10 cm (5:1 aspect) with 20 cm (8:1), extended aspect ratio. There was little bias (slope = 1.05) between these pairs and the scatter was much reduced with the higher aspect ratio.

When deployed in replicate, both the 10 cm and 20 cm traps showed good repeatability with paired t-test showing equal means ($P<0.05$) in all 4 comparisons (figure 3). The 10 cm traps replicate with an average difference between pairs of ± 11% and the 20 cm traps (with and 8:1 aspect ratio) replicate with an average difference between pairs of ± 14%. An intercomparison of capture efficiency between the 10 and 20 cm traps resulted in a design change from an aspect ratio of 5:1 to 8:1 for the larger diameter traps. A larger trap diameter results in a higher trap Reynolds number, with presumably lower collection efficiency. There was little bias (slope = 1.05) between the two types of traps and the scatter was much reduced with the extended aspect ratio, which became our standard for 20 cm diameter traps.

For its initial test deployment, the prototype sequencing trap was deployed at the 100m deep station (figure 1) at 5m above the bottom (within the BNL) in mid-July, 1990. Samples were collected at intervals of 18 hours (figure 4) in order to examine the variability within the BNL. Over the 18 day deployment, fluxes ranged over an order of magnitude. The first, and largest, peak in the mass flux corresponded directly with a large upwelling recorded at the water intake 20 km inshore of the trap location. The increased mass flux associated with the upwelling implies transport of large amounts of BNL materials from deeper regions further offshore. Upwellings are a regular feature in all of the Great Lakes and occur frequently during the stratified period. The mass transport associated with this process has not been quantified within the Great Lakes, but must be large. After the initial upwelling, fluxes were not related to intake temperatures or winds at a nearby airport; variability is most likely associated with advection. This complex pattern of mass flux reinforces the need for interdisciplinary approaches to sediment-water exchanges and coupling the deployment of traps with current meters and other time-series instruments.

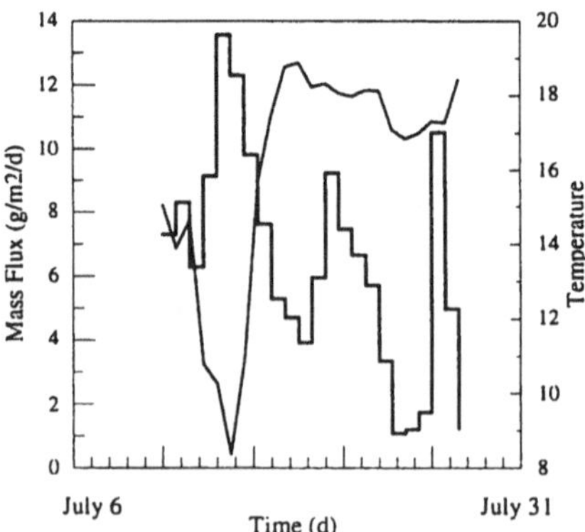

Figure 4. Mass flux (solid line) results from the first deployment of the sequencing traps at 5m above bottom at the 100m deep station. The traps sampled at 18 hour intervals and captured startling variability in flux compared with the long-term average (9.2 ± 5.9) for the summer period.

Interpretation of trap data has led to an increased understanding of particle (and associated constituent) behavior including water column residence times, settling velocities, sediment-water exchange via resuspension long-term removal rates via burial and geochemical properties of mobile materials. Analysis of trapped materials also provides information on sources, fluxes and degradation rates of natural and contaminant organic compounds in the water column. Throughout this long time series, approximately 80% of the primary production has been remineralized within the epilimnion; less than 5% reaches traps at 5m above bottom and approximately 2% is finally incorporated into surface sediments.

ACKNOWLEDGEMENTS

I would like to thank the crew of the RV Shenehon and all of the support technicians and engineers that have helped in the sediment trap program over the years. I would also like to thank Jurg Bloesch for his insightful comments on the draft manuscript.

REFERENCES

Baker, E.T., Milburn, H.B., and Tennant, D.A.: 1988, J. Mar. Res. 46:573-592.
Bloesch, J. and Burns, N.M.: 1980, Schweiz. Z. Hydrol. 42(1) 16-55.
Cahill, R.A.: 1981, Il. Geol. Survet Circular no 517, 96pp, Champaign, IL.
Chambers, R.L. and Eadie, B.J.: 1981, Sedimentology 28:439-447.
Eadie, B.J., Chambers, R.L., Gardner, WS., and Bell, G.L.: 1984, J. Great Lakes Res. 10(3):307-321.
Eadie, B.J. and Robbins, J.A.: 1987, Sources and Fates of Aquatic Pollutants, (R. Hites and S. Eisenreich, eds.), Advances in Chemistry Series No.216, American Chemical Society, Washington, D.C., p. 319-364.
EPA: 1987, The Great Lakes: An Environmental Atlas and Resource Book. EPA-905/9-87-002, Chicago, IL
Gardner, W.D.: 1980a, J. Mar. Res. 38:17-39.
Gardner, W.D.: 1980b, J. Mar. Res. 38:41-52.
Gardner, W.D.: 1996, Report on the JGOFS symposium in Villefranche sur Mer. May, 1995.
Hawley, N. : 1988, J. Great Lakes Res. 14:76-88.
Jannasch, H.W., Zafiriou, O.C., and Farrington, J.W.: 1980, Limnol. Oceanogr. 25:939-943.
Lee, C., Wakeham, S. and Hedges, J.: 1989,. JGOFs Report 10. WHOI. Woods Hole, MA.
Lerman, A.: 1979., Geochemical Processes Water arid Sediment Erivironments. John Wiley and Sons, Inc., New York, 481 pp.
Lineback, J.A. and Gross, D.L.: 1972, Il. Geol. Survey Environmental Notes no 58, 25pp, Champaign, IL.
Quinn, F.H.: 1992, J. Great lakes Res. 18:22-28.
Robbins, J.A. and Edgington, D.: 1975, Geochirn. Cosmochirn. Acta 39:285-304.
Robbins, J.A.: 1982, Hydrobiologia 92:611-622.
Robbins, J.A. and. Eadie, B.J: 1991 J Geophys Res. 96:17081-17104.
Wahlgren, M.A., Robbins, J. A. and Edgington, D.N.: 1980,: Transuranic Elements in the Environment. (Hanson, W.C.,ed), Technical Information Center/U.S. Department Energy, Washington, D.C.,

THE DYNAMICS OF SUSPENDED SEDIMENT TRANSPORT IN THE RIBBLE ESTUARY

MG Lyons
Westlakes Research Institute, Moor Row, Cumbria, CA14 4TP, UK.

Abstract. Velocity and turbidity were measured over a tidal cycle during both spring and neap tides in the Ribble estuary, England. The data was analysed to determine the relative magnitudes of the various components of the residual circulation and sediment flux. The Lagrangian residual circulation was found to be dominated by the influence of freshwater input to the estuary, the landward directed component of the circulation due to the Stokes drift being cancelled by the Eulerian flow induced by set-up. The dominant driving mechanism of the residual flux of suspended sediment was found to vary both spatially and over the spring-neap cycle. During the neap tide the flux in the mid estuary was negligable, however, during the spring tide landward transport of sediment by the mechanism of tidal pumping was found to be the dominant mechanism.

Key words. Estuarine, residual circulation, sediment transport, Stokes drift, tidal pumping, spring-neap cycle.

1. Introduction

The Ribble is a partially mixed shallow macrotidal estuary located in the north-west of England. Until the late-1970s the main channel of the estuary, which is enclosed within a training wall, was regularly dredged to allow passage to the upper estuary for shipping. Since the cessation of dredging the estuary has been subject to high rates of accretion (Beresford Hartwell *et al* 1995) and the partial decay of the training wall. The estuary is not expected to reach a new equilibrium state until well into the next century (O'Connor 1987).

Observations of the Ribble have revealed the existence of a turbidity maximum between the head of the estuary at Penwortham and the Ribble/Douglas confluence (Burton *et al* 1995). Substantial storage of intertidal mud has also been observed during neap tidal periods within the upper estuary (Burton 1994). A significant body of work now exists to show that tidal asymmetry, in which flood currents exceed ebb currents, is a principal mechanism in the generation of such deposits in shallow macrotidal estuaries (Allen *et al* 1980, Uncles *et al* 1985, Uncles *et al* 1992). Burton *et al* (1995) confirmed the importance of this mechanism in the Ribble by the application of a one-dimensional numerical model.

During December 1991, a dataset was collected for the purpose of calibrating a one-dimensional model of the estuary on behalf of British Nuclear Fuels Plc. (BNFL). Observations were made within the main channel at three locations during both spring and neap tidal cycles. The data is analysed in this paper to investigate the dynamics of sediment transport within the estuary.

Water, Air and Soil Pollution **99**: 141-148, 1997.

2. Observations

2.1 STUDY METHODOLOGY

Three locations in the estuary were manned during spring and neap tidal cycles. Station 1 was sited near the head of the estuary at Penwortham Bridge, from which the measurements were taken. Station 2 was located 6 km seaward in the centre of a straight section of the channel, still regulated by the training walls. Station 3 (5 km seaward of Station 2) was positioned just seaward of the confluence of the Ribble and the Douglas, its principal tributary. Observations at Stations 2 and 3 were taken from a bilge-keel boat. Velocity and salinity measurements were collected using a NBA DNC-3 current meter fitted with a conductivity cell. Turbidity readings were taken using an Alec PT-036 turbidity meter.

Measurements were made every half hour over a complete tidal cycle from the surface to the bed at one metre intervals. Where the water column was too shallow to allow this, measurements were taken near the surface and the bed. At depths of less than 0.5m a single mid-depth reading was taken to be representative of the entire water column.

River discharge data was obtained from the National Rivers Authority (NRA).

2.2 DATA

Throughout the study period the riverine discharge to the estuary remained below 11 m^3s^{-1}, low compared to the annual mean of 33 m^3s^{-1}. The landward limit of the salt water intrusion was in the vicinity of Station 1 during the spring tidal period and somewhat seaward of Station 1 during neap tides. At low tide the estuary was riverine as far seaward as Station 3 during both spring and neap periods.

In common with many macrotidal estuaries the duration of the flood tide was shorter than that of the ebb, high spring flood-tide velocities (up to 1.5 m s^{-1}) commonly resulting in the formation of a bore in the middle to upper estuary. Observations showed the bore to be associated with high levels of turbulence.

The turbulence produced by the interaction between with the high velocity flow observed during springs and the bed of the estuary acted to mix significant quantities of sediment vertically through the water column during the flood. During the longer ebb tide, peak velocities were typically somewhat lower and occurred near the surface. The lower velocities led to a reduction in vertical mixing of sediment during the ebb. In addition reduced vertical mixing of freshwater allowed stratification to develop during the ebb. Stratification of the water column also inhibits vertical mixing and vertical transport of sediment, thus reducing further the sediment concentrations in the water column (Dyer 1986).

3. Analysis of Residual Circulation and Sediment Fluxes

3.1 THEORY

The net transport of sediment within the water column when averaged over a tidal cycle is the product of the interaction of a range of mechanisms. Freshwater entering the estuary gives rise to a residual current which may transport seaward both sediment entering the estuary from riverine sources and sediment suspended within the estuarine water column, originating from local resuspension or marine sources. If the water column is stratified as was been observed during the ebb tide in the Ribble, sediment derived from riverine sources is more likely to be transported within the upper layer while resuspended bed-sediment and sediment of marine origin is likely to be transported near the bed. The differing velocities associated with each layer in combination with the differing sediment concentrations result in vertically sheared sediment transport. The net transport in such a situation may be either seaward or landward.

The combination of asymmetry in the duration of the flood and ebb tides and differences in sediment concentrations in the water column over the tidal cycle may result in a net transport of sediment up-estuary. The mechanism of tidal pumping of sediment within an estuary is, however, considerably enhanced by the action of the Stokes drift, a mass transport phenomenon resulting from the variation with time of estuarine cross-sectional area at a given location during the passage of the tidal wave. The result is a net up-estuary flux of water over a tidal cycle. Indeed Uncles and Jordan (1980) identified the flux of sediment due to the Stokes drift as being the main transport mechanism in a study of the Severn estuary. By volume conservation the set-up gradient induced by this flux must be balanced by a down-estuary Eulerian flow of water which is also capable of transporting sediment.

Analysis of the velocity and turbidity data allows the relative contributions due to freshwater, tidal pumping and vertical shear to the residual fluxes of water and suspended sediment at a given location, to be quantified. The method followed here (Uncles and Jordan, 1979, Uncles *et al*, 1985) is one-dimensional. H and U are the measured depth and axial flow velocity per unit width respectively at a given instant. Diamond brackets represent an average over a tidal cycle and overbars a depth average.

3.1.1 The Residual Circulation

$\overline{u}_E, \overline{u}_S, \overline{u}_L$ represent the depth-averaged Eulerian, Stokes and Lagrangian components of the residual current respectively, such that:

$$\overline{u}_L = \overline{u}_E + \overline{u}_S \qquad (1)$$

The depth-averaged Eulerian residual, the mean flow as measured at a given location, is given by

$$\overline{u}_E = \langle \overline{U} \rangle \qquad (2)$$

While the depth-averaged Stokes drift is represented by

$$\bar{u}_S = \langle \tilde{H}\tilde{U} \rangle / \bar{H} \quad (3)$$

where

$$\tilde{U} = \bar{U} - \langle \bar{U} \rangle \text{ and } \tilde{H} = H - \langle H \rangle \quad (4)$$

In addition riverine runoff induces a residual flow within the estuary, which is defined as $\bar{u}_F$. If $\langle A \rangle$ is taken to be the tidally averaged area of the estuarine cross-section at the measurement location and $\langle Q_F \rangle$ the rate of freshwater input up-estuary of the section, then:

$$\bar{u}_F = \langle Q_F \rangle / \langle A \rangle \quad (5)$$

Uncles and Jordan (1979) have shown that the residual rate of transport of water per unit width of water column is:

$$\langle Q \rangle = \langle H\bar{U} \rangle = \langle H \rangle \bar{u}_L \quad (6)$$

3.1.2 Residual Transport of Suspended Sediment

The residual transport of suspended sediment per unit width of water column G (units ppm m s^{-1}, where ppm = parts per million by weight of water) can be regarded as being composed of three components, G_L which is due to the Lagrangian residual flo w of water, G_{TP} due to tidal pumping and G_V due to vertical shear, such that:

$$G = G_L + G_{TP} + G_V \quad (7)$$

The suspended sediment concentration at a given instant is given by P, so:

$$G_L = \bar{u}_L \langle \bar{P} \rangle \quad (8)$$

$$G_{TP} = \langle \tilde{Q}\tilde{P} \rangle / \langle H \rangle \quad (9)$$

$$G_V = \langle H\overline{U'P'} \rangle / \langle H \rangle \quad (10)$$

where

$$P' = P - \bar{P} \quad (11)$$

3.2. RESULTS

3.2.1 Residual Currents

Figure 1(a) summarises the results of the calculations for $\overline{u}_S, \overline{u}_F, \overline{u}_E$ and $\overline{u}_L$ during the spring tidal period. The Stokes drift was directed up-estuary at all three Stations, reaching a maximum value of 30 cm s^{-1} at the most seaward Station (3), declining to a minimum of 10 cm s^{-1} at Station 2 before rising again to a level of approximately 25 cm s^{-1} at Station 1. Uncles and Jordan (1980) have shown that $\overline{u}_S$ is a product of the partially progressive nature of the tide within an estuary, the magnitude of $\overline{u}_S$ being dependant at any point in the estuary on the magnitude of friction at that location. It is probable that the minimum value of the Stokes drift is reached at Station 2 due to the topography of the channel at that point, which is located in the centre of a straight stretch which is still constrained by the old training walls. In contrast, at Station 3 the cessation of dredging has resulted in the formation of sandbanks within the main channel which are likely to increase the bed friction substantially. Station 1 is located up-estuary of the trained section within a narrow area of bends and mudbanks which are also likely to result in increased friction.

Figure 1. Components of the residual circulation during the spring (a) and neap (b) tidal periods. The x-axis measures distance below the tidal limit of the estuary moving through Stations 1 to 3 with x increasing.

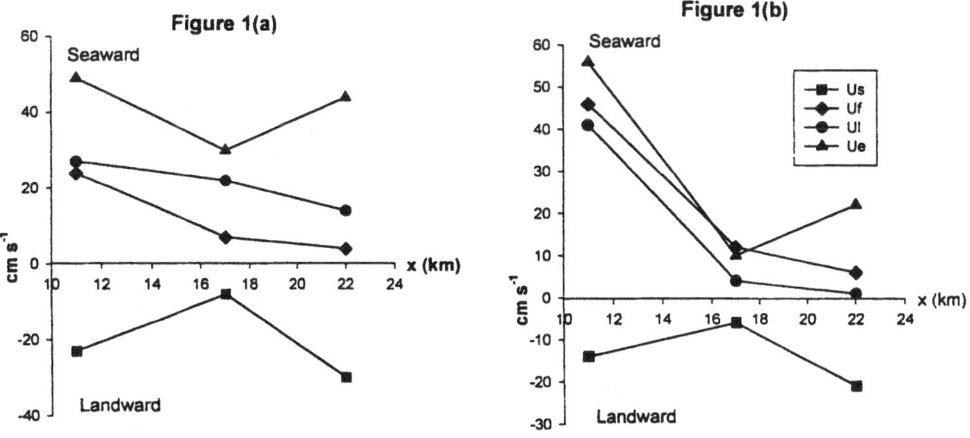

The Stokes drift results in a flow of water into the estuary which in turn results in a set-up gradient along the longitudinal axis. This gradient induces an Eulerian residual current which flows to oppose the Stokes drift. In the hypothetical case of a steady state existing within the estuary and if uniform conditions occurred at each location, then the component of the overall seaward Eulerian flow induced by the set-up would be in equilibrium with the landward Stokes drift, balancing it exactly. In such a case the Eulerian current resulting from the flow of freshwater upstream of the measurement point would be equal to the observed Lagrangian flow, such that $\overline{u}_L = \overline{u}_F$ (assuming constant depth across the channel and for a unit width). At the head of the estuary this relationship is approximately true, however, some discrepancy arises at the other locations although the trend towards reduction remains consistent. It is likely that main cause of the imbalance between the two terms is the result of cross-sectional variations in topography which act to introduce variations from the unit width approach taken here.

During the neap tide (Figure 1(b)) a similar regime was observed. The Stokes drift again reached a minimum at Station 2. Values were smaller due to the smaller frictional dissipation of the tide during neaps (Uncles and Jordan 1980), reaching a maximum of approximately 20 cm s^{-1} at Station 3. The $\bar{u}_L = \bar{u}_F$ relationship was improved at all locations with the pattern of variation consistent. The value of $\bar{u}_F$ at the head of the estuary (45 cm s^{-1}) was larger than that measured during the spring tide due to the reduced quantity of sea-water reaching the upper estuary, resulting in a smaller tidally averaged cross-sectional area for the channel.

3.2.2 Residual Fluxes of Sediment

From Equation 7 the residual flux of sediment is the sum of that due to the residual flow of water, tidal pumping and the vertical shear. If an equilibrium situation is again assumed, with uniform cross-section, so $\bar{u}_L = \bar{u}_F$, we obtain from equation 8 by substitution $G_L = G_F$. Equation 7 then reads

$$G = G_F + G_{TP} + G_V \qquad 15$$

allowing comparison of the relative magnitudes of the components of the flux due to freshwater, tidal pumping and vertical shear. Values of G_F, G_{TP}, G_V and the total flux G calculated for each location for the spring tidal period appear in Figure 2(a).

Transport through the mechanism of vertical shear was insignificant at all locations. At the head of the estuary landward transport by tidal pumping balanced the seaward transport by the freshwater induced current, resulting in no net transport of sediment during the spring period. At Station 2 up-estuary pumping of sediment was six times more significant than the seaward transport due to the freshwater flow. Although haline stratification was detected at this location during the ebb tide it appears to have had little influence on the transport of sediment when averaged over a tidal cycle. At Station 3 the role of tidal pumping in the transport of sediment was far smaller than that at Station 2, but still larger than that due to the freshwater induced residual flow.

During the neap tide (Figure 2(b)) transport of sediment by all pathways was reduced in comparison to transport during spring tides. Again transport by the vertical shear mechanism was negligible. At Station 1 both G_{TP} and G_V were approximately zero with the freshwater flow resulting in a small seaward flux. As this Station was above the landward limit of the saline intrusion during the neap tide and the influence of the flood tide was not

Figure 2. Components of the residual suspended sediment flux during the spring (a) and neap (b) tidal periods. The x-axis measures distance below the tidal limit of the estuary moving through Stations 1 to 3 with x increasing.

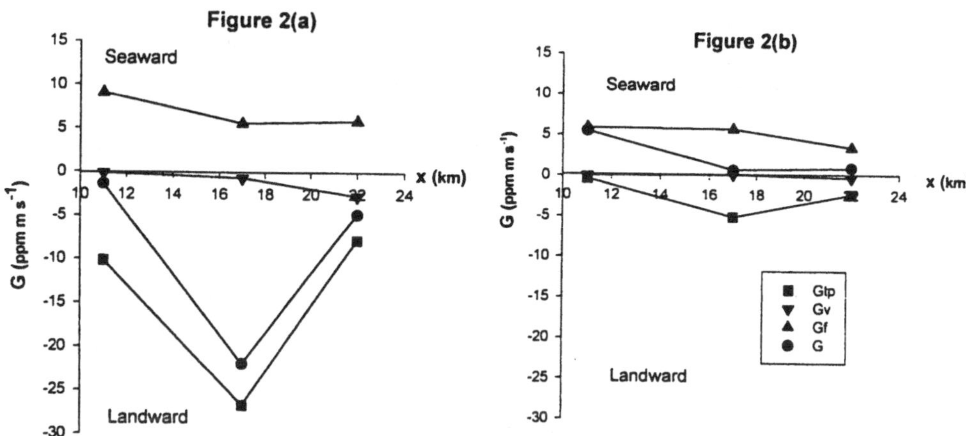

sufficient to reverse velocities, any influence from either G_{TP} or G_V would not be expected Tidal pumping approximately balanced the transport induced by the freshwater flow at Stations 2 and 3 with the maximum value of G_{TP} again occurring at Station 2

4. Discussion

Although performed in December, the low runoff conditions in the Ribble more closely approximated those of summer throughout the period of study. In the middle and upper estuary the Eulerian component of the residual circulation was directed seaward at all three stations during both spring and neap tides while the Stokes drift was directed landward, varying in intensity due to local friction and channel topography. Summation of the Eulerian and Stokes components shows the resulting Lagrangian residual to be principally driven by the input of freshwater to the upper estuary.

Haline stratification was observed during the ebb tide, however, its influence on the residual transport of sediment via the mechanism of vertical shear has been found to be negligible. Sediment transport within the Ribble can be regarded as being the product of the interplay between tidal pumping and transport by freshwater induced currents. Tidal pumping was found to be directed up-estuary at all locations, reaching a maximum at Station 2, while transport due to the input of freshwater to the head of the estuary was directed seaward throughout. Both the direction and magnitude of the resulting residual flux of sediment varied with the spring/neap tidal cycle as the sum of the two competing mechanisms. During the spring tide the residual transport of sediment was dominated by tidal pumping throughout the estuary with maximum levels occurring at Station 2, probably acting as a significant contributing factor to the maintenance of the turbidity maximum which exists between stations 1 and 3 and which is associated with the head of the saline intrusion (Burton 1994). During the neap tide, tidal pumping of sediment was much reduced resulting in an equilibrium situation with no net transport.

It appears therefore that, during the spring tide, sediment is pumped into the middle and upper estuary in significant quantities. This situation is not reversed during the neap tide, resulting in a net flux of sediment into the estuary. This landward flux will eventually result in a silting up of the estuary. Indeed, observations have shown the recent development of significant sandbanks within the main channel of the estuary between Stations 1 and 3.

Acknowledgements

The author would like to thank BNFL Plc. for allowing the use of the field data analysed in this paper.

References

Allen GP, Salomon JC, Bassoullet P, Du Perihoat Y, De Grandpre C: 1980 *Sedimentary Geology*, **26**, 69-90.

Beresford Hartwell PR, Horsington RW, Mamas CJV, Randle K, Taylor EW, West JR, Sokhi RS: 1995 *Coastal Zone Topics: Process, Ecology and Management*, **1**, 14-19.

Burton DJ: 1994 *Modelling of solute and sediment transport in the Ribble estuary*. PhD thesis. University of Birmingham.

Burton DJ, West JR, Horsington RW, Randle K: 1995 *Environment International*, **21**, 131-141.

Dyer KR: 1986 *Coastal and Estuarine Sediment Dynamics*. 342pp. Wiley.

O'Connor BA: 1987 *J. Geog. Soc. Lond.*, **144**, 187-195.

Uncles RJ, Jordan MB: 1979 *Est. and Coast. Mar. Sci.*, **9**, 287-302.

Uncles RJ, Jordan MB: 1980 *Est. and Coast. Mar. Sci.*, **10**, 39-60.

Uncles RJ, Elliot RCA, Weston SA: 1985 *Est., Coast. and Shelf Sci.*, **20**, 147-167.

Uncles RJ, Stephens JA, Barton ML: 1992 In: D. Prandle, ed. *Dynamics and Exchanges in Estuaries and the Coastal Zone*. New York, Springer Verlag, 255-276.

A STUDY OF TRANSPORT AND MIXING IN NATURAL WATERS USING ICP-MS: WATER-PARTICLE INTERACTIONS

SUSAN C. PAULSEN and E. JOHN LIST
California Institute of Technology, Pasadena, California 91101, USA

Abstract. Water-particle interactions often may result in non-conservative chemical behavior when waters from different sources mix with one another. The results presented in this paper address the role of these interactions in freshwater and estuarine mixing and support a larger study to develop a method to help resolve flow distribution and water quality questions in surface waters using a source water "fingerprinting" technique. Inductively coupled plasma-mass spectrometry (ICP-MS) is used to "fingerprint" each water source based upon the concentrations and relative proportions of elements in that source. Estimates can then be made of the fractions of various "fingerprinted" waters in water samples that contain a mixture of source waters. Such estimates depend upon the selection of tracers that behave conservatively during mixing; in this paper, results to establish the maximum particle exchange capacity and conservative mixing behavior are presented for samples collected from the Sacramento River-San Francisco Bay-Delta estuary. Elements likely to behave conservatively include boron, sodium, magnesium, potassium, calcium, strontium, and molybdenum.

Key words: tracer, fingerprint, particle, mixing, ICP-MS, Sacramento River, San Joaquin River, Delta

1. Introduction

A key scientific issue in the behavior of estuarine ecosystems is the distribution of fresh and saline waters in bays and estuaries, particularly where there are multiple freshwater inputs of varying water quality. Understanding this distribution requires detailed knowledge of the water circulation patterns within the estuary system. Although hydrodynamic models do exist to describe conditions within such systems, in most cases the application of the models is limited by the fact that the residual transport from the net freshwater outflow is small compared to daily tidal fluxes.

For this reason a method to determine the way in which salt and other chemical constituents reach a given location is needed. Tracer studies with similar goals have been performed using a variety of analytical techniques and tracers in air pollution studies (see, for example, Hopke *et al.*, 1976; Dzubay *et al.*, 1988; Heaton *et al.*, 1992; Rogge 1993). Receptor modeling has recently been used in groundwater studies to trace specific sources of contamination (Olmez *et al.*, 1994; Olmez and Hayes, 1990); others have used simpler mass balance techniques to investigate the spreading of released tracers that are distinct from surrounding groundwaters (Santschi *et al.*, 1987; Harvey *et al.*, 1989; Stauffer 1985; Deverel and Millard, 1988). ICP-MS has made possible the accurate measurement of many elements in the periodic table, and this study is focused on determining which of these elements can be used as tracers. Although previous tracer studies have not used ICP-MS, at least two studies have been conducted that have confirmed the potential of ICP-MS analysis to fingerprint waters (see Stetzenbach *et al.*, 1994, and Henshaw *et al.*, 1989).

Water, Air and Soil Pollution **99**: 149-156, 1997.

For a chemical constituent to function successfully as a tracer, three conditions must be met: (1) sources must have distinct chemical or elemental signatures; (2) source concentrations must not vary significantly on timescales shorter than the mixing timescales of a system; and (3) signatures of sources must not be altered to any great extent either by chemical and/or biological reactions or by physical changes that occur during mixing (i.e., mixing should be conservative, or very nearly conservative).

The work presented in this paper addresses primarily the third of these conditions. The question of whether mixing is conservative for any given element depends largely upon particle-water interactions; data collected in the Sacramento-San Joaquin Bay Delta system demonstrate that "conservativeness" is largely element-dependent, with mixing behavior more conservative for the alkali metals and alkaline earth elements and less conservative for the transition metals and rare earth elements.

2. Study Area and Methods

The Delta is located in northern California and consists of two large freshwater sources which flow into a brackish bay. The Sacramento River, which flows into the Delta from the north, provides roughly 80% of the freshwater that flows into the system; the San Joaquin River, flowing from the south, provides roughly 15% of the freshwater inflow. In this estuary, mixing between fresh and brackish waters of varying chemical compositions occurs over long distances. Figure 1 shows the study area, including the five sites where samples supporting this paper were collected.

Water samples were collected from rivers and brackish waters in Teflon sample containers that had been thoroughly cleaned in an ultrapure nitric acid bath. Samples were acidified upon collection with ultrapure nitric acid to a concentration of 2%. Filtered samples were passed through 0.45 μm Teflon syringe filters, which are also thoroughly cleaned prior to use. Field blanks were collected for all samples. Analysis was conducted quantitatively for forty-one elements using a Hewlett Packard 4500 ICP-MS. Internal standards were used for each sample, and check standards of known concentration were made from reagents and analyzed every five samples. Interpolation between check standards was used to correct for instrumental drift (less than 10%). Every tenth sample was reanalyzed during a separate run to confirm results. Instrumental drift has been minimal and data are highly reproducible. Further, standard addition testing has confirmed the ability of ICP-MS to measure accurately concentrations of many elements in the brackish waters of the study area. These waters are far less saline than ocean waters, and the measurement of major cations (including lithium (Li), sodium (Na), magnesium (Mg), calcium (Ca), rubidium (Rb), strontium (Sr), and barium (Ba)) and several trace elements (including boron (B), molybdenum (Mo), the rare earth elements (REEs), and uranium (U)) suffers from few interferences in these brackish waters.

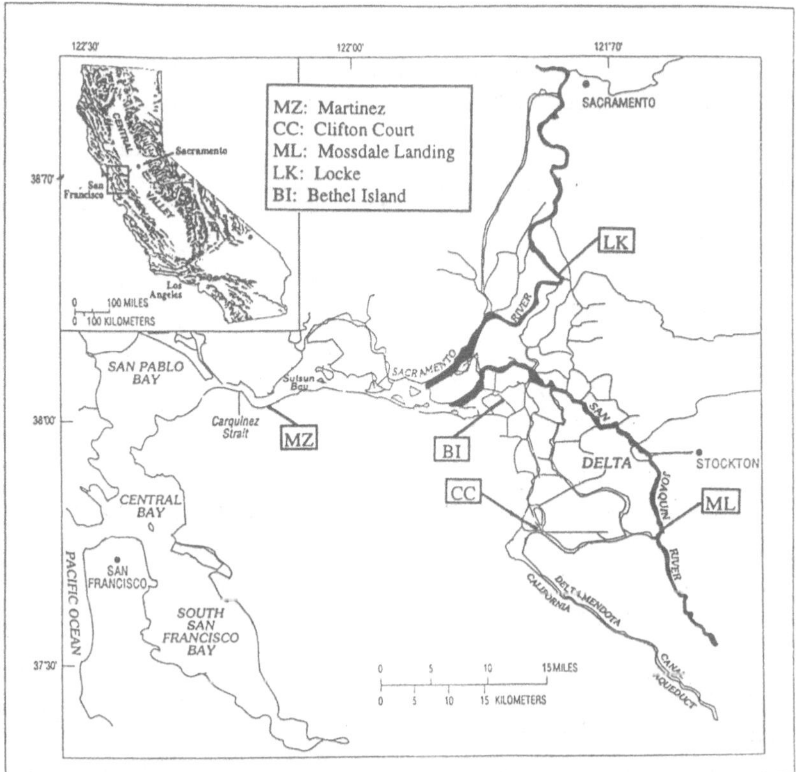

Figure 1. Study area and sample locations.

3. Results and discussion

3.1 VARIATION IN SOURCE SIGNATURES

Fingerprints, or signatures, of water sources are defined as the relative concentrations or proportions of measured concentrations of elements. Elements are chosen for inclusion in a fingerprint based upon significant and consistent variation in relative concentration between sources. To determine the signatures of the various sources contributing to the waters of the Delta, samples have been collected from the Sacramento and San Joaquin Rivers and from Carquinez Strait at Martinez, at the outlet of the Delta. The samples collected at Martinez are a mix of freshwater from the rivers and more saline Bay waters. It is difficult to define the "pure" Bay water end member due to mixing from various freshwater sources throughout the estuary and beyond the Delta. Because all water leaving the area of interest must pass through this strait, this collection point is used to define the "end member" of the study area. It is recognized that the composition of this end member will vary with time and freshwater outflow.

TABLE I
Measured elemental concentrations for sources to the San Francisco Bay-Delta estuary: Average concentrations and range of observed values

Element	Sacramento River at Locke (3/14/96 - 12/3/96)	San Joaquin River at Mossdale Landing (3/14/96 - 12/3/96)	San Pablo Bay at Martinez (4/13/96 - 8/29/96)
Li [ppb]	1.55 (0.74 - 3.16)	15.8 (3.37 - 41.4)	48.4 (8.30 - 99.5)
B [ppb]	36.2 (22.9 - 59.0)	304 (46.2 - 566)	880 (141 - 1550)
Si [ppm]	10.6 (6.23 - 19.3)	22.8 (3.69 - 60.6)	26.7 (10.3 - 50.5)
Ca [ppm]	4.40 (2.81 - 6.58)	13.4 (4.16 - 25.7)	33.4 (7.87 - 69.0)
Rb [ppb]	1.09 (0.53 - 2.33)	22.3 (0.96 - 103)	29.7 (5.99 - 57.7)
Sr [ppb]	63.7 (37.4 - 95.5)	328 (90 - 555)	1570 (180 - 3140)
Ba [ppb]	21.9 (12.5 - 44.4)	153 (32.9 - 430)	57.6 (28.7 - 98.6)
La [ppb]	0.199 (0.046 - 0.765)	7.82 (0.011 - 34.6)	2.53 (0.420 - 5.59)
U [ppb]	0.153 (0.050 - 0.290)	7.23 (0.790 - 13.5)	1.09 (0.455 - 1.87)

In the Delta, fingerprints of all water sources do vary seasonally and with changing flow conditions, particularly for the San Joaquin River. Concentrations of the alkali metals, alkaline earth elements, and halogens are generally higher in water from San Pablo Bay than in the river waters, while concentrations of transition metals and rare earth elements are significantly higher in San Joaquin River water than in Sacramento River water, as are concentrations of particulate matter. In Table I, average elemental concentrations and the range of observed concentrations are presented for daily composite samples collected from each of the sources. In addition, detailed data are presented for one representative sample date in Section 3.3 below; these data also show clearly the importance of adsorption to particulate matter in determining elemental concentrations.

The differences in elemental concentration between sources have been confirmed in grab samples collected on several different dates and for daily composite samples collected at these locations over a nine-month period; while concentrations in samples from individual sources vary (particularly the magnitude of concentrations in the San Joaquin River), the variations in concentration between sources follow the same patterns. These differences in elemental composition between freshwaters and saline waters should be similar in almost all estuarine environments and are confirmed in the literature for average river and ocean water concentrations (Burton and Liss, 1976; Aston, 1978; Martin and Meybeck, 1979; Bruland, 1983; Motekaitis and Martell, 1987; Bruland *et al.*, 1991; van den Burg, 1993).

3.2 Laboratory Mixing Studies and Mixing at Freshwater Confluences

Mixing studies have been conducted in the laboratory to determine the mixing characteristics of various elements. In these studies, waters collected from the San Joaquin River and Sacramento River were mixed in known proportions. Samples were agitated for several hours after mixing, and particulate matter was then allowed to settle to the bottom of the mixing containers. Because acidifying these samples after mixing would change the mixing behavior, samples were not acidified prior to analysis; the analysis results are therefore qualitative rather than quantitative. Concentration data were plotted against the known fraction of river waters, and conservative behavior was determined by a straight line fit through the data points.

Figures 2(a) and 2(b) show examples of conservative and non-conservative behavior for two different elements, Sr and vanadium (V). Elements which demonstrated conservative behavior (as determined by an r^2 correlation coefficient for a

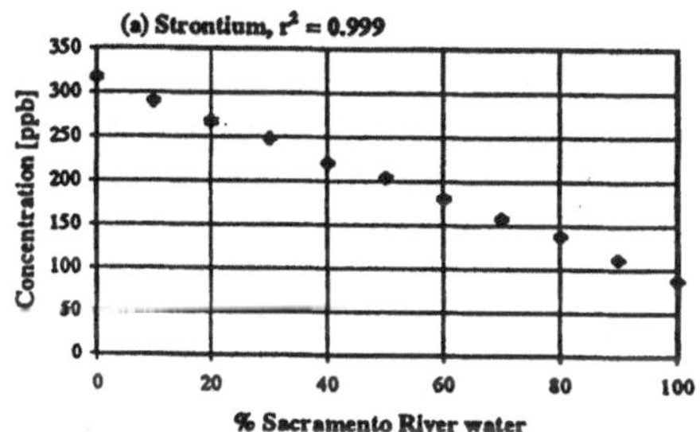

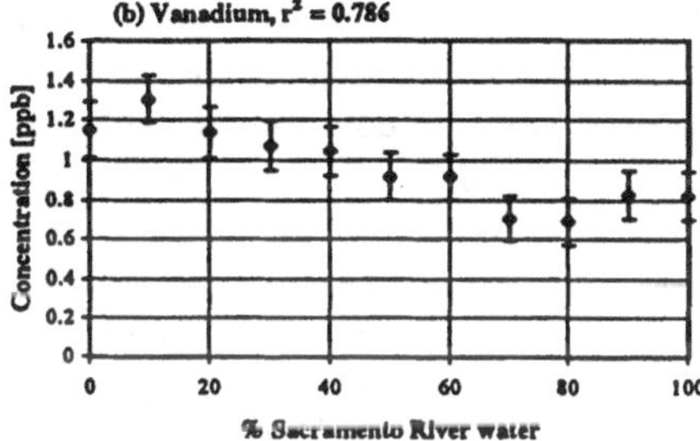

Figure 2. Results of laboratory mixing studies of Sacramento and San Joaquin River waters for two elements, Sr and V. Conservative mixing behavior is observed for Sr while non-conservative behavior is observed for V. All mixtures are composed of Sacramento River water and San Joaquin River water, so that a point falling at 80% Sacramento River water contains 20% San Joaquin River water.

straight line regression fit through the data points of 0.900 or greater) in this mixing series include: Li, B, Mg, silicon (Si), potassium (K), Ca, gallium (Ga), bromine (Br), Rb, Sr, Mo, iodine (I), barium (Ba), and uranium (U).

Additional studies were conducted at the confluences of the Sacramento River with the American River and the Feather River. These confluences were beyond the reach of tides when test samples were collected. The samples obtained were used to predict the fractions of water downstream of the confluence contributed by each of the upstream rivers; this calculation was based upon linear (conservative) mixing of elemental concentrations. Predicted fractions were then compared to the fractions of flow known from reservoir release data and measured flowrates. Unfiltered samples (whole water samples) that were acidified upon collection consistently predicted mixing fractions better than samples that were filtered and then acidified. Elements for which mixing was nearly conservative include: Li, Na, Mg, Si, K, Ca, Ga, Br, Rb, Sr, Mo, I, Ba, and U.

Both laboratory and field tests demonstrated that the same elements consistently predict conservative mixing when two freshwaters are mixed together. Some of these elements, however, may experience non-conservative behavior due to ion exchange and deposition of particulate matter when freshwaters mix with more saline waters. This effect is likely to be most pronounced at low to intermediate salinities, where sufficient river-borne particulate matter and high concentrations of Na^+ ions are available to participate in exchange reactions and enhance coagulation processes. In this salinity range, the concentrations of many elements can be measured accurately, and further testing was designed to address this question.

3.3 Determination of Maximum Cation Exchange Capacities

Because exchange reactions are a likely cause of non-conservative behavior during estuarine mixing, both filtered and unfiltered samples were collected to determine the maximum possible cation exchange capacities. Three sets of samples were collected at each of the five sampling locations: (1) whole water samples that were acidified upon collection; (2) filtered samples, filtered as described in Section 2 with the filtrate immediately acidified; and (3) samples that were acidified for a period of 24 - 48 hours and then filtered.

Whole water samples, acidified to 2% nitric acid upon collection, were analyzed to determine maximum cation exchange capacities, defined here as the amount of an element released to solution from particles upon acidification. Such strong acidification (to a pH between 0 and 1) will result in release of cations adsorbed to particle surfaces and possibly also in some dissolution of the particles themselves; anionic species adsorbed to particle surfaces will also be partially released due to complexation of hydrogen ions with competing ligands, which are less likely to complex with the surfaces of particles that are positively charged due to acidification. Because ICP-MS analysis analyzes small particles suspended in aqueous solution, samples were processed as described in (3) above. By comparison, these acidified, filtered samples provide a measure of the amount of any given element in the acidified solution which is associated with particles larger than 0.45 μm. Results for selected elements for each of the five sampling locations are presented in Figure 3.

The change in elemental concentration caused by filtration is negligible for many elements, including B, Na, Mg, K, Ca, Sr, and Mo. These elements are present primarily in the dissolved phase and should therefore behave conservatively during mixing. Filtration causes significant removal for many other elements, including the main components of most particulate matter (e.g., Fe) and elements which have a strong affinity for particulate matter, including the REEs (lanthanum (La) is a member of this group of elements), and thorium. Additionally, the transition metals showed significant removal in the freshwater samples; although these elements could not be accurately measured in more saline samples, removal was still observed. These elements are not likely to behave conservatively during estuarine mixing. Still a third class of elements

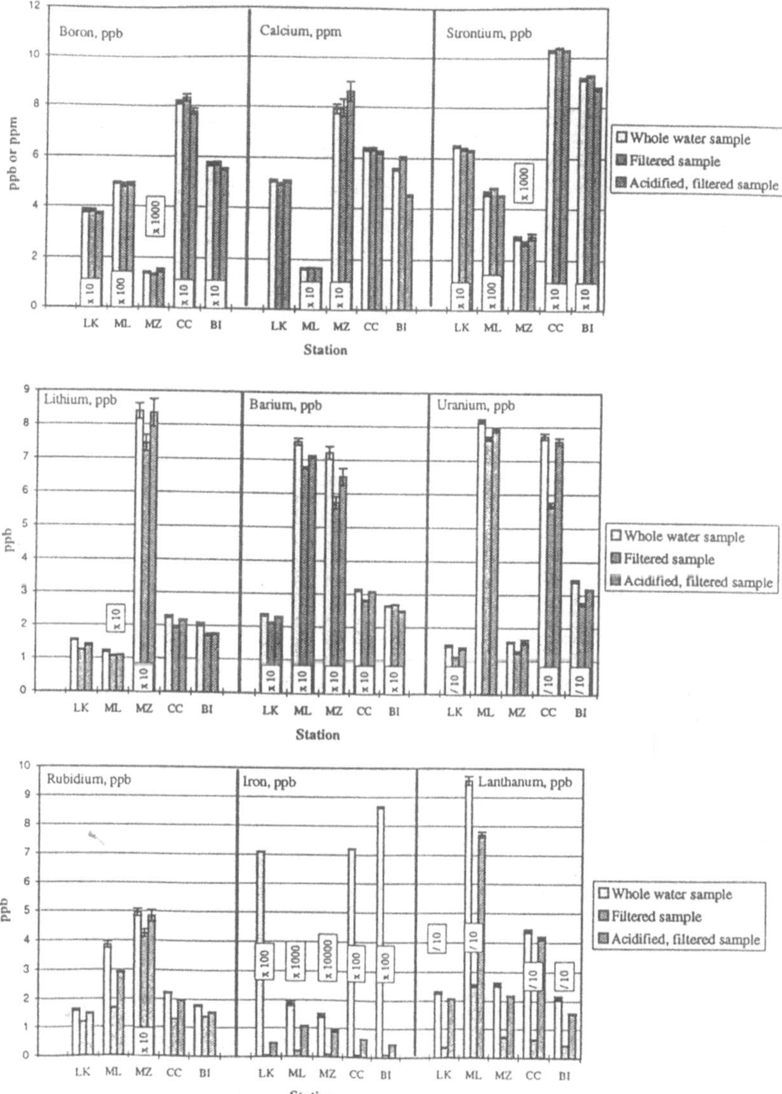

Figure 3. Results of filtration studies conducted to determine maximum exchange capacities. Panels show filtration results for samples collected at Locke (LK), Mossdale Landing (ML), Martinez (MZ), Clifton Court (CC), and Bethel Island (BI). Note that results for some stations are scaled; scaling factors are shown on the graph (i.e. value shown should be multiplied by scaling factor to obtain actual value). Absolute concentrations for iron are qualitative only due to interferences inherent in ICP-MS analysis for iron. While Fe concentration values are qualitative, relative differences between concentrations are quantitative.

exhibits near conservative behavior, with removal caused by filtration of roughly 30% or less. These elements include Li, Rb, Ba, and U.

Additional study will be required if these elements are to be used as tracers. Statistical analyses are being conducted for routinely collected field samples to determine if these elements do indeed behave conservatively during estuarine mixing. Behavior may be conservative if most of the mass of these elements that is associated with particulate matter is exchanged during estuarine mixing, or if these elements are associated with fine particulate matter which does not readily settle. The results for the samples which were acidified and then filtered show that this may be the case.

Acknowledgments

The authors gratefully acknowledge the financial support of the Mellon Foundation and the State of California, Department of Water Resources, through the Interagency Ecological Program. The continuing advice, counsel, and support of Professor James J. Morgan has been of great assistance to the project.

References

Aston, S. R.: 1978, *Estuarine chemistry*, in *Chemical Oceanography*, Volume 2, Academic Press, London, pp. 362-438.

Bruland, K. W.: 1983, *Trace Elements in Sea-water*, in *Chemical Oceanography*, Volume 8, Academic Press, London, pp. 157-220.

Bruland, K. W., Donat, J. R. and Hutchins, D. A.: 1991, *Limnology and Oceanography* **36**, 1555-1577.

Burton, J. D. & Liss, P. S.: 1976, *Estuarine Chemistry*, Page Bros. Ltd., Norwich, 229 pp.

Deverel, S. J. and Millard, S. P.: 1988 *Env. Sci. & Tech.* **22**, 697-702.

Dzubay, T. G., Stevens, R. K., Gordon, G. E., Olmez, I., Sheffield, A. E. and Courtney, W. J. : 1988, *Env. Sci. & Tech.* **22**, 46-52.

Harvey, R. W., George, L. H., Smith, R. L. and LeBlanc, D. R.: 1989, *Env. Sci. & Tech.* **23**, 51-56.

Heaton, R. W., Rahn, K. A. and Lowenthal, D. H.: 1992, *Atmos. Env.* **26A**, 1529-1543.

Henshaw, J. M., Heithmar, E. M. and Hinners, T. A.: 1989, *Anal. Chem.* **61**, 335-342.

Hopke, P. K., Gladney, E. S., Gordon, G. E., Zoller, W. H. and Jones, A. G.: 1976, *Atmos. Env.* **10**, 1015-1025.

Martin, J. M. and Meybeck, M.: 1979, *Marine Chem.* **7**, 173-206.

Motekaitis, R. J. and Martell, A. E.: 1987, *Marine Chem.* **21**, 101-116.

Olmez, I. and Hayes, M. J.: 1990, *Biological Trace Element Research* **26**, 7, 355-361.

Olmez, I., Pink, F. X. and Wheatcroft, R. A.: 1994, *Env. Sci. & Tech.* **28**, 8, 1487-1490.

Rogge, W.: 1993, *Molecular Tracers for Sources of Atmospheric Carbon Particles: Measurements and Model Predictions.* Ph.D Thesis, Cal. Inst. Technol. 505p.

Santschi, P. H., Hoehn, E., Lueck, A. and Farrenkothen, K.: 1987, *Env. Sci. & Tech.* **21**, 909-916.

Stauffer, R. E.: 1985, *Env. Sci. & Tech.* **19**, 405-411.

Stetzenbach, K. J., Amano, M., Kreamer, D. K. and Hodge, V. F.: 1994, *Ground Water* **32**, 976-985.

van den Berg, C. M. G.: 1993, *Estuaries* **16**, 512-520.

FILTRATION ENHANCES SUSPENDED SEDIMENT DEPOSITION FROM SURFACE WATER TO GRANULAR PERMEABLE BEDS

DAVID C.J.D. HOYAL[1], MARCUS I. BURSIK[1], JOSEPH F. ATKINSON[2], and JOSEPH V. DEPINTO[2]

[1]Department of Geology, [2]Department of Civil Engineering, State University of New York at Buffalo, Buffalo, NY, 14260, USA.

Abstract. We present results from an experimental study of suspended particle (4.5-36.5 μm silicon carbide powder) deposition from surface water to 'clean' equi-granular permeable beds in a small 12.5x12.5x15cm box and a re-circulating flume. Enhanced deposition rates of up to 5 times the accepted sediment deposition model (e.g., Einstein, *1968*) are explained by filtration of particles in the bed. Compared to this model deposition increases with increasing surface fluid speed, decreasing suspended particle size and increasing bed particle size. These results can be explained by an increased ability of particles to penetrate into the bed with the pore water which increases the effective filter thickness. The predominant deposition mechanism within the bed pores appears to be settling. Enhanced deposition, evident in Einstein's (1968) experimental data, was previously attributed to flocculation but may be explained better by filtration. These enhanced deposition rates drop off to close to the accepted model predicted rate after a certain volume of sediment has entered the bed, which may be due to the slumping of deposits from the top of bed elements. This reduction in deposition rate occurs long before the bed is filled with fine sediment.

Keywords: suspended sediment, deposition, granular substrate, gravel bed, filtration, rivers

1. Introduction

Although many natural surface waters flow over permeable beds of materials such as sand and gravel, very little is known about the influence of these substrates on suspended particle deposition. The assumption is commonly made that substrate properties make little difference to the deposition rate, but there have been few experimental studies to verify this assumption. One way that permeable beds might influence suspended particle deposition rates is by filtration in the upper regions of the bed. Surface-pore water exchanges and filtration of suspended particles has been suggested in a recent study of colloidal exchange between surface water and a sand bed by Packman and Brooks (1995).

The influence of bed permeability on suspended sediment deposition has application in alluvial streams for predicting the fate of toxic chemicals which may be sorbed to particle surfaces (Jobson and Carey, 1989; Packman and Brooks, 1995). Penetration of fine sediment particles into gravel beds is also a concern to fisheries managers because clogging of salmoniid redds by fine particles affects fish egg survival and emergence

Water, Air and Soil Pollution **99**: 157-171, 1997.

rates (Havis et al., 1993). The transport of sediment in overland flow is liable to deposit by similar mechanisms. At a fine scale the beds of all surface water systems are somewhat permeable and may trap fine particles. Flume investigations of bed clogging using sand sized suspended sediment have been carried out by Schalchi (1992), Cuningham et al., (1987), Diplas (1986), Carling (1984) and Beschta and Jackson (1979) but very few workers have investigated the influence of the bed on the sediment removal rate from the water column.

To investigate deposition to permeable beds a programme of laboratory experiments in a small scale laboratory mixing box and re-circulating flume is being conducted using silt sized silicon carbide powders. Sediment water mixtures are chemically dispersed to reduce coagulation effects which allows an analysis of physical deposition processes uncomplicated by the coagulation process. Although in natural systems the dynamics of fine particles may be strongly influenced by coagulation, particle clusters are likely to be influenced by the physical environment in a similar manner to discrete particles. Results from the experiments suggest that filtration in the surface of permeable beds may significantly enhance suspended particle deposition from flowing surface water.

2. Current Theory and Einstein's (1968) Experiments

The established sediment deposition model based on the advection-diffusion equation has been applied for deposition to many substrates including granular permeable beds (Dobbins, 1944; Einstein, 1968, 1970). In one dimension this model can be expressed as

$$\frac{\partial C}{\partial t} = \frac{\partial}{\partial z}\left(K \frac{\partial C}{\partial z}\right) + w_S \frac{\partial C}{\partial z} \tag{1}$$

where C is the concentration of sediment suspended in the fluid, z is the vertical coordinate, K is the diffusion coefficient for sediment, t is time, and w_s is the sediment fall velocity in still water. A convenient analytical solution to Equation (1) can be found for the case of a uniform concentration of particles with depth assuming no diffusion at the base of the fluid (Martin and Nokes, 1988). The decay in the mass (M) of particles in the fluid column is given by

$$\frac{dM}{dt} = -A w_S C(0) \tag{2}$$

where A is the horizontal crossectional area of the bed and C(0) is the concentration at the base of the fluid. Since the concentration is vertically uniform the concentration anywhere in fluid can be determined from

$$C(0) = \frac{M}{Ah} \tag{3}$$

where h is the thickness of the fluid layer. Substitution of Equation (3) into Equation (2) gives

$$\frac{dM}{dt} = -\frac{w_s}{h}M \qquad (4)$$

which when integrated and divided by the fluid volume gives the solution

$$C(t) = C_0 e^{\frac{-w_s}{h}t} \qquad (5)$$

One of the few experiments that measured suspended sediment deposition to granular permeable beds was performed by Einstein (1968) in a re-circulating flume with gravel beds using silica flour as the suspended sediment. The concentration decay over time was exponential in form and Einstein fitted the model with an exponential decay coefficient (λ) determined from the data. Comparing the average of his data with the uniform sediment deposition equation (Equation 5), Einstein found good agreement. However, the experimental deposition rates (λ), which are presented in Figure 1 normalized by the model (i.e., divided by w_s/h), show considerable scatter and some apparent systematic trends (lines on graph). Results in which λ falls below the model value w_s/h, (i.e., line at unity), could easily be explained by erosion, whereby velocities in the bed are high enough to resuspend deposited particles back into the water column.

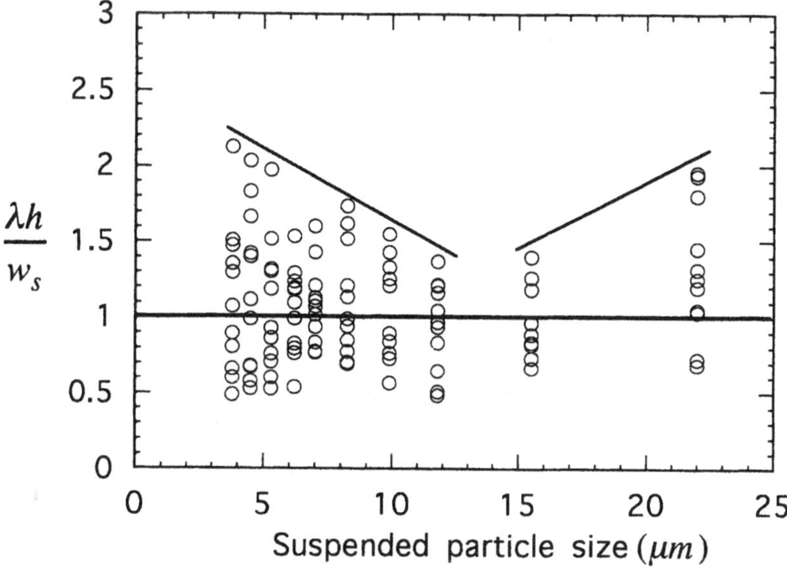

Figure 1 - Einstein's (1968) flume data plotted as non-dimensional deposition rate $\lambda h/w_s$ as a function of suspended particle size. Bed of gravel, suspended sediment is silica flour.

In this paper we are more interested in the two apparent trends of increasing deposition for smaller and larger particles. The increase in deposition (compared to Equation 5) with increasing particle size was attributed by Einstein (1968) to deviations from a uniform sediment concentration profile. The increase in normalized deposition rates with decreasing particle size was recognized by Owen (1969) and attributed to increasing coagulation of smaller particles. However, there is another explanation for the trend identified by Owen (1969) which appears equally plausible. It is well known that surface water and bed porewater are exchanged in the upper regions of the bed due to a penetration of the surface water velocity profile into the bed (Alonso and Mendoza, 1992; Zippe and Graf, 1983; Zagni and Smith, 1976). Filtration of this sediment-water mixture in the bed pores could also account for the enhanced deposition of finer particles which are able to penetrate into the bed deeper than coarser particles because of their higher likelihood of traveling with the fluid into the bed pore spaces.

Einstein justified his use of a settling boundary condition at the top of the bed (or just inside the bed), based on his hypothesis that once particles entered the bed they were only transported downwards into the bed by settling, so that the downwards flux is given by Aw_sC. However, the existence of advective currents within the bed suggests that the bed is part of the depositional system and particles can be redistributed to filtration sites allowing the bed to actively filter particles from the water. Filtration sites in water are likely to be areas in the bed onto which suspended particles can settle and rest. Thus, for permeable beds the total area of the bed, including filtration sites, may control deposition rates, rather than the simple horizontal crossectional area as in Equation 2. Einstein's boundary condition, however, does seem reasonable once all the filtration sites are full and the only effect of fluid motion in the bed is to exchange particles between the sites and the pore water.

3. Flume Experiment

To ensure that fine particle deposition is indeed enhanced when surface water flows over granular permeable beds similar to those studied in Einstein's (1968) experiments and not unlike many found in nature we performed four flume experiments. A suspended sediment mixture of 1g/l of 600 mesh silicon carbide (median diameter 9.3μm) in tap water, with 5g/l sodium hexa-metaphosphate to minimize flocculation, was recirculated over 5cm and 10cm thick beds of moderately sorted 2cm diameter gravel or medium sand in a 2.5m re-circulating flume. The vertical sediment distribution was near-uniform due to mixing within the pump. Comparison of deposition time curves, presented in Figure 2, shows that the deposition rates to the gravel are over twice the deposition rates to the sand. These results indicate that some characteristic of the bed substrate affects the deposition rate, contrary to what is predicted by the uniform model (Equation 5). Filtration seems the most obvious candidate for this increased deposition.

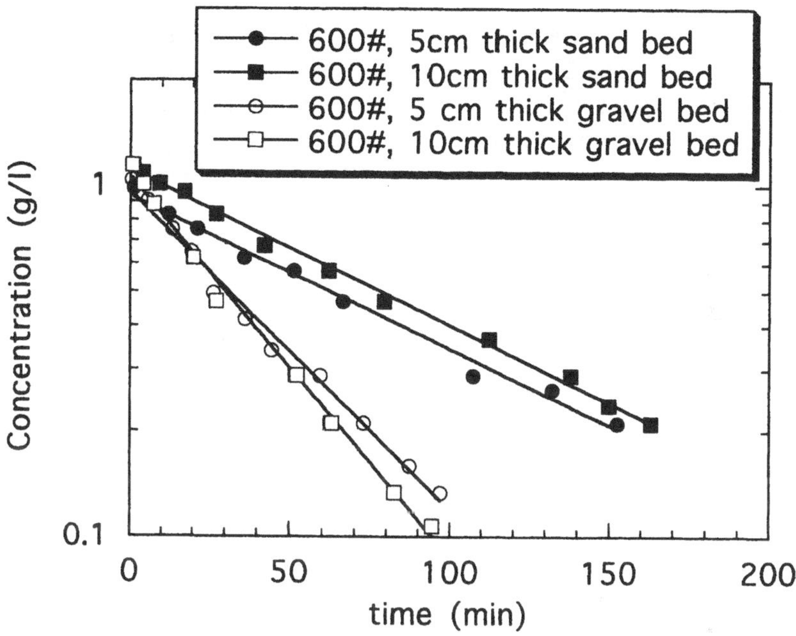

Figure 2 - Sediment concentration (g/l) as a function of time (minutes) in a 2.5m re-circulating flume. The beds are composed of medium sand (5, 10cm thick) and poorly sorted 2cm diameter gravel (5 and 10cm thick). Suspended particles are 600 mesh silicon carbide. Surface fluid depth (h) = 13cm.

4. Idealized Mixing Box Experiments

If the increased deposition recognized in both ours and Einstein's (1968) flume experiments is induced by filtration in the bed, then this deposition increase should be controlled by filtration variables. The study of granular filtration is relatively mature (see for example Tien, 1989) and important variables include (1) the size of suspended particles, (2) the size of collector elements and pores, (3) the velocity of fluid in the filter bed and (4) the amount of sediment already resident in the filter bed. Depositional mechanisms in filter beds include inertial impaction, interception, settling and Brownian motion (Tien, 1989). However, for larger particles, settling is the dominant depositional mechanism in water.

To test our settling-filtration hypothesis we performed a large number of experiments in a small box mixed by a rotating impeller with beds of glass beads, gravel, fibrous pads, flat beds and Perspex grids. These idealized experiments were designed to determine the dependence of particle deposition in permeable beds on the above-mentioned filtration variables.

4.1. GENERAL EXPERIMENTAL METHOD

Particle water mixtures were mixed by a thin wire impeller set at mid-depth over various bed surfaces in a 12.5 x 12.5 x 15cm Perspex box. The surface water depth was 11.5cm. All experiments had mixing speeds high enough to maintain a uniform sediment/water mixture in the surface layer so that results could be compared directly with Equation (5). Distilled water was used and 5g/l sodium hexa-metaphosphate was added to the water to reduce coagulation of particles. Sediment concentration was determined by measuring the attenuation of light from a small microscope lamp through the experimental box to a photographic lightmeter. Regressions of light vs. concentration all had correlation coefficients $r > 0.99$. Silicon carbide powders (specific gravity 3.2) were used because they are relatively well size graded and inexpensive. A list of particle size characteristics, as specified by the manufacturer, and the initial concentrations used in all experiments (unless otherwise specified) are presented in Table 1.

TABLE 1

Characteristics of the suspended silicon carbide particles

Grit	$d_{s50} \pm$ SD (μm)	Fall vel. (cm/min)	Initial Conc. (g/l)
280	36.5 ± 1.5	8.56	2.0
320	29.2 ± 1.5	5.47	2.0
400	17.3 ± 1.0	1.92	1.0
600	9.3 ± 1.0	0.55	0.66
800	6.5 ± 1.0	0.27	0.4
1000	4.5 ± 0.8	0.13	0.28

4.2. METHOD FOR SPECIFIC EXPERIMENTS

Because beds vary widely in the natural environment, a wide range of bed materials was tested. Specific experiments include:

Experiment 1 - The control experiment. Suspension of 280, 320, 400, 800 and 1000 mesh silicon carbide particles, stirred over a flat Perspex bed, (i.e., the bottom of the box). Results presented in Figure 3.

Experiment 2 - Suspension of 280, 320, 400, 800 and 1000 mesh silicon carbide particles stirred over a bed of 1.44±0.3cm angular crushed road gravel, 2.5cm thick. Results presented in Figure 4.

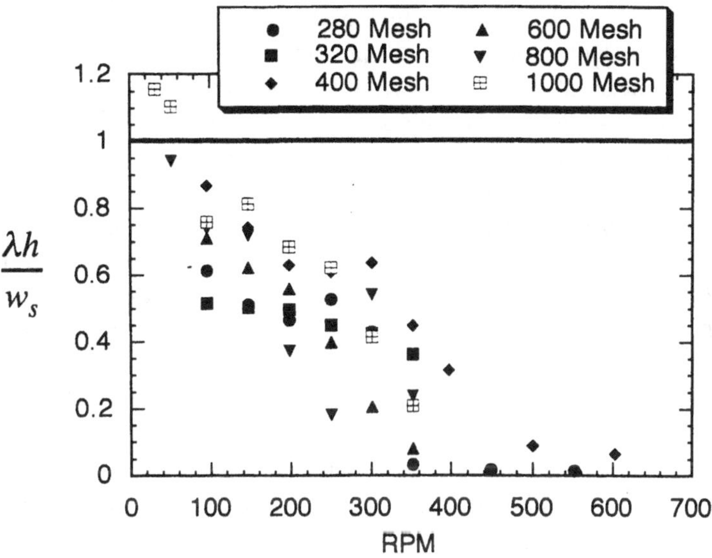

Figure 3 - Non-dimensional deposition rate $\lambda h/w_s$ as a function of mixer rotation speed (RPM), for suspended particles of 280,320,400,800,1000 silicon carbide stirred over a bed of flat Perspex.

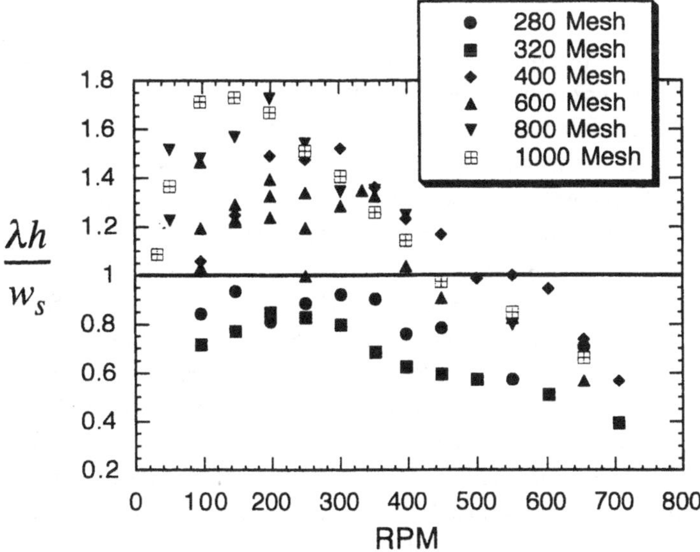

Figure 4 - Non-dimensional deposition rate $\lambda h/w_s$ as a function of mixer rotation speed (RPM), for suspended particles of 280, 320, 400, 600, 800, 1000 mesh silicon carbide stirred over a bed of 1.44±0.30cm angular road metal.

Experiment 3 - Suspension of 320, 400, 800 and 1000 mesh silicon carbide particles stirred over beds of 1, 3, 4, 5, 6, and 13mm glass beads. Bed thickness 2.5cm for 1mm beads and 4.5cm for all other sizes. Results presented in Figure 5.
Experiment 4 - Suspension of 280, 320, 400, 800 and 1000 mesh silicon carbide particles, stirred over a bed of 0.6cm fibrous air conditioner pad (Purolator, 3M brand). Results presented in Figure 6.
Experiment 5 - Suspension of 600 mesh silicon carbide particles stirred over a bed of stacked 6mm grids spaced 6mm apart vertically. One grid added, in each experiment, up to a stack of 5. The depth of surface water above the top of the upper grid is always 11.5cm. Each grid has a top surface area equal to half the horizontal crossectional area of the box (i.e., 0.5A). Mixing speed is the same for each experiment, (192 RPM). Results presented in Figure 7.
Experiment 6 - Suspension of 600 mesh particles, stirred over a bed of gravel (median diameter 0.69±0.23cm) of different thickness'. Data presented in Figure 8.
Experiment 7 - Suspension of 600 mesh silicon carbide particles with 5g/l initial concentration stirred over a bed of 13mm glass beads (marbles). Results presented in Figure 9.

5. Discussion of Results

5.1. LOW CONCENTRATION EXPERIMENTS

Many of the experimental data are presented normalized by the established model (Equation 5) with $\lambda h/w_s$ on the y axis. As in the previous figures, λ is the experimentally determined exponent fitted by least squares from the concentration time data. This allows direct comparison with the model predicted rates.

Deposition in the control, flat bed experiment (Experiment 1) (Figure 3) shows rates which quickly drop below the model (Equation 5) as mixing speed increases. This can be explained by erosion from areas of the flat bed due to high bottom shear stress. A comparison with the permeable beds, which show deposition at or above the model predicted rate for much higher stirring speeds, suggests that permeable beds may enable significant deposition at much higher velocities than a flat bed. This can be explained by a lowering of shear stress inside the bed pores compared to the surface of a flat bed.

Deposition to the permeable beds was generally greater than predicted by the established model and was affected by the size of the suspended particle, size of the glass beads in the bed and the stirring speed. Results for permeable beds presented on Figures 4, 5 and 6 appear to change systematically with changes in the independent variables and indicate the following general trends compared to Equation (5):

1) increasing deposition with increasing velocity across the granular substrate (impeller speed);

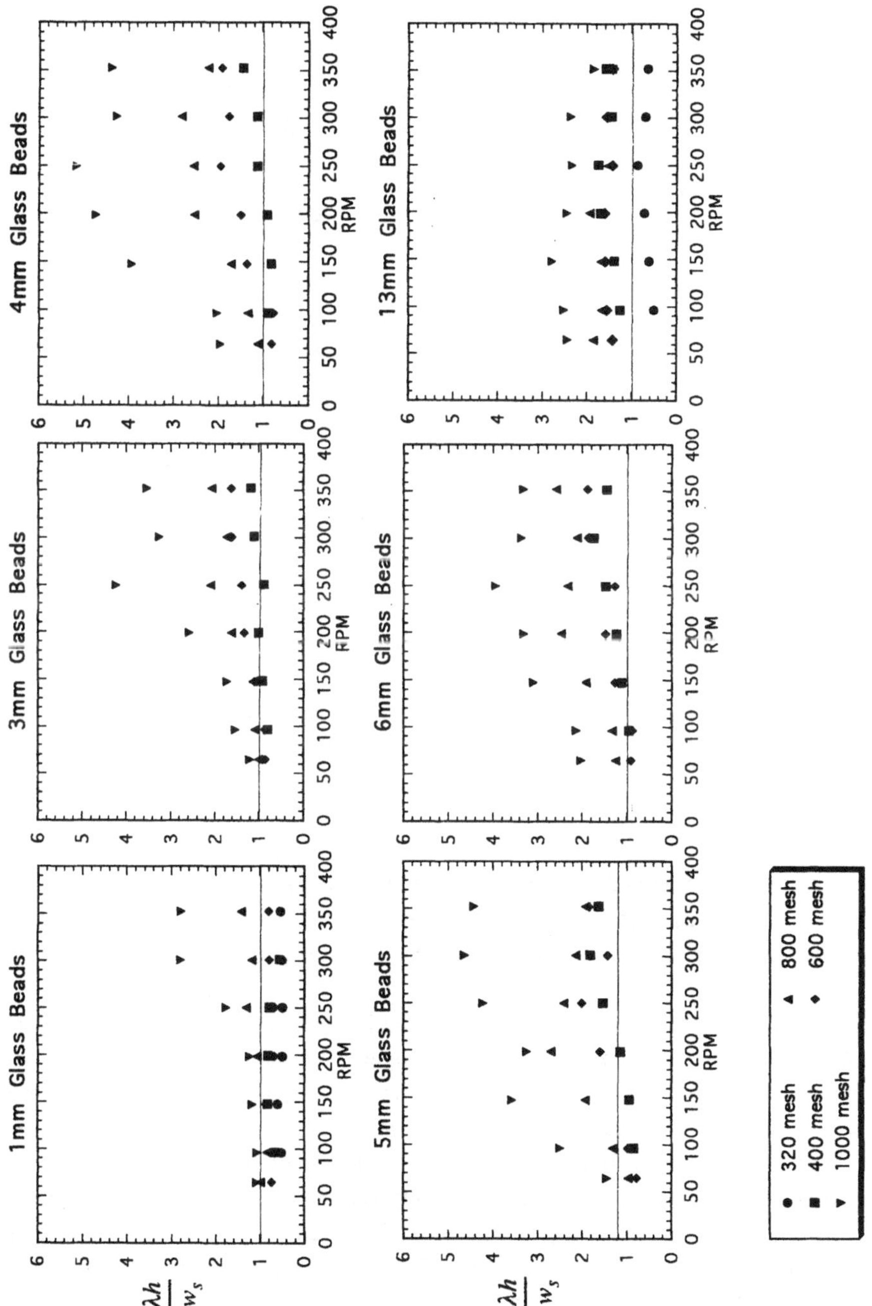

Figure 5 - Non-dimensional deposition rate $\lambda h/w_s$ as a function of mixer rotation speed (RPM) for suspended particles of 320, 400, 600, 800, 1000 mesh silicon carbide stirred over beds of 1,3,4,5,6,13mm glass beads.

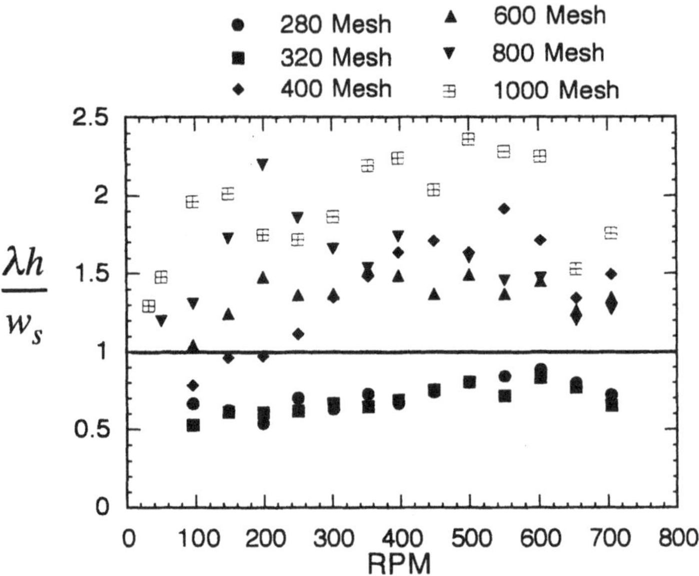

Figure 6 - Non-dimensional deposition rate $\lambda h/w_s$ as a function of mixer rotation speed (RPM), for suspended particles of 280, 320, 400, 800, 1000 mesh silicon carbide stirred over a bed of 0.6cm fibrous air-conditioner filter.

2) increasing deposition with decreasing suspended particle size;
3) for the glass beads (Figure 5) increasing deposition to beds of larger particles, up to 4mm (reduced rates for bed particles > 4mm are explained below).

Qualitatively all of these trends can be explained by the penetration of particles into the bed as surface water and porewater are exchanged. An increase in particle penetration depth could lead to a thicker filter for the removal of particles. Particle penetration would be expected to increase with increasing velocity in the bed, decreasing suspended particle size and increasing bed permeability, as exhibited by the data.

Experimental deposition curves in which deposition never rises above the established model, (i.e., below unity on Figure 5), for example 280 and 320 mesh, may indicate experimental situations where the suspended particles settle before they are able to be carried significantly into the bed with the pore fluid exchange. For smaller suspended particles with deposition rates which do rise steadily above unity, deposition invariably begins to flatten out and then decrease as mixing is increased further. These trends may indicate that the penetration depth has exceeded the depth of the experimental bed so that penetration is truncated and as mixing increases further, erosion may begin in

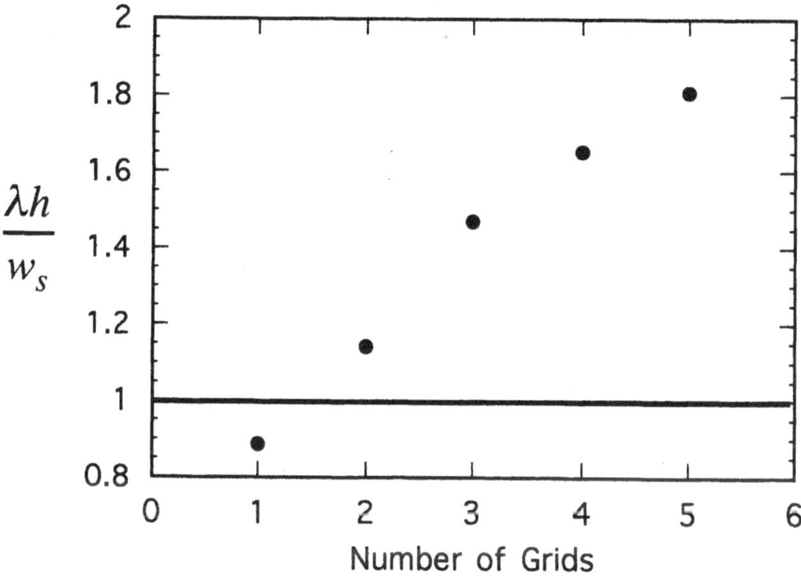

Figure 7 - Non-dimensional deposition rate $\lambda h/w_s$ as a function of mixer rotation speed (RPM), for suspended particles of 600 mesh silicon carbide stirred over a bed of 6mm Perspex grids spaced 6mm vertically (each grid covers half horizontal crossectional area of the box).

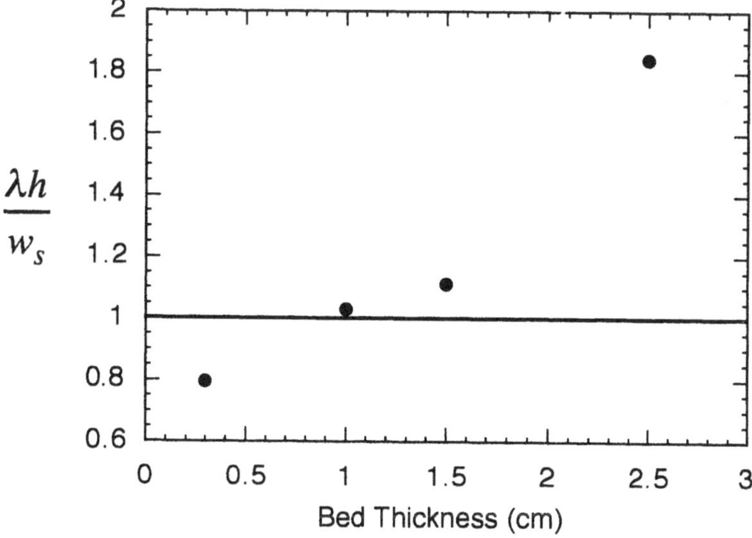

Figure 8 - Non-dimensional deposition rate $\lambda h/w_s$ as a function of mixer rotation speed (RPM), for suspended particles of 600 mesh silicon carbide mixed over beds of 0.69±0.23cm rounded gravel, of different thickness'.

certain portions of the bed. With even stronger stirring erosion may dominate so that the deposition rate drops below Equation 5. This effect is illustrated in Figure 4 for a 1.44cm median diameter gravel 2.5 cm thick. Erosion may also explain why the deposition rate begins to drop for the most permeable, 5, 6 and 13mm glass bead beds. Theoretically, if the permeable bed is of much greater thickness than the penetration depth, particles may be eroded from the upper bed layers, but will be filtered in lower regions of the bed, effectively moving the filter region down into the bed and deposition may still be enhanced.

Different materials show strikingly different filtration efficiencies even at similar permeability. For example, results from Experiment 4 (Figure 6), with a bed of 0.6cm fibrous air conditioner filter indicate that certain bed types (e.g., fibrous) may hold particles at much higher stirring velocities than do granular beds (Figure 5).

The mechanism of particle filtration within the bed appears to be settling because the observed deposition increase is always a small multiple of the established deposition model which is dependent on the Stokes settling velocity. This conclusion is consistent with the filtration literature which suggests that settling dominates for particles in this size range in water (Tien, 1989).

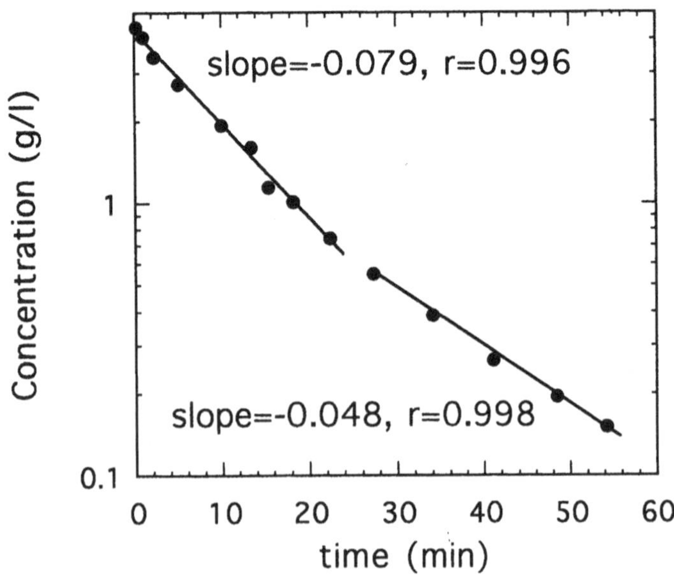

Figure 9 - Sediment concentration (g/l) as a function of time (minutes) for a high initial concentration (5 g.l) of 600 mesh silicon carbide stirred over a bed of 13mm glass beads. Model predicted rate for 600mesh (Equation 5) is 0.048.

The suggested deposition mechanism of settling in bed pores is also supported by the results from Experiment 6 (Figure 7) which indicate that deposition is enhanced even to an open framework of Perspex grids. Each of these grids has a horizontal surface area of 1/2 of the area of a horizontal crossection through the box. The increase in deposition with the addition of each grid is on the order of 0.2-0.3 reasonably close to 0.5 for the first 3 grids, indicating that for this geometry the increase in deposition might be explained by an increase in bed surface available for deposition. Of course, in reality the velocity of the fluid and the particle concentration must drop off with distance into the bed, which explains why the deposition rate begins to drop off with the addition of the last 2 grids. A similar pattern of increasing deposition with bed depth is evident in the results of Experiment 6 (Figure 8) with a bed of 1cm gravel and again the hypothesis of deposition by settling in the pores appears reasonable.

5.2. EFFECT OF SEDIMENT ALREADY DEPOSITED ON THE DEPOSITION RATE

The existence of a filtration effect to permeable beds appears to be controlled by the availability of filtration sites within the bed matrix. These sites will fill up as more sediment is deposited so that the deposition rate could be expected to decrease over time. The results from Experiment 7 (Figure 9) which had much higher initial particle concentrations in the water column than the previous experiments confirm this prediction. The results indicate that deposition occurs in two phases with a well defined inflection point. The first, enhanced deposition phase has been discussed in Section 5.1. Once the deposition sites are filled the deposition rate appears to revert to close to the established model (Equation 5). All of the deposition curves from the low concentration experiments (Experiments 1 - 6) are for a mass of particles in the bed less than the mass at the inflection point, and are representative of a bed with active filtration sites.

The loss of particle filtration potential does not appear to be a result of restricted flow into the bed since the deposition rate dropped off even when the pores for the larger bed materials were still very open. However, a plausible explanation for this phenomenon can be developed based on detailed observations of the sites where the fine particles deposit in the bed. Visual observations of the deposits showed that for the granular beds, fine particles deposit on the tops of the bed particles, which is to be expected since the adhesion of the particles was restricted by the use of the dispersant. When the volume of particles on top of these elements reached a certain height the particles began to slump off the bed particles and move downwards through the bed as particle falls or slumps. At this point the filtration capacity of the bed may be exceeded since particles on surfaces may be exchanged with suspended particles but there is no net flux to the bed. Once this capacity is exceeded the area for settling below the bed surface may revert to the horizontal crossectional area, A in Equation 2. Visual observations also suggest that slumping began at lower concentration for coarser sediment particles which appeared to have a lower angle of repose.

Natural permeable beds are unlikely to be 'clean' except after flushing events and may contain enough particles to exceed the point at which filtration is inhibited. Thus, the most important effect of permeable beds on natural sediment deposition may be to maintain a deposition rate close to the established model at much higher surface water velocities than are found for flat beds, by inhibiting erosion. The absorbing boundary used in this study demonstrates one situation in which boundary absorbency and the turbulent transport of particles can increase deposition rates as predicted by Hoyal et al (1995). In these experiments the deposition increase is only a small multiple of the model (Equation 5) suggesting that the deposition mechanism within the pores is predominantly by gravity.

6. Conclusions

Our experimental results indicate that deposition is commonly more rapid than predicted by the traditional suspended sedimentation model (Equation 5), which we explain by filtration within the bed pores. Deposition is affected by variables that influence granular bed filtration and, in general, increases with increasing surface fluid velocity, decreasing suspended sediment size and increasing bed particle size. These trends can be explained by the penetration of particles into the bed. In our experiments it appears that the deposition mechanism within the bed pores is by settling. Deposition rates drop off after a certain volume of sediment has entered the bed and we suggest that this may be due to a filling of all the surface space on top of bed elements so that sediment begins to slump off bed particles and move downwards in the bed. This reduction in deposition rate occurs long before the bed is filled with sediment. Thus, the greatest deposition rates are to clean beds and once slumping begins from the particle surfaces the deposition rate can be approximated by the established model (Equation 5). However, another important effect of bed permeability is to maintain high deposition rates at high surface water velocities where very little sediment would deposit on a flat bed. Based on our experimental results we suggest that the enhanced deposition for smaller particles, previously attributed to coagulation by Owen (1969), may be due to filtration in the surface of the bed. Although natural beds are not composed of equigranular materials, experiments on the materials we used are useful for gaining a fundamental understanding of the deposition mechanism. The effect of these bed filtration processes on deposition in environmental scale systems including gravel and sand bed rivers are yet to be demonstrated. These results may also have implications for turbidity currents driven by the density of suspended sediment. If these flows are poorly sorted, larger deposited particles may increase the deposition rate of suspended particles still in the water column and thus affect the dynamics driving the flow and the sedimentary structures in the resulting deposits.

Acknowledgments

This research was supported by grant EAR-9316656 from the National Science Foundation and by a subcontract from SAIC. Thanks also to Carl Renshaw for a useful discussion of the experimental results and to two anonymous reviewers for helpful comments. Special thanks to Cheryl Israel for her support and assistance.

References

Alonso, C.V. and Mendoza, C.: 1992, *Water Resour. Res.*, **28**, 9, 2459-2468

Beschta, R.L., and Jackson, W.L.: 1979, *Jour. of Fish. Res. Board Canada*, **36**, 204-210

Carling, P.A.: 1984, *Can. Jour. of Fish. and Aquatic Sci.*, **41**, 263-270, 1984

Cunningham, M., Anderson, C.J., and Bouwer, H.: 1987, *Jour of Irr. and Drain. Eng.*, ASCE, **113**, 1, 106-118

Diplas, P.: 1986, Ph.D Dissertation, University of Minnesota, 194 pp.

Dobbins, W.E.: 1944, *Trans. ASCE*, **109**, 629-656

Einstein, H.A.: 1968, *Jour. Hydraul. Div., ASCE*, **94**,HY5, 1197-1205.

Einstein, H.A.: 1970, *Jour. Hydraul. Div., ASCE*, **96**,HY2, 581-582

Havis, R.N., Alonso, C.V., King, J.G., and Thurow, R.F.: 1993, *Water Resour. Bull.*, **29**, 3, 435-444

Hoyal, D.C.J.D., Atkinson, J.F., DePinto, J.V., and Taylor, S.W.: 1995, *Jour. Hydraul. Res.*, **33**, 349-360

Jobson, H.E., and Carey, W.P.: 1989, *Water Resour. Res.*, **25**, 1, 135-140

Martin, D., and Nokes, R.: 1988, *Nature*, **332**, 7, 534-536

Owen, M.W.: 1969, *Jour. Hydraul. Div., ASCE*, **95**,HY3, 1085-1086

Packman, A.I., and Brooks, N.H.: 1995, *Marine and Freshwater Re..*, **46**, 233-236

Schalchli, U.: 1992, *Hydrobiologica*, **235/236**, 189-197

Tien., C., *Granular Filtration of Hydrosols and Aerosols*, Butterworths Series in Chemical Engineering, London, 365 pp., 1989

Zagni, A.F.E., and Smith, K.V.H.: 1976, *Jour. Hydraul. Div., ASCE*, **102**,HY2, 207-223

Zippe, H.J., and Graf, W.H.: 1983, *Jour. Hydraul. Res.*, **21**,1, 51-64

DIGITAL IMAGING CHARACTERIZATION OF THE KINEMATICS OF WATER-SEDIMENT INTERACTION

H. CAPART[1], H.-H. LIU, X. VAN CROMBRUGGHE[2] and D.-L. YOUNG

Dept. of Civil Engineering and Hydraulic Research Laboratory, National Taiwan University, Taipei, Taiwan, Republic of China
[1] Current affiliation: Fonds National de la Recherche Scientifique and Univ. Catholique de Louvain, Belgium. [2] Current affiliation: Univ. Libre de Bruxelles, Belgium.

Abstract. Digital imaging techniques especially geared towards the laboratory characterization of the kinematics of water-sediment interaction are presented. More specifically, the methods proposed apply to the motion of cohesionless spherical particles in transient water flow, with the aim of obtaining both particle velocity and concentration fields from sequences of digital images. A special particle identification algorithm is devised in order to deal with densely packed particles (in contrast to the sparse seedings of tracers used in studies of pure fluid kinematics) and to allow application of digital particle tracking velocimetry (DPTV). A procedure for extracting the concentration field from the knowledge of the discrete particle positions is then detailed. Finally, the various tools are illustrated for the laboratory case of a dambreak wave over a movable bed.

Keywords: sediment transport, transient flow fields, laboratory measurements, digital particle tracking velocimetry.

1. Introduction

The experimental characterization of the flow of sediment-laden water falls within the scope of experimental fluid mechanics as applied to multiphase flows. In this discipline, advances in instrumentation increasingly allow reliable point and field measurements of local concentration and flow velocities, in addition to aggregate measurements of quantities such as the sediment discharge across a channel section. Techniques such as laser Doppler velocimetry (LDV) and particle imaging velocimetry (PIV), for instance, have been applied successfully to a number of multiphase flows (see the review of Bachalo, 1994). With the rapid development of digital imaging methods, a technique beginning to find wide acceptance in fluid mechanics studies is digital particle tracking velocimetry (DPTV) (e.g. Haynes and Turner, 1992; Lloyd et al., 1995). This technique, based on the tracking of individual particles on a sequence of digital images, provides a powerful means of acquiring instantaneous velocity fields. In particular, it provides a promising tool for the characterization of the kinematics of water-sediment interaction, which constitutes the focus of the present report.

A number of specific challenges arise which make application of digital particle tracking to sediment transport cases quite different from application to pure fluid kinematics. Whereas for pure fluid flow characterization, sparse seedings of neutrally buoyant tracer particles are homogeneously dispersed in the fluid for the purpose of tracking, in sediment transport cases one has to deal with a highly nonhomogeneous dispersion of sediment particles presenting regions of very densely packed grains near the flow bed. This situation creates specific difficulties in the process of identifying and tracking individual particles on the digital images. A second difference is that for sediment kinematics, it is not only the velocity field, but also the instantaneous concentration field which is of interest. Methods should then be devised allowing to extract such concentration fields from the knowledge of the position of individual particles. In the present study, we develop a number of tools designed to meet those

Water, Air and Soil Pollution **99**: 173-177, 1997.

challenges and therefore work towards specialization of digital particle tracking techniques to the study of sediment motion. We first detail a special particle identification algorithm applicable to densely packed spherical grains, a necessary preliminary step to particle tracking and concentration field extraction. A procedure for evaluation of the concentration field is then presented. Finally, application of the different tools is illustrated for a case of transient sediment-laden wave flowing over an erodible bed.

2. Particle identification and tracking

Digital imaging involves the acquisition and interrogation of images manipulated in the form of two-dimensional arrays of discrete grey (or color) level data. Once sequences of digital images have been acquired, using for example a charge-coupled device (CCD) camera coupled with a frame grabber board and linked to a personal computer, the analysis process of digital particle tracking velocimetry involves the following steps: (1) particle identification, isolating particles from the surrounding fluid and locating them on the digital images; (2)particle tracking, following identified particles from one image to the next; and (3) data organization, constructing a velocity field representation from the individual particle trajectories. The first step presents specific difficulties in the case of sediment transport. If particles are loosely spaced, as is the case with the sparse seedings in DPTV studies of pure fluid flow, the method used is usually to insure a high luminosity contrast between tracer particles and the adjoining fluid, and simply apply a grey level thresholding to the resulting images (Lee and Hsu, 1994; Lloyd et al., 1995). When the particles to be tracked are close together, however, this is no longer sufficient for a reliable identification. An alternative method was developed in the present research, applicable to the identification of densely packed spherical particles, and which resorts to local measures of radial symmetry.

With reference to Figure 1, let $f(x,y)$denote the grey level at every pixel (x,y) of a digital image (Fig.1a), a transformed function $g(x,y)$ is obtained by applying a 'radial mask' to the original image in the form of the following weighted sum (in non-dimensional pixel units):

$$g(x_0,y_0) = \sum_{(x,y)\in\Omega} \omega(x-x_0,y-y_0) f(x,y) \tag{1}$$

where the neighborhood Ω is chosen to correspond roughly to the pixel size of individual particles (i.e. the approximate number of pixels of a typical particle diameter).

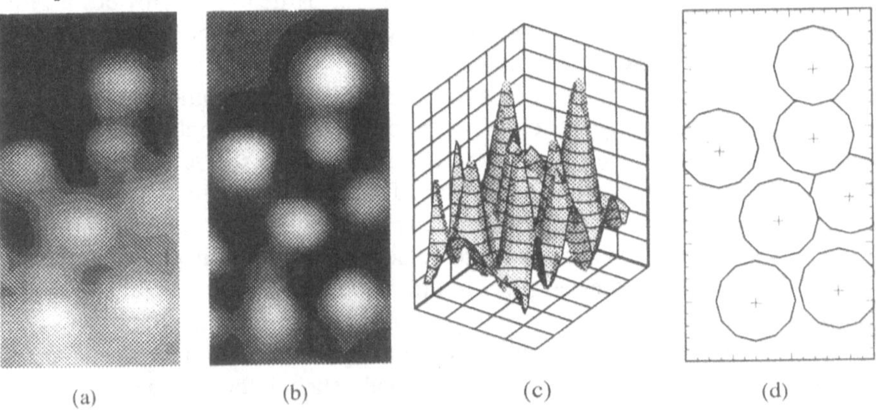

Figure 1: Particle identification procedure: (a) original image area; (b) image area as transformed by 'radial filtering'; (c) three-dimensional view of the transform; (d) maxima of the transform correspond to particle centroids.

The set of weighting coefficients ω sums to zero, rewards high luminosities close o the reference pixel (x_0,y_0) and penalizes them away from it, and vice-versa for low luminosities (assuming light-colored particles before a dark background). The effect of the mask (see Fig.1b) is to produce high positive values when the reference pixel corresponds to the centroid of a particle, close-to-zero values in empty regions, and high negative values when the reference pixel is in-between particles. Pixels that are even slightly off-center yield significantly lower g values than the ones that correspond to centroids (Fig. 1c) which means that particle centroids can be located by seeking the local maxima of the transformed function (Fig. 1d). The method was found to reliably identify particles even when the illumination was not uniform over the physical imaging window (some particles being embedded deeper into the flow than others), as illustrated below at Figure 3.

Once particles are identified, they can be tracked from one image to the next using for example the improved algorithm proposed by Haynes and Turner (1992; Lloyd *et al.*, 1995) which takes into account the relative positions of neighboring particles. In the present work, a simpler procedure is used, based on a predictor-corrector algorithm: the likely position of a particle on the next image is extrapolated from its past trajectory, then adjusted by exploring a local neighborhood in search of an actual particle centroid. This rather coarse method has to be manually supervised in the regions of steeper velocity gradients.

3. Concentration field extraction

An instantaneous concentration field can be derived from the knowledge of the positions (x_i,y_i) of the individual grains on two-dimensional images by using the following procedure: (1) a surface grain concentration $\eta(x,y)$, defined as the number of particles by unit surface observed on the digital images, is first obtained; (2) this surface grain concentration is then converted into the usual volume concentration C using an approach inspired by Bagnold's (1954) analysis. The local grain concentration is obtained throughout the plane by constructing a triangular mesh with nodes placed at every particle centroid, in addition to a few boundary nodes. The mesh is optimized using the usual mesh generation techniques in such a way as to obtain a set of triangles which are as close to equilateral as possible. With reference to Fig. 2a, surface concentration values η can then be evaluated easily for every element j of the mesh as

$$\eta_j = \frac{\sum_{k=1}^{3} \frac{\theta_k}{2\pi} \mu_k}{\mathbf{Area}_j} \tag{2}$$

where $\mu_k = 1$ for a particle centroid and 0 for a boundary node. Expression (2) presents the desirable property that integration of the surface concentration over any region yields back the exact number of particles which are included in the region, i.e. strict conservation is maintained. The surface concentration η then needs to be converted into the volumetric concentration C. Drawing inspiration from the approach of Bagnold (1954), a relationship can be established in terms of ratios of the actual concentrations to the concentrations of maximum packing, i.e.

$$C/C_0 = (\eta/\eta_0)^{3/2} \tag{3}$$

where C_0 and η_0 correspond to the tetrahedral-rectangular piling of rigid spheres of uniform diameter D associated with minimum pore space (see Fig. 2b). Simple geometric considerations lead to the expressions $C_0 = \pi/3\sqrt{2} \cong 0.74$ and $\eta_0 = \sqrt{2}/D^2$, from which eq. (3) can be evaluated. In this way, a local value of concentration can be obtained at every location of the imaging region.

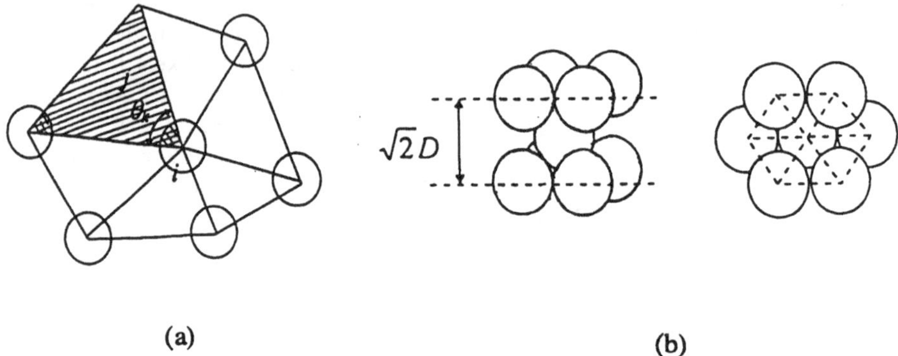

Figure 2: Concentration field extraction: (a) evaluation of the 'surface' particle concentration; (b) tetrahedral-rectangular piling of spheres used as reference for transformation from surface to volumetric concentration.

Other useful information can be deduced readily from the knowledge of velocity and concentration fields. In particular, the volumetric sediment transport rate across any interior boundary of the imaging region can be evaluated as the integral of the product of concentration with the component of velocity normal to the boundary. Provided visual access is possible, digital imaging constitutes a non-intrusive way of measuring the instantaneous sediment transport rate, a complicated measurement using classical approaches with sediment traps or probes.

4. Sample results for a fast flow event

The above methods can now be illustrated by applying them to the two-phase, transient two-dimensional flow which immediately follows the release of a dambreak wave over a movable particle bed. This case of fast transcritical, non-equilibrium sediment transport has been investigated by Capart and Young (1996). A volume of water of 10 cm initial depth is released by rapidly raising a sluice gate, resulting in sediment particles (artificial pearls of diameter 6 mm and specific gravity 1.05 chosen to magnify the phenomenon) being picked up by the developing shallow-water wave and convected downstream. Figure 3a shows one digital image of the sequence, taken 100ms after lifting of the gate (which was located at the image center), where the darker mass is pure water and the particles are colored white. Fig. 3b shows the optimized triangular mesh built upon the previously located particle centroids. Contours of the extracted concentration field are plotted at Fig. 3c. Finally, the individual vectors of the velocity field are shown at Fig. 3d. It is seen that a rather complete characterization of the transient two-phase flow can be obtained.

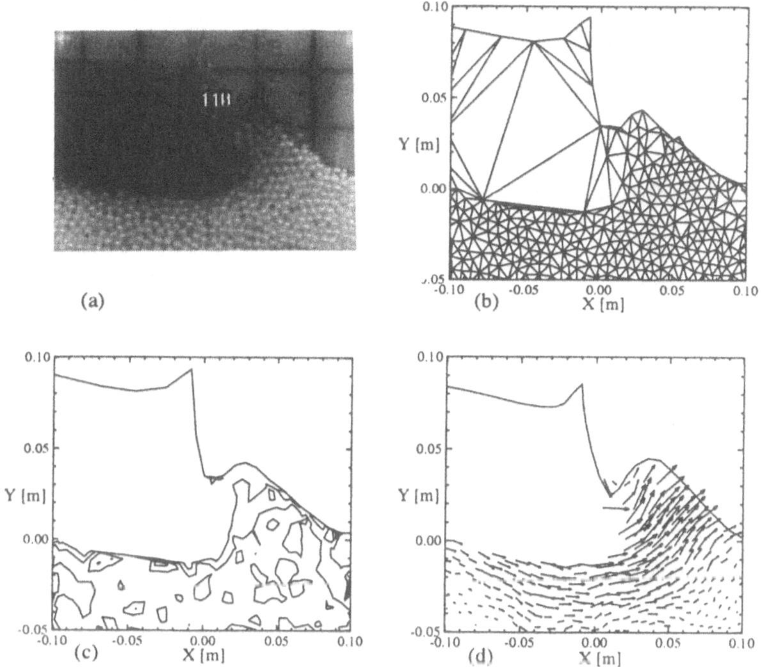

Figure 3: Image analysis process: (a) original frame; (b) triangular mesh with nodes at particle centroids; (c) concentration field; (d) particle velocity vectors.

5. Concluding remarks

Specialized tools for the application of digital imaging techniques to the characterization of two-phase flows have been developed. Fine-tuning of the algorithms as well as the continued development of more robust procedures are necessary. Nevertheless, the results obtained so far already demonstrate that application of digital imaging is feasible, yielding a very high information content as compared to more traditional approaches. Furthermore, since the digital image analysis process can be automatized to a large extent, the technique can deal readily with the large volumes of data associated with transient events. It is hoped that this kind of experimental characterization can complement fundamental studies of the physical mechanisms involved as well as attempts at their numerical modelling, and thus contribute to the study of water-sediment interactions in their mechanical aspects.

References

Bachalo, W.D.: 1994, *Int. J. Multiphase Flow* **20**, 261-295.
Bagnold, R.A.: 1954, *Proc. Roy. Soc. London* A **225**, 49-63.
Capart, H., Young, D.-L.: 1996 *Proc. Eighth Hydraulic Engineering Conf. of the Chinese Inst. of Civil Engineers*, Taipei, 379-386 (in Chinese).
Haynes, I.D., Turner, J.T.: 1992 *Sixth Int. Symp. on the Application of Laser Techniques to Fluid Mechanics*, 3.3.1-3.3.6.
Lee, H.-Y., Hsu, I.-S.: 1994, *J. Hydr. Engrg.* **120**, 831-845.
Lloyd, P.M., Ball, D.J., Stansby, P.K.: 1995, *J. Hydr. Res.* **33**, 519-533.

COMPARISON OF SEDIMENT PORE WATER SAMPLING FOR SPECIFIC PARAMETERS USING TWO TECHNIQUES

T.N. Angelidis
Aristotle University, Chemistry Department (BOX114), 54006 Thessaloniki, Greece

Abstract. The suitability of the two more extensively used sampling techniques, namely *in-situ* dialysis and centrifugation followed by filtration were compared. Field measurements derived from Lake Simbirizzi, Sardegna, Italy were used for the comparison. Two groups of substances were selected; the first one consists of the relatively inert in the aquatic environment chlorides and bromides, while the second one consists of substances involved in diagenetic processes (ammonium-nitrogen and soluble reactive phosphorus). The field measurements show that for the applied sampling techniques the measurements are comparable for the first group of substances, while measurements derived by centrifugation followed by filtration give considerably different concentrations using *in-situ* dialysis for the second group of substances. Since the second group is related to sediment-water interactions (adsorption-desorption and chemical processes), it seems that forced separation by centrifugation influences considerably the derived measurements. As a result *in-situ* dialysis is expected to give more accurate measurements of porewater concentrations for nutrients and other reactive substances.

keywords: pore water, sampling, dialysis, centrifugation-filtration, comparison

1.Introduction

Since an important fraction of the substances present in the aquatic environment is reversibly associated with surficial sediments (Berner, 1980; Forstner and Wittmann,1981), the study of substances dynamics in surficial sediments is prerequisite to the understanding of their behaviour in whole aquatic systems. Particular aspects to be addressed include the study of sediment-water exchanges, the chemical speciation of substances in surficial sediments, and their diagenetic modeling. Such studies may require reliable measurements of substances concentrations in the porewaters. Although several techniques such as squeezing (Reeburgh,1967; Bender *et al.*,1987), centrifugation followed by filtration (Elderfield *et al.*,1981) and *in-situ* dialysis (Hesslein,1976; Van Eck and Smits,1984) are available to sample sediment porewaters, their suitability for various substances has not been verified. Each technique is subject to several potential artifacts. For example, centrifugation and filtration methods are prone particularly to artifacts resulting from sample oxidation and temperature variations (Bischoff *et al.*,1970; Fanning and Pilson,1971; Bray *et al.*,1973; Lyons *et al.*,1979). For more substances the concentrations are often too small and contamination from equipment or from colloidal particulate matter passing through filters may be a problem. The dialysis approach is also subject to errors arising from incomplete equilibration, membrane breakdown, contamination, or from possible development of an electrical potential across the membrane which may interfere with the free diffusion of ions (Martens and Klump,1980; Carignan,1984).

For these reasons, no particular porewater sampling procedure can be, *a priori*, considered artifact-free. Concentration measurements should be regarded therefore as operationally defined (method dependent) until it is shown that independent procedures consistently yield similar results. Moreover, other aspects such as the reproducibility of the results or methological simplicity may become important factors when selecting

Water, Air and Soil Pollution **99**: 179-185, 1997.

among procedures that are otherwise equivalent. There is therefore a definite need for comparative studies of the performance of different porewater sampling methologies.

Three main factors are critical during porewater sampling especially for substances sensitive to oxidation-reduction, adsorption-desorption and precipitation-dissolution processes, number of manipulations, temperature and time. Too many manipulations will expose the samples to various atmospheres and temperatures affecting their concentrations. Time is critical especially in the case of field measurements where usually the sampling site and the analytical laboratory are separated geographically . The time elapsed between sampling and analysis may be connected with significant concentration changes. Usually comparison is made in laboratory where the conditions are more strictly controlled (Carignan *et al.*,1985). The main effort of the present research work was to compare the results derived from field measurements applying *in-situ* dialysis and centrifugation-filtration techniques. Two groups of substances were selected for the comparison; the first one consists of the relatively inert in the aquatic environment chlorides and bromides and the second one consists of the active and sensitive ammonium-nitrogen (NH_4^+-N) and soluble reactive phosphorus (SRP).

2.Methods

2.1.STUDY SITES

Lake Simbirizzi is an artificial lake located near Cagliari, Sardegna. The lake is part of the Cagliari potable water system. Simbirizzi serves as a deposit for excess water arriving at the city of Cagliari. The main problems of the lake are eutrophication and high salinity. Four stations (A, B, C and D) representative of the morphology of the lake were selected for sampling.

2.2.*In-situ* DIALYSIS

The dialyzers (peepers) used in this study consisted of 80x21x4cm plexiglass plates in which 18x1.2x0.6 cm chambers spaced one cm center to center were machined and covered with filtering membranes (METRICEL, DM-450) and a 0.2cm thick plexiglass sheet with apertures matching the chambers. The entire set-up is shown in Figure 1 (Van Eck and Smits,1984). The membrane and window were fastened to the main body of the dialyzer with stainless steel screws spaced 1.5cm apart. Prior to their insertion into the sediment, the dialyzers were bubbled with nitrogen for 12h in stainless steel cages filled with deionized water. The dialyzers were inserted into the sediment by SCUBA divers for one week. Upon retrieval they were rinsed quickly in two succesive baths of deionized water to dislodge adhering sediment particles and sampled within half an hour. The samples were stored in polystyrene vials and kept at 4°C.

2.3.CENTRIFUGATION-FILTRATION

To compare *in-situ* dialysis to centrifugation+filtration, short (25cm) cores were collected with 8cm diameter plexiglass tubes from the sediments near each dialyzer. The cores were sectioned into 1cm slices in a glove box under a nitrogen atmosphere, using a special sectioning device (Blomqvist and Abrahamsson,1987). Aliquots of each section were introduced into 50ml polycarbonate centrifuge tubes, the tubes were then tightly

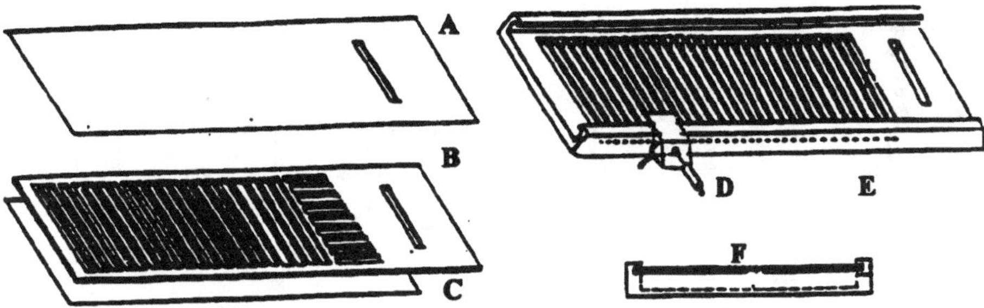

Figure 1. The set-up of the device used for the *in-situ* dialysis porewater sampling (A:protective cap, B: window cap, C: membrane, D: sampling device, E: main body, F: section of a sampling compartment).

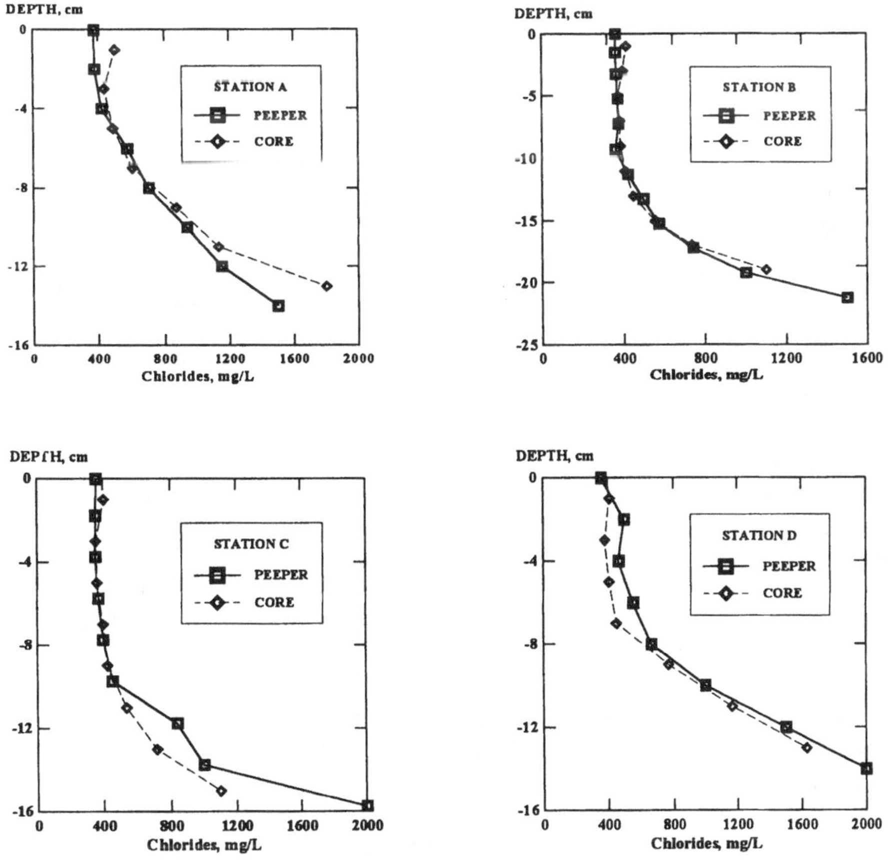

Figure 2. Comparison of chlorides profiles derived by in-situ dialysis and centrifugation-filtration at the four sampling stations (A, B, C and D).

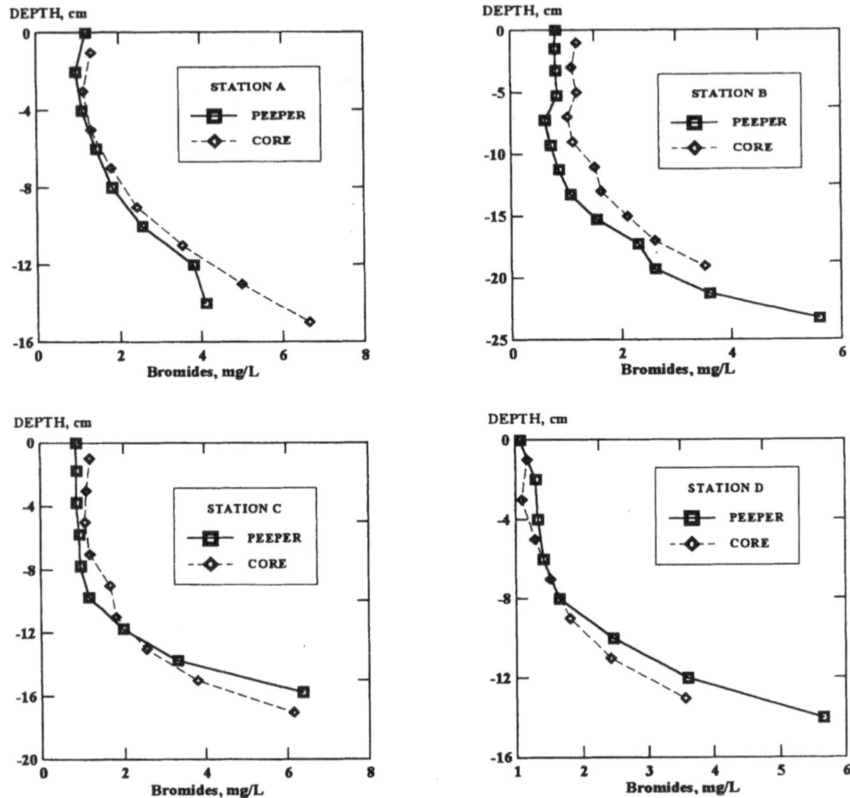

Figure3. Comparison of bromides profiles derived by *in-situ* dialysis and centrifugation-filtration at the four sampling stations (A,B,C and D).

capped, removed from the glove box, and centrifuged at 5000 rpm for 20 minutes. After centrifugation, the tubes were returned to the glove box and the supernatant collected in plastic syringes and filtered through 0.45 μm membrane filters using in line holders. The filtrates were collected in clean polystyrene vials and stored at 4°C. During manipulations under N_2 atmosphere, the oxygen partial pressure inside the glove box was continuously monitored with a dissolved oxygen electrode and kept below 10^{-3} atm.

2.4.ANALYSES

The overlying and porewater samples collected with both procedures were analyzed for chlorides, bromides, NH_4^+-N and SRP by high performance ion chromatography.

3.Results and Discussion

In Figures 2, 3, 4 and 5 the comparative results for each substance and station are summarized. Chlorides and bromides (Figures 2&3) gave similar profiles showing

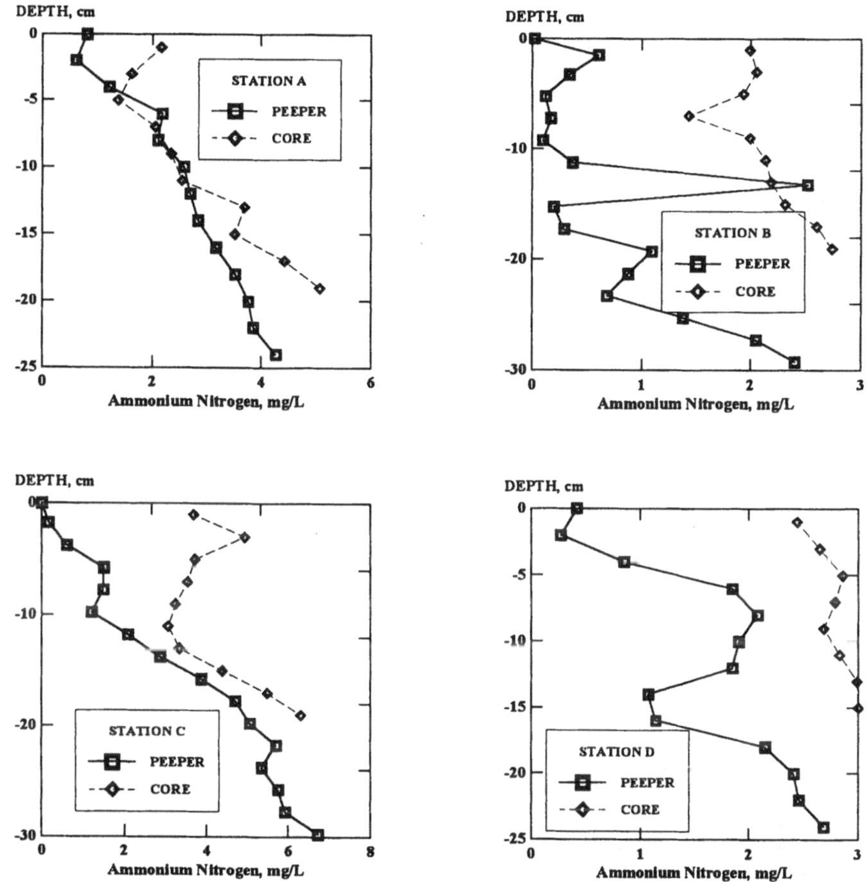

Figure 4. Comparison of NH_4^+-N profiles derived by *in-situ* dialysis and centrifugation-filtration at the four sampling stations (A,B,C and D).

that the measured concentrations are method independent. The high concentrations of these substances observed deep in the sediments were connected with the existence of a salt layer.

The situation is different with respect to NH_4^+-N and SRP. Centrifugation-filtration yields higher concentrations for both substances especially near the water-sediment interface (Figures 4&5). The respective substance concentrations in overlying water (0.1 to 0.4 mg/L for NH_4^+-N and 0.02 to 0.1 mg/L for SRP) are about equal to the measured porewater concentrations by *in-situ* dialysis at the first few centimeters of the sediment. Since, the centrifugation-filtration technique due to the high number of manipulations is more sensitive to oxidation-reduction processes, it is expected that the *in-situ* dialysis measurements are more accurate. Another factor that must not be ignored is that under the forced extraction of water during centrifugation an adsorbed quantity of these substances (Rosenfeld,1979; Furumai and Ohgaki,1989) is transferred to the water, due to changes of the nature of the solid material which becomes more dense after

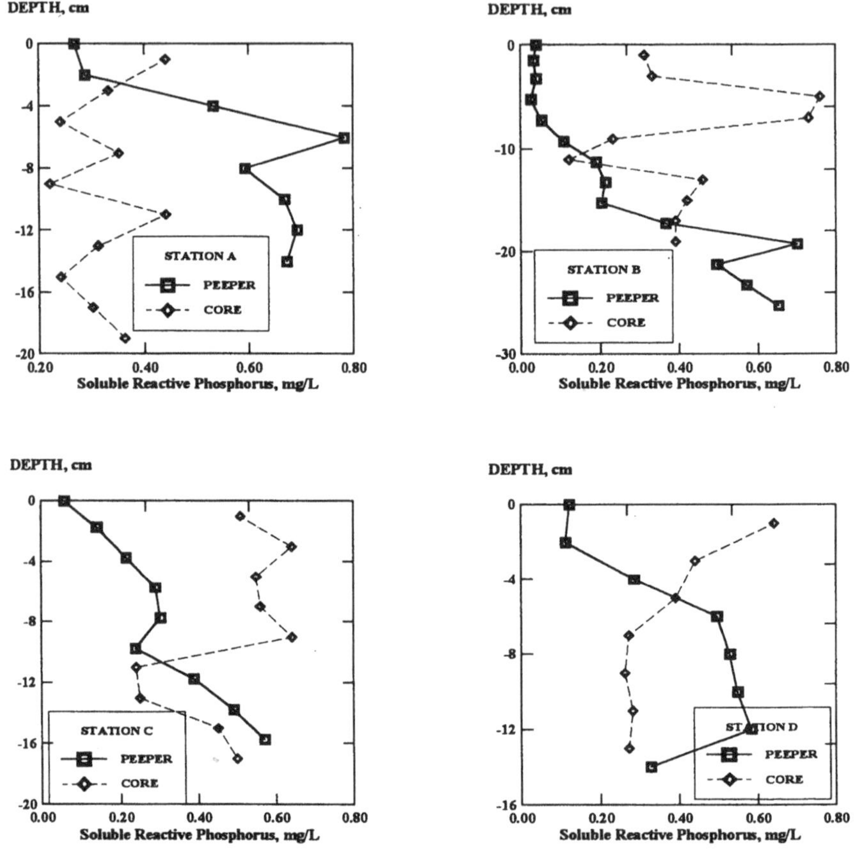

Figure 5. Comparison of SRP profiles derived by *in-situ* dialysis and centrifugation-filtration at the four sampling stations (A,B,C and D).

centrifugation. This observation is emphasized by the fact that the differences in measured concentrations between the two techniques are more pronounced at the highest part of the profiles where the sediment is less dense and has better adsorption properties, due to the higher organic matter content . The same observation was made by Bender *et al.*(1987) during the study of a special squeezing device, where the porewater is also force-extracted.

4.Conclusion

Although differences between the two techniques and their explanations require further research work, dialysis seems to prevail as a porewater sampling technique. In addition to the fact that it seems to be more accurate, it also has several practical advantages over the centrifugation-filtration methods. Because dialysis is performed *in-situ*, problems arising from temperature changes or exposure to atmospheric O_2 during porewater extraction by other methods are minimized. Dialysis requires less equipment and is less time-consuming than the other methods. Its relative simplicity also minimizes the chances of sample contamination from the equipment and makes it better adapted to field conditions.

Acknowledgments

The author is greatful for the valuable assistance of H.Muntau and the staff of the Environment Institute of JRC Ispra (Italy).

References

Bender M., Martin W., Hess J., Sayles F., Ball L. and Lambert C.: 1987, *Limnol. Oceanogr.* **32(6),** 1214-1225.
Berner R.A.: 1980, *Early Diagenesis*, Princeton University Press, Princeton, N.J., USA
Bischoff J.L., Greer R.E. and Luistro A.O.: 1970, *Science* **167**, 1245-1246.
Blomqvist S. and Abrahamsson B.: 1987, *Schweiz. Z. Hydrol.* **49/3**, 393-396.
Bray J.T., Bricker O.P. and Troup B.N.: 1973, *Science* **180**, 1362-1364.
Carignan R.: 1984, *Limnol. Oceanogr.* **29**, 667-670.
Carignan R., Rapin F. and Tessier A,: 1985, *Geochim. Cosmochim. Acta* **49**, 2493-2497.
Elderfield H., McCaffrey R.J., Luedtke N., Bender M. and Truesdale V.W.: 1981, *Amer.J.Sci.*, 1021-1055.
Fanning K.A. and Pilson M.E.Q.: 1971, *Science* **173**, 1228-1231.
Forstner U. and Wittmann G.T.W.: 1981 *Metal Pollution in the Aquatic Environment*, 2nd edn. Springer Verlag.
Furumai H. and Ohgaki S.: 1989, *Wat. Res.* **6**, 677-683.
Hesslein R.H.: 1976, *Limnol. Oceanogr.* **21**, 912-914.
Lyons W.B., Gaudette H.E. and Smith G.M.: 1979, *Nature* **277**, 48-49.
Martens C.S. and Klump J.V.: 1980, *Geochim. Cosmochim. Acta* **44**, 471-490.
Reeburgh W.S.: 1967, *Limnol. Oceanogr.* **12**, 163-165.
Rosenberg J.K.: 1979, *Limnol. Oeanogr.* **24(2),** 356-364.
Van Eck G.T.M. and Smits J.G.C.: 1984, *Proc. 3rd International Symposium, Interactions between sediments and water*, Geneva, 166-171.

SEDIMENT-WATER INTERACTIONS AFFECT ASSESSMENTS OF METALS DISCHARGES AT ELECTRIC UTILITIES

J. S. MATTICE, D. B. PORCELLA and R. W. BROCKSEN

Water Toxics Assessment and Watershed Management, Electric Power Research Institute, 3412 Hillview Ave, Palo Alto, CA, 94304-1395 USA

Abstract. We present three examples to show the importance of sediment-water interactions to electric utilities: 1) Selenium (Se), in ash pond effluents, has caused declines in fish populations in North Carolina. A biogeochemistry model appears to explain Se dynamics for several reservoirs. However, further work on sediment water interactions is needed to predict the speed of reservoir Se declines following cessation of inputs; 2) Mercury (Hg), volatilized in stack gases from coal fired power plants, is a public and wildlife health concern. Sediments play a major role in the biogeochemistry of Hg as documented in the Mercury Cycling Model (MCM); As with Se, questions about sediment water interactions limit the confidence in predictions about dynamics and effects of Hg; and 3) One of the recommendations from a recent Pellston Conference was to evaluate the use of a new paradigm as a basis for metals regulations. Under this new paradigm, effects of surface active metals (Ag, Al, Cd, Cu, Ni and Zn) on fish can be viewed as dependent on competition between the gill, a 'biotic ligand', and other environmental ligands for metals in discharges. Under this new paradigm, then, the mechanics of toxicity can be viewed as analogous to interactions at the sediment-water interface. It is clear from these three examples that fostering discussion among chemists and toxicologists, through joint participation at meetings and publication in journals used by both fields, is critical for development of accurate assessment capabilities and support of cost effective decision making.

Key Words: trace metals/metalloids; gills; ecosystem models; bioaccumulation; toxicity; mechanisms; bioavailability

1. Introduction

Electric utilities are dealing with operational problems that are related to sediment-water interactions. We will present three examples which have already been the focus of electric utility industry-sponsored research. Our primary purpose is to point out the need for chemists and toxicologists to work in concert or communicate extensively to enhance progress toward developing accurate assessment techniques to support cost effective decision making.

Selenium is the first example. Selenium, released to cooling reservoirs after coal ash is sluiced to ponds for disposal, has caused documented declines in fish populations in North Carolina (Cumbie and Van Horn 1978). The primary route of exposure appears to be uptake of selenium from water by lower trophic levels and bioaccumulation via the food web (Bowie et al. 1996). Food web transfer and biogeochemical cycling produce levels a thousand times higher in organisms and sediments (ppm) compared to water (ppb). Studies suggest that sediments can play a major role in ameliorating the immediate responses, but may extend the exposure of fish after termination of inputs of selenium.

A second example is mercury. Mercury concentrations in fish have raised questions about the potential health effects of releases into the atmosphere via coal burning and subsequent cycling through aquatic ecosystems (Porcella 1994). Methyl-mercury is the species of concern because it can have severe effects on neural function of humans, particularly the young, and it is the predominant form found in fish. Because most mercury species

Water, Air and Soil Pollution **99**: 187-199, 1997.

released from power plants are elemental or inorganic, methylation is required in the chain of transport to humans. Analyses show that sediment-water interactions play a major role in cycling and transformation of mercury (Hudson et al. 1994).

Finally, some of the potentially persistent metals (aluminum, cadmium, copper, nickel, silver, zinc) currently are receiving attention, because, like selenium, they are components of fly ash. A recent Pellston Conference organized by the Society for Environmental Toxicology and Chemistry convened a panel of international experts in metals fate and effects assessment to summarize current knowledge about these metals and propose a research agenda to support more scientifically defensible methods for deriving water quality criteria and effluent limits (Bergman et al. 1997).

In addition to concerns with metals released from sediments into the water column, electric utilities face regulation due directly to sediment metals concentrations *per se*. For example, Fenner (1992) and Howe (1992) independently indicated that Region V (EPA) already specifies sediment monitoring and toxicity testing in some permits. The State of Washington also has issued Sediment Management Standards for marine sediment in Puget Sound. USEPA draft publications dealing with sediment criteria for metals (USEPA 1993a, 1993b) and a two volume users guide for the Equilibrium Partitioning Method that is being developed also indicate that sediments will play a larger role in regulating effluents in the future.

Given the implications of sediment-water interactions and their potential effects on electric utility operations, EPRI and individual utilities are conducting studies of potential ecological effects of toxicants in sediment water systems. We summarize some of the studies below indicating how individual investigators interested in sediment-water interactions might focus their work to reduce the uncertainty in risk assessment and management of toxicants. Specific details of the work may be obtained in the cited references.

2. Selenium - role of sediments

A model framework is being developed (Porcella et al. 1991; Bowie and Grieb 1991; Cutter 1991; Cutter et al. 1991; Bowie et al. 1996) to evaluate the effects of Se released from ash pond or coal pile runoff on aquatic ecosystems. The five modules that comprise the Se model, include biogeochemistry, pharmacokinetics, food web transfer, toxic effects, and ecosystem effects. The model is being developed in concert with laboratory experiments and field studies in three reservoirs that receive Se inputs related to fossil fuel combustion and two reference reservoirs that do not receive such inputs. Earlier work showed that different Se species have different toxicological properties. Thus, the field work and laboratory experiments measure speciation of Se in solution and in inorganic and

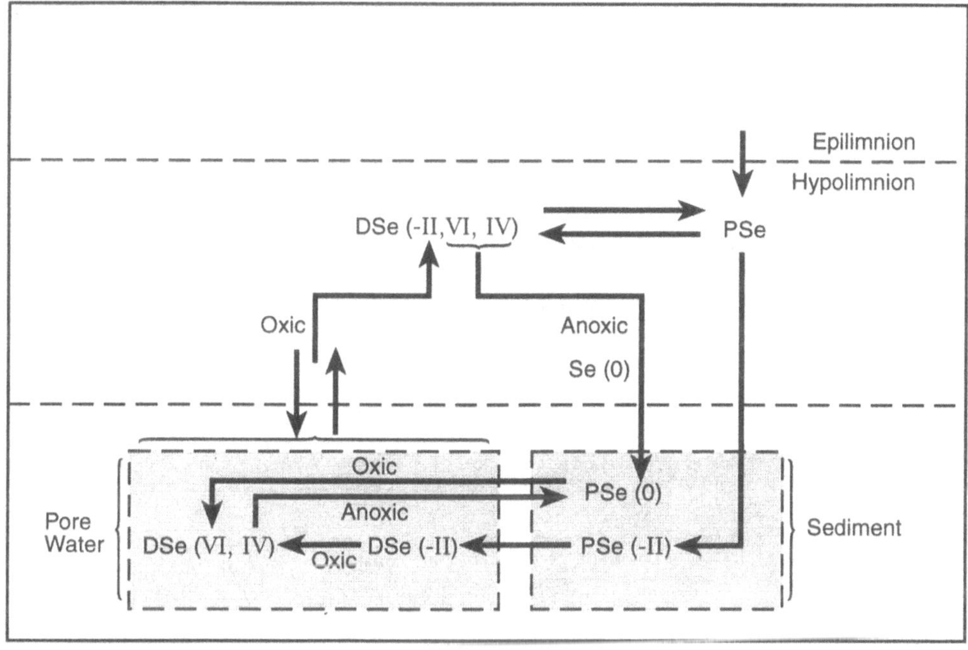

Figure 1. Diagram of the dynamics of Se exchanges and transformations between sediment, pore water and water column. Se species as defined in the text. D= dissolved, P = particulate (modified from Cutter 1991a).

organic particles in the water column, sediments, and pore water, as well as uptake, biotransformation, and excretion rates by microorganisms, phytoplankton, zooplankton, benthic invertebrates, and fish. These studies permit iterative model formulation and evaluation of Se processing and effects.

In this presentation we focus on Se biogeochemistry in Se sediment-water interactions as part of effects assessment. The Se biogeochemistry module is based on a dynamic mass balance approach. The slow rates of Se oxidation reactions and the continuous biological regeneration of soluble reduced organic chemical species prevent use of thermodynamic equilibrium models. The model describes the reservoir inputs, transformations, and outputs of the major species of Se [inorganic and organic selenides (-II); selenite (IV); selenate (VI); and elemental selenium (0)] and distribution of these species as particulate or dissolved Se in the water column, sediments and pore waters (Figure 1).

The model can treat the sediments as 2 or 3-dimensional, simulating sediments as two vertical compartments, an upper layer in which biotic activity takes place and in which diffusion and particle settling result in exchanges with the hypolimnion, and a lower layer which acts as a sink for particulate Se. Resuspension is not modeled, but direct sedimentation of Se is modeled as the net of deposition and resuspension. Within the

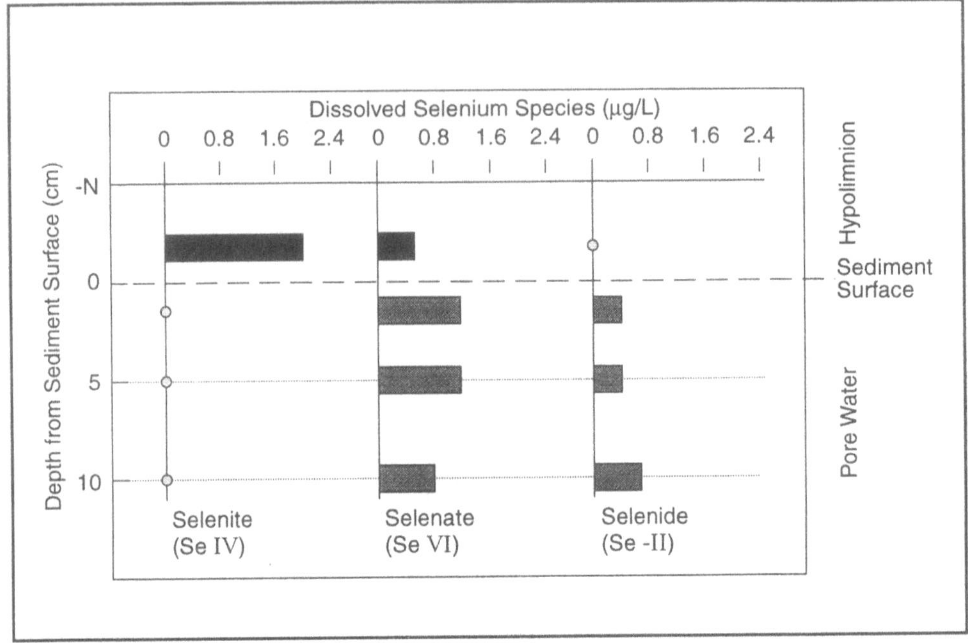

Figure 2. Relative dissolved Se concentrations in the hypolimnion and pore water indicate that Se (IV) will diffuse into the pore water, while Se (VI) and Se (-II) diffuse into the hypolimnetic waters (modified from Cutter 1991a).

water column, biological uptake, adsorption and reduction reactions transform dissolved Se to particulate Se, while excretion and particle decomposition transform particulate Se to dissolved Se. Particulate organic Se in the hypolimnion can accumulate in the sediments via deposition (settling) of inorganic particles, organic detritus, and algae. Organic selenides (Se-II), with lesser percentages of selenite (Se IV) and selenate (Se VI) form the major share of the organic particulate Se in the sediments. Inorganic particulate Se settling to the sediments consists of elemental Se (Se (0)) and Se (IV) and Se (VI) adsorbed to clays. Se (0) increases if the hypolimnion becomes anaerobic during summer. Within the sediments, organic decomposition releases the organic Se (-II), Se (IV) and Se (VI) into the pore water in proportion to their concentrations in the organic particles. From the pore water, redox reactions and diffusion pressures between pore water and the overlying hypolimnion determine the primary Se species and net fluxes. However, except for a thin oxidized layer at the sediment water interface during oxic periods in the overlying water, the sediments are assumed to be anaerobic. Thus, reduction of Se (VI) and Se (IV) to Se (0) from either organic or inorganic sources appears likely to predominate. Within the microlayer of oxygenated sediment at the sediment water interface, Se (0) and Se (-II) may be oxidized to Se (IV) and Se (VI). These reactions appear to be slow, but could have long term consequences.

Se studies and model application can help predict the dynamics of the decline in water column Se after cessation of ash pond discharges. The relative concentrations in pore water and the water column directly over the sediment in Hyco Reservoir provide an estimate of the potential net flux across the sediment water interface for the different species of Se (Figure 2). For Se (IV), the concentration in the water directly above the sediment far exceeded that in the sediment pore water even as shallow as one centimeter below the sediment surface. However, Se (VI) and Se (-II) occurred at higher concentrations in the sediment pore water than in the water directly above it. Thus, Se (IV) would be expected to diffuse from the water column into the pore water, while Se (VI) and Se (-II) would be expected to diffuse from the pore water into the water column. The decrease of Se (IV) would exceed the increases due to Se (VI) and Se (-II). These fluxes would change with seasonal variations in the Se concentrations of the overlying water. The Se in the pore water and sequestered in the sediments could yield a source of Se influx into the water column and food chain long after inputs from the ash pond stop. However, the rates of release are not well characterized, and further biogeochemical studies could provide less conservative estimates of these rates.

Although model testing has been limited to the three reservoirs in the study that have received Se inputs from ash ponds, the Se model has provided impressive matches to real data (Bowie et al. 1996). The rates for model processes in the biogeochemical and food web modules were derived from laboratory studies and applied to data collected from Hyco Reservoir over the period from 1985 to 1992 (Figure 3). For the periods for which samples were taken, the models simulated field data well. Testing of the modules and the full model at a broader range of reservoirs would provide stronger evaluation of their completeness and generality.

3. Mercury - role of sediments

Like Se, speciation is important to understanding Hg cycling and effects. In surface waters, Hg occurs in three forms: elemental (Hg(0)), inorganic (Hg(II)), and MeHg (CH_3Hg^+). These forms differ in residence time in the water, particulate distribution, sediment, and biotic compartments, in transformation and accumulation rates within these same compartments, and in their effects. Under chronic exposure conditions (long-term, low loading) MeHg is the major concern because of the ability of fish to accumulate large quantities in their muscle tissue -- binding to sulfhydryl groups -- with little or no direct deleterious effects on the health of the fish. However, risk assessment has identified potential concerns with fish consumers (humans and wildlife) who may suffer neurological effects from high consumption of MeHg (Clarkson 1990).

Direct uses/discharges of MeHg largely have ceased, and bacteria produce MeHg *in situ* from inorganic Hg(II) (Zillioux et al. 1993). In lakes, most of the mercury can be found in the sediments, where the bulk of methylation takes place (Watras et al. 1994). In fact, as shown in Figure 4, investigators have identified the sediment-water interface as the site where most MeHg is formed. Sulfate reducing bacteria appear to be the major producers of

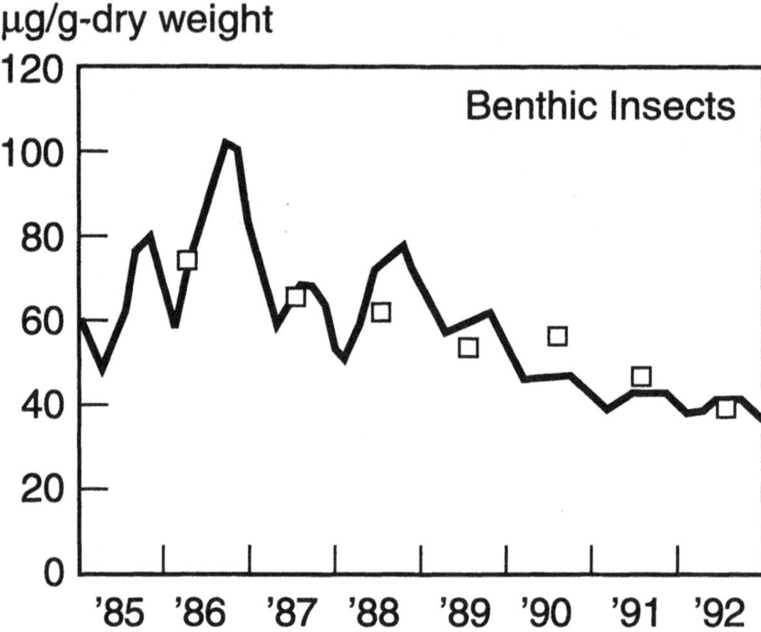

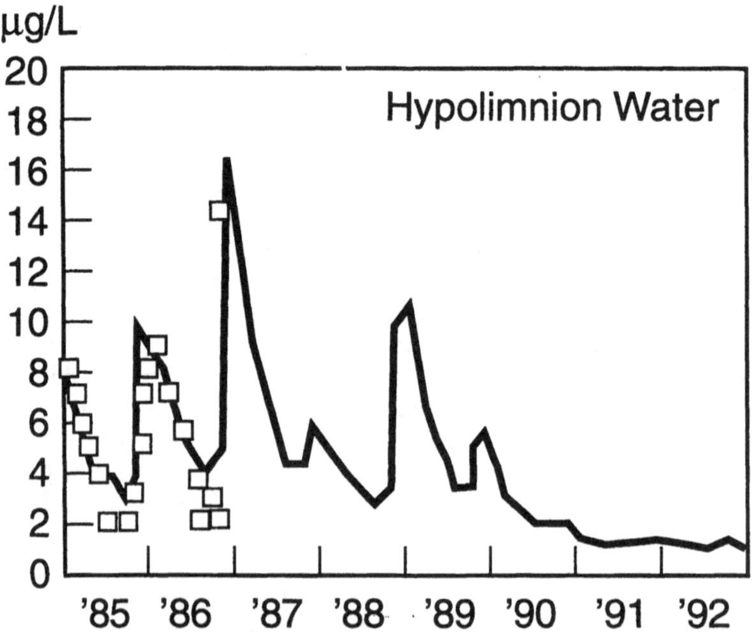

Figure 3. Comparison of total Se model predictions and data for the hypolimnion and benthic insects in Hyco Reservoir from 1985 to 1992 (modified from Bowie et al. 1996).

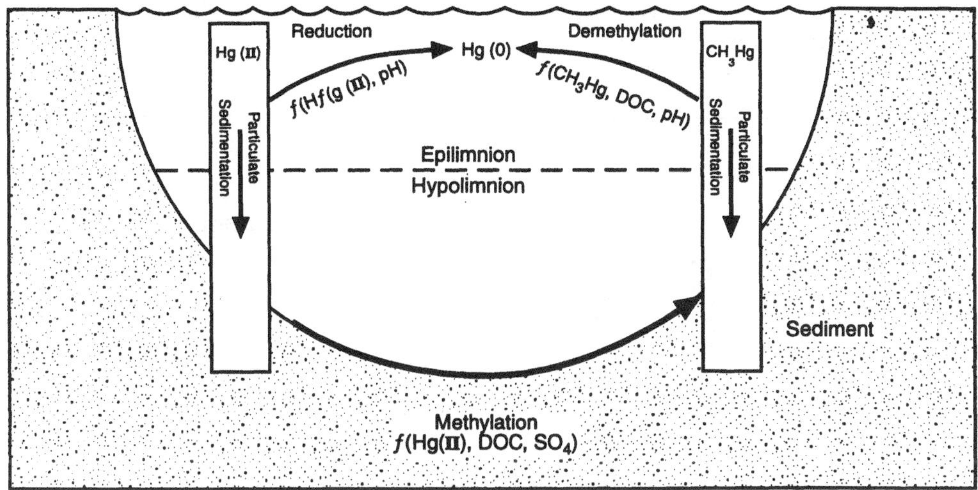

Figure 4. Sediment-water interface reactions affect Hg transformations in surface waters.

MeHg (Gilmour et al. 1992). A competition between Hg(II) availability, sulfide precipitation, and methylation processes seems to govern the rate of MeHg production. Factors involved with Hg(II) availability include: Hg reduction to Hg(0), a photochemical/biochemical reaction which is a function of pH; the presence of dissolved organic compounds that bind Hg(II) and sometimes enhance its delivery to the surface water but then compete with bacteria for the substrate Hg(II); and the biotic production rate (Hudson et al. 1994). Processes like demethylation can greatly influence MeHg concentrations available for fish bioaccumulation (Porcella 1994).

Because most (90 percent, Porcella 1994) Hg(II) transfers to the sediment in seepage lakes largely via particle sedimentation (the net of resuspension and settling), the sediment represents a long-term reservoir of potential Hg methylation. In model analyses performed with the Mercury Cycling Model applied to seepage lakes, changes in Hg input have quantifiably small effects on the accumulation of mercury in fish (Porcella 1994). Further model studies show that removal of Hg contaminated sediments or covering of contaminated sediments with low Hg soils have substantial effects on fish mercury for seepage lakes where all the Hg comes from the atmosphere. However, surface water systems dominated by highly contaminated sediments (such as Superfund sites) might undergo different patterns than do relatively simplistic seepage lakes. Further, the rates of change in fish Hg are slow, taking up to decades to reach new steady states.

Clearly, the interactions at the sediment-water interface are extremely important to the availability of substrate (Hg(II)) and its subsequent methylation. Future research requires investigation of: the bioavailability of soluble and particulate forms of Hg(II); processes governing MeHg production including sulfate reduction and the role of organic carbon;

how sulfides from sulfate reduction might compete with Hg(II) availability; and how vertical and spatial location of the processes of methylation/demethylation govern the amount of Hg accumulated in fish.

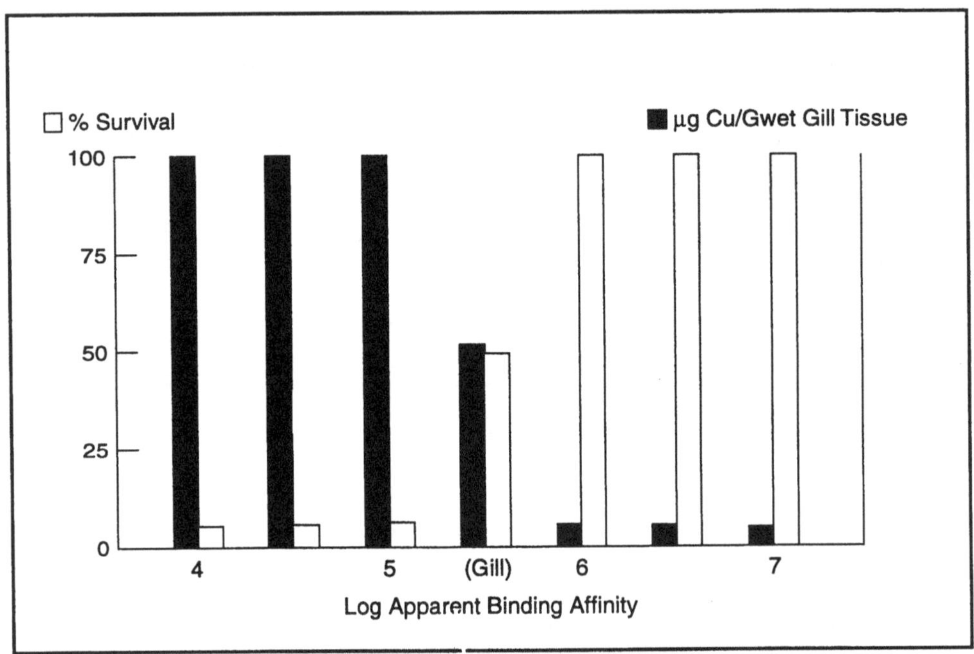

Figure 5. Hypothetical expectation of toxicity and gill Cu in response to equimolar exposures of copper and organic acid ligands of different apparent binding affinities. The log gill binding affinity is between 5 and 6. Organic acid ligands with higher binding affinities than the gill bind the copper and keep it from depositing on the gill. When the gill binding affinity is greater than that of the organic acid ligand Cu deposits on the gill and causes toxicity. Toxicity (1/survivorship) and gill copper are directly correlated.

4. Application of sediment-water chemistry to metals toxicity[1]

The draft report from a Society of Environmental Toxicology and Chemistry Pellston Conference held in Pensacola Beach, Florida, on February 10-14, 1996 (Bergman, et al. 1997) discusses a new paradigm for assessing toxicity of metals and regulating discharges. This new paradigm is based on current knowledge of mechanisms of metal toxicity to fish and treats the fish gill as a biotic ligand, in competition with other environmental ligands, e.g., dissolved organic carbon (DOC), for metals released in discharges.

[1]The discussion in this section applies to metals such as Ag, Al, Cd, Cu, Ni and Zn, whose major effects appear to be ecological, mediated via interactions with gills or other osmoregulatory, ionoregulatory or respiratory surfaces i.e., surface-active toxicants

Researchers on the Lake Acidification and Fisheries (LAF) project (see papers in the Canadian Journal of Fisheries and Aquatic Sciences 45:1561ff; 47:1578ff; and 48:1987ff) conducted some early work leading to development of this new paradigm. LAF researchers found that aluminum toxicity to fish is mediated via decreases in respiratory efficiency and/or impairment of osmotic and ionic regulation (Wood & McDonald 1987; Freda & McDonald 1988; Mount et al. 1988; Playle & Wood 1989; McDonald et. al. 1989). However, Fernandez et al. (1988) suggested that none of the existing methods for Al correlated well with toxicity to fish, and described a way to use organic acids of known binding affinities for Al to estimate the fraction of Al that is toxic. This method assumes that the first step in the Al toxicity mechanism is binding to the gill. In fact, Wood and McDonald (1987) had found earlier that sensitivity of brook trout, rainbow trout and smallmouth bass exposed to identical Al concentrations was directly correlated with Al accumulation on the gill. Fernandez et al. (1990) compared survival of adult brook trout exposed for 7 days to Al alone or Al accompanied with equimolar concentrations of a series of organic acids with known Al binding constants. The expectation was that organic acid ligands that bound Al more strongly would protect the fish from toxicity, while those that bound Al less strongly would confer less or no protection (Figure 5). In fact, organic acids with binding affinities of log 5.24 or less did not protect the fish (0% survival), whereas those with binding affinities of log 5.32 or more did protect the fish (> 90% survival). Kline (1992) found the same general relationship for rainbow trout. However, both researchers found that various water quality parameters (pH, temperature, etc.) affected this relationship. This led to use of terms such as "apparent binding affinity", "apparent binding constants" or "conditional stability constants", all to indicate that the binding affinity value even for the gill of a single fish varies within a range of values.

Other researchers have extended these conclusions about acute toxicity to other metals (Playle et al. 1993a and b; MacRae et al. in press) and found causal relationships between gill functions and toxicity (Bergman et al. 1997). Metals such as Ag, Cd, Cu, and Zn are acutely toxic to fish because they interfere with the active uptake of Na^+ or Ca^{2+} (Table 1).

TABLE I
Physiological mechanisms of metal toxicity in freshwater fish at concentrations near the appropriate acute criteria (modified from Bergman et al.1996).

Metal	Toxic Mechanism[1]
Ag	Blockage of Na^+ Uptake
Al	Blockage of Na^+ Uptake[2]
Cd	Blockage of Ca^{2+} Uptake
Cu	Blockage of Na^+ Uptake
Zn	Blockage of Ca^{2+} Uptake

[1] Sources cited in Bergman et al. (1996).

[2] Also can cause edema of gills and respiratory distress

This uptake is important because electrochemical gradients cause both ions to diffuse from the gills of fresh water fish into the surrounding water. Such losses, depending on severity, can impair health or cause death. Al has additional effects because it can cause edema of the gill that, in turn, reduces gas (O_2, CO_2, NH_4^+) exchange. The metals are surface active toxicants and initially bind to anionic sites on or in the gills. This is probably why the free metal divalent cations appear most often to be the most toxic metal species. Evidence suggests that there are a certain number of negatively charged binding sites, and that the metals compete with other ions for those sites. Examples of these competing ions are Ca^{2+}, Na^+ and H^+ ions. All reduce toxicity of metals via competition for the sites. Furthermore, toxicity is reduced if natural or synthetic ligands such as dissolved organic carbon, HCO_3^-, Cl^-, ethylenediamine tetraacetic acid (EDTA), etc., complex the metals, reducing the free metal ion concentration.

Models support the validity of the concepts sketched above. The binding process is subject to a limited set of rules, and thermodynamic models (MINEQL+, Schecker and Mc Avoy 1992; MINTEQA2, Allison et al. 1991) have proven useful in predicting binding of metals to gills under a range of water quality conditions. In addition, the gill surface interaction model (Pagenkopf 1983; Morel 1983; Morel and Hering 1993) provides reasonable explanations of the interactions of the metals with gills and water quality parameters.

Much work will be required, however, before the model can be validated for use in regulation of effluents or management of metals. Generality of application to all metals under broad ranges of water quality conditions will be required. However, the discussion above calls attention to the parallels between the information needed to validate this proposed method and the identification and process rate estimation that is the focus of chemists interested in processes that occur at the sediment-water interface. These observations suggest that increased interactions between chemists and toxicologists could be essential to optimize progress toward developing the required decision tools to regulate these 'surface-active' metals.

5. Conclusion

Progress in environmental biogeochemistry and toxicology of metals over the past decade has called attention to the critical need for collaboration of toxicologists and chemists. Current information indicates the importance of sediment-water interactions in controlling the effective water column and pore water concentrations of metals and metalloids. Se may be available from contaminated sediments over relatively long periods of time depending on the types and rates of processes that occur between sediments, pore waters and the water column. Hg availability and methylation depend on processes at or near the sediment-water interface. Finally, the processes that control reactions where solids and liquids meet at the surface of lake sediments appear to provide useful analogies to what occurs at the surfaces of fish gills and to help explain the rates and mechanisms of toxicity. Thus, interactions among sediment-water chemists and toxicologists appear necessary to support optimal progress toward environmental assessment of metals and cost effective effluent regulation.

Acknowledgments

The Electric Power Research Institute provided support for the authors and much of the work that is cited. We are also thankful to the researchers on the projects and at the Pellston Conference for use of information and data, some of it made available in prepublication form.

References

Allison, J.D., Brown, D.S., and Novo-Gradac, K.J.: 1991, USEPA, MINTEQA2/PRODEFA2, a geochemical assessment model for environmental systems: Version 3.0 Users manual. U. S. Environmental Protection Agency, Washington, D.C.

Bergman, H.L., Dorward-King, E.J. Allen, H.E., DiToro, D., Erickson, R.J., Mattice, J.S. and Reiley, M.C.: 1997, Proceedings of the 26th Pellston Workshop, SETAC Foundation for Environmental Education, 1010 North 12th Avenue, Pensacola, Florida, 32501-3370.

Bowie, G. L. and Grieb, T. M.: 1991, Water, Air, and Soil Pollution **57-58**, 13-22.

Bowie, G. L., Sanders, J. G., Riedel, G. F., Gilmour, C. C., Breitburg, D. L., Cutter, G. A. and Porcella, D. B.: 1996, Water, Air, and Soil Pollution **90**: 93-104.

Clarkson, T. W. 1990. Human Health Risks From Methylmercury in Fish. *Environ. Toxicol. Chem.* **9**:821-823.

Cowan, C. E. 1988. Review of Selenium Thermodynamic Data. EPRI Report EA-5655. Electric Power Research Institute. Palo Alto CA 94304. 73 p.

Cumbie, P. M. and Van Horn, S. L.:1978, Proc. Ann. Conf. SE Assoc. Fish Wildl. Agencies **32,** 612-624.

Cutter, G. A.: 1991a, EN-7281, Volume 1, Electric Power Research Institute, 3412 Hillview Ave., Palo Alto, CA 94304-1395.

Fenner, K. A.: 1992, TR-100434, Electric Power Research Institute, 3412 Hillview Ave., Palo Alto CA 94304-1395, KA23-24.

Fernandez, J.D., Mount, D.R., Hockett, J.R., and Bergman, H.L.: 1988, Fernandez, J.D. Proceedings 9th Annual Meeting of the Society of Environmental Toxicology and Chemistry, Arlington, VA, November, 1988.

Fernandez, J.D., Kline, E.R., and Bergman, H.L.: 1990, Proceedings 11th Annual Meeting of the Society of Environmental Toxicology and Chemistry, Arlington, VA, November, 1990.

Freda, J. and McDonald, D. G.: 1988, J. exp. Biol. **136**: 243-258.

Gilmour, C. C., Henry, E. A. and Mitchell, R.: 1992, Env. Sci. Tech. **26**:2281-2287.

Hudson, R. J. M., Gherini, S. A., Watras, C. J. and Porcella, D. B.: 1994, A mechanistic model of the biogeochemical cycle of mercury in lakes. In "Mercury Pollution: integration and synthesis", Watras, C. J. and J. W. Huckabee, eds. Lewis Publishers (CRC Press, Boca Raton, FL). pp. 473-523.

Howe, P., Masters, E., Atteberry, R., and Redmon, P.: TR-100434, Electric Power Research Institute, 3412 Hillview Ave., Palo Alto CA 94304-1395, 3-1 to 3-10.

Hudson, R. J. M., Gherini, S. A., Watras, C. J., Porcella, D. B.: 1994, Chapter V.1, In: Watras, C. J. and Huckabee, J. W., *Mercury Pollution integration and Synthesis*, Lewis Publishers, Chelsea, MI, p473-523.

Kline, E.: 1992, M.S. thesis, University of Wyoming, Laramie, WY.

McCrae, R.K., Smith, D.E., Swoboda-Colberg, N., Meyer, J.S. and Bergman, H.L.: in press, Environmental Toxicology and Chemistry.

McDonald, D. G., Reader, J. P. and Dalziel, T. K. R.: 1989, p 221-242. In R. Morris, D. J. A. Brown, E. W. Taylor and J. A. Brown (ed.) Acid Toxicity and Aquatic Animals, Society for Experimental Biology Seminar Series. Cambridge University Press, UK.

Morel, F.M.M.: 1983, Principles of Aquatic Chemistry, John Wiley and Sons, Toronto, Canada. 44gpp.

Morel, F.M.M. and Hering, J.G.: 1993, Principles and Applications of Aquatic Chemistry, John Wiley and Sons, New York, NY. 588pp.

Mount, D.R., Ingersoll, C.G., Gulley, D.D., Fernandez, J.D., LaPoint, T.W., and Bergman, H.L.: 1988, Canadian Journal of Fisheries and Aquatic Sciences **45**: 1623-1632.

Pagenkopf, G.K: 1983, Environmental Science and Technology **17**: 342-347.

Playle, R.C., and C.M. Wood: 1989, J. Comp. Physiol. **B159**:527-53.

Playle, R.C. Dixon, D.G., and Burnison, K.: 1983a, Can. J. Fish. Aquat. Sci. **50**: 2667-2677.

Playle, R.C., Dixon, D.G. and Burnison, K: 1993b, Can. J. Fish Aquat. Sci. **50**: 278-287.

Porcella, D. B.: 1994, Mercury in the environment: Biogeochemistry. In "Mercury Pollution: integration and synthesis". Watras, C. J. and J. W. Huckabee, eds. Lewis Publishers (CRC Press, Boca Raton, FL). pp. 3-19.

Porcella, D. B.: 1994, Chapter I.1, In: Watras, C. J. and Huckabee, J. W., *Mercury Pollution integration and Synthesis*, Lewis Publishers, Chelsea, MI, p3-19.

Porcella, D. L., Bowie, G. L., Sanders, J. G. and Cutter G. A.: 1991, Water, Air and Soil Pollution **57-58**, 3-11.

Schecher, W.D. and McAvoy, D.C.: 1992, Computers, Environment and Urban Systems **1b**: 65-76.

USEPA : 1993a, EPA-822-R-93-011, Water Resource Center, USEPA, 401 M Street, SW., Washington, DC 20460.

USEPA : 1993b, EPA-822-R-93-017, Water Resource Center, USEPA, 401 M Street, SW., Washington, DC 20460.

Watras, C.J., Bloom, N.S., Hudson, R.J.M., Gherini, S., Munson, R., Class, S.A., Morrison, K.A., Hurley, J., Wiener, J.G., Fitzgerald, W.F., Mason, R., Vandal, G.,

Powell, D., Rada, R., Rislov, L., Winfrey, M., Elder, J., Krabbenhoft, D., Andren, A.W., Babiarz, C., Porcella, D.B., Huckabee, J.W. .: 1994, Chapter I.12, In: Watras, C. J. and Huckabee, J. W., *Mercury Pollution Integration and Synthesis*, Lewis Publishers, Chelsea, MI, p153-177.

Wood, C.M.: 1989, In: R. Morris and D.J.A. Brown, Eds., Acid Toxicity and Aquatic Animals, Cambridge University Press, U.K. 125-152.

Wood, C.M. and McDonald DG: 1987, In: Ecophysiology of Acid Stress in Aquatic Organisms, Annls. Soc. Zool. Belg. Supp. 399-410.

Zillioux, E. J., Porcella, D. B. and Benoit, J. M.: 1993, Mercury cycling and effects in freshwater wetland ecosystems. Environ. Toxicol. Chem. **12**:2245-2264.

EFFECT OF IONIC COMPOSITION AND TEMPERATURE ON THE RADIOCAESIUM FIXATION IN FRESHWATER SEDIMENTS

M. J. MADRUGA[1], A. CREMERS[2]

[1] *DGA, Radiological Protection and Safety Department, E.N. 10, 2685 Sacavém, Portugal*
[2] *Laboratory for Colloid Chemistry, Catholic University of Leuven, Belgium*

Abstract. The objective of this study was to verify in which way the ionic composition of the sediment and that of the overlying water column may have an effect on radiocaesium fixation, through possible structural modifications of the frayed edge sites (FES) pools of the sediments. Two experimental protocols have been considered: i) a condition in which sediments were homoionically saturated with either potassium, ammonium, calcium, magnesium or sodium ions, and ii) a mixed potassium-calcium scenario in the liquid phase. Nine freshwater sediments from four different locations were used in this study. For homoionic potassium and ammonium saturated sediments a nearly quantitative radiocaesium desorption (90-100%) was observed, whereas for calcium and magnesium the desorption yields were about 20%. It appears that the action of strongly hydrated ions (Na^+, Mg^{2+}, Ca^{2+}) leads to a pronounced enhancement of radiocaesium fixation in the solid phase, whereas poorly hydrated ions (K^+, NH_4^+) have the opposite effect and promote sorption reversibility.

Another issue considered in this study concerns the effect of temperature and sediment drying on the radiocaesium fixation. Drying the sediments at 110°C leads to a significant increase in radiocaesium fixation levels, while drying the sediments at room temperature (25°C) has a very limited effect on radiocaesium fixation and appears to put a brake on the aging effects.

Key words: freshwater sediments, radiocaesium, fixation, ionic composition, temperature

1. Introduction

In the studies on the environmental impact of radiocaesium in freshwater bodies two processes can be considered: the "short-term" or instantaneous equilibrium effects that are related to the sorption potential of sediments and suspended particles, and the "long-term" effects that refer to the potential of the sediment to act as a future source leading to the remobilization of radiocaesium to the overlying water column. It had been reported (Evans *et al.*, 1983; Comans *et al.*, 1989) that the amount of water soluble ^{137}Cs is significantly higher under anaerobic redox conditions due to the radiocaesium remobilization from sediments by ammonium, released during anaerobic conditions. However, other studies (Evans *et al.*, 1983; Comans *et al.*, 1991) provide convincing evidence that over the longer periods of time, radiocaesium sorption is only partially reversible.

There is a general consensus that the selective sorption of radiocaesium in soils and sediments is directly related to the action of micaceous clays (Sawhney, 1972; Brouwer *et al.*, 1983). It is also well established that the sorption sites in illite can be divided into surface exchange sites, interlattice positions near the edges of the clay particles (frayed edge sites) and interlattice positions in the interior of the particles (Bolt *et al.*, 1963;

Water, Air and Soil Pollution **99**: 201-208, 1997.

Brouwer *et al.*, 1983). Moreover, ions with low hydration energy (Cs^+, Rb^+, K^+, NH_4^+) induce interlayer dehydration and layer collapse resulting in the fixation of cations in interlayer positions. It is now accepted that this fixation takes place at the edge of the clay particles (frayed edge sites -FES) generated by the effect of weathering and possibly by the action of large hydrated ions such as Ca^{2+} and Mg^{2+} (Comans *et al.*, 1991). The frayed edge sites capacity such as the caesium selectivity in these sites can be determined experimentally (Cremers *et al.*, 1988). On the basis of the results obtained for sediments and soils (De Preter, 1990; Valcke, 1993) which show a rather consistent and coherent caesium selectivity pattern in the frayed edge sites it can be postulated that these sites have similar structures. However, this is an oversimplification considering the irreversible pattern of radiocaesium sorption obtained in various systems (Madruga, 1993; Valcke, 1993). These irreversibility effects can be discussed in terms of possible modifications of the structural properties of the FES during the adsorption phase.

The objective of this study is to verify in which way the ionic composition of the sediment, and that of the overlying water column, may have an effect on radiocaesium fixation through possible structural modifications of the FES pool. Two scenarios have been considered : a) a condition in which sediments are homoionically saturated with various ions, and b) mixed potassium-calcium scenario in the liquid phase.

In sediment sampling methodology is important to know if a sediment (or soil) should be dried and if so, at what temperature and what is the possible effect on the desorption behaviour of radiocaesium. To answer to these questions the effect of temperature and sediment drying on the radiocaesium fixation has already been studied.

2. Materials and Methods

2. 1. EFFECT OF SUBSTRATE IONIC COMPOSITION

Nine freshwater sediments from four different locations [T1 and T2 from Tejo River (Portugal); A and S from Tejo estuary (Portugal); D3 and D4 from Devoke Water (England); KR4, KR5 and KR6 from Kiev reservoir (Ukraine)] were used in these experiments.

2.1.1. *Homoionic scenarios*

Two sets of desorption tests were studied using a protocol for which either the sediment (set 1) or the adsorbent Giese-granulate (ammonium copper hexacyanoferrate) (set 2) were enclosed in a dialysis membrane. The Giese-granulate is an ion-exchanger, characterized by exceedingly high K_d (Cs) values. For example in a 10^{-3} mol dm^{-3} NH_4^+ solution a value of about $5x10^5$ cm^3 g^{-1} was determined.

In set 1, a comparison of desorption behaviour was made in homoionic potassium, ammonium, calcium and magnesium saturated sediments. Sediment samples [T2, A] were enclosed in dialysis membranes (1 g/10 cm^3) and homoionically saturated with different solutions: $5x10^{-3}$ mol dm^{-3} K^+, 10^{-3} mol dm^{-3} NH_4^+, $5x10^{-3}$ mol dm^{-3} Ca^{2+} and 10^{-2} mol dm^{-3} Mg^{2+}. Systems were labelled with ^{137}Cs in these same solutions, and allowed to age for two days. The amount of radiocaesium adsorbed in the sediments was determined through the measurement of radioactivity in the liquid phase by gamma spectrometry. The dialysis membranes with the sediments were finally transferred to vessels containing 200 cm^3 of 10^{-3} mol dm^{-3} NH_4Cl and 15 g (dry) of Giese granulate. Desorption progress was monitored by intermittent counting of the adsorbent by gamma spectrometry.

For set 2, for all the sediment samples [T1, T2, A, S, D3, D4, KR4, KR5, KR6] a comparison was made between the potassium, sodium, and calcium forms of the sediment and their natural conditions (where the ionic composition of the sediment was unaltered). Homoionic systems (in K^+ $2x10^{-3}$ mol dm^{-3}, Na^+ $2x10^{-3}$ mol dm^{-3}, Ca^{2+} 10^{-3} mol dm^{-3}), at solid/liquid ratio of 1 g/10 cm^3, were labelled with radiocaesium in a 10^{-5} mol dm^{-3} KCl ^{137}Cs labelled solution, and allowed to age for either 4 or 52 days. The radiocaesium adsorption in the solid phase (sediments) was measured by gamma spectrometry, after ultracentrifugation. The sediment samples were subsequently submitted to the desorption membrane protocol (Wauters *et al.*, 1992). Samples were dispersed in 200 cm^3 of 10^{-3} mol dm^{-3} NH_4Cl containing a dialysis membrane with 1.5 g (dry) Giese granulate. Desorption progress was monitored counting the adsorbent (fresh adsorbent is used after each sampling) by gamma spectrometry.

2.1.2. *Biionic scenarios*

Sediment samples [T2] were exhaustively preconditioned in three K-Ca mixtures of the following composition (meq dm^3): K= 0.1, Ca= 3, (PAR) = 0.08 [PAR= potassium adsorption ratio= $m_K/(m_{Ca}+m_{Mg})^{1/2}$]; K= 0.5, Ca= 3, (PAR) = 0.41; K= 1, Ca= 3, (PAR) = 0.82, i.e. they were submitted to a range of potassium adsorption ratios (PAR), keeping total concentration at low level and nearly constant (3.1 - 4 meq dm^{-3}). ^{137}Cs labelling was made in these conditions and systems were allowed to age for 2, 129 and 165 days. ^{137}Cs desorption was studied using the membrane protocol described in paragraph 2.1.1. (set 2).

2.2. EFFECT OF SEDIMENT DRYING AND TEMPERATURE

Sediment samples [T1, T2, A, D3 and KR4] (~1 g) were weighed in aluminium plates and contaminated using 2.5 cm^3 10^{-5} mol dm^{-3} KCl ^{137}Cs labelled solution. These systems were submitted to five drying-wetting cycles at 25°C using 2.5 cm^3 bidistilled water. The entire procedure took 4 days.

Another set of samples were labelled with ^{137}Cs, using the same procedure, dried at 25°C and left to age for 48 days. The effect of temperature was studied by oven-drying the ^{137}Cs contaminated samples at 110°C and left to age for 4 and 48 days. In parallel, reference samples were contaminated with ^{137}Cs and left to age at room temperature (25°C) in wet conditions for 4 and 48 days. For all these experiments the amount of radiocaesium adsorbed was determined counting the sediments by gamma spectrometry, after ultracentrifugation.

For the radiocaesium desorption evaluation the samples were submitted to the membrane protocol already described (paragraph 2.1.1 (set 2)).

3. Results and Discussion

3.1. EFFECT OF SUBSTRATE IONIC COMPOSITION

3.1.1. *Homoionic scenarios*

Figure 1 shows the radiocaesium desorption yields, in terms of desorption levels versus desorption time, for set 1. It is seen that very well defined plateaus are obtained for all the sediment saturated forms. However, a completely different behaviour in the potassium, ammonium, calcium and magnesium sediment saturated forms was observed. For homoionic potassium and ammonium saturations, a nearly quantitative desorption (90-100%) was observed, whereas in the case of calcium and magnesium, the desorption yields were about 20%. This means that the magnesium and calcium appear to modify the structural configuration of the sediments in such a way, that the radiocaesium desorption will be significantly hindered. These effects are confirmed by the results of the second set.

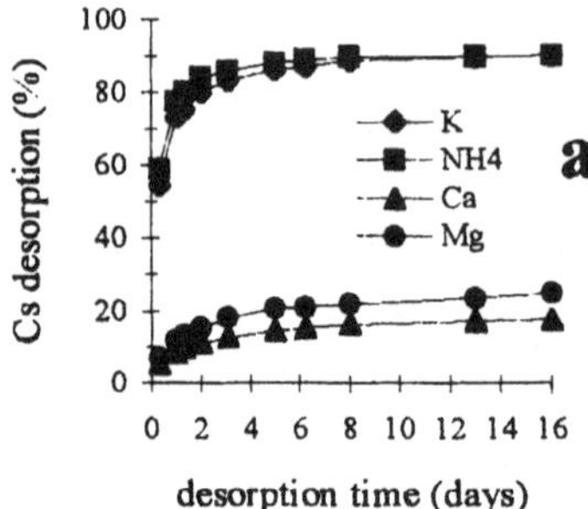

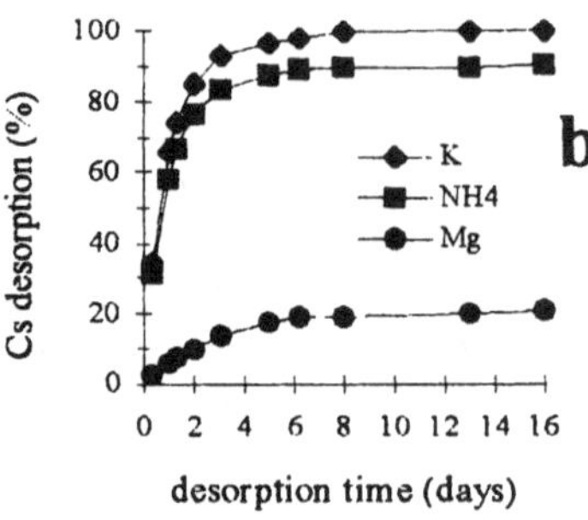

Fig. 1. Time dependence of radiocaesium desorption for T2 (a) and A (b) sediments in 5×10^{-3} mol dm^{-3} K^+, 10^{-3} mol dm^{-3} NH_4^+, 5×10^{-3} mol dm^{-3} Ca^{2+}, 10^{-2} mol dm^{-3} Mg^{2+} solutions and Giese granulate. Aging time 2 days.

Figure 2 summarizes the results of the second set, for the two aging times, expressed directly in terms of a comparison of the maximum fixation levels in the various systems studied for the natural, potassium, sodium and calcium states. Overlooking some minor differences in the behaviour, a very consistent pattern was observed. For short aging time (Figure 2a), there is a significant decrease of fixation level upon changing from the natural conditions to the K-state. There is also a very significant increase of radiocaesium fixation in the sodium condition, but the highest fixation level was found in the Ca-state. The nature of the effect is obvious: it appears that the action of strongly hydrated ions (Na^+, Mg^{2+}, Ca^{2+}) leads to (structural) configurations unfavourable for radiocaesium desorption within a very short time scale (days). This kind of effect (although less pronounced) has also been demonstrated in illite (presaturated with K^+, Ca^{2+} and Mg^{2+}), using the desorption membrane protocol (Wauters *et al.*, 1992), and using another approach (comparison of sorption and desorption distribution coefficient values); (Comans *et al.*, 1991). As a explanation we suggest that the presence of strongly hydrated ions in the FES pool may lead to a wedge-effect, allowing a deeper penetration of radiocaesium into the solid, resulting in a dramatic drop in subsequent desorption levels. The different behaviour between the natural conditions and the Ca-state is also of some interest. In the case of Kiev [KR4] reservoir sediments, both conditions lead to nearly identical fixation levels, due to the fact that, in the natural state, these sediments are practically in the homoionic Ca-state (Madruga, 1993).

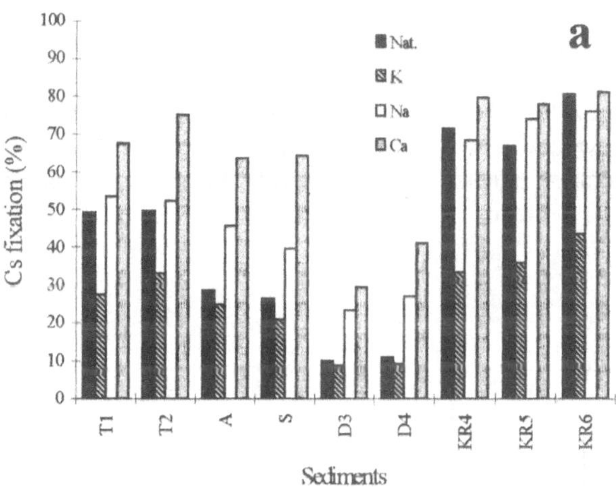

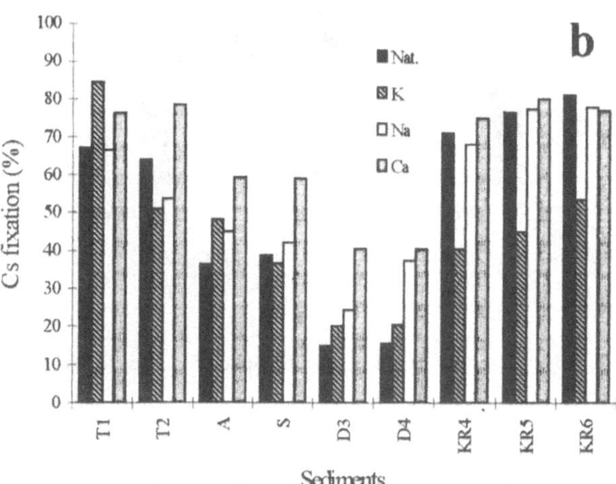

Fig. 2. Effect of K, Na and Ca ions on radiocaesium fixation for all selected sediments and 4 (a) and 52 (b) days aging times.

When the various sediments are submitted to an aging process of 52 days (Figure 2b), it appears that the differences in fixation behaviour become less pronounced and the Ca-saturated sediments barely show any aging effects. The difference in fixation behaviour between the freshwater sediments [T1 and T2] and estuarine sediments [A and S] confirms results obtained (Madruga, 1993) which show a higher fixation level in the systems with lower FES capacity. However, on the basis of the results presented here, the differences on the fixation levels are very likely related to differences in the ionic composition of the sediments.

3.1.2. *Biionic scenarios*

Figure 3 summarizes the results in terms of fixation levels, obtained from the plateau desorption values, versus PAR data. Aging effects are clearly evidenced. However, the most important point is that the fixation tendency is promoted at low PAR values, i.e. the calcium effect, as evidenced in the homoionic scenario, is also shown in the biionic scenario. These calcium effects on radiocaesium fixation are also reported for soils (Wauters, *et al.*, 1992; Valcke, 1993).

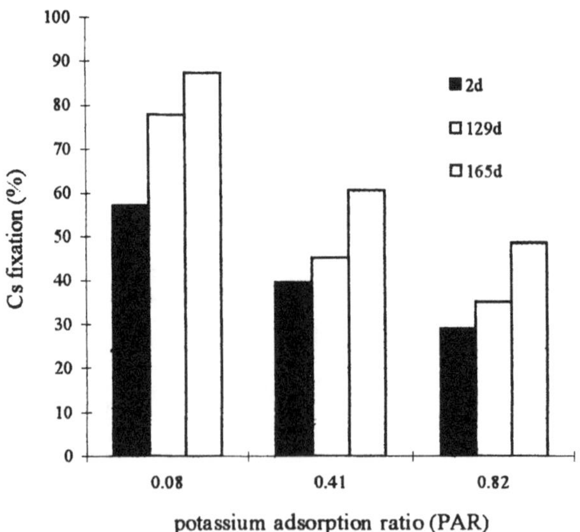

Fig. 3. Effect of potassium adsorption ratio on radiocaesium fixation for T2 sediment, at different aging times (2, 129 and 165 days).

These findings may have several practical implications. First of all, they may possibly explain the erratic fixation behaviour in systems (soils and sediments) in terms of differences in ionic composition (for example, low and high potassium levels in lakes). Secondly, they may possibly suggest a potential countermeasure after a nuclear accident in terms of calcium treatments. Finally, the favourable effect found for sodium may lead to an enhancement of fixation in a marine scenario.

3.2. EFFECT OF SEDIMENT DRYING AND TEMPERATURE

The results are summarized in Table I in terms of fixation levels, as obtained from the desorption plateaus, for 4 and 48 days aging times. For short aging times (4 days) five wetting-drying cycles have a limited effect on radiocaesium fixation (except for sediment A), as compared to the reference samples left to age for 4 days under wet conditions. This is in agreement with results obtained on illite clay (De Preter, 1990). A slight increase is in fact expected on the basis of a similar and well known effect of wetting-drying on potassium fixation in soils. The drying procedure at 110°C leads to a significant increase in fixation levels.

Very important aspects of radiocaesium behaviour are related to aging effects. It is clear that, for wet systems, the aging causes an increase in fixation levels by a factor of about 2 (except for the KR4 sediment). Most important is the finding that in the dried systems, aging leads to minimal changes in fixation levels (both at 25°C and 110°C). This is a rather important information for sample preparation. It means that, when sediments (or soils) are air-dried after sampling, aging effects are interrupted and, if one is interested in studying radiocaesium desorption yields, the picture obtained relates in fact to the date of sampling. Moreover, these data show that drying the system at high temperature (110°C) should be avoided since leads to an increase of radiocaesium fixation. If, as demonstrated above, the aging effects are interrupted in dry conditions, then perhaps it would be expected aging effects in soils (as obtained from time-dependence of transfer factors) to be more pronounced for years characterized by high rainfall.

TABLE I

Effect of sediment drying and temperature on radiocaesium fixation levels (%) for 4 and 48 days aging times.

Aging time (days)	Sediments	Wet	Dry (25°C)	Dry (110°C)
4	T1	41.5	44.6*	66.7
	T2	41.4	54.3*	-
	A	26	47.8*	59.6
	D3	7.5	8.9*	-
	KR4	68.8	71.8*	-
48	T1	75.9	49.3	63
	T2	74.7	50.7	-
	A	39.7	33.1	55.5
	D3	14.5	8.7	-
	KR4	70.3	65.5	-

* 5 wetting/drying cycles

4. Conclusions

It can be concluded that the ionic composition of the liquid phase (and therefore of the solid) plays a key role in the fixation behaviour of radiocaesium: strongly hydrated cations (Ca^{2+} , Mg^{2+}, Na^{+}) lead to a pronounced enhancement of fixation in the solid phases; poorly hydrated ions such as K^{+} and NH_4^{+} have the opposite effect and promote sorption reversibility. Drying a sediment at room temperature has a very limited effect on fixation and appears to put a brake on the aging effects.

References

Bolt, G., Sumner, M., and Kamphorst A.: 1963, *Soil Sci. Soc. Am. Proc.* **27**, 294-299.

Brouwer, E., Baeyens, A., Maes, A. and Cremers, A.: 1983, *J. Phys. Chem.* **87**, 1213-1219.

Chhabra, R., Pleysier, J., and Cremers, A.: 1975, *Proc. Int. Clay Conf.*, Mexico: Appl. Publs. Lim., 439-449.

Comans, R., Middelburg, J, Zonderhuis, J., Woittiez, J., De Lange, G., Das, H., and Van der Weijden, C.: 1989, *Nature* **339**, 367-369.

Comans, R., Haller, M., and De Preter, P.: 1991, *Geochim. Cosmochim. Acta* **55**, 433-440.

Cremers, A., Elsen, A., De Preter, P., Maes, A.: 1988, *Nature* **335**, 247-249.

De Preter, P.: 1990, Radiocaesium retention in the aquatic, terrestrial and urban environment: a quantitative and unifying analysis, PhD Thesis, Katholieke Universiteit Leuven, Belgium, 93 pages.

Evans, D., Alberts, J. and Clark, R.: 1983, *Geochim. Cosmochim. Acta* **47**, 1041-1049.

Madruga, M.J.: 1993, Adsorption-desorption behaviour of radiocaesium and radiostrontium in sediments, PhD Thesis, Katholieke Universiteit Leuven, Belgium, 121 pages.

Sawhney, B.: 1972, *Clays & Clay Min.* **20**, 93-100.

Valcke, E.: 1993, The behaviour dynamics of radiocaesium and radiostrontium in soils rich in organic matter, PhD Thesis, Katholieke Universiteit Leuven, Belgium, 135 pages.

Wauters, J., Sweeck, L., Valcke, E., Elsen, A., and Cremers, A.: 1992, *Sci. Total Environ.* **12**, 79-87.

RELATIONSHIPS BETWEEN RADIONUCLIDE CONTENT AND TEXTURAL PROPERTIES IN IRISH SEA INTERTIDAL SEDIMENTS

J. CLIFTON [1,2], P. McDONALD [1], A. PLATER [2] and F. OLDFIELD [2]

[1] Princess Royal Building, Westlakes Research Institute, Moor Row, Cumbria CA24 3LN, U.K. [2] Department of Geography, University of Liverpool, PO Box 147, Liverpool L69 3BX, U.K.

Abstract. Intertidal sediments from a range of depositional environments in the eastern Irish Sea have been analysed with regard to their radionuclide content, particle size distribution and magnetic properties. Concentrations of ^{241}Am and ^{137}Cs are highly influenced by the abundance of sediment finer than 32μm in fine-grained sedimentary environments, whilst radionuclide activity in coarser sediments is less dependent on particle size. Investigation of the magnetic properties of these sediments highlights a similar association between this size fraction and magnetic remanence, the latter being shown to offer strong potential as a grain size proxy in monitoring ^{137}Cs concentrations in fine-grained sediments. Analysis of particle size fractions indicates the extent to which ^{137}Cs may be enriched in the clay size fractions of both fine and coarse sediments.

Key words: americium-241, caesium-137, fractionation, grain size, magnetic properties, normalisation, Sellafield

1. Introduction

Low-level radioactive liquid waste has been discharged to the Irish Sea from the British Nuclear Fuels (BNFL) nuclear reprocessing plant at Sellafield in Cumbria since 1952 (Gray *et al.*, 1995). Particle-reactive isotopes such as americium-241 (^{241}Am, half-life=432.7y) tend to adsorb rapidly onto sediment following discharge, whilst isotopes of the alkali metals such as caesium-137 (^{137}Cs, half-life=30.2y) remain predominantly in the dissolved ionic state in saline water and undergo wider dispersion by tidal currents. Both radionuclides, in common with trace metals, exhibit preferential sorption with respect to fine grained sediment (Carpenter *et al.*, 1991; Livens & Baxter, 1988); the greater surface area of fine particles enhancing the extent to which radionuclide adsorption may take place. Micaceous minerals such as illite and coatings of organic matter and sesquioxides in fine sediment fractions serve to enhance the sorption of ^{137}Cs and ^{241}Am (Comans & Hockley; 1992, McDonald *et al.*, 1990).

A range of techniques have been developed to minimise particle size-induced variations in radionuclide content in order to elucidate the underlying spatial and temporal trends in activity. These have traditionally included normalising radionuclide data to the abundance of a specific particle size class such as the fraction finer than 63μm (Assinder *et al.*, 1993), although fractions such as the percentage less than 75μm and 50μm have been used (Aston & Stanners, 1982; Clifton & Hamilton, 1982). Physical separation of particle size fractions by settling techniques has been employed to identify the dominant fraction contributing to bulk sample activity (Ramsay & Raw, 1987).

Water, Air and Soil Pollution **99**: 209-216, 1997.

Alternative methods make use of sediment properties which are dependent upon particle size. The abundance of elements such as aluminium (Al) or lithium (Li), which are structurally incorporated within clay minerals, has been utilised to reflect the presence of fine sediment in studies of trace metal pollution (Daskalakis & O'Connor, 1995; Loring, 1991). Investigation of the magnetic properties of sediments has indicated significant associations with the presence of fine mineral sediments (Bonnett *et al.*, 1988; Oldfield *et al.*, 1993).

This study undertakes detailed analysis of the composition and radionuclide content of a wide variety of sediments from the eastern Irish Sea in order to identify the correct grain size class to utilise in normalisation procedures. The potential of magnetic properties to act as a grain size proxy measurement in radionuclide monitoring will be investigated. The effectiveness of particle size separation techniques and the nature of radionuclide concentration in size fractions will be examined.

2. Methodology

Surface intertidal sediments (0-1cm depth) were collected using a plastic trowel from 57

Fig. 1. Location map of the central Irish Sea

locations along the eastern Irish Sea coastline of south west Scotland, England and north Wales (Figure 1). These sites were identified from map evidence as representative of distinctive stretches of coastline. Sampling was carried out over a four week period and samples frozen within 48 hours of collection.

All particle size analyses utilised 0.45μm filtered sea water in order to reduce the potential for deflocculation, thereby obtaining size data approximating to those prevailing under natural conditions. Following separation of sediment by wet sieving through a 1000μm sieve, the finer fraction was analysed using a Malvern Mastersizer laser granulometer. Replicate analyses revealed the calculated mean grain size to vary by less than 5% in the majority of samples. Sediment coarser than 1000μm was dry sieved to 4000μm and the results combined with laser granulometry data to produce complete particle size distribution information.

Radionuclide activity was determined by counting freeze-dried bulk samples in calibrated geometries on a planar high-purity germanium gamma spectrometer for periods of up to 8 hours. Spectral calibration was performed using commercially available software and 95% confidence limits applied through counting standard IAEA sediments. Detection limits were in the order of 1 Bq kg^{-1}. Low and high frequency magnetic susceptibility (χ_{LF} and χ_{HF}), anhysteretic remanent magnetisation normalised to the d.c. biasing field (χ_{arm}), and saturated isothermal remanent magnetisation (SIRM) were measured following the procedure described by Oldfield & Yu (1994).

Ten representative samples were identified for particle size fractionation through analysis of the textural parameters of the entire sample suite. Sediment coarser than 63μm was isolated by wet sieving, air dried and sieved to 1000μm. The finer fraction was suspended in a settling tube filled with distilled water made up to pH 7.7 with NH_4HCO_3, replicating the measured pH of filtered sea water. Particle size fractions were obtained through calculating settling times using the measured density of the fraction finer than 63μm. Analysis of radionuclide activities, magnetic properties and particle size distributions were carried out on each size fraction following the procedures outlined above.

3. Results

3.1 PARTICLE SIZE AND RADIONUCLIDE ACTIVITY RELATIONSHIPS IN BULK SAMPLES

Tables I to IV detail rank correlation coefficients between radionuclide activity and the abundance of successive particle size classes derived from size analysis of bulk samples, grouped according to type of depositional environment. Caesium-137 is seen to display stronger correlations than ^{241}Am with respect to all particle size fractions. The significance of mean grain size as an indicator of radionuclide content in all sites derives solely from the strength of the correlation in mudflat environments, whilst coastal and salt marsh sediments display little or no correlation. With regard to the latter, the strong negative correlation between radionuclide content and the abundance of sediment in the 32-125μm size class significantly alters the degree of correlation with mean grain size, this trend being unique to salt marsh sediments.

Radionuclide activities in mudflat sediments exhibit a more gradual shift towards a negative particle size association in the 32-63μm size range. Open coastal sediments display weaker yet significant correlations between radionuclide content and particle size

TABLE I
Radionuclide content and size class abundance correlations for all sites
^{241}Am n=50, ^{137}Cs n=57; *italic*=significant at 5% level, all others significant at 1% level.
NS=not significant at 5% level

	Mean	% <2μm	% 2-8μm	% 8-16μm	% 16-32μm	% 32-63μm	% 63-125μm	% 125-250μm	% >250 μm
^{241}Am	-0.52	0.71	0.64	0.64	0.7	0.47	NS	*-0.27*	*-0.27*
^{137}Cs	-0.74	0.76	0.81	0.82	0.84	0.64	0.35	*-0.5*	-0.51

TABLE II
Radionuclide content and size class abundance correlations for salt marsh sites
n=17; *italic*=significant at 5% level, all others significant at 1% level. NS=not significant at 5% level

	Mean	% <2μm	% 2-8μm	% 8-16μm	% 16-32μm	% 32-63μm	% 63-125μm	% 125-250μm	% >250 μm
^{241}Am	NS	*0.58*	0.65	0.69	0.55	-0.78	-0.81	NS	*0.52*
^{137}Cs	NS	*0.54*	0.63	0.73	0.62	-0.86	-0.81	NS	*0.49*

TABLE III
Radionuclide content and size class abundance correlations for mudflat sites
n=11; *italic*=significant at 5% level, all others significant at 1% level. NS=not significant at 5% level

	Mean	% <2μm	% 2-8μm	% 8-16μm	% 16-32μm	% 32-63μm	% 63-125μm	% 125-250μm	% >250 μm
^{241}Am	-0.7	0.82	*0.68*	*0.67*	0.84	NS	*-0.52*	*-0.57*	NS
^{137}Cs	-0.8	0.91	0.85	0.85	0.95	NS	*-0.62*	*-0.69*	NS

TABLE IV
Radionuclide content and size class abundance correlations for open coastal sites
^{241}Am n=22, ^{137}Cs n=28; *italic*=significant at 5% level, all others significant at 1% level.
NS=not significant at 5% level

	Mean	% <2μm	% 2-8μm	% 8-16μm	% 16-32μm	% 32-63μm	% 63-125μm	% 125-250μm	% >250 μm
^{241}Am	NS	NS	NS	NS	0.57	*0.52*	*0.48*	NS	NS
^{137}Cs	*-0.44*	NS	*0.45*	*0.45*	0.58	0.63	0.57	NS	*-0.44*

across the 16-125μm size range. This reflects the association of radionuclides with the finest fraction of sediments characterised by a wide range of grain size distributions rather than any association with a specific size class as in the case of low energy depositional environments. The importance of the abundance of material finer than 32μm with respect to radionuclide concentrations in fine intertidal sediments points towards its use as a granulometric approach to correct for the grain size effect. In addition, the strong negative association between radionuclide content and sediment in the 32-125μm size range in salt marsh sediments offers similar potential in environments where material of this size is present.

3.2 BULK PARTICLE SIZE, RADIONUCLIDE ACTIVITY AND MAGNETIC PARAMETERS

Comparison of magnetic parameters with published data (Maher, 1988) indicates that fine-grained sediments are characterised by magnetic grains between 0.02 and 0.05μm diameter, whilst magnetic grains approaching 1μm diameter are present in association with coarse-grained sediments. Correlation of magnetic properties with the abundance of mineral size fractions highlights a significant association between magnetic remanence expressed as χ_{arm} and sediment finer than 32μm in low energy environments; a minimal correlation with particle size fractions being found in open coastal sediments.

Previous studies (Oldfield & Yu, 1994) suggest that such a relationship reflects the presence of bacterial magnetite, formed authigenically and dispersed with the associated fine-grained sediments. Figures 2a to 2d are plots of the regression of radionuclide content against magnetic remanence, the strong linear correlation with regard to ^{137}Cs indicating the potential of χ_{arm} to serve as a grain size proxy in the monitoring of ^{137}Cs in fine intertidal sediments. A less significant correlation between remanence and ^{241}Am is present, resulting from the weaker particle size dependence of ^{241}Am, as illustrated in Tables I to IV.

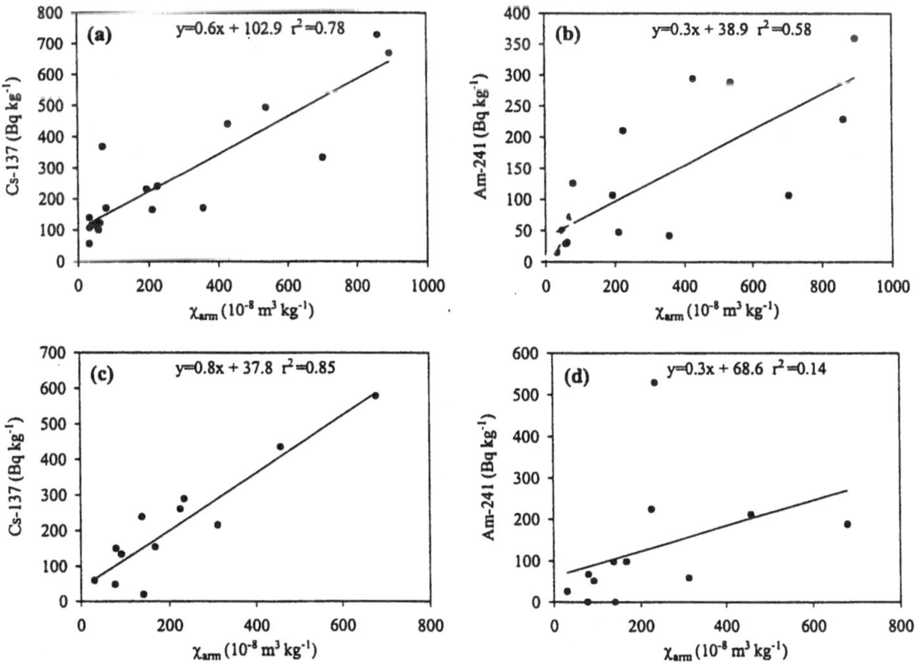

Fig. 2. Radionuclide activity as a function of magnetic remanence in bulk salt marsh (a-b) and mudflat (c-d) sediments.

3.3 PARTICLE SIZE FRACTIONATION

Table V indicates the percentage by volume of sediment in each particle size fraction found to lie within the required size range as determined by laser granulometry. The abundance of sediment within the required particle size range exceeds 50% of the total volume in the <2μm and 32-63μm particle size fractions, falling to less than 30% in the 8-16μm fractions.

It is clear that certain factors restrict the extent to which particle size separation can be carried out accurately. These may include the assumption in Stokes' Law of spheres falling undisturbed, which clearly may not apply in the case of natural sediments in a settling tube. In addition, the range of dry density values between samples for the fraction finer than 63μm may indicate the extent to which density vary for individual size fractions, which will limit the extent to which Stokes' Law can predict settling velocity. An inverse relationship is present between the density of the fraction finer than 63μm and the percentage in the required range in size fractions below 8μm, indicating that separation of the finer fractions may be adversely affected by differences between the specific gravity of individual fractions and that of the fraction finer than 63μm.

TABLE V
Radionuclide content and size class abundance correlations for open coastal sites

Sample site	<63μm fraction density (g cm^{-3})	Percentage of sediment within required particle size range				
		<2um fraction	2-8μm fraction	8-16μm fraction	16-32μm fraction	32-63μm fraction
4	2.5	-	-	-	-	54.1
5	2.53	-	-	-	-	54.5
10	2.35	57.9	36.8	25.6	38.4	54.8
13	2.72	41.6	30.9	24.4	33.8	46.1
14	2.48	71.7	43.4	24.4	29.1	47.8
24	2.07	70.2	48.9	33.8	45.7	56.7
31	2.36	72.8	43.3	26.3	30.9	51.5
43	2.99	46.7	44.8	29.7	36	48.6
46	1.66	87	50.7	22.9	22.5	42.9
51	1.8	65.2	43.5	24.7	22.5	19

Figures 3a and 3b express χ_{arm} values and ^{137}Cs activities in selected particle size fractions as enrichment factors relative to bulk sample data. Samples from salt marsh and open coastal sediments are plotted, mudflat sediment fractions displaying a similar trend to salt marsh samples. Figure 3a illustrates a gradual increase in χ_{arm} and ^{137}Cs in successively finer fractions, most notably the fractions finer than 16μm. This gives rise to a linear relationship between the two variables similar to that resulting from bulk sample analysis, despite the inherent problems of carrying out accurate particle size fractionation. An outlying clay-sized fraction representing an enhanced degree of ^{137}Cs enrichment implies some potential variance from the general relationship. A reduction in activity by 10% would eliminate this deviation, indicating that the magnitude of this variation in terms of ^{137}Cs activity is not very significant.

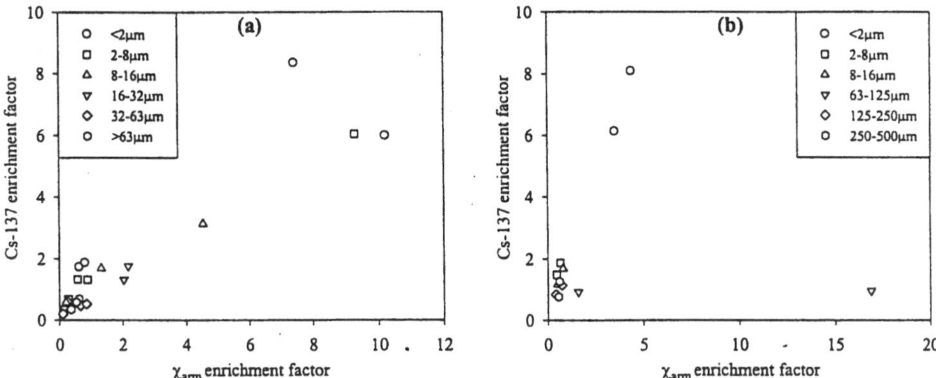

Fig. 3. Enrichment factors of ^{137}Cs and magnetic remanence in selected particle size fractions from (a) salt marsh and (b) open coastal environments. Enrichment factor = size fraction concentration/bulk sample concentration.

Figure 3b, by contrast, highlights the concentration of ^{137}Cs activity and χ_{arm} in the finest size fractions of coastal sediments, with relatively little variation in enrichment factors amongst the majority of other particle size fractions in this depositional environment. The only exception to this trend results from a highly enriched magnetic remanence relative to ^{137}Cs activity. This is due to the magnetic influence of coal mine tailings found along the Cumbrian coastline from where this sample derives; these tailings being concentrated in one size fraction by the sorting effect of wave action. A similar degree of ^{137}Cs activity enrichment amongst the majority of coastal sediment particle size fractions indicates the availability of sorption sites, possibly associated with organic particles or fine inclusions in the coarse matrix, which permit measurable ^{137}Cs concentrations in the coarsest size fractions.

4. Conclusion

The relationships between particle size, radionuclide content and magnetic properties have been explored in bulk samples and particle size fractions representing a heterogeneous suite of Irish Sea intertidal sediments. Although ^{137}Cs displays greater particle size dependence than ^{241}Am, both radionuclides are significantly associated with the abundance of sediment finer than 32μm in low energy environments, indicating the optimum grain size fraction to utilise when correcting for textural differences in radionuclide monitoring procedures. The grain size dependence of magnetic remanence is seen to result in highly significant correlations with ^{137}Cs in these environments, underlining its potential as a grain size proxy, whilst a poorer correlation between remanence and ^{241}Am activities is evident.

Potential difficulties affecting the accuracy of particle size fractionation procedures have been identified. Size fractionation of a representative group of samples has shown the extent to which the enrichment of ^{137}Cs in smaller particle size fractions of fine-grained sediments is mirrored by χ_{arm}. Sediments from higher energy environments are characterised by preferential enrichment of ^{137}Cs and χ_{arm} in the finest particle size fractions only, thereby reducing the extent to which a consistent relationship with grain size can be observed. This highlights the need to identify alternative grain size proxies to aid the monitoring of coarse-grained sediments which may contain substantial inventories of anthropogenic radionuclides.

Acknowledgements

The authors would like to acknowledge BNFL for their support of this project.

References

Assinder, D. J., Yamamoto, M., Kim, C.K., Seki, R., Takaku, Y., Yamauchi, Y., Igarishi, S., Komura, K., Ueno, K.: 1993, *J. Rad. Nucl. Chem.* **170**, 333-346.

Aston, S. R. and Stanners, D. A.: 1982, *Est. Coast Sh. Sci.* **14**, 167-174.

Bonnett, P. J. P., Appleby, P.G., Oldfield, F.: 1988, *Sci. Tot. Env.* **70**, 215-236.

Carpenter, R. C., Burton, P.J., Strange, L.P., Pratley, F.W.: 1991, *Radionuclides in intertidal sands from Morecambe Bay to the Dee Estuary,* UKAEA Technology Report AERE-R 13803, 14pp.

Clifton, R. J. and Hamilton, E. I.: 1982, *Est. Coast Sh. Sci.* **14**, 433-446.

Comans, R. N. J. and Hockley, D. E.: 1992, *Geo. Cos. Acta* **56,** 1157-1164.

Daskalakis, K. D. and O'Connor, T. P.: 1995, *Env. Sci. Tech.* **29**, 470-477.

Gray, J., Jones, S.R., Smith, A.D.: 1995, *J. Rad. Prot.* **15**, 99-131.

Livens, F. R. and Baxter, M. S.: 1988, *Sci. Tot. Env* **70**, 1-17.

Loring, D. H.: 1991, *I.C.E.S. J. Mar. Sci.* **48**, 101-115.

Maher, B.: 1988, *Geophys. J.* **94**, 83-96.

McDonald, P., Cook, G.T., Baxter, M.S., Thomson, J.C.: 1990, *J. Env. Rad.* **12**, 285-298.

Oldfield, F., Richardson, N., Appleby, P.G.: 1993, *J. Env. Rad.* **19**, 1-24.

Oldfield, F. and Yu, L.: 1994, *Sedimentology* **41**, 1093-1108.

Ramsay, J.D.F. and Raw, G.: 1987, *Studies of environmental radioactivity in Cumbria, part 9: Physical characteristics of colloids and particulates in coastal sediments and seawater,* UKAEA Technology Report AERE-R 12086, 10pp.

COMPARISON OF TWO REAGENTS, SODIUM PYROPHOSPHATE AND SODIUM HYDROXIDE, IN THE EXTRACTION OF LABILE METAL ORGANIC COMPLEXES

G.E.M. HALL and P. PELCHAT

Geological Survey of Canada, 601 Booth St., Ottawa, ON, Canada, K1A 0E8

Abstract. The international reference lake sediment, LKSD-4, was used to compare Hg, organic C and Zn extracted from its 'soluble organic' phase by two commonly used reagents: 0.1 M $Na_4P_2O_7$ solution at pH 10 and 0.5 M NaOH solution at pH 12. While recoveries of Hg and Zn by 0.1 M $Na_4P_2O_7$ are not affected by changes in sample weight to reagent volume ratio (W/V) or contact time, those by NaOH show a marked dependency. In general, the NaOH leach extracts more organic C and Hg from LKSD-4 but less Zn. Over the range of conditions studied, the NaOH-based method extracted 4.7-9.8% C, 27-103 ng g^{-1} Hg and 19-69 μg g^{-1} Zn from LKSD-4, compared to 2.3-2.8% C, 17-24 ng g^{-1} Hg and 64-72 μg g^{-1} Zn by the $Na_4P_2O_7$ leach. Clearly, different groups of organic substances are being dissolved by these two reagents and therefore a comparison of data from different laboratories becomes meaningless. This paper suggests that more research is needed into the exact nature of metal-organic associations extracted by selective leaches and into associated artifacts of extraction such as readsorption phenomena.

Keywords: selective leach; organic; humic; fulvic; analysis; mercury; zinc

1. Introduction

The efficiency of sodium pyrophosphate solution in extracting metals bound to humic and fulvic complexes in humus, soil and sediment samples has been studied recently at the Geological Survey of Canada (Hall *et al.*, 1995; Hall *et al.*, 1996). This work was stimulated by interest amongst the mineral exploration community in distinguishing that portion of metal held in a relatively labile form, in the surficial environment, from that portion present in mineralogical structures such as crystalline oxides or silicates (e.g. Antropova *et al.*, 1992). The high scavenging capacity of humic and fulvic complexes is usually attributed to their carboxylic acid functional groups, with contributions from other groups such as -NH_2 (amine) and -SH (thiol). As metals can be adsorbed, complexed or chelated by these organic substances, design of a phase-specific leach is challenging. Oxidative destruction of the organic matter, by reagents such as H_2O_2 or NaOCl, is undesirable as other phases such as sulphides and Mn oxides may also be dissolved and precipitation can occur at high pH. Dissolution of organic matter by sodium pyrophosphate, $Na_4P_2O_7$, has been adopted by soil scientists (e.g. McKeague, 1968; Bascomb and Thanigasalam, 1978; Jeanroy and Guillet, 1981) and, moreover, the ratio of C in humic to C in fulvic components of the extract has been used to classify A and B horizons. The normal procedure employed by soil scientists is that described by McKeague *et al.* (1971) where 1 g of sample is stirred for 16 h in 100 ml of 0.1 M $Na_4P_2O_7$ at room temperature. Application of this procedure to 70 soil and sediment samples and the international lake sediment reference material, LKSD-4, demonstrated that the commonly used period of 16 h for the 0.1 M $Na_4P_2O_7$ leach could be shortened to 1 h without significant decrease in extraction efficiency (<5%); indeed, the shorter contact time is of benefit to elements such as Ni which are prone to readsorption (Hall *et al.*, 1996). Analyses of 28 1-g samples of LKSD-4 over a two year period

Water, Air and Soil Pollution **99**: 217-223, 1997.

demonstrated relative standard deviations in the range 2-6% for elements such as Zn, Pb, Cu, Fe, Mn and C.

While the $Na_4P_2O_7$ leach is being adopted widely in exploration geochemistry, the NaOH reagent remains that suggested by the International Humic Substances Society (Kersten and Forstner, 1989) and is prominent in environmental studies in sediment analysis (e.g. Dmytriw *et al.*, 1995). Kersten and Forstner (1989) warn against the use of NaOH for two main reasons: it causes unacceptable dissolution of clay minerals; and metals tend to hydrolyze and readsorb onto reactive surface sites following removal of their organic coatings. This paper compares the efficacy of extraction of Hg, Zn and C bound to 'soluble' organic matter in the geological reference material (GRM), LKSD-4, using these two popular leach solutions, 0.1 M $Na_4P_2O_7$ and 0.5 M NaOH.

2. Methods

The GRM, LKSD-4 (from the Canadian Certified Reference Materials Project, CCRMP), was used to study the effects of (a) the ratio of sample weight to volume of reagent and (b) contact time in both the 0.1 M $Na_4P_2O_7$ (at pH 10) and 0.5 M NaOH (at pH 12) leaches. This GRM contains 17.7% C, 190 ng g^{-1} total Hg and 194 µg g^{-1} Zn (Lynch, 1990). Each test was performed in triplicate. In the first half of the study, contact times of 1 h ($Na_4P_2O_7$) and 18 h (NaOH) were used, 18 h having being recommended by Schnitzer and Skinner (1968) in the development of the NaOH procedure. Ratios of sample weight:reagent volume examined were 0.005, 0.01, 0.02, 0.04 and 0.1 g ml^{-1}, the minimum sample weight taken being 0.25 g. The NaOH leach was carried out under N_2 (to minimise auto-oxidation and hydrolysis) and constant shaking was employed for both leaches.

Contact times of 1, 2, 3, 4 and 18 h were tested, using two ratios of sample weight:reagent volume: 0.005 and 0.01 for the $Na_4P_2O_7$ procedure; and 0.04 and 0.06 for the NaOH extraction. These ratios reflect those commonly used in the literature. Each solution was centrifuged at 2800 rpm for 10-15 min for separation from the residue prior to analysis.

Each leach solution was diluted with water and, for Hg determination, the solution immediately made 4 M in HCl and 0.5% in BrCl (to maintain Hg in solution as Hg^{2+}). The dilution factor was variable, the criterion being that the final solution contained less than 400 ng l^{-1} Hg, the uppermost calibration standard. Mercury was determined in the resulting solutions by ICP-MS with $NaBH_4$ as reductant in-line to form Hg^0 vapour which was swept into the central channel of the ICP torch (for details, see Hall and Pelchat, 1997). Zinc was determined by flame AAS at 213.9 nm. Organic C in solution was determined using the Schimadzu Model 5000 TOC instrument, based on combustion at 680 °C and non-dispersive infra-red detection.

3. Results and Discussion

The influence of changing the sample weight to reagent volume ratio on the extraction of Hg and organic C in LKSD-4 is illustrated in Fig. 1 for both methods. Change in this ratio, from 0.005 to 0.1, has little effect on the Hg or C concentrations extracted by $Na_4P_2O_7$ which are about 18 ng g^{-1} and 2.5%, respectively (Fig. 1a). However, there is a marked increase in Hg (exponential?) extracted by NaOH solution when the sample weight:reagent volume ratio decreases below 0.04, resulting in a maximum average Hg concentration of 102 ng g^{-1} as compared to a minimum of 55 ng g^{-1} (Fig. 1b). This increase is not evident in the behaviour of C which is constant at about 8.4% except for the increase to 9.8% at the minimum sample weight:volume ratio of 0.005. Clearly, a much greater component of the organic C is being extracted by the NaOH leach (8.4 vs. 2.5% C) but the lack of coherence between the Hg and C graphs for this leach suggests that much of the corresponding Hg is not associated with organics. The results for Zn are presented in Fig. 2.

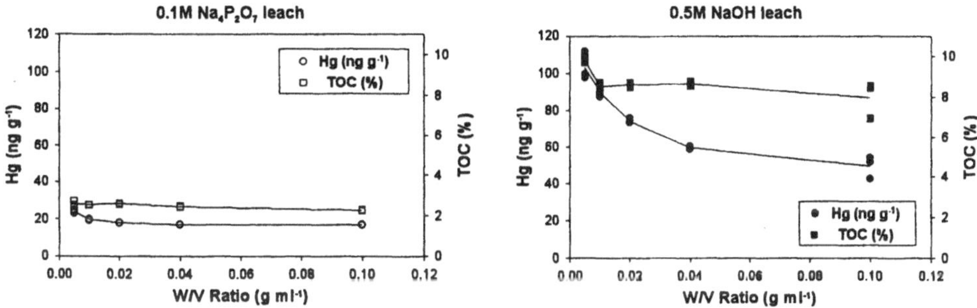

Fig. 1 Effect of sample weight to reagent volume ratio (W/V) on Hg and organic C extracted from LKSD-4 by (a) 0.1 M $Na_4P_2O_7$ and (b) 0.5 M NaOH solutions.

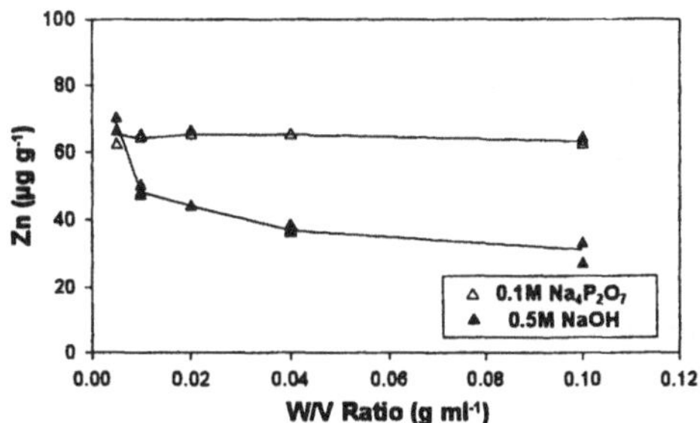

Fig. 2 Effect of sample weight to reagent volume ratio (W/V) on Zn extracted from LKSD-4 by 0.1 M $Na_4P_2O_7$ and 0.5 M NaOH solutions.

Note the plateau again evident for the $Na_4P_2O_7$ graph and a sharp increase in Zn extracted by NaOH at the minimum sample:volume ratio which coincides with the increase in C. Here we see the converse in trends: Zn in the 'labile organic' component of LKSD-4 is about 65 µg g^{-1} by the $Na_4P_2O_7$ extractant, whereas it ranges from a low of only 31 µg g^{-1} to 69 µg g^{-1} using NaOH. The Hg and Zn graphs suggest that more than one 'phase' is being dissolved by NaOH and that readsorption could be occurring. If 0.1 M $Na_4P_2O_7$ is extracting the humic and fulvic component of LKSD-4 with adequate specificity (as indicated by McKeague, 1968; McKeague *et al.*, 1971; Bascomb and Thanigasalam, 1978), then the 0.5 M NaOH leach is over-estimating the associated labile Hg and under-estimating Zn. Clearly it is undesirable to select conditions of extraction where slight deviations cause major changes in recovery, as would be the case when operating below a ratio of 0.04 in Fig. 1b (see Dmytriw *et al.*, 1995).

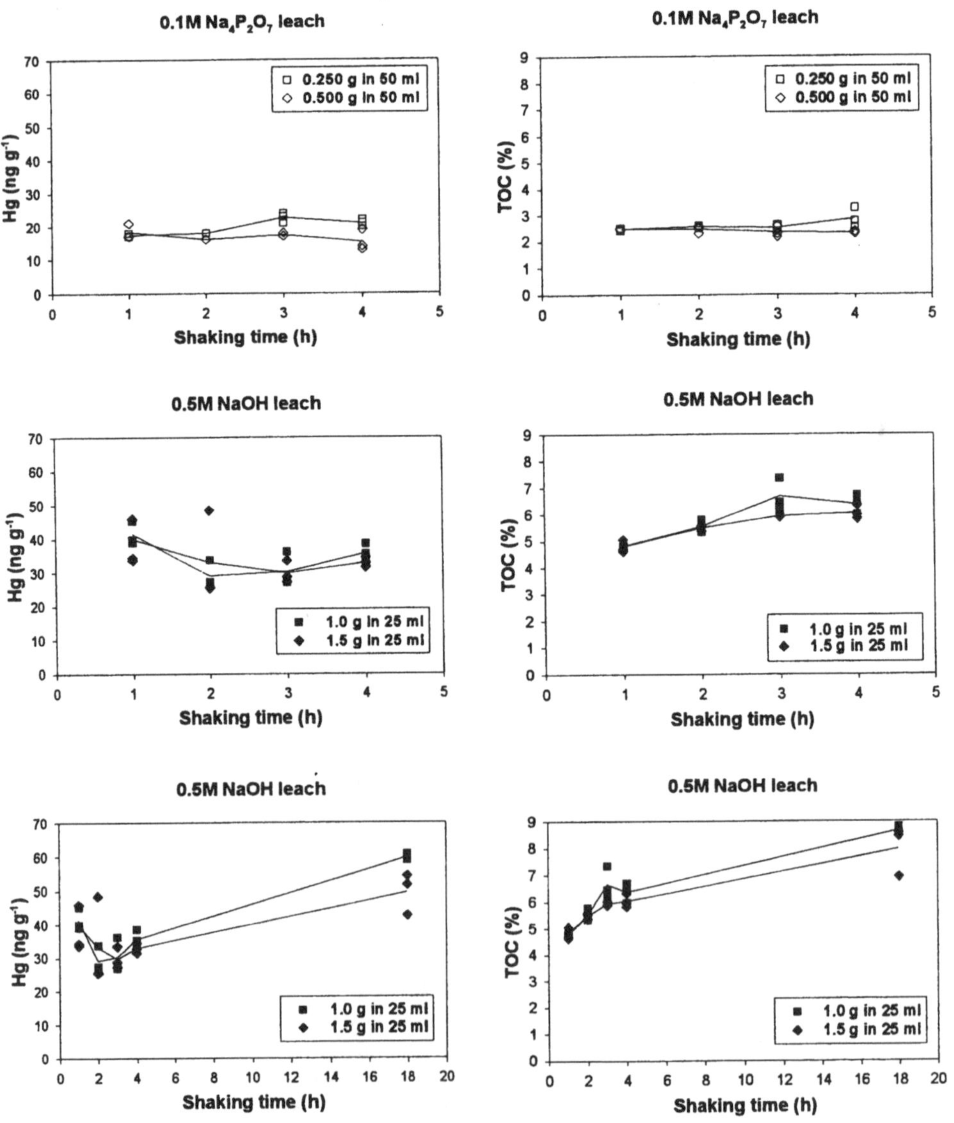

Fig. 3 Effect of contact time on Hg and organic C extracted from LKSD-4 by (a) 0.1 M $Na_4P_2O_7$ and (b)-(c) 0.5 M NaOH solutions.

Contact time has a slight effect on the extraction of Hg and C by 0.1 M $Na_4P_2O_7$ at the lower sample weight of 0.25 g of LKSD-4 which may be a reflection of heterogeneityand noisier measurements at the lower concentrations (Fig. 3a). Extraction of Hg and C by 0.5M NaOH differs over the 4 h period examined: Hg concentrations fall at hour 2 and 3 compared to that at hour 1 and rise again slowly whereas C concentration increases continuously with time from 1 h on (Figs. 3b, c). Triplicate analyses indicate inferior reproducibility by the NaOH leach (Fig. 3).

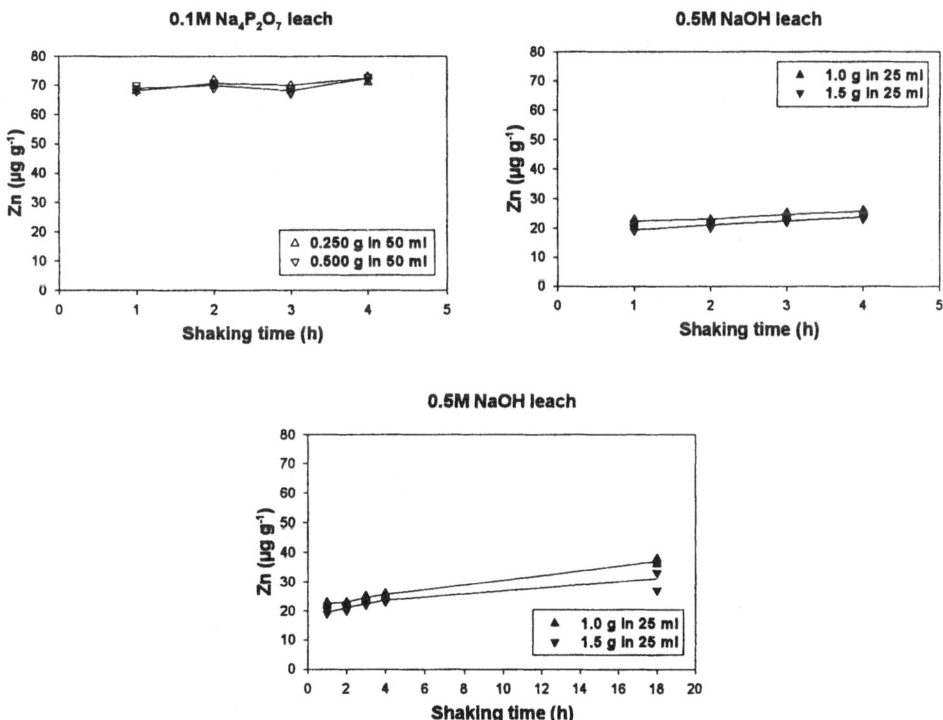

Fig. 4 Effect of contact time on Zn extracted from LKSD-4 by 0.1 M $Na_4P_2O_7$ and 0.5 M NaOH solutions.

The effect of contact time on Zn extraction by both methods is shown in Fig. 4: Zn levels are unchanged (70±2 µg g^{-1}) with time using $Na_4P_2O_7$ but increase gradually with NaOH reagent, from 24-26 µg g^{-1} at 1 h to a concentration of 30-37 µg g^{-1} at 18 h.

Relatively little is known about the extent to which the organic matter extracted by 0.5 M NaOH differs from that dissolved by 0.1 M $Na_4P_2O_7$. In a comparison of the two reagents applied to several different soil types, Schnitzer and Schuppli (1989a) found there to be a tendency for the humic acids derived from the 0.1 M $Na_4P_2O_7$ leach to be more aromatic and of higher molecular weight but did not find the significant difference in recoveries of organic C we find in LKSD-4 (2.5 vs. 8.4% C). They recommended the 0.1 M $Na_4P_2O_7$ leach as an attractive alternative to 0.5 NaOH as the probabilities for auto-oxidation and hydrolysis are greatly diminished. A sequential extraction procedure then developed by Schnitzer and Schuppli (1989b), to define various organic fractions in soil, placed 0.1 M $Na_4P_2O_7$ before 0.5 M NaOH to specifically dissolve organic matter complexed to metals and clays as opposed to "free" organic matter including less decomposed material which would finally be leached with the latter reagent. In outlining knowledge gaps yet to be addressed, Schnitzer (1991) calls for future research into metal, clay and humic sustance (humic acids, fulvic acids and humin) interactions. Given the complexity of these interactions (via adsorption, chelation and complexation) and the range of chemical structures involved (e.g. aliphatic-, aromatic-, phenolic-, carbohydrate-C), the classification of the 'soluble' or 'poorly soluble organic' phases liberated in commonly used 'selective' leaches is very oversimplified. It is hardly surprising, therefore, that dissimilar metal concentrations are obtained by the application of these different reagents. Further attention should be given to exactly what group(s) of metal-organic substances are of interest, to, for example, distinguish between diagenesis in a sediment core and a process taking place prior to deposition.

4. Conclusion

The 0.1 M $Na_4P_2O_7$ leach at pH 10, designed to dissolve the 'soluble organic' phase of a soil or sediment, shows minimal change in the amount of organic C, Hg or Zn leached from LKSD-4 when parameters of sample weight to reagent volume (W/V) and contact time are altered. This is in contrast to the 0.5 M NaOH leach at pH 12, where there appears to be an exponential increase in Hg (and Zn) extracted with decrease in W/V which does not correspond to the behaviour of organic C. Contact time has a less dramatic effect on recoveries by NaOH which are seen to increase generally with time for all three elements, though Hg does show a slight decrease initially after 1 h. Thus, the ratio of W/V selected in the NaOH procedure will have a significant effect on the results. Over the range of conditions studied, the NaOH-based method extracted 4.7-9.8% C and 27-103 ng g^{-1} Hg from LKSD-4, compared to 2.3-2.8% C and 17-24 ng g^{-1} Hg by the $Na_4P_2O_7$ leach. Unlike Hg, extraction of Zn by the $Na_4P_2O_7$ leach was consistently higher, at 64-72 $\mu g\ g^{-1}$ over this range, compared to 19-69 $\mu g\ g^{-1}$ by NaOH. Clearly the two reagents are extracting different groups of organic substances and results for both cannot be equated to the same nominal phase.

References

Antropova, L.V., Goldberg, I.S., Voroshilov, N.A., and Ryss, Ju.S.: 1992, *J. Geochem. Explor.* **43**, 157-166.
Bascomb, C.L. and Thanigasalam, K.: 1978, *J.Soil Sci.* **29**, 382-387.
Dmytriw, R., Mucci, A., Lucotte, M., and Pichet, P.: 1995, *Water Air Soil Pollut.* **80**, 1099-1103.
Hall, G.E.M., MacLaurin, A.I, and Vaive, J.E.: 1995, *J. Geochem. Explor.* **54**, 27-38.
Hall, G.E.M., Vaive, J.E., and MacLaurin, A.I.: 1996, *J. Geochem. Explor.* **56**, 23-36.
Hall, G.E.M. and Pelchat, J.C.: 1997, *J. Anal. At. Spectrom.* **12**, *page info coming (Jan issue)*

Jeanroy, E. and Guillet, B.: 1981, *Geoderma* **26**, 95-105.
Kersten, M. and Forstner, U.: 1989, *Trace Element Speciation: Analytical Methods and Problems*, G. Batley [ed], CRC Press, 245-317.
Lynch, J.J.: 1990, *Geostand. Newsl.* **14**, 153-167.
McKeague, J.A.: 1968, *Can. J. Soil Sci.* **48**, 27-35.
McKeague, J.A., Brydon, J.E., and Miles, N.M.: 1971, *Soil Sci. Soc. Amer. Proc.* **35**, 33-38.
Schnitzer, M.: 1991, *Soil Sci.* **151**, 41-58.
Schnitzer, M. and Schuppli, P.: 1989a, *Can. J. Soil Sci.* **69**, 253-262.
Schnitzer, M. and Schuppli, P.: 1989b, *Soil Sci. Soc. Am. J.* **53**, 1418-1424.
Schnitzer, M. and Skinner, S.I.M.: 1968, *Soil Sci.* **105**, 392-396.

MODELING THE DYNAMICS OF THE SORPTION OF HYDROPHOBIC ORGANIC CHEMICALS TO SUSPENDED SEDIMENTS

WILBERT LICK, ZENITHA CHRONEER, and VENKATRAO RAPAKA

Department of Mechanical and Environmental Engineering, University of California, Santa Barbara, CA 93106

Abstract. The adsorption and desorption of hydrophobic organic chemicals to suspended sediments are often quite slow and, because of this, these finite rates of sorption must be included for accurate predictions of the transport of contaminants in surface waters in many cases. A theoretical model of adsorption and desorption which includes finite sorption rates has previously been developed and shown to be in good agreement with experimental results. However, this model generally requires quite large amounts of computer time. In the present study, a simplified but reasonably accurate version of this model is developed which requires much less computer time and is therefore suitable for large-scale calculations of the transport of contaminants in surface waters. Applications of this model are presented in order to illustrate the effects of finite sorption rates on this transport.

Key words: sorption, h· drophobic organic chemicals, suspended sediments, finite sorption rates, Green Bay, transport, numerical model

Introduction

Recent long-term adsorption and desorption experiments with four hydrophobic organic chemicals (hexachlorobenzene and three PCB congeners) have clearly demonstrated that the rates of adsorption and desorption for these chemicals to natural suspended sediments can be relatively slow (days to more than a year) and therefore that the assumption of chemical equilibrium may not be valid in many realistic situations (Jepsen et al, 1995; Tye et al, 1996; Jepsen and Lick, 1996; Borglin et al, 1996). Based on this experimental work, a theoretical model of the adsorption, desorption, and partitioning of hydrophobic organic chemicals (HOCs) to suspended sediments has been developed and used to compare with experimental results (Lick and Rapaka, 1996). The model is relatively simple with retarded chemical diffusion within the particle/floc as the basic transport process. A complicating factor is the variations in the particle/floc size and density distributions. These distributions are dependent on local conditions and may also vary with time. In the modeling, they were chosen on the basis of measurements taken during the sorption experiments. In general, good agreement between the numerical and experimental results was obtained for all cases considered thus helping to substantiate the model and the processes implicit in the model.

This sorption model was developed primarily to understand and quantify the results of laboratory sorption experiments. However, it is unsuitable for use in large-scale

Water, Air and Soil Pollution **99**: 225-235, 1997.

contaminant transport calculations because of the large amounts of computer time required for the diffusion calculations. In typical calculations of the transport of contaminants in surface waters, several thousand grid points are needed and the calculations generally need to be performed for several thousand time steps or more. The solution of the diffusion equation for each class of particles at each of these grid points for long periods of time requires excessive computer time and makes the overall transport calculation impracticable. In order to greatly reduce the required computer time while still maintaining reasonable accuracy, an approximate sorption model has been developed. By means of this model, calculations have been made in order to compare with the results of the diffusion model and to illustrate the effects of finite sorption rates on the transport of contaminants in surface waters.

In the following section, recent experimental and modeling work on sorption is briefly reviewed. The modified and computationally more efficient sorption model is then described in the third section. Applications of this model to the transport of contaminated sediments are presented in the fourth section. A summary and concluding remarks are given in the final section.

Previous Work

EXPERIMENTAL WORK

In order to obtain a quantitative understanding of the processes governing adsorption, desorption, and partitioning of HOCs to suspended sediments, long-term adsorption and desorption experiments are necessary. Particularly relevant to the present study are the long-term adsorption experiments done by Jepsen et al (1995), Tye et al (1996), and Jepsen and Lick (1996) and the long-term desorption experiments done by Borglin et al (1996). The adsorption experiments were done by means of a batch-mixing procedure. The HOCs used were HCB and three PCB congeners (a monochlorobiphenyl, MCB; a dichlorobiphenyl, DCB; and a hexachlorobiphenyl, HPCB). The sediments were from the Detroit River in Michigan and were mixed with either filtered tap water or Optima pure water. Experiments were conducted for different periods of time up to six months in order to insure that a steady state and complete equilibration between the chemical, sediment, and water were eventually attained. The results of the experiments were generally reported as the logarithm of the partition coefficient as a function of time. For a mixture of sediment and water, the partition coefficient is defined as $K_p = C_s/C_w$ where C_s is the mass of the HOC sorbed to the sediment divided by the mass of the sediment, and C_w is the mass of the HOC dissolved in the water divided by the volume of water.

From these experiments, equilibrium partition coefficients and adsorption rates for the four HOCs were determined. In addition, it was found that the adsorption process was slow (from less than one day to as long as 30 days), that the sorption isotherms were linear over a wide range of contaminant concentrations, and that the most significant factors affecting the adsorption and partitioning of HOCs were colloids from the water (and especially their flocculation), colloids from the sediments, particle and floc size and density distributions, the organic content of the sediment, and properties of the chemical.

The long-term desorption experiments were done by means of a purge-and-trap procedure. HOCs used were HCB, MCB, and HPCB. As with the adsorption experiments, natural sediments from the Detroit River and either filtered tap water or Optima pure water were used. The experiments demonstrated the dependence of the desorption rates on the particle/floc size and density distributions, the type of water, and the organic content of the sediments. It was also demonstrated that (a) desorption was more rapid for sediments which were only partially equilibrated than for those which were fully equilibrated, and (b) desorption rates depended on the equilibrium partition coefficient of the chemical, i.e., the larger the partition coefficient the slower the desorption. For HCB, 90% desorption times were on the order of 60 days. For HPCB, which has a much larger partition coefficient, this desorption time was on the order of a year.

A MODEL OF THE SORPTION PROCESS

In order to more quantitatively understand and describe the experimental results and also to be able to accurately predict the sorption, transport, and fate of HOCs, a quantitative model of the sorption process was developed (Lick and Rapaka, 1996) and is summarized here. The model includes effects of (a) chemical diffusion through the pores within a particle/floc, (b) quasi-equilibrium chemical sorption within a particle/floc, (c) approximate particle and floc size and density distributions and changes with time of these quantities, and (d) effects of organic content and chemical properties. Because of the assumption of quasi-equilibrium chemical sorption, the transport of the chemical within the particle can be described by a single time-dependent diffusion equation with no reaction term but with an effective diffusion coefficient, D, given by

$$D = \frac{D_m}{1 + \left(\frac{1-\phi}{\phi}\right) \rho_p K_p} \tag{1}$$

where D_m is the molecular diffusion coefficient (cm^2/s) of the chemical in the fluid within the particle without consideration of any reactions but corrected for tortuosity, ϕ is the porosity of the particle, and ρ_p is the mass density of the solid particle (approximately 2.6 kg/L). For a natural suspension of sediments, the particles and flocs have a distribution of sizes and densities. These quantities change with time. These distributions and their changes with time have a significant effect on sorption dynamics and were included in the modeling. The particle/floc size distributions can be measured directly, while the density distributions can be inferred from measurements of the settling speeds of the particle/flocs.

Calculated results of the model were compared to results from both adsorption and desorption experiments. In all cases, good agreement with available experimental data was obtained. An example of this agreement is shown in Figure 1, where the partition coefficients for HCB, MCB, DCB, and HPCB as calculated from the model and as determined experimentally are plotted as a function of time for the case of adsorption to suspended sediments. It can be seen that there is good agreement for all four HOCs and for all times.

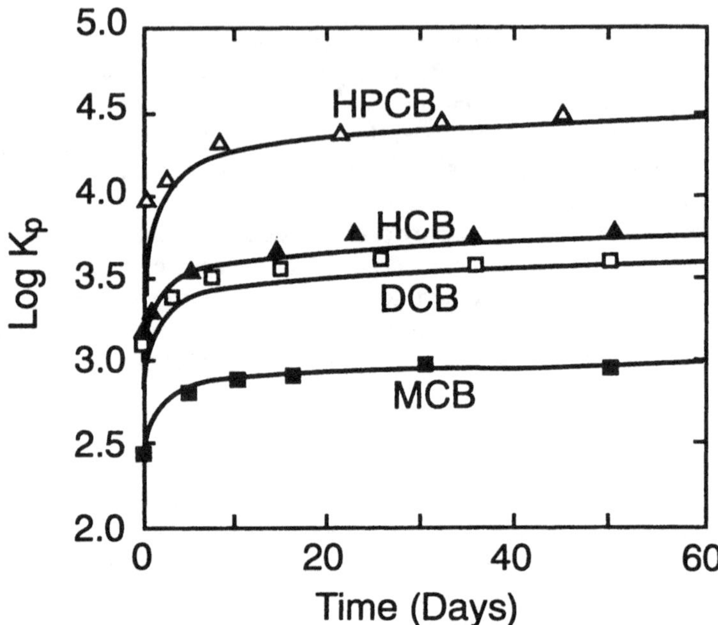

Fig. 1. Partition coefficients for the adsorption of HCB and three PCB congeners (MCB, DCB, and HPCB) to sediments from the Detroit River. Log K_p is shown as a function of time. Experimental data is shown as open and closed symbols while the modeling results are shown as solid lines.

A Computationally Efficient Model

The sorption model described above (hereafter referred to as the diffusion model), when used as part of a general calculation of the transport of contaminants in surface bodies of water, requires considerable computer time and makes the overall transport calculation impracticable. In order to greatly reduce the required computer time while still maintaining reasonable accuracy, an approximate sorption model has been developed and is described here. Consider sediments consisting of single-size particles first. In this case, it is assumed that the time-dependent variation of C_s, the average contaminant concentration in the particle/floc, due to the transfer of the contaminant from the water to the solid particle is given by

$$\frac{dC_s}{dt} = k\left(K_p C_w - C_s\right) \tag{2}$$

where k is a mass transfer coefficient (s^{-1}). The quantity ($K_pC_w - C_s$) is a forcing function which is proportional to the difference between C_s and its equilibrium value, K_pC_w, and is zero when the contaminant sorbed to the particle/floc is in chemical equilibrium with the contaminant dissolved in the water. This model will be referred to as the mass transfer model.

The coefficient k is not known from basic principles or directly from experimental results but is chosen so as to obtain good agreement between the solutions of the above equation and the diffusion equation. Good agreement with experimental results is therefore indirectly obtained. In order to determine k, consider the case of desorption from a sorbed concentration of C_s to a dissolved contaminant concentration in the water of zero. Integration of the above equation then gives

$$C_s = C_o \, e^{-kt} \tag{3}$$

where C_o is the initial value of C_s. The time for 75% desorption of C_o is given by

$$t^* = \frac{1.386}{k} \tag{4}$$

For the diffusion equation, the time for 75% desorption is given approximately by (Carslaw and Jaeger, 1959)

$$t^* = 0.0225 \, \frac{d^2}{D} \tag{5}$$

where d is the diameter of the particle. For the results of the diffusion and mass transfer models to agree at t^*, Eqs. (4) and (5) must agree and therefore k is determined by

$$k = \frac{D}{0.0165 d^2} \tag{6}$$

In order to determine the accuracy of this approximation, desorption calculations were made with both the diffusion and the mass transfer models. Results are shown in Figure 2 for the desorption of HCB ($K_p = 10^4$, $D = 2 \times 10^{-14}$ cm^2/s) for particle/flocs with a diameter of 14.56 μm. It can be seen that there is reasonable agreement between the results of the two models for all time and that they coincide for 75% desorption. For a mixture of particles of different size classes, Eqs. (2) and (6) are assumed to be valid for each size class. For this case, there is again good agreement between the results of the two models for all time; the two models coincide at a percent desorbed slightly greater than 75% due to the fact that the rates of desorption are different for the three sizes. Calculations have also been made for the case of adsorption where the sorbed concentration was initially zero but increased with time while the dissolved concentration was constant. For both the single-size and multi-size particles, the results of the mass transfer and diffusion models agree well for all time and coincide at approximately 75% adsorption.

However, in the adsorption experiments described above, the contaminant concentrations in the water were not constant and often changed by large amounts (often by more than an order of magnitude) and quite rapidly with time. The reason for this is that, since the total amount of contaminant was constant, the dissolved contaminant

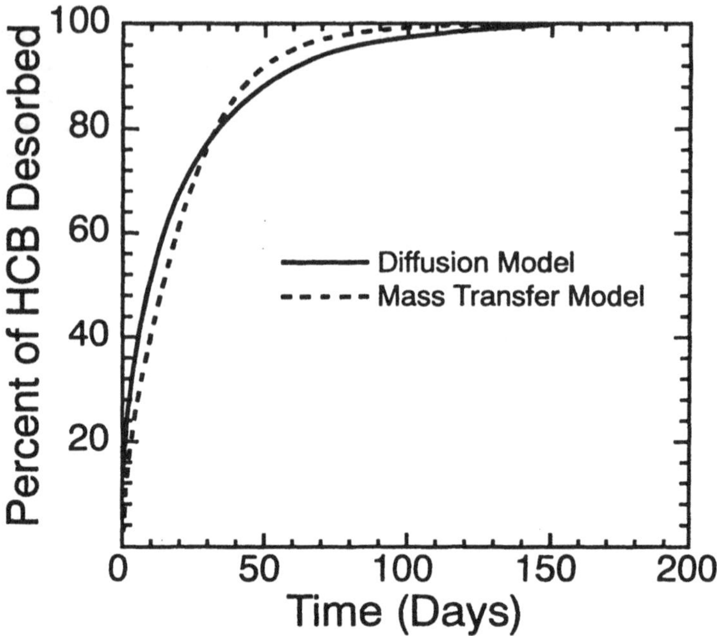

Fig. 2. Desorption of HCB from suspended sediments as calculated by the diffusion and mass transfer models. Shown is the percent of sorbed HCB which has desorbed as a function of time. For this calculation, $D = 2.0 \times 10^{-14}$ cm^2/s, log $K_p = 4.0$, and the particle diameter was 14.56 μm.

concentration must decrease as the contaminant is sorbed to the solid. For large values of K_pC where most of the contaminant was eventually sorbed to the solid, the variations of C_w were quite large, especially at small time. In this case, the mass transfer model with k given by Eq. (6) did not give good results.

In order to obtain a better approximation for k for the single particle size, adsorption case, the expression for k given by Eq. (6) was modified to

$$k = \frac{D}{0.0165\, d^2}\left(1 + a\left|\frac{dC_w}{dt}\right|\right) \tag{7}$$

where a is a parameter to be determined by comparison of the results of the diffusion and mass transfer models. It is convenient to write a as

$$a = \frac{a_0}{\left(\frac{m}{V}\right) K_p\, C} \tag{8}$$

where m/V is the total mass of the chemical per unit volume and C is the sediment concentration. Much of the dependence of a on m/V, K_p, and C has been included in the denominator and, as a result, a_o is a slowly varying parameter. For the general case considered here of particles of a single size, calculations show that a_o is generally between

0 and 2.4 and is a slowly varying function of K_p and C, but is not a function of particle size or D. With this modification, generally good agreement was obtained between the results of the diffusion and mass transfer models.

For a sediment consisting of a mixture of different particle sizes, Eqs. (2), (7), and (8) are assumed to be valid for each size class. Again, the coefficients a_{oi} must be determined by a comparison of the results of the diffusion and mass transfer calculations. This has been done for HCB, MCB, DCB, and HPCB and for a wide range of conditions. The resulting values of a_{oi} can be approximated as simple polynomials (Rapaka and Lick, 1996).

A Transport Calculation

In order to illustrate the effects of finite sorption rates on the transport of HOCs in surface waters, calculations have been done for the flow of sediments and sorbed HCB from the Fox River into Green Bay. Green Bay (Figure 3) is a narrow extension of Lake Michigan and is 188 km long with a maximum width of 37 km. The southern part of the bay is relatively shallow, with an average depth of approximately 10 m and a maximum depth of 33 m, while the northern part of the bay is deeper, with an average depth of 17 m and a maximum depth of almost 45 m. The northeastern part of the bay opens into Lake Michigan. The Fox River flows into the southern part of the bay and is the major contributor of contaminants to the bay. Calculations have previously been made of the hydrodynamics and sediment transport in the bay (Chroneer et al, 1995), and these same models are used in the present calculations.

In order to be specific, consider the following conditions. The water in the bay has a dissolved HCB concentration of 10 ng/L. The water from the Fox flows into the bay at a rate of 500 m^3/s and has a sediment concentration of 100 mg/L. In order to illustrate the effects of finite sorption rates without the complicating effects of sediment flocculation, it is here assumed that no flocculation occurs. Three different sediment size classes are present with diameters of 4, 15, and 30 μm; settling speeds of 20, 250, and

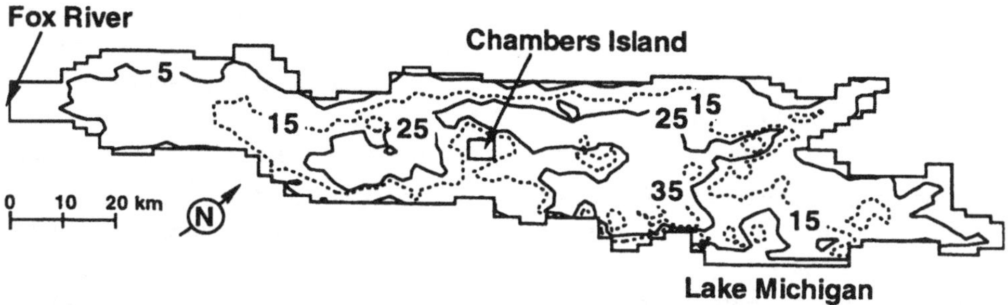

Fig. 3. Green Bay bathymetry. Depth contours of 5, 15, 25, and 35 m are shown.

1000 μm/s; and size fractions of 10, 60, and 30% respectively. The coefficient k depends on the diffusion coefficient and diameter of the particles is shown in Eq. (6). In the present calculations, it is assumed that $D = 2 \times 10^{-14}$ cm^2/s, a reasonable value from experimental results. The sediments from the Fox are free of HCB. As these sediments flow into the bay, they adsorb HCB from the waters in the bay at a rate depending on their mass transfer coefficient, k, and the difference between the contaminant concentration of the sediment and its equilibrium value with the water (see Eq. (2)). As a result of this mass transfer, the concentrations of HCB in the water, C_w, and on the sediments, C_s, change as a function of time throughout the bay. However, the major effect initially is near the mouth of the Fox where the sediments enter the bay.

In order to illustrate these effects, consider the changes in C_s as a function of time at a distance of 20 km from the Fox (Figure 4a) and at 10 days as a function of distance from the Fox in the direction of the long axis of the bay (Figure 4b). Figure 4a shows the equilibrium value of C_s and the non-equilibrium values of the average C_s for the suspended sediments of all sizes and the C_s for each of the three particle size classes. For the equilibrium case, $C_s = K_p C_w$ and C_s is therefore proportional to C_w. Since $K_p = 10^4$ and C_w is initially 10 ng/L, C_s is 100 μg/kg at zero time. At any location, C_w decreases with time as sediments from the Fox (which are initially clean) are transported into the Bay and continually adsorb contaminant along the way. Since C_s is proportional to C_w, C_s also decreases with time at any specific location. For the non-equilibrium case, C_s is zero initially for all size classes. As time increases, it can be seen that the fine sediments adsorb relatively rapidly and approach the equilibrium value. However, the medium and coarse sediments adsorb slowly and are far from equilibrium even after ten days. The average value of C_s is also far from equilibrium.

The changes in C_s with distance along a line from the Fox (see Figure 3) at a time of 10 days are shown in Figure 4b for both the equilibrium and non-equilibrium cases. Due to the finite rates of adsorption, the non-equilibrium values for C_s are generally below the equilibrium value of C_s and always below $K_p C_w$. Close to the Fox, because of the slow adsorption by the medium and coarse particles, the average contaminant concentration of the particles is far from equilibrium. Further out in the bay, the average value of C_s approaches that of the fine particles and the equilibrium value of C_s, even though the medium and coarse particles are still far from equilibrium. The reason for this is that most of the medium and coarse particles have settled out of the water column and the resulting sediments in suspension are primarily fine particles. The average value of C_s is thus dominated by C_s for the fine particles.

The spatial variations of the average C_s for both the equilibrium and non-equilibrium cases at a time of 10 days are compared in Figure 5. The general increase in C_s with distance from the Fox is evident. The maximum differences in C_s between the equilibrium and non-equilibrium cases occur approximately 20 km from the Fox. As time increases, C_s will decrease as sediments from the Fox continually adsorb contaminants dissolved in the Bay waters.

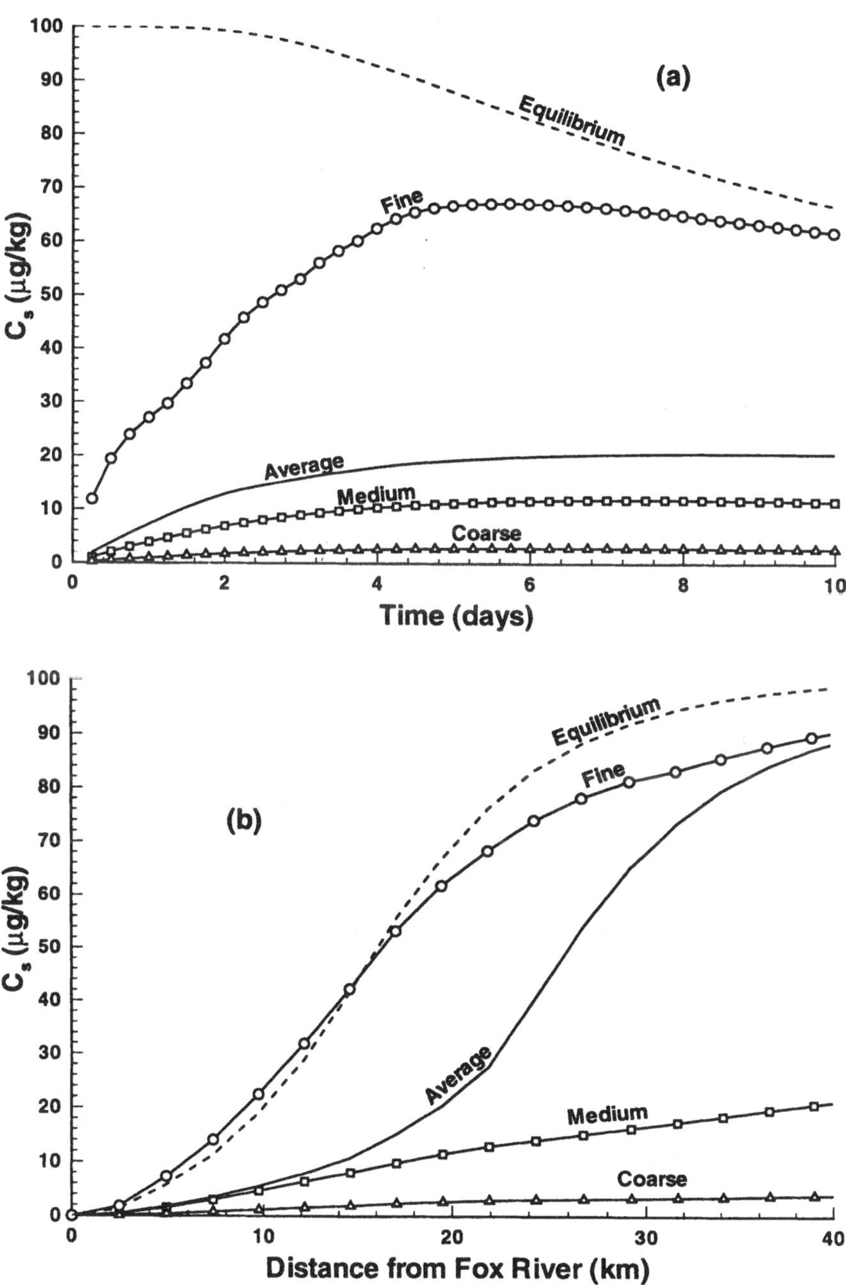

Fig. 4. Contaminant concentrations for the case of clean sediments from the Fox River entering into Green Bay. The waters in the bay initially have a dissolved HCB concentration of 10 ng/L. The sediments from the Fox are free of HCB. (a) C_s as a function of time at a distance of 20 km from the Fox. Contaminant concentrations for each size fraction and the average for all sediments (both non-equilibrium and equilibrium) are shown. (b) C_s as a function of distance from the Fox at a time of 10 days. Contaminant concentrations for each size fraction and the average for all sediments (both non-equilibrium and equilibrium) are shown.

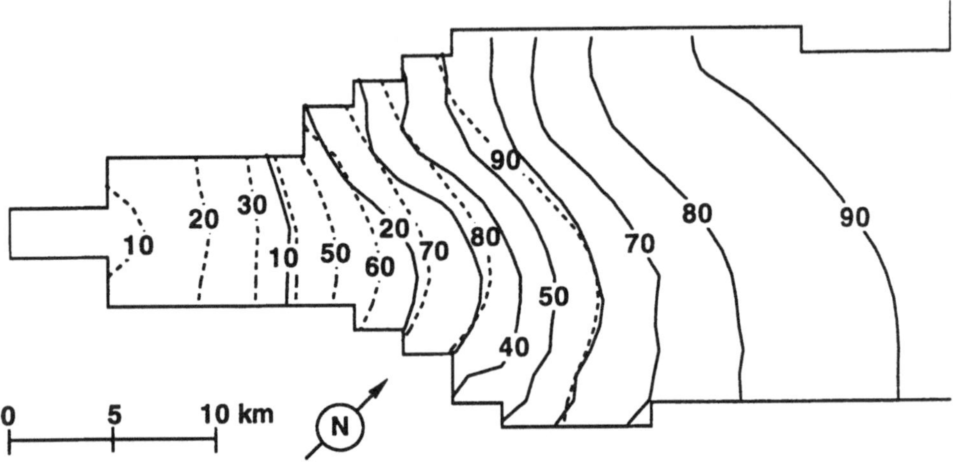

Fig. 5. The spatial distribution of C_s (μg/kg) in Green Bay for the equilibrium and non-equilibrium cases. Solid lines are the non-equilibrium case, while the dashed lines represent the equilibrium case.

Summary and Concluding Remarks

For many realistic conditions in the transport of contaminants in surface waters, the times required for the sorption of hydrophobic organic chemicals to suspended sediments may be comparable to or longer than transport time scales. Because of this, these sorption rates must be explicitly considered in attempting to accurately predict the transport and fate of contaminants. A chemical equilibrium assumption may not be valid. In the present study, a simplified sorption model was presented and also used to illustrate the effects of finite sorption rates. Values of parameters were based on experimental results.

Only the non-equilibrium transfer of contaminants to suspended sediments was considered here; the bottom sediments were assumed to be in chemical equilibrium. Recent work has demonstrated that the sorption of hydrophobic organic chemicals to and from bottom sediments requires even longer times than those for suspended sediments. This may significantly affect the flux of contaminants from the bottom sediments to the overlying water and the availability of these contaminants to benthic organisms.

Acknowledgements

This research was funded by the U.S. Environmental Protection Agency and by the University of California Campus Laboratory Collaboration Program.

References

Borglin, S., Wilke, A., Jepsen, R., and Lick, W.: 1996, *Environ. Toxicol. Chem.*, **15**, in press.

Carslaw, H.S., and Jaeger, J.C.: 1959, *Conduction of Heat in Solids*, Oxford University Press, London, Great Britain.

Chroneer, Z., Cardenas, M., Lick, J., and Lick, W.: 1995, *Estuarine and Coastal Modeling*, edited by M. Spaulding and R. Cheng, 313-324.

Jepsen, R., Borglin, S., Lick, W., and Swackhamer, D.: 1995, *Environ. Toxicol. Chem.*, **14**, 1487-1497.

Jepsen, R. and Lick, W.: 1996, *J. Great Lakes Res.*, **22**, 341-353.

Lick, W. and Rapaka, V.: 1996, *Environ. Toxicol. Chem.*, **15**, 1038-1048.

Rapaka, V. and Lick, W.: 1996, *The Sorption of Hydrophobic Organic Chemicals to Suspended Sediments*, University of California at Santa Barbara Report.

Tye, R., Jepsen, R., and Lick, W.: 1996. *Environ. Toxicol. Chem.*, **15**, 643-651.

ANALYSIS OF ORGANIC MICROPOLLUTANTS IN SEDIMENT SAMPLES OF THE VENICE LAGOON, ITALY

E. FATTORE, E. BENFENATI, G. MARIANI, E. COOLS, G. VEZZOLI and R. FANELLI

Istituto di Ricerche Farmacologiche "Mario Negri"
Via Eritrea 62, 20157 Milano, Italy.

Abstract. Surface sediment samples were collected at six locations of the Lagoon of Venice reflecting potential different contamination sources and representative of different hydrological situations. Analysis of polychlorinated dibenzo-*p*-dioxins and dibenzofurans (PCDD/PCDFs), polychlorinated biphenyls (PCBs), polycyclic aromatic hydrocarbons (PAHs), linear alkylbenzenes (LABs) and coprostanol have been carried out by gas chromatography-mass spectrometry (GC-MS) to assess the influence of various pollution sources: urban, industrial and combustion processes. PAHs, PCBs, coprostanol, and LABs showed the highest levels in the sample collected within the city of Venice (Canal Grande) indicating a very heavy contamination due to combustion sources and to the domestic waste waters directly entering the canals of the city. The highest levels of PCDD/PCDFs were found in samples collected near the industrial area of Porto Marghera. The investigation on the PCDD/PCDFs homologue profiles suggested the presence of two different sources for these pollutants, one due to combustion processes and another one due to the chemical processes of Porto Marghera.

Key words. Venice Lagoon, sediments, PAHs, PCDD/PCDFs, PCBs, LABs, coprostanol.

1. Introduction

The Venice Lagoon (North-East Italy) has a surface area of 550 km^2 and an average depth about 1 m (Figure 1). It is affected by different sources of pollution mainly due to the untreated domestic waste waters coming from city of Venice, to the presence of the big industrial district of Porto Marghera and to the exhausts of the motorboats crossing the lagoon. They contribute to deteriorate the lagoon system (Raccanelli *et al.*, 1989; Capodaglio *et al.*, 1990 a; b; Pavoni *et al.*, 1990 a; b; Calvo *et al.*, 1991 a; b; Pavoni *et al.*, 1992) that is particularly vulnerable because of his shallow waters and poor water exchange in many areas.

The selected compounds analysed were polychlorinated dibenzo-*p*-dioxins and polychlorinated dibenzofurans (PCDD/PCDFs), polychlorinated biphenyls (PCBs), polycyclic aromatic hydrocarbons (PAHs), linear alkylbenzenes (LABs) and coprostanol (5-β-cholestan-3β-ol). The environmental concern of some of these, like PCDDs, PCDFs, PCBs and PAHs is due to their persistence and their carcinogenic potential. Coprostanol is a specific product of biohydrogenation of cholesterol in the intestine of higher animals including man and it is persistent in anoxic sediments. For this reason it has been suggested as a marker compound to identify fecal contamination due to domestic waste waters in the marine environment (Grimalt *et al.*, 1990; Venkatesan and Kaplan, 1990). LABs are domestic waste waters tracers too, since they are the starting

Water, Air and Soil Pollution **99**: 237-244, 1997.

compounds for the synthesis of the linear alkylbenzenesulfonates and persist as unsulfonated residues in the common synthetic detergents (Eganhouse *et al.*, 1983; Ishiwatari *et al.*, 1983).

Our study was a part of a more extensive research program (Progetto Sistema Lagunare Veneziano) on the pollution in the Venice Lagoon, some results of which were already published (Benfenati *et al.*, 1994 a; Turrio Baldassarri *et al.*, 1994). The purpose was to investigate the degree of contamination and the sources of these pollutants within the lagoon.

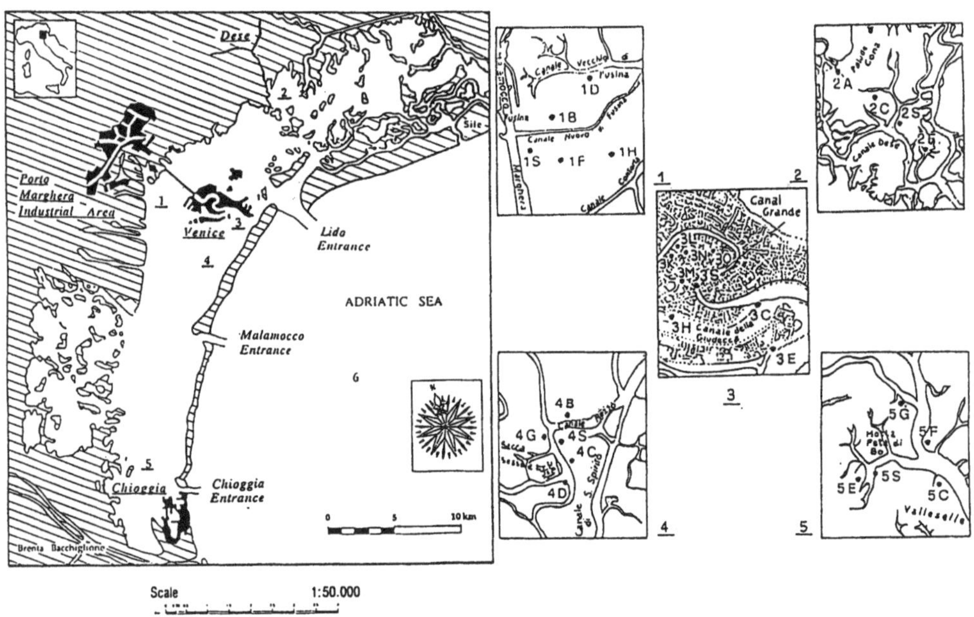

Fig. 1. The Venice Lagoon and the sampling sites.

2. Materials and Methods

2.1. SAMPLING CAMPAIGNS

Surface sediments samples (about the top 10 cm) were collected by a steel dredge during July 1992 in five sampling stations inside the lagoon, representative of different kinds of pollution (Table I), and one outside (Figure 1). Further samples were collected during two subsequent campaigns: in March 1994 five samples were collected in the internal canals of Venice and were analysed for PAHs and in July 1994 sediment samples were collected in the previously investigated stations of the lagoon and analysed for PCDDs and PCDFs (Table I). The surface sediments of the Venice Lagoon are rich of carbonate and contain mainly fine grained sized fractions (<62 μm) (Calvo *et al.*, 1991 b).

TABLE I.
Sampling stations and sample codes

SAMPLING STATION	CHARACTERISTICS OF THE AREA	SAMPLE CODE July '92	March '94*	July '94**
Station 1 (Porto Marghera)	Industrial district of Marghera, slow water exchange.	1S		1B, 1D, 1F, 1H
Station 2 (Cona Marsh)	Mouth of the Dese river; slow water exchange and considerable amounts of fresh water draining a largely agricultural basin.	2S		2A, 2C, 2F
Station 3 (Venice)	Canals of Venice, heavy pollution due to untreated sewage. Moderate water exchange.	3S	3K, 3L, 3M, 3N, 3O	3C, 3E, 3H
Station 4 (Sacca Sessola)	Relatively small load of urban and industrial contaminants, quite good water exchange.	4S		4B, 4C, 4D, 4G
Station 5 (Chioggia Basin)	Quite far from any industrial source but close to the town of Chioggia; ample water exchange.	5S		5C, 5E, 5F, 5G
Station 6 (Adriatic Sea)	Landmark outside the lagoon, about 5.5 km from the cost.	6S		

*Only PAH determination
**Only PCDD/PCDFs determination

2.2. ANALYSIS

The samples were kept frozen until analysis. Then they were freeze-dried, spiked with internal standards for each class of compounds and different amounts of dry weight (1 g for coprostanol, 10 g for PAHs and 100 g for PCDDs, PCDFs and PCBs) were extracted in a Soxhlet apparatus using an acetone/*n*-hexane mixture 1/1 v/v (Berlincioni *et al.*, 1989). The internal reference standards used were: ^{13}C-labelled 2,3,7,8-substituted analogues (Cambridge Isotope Laboratories, Woburn, MA, USA) for each class of PCDDs and PCDFs; ^{13}C-labelled pentachlorobiphenyl (IUPAC #101) (Cambridge Isotope Laboratories) for PCBs; acenaphthene-d_{10}, benzo[*a*]anthracene-d_{12} (Supelco, Inc., Bellefonte, PA, USA) and benzo[*ghi*]perylene $^{13}C_{12}$ (Cambridge Isotope Laboratories) for PAHs; 1-phenyldodecane (Aldrich-Chemie, Steinheim, Germany) for LABs; coprostanol-d_4 for sterols (Benfenati *et al.*, 1994 b). Analytical procedure for PCDDs/Fs and PCBs was previously described (Gizzi *et al.*, 1982, Benfenati *et al.*, 1992). Samples relative to the campaign of March 1994 were submitted to a further purification by activated carbon AMOCO PX-21 (Lindstrom and Rappe, 1988) to improve the specificity of the methodology. Blanks were carried out and samples relative to the campaign of July 1992 followed the entire analytical procedure in triplicate. Instrumental analyses were performed by gas chromatography-mass spectrometry (GC-MS) with a HP 5980 MSD 5971 for PAHs, coprostanol and LABs and a HP 5890 VG-TS 250 for PCDD/Fs and PCBs. The gas chromatographic conditions are shown in Table II. Sterols were silylated just before injection (Benfenati *et al.*, 1994 b).

TABLE II

Instrumental conditions for the analysis of PCDDs, PCDFs, PCBs, PAHs, coprostanol and LABs.

Compound	Capillary Column	T programme	T inj.	head pressure
PCDDs PCDFs	Chrompack CP Sil 8 CB 50mx0.25mm film thickness 0.25 μm	160°C x 1min 20°C/min 300°C x 20min	280°	140 kPa
PCBs	Chrompack CP Sil 8 CB 25mx0.25mm film thickness 0.25 μm	120°C x 1min 10°C/min 300°C x 10 min	280°C	80 kPa
PAHs	Chrompack CP Sil 8 CB 25mx0.25mm film thickness 0.25 μm	70°C x 1min 10°C/min 300°C x 10 min	280°C	30 kPa
Coprostanol	Chrompack CP Sil 8 CB 25mx0.32mm film thickness 0.12 μm	120°C x 2 min 12°C/min 285°C 20°C/min 300°C x 4 min	300°C	30 kPa
LABs	Chrompack CP Sil 8 CB 25mx0.25mm film thickness 0.25 μm	80°C x 2 min 10°C/min 130°C 4°C/min 250°C x 4 min	260°C	40 kPa

3. Results and Discussion

3.1. COPROSTANOL AND LABs

Coprostanol concentrations are shown in Figure 2. The lowest values, 0.04 μg/g of dry weight, were detected in sample 6S, collected in the open sea, while the highest, 4.41 μg/g, in sample 3S, relative to the Canal Grande, the largest canal within the city of Venice. This value is lower than levels detected in a previous investigation on the internal canals of Venice (Sherwin *et al*, 1993). The levels in the other samples of the lagoon resulted at least one order of magnitude lower than inside Venice. A detailed investigation on sterols of different origins in the Venice Lagoon will be published elsewhere.

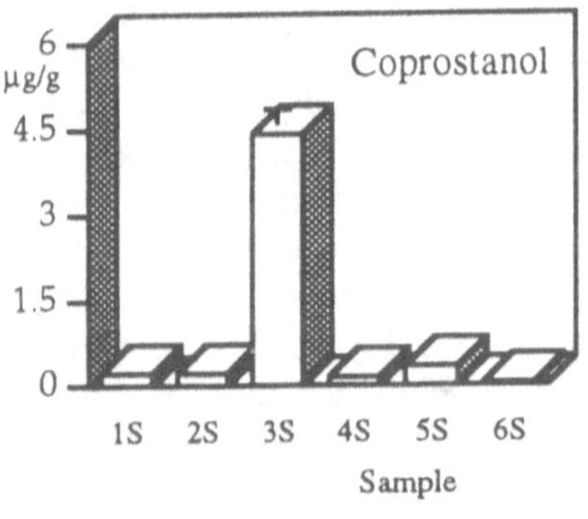

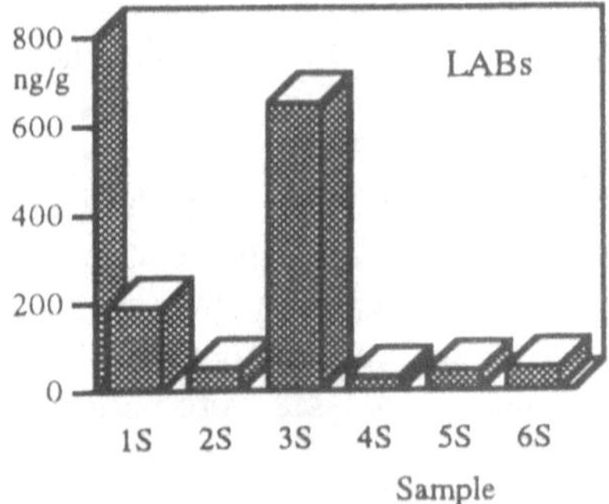

Fig. 2. Average concentrations of coprostanol and LABs in sediment samples 1S-6S.

LABs concentrations varied from 31 to 650 ng/g of dry weight (Figure 2). Excluding sample 3S and 1S, no meaningful differences were found between the open sea (6S) and the others samples of the lagoon. The highest values, as expected, were detected at sample 3S and they were similar to the least bound of the range present in the polluted Tokyo Bay (Ishiwatari *et. al.*, 1983). Meaningful higher LAB levels were found at sample 1S too, close to Porto Marghera, where no elevated values of coprostanol were detected. Thus LABs could be not only tracers of domestic wastes but indicate pollution of industrial origin too.

3.2. PAHs, PCBs AND PCDD/PCDFs

Concentration of total PAHs ranged between 184, in the open sea, and 201678 ng/g dry weight in sample 3S (Table III). Within the lagoon, sample 5S showed the lowest levels and they were not meaningfully different from those detected in the open sea. These levels are comparable to the concentrations found in the surface sections of a sediment core collected in the Venice Lagoon (Pavoni *et al.*,1987). Except sample 3S the highest concentrations were in sample collected near Porto Marghera. The PAH contamination within Canal Grande was similar to that found in very polluted areas such as some sites of the New York Bight and Rio Santiago, Argentina (Colombo *et. al.*, 1989). Unlike others pollutants of this study, PAH concentrations at sample 2S resulted quite high compared to the other samples of the lagoon.

TABLE III
Average concentrations of total PAHs and PCBs with relative standard deviations in sediment samples 1S-6S.

Sample	Total PAHs	Total PCBs
1S	742±159	52±21
2S	508±79	48±7
3S	201678±50143	1583±1
4S	248±31	42±4
5S	189±23	18±4
6S	184±43	39±12

The distribution of the individual PAHs (Figure 3) showed that the differences between the sample collected at station 3 and the other samples was mainly due to the high amounts of phenanthrene, anthracene, fluoranthene and pyrene that are dominant in comparison to the other PAHs. The concentrations related to the five samples collected into the internal canals of Venice confirmed only partially the results of sample 3S. Total PAHs ranged between 6570 ng/g in sample 3M (Rio Novo) to 111215 ng/g of dry weight in sample 3L (Rio Marin). In most of the samples, the PAH profiles are dominated by fluoranthene, pyrene, phenanthrene and benzo[*b*+*k*]fluoranthene, while anthracene is quite low in comparison to the others. Fluoranthene, pyrene and benzo[*b*+*k*]fluoranthene often dominate the profile of air samples from urban areas (Halsall *et al.*, 1993), so that the PAH contamination in the sediments of the lagoon is probably arising from combustion processes due to the motor boats rather than from oil leakage.

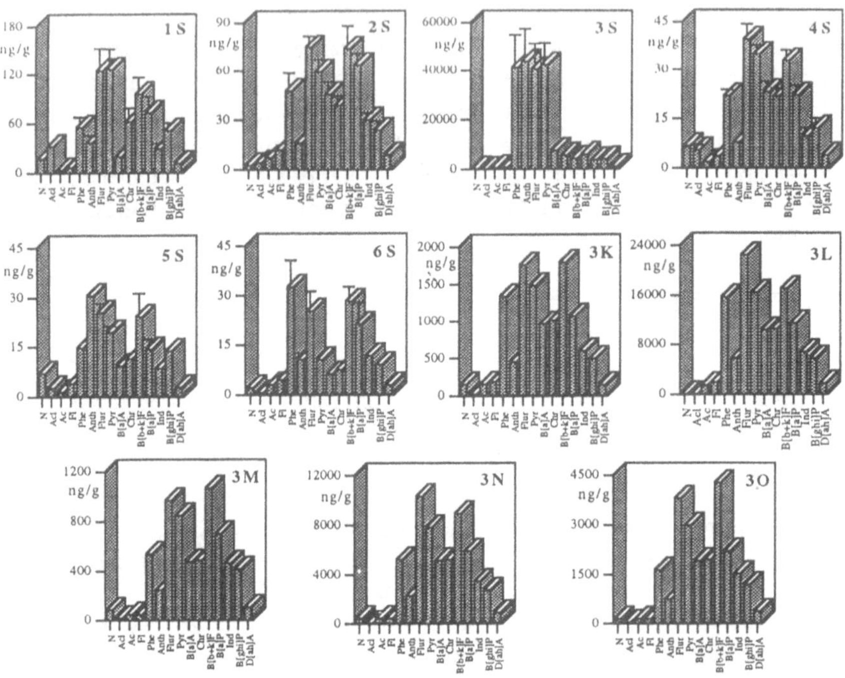

Fig. 3. PAH concentrations in sediment samples 1S-6S and 3K-3O of the Venice Lagoon.

Legend: Naphtalene = N; Acenaphthylene = Acl; Acenaphthene = Ac; Fluorene = Fl; Phenanthrene = Phe; Anthracene = Anth; Fluoranthene = Flur; Pyrene = Pyr; Benzo[*a*]anthracene = B[*a*]A; Chrysene = Chr; Benzo[*b*]fluoranthene+Benzo[*k*]fluoranthene = B[*b*+*k*]F; Benzo[*a*]pyrene = B[*a*]Pyr; Indeno[1,2,3-*cd*]pyren = Ind; Benzo[*ghi*]perylene = B[*ghi*]Per; Dibenzo[*a,h*]anthracene = D[*ah*]A.

Total PCB concentrations (Table III) varied from 18 to 1583 ng/g of dry weight. Except sample 3S, these values are similar to levels detected in sediments of the Adriatic Sea (Galassi *et al.*, 1993), and lightly higher or comparable to levels found in the Venice Lagoon (Donazzolo *et al.*, 1983; Pavoni *et al.*, 1987; Raccanelli *et al.*, 1989; Turrio Baldassarri *et al.*, 1994). The high concentrations in sample 3S, as for PAHs, could be due to the exhausts from the boat engines. The presence of PCB species in the emissions of combustion process has been reported (Benfenati *et al.*, 1992).

The results of the analysis of the PCDD/PCDFs showed a contamination within the lagoon of one order of magnitude higher than in the open sea. The PCDD/PCDF total concentrations detected in the singular samples of the lagoon ranged between 56 and 2370 pg/g of dry weight. The highest values were found in some samples collected near Porto Marghera and in the sample 3S and they are similar to the least concentrations detected in sediments from heavy industrialised and urbanised areas (Evers *et al.*, 1988; Wenning *et al.*, 1992). The average homologue profile (sum of the isomers with the same number of chlorine atoms), calculated with all the samples analysed for PCDDs and PCDFs in each sampling station (Figure 4) pointed out some differences in the sampling sites, probably referable to different origins of these pollutants (the chemometric elaboration of the singular homologue profiles will be presented elsewhere). In particular at least two kinds of profiles can be identified: the first, relative to station 1 and 4, characterised from high levels of OCDF and a high ratio

PCDFs/PCDDs and another with higher levels of OCDD, TCDFs, PnCDFs and HxCDFs. This has suggested the presence of a point source of these pollutants probably arising from the chemical processes occurring in Porto Marghera and mixing with the background or another source linked to the combustion.

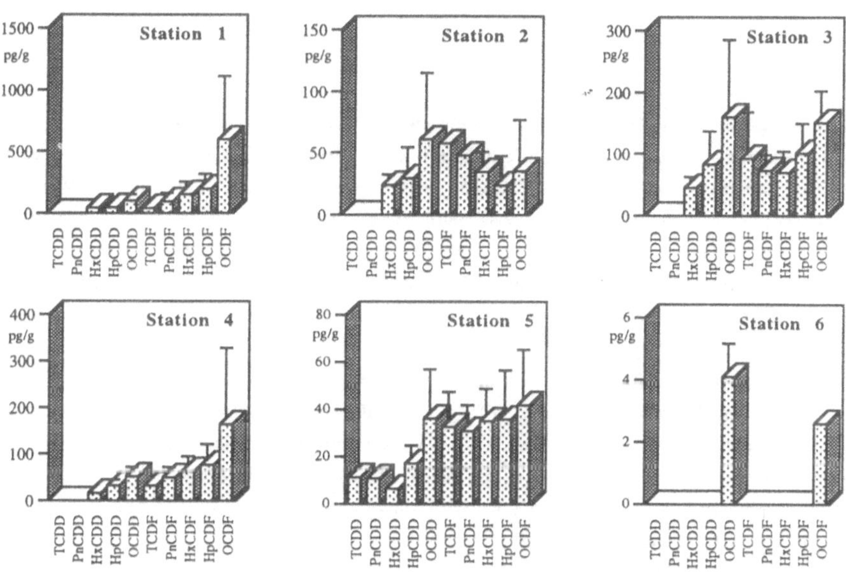

Fig. 4. Average homologue profiles of PCDDs and PCDFs in samples collected during the July '92 and July '94 campaigns in the six sampling stations of the Lagoon of Venice.

Legend: TCDD = tetrachloro dibenzo-*p*-dioxins; PnCDD = pentachloro dibenzo-*p*-dioxins; HxCDD = hexachloro dibenzo-*p*-dioxins; HpCDD = heptachloro dibenzo-*p*-dioxins; OCDD = octachloro dibenzo-*p*-dioxin; TCDF = tetrachloro dibenzofurans; PnCDF = pentachloro dibenzofurans; HxCDF = hexachloro dibenzofurans; HpCDF = heptachloro dibenzofurans; OCDD = octachloro dibenzofuran.

4. Conclusions

Excluding the sediment of the city of Venice, this study has shown a moderate contamination in most of the samples analysed. The concentrations of PAHs and PCBs were below the believed thresholds of toxicity for most of the marine benthic organisms (Kennicutt *et al.*, 1994). The sample relative to Venice showed very high level for nearly all the pollutants analysed. Coprostanol and LABs indicate a strong contamination due to the untreated domestic waste waters, while the high levels of PAHs point out a strong impact of the motor boat traffic. Also for PCBs and PCDD/PCDFs in the Canal Grande the source is probably due to the combustion in the motor boats' engines even if a contribute of the urban wastes discharged in the canal can not be ruled out. Higher concentrations of PCDDs and PCDFs were found in some samples collected near Porto Marghera; probably these compounds rise from a punctual source in this area.

Acknowledgements

This work was supported by the National Project MURST, Progetto Sistema Lagunare Veneziano. Elena Fattore is recipient of a fellowship from Fondazione Lombardia per l'Ambiente, Milan.

References

Benfenati, E., Valzacchi, S., Mariani, G., Airoldi, L. and Fanelli, R.: 1992, *Chemosphere* **8**, 1077-1083.

Benfenati, E., Fattore, E., Mariani, G., Cools, E. and Fanelli, R.: Mass spectrometric analysis of organic compounds in surface sediments from the Venice Lagoon. International Symposium Chromatography and Mass-Spectrometry in Environmental Analysis. S. Petersburg, Russia, October 3-7, 1994 a.

Benfenati, E., Cools, E., Fattore, E. and Fanelli, R.: 1994, *Chemosphere* **29**, 1393-1405 b.

Berlincioni, M., Di Domenico, A., Fanelli, R., Palma, S. and Zapponi, G.: 1989, *Chemosphere* **19**, 501-506.

Calvo, C., Donazzolo, R., Guidi, F. and Orio, A. A.: 1991, *Wat. Res.* **25**, 1295-1302 a.

Calvo, C., Grasso, M. and Gardenghi, G.: 1991, *Mar. Pollut. Bull.* **22**, 543-547 b.

Capodaglio, A., Novotny, V. and Zheng, S.: 1990, *Ingegneria Ambientale* **19**, 563-573 a.

Capodaglio, A., Novotny, V. and Zheng, S.: 1990, *Ingegneria Ambientale* **19**, 638-643 b.

Colombo, J. C., Pelletier, E., Brochu, C. and Khalil, M.: 1989, *Environ. Sci. Technol.*, **23**, 888-894.

Donazzolo, R., Menegazzo Vitturi, L., Orio, A. A. and Pavoni, B.: 1983, *Environ. Technol. Lett.* **4**, 451-462.

Eganhouse, R. P., Blumfield, D. L. and Kaplan, I. R.: 1983, *Environ. Sci. Technol.* **17**, 523-530.

Evers, E. H. G., Ree, K. C. M. and Olie, K.: 1988, *Chemosphere* **17**, 2271-2288.

Galassi, S., Guzzella, L. and De Paolis, A.: 1993, *Fresenius Envir. Bull.* **2**, 25-30.

Gizzi, F., Reginato, R., Benfenati, E. and Fanelli, R.: 1982, *Chemosphere* **11**, 577-583.

Grimalt, J. O., Fernandez, P., Bayona, J. M. and Albaigés, J.: 1990, *Environ. Sci. Technol.* **24**, 357-363.

Halsall, C., Burnett, V., Davis, B., Jones, P., Pettit, C. and Jones, K. C.: 1993, *Chemosphere* **26**, 2185-2197.

Ishiwatari, R., Takada, H., Yun, S-J. and Matsumoto, E.: 1983, *Nature* **301**, 599-600.

Kennicutt, M. C., Wade, T. L., Presley, B. J., Requejo, A. G., Brooks, J. M. and Denoux, G. J.: 1994, *Environ. Sci. Technol.* **28**, 1-15.

Lindstrom, G. and Rappe, C.: 1988, *Chemosphere* **17**, 921-935.

Pavoni, B., Sfriso, A. and Marcomini, A.: 1987, *Mar. Chem.* **21**, 25-35.

Pavoni, B., Calvo, C., Sfriso, A. and Orio, A. A.: 1990, *Sci. Total Environ.* **91**, 13-21 a.

Pavoni, B., Sfriso, A., Donazzolo, R. and Orio, A. A.: 1990, *Sci. Total Environ.* **96**, 235-252 b.

Pavoni, B., Marcomini, A., Sfriso, A., Donazzolo, R. and Orio, A. A.: 1992, *The Science of global change*, D. A. Dunnette, R. J. O'Brien (*eds.*), American Chemical Society, Washington, pp. 287-305.

Raccanelli, S., Pavoni, B., Marcomini, A. and Orio A. A.: 1989, *Sci. Total Environ.* **79**, 111-123.

Sherwin, M. R., van Vleet, E. S., Fossato, V. U. and Dolci, F.: 1993, Mar. Pollut. Bull. **26**, 501-507.

Turrio Baldassarri, L., Di Domenico, A., Iacovella, N., La Rocca, C. and Rodriguez, F.: 1994, *Organohalogen Compounds* **20**, 183-186.

Venkatesan, M. I. and Kaplan, I. R.: 1990, *Environ. Sci. Technol.* **24**, 208-214.

Wenning, R. J., Harris, M. A., Ungs, M. J., Paustenbach, D. J. and Bedbury, H.: 1992, *Arch. Environ. Contam. Toxicol.* **22**, 397-413.

PCBs AND ORGANOCHLORINE PESTICIDES IN LAKE ORTA (NORTHERN ITALY) SEDIMENTS

L. GUZZELLA
Istituto di Ricerca sulle Acque, C.N.R., via Mornera 25, 20047 Brugherio (MI), Italy

Abstract - PCB congeners and organochlorine pesticides (DDT, lindane and HCB) distribution were studied in Lake Orta sediments. The results indicated a contaminated area in the northern part of the sub-basin. The observed high levels of organochlorine compounds (OCs) may be explained by the focusing phenomenon, ie. the preferential transport of lighter and smaller particles from the emission sources to this area. The PCBs and DDT values were correlated with the organic carbon content and the heavy metal contamination. The toxicity of the sediment samples was related also to PCB content. PCBs and OCs pollution of Lake Orta was of the same order of magnitude as in Lake Como, which is the most contamined lake in Northern Italy.

Keywords: PCB, organochlorine pesticide, sediment, organic carbon content, toxicity.

1. Introduction

Since PCBs and organochlorine pesticides have a high affinity for particulate matter, one of their primary sinks is thought to be the bottom sediments. Sediments act, in fact, as one of the main sinks and potential sources for chlorinated biphenyls (PCBs) and organochlorine pesticides (HCB, DDT, lindane) pollution. Sediment analysis has therefore been used widely to detect PCB and pesticide contamination and to assess ecological risks (Oliver *et al.*, 1989). The contamination of lake sediments may be related to point sources (e.g., industrial discharges and waste plant effluents) but more frequently, as in the case of pesticide pollution, it is attributed to diffuse sources (precipitation, agriculture runoff, particle transport). The investigation of pollutant distribution in sediments provides a valuable record of contamination in the lake (Eisenreich *et al.*, 1989).

Most studies aimed at the reconstruction of the historical inputs of PCBs and organochlorine pesticides (OCs) to lacustrine environments are based on profile descriptions of contamination in core samples (Eisenreich *et al.*, 1989). Many investigations have been undertaken in North America and Europe (Oliver *et al.*, 1989; Hermanson *et al.*, 1991; Sanders et al., 1992; Hornbuckle et al., 1993), but few data are available on the Northern Italian Lakes. Previous studies on Lakes Maggiore, Como and Garda had very limited sampling points and do not allow the determination of point source pollution (Bossi *et al.*,1992;Galassi *et al.*, 1995; Provini *et al.*,1995).

Recently, in a study conducted after the Lake Orta remediation operation, sediments were sampled and analysed for organic carbon content, heavy metals, salts and organic micropollutants concentrations and also for toxicity tested with bacteria, plants and crustacea assays (Ross *et al.*, 1996). In the present paper the results obtained with PCB and OCs analyses of the same sediment samples are reported. Lake Orta was chosen for the study because of the lack of information on sediment toxicity and organic pollutant contamination in contrast to the great amount of data existing on inorganic contaminants (heavy metals and ammonia distribution). The PCBs and OCs concentrations were

Water, Air and Soil Pollution **99**: 245-254, 1997.

compared also to heavy metal pollution (Baudo *et al.*, 1989, 1993) and to sediment toxicity tested with Microtox *Solid Phase* (Ross *et al.*, 1996).

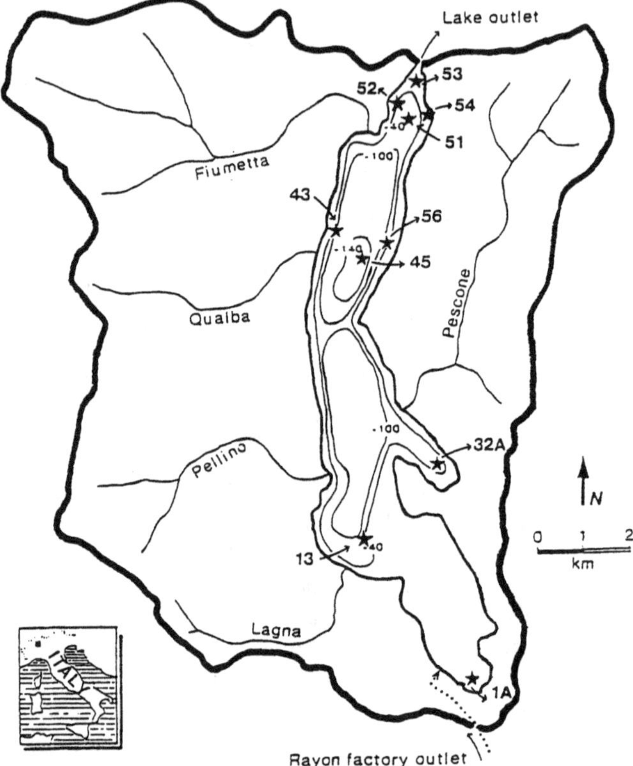

Figure 1. The location of the sediment sampling stations in the Lake Orta. The main physical characteristics of the lake are: lake area = 18.14 km^2, volume = 1.29 km^3, mean depth = 70.9 m, maximum depth = 143 m, watershed area = 116 km^2, actual mean residence time = 10.7 years, theoretical renewal time = 8.5 years.

2. Materials and Methods

2.1 STUDY AREA

Lake Orta is a warm monomictic lake located near Novara in the subalpine region of Northern Italy. Chemical contamination of the lake began in 1926 with massive discharges of ammonium sulphate and copper from a local rayon factory. All forms of aquatic life were greatly affected by this chemical pollution and by 1929 the pelagic and benthic communities were virtually absent. Oxidation of ammonia led to a progressive accumulation of nitrates and a drastic acidification of the water column (Provini and Gaggino, 1986). The introduction of a waste treatment system at the local rayon factory in 1958, and improvements to it in 1980, have greatly reduced the ammonium and copper concentrations in the waste stream discharged into the lake. In 1989 a recovery operation began, and 14,500 metric tons of calcium carbonate were added to the lake water, increasing the pH from 4 to 6 and decreasing ammonia nitrogen levels (Calderoni et al., 1992).

Since the early 1950's, discharges from several electroplating plants on a tributary, the Lagna River, have increased the contamination of the lake through discharges of nickel and chromium. Other industrial activities were developed intensely in 1950-60's. Several industrial companies (textiles, mechanics, electromechanics, galvanics, wood, paper and leather manufactures) were concentrated in the north-eastern part of the lake basin and contributed to the lacustrine pollution. Non-point pollution sources, horticulture and floriculture were intensively widespread in the south-western and north-eastern part of the basin.

2.2 SAMPLING AND ANALYTICAL PROCEDURE

The locations of the sediment sampling stations in Lake Orta are shown in Figure 1. In July 1992, surface sediment samples were collected with a Ponar grab from stations 52, 53, 54, 43, 56, 32A and 1A, and vertical cores were taken from stations 51, 45 and 13 (Table 1). The cores were divided into sections and three strata were selected for analysis: 0-10 cm (recent deposits), 10-20 cm (intermediate-age deposits) and another 10 cm interval from a deeper layer (ancient deposits). The sediment samples were stored in glass vessels in the darkness at -20°C until analysis.

Homogenised subsamples were freeze-dried and 5 g freeze-dried sediment aliquots were extracted for 8 h in a Soxhlet apparatus with 150 mL n-hexane. Concentrated extracts (1-2 mL) were passed through a Florisil column (4 x 0.7 cm) with a layer of Cu powder on top (0.1 g). For the detection of sulphur compounds, Cu powder, previously activated with HCl (18%) and washed with acetone and n-hexane, was added. The Florisil column was eluted with 25 mL of n-hexane and the eluate was concentrated to 1 mL. Cleaned-up extracts were injected into a gas chromatograph (Carlo Erba 8160) equipped with a ^{63}Ni electron capture detector (ECD). The on-column system was adopted for injection. Instrumental parameters and operational conditions were as follows: a fused silica capillary column (CP-Sil-8 CB, 50 m x 0.25 mm i.d.); helium as carrier gas at 1 mL/min and nitrogen as auxiliary gas detector at 60 mL/min; the oven temperature program was set up at 100 °C for 1 min, from 100 to 180 °C at 20°C/min, from 180 to 250 °C at 1.5 °C/min, from 250 to 270°C to 10°C/min, at 270 °C for 15 min.

Aroclor 1260 (Alltech) was used as a reference standard for PCB quantification. The 18 PCB congeners in the reference mixture were identified and measured against reference-pure PCBs (BCR, Brussels). The sum of PCB congeners determinated in this way amounted to about 85% of the standard Aroclor 1260 mixture by weight. Reference pure standard (Alltech) were used for pp'-DDT and relative metabolites, HCB and HCH isomers quantification. Each analysis was performed in duplicate/triplicate and analytical variability was 10%. The detection limits of the adopted procedure are: 0.1 ng/g for each PCB congener and 0.05 ng/g for OC pesticides. Blanks were analysed by the same procedure as for the samples and did not reveal any PCBs/OCs contami-nation. Good laboratory practices were verified by testing a freeze-dried reference sludge sample (BCR, Brussels).

Sediment core sections were dated on the basis of sedimentation rates measured by Provini and Gaggino (1985) using Cs^{137}. Organic carbon content was determined on dry samples by chromic acid oxidation according to Jackson (1958), modified by Gaudette et al. (1974).

TABLE I
Main characteristics of the sampling stations and of the sediment samples

Stations and Samples	Sampling station			Sediment samples		
	Sub-basin	Depth of station (m)	Type of sample	Dating of samples	Water (%)	Organic carbon (%)
51 (0-10cm)	North	71	core	1970-92	65.4	9.96
51(10-20cm)	North	71	core	1948-70	64.4	6.15
51(38-48cm)	North	71	core	1885-1907	58.5	4.15
52	North	33	grab	1970-92	59.6	4.87
53	North	19	grab	1970-92	75.1	11.65
54	North	30	grab	1970-92	67.0	9.84
43	North	89	grab	1970-92	39.0	7.44
56	North	58	grab	1970-92	42.5	4.53
45 (0-10cm)	North	143	core	1970-92	25.0	3.20
45(10-20cm)	North	143	core	1948-70	31.0	4.11
45(32-42cm)	North	143	core	1900-22	24.0	3.20
32A	Central	90	grab	1961-92	42.2	5.83
13 (0-10cm)	South	47	core	1950-92	49.0	6.84
13(10-20cm)	South	47	core	1908-50	41.0	4.78
13(50-60cm)	South	47	core	1742-84	28.0	5.80
1A	South	11	grab	1950-92	74.6	4.97

3. Results and Discussion

3.1 PCB AND OC LAKE CONTAMINATION

The main characteristics of the sampling station (sub-basin, depth and type of samples) and the water and organic content of sediment samples are reported in Table I. The selection of the sampling sites was conducted on the basis of the results obtained by a detailed monitoring study on organic and inorganic composition of Lake Orta sediments (Baudo *et al.*, 1989). The most critical lake zones were chosen in the present investigation in order to characterise the three sub-basins and the different bottom platforms.

According to Provini and Gaggino (1985), all the superficial core and grab samples belong to the industrial period and include 1960-70's, the most important period of organochlorine pesticide intensive application and PCB discharges. On the contrary, the deepest layers of the three selected cores belong to the pre-industrial period.

The organic carbon distribution in the lake sediment samples (Table I) showed an enrichment in two well circumscribed zones of the north sub-basin: the first one near the lake outlet (station 51, 53 and 54) and the second one in the bottom platform of the same sub-basin (station 45). The results were in good agreement with the values reported in Baudo *et al.* for Lake Orta. In their study, Baudo *et al.* (1989) also reported a high content of heavy metals and organic carbon in the depositional area near the outlet of the lake. Since no known pollution source is present in this region, the contamination was

TABLE II

PCB congeners content in Lake Orta sediment samples (ng/g)

PCBs	Lake Orta sediment samples															
	1A	13	13	13	32A	45	45	45	43	56	51	51	51	52	53	54
		0-10	10-20	50-60		0-10	10-20	32-42			0-10	10-20	38-48			
101	1.79	1.25	0.29	0.49	1.20	0.28	0.19	0.45	4.32	0.34	7.81	1.02	0.73	9.89	22.91	3.59
110	2.14	1.74	0.40	0.46	1.54	0.28	0.26	0.50	5.07	0.34	5.52	1.12	0.65	11.27	19.57	4.14
151	1.08	0.60	0.13	0.23	0.71	0.14	0.07	0.17	2.25	0.13	3.71	0.31	0.02	3.53	9.08	1.92
149	2.71	2.03	0.36	0.68	2.28	0.29	0.22	0.40	5.13	0.29	10.16	1.04	0.66	14.33	29.10	7.05
118	2.88	1.51	0.21	0.36	1.93	0.00	0.00	0.00	6.77	0.14	5.59	0.48	0.19	13.25	41.07	7.53
153	5.33	3.62	0.62	1.11	2.80	0.43	0.24	0.53	10.30	0.52	19.12	1.65	0.50	16.48	43.04	13.86
141	1.10	0.62	0.13	0.26	0.76	0.22	0.10	0.18	1.61	0.14	3.68	0.33	0.26	3.30	7.39	1.94
138	3.98	2.78	0.37	0.56	2.19	0.61	0.10	0.31	8.85	0.34	13.41	0.87	0.42	14.16	30.84	9.47
187	2.23	1.60	0.25	0.52	1.14	0.51	0.14	0.23	3.29	0.24	7.88	0.87	0.34	4.69	17.78	5.75
183	1.24	1.64	0.00	0.15	0.91	0.35	0.28	0.27	1.96	0.26	9.46	1.74	0.44	4.46	8.36	3.20
185	0.22	0.18	0.09	0.07	0.14	0.04	0.03	0.04	0.52	0.05	0.64	0.12	0.04	0.61	1.93	0.59
174	2.05	1.26	0.29	0.50	1.49	0.07	0.24	0.17	2.67	0.21	6.30	0.66	0.21	4.43	14.30	4.45
177	1.14	0.98	0.33	0.35	1.13	0.00	0.29	0.38	2.95	0.26	5.09	0.97	0.53	3.40	6.44	2.82
180	4.10	2.62	0.44	0.76	2.46	0.33	0.18	0.34	6.54	0.37	16.28	1.37	0.36	9.50	35.50	10.53
170	1.26	1.07	0.14	0.23	0.90	0.10	0.05	0.04	2.22	0.12	3.44	0.27	0.11	3.20	7.56	3.08
201	0.91	0.90	0.14	0.14	0.57	0.05	0.03	0.07	1.63	0.13	4.36	0.34	0.00	1.95	7.39	3.16
196	0.76	0.92	0.07	0.15	0.79	0.08	0.08	0.06	1.83	0.12	4.57	0.55	0.00	2.48	8.63	4.10
194	0.57	0.63	0.12	0.07	0.75	0.09	0.07	0.20	1.24	0.12	2.96	0.26	0.12	1.80	6.16	3.11
Total	35.49	25.94	4.38	7.09	23.69	3.86	2.57	4.34	69.15	4.13	129.98	13.97	5.78	122.73	317.06	90.29

attributed to the focusing phenomenon, i.e. the preferential transport of lighter and smaller particles to this area from the real emission sources located farther southwards. The particulate matter which escaped the sediment trap caused by the flat bottom at 120 m of depth, can settle easily in the north part of the lake. Because of the lower depth in this part of the lake and the prevailing water current from south to north, the smaller particles tend to settle easily. As a consequence, it is not surprising that all the heavy metal contaminants, phosphorus, nitrogen, sulphur, data reported in Baudo *et al.* (1989), and organic carbon were more concentrated in this accumulation area in respect to the erosion area located in the south site of the lake.

The PCB congener content of the sediment samples is reported in Table II. The highest PCB concentration was observed in the more superficial sediment samples with the exception of station 45. This fact is in agreement with the commercial use of these chemical compounds. The production and usage of PCBs in Italy, in fact, began in the 1940s but became very important only between 1960 and 1970 (Maroni *et al.*,1991). As concerns only the superficial sediment samples, the least contaminated lake area is located in the flat bottom and in the eastern part of the north sub-basin (station 43 and 56). This part of Lake Orta received the input from a residential area not particularly industrialised. On the contrary sampling stations 51, 52, 53 and 54 are influenced by the previously described focusing phenomenon and are characterised by PCB total concentrations with values higher than 100 ng/g.

The profile of PCB congener distribution is very similar to the commercial mixtures of Aroclor 1254 and 1260 with an enrichment of penta and hexa-chlorobiphenyls (101,110,149) and a shortage of hepta and octa-chlorobiphenyls (180,170,194) in the deeper horizons of the three cores (Figure 2). If the results of the single congeners are expressed in percent of the total PCB content, the PCBs distribution becomes clear in station 51 and 43 where the first layer was dated 1970-1992 and the second one at 1948-70. The observed different pattern of PCB congener distribution may be explained either by changes in commercial Aroclor mixtures in the last decades in Italy (hypothesis quite unprobable because Aroclors with higher chlorine content were used in the past years) or by a preferential rain transport of the less chlorinated PCBs. In fact, because of the different volatility capacity of PCB congeners, penta and hexa chlorobiphenyls can be more easily transported at a longer distance than hepta and octa-chlorophenyls. Swackhamer *et al.* (1988) measured PCB contents in dry particulate matter, snow and rain of the Siskiwit Lake and demonstrated that octa-chlorobiphenyls came from particulate matter while lighter PCBs from snow and rain.

Organochlorine pesticides results are reported in Table III. The OCs pollution of Lake Orta sediments showed a similar pattern to the PCBs (higher concentration in superficial sediment samples, accumulation of OCs in stations 51, 52, 53 and 54 due to the focusing phenomenon, less contaminated area in the flat bottom and in the east part of the north sub-basin) with the exception of station 43. The highest total DDT and total HCH concentrations (pp'-DDT and its metabolites reached a concentration of 120 ng/g while lindane - gamma HCH - and its isomers values of 25 ng/g) were observed in station 43 probably because of local discharge inputs. In Italy the use of pp'-DDT and lindane was forbidden in the 1970s, the application of lindane is actually restricted to agricultural use and the application of pp'-DDT to floriculture. The presence of pp'-DDT metabolites indicates that the pesticide is no longer used in the catchment basin and that the DDT contamination is due to the past usage of this pesticide.

TABLE III
Organochlorine pesticides content in Lake Orta sediment samples (ng/g)

Pesticides	Lake Orta sediment samples															
	1A	13	13	13	32A	45	45	45	43	56	51	51	51	52	53	54
		0-10	10-20	50-60		0-10	10-20	32-42			0-10	10-20	38-48			
pp'-DDE	3.50	6.79	0.29	0.41	4.54	0.69	0.15	0.32	25.71	0.78	28.75	0.68	0.61	13.85	13.30	16.62
pp'-DDD	7.00	10.28	0.31	0.03	6.96	0.84	0.10	0.16	58.38	1.20	43.75	0.49	0.32	18.38	23.52	25.80
pp'-DDT	1.45	1.18	0.14	0.27	1.00	1.03	0.28	0.37	9.02	0.27	5.93	0.42	0.26	5.77	3.90	4.32
op'-DDE	0.09	0.45	0.00	0.00	0.46	0.08	0.00	0.00	1.47	0.54	1.90	0.00	0.00	0.43	0.49	1.14
op'-DDD	2.74	4.00	0.00	0.00	2.71	0.39	0.00	0.00	24.28	0.21	13.65	0.00	0.00	15.46	22.11	16.16
op'-DDT	0.21	0.18	0.00	0.00	0.26	0.08	0.00	0.00	0.88	0.12	0.83	0.00	0.00	0.69	0.48	0.46
DDT total	14.99	22.88	0.74	0.71	15.93	3.11	0.53	0.85	119.74	3.12	94.81	1.59	1.19	54.58	63.80	64.50
alfa HCH	0.21	0.17	0.27	0.06	0.14	0.27	0.04	0.06	0.50	0.08	0.61	0.08	0.06	0.57	0.46	0.40
beta HCH	1.39	0.44	0.10	0.17	0.44	0.12	0.14	0.07	11.03	0.99	0.45	0.12	0.15	2.07	3.09	1.21
gamma HCH	3.22	0.32	0.42	0.24	0.94	0.33	0.29	0.33	13.73	2.07	1.78	0.63	0.41	2.83	3.72	2.11
HCH total	4.82	0.93	0.79	0.47	1.52	0.72	0.47	0.46	25.26	3.14	2.84	0.83	0.62	5.47	7.27	3.72
HCB	0.86	0.61	0.06	0.10	0.19	0.15	0.07	0.08	0.62	0.07	0.69	0.18	0.09	2.42	0.69	0.77

The presence of lindane isomers (alfa and beta HCH) indicate a previous use of HCH commer-cial mixtures while HCB con-tamination is limited to an en-riched area located in an accu-mulation zone, near station 52.

3.2 COMPARISON WITH HEAVY METAL CONTAMINATION AND TOXICITY

The PCB and OC concentrations in Lake Orta sediments (Table II and III) were compared to organic carbon content (Table I), to heavy metal contamination (Baudo *et al.*, 1993) and to toxicity tested with Microtox *Solid Phase* (Ross *et al.*, 1996), using the Statgraphic Software. Organic carbon content was positively correlated with total PCBs ($r = 0.80$) and total DDT ($r=0.72$) ($P<0.01, n=16$), validating the hypothesis of a contaminated accumulation area near the outlet of the lake.

For heavy metals, the results showed a good correlation between total PCBs, lead ($r = 0.83$) and zinc ($r=0.92$) sediment content and a weaker correlation with copper ($r = 0.64$) values ($P<0.01$, $n=16$).

Similar correlations were observed between total DDT, lead ($r = 0.69$), zinc ($r = 0.70$) and nickel ($r = 0.68$) and also total HCH and nickel ($r = 0.79$) ($P<0.01$, $n=16$). Because Pb, Zn and Ni metal contamination followed the same distribution of organic carbon content in Lake Orta sediments, the observed correlations are justified.

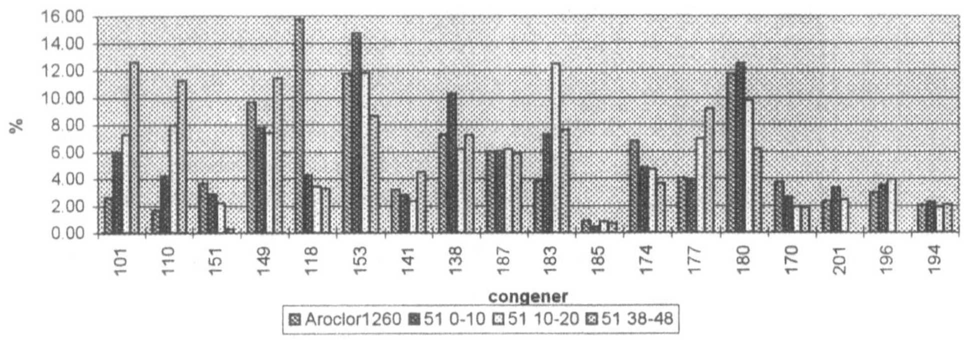

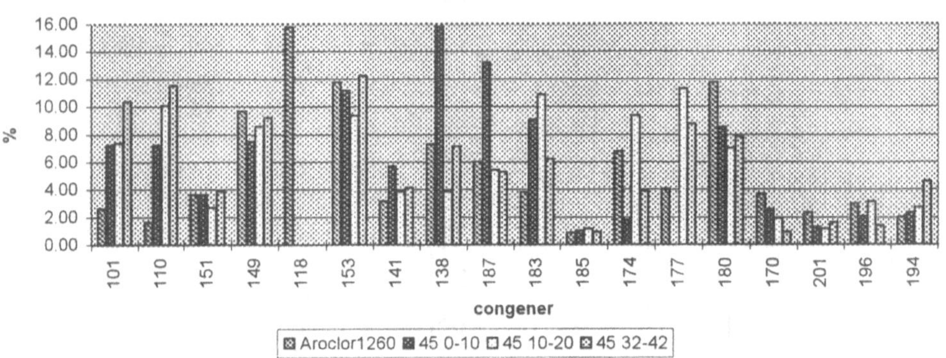

Figure 2 . PCB congeners profile in Aroclor 1260 and Lake Orta stations 51 and 45.

The positive correlation obtained comparing PCB with *Solid Phase* Microtox test results is very interesting. In fact, computing the data in a Multivariate Statistical Analysis with the CSS Statistic Software, the obtained toxicity by the lake sediments (expressed in g/l of solid sample that determine the EC50 value) can be explained by the following equation:

Solid Phase EC50 (T.U.) = 2.8+1.1 Cu (ng/g) -0.85 Pb (ng/g) +0.88 total PCBs (ng/g) - 0.33 K (ug/g) - 0.57 Org. Carbon (%) (1)
$r = 0.994$; $n=16$; $P < 0.001$, Cu=0.00001, Pb=0.00088, PCB=0.00034, k=0.0009, OC= 0.0066.

In equation (1) it is demonstrated that PCBs content of the lake sediments contribute to the toxic response of the samples and, therefore, they could play an important role in the risk assessment evaluation of Lake Orta.

3.3 COMPARISON WITH OTHER LAKE SEDIMENTS

Comparison of the present data with previous analyses carried out in other Northern Italian lakes is feasible for Lakes Maggiore, Como and Garda. Bossi *et al.* (1992)

analysed sediment cores from the three main sub-basins of Lake Garda and found a PCB content similar to the one of the Orta south and central sub-basin sediments while the north basin of Orta is ca ten times more polluted (station 53) than the most contaminated Lake Garda core (Toscolano station). Similar results are evident also for DDT and HCH contamination: Lake Orta station 43 is four and ten times respectively more polluted than the same Toscolano station.

For Lakes Maggiore and Como, the comparison is possible only making an average of the 1992-1970 sediment core layer concentrations in the two lakes. The comparison showed that sediment PCB content of the north part of the Lake Orta is very similar to that in Lake Como, the most polluted lake investigated (Provini *et al.*, 1995) while the total DDT contamination of Lake Orta sediments is greater than Maggiore and Como (Galassi *et al.*, 1995). The contamination of station 43 is, in fact, of the same order of magnitude as the most polluted Como sediment layer dated to the middle of the1960s.

4. Conclusions

The PCB and OC sediment results permitted us to conclude that the least contaminated lake area is located in the central sub-basin and in the eastern part of the north sub-basin while the depositional area near the lake outlet is particularly enriched in organic micropollutants. The profile of PCB congener distribution is very similar to the commercial Aroclor 1254 and 1260 with an enrichment of the lighter chlorobiphenyls and a scarcity of the heavier ones in the deeper sediment samples probably because of the different volatility capacities of PCB congeners. The PCB and OC sediment concentrations are significantly correlated to the organic carbon content, to lead, zinc and nickel heavy metal pollution and to toxicity tested with Microtox *Solid Phase.*
In addition, the PCB and OC sediment contamination of the Lake Orta south-central basins is of the same order of magnitude as in Lakes Maggiore and Garda while the sediment contamination of the northern lake basin showed similar concentrations to those measured in Lake Como.

Acknowledgements

The authors are greateful to A. Lami and R. Baudo of the Italian Institute of Hydrobiology (CNR-Verbania-Pallanza, Novara, Italy) and to P. Ross of The Citadel Charleston South Carolina, USA, for their assistance in collecting and processing the sediment samples.

References

Baudo, R., Amantini, L., Bo, F., Cenci, R., Hannaert, P., Lattanzio, A., Marengo, G. and Muntau, H. :1989, *Sci. Total Environ.* **87/88**, 117-128

Baudo, R., Tartari G.A., Lami, A., Ross P., Bo F., Cenci R., Vivian R. and Muntau, H: 1993,*14th Annual Meeting SETAC Ecological Risk Assessment:Lessons Learned?* Houston. TX, USA, 14-18 November 1993. Abstract Book:141.

Bossi , R., Larsen, B. and Premazzi, G. :1992, *Sci. Total Environ.* **24**,77-93.

Calderoni, A., Mosello, R., and Ruggiu, D.:1992, *Mem. dell'Ist. Ital. Idrobiol.* **50**, 201-224.

Eisenreich, S.J., Capel, P.D., Robbins, J.A., and Bourbonniere, R.:1989, *Environ. Sci. Tech.* **23**, 1116-1126.

Galassi, S., Provini, A., Guzzella, L. and De Paolis, A. : 1995, *SITE Atti 16*, Venice, 22-29 September 1994, 341-343.

Gaudette, H.E., Flight, W.R. and Toner, L. :1974, *J. Sediment Petrol* 44, 249-253.

Hermanson, M.H., Christensen, E.R., Buser, D.J. and Chen, L.:1991, *J. Great Lakes Res.*,**17**,94-108.
Hornbuckle, K.C., Achman, D.R., and Eisenreich, S.J.: 1993, *Environ. Sci. Tech.* **27**, 87-98.
Jackson, M.L. :1958, *Soil Chem. Anal.*, Prentice Hall, Englewood Cliffs, N.J. 205-223.
Maroni, M., Colombi A., Carrer., P., Barbieri, F., Foa, V., del Frate, G., and Bossi, C.:1991, *Acqua Aria*,**8**, 761-777.
Oliver, B.G., Charlton, M.N. and Durham, R.W.:1989, *Environ. Sci. Tech.* **23**, 200-208.
Provini , A. and Galassi S.:1993, *Handbook of Hazadous Materials.* , Academic Press Inc., 491-504.
Provini, A. and Gaggino, G.F. :1985, *Verh. Internat. Verein. Limnol.* **22**, 2390-2393.
Provini, A., Galassi, S., Guzzella, L. and Valli, G. : 1995, *Mar. Freshwater Res.* **46**, 129-36.
Ross, P., Baudo, R., Bartone, C.: 1996, manuscript sent to *Environ. Toxicol. Chem.*
Sanders,G., Jones, K.C. Hamilton-Taylor, J. and Dorr,H.:1992, *Environ. Science Tech.* **26**, 1815-1821.
Swackhamer, D.L., McVeety, B.D. and Hites R.A.:1988, *Environ. Sci. and Tech.* **22**, 664-672.

POLLUTANT EXCHANGE AT THE WATER/SEDIMENT INTERFACE IN THE VENICE CANALS

E. ARGESE[1], E. RAMIERI[2], C. BETTIOL[2], B. PAVONI[2], E. CHIOZZOTTO[3] and A. SFRISO[2]

[1]*Centro di Studio sulla Chimica e le Tecnologie per l'Ambiente (CNR), Dipartimento di Scienze Ambientali,* [2]*Dipartimento di Scienze Ambientali, Università Ca'Foscari di Venezia, Calle Larga S. Marta 2137, 30123 Venezia, Italy.* [3]*Assessorato all'Ecologia, Comune di Venezia, S.Marco 4136, 30124 Venezia, Italy*

Abstract. The space-time distribution of some pollutants (Cu, Pb, Zn, Cd, Fe, Mn, V, Ni, Cr) in the sludge of the canals of Venice was studied. The contamination levels were comparable to, or higher, than those measured in the most polluted sediments of the Lagoon of Venice. Sediments were collected by two different sampling techniques: 1) collection of sediment cores (upper 5 cm) by a syringe-type corer; 2) collection by traps, placed on the bottom of the canal. Traps permitted the sampling of sediments essentially resuspended by overlying water turbulence. This sediment fraction is subjected to variations of its physicochemical parameters (principally change of redox conditions) and therefore to pollutant exchange at the water/sediment interface.

The metals principally exchanged during sediment resuspension were Cd, Pb, Zn and Cu. These metals have principally an anthropogenic origin and are bound to the most labile geochemical phases of the sediment (such as sulphides), which can be oxidised during sediment resuspension, releasing metals into the water. Fe, Cr and Ni were only partially exchanged, while Mn and V were generally not exchanged; a significant fraction of these metals is of natural origin and is bound to the most refractory phases of the sediment.

Key words: heavy metals, sediments, pollution, resuspension, release processes, bioavailability, anthropogenic metal, residual metal, geochemical phases

1. Introduction

The city of Venice is crossed by a complex network of canals, where high amounts of organic matter and pollutants, such as heavy metals and nutrients, are vehiculated. The main pollution sources are represented by urban waste water, sewage, commercial and artisan activities and boat traffic (Zonta *et al.*, 1985).

Pollutants and suspended particles are more or less retained in the system, settling on the bottom and producing a sludge which is greatly reduced and rich in organics (in the most degraded sites even the superficial layer of the sludge and the water column can be reduced). This process, which is particularly evident in sites characterised by scarce hydrodynamics, produces barriers of sludge that favour the settling of suspended particles and limit the water exchange with other canals or with the Lagoon.

Hydrodynamics in the canals of Venice is complex. Many canals are characterised by small tide differences at the two extremities (Dorigo, 1966; Alberotanza and Dazzi, 1970). During flood-tide, two currents, produced by the tide, enter the canal in opposite directions. They meet in the inner part of the canal, producing in this zone a vertical water movement, whereas horizontal movement is strongly limited. In these conditions of limited water circulation, pollutants can accumulate easily on the bottom of the canal.

Metals can be exchanged at the water/sediment interface when variations of physicochemical conditions occur, for example when fresh water mixes with salt water (Salomons, 1985). In the reduced environment of sludge, bacterial degradation of organic matter causes sulphate reduction to insoluble sulphides (Morse *et al.*, 1987). Metals can be adsorbed by authigenic sulphides or coprecipitated with them on the bottom of the canals. During sediment resuspension, sulphides can be oxidised to sulphates and associated metals can be released to the water (Morse *et al.*, 1987; Morse, 1994). Sediments therefore act both as sinks and as sources of metals (Förstner and Wittmann, 1981).

In this work, the pollution level and the space-time distribution of heavy metals in three sites of a Venice canal were studied. Sediment samples were collected using two different techniques. A comparison between results obtained with the two sampling methods was utilised to evaluate the pollutant exchange efficiency at the water/sediment interface and to determine the parameters that mainly influence these processes.

1.1. STUDY AREA

The canal studied in the present work was the Misericordia canal (Figure 1). It is 1079 m long and has a width ranging between 7 and 17 m. At the two ends it is connected to the northern part of the Lagoon of Venice; at the eastern end it is connected also to the Grand Canal, the main canal of the city.

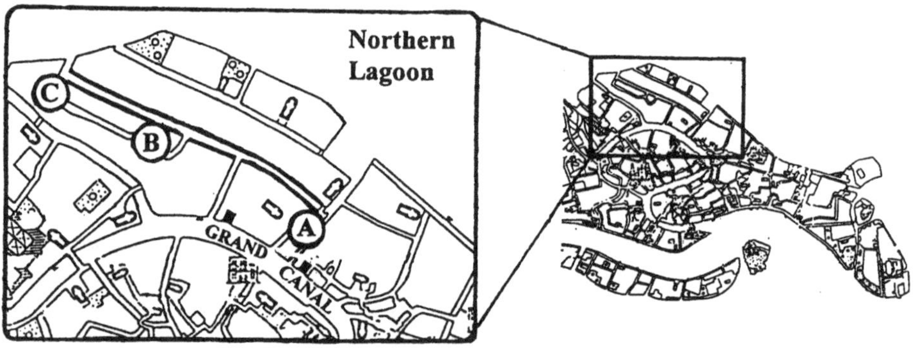

Fig. 1. Map of the city of Venice, showing the Misericordia canal with the three sampling sites,

The Misericordia canal can be considered representative of the network of canals of Venice, since it is characterised by boat traffic, hydrodynamic and pollution conditions similar to those found in other canals of the city.

Along the canal, three sampling sites with different characteristics were considered (Figure 1). Sites A and C, located at the opposite ends of the canal, were respectively 150 cm and 80 cm deep. These sites are characterised by good hydrodynamics, in particular site A, which actively exchanges water with the Northern Lagoon. Site B was located in the central part of the canal and was shallower (50 cm). In this site, characterised by scarce hydrodynamics, sediments tend to accumulate on the bottom of the canal.

2.Materials and methods

In the period March - August 1995, 5 cm sediment cores were collected monthly using a plexiglas syringe type corer; in the same period samples were taken also monthly using 10 cm high plexiglas traps placed on the sediment surface (Sfriso *et al.*, 1992). These traps were used to sample sediments which were resuspended by overlying water turbulence, principally due to boat traffic. Suspended sediments are subjected to variations of environmental physicochemical parameters (principally variations of redox conditions) and, therefore, to possible pollutant exchange at the water/sediment interface. Collected sediments were frozen, lyophilised and carefully homogenised. Samples were preserved at 4 ° C. Sample bottles and other laboratory tools were leached in 0.1 N HNO_3 for 48 hours prior to use. The extraction solutions were prepared using Milli-Q water and BDH Aristar acids (HNO_3 69%, HCl 35%). For total dissolution, Merck Ultrapur acids (HCl 30%, HNO_3 60%, HF 48%) and Merck Suprapur acids (H_3BO_3) were used.

To evaluate the fraction of metal bound to the most labile sediment components and, therefore, more easily exchangeable, fractions of the samples were digested by cold 1 N HCl and by hot 8 N HNO_3. Digestion in Teflon bombs with an acid mixture was used to completely solubilise the sediment and to estimate the total metal concentration (Guerzoni *et al.*, 1987). The first two methods are commonly used to evaluate the anthropogenic fraction of metals (Zonta *et al.*, 1994), even if they don't permit a precise estimation of such component. In fact, the strong oxidising effect and the high extraction temperature of HNO_3, besides permitting the solubilisation of the anthropogenic heavy metal bound to the principal geochemical phases (Fe and Mn oxides/hydroxides, organic matter, carbonates and sulphides), determine also a partial extraction of residual heavy metal by partially solubilising the silicate matrix. Cold extraction with HCl generally permits the evaluation of the anthropogenic fraction, but it can lead to an underestimation of this component for some metals, since it is not fully efficient in the solubilisation of authigenic pyrite (Zaggia *et al.*, 1996).

The solutions obtained were analysed for 9 heavy metals (Cd, Pb, Zn, Cu, Fe, Mn, Ni, Cr, V) by flame or flameless atomic absorption spectrometry (Varian spectrometer model Spectra 250 Plus, equipped with a Varian GTA-96 graphite furnace). Calibration was carried out by using the standard addition technique. Standards were prepared using Milli-Q water and Merck standard solutions (1000 μg/g).

The reproducibility and reliability of data obtained with each extraction solution were tested by applying Student's distribution to the experimental values obtained by ten replicates on the same sample. The reproducibility of the three extraction procedures was for all metals within a 95% C.I. of 10%.

Grain size distribution was determined for representative samples, obtained by mixing aliquots of sediments collected at the same site in different months. The samples were homogenised and sieved at 300 μm. The particle grain size was measured in the range 0.9-300 μm, using a Microtrac particle-size analyser (Leeds and Northrup mod. 7995, USA) which recorded volumetric percentages in 15 diameter size classes (Plantz, 1984).

3. Results and discussion

Mean, maximum and minimum concentrations of total heavy metals in sediments collected by traps and corers at the three sites of the Misericordia canal are shown in Tables 1 and 2.

Concentrations were comparable to those measured in other canals of the city (Chiozzotto and Scattolin, 1994). The recorded mean levels of most heavy metals were either higher (1.5-3 fold) or comparable to those measured in the most polluted areas of the Lagoon of Venice, such as the industrial area of Porto Marghera (Donazzolo *et al.*, 1984).

TABLE I

Mean, maximum and minimum concentrations of total metals (μg/g) in sediments collected by traps

	Site A			Site B			Site C		
	mean	min	max	mean	min	max	mean	min	max
Fe	25700	24900	26700	29600	27900	31900	28600	27100	29800
Zn	919	843	967	1050	911	1280	748	698	830
Cu	241	227	260	323	267	376	218	185	289
Pb	177	160	203	193	181	239	152	146	159
Cd	4.61	4.39	5.06	5.94	5.25	7.42	4.25	3.31	5.27
Mn	371	343	396	388	347	418	404	363	448
Cr	76.9	60.6	91.4	91.2	75.8	128	87.8	77.5	107
Ni	31.7	30.1	33.9	38.0	33.3	43.5	37.6	33.4	42.9
V	59.5	49.5	68.4	69.1	58.0	78.9	67.9	55.8	77.7

TABLE II

Mean, maximum and minimum concentrations of total metals (μg/g)) in sediments collected by corer.

	Site A			Site B			Site C		
	mean	min	max	mean	min	max	mean	min	max
Fe	28500	26000	30200	30100	28200	31200	29700	28500	31300
Zn	1640	1180	2090	990	874	1130	827	728	992
Cu	333	247	428	315	276	354	240	213	301
Pb	246	185	280	197	186	213	200	149	296
Cd	7.76	5.15	9.63	5.53	4.83	6.23	3.68	3.42	3.96
Mn	371	337	418	378	368	390	381	369	403
Cr	80.4	68.3	90.0	90.0	79.9	103	81.9	75.7	88.7
Ni	34.1	27.8	38.5	38.7	33.4	42.5	41.4	39.4	44.6
V	58.7	52.5	69.6	73.9	62.2	86.1	63.8	55.1	76.1

The mean percentages of metal extracted by cold 1 N HCl and hot 8 N HNO_3 were referred to the total concentration (Table 3). The extraction by HNO_3 shows a high efficiency for all the analysed metals. It completely solubilised Cu, Pb, Cd and most of Fe (97%), Zn (97%), Mn (94%) and Ni (88%), and extracted a high amount of Cr (69%) and V (70%). Therefore the extraction by HNO_3 can be a suitable procedure to estimate the total concentration of most of the analysed metals (Cu, Pb, Zn, Cd, Fe, Mn) present in the sludge of the canals of Venice.

It has been demonstrated that hot HNO_3 solubilises the principal mineralogical geochemical phases present in the sediment (Zaggia *et al.*, 1996). In particular, it is able to solubilise carbonates, sulphides, organic matter, Fe-Mn oxides/hydroxides and also some clay minerals, such as smectites and chlorites, leaving only the most refractory minerals (quartz, feldspars and micas), which are characterised by a low amount of associated heavy metals. The very high extraction efficiency shown by HNO_3 for heavy metals associated with the sludge of the canals of Venice is very likely due to their particular mineralogical composition. Such sludges are richer in organic matter and sulphides than sediments collected in the open areas of the Lagoon of Venice, whereas they are poorer in the HNO_3 resistant mineralogical component.

TABLE III

Efficiency of HCl and HNO_3 extractions for heavy metals present in the sludge of the canals of Venice

	Cu	Pb	Zn	Cd	Fe	Cr	Mn	Ni	V
HNO_3	99.9%	100%	96.6%	101%	96.5%	69.3%	93.7%	87.7%	71.6%
HCl	78.5%	93.0%	90.1%	96.3%	39.9%	29.4%	67.5%	35.9%	34.5%

The extracting power of HCl was lower than that of HNO_3. This acid generally extracts metals bound to organic matter, calcite, dolomite, Fe-Mn oxides/hydroxides, and only a limited part of metals bound to sulphides (it is not able to solubilise authigenic pyrite) (Zaggia *et al.*, 1996). However it is not suitable for solubilising other resistant authigenic minerals. HCl extracted a high amount of Cd (96%), Pb (93%), Zn (90%) and Cu (79%). These metals are bound mainly to the most labile geochemical phases of the sediment. HCl also extracted 68% of Mn, 40% of Fe, 35% of V and Ni and 29% of Cr. It has been demonstrated that the metal extracted by HCl corresponds approximately to the total metal fraction bound to four geochemical phases, i.e. exchangeable metal, metal bound to carbonates, metal bound to Fe-Mn oxides/hydroxides, metal bound to organic matter and sulphides. This total metal fraction is considered generally to be representative of the anthropogenic metal (Chester an Voutsinou, 1981; Agemian and Chau, 1976).

With respect to sediments collected in open areas of the Lagoon (Zaggia *et al.*, 1996), HCl demonstrated a higher extraction efficiency for Cu, Zn, Fe and Cr present in the sludge of the canals. These differences, as in the case of the extraction with HNO_3, could be ascribed to the different mineralogical compositions of the collected sediments. The sediments of the canals of Venice are less rich in the HCl refractory components (typically pyrite) than the sediments of the Venice Lagoon. In the canals of the city sulphides are mainly represented by greigite, mackinavite and amorphous iron sulphide (Zaggia and Zonta, 1996), which can be solubilised by HCl. In conclusion, the extraction by HCl enables a good evaluation of the anthropogenic component of metals present in the sludge of the canals.

The data reported in Table 3 underline that in the canals of Venice almost 100% of Cd, Pb and Zn have an anthropogenic origin, as well as Cu. This fact, and the very high levels of contamination found, point out the potential toxicity of the sludge. When variations in physicochemical conditions occur, metals bound to the most labile geochemical phases can be released in the water, becoming in this way bioavailable (Morse, 1994).

3.1 SPATIAL DISTRIBUTION OF HEAVY METAL CONCENTRATIONS IN THE MISERICORDIA CANAL

In general, the spatial distribution of Mn, Fe, Ni, Cr and V in the cores from the three sites can be considered sufficiently homogenous (see Table 2). In contrast, cores collected at site A showed higher levels of Zn, Cd, Pb and Cu, with respect to cores collected from the other sites. These metals have principally an anthropogenic origin and are bound to the most labile component of the sediment.

For sediments collected by traps, site B showed the highest heavy metal levels. This was particularly true for Cu, Pb, Zn and Cd. For the other metals, the differences in concentrations at the three sites were less evident, especially at sites B and C.

3.2 TIME DISTRIBUTION OF HEAVY METAL CONCENTRATIONS IN THE MISERICORDIA CANAL

In general, heavy metal concentrations in core samples extracted with HCl and HNO_3 presented similar time distributions (Figure 2); similar time trends were obtained for samples collected with traps. The variations in concentration with time were dependent on the metal considered.

Generally, the metals characterised by the greater time variability were Cd, Cu, Zn and Pb, i.e. the metals bound to the most labile geochemical phases of sediments and therefore the ones more easily exchanged with overlying water.

3.4. COMPARISON BETWEEN HEAVY METAL CONCENTRATIONS MEASURED IN CORE AND TRAP SAMPLES

Heavy metal concentrations, measured in sediments collected by corer, were compared with those measured in sediments collected by traps. Table 4 reports differences in percentages between heavy metal concentrations in core and trap samples from the three sites. For site B, differences were not significant. In fact, the mean differences for all the metals were lower than 10%. On the contrary, for site A, these differences were evident. In particular some metals showed greater concentrations in the cores. The differences, for the three extracting procedures, were about 42% for Pb and 67% for Cd. Mean differences were about 39% for total Cu and 18% for anthropogenic Cu. For Zn, mean differences were 98% for the anthropogenic fraction, 86% for the HNO_3 extracted fraction and 78% for total metal. Ni and Cr showed evident differences between core and trap only for the anthropogenic fraction (on average 24% for Cr and 25% for Ni), whereas total Ni and Cr in cores and in traps, except for some months, were similar. No significant difference was found for Fe, Mn and V. At site C, differences between heavy metal concentrations in cores and traps were significant only for Pb and, to a lesser extent, for Cd.

At site A the higher differences were shown by Cd, Pb, Zn and Cu, i.e. by the metals which are mostly of anthropogenic origin and are bound mainly to the most labile geochemical phases of sediments. On the contrary, metals that showed slight differences between cores and traps are prevalently bound to the more resistant phases. These findings indicate that differences between heavy metal concentrations in core and trap samples are

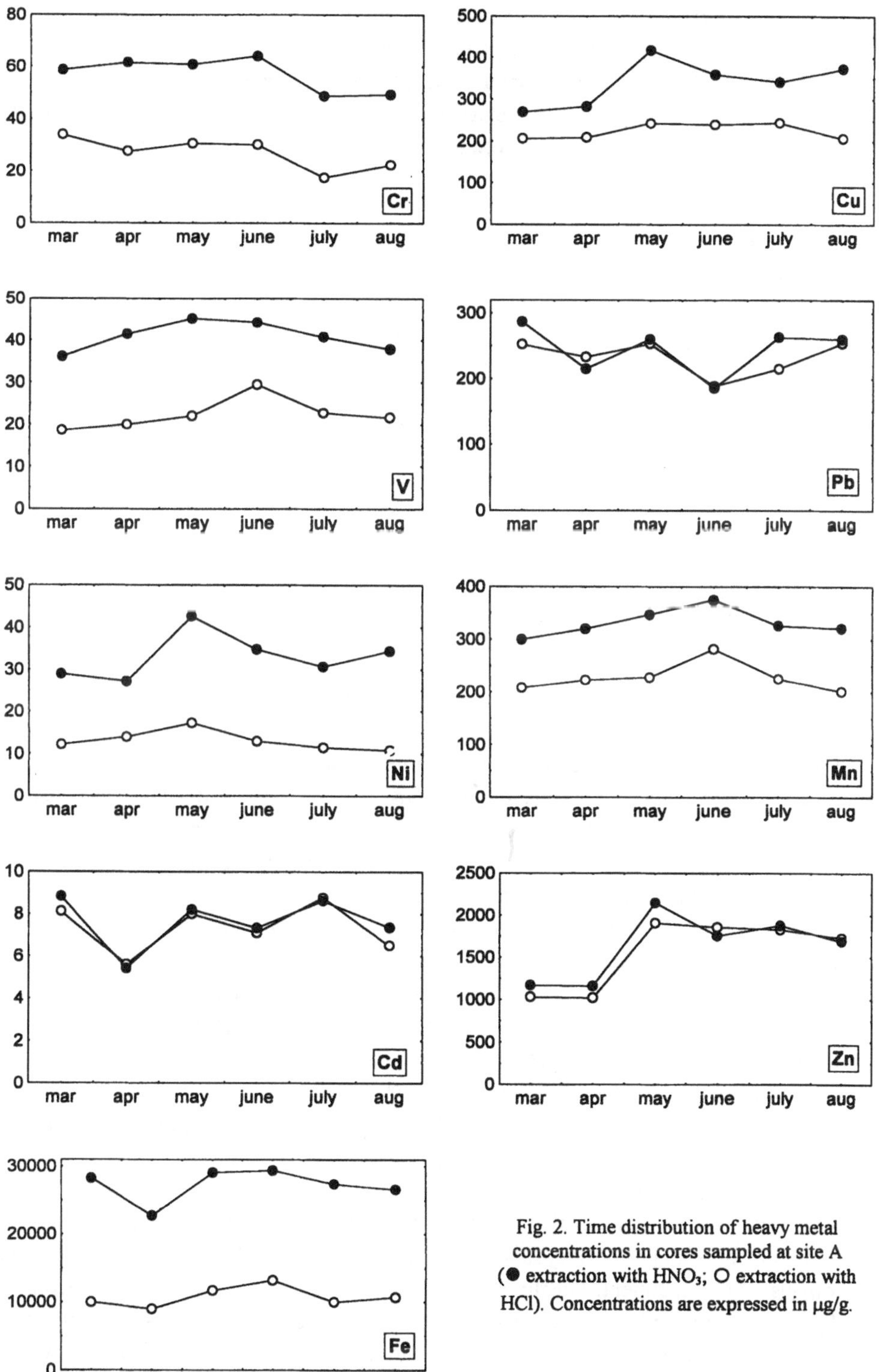

Fig. 2. Time distribution of heavy metal concentrations in cores sampled at site A (● extraction with HNO_3; ○ extraction with HCl). Concentrations are expressed in μg/g.

due to the metal exchange at the water/sediment interface. During resuspension, the most labile phases (such as sulphides and organic matter), passing from a very reduced (in situ sediment) to a more oxidised environment (overlying water column), are partially solubilised, releasing the associated metals into water. The results obtained demonstrated that exchange processes at site A are more efficient than in the other two sites.

TABLE IV

Mean values of the monthly percentage differences between the heavy metal concentrations measured in the sediments sampled by corer and trap in sites A, B and C. The percentage differences were calculated by the formula (C-T)/T × 100, where C and T are the concentrations of metals in core and trap, respectively.

	Site A			Site B			Site C		
	HCl	HNO_3	Tot	HCl	HNO_3	Tot	HCl	HNO_3	Tot
Cu	18	39	38	2.4	-3.8	-1.9	-2.5	7.1	11
Pb	42	44	40	-2.8	-1.2	2.5	19	32	31
Zn	98	86	78	-3.1	-6.5	-5.9	3.4	1.3	9.1
Cd	68	67	67	-3.5	-5.5	-6.7	-10	-13	-12
Fe	7.7	12	11	-8.3	2.9	1.6	-6.5	0.1	3.7
Cr	24	11	6.4	-15	-3.5	0.3	-1.6	-9.6	-7.4
Mn	-9.9	-3.3	0.3	-8.9	-4.2	-2.1	-16	-7.9	-3.3
Ni	25	14	7.6	-8.9	-4.0	2.4	-5.7	6.1	11
V	11	-1.3	-1.2	10	6.2	7.1	-0.5	-3.5	-3.8

This can be explained by considering the different environmental characteristics of the three sites. Site A, which is deeper (150 cm) than the other sites, is characterised by good hydrodynamics and by an active exchange with the more oxygenated water of the Northern Lagoon and Grand Canal. Under these conditions, metal exchange during resuspension is favoured. On the contrary, site B is shallower (50 cm), has very limited hydrodynamics and water exchange with the Northern Lagoon. Finally, site C (80 cm deep) shows intermediate environmental characteristics.

Another important difference between sites A and the other ones depended on the grain size of the resuspended material. At the deeper site A, resuspension principally involves the finer and superficial fraction of the sediment. Grain size distribution of sediments collected by trap was shifted toward finer particles (mean diameter 55.4 μm), whereas larger particles prevailed in the core (mean diameter 92.4 μm). At site B, due to the lower depth, the sediment is resuspended mainly by boat traffic, involving even coarser particle aggregates. In fact, at this site, sediments collected by trap and core showed similar grain size distribution and similar mean diameters (about 83.0 μm).

At site A, the greater depth, the higher hydrodynamics and the grain size characteristics of resuspended particles, keep the sediment in suspension in the overlying oxygenated water column for a longer period, favouring the release of metals bound to the oxidisable components of the sediment. On the contrary, at site B, because of the lower depth, the poor hydrodynamics and the presence in the resuspended sediment of coarse particles, the sediment tends to settle rapidly on the bottom. Moreover, the absence of exchange with the oxygenated water of the Northern Lagoon and the limited hydrodynamics determine poorly oxygenated conditions in the overlying water column, minimising the metal release due to the variation of redox conditions during sediment resuspension.

4. Conclusions

The results of this study indicate that the sludge of the canals of Venice presents levels of contamination comparable to, or higher, than those measured in the most polluted areas of the Lagoon of Venice.

The two different techniques used to collect sediment samples, besides giving rise to reproducible and reliable results, have shown themselves suitable to investigate space-time distributions of various pollutants present in the sludge of the canals. Moreover, the contemporary analysis of samples collected with the two techniques has proved useful in evaluating the intensity of pollutant exchange at the water/sediment interface, and in determining the parameters that mainly influence these processes. In fact, the results show that hydrodynamics, depth, oxygenation levels and the different associative forms between metals and suspended particles play an essential role in the bioavailability of heavy metals.

The most available metals were those of prevalent anthropogenic origin (Cu, Cd, Pb and Zn) and, in particular, the fraction bound to the most labile geochemical phases. These metals also present the highest space-time variability.

References

Agemian, H. and Chau, A. S. Y.: 1976, *Analyst* **101**, 761-767.

Alberotanza, L. and Dazzi, R.: 1970, *Techn. Report n.12, National Council of Research (CNR)*, (in Italian).

Chester, R. and Voutsinou, F. G.: 1981, *Mar. Pollut. Bull.* **12**, 84-91.

Chiozzotto, E. and Scattolin, M.: 1994, *Final Report for the Venice Municipality* (in Italian).

Donazzolo, R., Orio, A. A., Pavoni, B. and Perin, G.: 1984, *Oceanol. Acta* **7**, 25-32.

Dorigo, L.: 1966, *Istituto Veneto di Scienze, Lettere ed Arti, Rapporti e Studi*, vol.3, 129-151 (in Italian).

Förstner, U. and Wittmann, G. T. W.: 1981, in: *Metal Pollution in the Aquatic Environment* (2nd ed.), Springer-Verlag, Berlin, pp.190-227.

Guerzoni, S., Rovatti, G., Molinaroli, E. and Rampazzo, G.: 1987, *Chemistry and Ecology* **3**, 39-48.

Morse, J. W.: 1994, *Mar. Chem.* **46**, 1-6.

Morse, J. W., Millero, J. F., Cornwell, S. C. and Richard, D.: 1987, *Earth Sci. Rev.* **24**, 1-42.

Plantz, P. E.: 1984, in: *Moderm Method of Particle Size Analysis*, J.Wiley & Sons, New York, p.173.

Salomons, W.: 1985, *Environ. Technol. Letters* **6**, 315-326.

Sfriso, A., Pavoni, B., Marcomini, A., Racanelli, S. and Orio, A. A.: 1992, *Environ. Technol.* **13**, 473-483.

Zaggia, L., Argese, E. and Zonta, R.: 1996, *Toxicol. Environ. Chem.* **54**, 11-21.

Zaggia, L. and Zonta, R. : 1996, *Appl. Geochem.*, submitted.

Zonta, R., Argese, E., Costa, F., Racanelli, S., Scattolin, M. and Zaggia, L.: 1985, in: *Proc. Int. Conf. Heavy Metals in the Environment*, Hamburg, pp.439-442.

Zonta, R., Zaggia, L. and Argese, E.: 1994, *Sci. Total Environ.* **151**, 19-28.

DIFFERENCES IN IRON, MANGANESE, AND PHOSPHORUS BINDING IN FRESHWATER SEDIMENT VEGETATED WITH *LITTORELLA UNIFLORA* AND BENTHIC MICROALGAE

KASPER KJELLBERG CHRISTENSEN

Odense University, Institute of Biology, Campusvej 55, DK-5230 Odense M, Denmark

Abstract. Undisturbed sediment cores from an oligotrophic lake were percolated with artificial porewater to examine the effects of isoetid macrophytes, *Littorella uniflora*, and benthic microalgae on daily dynamics of sediment retention of phosphorus (P) by either iron (Fe) or manganese (Mn). Retention of Fe and Mn was observed due to oxidation processes mediated by oxygen release from *L. uniflora* roots and benthic microalgae. Therefore increased retention of P was observed because of P precipitation with oxidized Fe- and Mn-compounds. During light periods, the ratio between Fe and P precipitation in the sediment was positively correlated with the P uptake by *L. uniflora* ($p < 0.001$, $r^2 = 0.984$). The atomic precipitation ratio between Fe and P was between 1 and 2. The ratio between oxidized Fe-compounds and Fe-bound phosphate in the sediment was positively correlated with the root density of *L. uniflora* ($p < 0.001$, $r^2 = 0.995$). The ratio between Mn and P precipitation was higher (≥ 6) than the ratio between Fe and P. The role of benthic primary producers on P retention in freshwater littoral sediments is discussed.

Key words: isoetid, macrophyte, oligotrophic lake, redox potential, Fe, Mn, and P cycling.

1. Introduction

Growth of vascular macrophytes and benthic microalgae can increase phosphorus (P) retention in sediments both by direct phosphate (PO_4^{3-}) uptake and by precipitation of PO_4^{3-} with oxidized iron (Fe(III)) and manganese (Mn(IV)) compounds which form due to oxygen release from the primary producers (Jaynes and Carpenter, 1986; Carlton and Wetzel, 1988; Christensen and Andersen, in press; Christensen *et al.*, in press). Macrophytes can increase redox potential (E_h) in the rhizosphere by root oxygen release (Wium-Andersen and Andersen, 1972; Sand-Jensen *et al.*, 1982; Pedersen and Sand-Jensen, 1995; Christensen *et al.*, in press) whereas benthic microalgae only oxidize the surface of the sediment (Carlton and Wetzel, 1987, 1988; Christensen *et al.*, in press). The degree by which Fe(III) and Mn(IV) precipitates with PO_4^{3-} in the sediment, influences the capacity of the sediment to retain P. Low metal to P precipitation ratios indicate high P retention capacities in sediments (Christensen *et al.*, in press) and high metal concentrations relative to P concentrations in sediments indicate high P binding capacity. Jensen *et al.* (1992) found that the total phosphorus (TP) concentration in Danish shallow lake water was negatively correlated to the ratio between total iron (TFe) and TP in the oxidized surface sediment. When the atomic ratio between TFe and TP was > 8, P binding in the surface sediment would control P release from sediments.

The atomic ratios by which Fe(III) and PO_4^{3-} (Fe:P) and Mn(IV) and PO_4^{3-} (Mn:P) precipitate in sediments are affected by the porewater concentrations of the dissolved ions. Under slightly acidic conditions pure $FePO_4 \cdot 2H_2O$ (strengite) will be precipitated (Fe:P = 1) due to Fe(II) oxidation (if PO_4^{3-} is present), whereas in the neutral and slightly alkaline pH ranges metastable Fe(III)-compounds containing both PO_4^{3-} and hydroxide ions (OH^-) in

Water, Air and Soil Pollution **99**: 265-273, 1997.

variable proportions, depending upon the pH, will be formed (Fe:P > 1). The affinity of Mn(IV) for OH^- is much larger within the pH range of natural waters and therefore Mn(IV)-oxyhydroxides prevalently are formed. When Fe(III)- and Mn(IV)-oxyhydroxides are formed, adsorption of PO_4^{3-} on the surface of the oxyhydroxides is possible, however, the capacity for the metal-oxyhydroxides to bind PO_4^{3-} is much less than for the pure Fe(III)- and Mn(IV)-phosphates (Stumm and Morgan, 1970; Lijklema, 1980; Christensen *et al.*, in press).

The experimental results presented in this paper were obtained in conjunction with a long-term analysis of the influence of the isoetid macrophyte, *Littorella uniflora* (L.) Ascherson, and benthic microalgae on Fe, Mn, and P retention (Christensen and Andersen, in press; Christensen *et al.*, in press). The long-term analysis showed that *L. uniflora* could increase P retention both by P uptake and P precipitation with oxidized Fe- and Mn-compounds in and below the rhizosphere sediment. Also, sediment covered with benthic microalgae showed increased sediment P retention, although the capacity was lower than for sediment with *L. uniflora*. Both for sediment with *L. uniflora* and microalgae, Fe showed higher P binding capacity than Mn due to lower Fe:P precipitation ratios and more efficient Fe retention than Mn:P precipitation ratios and Mn retention. In the present study P, Fe, and Mn retentions were investigated in short-term porewater percolation experiments. By comparison of P retention with Fe and Mn retentions, the Fe:P and Mn:P precipitation ratios in the sediment were calculated and compared with the P uptake by *L. uniflora* and benthic microalgae. This comparison was not possible in the long-term studies. In this paper it is shown, that P uptake by benthic primary producers inhabitating littoral zones, can probably increase the Fe:P precipitation ratio in the surface sediments and thereby increase the P binding capacity.

2. Material and methods

Undisturbed sediment cores with *L. uniflora* and bare sediment with a microalgal layer were collected in the oligotrophic/dystrophic shallow softwater Danish Lake Kvie. A description of Lake Kvie is available in Olsen and Andersen (1994). In each experiment 5 cores (32 cm long, 8 cm inner diameter) with *L. uniflora* and 5 cores with microalgae were collected. After returning to the laboratory, the cores were adjusted to a sediment height of 12 cm and inserted in core-holders (see Christensen and Andersen, in press, for description of the setup). The core-holders were connected separately at the bottom to a reservoir of artificial porewater supplied by a peristaltic pump that percolated the artificial porewater up through the sediment (~6 l m^{-2} h^{-1}). The experiments were performed at 15°C, 12/12 light/dark cycles, and with an irradiance of ~200 µmol photons m^{-2} s^{-1} at the sediment surface during light periods. Anoxic artificial porewater with nutrients and a pH of 6.15 was produced in a gas proof plastic bag as described by Christensen and Andersen, in press. Three time-course experiments were carried out in the period from July to December 1994: one for examination of P retention, one for examination of P and Fe retentions, and one for examination of P and Mn retentions. The artificial porewater contained 100 µM PO_4^{3-} and because Mn compared with Fe has lower capacity for PO_4^{3-} binding (Stumm and Morgan, 1970), a higher concentration of Mn (300 µM) compared to Fe (50 µM) was added to the porewater. Retention of P, Fe, or Mn over time was calculated as the difference between the amount of PO_4^{3-}, Fe(II), or Mn(II) added with the porewater and the amount of total phosphorus (TP), total iron (TFe), or total manganese (TMn) recovered in the overlying water.

Before the time-course measurements of P, Fe, and Mn retentions were carried out, the sediment was preincubated with porewater percolation for 30 to 46 days to obtain steady-state between daily retention relative to daily addition (see Christensen and Andersen, in press; Christensen *et al.*, in press). This design ensured that the P retention measured in the experiment with only P in the porewater was due to P uptake by the primary producers because the sediment has been physically/chemically saturated with P in the preincubation period. In the experiment with both P and Fe in the porewater, Fe retention was caused by precipitation of solid Fe(III)-compounds in the sediment, and therefore the P retentions were both due to P uptake by the primary producers and P precipitated with Fe(III)-compounds. The P precipitated with Fe(III)-compounds can be calculated by subtraction of the P retention measured in the experiment with only P in the porewater from the P retention measured in the experiment with both Fe and P in the porewater. This calculation assumes that P uptake by plants in the two experiments was identical and that the Fe uptake by plants was negligible. In the experiment with P and Mn in the porewater, precipitation of Mn(IV)-compounds with P was analyzed. When retentions were measured there were no significant differences in shoot biomass of *L. uniflora* among the three treatments (one way ANOVA: $P >> 0.05$, biomass = 430 ± 16 g dry weight m^{-2}, mean ± SE, n = 15).

Measurement and calculations of redox potential were performed according to Hargrave (1972) using a platinum electrode with a calomel electrode as a reference. Oxidized Fe and Fe-bound PO_4^{3-} in the sediment were extracted with a bicarbonate-dithionite solution (Jensen and Thamdrup, 1993). For a closer description of the chemical methods used see Christensen *et al.*, in press.

Differences in Fe, Mn, and P retentions between the two sediment types over time was tested with two-way ANOVA with repeated measures. For comparison of retention efficiencies between sediment type over time multiple unplanned comparison by the T-method (equal sample size) was used. Two-tailed t-test was used for comparison of E_h between the two sediment types (Sokal and Rohlf, 1995).

3. Results and discussion

3.1. SEDIMENT CHARACTERISTICS

The density of the wet, sandy sediment was 1.8-1.9 g cm^{-3} in both sediment types with loss on ignition of 0.5-1.5% of dry weight. In all experiments *L. uniflora* increased their biomass during the incubation period by more than 100% the initial biomass and oxygen fluxes between water and sediment confirmed high productivity in both cores with *L. uniflora* and microalgae in light periods. Redox potential (E_h) measured during the light in the experiment with only P in the porewater, showed significantly ($P < 0.001$) higher E_h in cores with *L. uniflora* than in cores with microalgae except at the sediment surface and in the overlying water (Figure 1) confirming that macrophytes can increase E_h in the entire rhizosphere by root oxygen release whereas benthic microalgae can only oxidize the surface of the sediment.

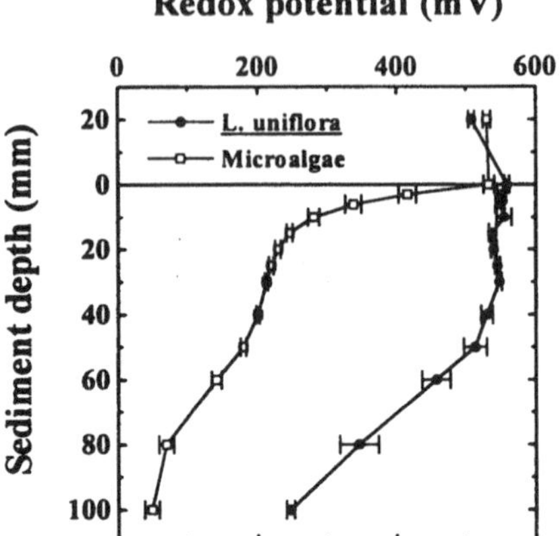

Fig. 1. Redox potential (E_h) in cores *L. uniflora* in cores with benthic microalgae. Average values ± SE for 5 cores with *L. uniflora* and 5 cores with microalgae. The figure is redrawn from Christensen and Andersen, in press.

3.2. P UPTAKE BY *L. UNIFLORA* AND MICROALGAE

Sediment inhabited with *L. uniflora*, showed always significantly ($P < 0.01$) higher P retention than sediment with microalgae (Figure 2). The measured P retention is due to P uptake by *L. uniflora* and microalgae because the sediment was saturated with P in the preincubation period and there was no Fe or Mn in the porewater. P retention by *L. uniflora* increased from 230 to 320 μmol m^{-2} h^{-1} by transition from darkness to light. However, after 4 h in light P retention was reduced which was probably caused by reduced P assimilation by *L. uniflora* and porewater advection due to oxygen bubbles ascending from the sediment in light (Christensen and Andersen, in press). For microalgae there also was an increase in P retention from darkness (-20 μmol m^{-2} h^{-1}) to light (230 μmol m^{-2} h^{-1}) followed by a gradual decrease over time in light indicating saturation of algae P assimilation.

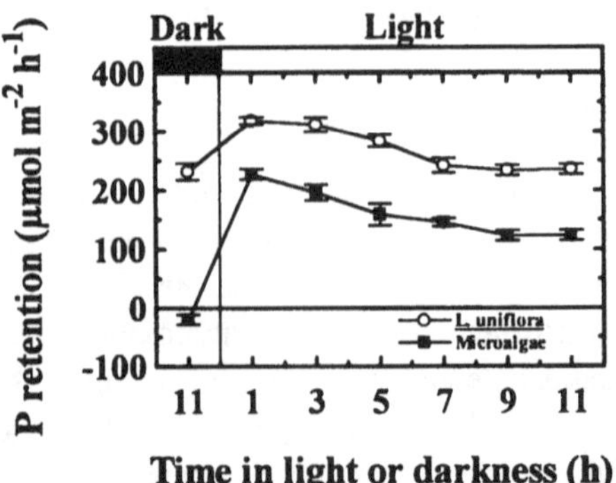

Fig. 2. P retention by *L. uniflora* and benthic microalgae. The cores received artificial porewater containing 100 μM PO_4^{3-}. The data represents average values ± SE of five cores for 2 h periods. The negative value indicates a release of P from the sediment/microalgae higher than the amount P added by the artificial porewater.

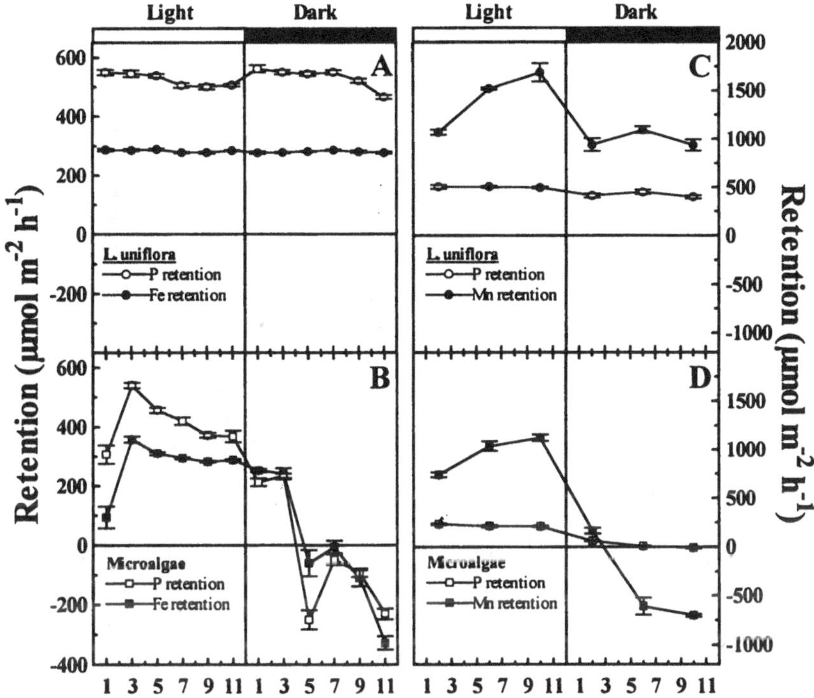

Fig. 3.
Left panel: Fe and P retentions analyzed in a 12 h light and a 12 h dark period in cores with *L. uniflora* (A) and microalgae (B). The cores received artificial porewater containing 50 µM Fe(II) and 100 µM PO_4^{3-}. The data represents average values ± SE of five cores for 2 h periods.
Right panel: Mn and P retention analyzed in a 12 h light and a 12 h dark period in cores with *L. uniflora* (C) and microalgae (D). The cores received artificial porewater containing 300 µM Mn(II) and 100 µM PO_4^{3-}. The data represents average values ± SE of five cores for 4 h periods.
Note the different retention scales between left and right panel. Negative values indicate a release of Fe, Mn, or P from the sediment/microalgae higher than the added amounts of Fe, Mn, or P by the artificial porewater.

3.3. Fe RETENTION

Sediment with *L. uniflora* and sediment with microalgae showed high Fe retention in light (Figure 3AB). However, in darkness Fe retention was reduced in sediment with microalgae whereas Fe retention remained constant in sediment with *L. uniflora*. High Fe retention in light was due to Fe(II) oxidation mediated by oxygen release from *L. uniflora* roots and microalgae. However, in sediment with *L. uniflora*, high Fe retention was also measured in darkness indicating that oxygen release from *L. uniflora* roots was sufficient to maintain Fe(II) oxidation. This is in agreement with observations by Christensen *et al.* (1994) who showed oxygen release from *L. uniflora* roots in both light and darkness. Carlton and Wetzel (1987) showed that benthic microalgae were only able to keep the upper part of the sediment

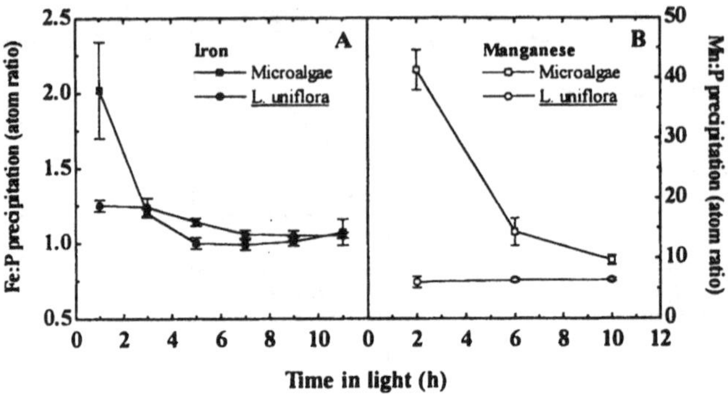

Fig. 4. Ratios of Fe and P (A), and Mn and P (B) precipitation in the sediment over time in light in the experiments with Fe or Mn in the porewater, respectively. The ratios were corrected for P uptake by *L. uniflora* and microalgae. The data represents average values ± SE of five cores for 2 h (Fe:P) and 4 h (Mn:P) periods.

oxidized in light. The P retention in sediment percolated with Fe and P with *L. uniflora* (Figure 3A), was higher in both light and darkness (~500 μmol m^{-2} h^{-1}) when compared to the experiment with only P in the porewater (Figure 2) due to P precipitation with oxidized Fe species. For cores with microalgae, P retention in presence of Fe was also higher in light when compared to the experiment with only P in the porewater, however, in darkness P release from the sediment was observed due to low oxidizing capacity by microalgae in darkness (Figure 3B).

3.4. Mn RETENTION

The chemical oxidation of Mn(II) is much slower and requires higher E_h than the oxidation of Fe(II) (Patrick and Henderson, 1981; Luther, 1990) and therefore lower Mn retention compared to Fe retention was expected. Mn retention reached higher values than Fe retention because the concentration of Mn in the porewater was higher than the concentration of Fe (Figure 3CD). Mn retention was more sensitive to changes from light to dark condition than Fe retention. Thus, in sediment with both *L. uniflora* and microalgae, Mn retention increased with time in light ($P < 0.01$) and decreased significantly ($P < 0.001$) immediately after transition to darkness. This indicates the oxidizing capacity in darkness was lower than in light because of the decrease in oxygen release from the *L. uniflora* roots (Christensen *et al.*, 1994) and benthic microalgae (Carlton and Wetzel, 1987). Thus, higher dependency of Mn retention compared to Fe retention by changes in light/dark condition is in agreement with much easier oxidation of Fe(II) when compared to Mn(II). The change in P retention for cores with *L. uniflora* with Mn was smaller (Figure 3C), however, P retention was significantly ($P < 0.05$) higher in light (~505 μmol m^{-2} h^{-1}) than in darkness (~420 μmol m^{-2} h^{-1}). The small change in P retention indicates a low capacity for Mn(IV)-compounds to bind PO_4^{3-} (see below). Also for sediment with microalgae, the relation between Mn and P retention was much less pronounced than the relation between Fe and P retention (Figure 3D).

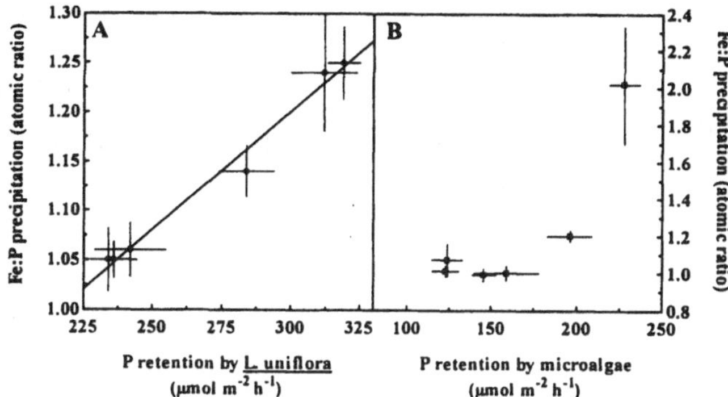

Fig. 5. Correlation between P retention by *L. uniflora* (A) and microalgae (B) and the precipitation ratio between Fe and P in the sediment. The data represents average values ± SE of five cores.

3.5. Fe:P AND Mn:P PRECIPITATION RATIOS

The Fe:P and Mn:P precipitation ratios in the sediment were calculated in the light periods by correcting for P assimilated by *L. uniflora* and microalgae (Figure 4). In cores with *L. uniflora*, the Fe:P precipitation ratio was ~1.25 at the start of the light period but was reduced to ~1.05 at the end of this period. In cores with microalgae, the Fe:P precipitation ratio was high (~2) initially after the dark period but stabilized to about 1.0 after 4 h in light. The Fe:P precipitation ratio in the sediment was positively correlated with the P uptake by *L. uniflora* ($p < 0.001$, $r^2 = 0.984$, Figure 5A). Also, high microalgae P uptake was coincident with high Fe:P precipitation ratios, however, microalgae P retentions below 175 µmol m^{-2} h^{-1} resulted in Fe:P precipitation ratios of ~1 (Figure 5B). Thus, PO_4^{3-} uptake by benthic primary producers may influence the ratio by which Fe(III) precipitates with PO_4^{3-} in sediments. When P uptake by *L. uniflora* or microalgae increase, the concentration of PO_4^{3-} in the porewater will decrease and consequently less PO_4^{3-} is available for precipitation with oxidized Fe or Mn species and the precipitation ratio between Fe:P and Mn:P will increase. These results are in agreement with both crystallization experiments (Lijklema, 1980; Hsu, 1982) and geochemical models (Fox, 1989; Griffloen, 1994) which show that the ratio between Fe(II) and PO_4^{3-} in solution will influence the Fe:P ratio in the precipitated Fe(III)-phosphates.

The Mn:P precipitation ratio was constant at ~6 during the entire period in cores with *L. uniflora* (Figure 4B). In cores with microalgae, the Mn:P precipitation ratio was high (~40) just after the dark period but after 4 h in light the ratio was reduced to ~12. Higher Mn:P precipitation ratios compared to Fe:P precipitation ratios are in agreement with the solubility theory concerning Fe, Mn, and P at slightly acidic conditions (Stumm and Morgan, 1970). Thus, Mn has much higher affinity for OH^- ions than Fe(III), and therefore precipitation of metastable Mn(IV)-compounds containing PO_4^{3-} and OH^- is predicted. The Mn:P precipitation ratio was higher for sediment with microalgae when compared to sediment with *L. uniflora*. Also, the Fe:P precipitation ratio in sediment with microalgae was higher than the Fe:P precipitation ratio in the sediment with *L. uniflora*. This might be an effect of increased pH in the surface sediment in light in cores with microalgae due to photosynthetic activity (Carlton and Wetzel, 1987, 1988). An increase in pH will increase the competition between

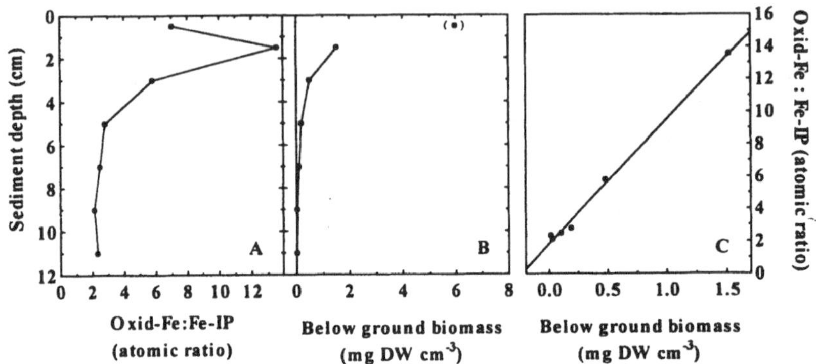

Fig. 6. A: Ratios of the concentrations between oxidized Fe-compounds and Fe-bound PO_4^{3-} (Oxid-Fe:Fe-IP) in the sediment at the end of the experiment with Fe in the porewater. B: The below ground biomass of *L. uniflora* decreased exponentially with sediment depth. C: There was a significant ($p < 0.001$, $r^2 = 0.995$) positive correlation between the root biomass and the Oxid-Fe:Fe-IP ratio in the end of the experiment. All points represent a pooled sample from five cores. The sediment Fe and P pools are recalculated from Christensen *et al.*, in press.

OH^- and PO_4^{3-} for binding sites to Mn(IV)- and Fe(III)-compounds (Stumm and Morgan, 1970) and therefore the precipitation ratio is increased.

The increased Fe:P ratio due to *L. uniflora* P uptake was also indicated by an analysis of the *L. uniflora* root biomass compared with the ratio between the concentration of oxidized Fe and Fe-bound PO_4^{3-} in the sediment (Figure 6). At the end of the experiment with Fe in the porewater, the ratio between the concentration of oxidized Fe and Fe-bound PO_4^{3-} and the root biomass was highest in the surface of the sediment (except in the 0-1 cm interval) and decreased with sediment depth (Figure 6A,B). Therefore a positive correlation between concentrations of oxidized Fe and Fe-bound PO_4^{3-} in the sediment and the root biomass was found ($p < 0.001$, $r^2 = 0.995$, Figure 6C). Data from the upper 1 cm were not included in this correlation analysis because the below ground biomass constituted mainly of stems which probably have much lower P uptake capacity than roots.

4. Conclusions

Benthic primary producers can increase P retention in sediments by direct PO_4^{3-} uptake and by mediating precipitation of PO_4^{3-} with Fe(III)- and Mn(IV)-compounds. However, P uptake by benthic primary producers may also influence the ratio by which PO_4^{-3} precipitates with Fe(III). Thus, areas with high benthic primary production can probably increase the Fe:P precipitation ratio in the sediments, and thereby increase the PO_4^{3-} adsorption capacity. Areas with isoetid growth may be major sites for increased PO_4^{3-} adsorption capacity due to their evergreen life strategy, P assimilation, and oxygen release deeper in the sediment than benthic microalgae. However, increased PO_4^{3-} adsorption capacity created in the growth period of sites with non-evergreen macrophytes may also be important due to reduced P release from these sediments in the unproductive period of the year.

Acknowledgements

This study was supported by The Danish Environmental Research Programme. I thank F.Ø Andersen, H.S. Jensen, S. Findlay, and C. Wigand for helpful comments on the manuscript.

References

Carlton, R.G. and Wetzel, R.G.: 1987, *Can. J. Bot.* **65,** 1031-1037.

Carlton, R.G. and Wetzel, R.G.: 1988, *Limnol. Oceanogr.* **33,** 562-570.

Christensen, K.K. and Andersen, F.Ø.: in press. Influence of *Littorella uniflora* on phosphorus retention in sediment supplied with artificial porewater. *Aquat. Bot.*

Christensen, K.K., Andersen, F.Ø. and Jensen, H.S.: in press. Comparison of iron, manganese, and phosphorus retention in freshwater littoral sediment with growth of *Littorella uniflora* and benthic microalgae. *Biogeochemistry.*

Christensen, P.B., Revsbech, N.P. and Sand-Jensen, K.: 1994, *Plant Physiol.* **105,** 847-852.

Fox, L.E.: 1989, *Geochim. Cosmochim. Acta* **53,** 417-428.

Griffloen, J.: 1994, *Environ. Sci. Technol.* **28,** 675-681.

Hargrave, B.T.: 1972, *Oikos* **23,** 167-177.

Hsu, P.H.: 1982, *Soil Sci. Soc. Am. J.* **46,** 928-932.

Jaynes, M.L. and Carpenter, S.R.: 1986, *Ecology* **67,** 875-882.

Jensen, H.S., Kristensen, P., Jeppesen, E. and Skytthe, A.: 1992, *Hydrobiologia* **235/236,** 731-743.

Jensen, H.S. and Thamdrup, B.: 1993, *Hydrobiologia* **253,** 47-59.

Lijklema, L.: 1980, *Environ. Sci. Technol.* **14,** 537-541.

Luther, G.W. : 1990, 'The frontier-molecular-orbital theory approach in geochemical processes', in Stumm, W. (Ed.), *Aquatic chemical kinetics: reaction rates of processes in natural waters,* pp. 173-198. Wiley, New York.

Olsen, K.R. and Andersen, F.Ø.: 1994, *Hydrobiologia* **275/276,** 255-265.

Patrick, W.H., Jr. and Henderson, R.E.: 1981, *Soil Sci. Soc. Am. J.* **45,** 855-859.

Pedersen, O. and Sand-Jensen, K.: 1995, *Ecology* **76,** 1536-1545.

Sand-Jensen, K., Prahl, C. and Stokholm, H.: 1982, *Oikos* **38,** 349-354.

Sokal, R.R. and Rohlf, F.J.: 1995, *Biometry,* W.H.Freeman and company, New York.

Stumm, W. and Morgan, J.J.: 1970, *Aquatic chemistry. An introduction emphasizing chemical equilibria in natural waters,* Wiley, New York.

Wium-Andersen, S. and Andersen, J.M.: 1972, *Limnol. Oceanogr.* **17,** 948-952.

SELENIUM ASSOCIATIONS IN ESTUARINE SEDIMENTS: REDOX EFFECTS

GREGORY M. PETERS[1], WILLIAM A. MAHER[2], JOHN P. BARFORD[1], VINCENT G. GOMES[1]
[1]*Chemical Engineering Department, University of Sydney 2006, Australia,* [2]*Faculty of Applied Science, University of Canberra, PO Box 1, Belconnen ACT 2616, Australia*

Abstract. Selenium (Se) is a contaminant of concern in environments affected by discharges from smelting and coal-burning industries. Experiments have been performed to investigate the phase associations of selenium in contaminated sediments under a range of controlled redox conditions. In this study, Se sediment associations were examined using the BCR sequential extraction technique after stabilisation at different redox states. It was shown that although most of the sediment-bound Se is associated with the operationally-defined "organic/sulfide" fraction, as the measured redox potential of the system is increased, more Se moves into the "exchangeable" and "iron/manganese oxyhydroxide" fractions. In these fractions, contaminants can be expected to be more bioavailable. As the mass of Se absorbed to sediments is typically at least an order of magnitude higher than the mass dissolved in porewaters, significant Se exposure may result from oxidative shifts in Se associations.

Keywords: selenium, sediment, estuaries, sequential extraction, redox potential

1. Introduction

1.1 LAKE MACQUARIE

Lake Macquarie is the largest barrier lagoon in eastern Australia and is located 100 km north of Sydney on the New South Wales coast. This estuary is an important ecosystem for commercial and recreational fishing and has a history of localised industrial use focused on Cockle Creek, which flows into its northern reaches.

Total heavy metal analysis of sediments has shown Cockle Creek to be the primary source of heavy metal pollution (Roy & Crawford, 1984; Batley, 1987). A lead-zinc smelter owned by CRA/Pasminco has released Se, Pb, Zn, Cd and Cu to the sediment of the northern Lake since 1898. Near the mouth of Cockle Creek, Batley (1987) found above-background concentrations of these contaminants, including Se at 14 mg/kg. The coal-fired power stations of Delta Power are responsible for Se inputs via ash-dam runoff and atmospheric emissions (Woodford, 1995).

Roberts (1994) reported high levels of Se in commercially-valuable fish taken from Lake Macquarie. Mullet (*Mulgil cephalus*) and silverbiddy (*Gerres ovatus*) had average muscle tissue Se concentrations of 10 and 8.2 $\mu g.g^{-1}$ DW. These values correspond to 2.0 and 1.6 $\mu g.g^{-1}$ on a wet basis. Since the Australian National Health and Medical Research Council (NHMRC) allowable limit for human consumption is 1 $\mu g.g^{-1}$, the fish are significantly contaminated. The maximum concentration of Se in mullet was 64 mg/kg (dry basis), twelve times the NHMRC limit (Roberts, 1994).

1.2 SEQUENTIAL EXTRACTION

Contaminants can be bound to different components of mineral phases and organic matter present in sediment. Sediment organisms are less likely to be exposed to Se held within the crystal lattice of sediment grains than dissolved Se in the porewater. Between

Water, Air and Soil Pollution **99**: 275-282, 1997.

these two extremes, adsorbed contaminants may become bioavailable at different rates due to rate changes in processes like bioturbation, ingestion, and variations in dissolved oxygen, nutrients and salinity in the overlying seawater (Tessier *et al.*, 1979) To examine the associations between contaminants and sediments, sequential extraction schemes have been developed. For many years, the unofficial benchmark has been the scheme of Tessier *et al.* (1979), giving rise to that of Maher (1984), Salomons and Förstner (1985), Ma and Uren (1995), Pickering (1986) and Ure *et al.* (1993). Differences between these schemes are discussed in the last two references. Indications are that Tessier's second and relatively strong third extraction steps remove a significant proportion of the heavy metals bound to mackinawite (FeS) (Wallmann *et al.*, 1993). In the presence of such a diversity of approaches to sequential extraction, the European Community Bureau of Reference has made an intensive attempt to develop a standard, resulting in the "BCR Scheme" (Ure *et al.*, 1993 - see Table 1).

Table 1: Comparison of the "Tessier" and "BCR" sequential extraction schemes

Nominal phase	Tessier *et al.* (1979)	Ure *et al.* (1993)
"Exchangeable"	1 M $MgCl_2$	0.11 M CH_3COOH
"Carbonate"	1 M $NaOAc/CH_3COOH$ pH5	
"Mn/Fe Oxyhydroxides"	0.4 M $NH_2OH.HCl$ in 25% CH_3COOH	0.1 M $NH_2OH.HCl$, pH2
"Organic + Sulfide"	30% H_2O_2/HNO_3 pH2 then 3.2 M NH_4Ac in 20% HNO_3	8.8 M H_2O_2, 1 M CH3COONH$_4$, pH2
"Residual"	Conc. $HF.HClO_4$	

The names of phases listed above are used here for reasons of simplicity. The actual phases extracted can only be defined "operationally", that is, in terms of the extractants used. Much effort has been expended by the various workers in order to improve the effectiveness and selectivity of the extractions and this will undoubtedly continue.

1.3 SELENIUM

Se is a trace element present in many different compounds in various livingorganisms. Found naturally as Se(II-), Se(0), Se(IV) and Se(VI), the metalloid undergoes microbially catalysed transformations between these forms. In anoxic sediments, it is an electron acceptor along with iron, nitrate, sulfate and other compounds (Oremland *et al.*, 1989). Since it is usually present in nanomolar or lower concentrations, Se is relatively insignificant geologically but the mineralisation process is an important regulator of concentrations of toxic selenooxyanions in sediments. Biological transformations from Se(VI) to Se(IV) and Se(0) have been observed in different environments by several workers (Gerhardt *et al.*, 1991; Maiers *et al.*, 1988; Oremland *et al.*, 1989; Steinberg & Oremland, 1990) over a range of redox conditions (Masscheleyn *et al.*, 1991).

1.3.1 BIOACCUMULATION AND TOXICITY

Despite its status as an essential trace element, Se is a contaminant capable of being bioaccumulated, biomagnified and causing great damage to avifauna (Ohlendorf, 1989).

Of the two oxidised forms, selenite is generally the more toxic form (Maier & Knight, 1993) and the more easily bioaccumulated, although organic forms such as seleno-DL-methionine may be even more toxic and readily bioaccumulated (Besser *et al.*, 1993). About 5 mg Se per day is considered toxic in the human diet, making Se the third most toxic trace element after mercury and lead (Alloway & Ayres, 1993). Swedish research has shown fish consumption to be a significant route of human Se exposure (Svensson *et al.*, 1991). The NHMRC standard for Se in fish muscle tissue is 1.0 $\mu g.g^{-1}$ (wet basis). Liu *et al.* (1987) performed extensive analyses of an aquatic food web. The researchers determined that biomagnification was most significant in benthic environments. Significant bioaccumulation can occur in aquatic food chains at water concentrations as low as 1 $\mu g.L^{-1}$ (Peterson & Nebeker, 1992).

1.3.2 SOLUBILITY AND SORPTION - PH AND EH DEPENDENCE

In sediment systems, the mobility of Se is highly dependent on its redox state. Elemental Se is relatively insoluble as are the metal salts of selenide, and these forms of Se can therefore be expected to remain immobilised in anoxic sediments (Elrashidi *et al.*, 1987). Selenite and selenate are much more soluble. A massive remobilisation of Se can be expected when selenides and elemental Se are oxidised to these forms. Based on experiments, Masscheleyn & Patrick (1993) stated that the redox potential at which the transformation between selenite and the less-mobile forms of elemental Se occurs is around -20 mV, a level between the reduction of iron oxides and sulfates. (They did not distinguish between Se(-II) and Se(0).) The selenite-selenate transformation occurs at about 270 mV, about the level at which denitrification occurs in environmental systems. Steinberg *et al.* (1992) confirmed this in a system moving to anoxia, showing that nitrate reduction precedes selenate reduction by selenate-respiring bacteria and the presence of nitrate precludes selenate reduction.

Chemically reducing conditions, expected in sediments below the redox discontinuity level (a depth at which a sudden negative swing in E_h occurs), result in Se solubility being controlled by the formation of elemental Se or insoluble metal selenides such as FeSe (Masscheleyn & Patrick, 1993). In moderately reduced sediments, adsorption mechanisms dominate Se behaviour. Several workers have examined the influence of pH and competing ions on Se adsorption. Iron oxyhydroxides are thought to be the most significant scavengers of Se oxyanions (Masscheleyn & Patrick, 1993). Selenite, which competes with phosphate for adsorption sites in soil matrices, is adsorbed to a greater extent than selenate which competes with sulfate. Above pH 7.5, selenate adsorption is negligible. Neal *et al.* (1987) found that selenite sorption dropped from 50 to less than 5% as soil pH was raised from 4 to 9. Due to adsorption and/or precipitation mechanisms, sediments contain the major part of the total estuarine Se inventory.

1.4 RATIONALE FOR THIS WORK

Maher *et al.* (1992) have reviewed current literature on Se in Australian marine environments and were concerned that Se has not been sufficiently characterised in Australian ecosystems for regulators to know whether it poses a significant environmental hazard. Currently, a major study involving public and private collaboration coordinated by the Lake Macquarie Health Unit is collecting data on the level of contamination of commercially harvested fish and invertebrate species (Farmer, 1995). While this will provide useful information on the distribution of Se in seafood from the Lake, there is a lack of data on current or potential contributions to Se in the food-web from contaminated Lake sediments. It is not known whether previously contaminated sediments, via bioturbative or human disruption of normal diagenetic processes, could result in significant Se fluxes compared to current industrial and municipal inputs.

In Masscheleyn *et al.* (1991), detailed attention was paid to the speciation of soluble and volatile Se compounds in a redox-controlled freshwater slurry. Masscheleyn *et al.* (1990) describes similar work on a brackish slurry. Experiments have shown that the bulk of Se in a sediment-water system is bound to sediments. In these two papers and in our own experiments (Peters *et al.*, 1996), equilibrium porewater Se concentrations at natural pHs are typically much lower than the sediment concentrations. They were closest in Masscheleyn *et al.* (1990) when at pH 7.5 and E_h 450 mV, 257 $\mu g.L^{-1}$ dissolved from sediments containing 9.1 $\mu g.g^{-1}$ Se. If adsorbed Se changes its sediment association to the point where it is almost as bioavailable as the dissolved Se, it may represent a greater contribution to total bioavailable Se in contaminated sediments. Therefore, while it is important from a regulatory perspective to determine the amount of soluble Se in a contaminated sediment ecosystem, we decided to examine the influence of redox conditions on the phase associations of sediment Se.

2. Method

Sediments were taken from a location near Nord's Wharf, an area of the Lake relatively well characterised with respect to Se levels and particle size distribution (Scholz *et al.*, 1995). The sediment was wet sieved (2 mm) to remove macrofauna and large inhomogeneities.

Figure 1: Schematic of apparatus

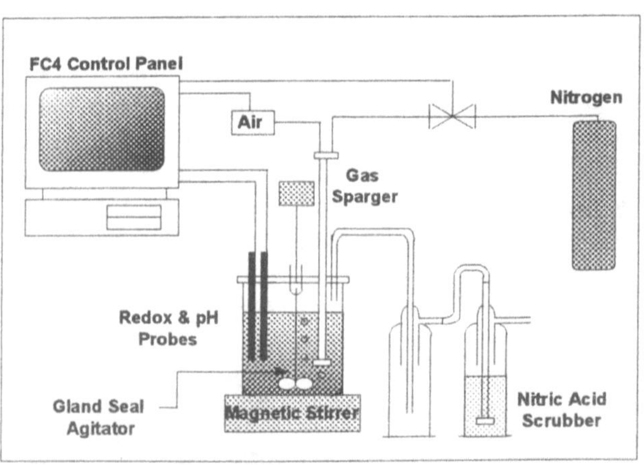

The laboratory reactor system used a modification of the E_h-pH control system of Masscheleyn *et al.* (1990) and is shown in Figure 1. It consists of a stirred reactor linked to a computer via a data aquisition and control device (FC4: Real Time Engineering, Sydney). This converts the potentials generated by the pH and E_h electrodes into a digital signal and compares them with the operator's setpoints. If the measured E_h is too low, the FC4 opens a solenoid valve to begin nitrogen sparging. An air pump is activated if the E_h is too low. To the 800 mL reactor, 105 g of sediment (dry basis) was added. The reactor was then filled with water collected from Lake Macquarie to produce a slurry mass ratio of approximately 1:7. The slurry was initially oxidised to 400 mV. We found air sparging inefficient in raising the E_h above 350 mV and therefore employed oxygen for further oxidation to 400 mV. In this oxidised state 3.78 mg of selenious acid was added (2.47 mg Se(0)), resulting in a 23.3 $\mu g.g^{-1}$ Se in the sediment on a dry basis. Four E_h setpoints were used: -250, -0, 200 and 400 mV, the lowest of these being reached by the combination of microbial reduction and nitrogen sparging. Stirring via a magnetic bar was used in initial trials but rejected in favour of overhead stirring on the basis that the average particle size and the bar itself were being reduced by abrasion.

2.3 ANALYSIS

The BCR sequential extraction scheme was chosen to examine Se-sediment associations. All sediment manipulations were carried out in a glove box. In order to maintain anoxic conditions in sediment samples removed from the reactor for $E_h \leq 0$ mV, water for making up the BCR extractant solutions was boiled and then cooled under high-purity nitrogen sparging. After five day's stabilisation at each setpoint, sediment slurry samples were removed from the reactor and centrifuged at 3500 rpm for 25 minutes. The sediment was divided into 3 equal parts: one to determine the dry weight of the sample by evaporation at 100°C and two for duplicate sequential extraction and Se analysis. The total Se concentration in extracts was determined according to the hydride generation atomic absorption spectroscopic method of Fio and Fujii (1991). Iron was determined by flame atomic absorption spectroscopy.

3. Results and discussion

The results of the analyses of sequential extracts are shown in Figures 2 and 3. Figure 2 shows that at each redox potential, over 85% of the Se resided in the "organic/sulfide" fraction of the sediment. There is a trend towards increasing Se in the other two fractions, particularly the "exchangeable" fraction as E_h increases. At 200 mV, 11% of the sediment-bound Se was in the "exchangeable fraction, with a further 1.5% in the "iron/manganese oxyhydroxide" fraction. Volatilisation losses to the atmosphere were minimal. Over the course of the entire experiment, a total of 26 μg Se was volatilised, representing 1% of the total Se inventory.

The significance of these results lies in the potential for migration and bioaccumulation of sediment-bound Se. It has been shown in laboratory trials that invertebrates are capable of bioaccumulating Se from a variety of sources including sediments, diatoms and water (Luoma, 1992). Data on the relative toxicity of different Se redox species has been generated (eg Davis *et al.*, 1988), yet to our knowledge, the impact of variation in the E_h of sediments on the bioavailability of adsorbed Se has not been investigated. Tessier *et al.* (1979) listed several experiments in which changes in the ionic composition of waters, as are normal in estuaries, affected the adsorption or desorption of sediment contaminants in the "exchangeable" fraction. Additionally, swings in estuarine pH could have an important impact on Se absorption (Hamdy & Gissel-Nielsen, 1977; Neal *et al.*, 1987). If sediment-bound Se moves into the "exchangeable" fraction as a result of changes to the E_h of the sediment, the potential for higher porewater concentrations is created and organisms may accumulate Se at a greater rate. Additionally, increased porewater concentrations would create a concentration driving force favouring efflux of soluble Se into the water column after the disappearance of external loads (Bowie *et al.*, 1996). Masscheleyn *et al.* (1990) noted that increasing E_h raised porewater Se concentrations in the $\mu g.L^{-1}$ range, but in our experiment we found sediment fractionation changes with potentially increased bioavailability in the $\mu g.g^{-1}$ range under similar changes in redox potential. It is possible that bioturbation by organisms such as callianassids (Aller, 1988) or human engineering activities could increase sediment E_h to levels at which Se bioavailability could become a concern.

Figure 2: Sediment-bound selenium

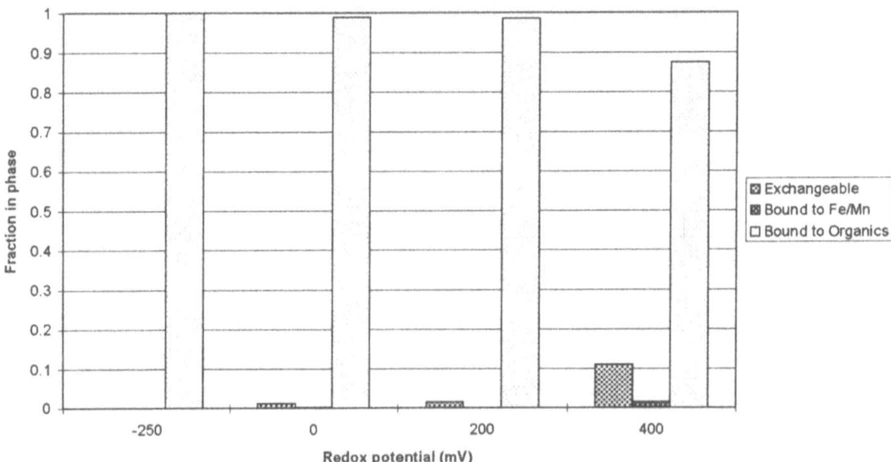

The associations of iron in the experimental slurry are shown in Figure 3. Whereas the proportion of the iron bound in the "organic" phase remains relatively constant at 33± 3% of the total over the entire range of redox potential, the proportion of easily removed "exchangeable" iron fell from 60 to 14% of the total with a corresponding increase in the "iron/manganese oxyhydroxide" fraction from 10 to 52% as the measured E_h was increased by 650 mV. Based on the literature (eg Masscheleyn *et al.*, 1993; Manning & Burau, 1995) it can be expected that Se is scavenged by iron. This is believed to relate to the increase in oxidised (ferric) iron at higher E_h. Under these conditions, iron is expected to persist in solid forms such as FeOOH, $Fe(OH)_3$ and Fe_2O_3 and scavenge selenite by adsorption (Masscheleyn, *et al.*, 1993). At lower E_hs, ferrous iron is more significant and is likely to become insoluble by bonding ionically with selenide. Iron unable to bind with selenide or sulfide would be detected as soluble and exchangeable iron. Under this interpretation, the observed movement of Se into the "exchangeable" and "iron/manganese oxyhydroxide" fractions with rising E_h may be related to a decrease in insoluble FeSe, an increase in the proportion of ferric iron and the adsorption of released Se into the "exchangable" fraction.

Figure 3: Sediment-bound iron

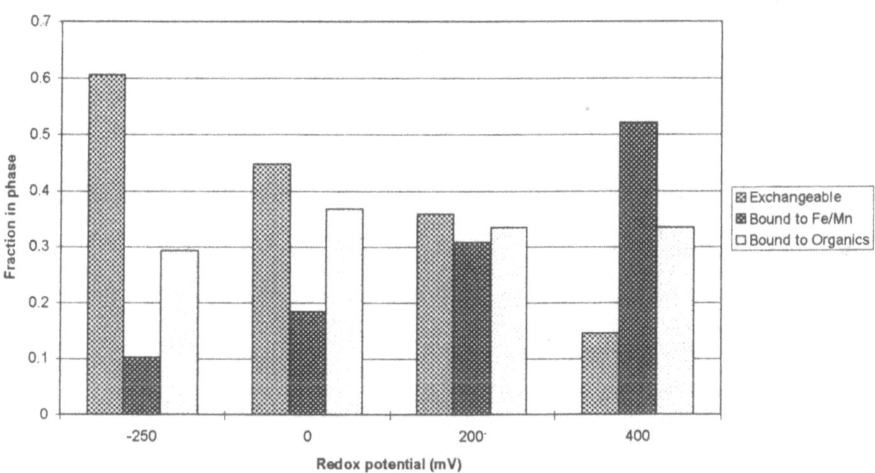

By comparison with the experiments of Masscheleyn *et al.* (1990, 1991), the pH remained relatively constant throughout this experiment because of the buffering capacity of seawater, staying between 8 and 8.5. This rules out significant influence of acidity in the degree of selenite sorption (Neal *et al.*, 1987). By comparison, the pH ranged from 4 to 6.9 in experiments by Masscheleyn *et al.* (1991) using a freshwater slurry and 7.5 to 8.1 in a brackish slurry (Masscheleyn *et al* (1990).

4. Conclusions

In a redox-stated slurry system, increases in the operationally-defined "ion exchangeable" and "iron/manganese oxyhydroxide" fractions of the sediment-bound selenium were observed. Selenium bound to the "organic/sulfide" fraction diminished from 100% to 87% of total sediment selenium. Most of the losses from the "organic/sulfide" fraction were to the "ion-exchangeable" fraction. It is worthwhile investigating the environmental implications of this change in areas where sediment selenium contamination may be exposed to fluctuations in salinity, dissolved oxygen, bioturbator populations and human activities.

Acknowledgements

Amber Swain and Denis Nobbs provided practical assistance throughout the work. Financial support from the Lake Macquarie City Council, Power Coal, the Australian Research Council and the Loxton Foundation is gratefully acknowledged.

References

Aller, R. C.: 1988, in T. H. Blackburn and J. Sørensen (eds) *Nitrogen Cycling in Coastal Marine Environments.*, John Wiley & Sons, New York.

Alloway, B. J. and Ayres, D. C.: 1993, *Chemical Principles of Environmental Pollution*, Blackie Academic & Professional, London.

Batley, G. E.: 1987, *Aust. J. Freshw. Res.* **38**, 591-606.

Besser, J. M., Ganfield, T. J., La Point, T. W.: 1993, *Environ. Toxicol. Chem.* **12**, (1), 57-72.

Bowie, G. L., Sanders J. G., Riedel F. F., Gilmour, C. C., Breitburg, D. L., Cutter, G. A., Porcella, D. B.: 1996, *Wat., Air, Soil Pollut.*, **90**, 93-104.

Davis, E. A., Maier, K. J., Knight, A.: 1988, *Calif. Agric.*, **1**, 18-20.

Elrashidi, M. A., Adriano, D. C., Workman, S. M., Lindsay, W. L.: 1987, *Soil Science* **144**, 141-152.

Farmer, N.: 1995, (Lake Macquarie City Council Catchment Management Officer) pers. comm, (memo 19951212).

Fio, J. L. and Fujii, R.: 1990, *Soil Sci. Soc. Am. J.*, **54**, 363-369.

Gerhardt, M. B., Green, F. B., Newman, R. D., Lundquist, T. J., Tresan, R. B., Oswald, W. J.: 1991, *Res. J. WPCF.* **63**, (5), 799-805.

Hamdy, A.A. and Gissel-Nielsen, G.: 1977, *Z. Pflanzenernaehr. Bodenkd.* **140**, 63-70.

Liu, D. L., Yang, Y. P., Hu, M. H., Harrison, P. J., Price, N. M.: 1987, *Mar. Environ. Res.* **22**, (2), 151-165.

Luoma, S. N., Johns C., Fisher, N. S., Steinberg, N. A., Oremland, R. S. and Reinfelder, J. R.: 1992, *Env. Sci. Tech.* **26**, 485-491.

Ma, Y. R. and Uren, N. C.: 1995, *Commun. Soil Sci. Plant Anal.* **26**, 3291-3303.

Maher, W. A.: 1984, *Bull. Environ. Contam. Toxicol.* **32**, 339-344.

Maher, W.A., Baldwin, S., Deaker, M., Irving, M.: 1992, *Applied Organometallic Chemistry*, **6**, 103-112.

Maier, K. J. and Knight, A. W.: 1993, *Arch. Environ. Contam. Toxicol.* **25**, (3), 265-370.

Maiers, D. T., Wichlacz, P. L., Thompson, D. L., Bruhn, D. F.: 1988, *Appl. Environ. Microbiol.* **54**, (10), 2591-2593.

Manning, A. B., Burau, R. G.: 1995, *Environ. Sci. Technol.* **29**, 2639-2646.

Masscheleyn, P. H., Delaune, R. D., Patrick, W. H.: 1990, *Environ. Sci. Technol.* **24**, 91-96.

Masscheleyn, P. H., Delaune, R. D., Patrick, W. H.: 1991, *J. Environ. Qual.* **20**, (3), 522-527.

Masscheleyn, P. H. and Patrick, W. H. Jr.: 1993, *Env. Toxic. Chem.* **12**, 2235-2243.

Neal, R. H., Sposito, G., Holtzclam, K. M.: 1987, *Soil Sci. Soc. Am. J.* **51**, 1161-1165.

Ohlendorf, H.M.: 1989, in *Selenium in Agriculture and the Environment*, Soil Science Society of America, Madison WI, 133-179.

Oremland, R. S., Hollibaugh, J. T., Maest, A. S., Presser, T. S., Miller, L. G., Culbertson, C. W.: 1989, *Appl. Environ. Microbiol.* **55**, (9), 2333-2343.

Peters, G. M., Barford, J. P., Gomes, V. G., Reible, D. D., Maher, W. A.: 1996, Bioturbator interaction with selenium contaminated sediment. 7th IASWS Symposium on Interactions Between Sediments and Water. Baveno, Italy, 22-25 September.

Peterson, J. A., Nebeker, A.V.: 1992, *Arch. Envir. Contam. Toxicol.* **23**, (2), 154-162.

Pickering, W. F.: 1986, *Ore Geol. Reviews* **1**, 99.

Roberts, B.: 1994, *The accumulation and distribution of selenium in fish from Lake Macquarie, NSW*, Honours thesis, Faculty of Applied Sciences, University of Canberra.

Roy, P. S. and Crawford, E. A.: 1984, *Est. Coast. Shelf Sci.* **19**, 341-358.

Salomons, W. and Förstner, U.: 1985, *Environ. Technol. Lett.* **1**, 506.

Scholz, D., Maher, W.A. Krikowa, F.: 1995, *Heavy metals in soil and sediment from Lake Macquarie, New South Wales*. (unpublished report), University of Canberra Faculty of Applied Science.

Steinberg, N. A., Blum, J. S., Hochstein, L., Oremland, R. S.: 1992, *Appl. Environ. Microbiol.* **58**, (1), 426-428.

Steinberg, N. A. and Oremland, R. S.: 1990, *Appl. Environ. Microbiol.* **56**, (11), 3550-3557.

Svensson, B., Schuetz, A., Nilsson, A., Aakesson, I., Aakesson, B., Skerfving, S.: 1992, *Sci. Tot. Envin.* **126**, 61-74.

Tessier, A., Campbell, P. G. C., Bisson, M.: 1979, *Anal. Chem*, **51**, 844.

Ure, A., Quevauviller, Ph., Muntau, H., Griepink, B.: 1993, *Report EUR 14763 EN*, CEC, Brussels.

Wallmann, K., Kerten, M., Gruber, J., Förstner, U.: 1993, *Int. J. Environ. Anal. Chem.* **51**, 187-200.

Woodford, J.: 1995, *Sydney Morning Herald*, September 13, p1.

Fe and Al Sedimentation and Their Importance as Carriers for P, N and C in a Large Humic Lake in Northern Sweden

ANDERS JONSSON
Institute of Physical Geography Umeå University S-901 87 Umeå Sweden

Abstract. The sedimentation of organic, amorphous oxides, crystalline oxides and crystalline silicate species of Fe and Al was investigated in humic Lake Örträsket. The covariation with C, N and P sedimentation also was studied. The results showed a strong temporal variation in the occurrence of different Fe species. During high discharge, such as in spring, Fe mostly occurred as crystalline Fe-silicate species, while the amorphous form dominated at other times of the year. Al generally sedimented as crystalline Al-silicate species (95%). Amorphous Fe species showed a strong relation with organic C and N, suggesting an association between organic material and amorphous Fe oxides. P was not as strongly related to amorphous Fe, probably as a result of apatite or clay bound-P species sedimentation.

Keywords: Sedimentation, Iron, Aluminum, Fractionation, Organic material, Phosphorus, Humic lakes

1. Introduction

Humic lakes have high allochthonous input of, e.g. iron (Fe), aluminium (Al) and organic matter (Arvola *et al.*, 1990; Lahermo *et al.*, 1995). Most carbon (C), nitrogen (N) and phosphorus (P) are bound in, or associated with, dissolved organic matter and the inorganic nutrient concentrations are low (Whalen and Cornwell, 1985; Salonen *et al.*, 1994). Crystalline and amorphous Fe oxides and different Al species have a high complexation capacity with organic material (Tipping, 1981; Davis, 1982; Ledin 1993), especially large organic molecules of a hydrophobic nature (Gu *et al.*, 1995). Sedimentation of these Fe and Al species may therefore influence the sedimentation of C and N. It is well documented that Fe adsorbs P (Koenings and Hooper, 1976; Buffle *et al.*, 1989) and thereby regulates P-sedimentation (Ostrofsky, 1987; Löfgren and Boström, 1989). In humic lakes, however, P sedimentation is complicated by the fact that Fe-P complexes are associated with organic colloids in Fe oxide-P-humic complexes (deHaan and deBoer, 1986; Jones *et al.*, 1988; deHaan, 1990), with a low tendency to settle. Since Fe and Al oxides in lakes occur largely in colloidal forms (Tipping *et al*, 1981; deHaan and deBoer, 1986; Ledin, 1993), aggregation of the colloids has to occur before sedimentation takes place. The colloids also can become associated with larger particles, e.g. clay particles (Deng and Stumm, 1994) and sediment.

The distribution of Fe and Al forms in humic lakes could influence the sedimentation of these metals and probably to a large extent also the sedimentation of C, N and P. The latter elements are critical to biological processes in lake water and sediments. Interaction between C, N and P and different forms of Fe and Al may therefore have a significant impact on the turnover of nutrients and biological activity in the water and the sediments.

Studies on the occurrence of Fe and Al species in sedimenting material and the temporal variability of the sedimentation are few. Therefore this study was designed to characterise sedimenting material with respect to organic Fe and Al, amorphous oxides, crystalline oxides and silicate crystalline forms of Fe and Al. The aim was to assess sedimentation rates of the various species, the influence of hydrological conditions on sedimentation, and possible relationships between the sedimentation of metal species and C, N and P sedimentation.

Water, Air and Soil Pollution **99**: 283-295, 1997.

2. Material and Methods

2.1. STUDY AREA

The investigation was carried out in Lake Örträsket, northern Sweden, which is a large and deep humic lake with a high, and seasonally variable input of Fe and Al. Lake Örträsket is the only lake along the main course of the River Öre (Figure 1). Two major streams, River Vargån and River Öràn, drain into Lake Örträsket. The annual mean discharge at Torrböle (Figure 1) for the years 1971-1991 is shown in Figure 2. Table I and Figure 3 present chemical and physical properties of the lake and the catchment basin.

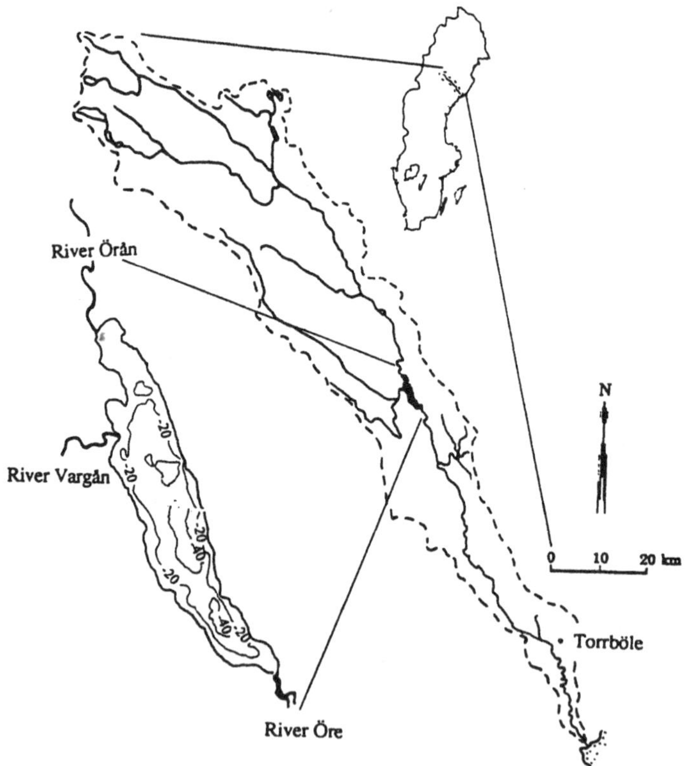

Fig. 1. The river Öre catchment basin and Lake Örträsket.

2.2. SAMPLING

Sediment traps were used to collect sedimenting material in the ice-free season of 1993 and 1994. On each collection date all traps were recovered and set out again. In 1993, traps were deployed on June 2 and emptied on July 7 and 28 and on August 12. In 1994 traps were deployed on April 12 and emptied on May 13, June 3 and 15, July 14 and 28, August 9 and 25, September 9 and 22 and on October 19. In 1993 the number of trap stations varied between 10 and 5 and in 1994 between 12 and 8. Figure 3 shows the distribution of the sediment traps in both years. A trap station consisted of two parallel cylinders with an aspect ratio of 7:1 and a diameter of 8.4 cm. The traps were deployed 5m above the lake bottom to avoid resuspended material.

TABLE I

Average lake water chemistry in Lake Örträsket (n=17) from 17 April - 21 September, 1994, and 31 May - 19 September 1995 (unpublished, Jansson *et al.* and unpublished, Bergström). Fe and Al (n=6, for each element) were analysed on 0.7 μm filtered water from 1994 only (unpublished, Jonsson). pH measurements (n=10) are also from the year 1994 (unpublished, Jonsson).

TOC $mg \cdot L^{-1}$	Tot N $\mu g \cdot L^{-1}$	Tot P $\mu g \cdot L^{-1}$	Diss Fe $\mu g \cdot L^{-1}$	Diss Al $\mu g \cdot L^{-1}$	pH
10.4	410	21	550	104	6.4

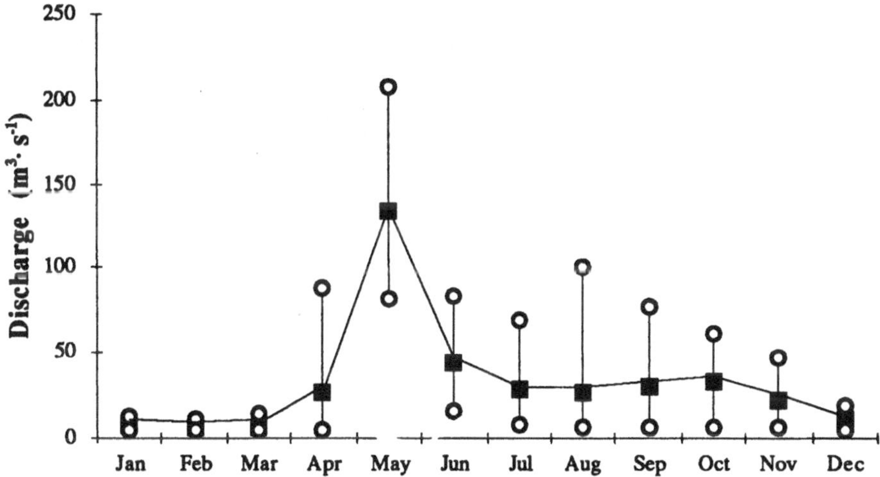

Fig. 2.. Discharge of the River Öre at Torrböle. Monthly mean, maximum and minimum values over the period 1971-1991.

On May 6, 1994, when the spring high flow episode was close to its maximum, 40 L of water was sampled from each of the two inlets (R. Öràn and R. Vargån). Trap-collected material and river water was centrifuged at 2600 rpm for 20 minutes to retrieve the particulate material. The recovered material was stored in a freezer. Samples from 1993 were also freeze-dried in 1993 and dried material was stored at room-temperature in the dark, while trap material from 1994 was not freeze-dried until 1995 when the extractions were made.

2.3. EXTRACTION PROCEDURES

The extraction procedures used in this study are commonly used in soil investigations, and were first developed for soil classification (McKeague and Day, 1966). Three separate extraction procedures were run according to the following protocol (Soon 1993):

(i) 20 mL of 0.1M copper chloride ($CuCl_2$), 24 hours shaking time (Soon, 1993).

(ii) 25 mL of 0.23mM ammonium oxalate-oxalic acid solution (pH 3.0), shaking time 4 hours in the dark (van Lagen, 1993).

(iii) 20 mL of 0.17M citrate dithionite solution, shaking time overnight (van Lagen, 1993).

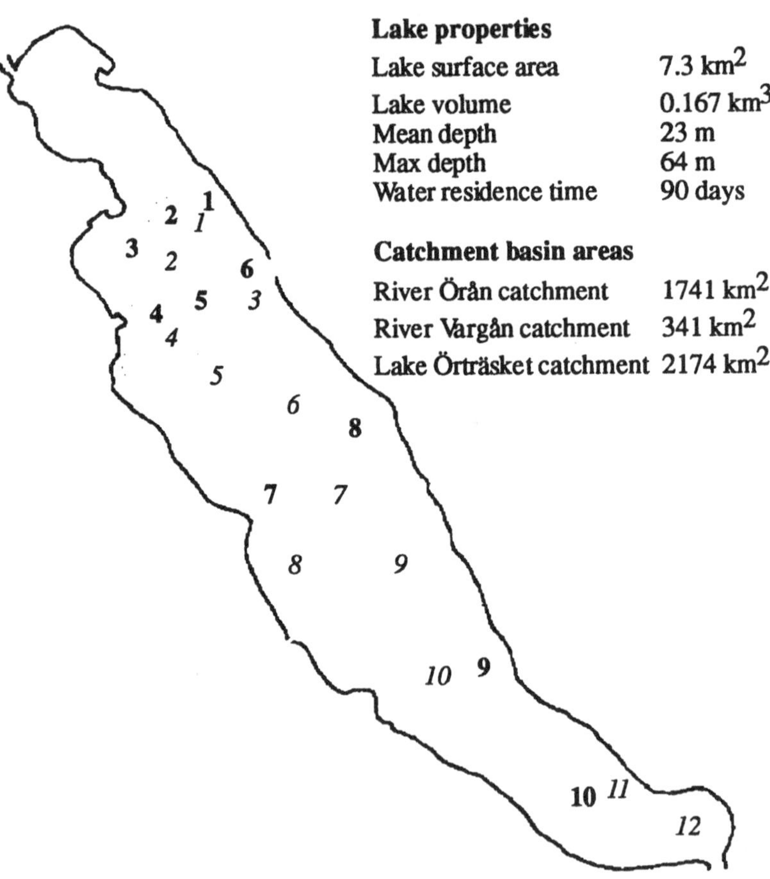

Fig. 3. Characteristics of Lake Örträsket and the distribution of trap stations in the lake in 1993 (bold) and 1994 (italics).

The same ratio of solution volume:sample weight was used in all extractions. Sample weights varied between 60 and 200 mg (mostly 100 mg). Extractions were made by adding the extraction solution to dry samples in 50 mL vials and shaking the suspension using a horizontal shaking apparatus. After shaking, the samples were centrifuged at 5600 rpm for 15 minutes. Samples from extraction (i) were filtered (Whatman GF/F filter) after centrifugation, while samples from extractions (ii) and (iii) had the coagulation agent "super floc" added before centrifugation (van Lagen, 1993). Analyses of various Fe and Al fractions were made on the supernatant obtained from each extraction. Analyses of total-Fe, -Al, -P, organic-C and -N were made on non-extracted solid samples.

2.4. CALCULATIONS

Organic species contents were derived directly from the $CuCl_2$ extraction (Juo and Kamprath, 1979; Hargrove and Thomas, 1981). Amorphous species were obtained by subtracting the amount of $CuCl_2$-extracted forms from those obtained with ammonium oxalate (Soon, 1993). The content of crystalline oxides was calculated by subtracting the ammonium oxalate extracted fractions from the dithionite citrate extracted Fe and Al (Parfitt and Childs, 1988; Borggaard *et al.*, 1990). Crystalline Fe- and Al-silicates (silicate crystalline species) were calculated by subtracting the dithionite-citrate extracted species from the total metal content of dry sediment (Jackson *et al.*, 1986). The terms organic species, amorphous oxides, crystalline oxides and silicate crystalline species, calculated as described above, will be used in the following.

Trap samples from the periods June 2 - July 7, 1993, and from April 12 - May 13, May 13 - June 3 and June 15 - July 14, 1994, were extracted separately (n=41). On each of the other nine occasions, samples from the various stations were pooled before analyses. Pooling was done in order to obtain sufficient material for the extraction procedures and also to obtain one sample that represented total sedimentation in the lake. Pooling was done by relating the trap yield to the area of the concurrent subarea. The whole trap sample was used from the trap with the least amount of dry matter, and material from the traps of the other subareas was added in proportion to the different amount:area ratios.

2.5. ANALYSES

All analyses were made at Umeå Marine Sciences Centre. Metals were analysed with an ICP-AES (ARL 3410). Solid samples were digested with a $LiBO_2$- method, where samples are heated to 900_C and dissolved with 10% HNO_3 before analyses of P, Fe and Al. Certified sediments from National Research Council, Canada, were used for standardization. Analyses of C and N were done with a Carlo Erba (1106) elemental analyzer. Precision for all measurements was better than ± 3%, provided the sediments were homogeneous.

3. Results

Table II presents the results of the extraction. Fe was more readily extracted than Al. Generally 74% and 5% of the total content was extracted for Fe and Al, respectively. Somewhat more Al was extracted with ammonium oxalate than with dithionite citrate at 4 extractions (cf. McKeague *et al.*, 1971), from material collected between April 12 - May 13, 1994. In these cases, the dithionite citrate fraction was set equal to the ammonium oxalate fraction, i.e. all Al oxides were considered as amorphous.

TABLE II

Arithmetic mean concentrations (n=50), standard deviation within parenthesis, of Fe and Al species extracted from the trap material collected in 1993 and 1994, The percentages of individual fractions in relation to total concentrations for each species are also shown.

	Concentration		**Percentages**	
	Fe	**Al**	**Fe**	**Al**
Species	(% of dw)	(% of dw)	(% of total)	(% of total)
Organic	0.00 (0.01)	0.07 (0.03)	0	1
Amorphous oxides	5.77 (2.72)	0.18 (0.12)	60	3
Crystalline oxides	1.36 (3.52)	0.05 (0.15)	14	1
Silicate crystalline	2.57 (3.04)	5.18 (1.18)	26	95
Total	9.70 (3.04)	5.48 (1.18)		

In Table III the concentrations of the extracted Fe and Al species in the inlet material collected during the spring high flow, May 6 1994, are compared with the concentration in the trap material collected between April 12 and May 13. The concentration of amorphous Fe was higher and the concentration of crystalline Fe species was somewhat lower in the inlet material than in the trap material.

TABLE III

The sum of the flow weighted concentration (% of dw) of Fe and Al species extracted in material from the two inlets, R. Vargån and R. Örån, at spring high flow, May 6, 1994 compared with the arithmetic mean of trap data from 6 stations in Lake Örträsket between April 12 and May 13, 1994. Trap data from station nos. 9 and 11 (Figure 1) have been excluded (see text). The percentages of individual fractions in relation to total concentrations for each species are also shown.

	Concentration (% of dw)		**(%) of total metal concentration**	
Fe	Inlets	Lake	Inlets	Lake
Silicate crystalline	4.81	3.29	44	59
Crystalline oxides	0.16	0.36	2	6
Amorphous oxides	3.32	1.96	54	35
Organic	0.01	0.00	0	0
Al				
Silicate crystalline	6.06	6.50	97	98
Crystalline oxides	0.01	0.00	0	0
Amorphous oxides	0.11	0.08	2	2
Organic	0.04	0.03	1	0

Since almost no organic Fe was obtained, this fraction is not considered when presenting further results.

The sedimentation of Al and Fe successively decreased after the peak in spring (Figure 4), suggesting that the sedimentation is highly dependent on discharge. The discharge during the investigated period in July - October, 1994 was very low, with a minimum in August. The discharge was high on the three first sampling periods (June 2-July 7-July 28-August 12, 1993), which also resulted in high sedimentation.

The concentration of the different species in the trap material was significantly related to discharge (Table IV). Concentration of silicate crystalline material was positively related to discharge (p^{2}0.01) and other species were negatively correlated ($p<0.01$). Two different relationships were found for amorphous Fe depending on whether runoff was higher or lower than about 10 $L \cdot s^{-1} \cdot km^{-2}$ (Figure 5). The Tot-Fe:Tot-Al ratio varied significantly (unpaired two way t-test, $p<0.001$) from 3.2 (average, n=10) to 1.6 (average, n=6) when runoff increased from less than 18 $L \cdot s^{-1} \cdot km^{-2}$ to more than 24 $L \cdot s^{-1} \cdot km^{-2}$ (Figure 6). At higher discharges the Tot-Fe:Tot-Al ratio decreased to about 0.8.

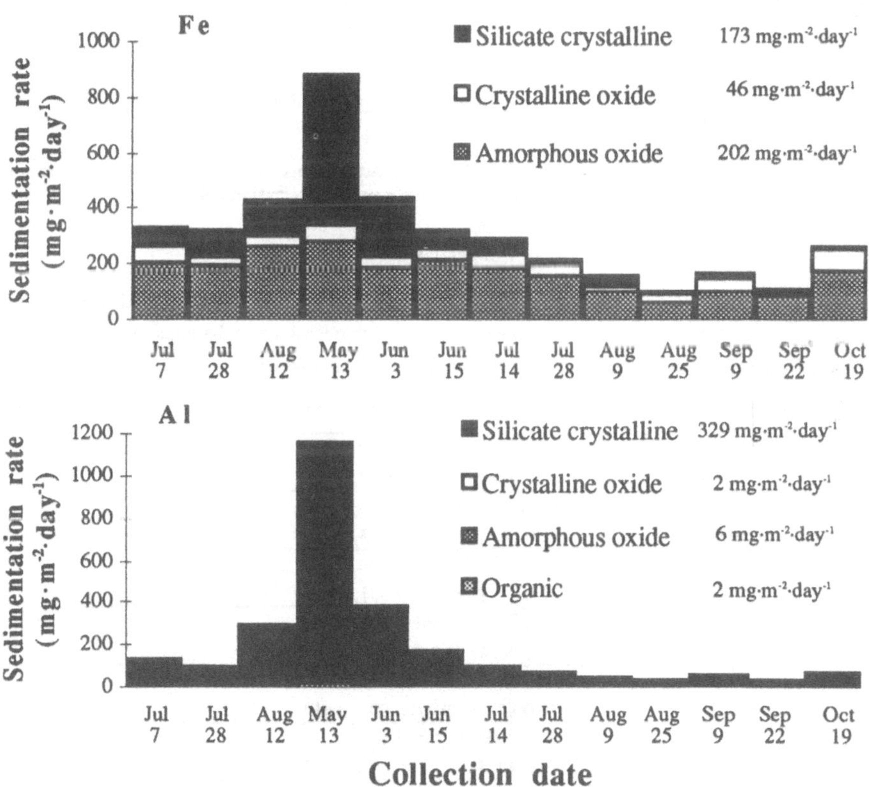

Fig. 4. Sedimentation rates of Fe and Al species in Lake Örträsket. The arithmetic mean sedimentation rates for each species extracted are also reported.

The concentrations of C, N and P in the sedimenting material showed a strong covariation with the amorphous Fe concentrations (Figure 7a-c). During the early spring period (April 12 - May 13, 1994) material caught in traps no. 9 and 11 (Figure 3) were richer in amorphous Fe than material caught in other traps (cf. Fig 7). These traps were deployed at large depths (60 and 42m, respectively) in the lower end of the lake. Material transported during the spring might not have reached these traps to the same extent as the other traps.

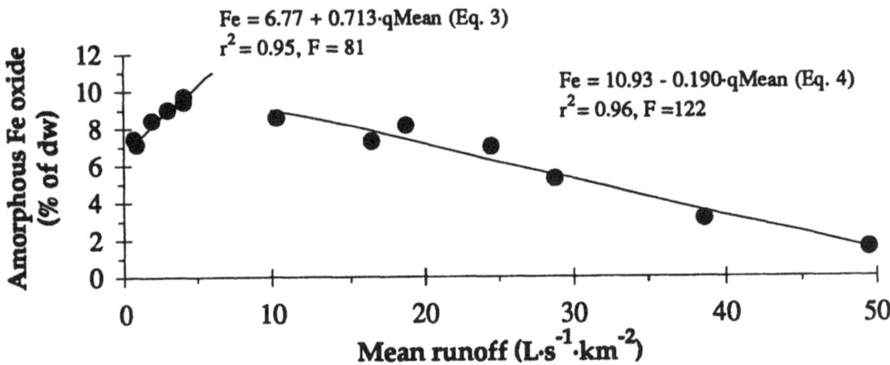

Fig. 5. The whole lake mean concentration of sedimenting amorphous Fe oxides related to the mean water discharge (qMean) into the lake for the period between two emptying occasions of the traps.

TABLE IV

Relations between whole lake mean concentration (% of dw) of sedimenting Fe and Al species and mean runoff (qMean, $L \cdot s^{-1} \cdot km^{-2}$) of the period between two emptying occasions of the traps. Amorphous Fe oxides has two relations, one when runoff is higher than 10 $L \cdot s^{-1} \cdot km^{-2}$ and one when runoff is lower.

	Species	Equation	
Fe	Silicate crystalline	Fe= 1.51 + 0.044·qMean r^2=0.49, F=10, p<0.008, n=13	(1)
	Crystalline oxides	Fe= 2.74 - 0.056·qMean r^2=0.56, F=14, p<0.003, n=13	(2)
q< 10	Amorphous oxides	Fe= 6.77 + 0.713·qMean r^2=0.95, F=81, p<0.001, n=6	(3)
q> 10	Amorphous oxides	Fe= 10.93 - 0.190·qMean r^2=0.96, F=122, p<0.001, n=7	(4)
Al	Silicate crystalline	Al= 3.25 + 0.070·qMean r^2=0.88, F=79, p<0.001, n=13	(5)
	Crystalline oxides	Al= 0.129 - 0.003·qMean r^2=0.43, F=22, p<0.016, n=13	(6)
	Amorphous oxides	Al= 0.322 - 0.005·qMean r^2=0.81, F=48, p<0.001, n=13	(7)
	Organic	Al= 0.109 - 0.002·qMean r^2=0.90, F=95, p<0.001, n=13	(8)

The material in these two traps also had higher amorphous Fe:N (13.1, 10.9) and amorphous Fe:P (46.7, 35.7) ratios than material in the other traps during spring. The ratios are similar to the ratios found in the material sedimenting in summer (Table V). The C:P and C:N ratios of the sedimenting material differ significantly ($p^{2}0.004$) between spring and summer (Table V). Spring material was relatively richer in N and P than the summer material.

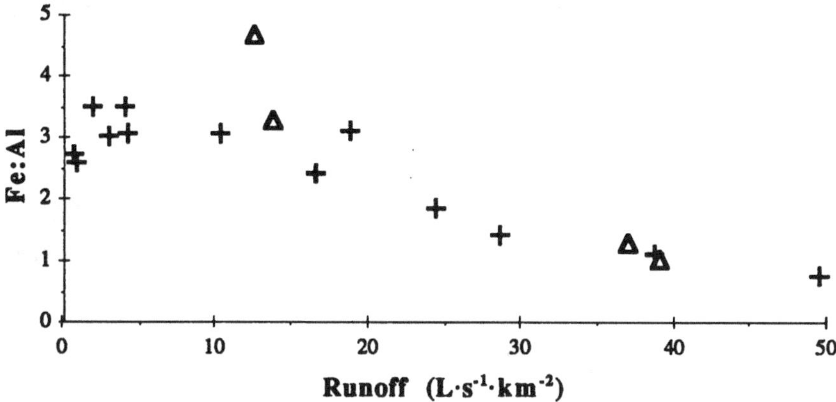

Fig. 6. The Tot-Fe:Tot-Al ratio related to the mean water discharge into the lake for the period between two emptying occasions of the traps. Data used in this investigation are marked with (+) and (Æ) denote unpublished data from Lake Örträsket..

TABLE V

Arithmetic mean concentrations (% of dw) of Fe, Al, C, N and P in the trap material divided into spring 12 April - 3 June, 1994, (n= 19) and summer - autumn data (n = 31) including data both from 1993 and 1994. Mass ratios calculated using the same data are also shown. The ratios are all significantly different ($p \leq 0.0004$, unpaired two way t-test) except for the Am Fe:C ratio. The abbreviation Am Fe represents amorphous Fe.

Period	Fe	Al	C	N	P
Spring	7.12	6.53	4.81	0.40	0.147
Summer	11.42	4.70	11.60	0.79	0.186

Period	C/N	C/P	Am Fe/C	Am Fe/N	Am Fe/P
Spring	12.3	32.1	0.64	7.9	21.0
Summer	14.8	61.2	0.65	9.6	39.1

4. Discussion

It is assumed that the extractions were specific for each of the chemical fractions. This is probably not entirely correct, since the extractions are only operational and they are more specific for Fe than Al (McKeague *et al.*, 1971; Jackson *et al.*, 1986; Parfitt and Childs, 1988; Paterson *et al.*, 1993), except for the organic $CuCl_2$ extracted fraction for which the procedure was developed for Al (Soon, 1993). However, it has been found that the distribution of different fractions in the extraction procedure compares well with corresponding distributions given by direct methods such as Moessbauer spectra, X-ray diffraction (XRD) and differential X-ray diffraction (DXRD) (Parfitt and Childs, 1988; Wang *et al.*, 1993). Any major misinterpretations are therefore considered unlikely.

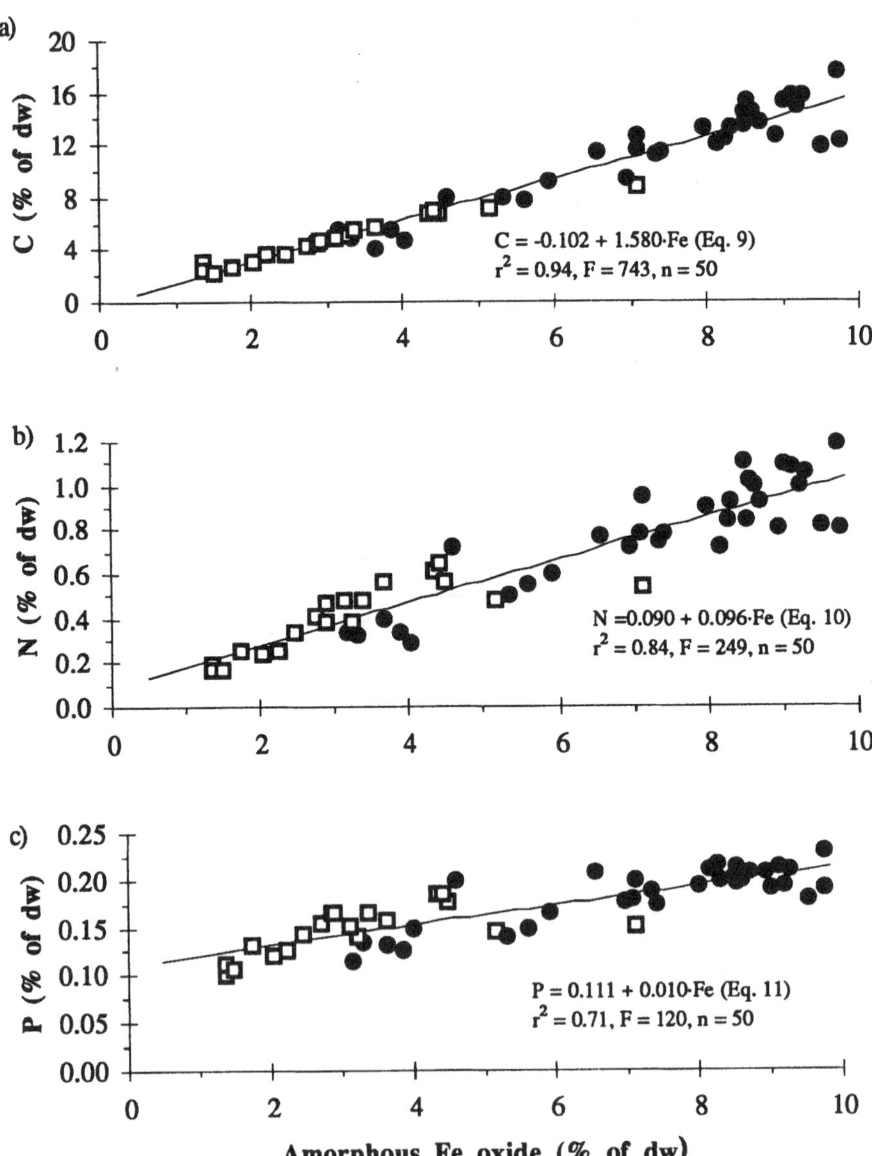

Fig. 7. The relation between amorphous Fe oxide and C (a), N (b) and P (c). All available data have been used, both weighted whole lake data (9 samples) and data from separate traps (41 samples). Open squares represent material in spring 1994 and closed circles material from summer and autumn 1993 and 1994.

Even though the data presented in Table III are not directly comparable, since the river data are from a specific date and the trap data are integrated over several days, they indicate that crystalline Fe species sediment separately, and that amorphous Fe species have a lower tendency to sediment. Probably this is a result of the high turbulence of the lake water during spring high flow. Crystalline Fe particles then occur as dense particles with higher sedimentation rates than amorphous Fe particles.

The sedimentation of silicate crystalline material, e.g. clay and silt, increases with increasing discharge (Eq. 1 and 5, Table IV). Thus, silicate crystalline material becomes an important component of the sedimenting material during high flow situations (Figure 4), especially during spring. It is shown also that the sedimenting material changes character towards a more crystalline nature when discharge increases above 18 - 24 $L{\cdot}s^{-1}{\cdot}km^{-2}$, inferred by the Tot-Fe:Tot-Al ratio, (Figure 6). Thus, clay and silt probably begin to be eroded and transported in large amounts in the river when discharge increases above 18-24 $L{\cdot}s^{-1}{\cdot}km^{-2}$.

Crystalline oxides, e.g. goethite and hematite, appear at their highest concentrations in the trap material during low flow situations (Eq. 2 and 6, Table IV), indicating that they occur as particles with low settling rates. In general, there was a rather small fraction extracted as crystalline oxides (Table II), especially for Al. Most likely crystalline oxides form complexes with organic material (Tipping, 1981), resulting in most Fe oxides (probably also Al oxides) also occurring in a colloidal state (Koenings and Hooper, 1976). These can not be expected to sediment. They are probably included in the dissolved Fe and Al fraction, which is found in rather high concentrations in Lake Örträsket (Table I).

Organic species of Fe and Al are a small fraction of the sedimenting material (Table II). Only Al seemed to be chelated with organic particulate material in any measurable amounts. Most Fe and Al bound to organic material probably exists in the dissolved phase (Ivarsson and Jansson 1994) and the organic Fe and Al are thus not recovered in the sedimenting material.

Amorphous Fe and Al oxides are the most abundant fractions besides silicate crystalline material (Table II), similar to what was found in meso-humic Toolik Lake, Alaska (USA) (Cornwell, 1987). Amorphous Fe oxides in particular dominate the sedimentation during low flow (Figure 4). Amorphous oxides have been found to appear as a gel around clay particles in soils (Jones and Uehara, 1973), and it has been shown that amorphous oxides may adsorb to clay particles in natural waters (Deng and Stumm, 1994). The sedimentation of amorphous oxides could thus, at least to some extent, be controlled by the sedimentation of clay material. A higher transport and sedimentation of clay particles would then result in a higher sedimentation of amorphous Fe. The concentration of amorphous Fe in sedimenting particles increases also with discharge (Eq. 3, Figure 5); a relation that no other species showed at discharges below 10 $L{\cdot}s^{-1}{\cdot}km^{-2}$. At discharges above 10 $L{\cdot}s^{-1}{\cdot}km^{-2}$ there is a negative relation between the concentration of amorphous Fe and discharge (Eq. 4, Figure 5) similar to what was found for other species except silicate crystalline material (Table IV). This may be an effect of the lake water becoming more turbulent during high flow events, and thus preventing slowly settling particles from sedimenting. The average transit time for water (Tw) during high flow situations ($q>18$ $L{\cdot}s^{-1}{\cdot}km^{-2}$) is also very short in Lake Örträsket with an average of 19 days during 1993 and 1994 high flow situations. The short transit time for water is a result of inflowing water passing through the epilimnion of the lake that does not mix with the total lake volume. The decreasing concentration of amorphous Fe also may be an effect of dilution with material that does not carry amorphous Fe, e.g. silicate crystalline material; the sedimentation of which also showed a positive relation with discharge (Eq. 1 and 5, Table IV). These particles are probably silty material eroded from the marine and glacial deposits reaching 5 km upstream of the R. Örån´s inlet to the lake (Ivarsson and Forsgren, 1988).

The circulation of amorphous Fe may be important for the distribution of other substances through complexation reactions (Tipping, 1981; Tipping *et al.*, 1981). The good correlations between C, N, P and amorphous Fe (Eq. 9-11, Figure 7a-c) suggest that amorphous Fe is a carrier for these elements. This particularly applies to C and N, for which the regression equations almost intercept the origin (no sedimentation of C and N occurs when there is no amorphous Fe sedimenting). C has a similar association with amorphous Fe in spring and summer, as shown by the linear correlation in Figure 7a (Eq. 9) and the similar amorphous Fe:C ratios in spring and summer (Table V). Organic material sedimenting in summer is probably more refractory, indicated by the C:N ratio being significantly ($p=0.004$) higher in summer than in spring (Table V). The relatively higher N content of organic material sedimenting in spring also results in a lower amorphous Fe:N ratio (Table V) and a steeper relation between amorphous Fe and N in spring (Figure 7b) than in summer and autumn. Thus, the organic material sedimenting in spring is probably less decomposed and therefore should be more suitable as a nutrient source for benthic organisms in the lake than organic material sedimenting in summer. This is further emphasised by the fact that suspended particulate organic material (> 0.7 μm) in the lake water has a successively increasing C:N ratio ranging from 11 in spring to 24 in the fall (unpublished data 1995, Bergström).

Several investigations have shown that P in the form of phosphate is complex-bound by Fe oxides (Tipping, 1981; Nilsson *et al.*, 1992). A relation intercepting zero between the P sedimentation and the amorphous Fe sedimentation could be expected, if all P sedimenting in Lake Örträsket were in the form of P-Fe oxide complexes. The large intercept of the regression equation (Eq. 11, Figure 7c), however, indicates that a large percentage of sedimented P in Lake Örträsket is not associated with amorphous Fe. Earlier studies have shown that there is a large fraction of apatite-P in sedimenting material in Lake Örträsket (Forsgren and Jansson, 1993), which is consistent with this result. The apatite-P fraction made up 26-62% of total-P in the sediment of Lake Örträsket (Forsgren and Jansson, 1993). It can also be seen that the amorphous Fe:P ratio is much higher in spring, when the transport of crystalline material is high (Fig 4), than in the summer (Table V). In contrast the amorphous Fe:C ratio does not differ much between spring and summer. Thus, the lower C:P ratio in spring (Table V) is probably a result of a higher sedimentation of crystalline material in high flow situations resulting in a higher sedimentation of inorganic P.

5. CONCLUSIONS

Crystalline Fe- and Al-silicates are important components of sedimenting material during high flow events and are thus important when calculating annual mass balances of the two elements. This is true especially for Al, 95% of which occurred as crystalline Al-silicates. Amorphous Fe (e.g. ferrihydrite) occurred in high concentrations and was the most common Fe species sedimenting in the lake. Crystalline Fe oxides (e.g. goethite) only represented a small share of settling Fe. The sedimentation of amorphous Fe oxides was strongly regulated by the discharge of water into the lake. Amorphous Fe oxides were important for the sedimentation of C, N and P.

ACKNOWLEDGEMENTS

This study was carried out with financial help from the Swedish National Science Research Council, Swedish Society of Anthropology and Geography (SSAG) and Kempe Memorial Foundation. I thank Professor Mats Jansson for valuable comments on the manuscript and Lars-Inge Karlsson and Ann-Kristin Bergström for technical assistance, and Dr Erik Lundberg and Tommy Olofsson who prepared the chemical analyses.

References

Arvola, L., Salonen, K. and Rask, M.: 1990, Limnologica. **20**, 243-251.
Borggaard, O. K., Jørgensen, S. S., Møberg, J. P. and Raben-Lange, B.: 1990, J. Soil Sci. **41**, 443-449.
Buffle, R., De Vitre, R. R., Perret, D. and Leppard, G. G.: 1989, Geochim Cosmochim Acta. **53,** 399-408.
Cornwell, J. C.: 1987, Arch. Hydrobiol. **109,** 161-179.
Davis, J. A.: 1982, Geochim. Cosmochim. Acta. **46,** 2381-2393.
deHaan, H. and deBoer, T.: 1986, Freshwat. Biol. **16**, 661-672.
de Haan, H.: 1990, Limnol. Oceanogr. **35,** 491-497.
Deng, Y. and Stumm, W.: 1994, Appl. Geochem. **9**, 23-26.
Forsgren, G. and Jansson, M.: 1993, Hydrobio. **253**, 233-248.
Gu, B, Schmitt, J., Chen, Z., Liang, L. and McCarthy, J.F.: 1995, Geochim. Cosmochim. Acta. **59**, 219-229.
Hargrove, W. L. and Thomas, G. W.: 1981, Soil Sci. Soc. Am. J. **45**, 151-153.
Ivarsson, H. and Forsgren G.: 1988, SGÅ 64. Lund, Sweden.
Ivarsson, H. and Jansson, M. 1994, Arch. Hydrobiol. **132**, 45-55.
Jackson, M. L., Lim, C. H. and Zelazny, L.: 1986, '*Oxides, Hydroxides, and Aluminosilicates*', in Klute. A (ed.) '*Methods of soil analysis, part 1*'. Physical and Mineralogical Methods- Agronomy Monograph no. 9, Second edition. American Society of Agronomy- Soil Sci. Soc. Am, USA. pp. 101-149.
Jones, R. C. and Uehara, G.: 1973, Soil Sci. Soc. Amer. Proc. **37**, 792-798.
Jones, R. I., Salonen, K. and deHaan, H.: 1988, Fresh. Wat. Biol. **19**, 357-369.
Jou, A. S. R. and Kamprath, E. J.: 1979, Soil Sci. Soc. Am. J. **43**, 35-38.
Koenings, J. P. and Hooper, F. F.: 1976, Limnol. Oceanogr. **21**, 684-696.
Lahermo, P., Mannio, J. and Tarvainen, T.: 1995, Appl. Geochem. **10**, 45-64.
Ledin, A.: 1993, Ph. Dr. Thesis, no. 91. Lindköping Studies in Arts and Science, Lindköping, Sweden.
Löfgren, S. and Boström, B.: 1989, Wat. Res. **23**, 1115-1125.
McKeague, J. A. and Day, J. H.: 1966, Can. J. Soil Sci. **46**, 13-22.
McKeague, J. A., Brydon, J. E. and Miles, N. M.: 1971, Soil Sci. Soc. Amer. Proc. **35**, 33-38.
Nilsson, N., Lövgren, L. and Sjöberg, S.: 1992, Chemical Speciation and Bioavailability. **4**, 121-130.
Ostrofsky, M. L.: 1987. Can. J. Fish. Aquat. Sci. **44**, 960-966.
Parfitt, R. L. and Childs, C. W.: 1988, Aust. J. Res. **26**, 121-144.
Paterson, E., Clark, L. and Birnie, A. C.: 1993, Commun. Soil Sci. Plant Anal. **24**, 2015-2023.
Salonen, K., Keskitalo, J. and Arvola, L.: 1994, Arch. Hydrobiol. **129**, 425-441.
Soon, Y. K.: 1993, Commun. Soil Sci. Plant Anal. **24**, 1683-1708.
Tipping, E.: 1981, Geochim. Cosmochim. Acta. **45**, 191-199.
Tipping, E., Woof, C. and Cooke, D.: 1981, Geochim. Cosmochim. Acta. **45**, 1411-1419.
Van Lagen, B.(ed.): 1993, '*Manual for chemical soil analyses*'. Department of Soil Science and Geology, Agricultural University Wageningen.
Wang, H. D., White, G. N., Turner, F. T. and Dixon, J. B.: 1993, Soil Sci. Soc. Am. J. **57**, 1381-1386.
Whalen, S. C. and Cornwell, J. C.: 1985, Can. J. Fish. Aquat. Sci. **42**, 797-808.

BEHAVIOUR OF Co, Fe, Mn AND Ni IN THE PO ESTUARY (ITALY).

MARINA CAMUSSO[1], SONIA CRESCENZIO[1], WALTER MARTINOTTI[2], MAURIZIO PETTINE[3] and ROMANO PAGNOTTA[3]

[1]*Water Research Institute, C.N.R., via della Mornera 25, 20047 Brugherio, Italy,* [2]*ENEL S.p.A.-Material and Environment Research Centre, Via Rubattino 54, 20134 Milan, Italy;* [3]*Water Research Institute, C.N.R., via Reno 1, ,00198 Rome, Italy.*

Abstract. The behaviour of Co, Fe, Mn, and Ni was investigated in the mixing area of the Po river in six surveys over the period March 1992 to June 1995 during low-to-medium solid load and flow conditions. The concentrations (nM) of dissolved elements for the riverine and marine end-members respectively were in the ranges 0.09 to 0.93 and 0.24 to 0.98 for Co, 10 to 53 and 5 to 20 for Ni, 12 to 87 and 5 to 33 for Fe, and 10 to 155 and 26 to 105 for Mn. Co and Mn behave non-conservatively in the Po estuary: an addition to the dissolved phase occurred for these elements that was more marked in the summer period. The concentrations of dissolved and particulate Ni decreased almost linearly with increase in salinity, with more marked variations in dissolved Ni concentration in winter than summer. The concentrations of dissolved and particulate Fe also varied conservatively with salinity. This unusual behaviour for dissolved Fe is attributed to the analytical procedure which excluded kinetically inert colloidal species.

Key-words: Estuary, Trace metals, Suspended matter

1. Introduction

When riverwater mixes with seawater a large number of physical and chemical processes take place which affect the distribution of trace metals over the particulate and dissolved phases and hence the composition of the deposited sediments (Martin and Whitfield, 1983; Salomons and Förstner, 1984; Millward and Turner, 1995). The rapid removal from riverwater of Fe and the non-conservative behaviour of Mn in most estuaries have been well documented, and processes affecting their fate in estuaries have been thoroughly investigated (Boyle *et al.*, 1977; Morris *et al*, 1982; Boughriet *et al.*, 1992). Fe and Mn are also important because their oxides may influence the behaviour of other metals (Millward and Moore, 1982; Bourg, 1983). The behaviour of Co and Ni has been studied in a number of estuaries (Yeats and Bewers, 1982; Duinker *et al.*, 1982; Chiffoleau *et al.*, 1994), but observations on the behaviour of trace metals for one estuary cannot be extrapolated to others, since the relative importance of biogeochemical processes may differ from one estuary to another. The Po river flows through a delta system to the Northern Adriatic basin, bringing about 50% of the total freshwater discharge and about 50% of the total nutrient load to the basin, making it one of the most productive in the Mediterranean (Degobbis and Gilmartin, 1990). Trace metal bioaccumulation and variability in the dissolved and particulate phases in the lower stretch of the Po have been investigated recently (Pettine *et al.*, 1994; Camusso *et al.*, 1994), but information on processes affecting the conservation of trace elements in the river-sea transition zone is limited (Ferrari and Ferrario, 1989; Dorten *et al.*, 1991).

As part of a larger project, started in 1992 to examine how physical, chemical and biological variables influence nutrients, organics and micro elements in the deltaic mixing area of the Po, the behaviour of dissolved and particulate Co, Ni, Fe and Mn along the salinity gradient were studied in six surveys (March 1992-June 1995) during low-to-medium solid transport (11-44 mg/l) and flow conditions (780-2000 m^3/s). These metals

Water, Air and Soil Pollution **99**: 297-304, 1997.

are of great interest because they are bioactive and their distribution may influence primary production in the coastal Adriatic. The main objectives of the study were: (1) to examine the influence of Po discharge on Co, Ni, Fe and Mn concentrations in the coastal Adriatic area; (2) to investigate processes affecting their dissolved and particulate concentrations as a function of mixing; and (3) to compare metal behaviours and concentrations to those reported from other European estuaries.

2. Material and Methods

2.1 STUDY AREA AND SAMPLING

The Po is the largest river in Italy with a drainage basin extending over one quarter of the national territory. The river terminates in a delta of 400 km^2 that has six main channels. Po di Pila, the largest channel, conveys 40-50% of the flow as measured at Pontelagoscuro — the closing section of the river, 90 km from the mouth. The delta area is subject to variations in chemical and physical conditions due to variations in the hydrological regime, the influence of the tides and the episodic presence of a salty wedge (12 km maximum). Between March 1992 and June 1995 grab samples were collected in the lower Po upstream of the delta area (at Pontelagoscuro), in Po di Pila, and in the river plume on the sea surface at various salinities. Six surveys took place: on March 18, 1992, February 10 and July 21, 1993, July 6, 1994 and March 1 and June 21, 1995. Water was collected with a Niskin bottle or pumped from 0.5 m depth (80% of cases). On average of 15 samples were collected at each survey (Fig. 1).

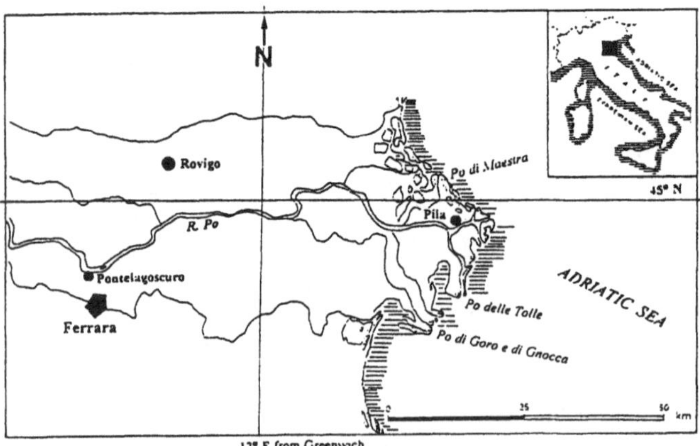

Fig. 1 Map of the delta area of the Po River showing sampling area

2.2 ANALYTICAL PROCEDURES AND METHODS

Water samples (5-7 L for low salinity and 18-20 L for high salinity) were filtered, under nitrogen pressure, through 0.4 μm polycarbonate membranes (Nuclepore) using Teflon covered apparatus (Sartorius) connected to HDPE vessels. The first 0.5 L of filtrate was discarded. Filtration began 2-3 h after collection and was completed within 2 days using five apparatuses in order to minimize changes due to biological processes. To determine dissolved metals, 1 L of each sample was placed, immediately after filtration, in a cleaned polyethylene bottle containing NH_4^+-charged Chelex 100 resin and 1M ammonium acetate buffer (pH 6.1) and shaken for 24 h. The metals were eluted from the resin with 2.5 M ultrapure nitric acid, achieving a 100-fold concentration (Boniforti *et al.*, 1984). Procedural blanks were processed and analyzed with the samples. Dissolved Fe and Mn were assayed on the acid eluates by emission spectrometry (ICP AES, Jobin Yvon 38); dissolved Co and Ni were analyzed by electrothermal atomic absorption spectrometry (EAAS), using a Varian AA 1475 equipped with a graphite furnace (GTA 95). The analytical detection limits were 0.1 nM for Co, 7.5 nM for Fe, 0.2 nM for Mn, and 0.3 nM for Ni. Precision at lowest concentrations was ±0.1% for Fe and Mn and ±5% for Co and Ni. Dissolved Na (used as a tracer for the fresh/saltwater mixing) was determined by ICP emission spectrometry (Perkin Elmer, Plasma 40). Water quality parameters (T, pH, DO, conductivity or salinity, and chlorophyll α) were measured *in situ* by means of a multiparameter probe (Idronaut Srl, Italy).

For analysis of particulate phase metals, preweighed filters were washed with 200 mL deionized water to remove seasalts and dried inside a laminar air flow hood before reweighing to determine total suspended solids (TSS). Particulate matter was leached from the filters by shaking with 0.3 M HCl (BDH, Aristar) for 2 h, and metal concentrations determined by ICP-AES and EAAS techniques as described for dissolved metals. This procedure determines the non-detrital, available fraction of metals that are associated with organic and mineral suspended matter (Malo, 1977). After leaching by the same procedure metal concentrations were also determined on certified estuarine material (NIST 1646). Metal concentrations were also determined after nitric acid digestion of the some particulate samples (and certified material) in order to determine leachable metals as fractions of the material removed by more complete digestion. The leachable fraction of Mn in Po suspended matter exceeded 80% of the nitric acid digested fraction; corresponding figures for Ni, Co and Fe were 60%, 50% and 15% respectively. All chemicals used were ultrapure grade; deionized water for solution preparation and cleaning was from a Millipore Milli-Q2 system.

3. Results

3.1 DISSOLVED METALS

The concentrations (nM) of dissolved elements in the riverine end-members were in the ranges 0.09 to 0.93 for Co, 10 to 53 for Ni, 12 to 87 for Fe, and 10 to 155 for Mn (Table 1). Average riverine concentrations of dissolved Co and Mn (n=12, 0.40±0.29 and 52±52 nM respectively) did not differ from median values of 0.51 and 51 nM for the lower Po measured over the period 1988-1990 (Pettine *et al.*, 1994) and are only slightly lower than the means determined in that period (n=35, 0.68±0.42 and 76±73 nM); our values can therefore be considered representative of lower Po concentrations. The highest estuarine values of Mn were measured on March 1, 1995 when the Po waters were characterized by a high solid load (109 mg/l): the day before, the load at Pontelagoscuro was 352 mg/l, illustrating the wide variability of the suspended matter load transported by the river.

Table 1 Flow ($m^3 s^{-1}$) and total suspended solids (TSS, mg l^{-1}) in the River Po, and end-member concentrations in the dissolved (nM) and particulate (µmol/g) phases in the six surveys. End-member concentrations are averages of 2-3 samples with similar salinity.

Dissolved		Po River				Adriatic Sea			
Date	Flow	Co	Fe	Mn	Ni	Co	Fe	Mn	Ni
March 18, 1992	776	0.57	42	42	49	0.98	5	52	11
February 10, 1993	1410	0.93	18	54	53	0.82	29	39	5
July 21, 1993	960	0.22	12	10	16	0.26	7	68	5
July 6, 1994	1110	0.41	20	23	10	0.63	6	28	8
March 1, 1995	2090	0.20	87	155	24	0.37	33	105	20
June 21, 1995	1810	0.09	16	27	11	0.24	17	26	13
Particulate		**Po River**				**Adriatic Sea**			
Date	TSS	Co	Fe	Mn	Ni	Co	Fe	Mn	Ni
March 18, 1992	11.5	0.03	47	15	0.32	0.02	37	4	0.21
February 10, 1993	11.4	0.13	148	25	0.74	0.10	81	11	0.29
July 21, 1993	19.6	0.09	80	22	0.92	0.04	77	26	0.50
July 6, 1994	34.7	0.17	107	22	0.61	0.17	39	21	0.55
March 1, 1995	109	0.07	63	16	0.33	0.04	25	4	0.26
June 21, 1995	44.4	0.14	90	18	1.60	0.04	23	13	0.60

Riverine concentrations of Ni were significantly higher in winter (average 42 nM) than summer (average 12 nM). Average river concentrations of dissolved Ni and Fe (23±18 and 33±40 nM) agreed well with previous mean values (Pettine *et al.*, 1994) measured in the lower Po (35±19 and 36±45 nM respectively).

The concentrations (nM) of dissolved elements in the marine end-members were in the ranges 0.24 to 0.98 for Co, 5 to 20 for Ni, 5 to 33 for Fe, and 26 to 105 for Mn (Table 1). Average marine Co and Mn concentrations were 0.50±0.31 and 52±46 nM (n=14) respectively, substantially similar to those of the riverine end-member; while Ni and Fe concentrations in coastal samples were consistently lower than in riverine samples, with average values of 10±6 and 15±12 nM respectively.

3.2 PARTICULATE METALS

Particulate metal concentrations, expressed here as µmol/g of suspended matter, represent only the leachable and more reactive fraction of these elements. Concentrations of leachable Co in the solid phase ranged between surveys from 0.03 to 0.17 and 0.02 to 0.17 µmol/g in the riverine and saline end-members respectively (Table 1), showing similar ranges but slightly higher averages (0.11±0.05 µmol/g) in riverine than marine samples (0.07±0.05 µmol/g). Average Mn and Fe concentrations in the solid phase were slightly higher in riverine samples (20±4 and 89±37 µmol/g) than marine end-members (13±9 and 50±34 µmol/g), with values ranging from 15 to 25 for Mn and 47 to 148 for Fe in the river, and from 4 to 26 (Mn) and 23 to 81 (Fe) in the sea. Particulate Ni concentrations varied from 0.32 to 1.60 µmol/g in river samples and from 0.21 to 0.60 µmol/g in marine samples with average values higher (0.75±0.46) in river than marine samples (0.40±0.22). Leachable metal concentrations in river suspended matter were consistently lower than previously found average values in suspended particulate matter after total acid digestion (Pettine *et al.*, 1994); these values were: 20% lower for Mn, 50% lower for Co, 60% lower for Ni, and 78 % lower for Fe.

3.3 TRACE METAL CHANGES WITH SALINITY

Along the salinity gradient dissolved cobalt showed non-conservative behaviour in the summer surveys (Fig. 2a), with addition to the dissolved phase occurring in estuarine samples. This deviation from linearity corresponded to a depletion in the particulate phase (Fig. 2b), particularly evident on samples taken on July 6, 1994. The same pattern was observed for dissolved manganese in summer (Fig. 2c) with addition in the dissolved phase

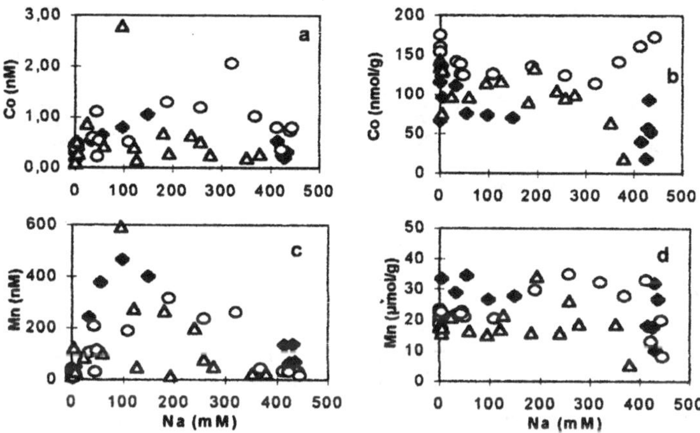

Fig. 2 Concentrations of dissolved Co (a) and Mn (c), particulate Co (b) and Mn (d) as a function of Na+ concentration (used as tracer for fresh/salt water mixing) during summer surveys. ◆ = July 21, 1993; O = July 6, 1994 and Δ = June 21, 1995

in all the summer surveys. The enrichment of dissolved manganese on June 21, 1995 also corresponded to an enrichment in the particulate phase for mid-salinity samples, suggesting a release of the element from the sediment (Fig 2d).

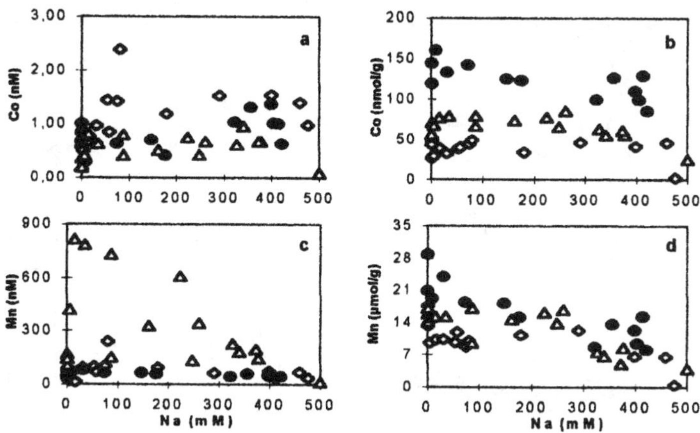

Fig. 3 Concentrations of dissolved Co (a) and Mn (c), particulate Co (b) and Mn (d) as a function of Na+ concentration during winter surveys. ◊ = March 18, 1992; ● = February 10, 1993 and Δ = March 1, 1995

In two winter samplings (March 18, 1992 and March 1, 1995) dissolved Co concentrations showed a positive deviation from the dilution line (Fig. 3a); however in the February 10,

1993 sampling a clear trend was not discerned. The corresponding particulate Co concentrations were fairly uniform along the salinity gradient (Fig. 3b). Manganese showed dissolved phase enrichment on two winter occasions: on March 1, 1995, (when the Po was characterized by high solid transport) dissolved Mn concentrations showed the highest values in the low and mid salinity ranges; on March 18, 1992 enrichment was less marked (Fig. 3c). The corresponding particulate Mn concentrations showed an almost linear decrease with salinity suggesting a prevalence of dilution processes (Fig. 3d).

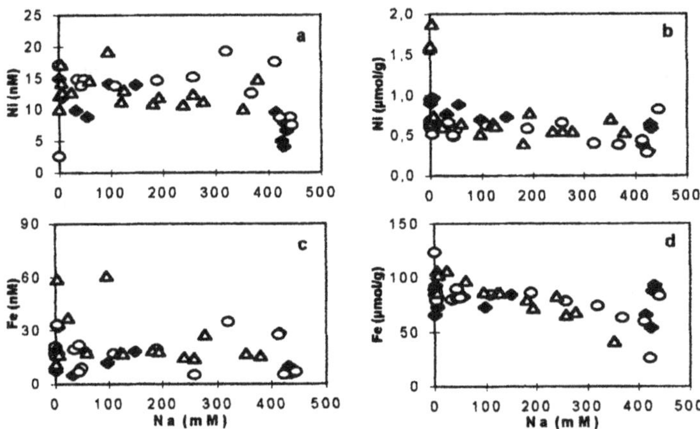

Fig. 4 Concentrations of dissolved Ni (a) and Fe (c), particulate Ni (b) and Fe (d) as a function of Na+ concentration during summer surveys. ◆ = July 21, 1993; O = July 6, 1994 and Δ = June 21, 1995

Dissolved Ni concentrations in the summer surveys were fairly uniform along the salinity gradient, showing small variations between riverine and estuarine samples (Fig. 4a); the corresponding particulate concentrations were also fairly uniform along the salinity gradient (Fig. 4b).

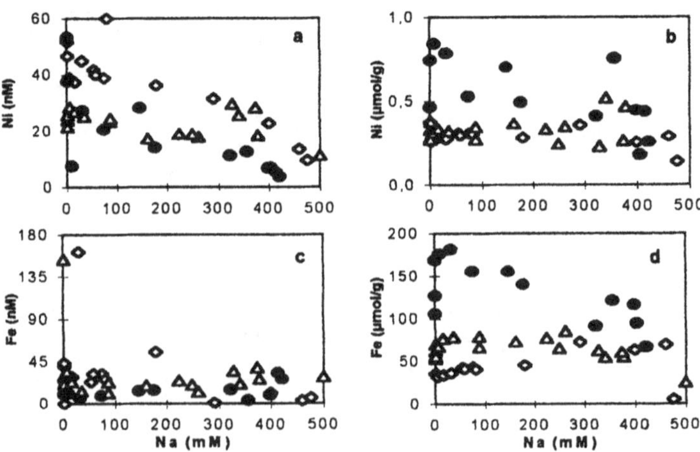

Fig. 5 Concentrations of dissolved Ni (a) and Fe (c), particulate Ni (b) and Fe (d) as a function of Na+ concentration during winter surveys. ◊ = March 18, 1992; ● = February 10, 1993 and Δ = March 1, 1995

In winter, dissolved Ni concentrations tended to decrease with increasing salinity (Fig. 5a); the corresponding particulate Ni concentrations were fairly uniform along the salinity gradient on March 18, 1992 and March 1, 1995, while the February 10, 1993 data were more scattered in the riverine and estuarine zones. On the latter occasion Ni had a higher affinity for the particulate phase than in two other surveys (Fig. 5b).
Dissolved iron concentrations in the summer surveys tended to decrease along the salinity gradient, but marked removal was not observed at low salinity in the estuary (Fig. 4c). The corresponding Fe concentrations in suspended particulate matter were nearly constant or tended to decrease with increase in salinity (Fig. 4d). Dissolved iron concentrations in winter were fairly uniform across the mixing zone, the only high values were found in the riverine end member (Fig. 5c). In winter particulate Fe showed the same pattern as particulate Ni, with a linear behaviour along the salinity gradient and a higher affinity for the particulate phase on February 10, 1993 than in the other two surveys (Fig. 5d).

4. Discussion and Conclusions

The concentrations of dissolved cobalt we found in the River Po (0.40 nM) are lower than the average for world rivers (3.4 nM), while concentrations in the coastal Adriatic (0.50 nM) agreed well with the average (0.85 nM) reported for ocean waters (Martin and Whitfield, 1983 and references cited). Dissolved manganese concentrations in the Po (52 nM) were lower than the average for world rivers (149 nM), while concentrations in the coastal Adriatic (52 nM) were 10 times higher than the average (3.6 nM) reported for ocean waters (Martin and Whitfield, 1983 and references cited). Dissolved nickel (23 nM) was higher than the average for pristine rivers of the Northern Adriatic basin (2.47 nM, Elbaz-Poulichet *et al.*, 1991) and the average for world rivers (8.5 nM), but similar to levels in the River Rhone (20.5 nM, Dai *et al.*, 1995). Ni concentrations in the coastal Adriatic (10 nM) were similar to those in the Venice lagoon (10-20 nM) and higher than the average (3.5 nM) for Western Mediterranean waters (Martin *et al.*, 1994 and references cited).
Our results show that Co and Mn behave non-conservatively in the Po estuary while Ni and Fe varied conservatively with salinity. The conservative behaviour of Fe requires comment: this metal is strongly influenced by redox processes (as is Mn), although it is generally removed from most estuaries by precipitation of Fe(III) hydroxides. In the Po conservative behaviour may reflect a balance of release and removal processes. However, it should also be stressed that we measured labile Fe species since our preconcentration procedure did not collect the colloidal Fe fraction which is mainly responsible for its estuarine removal. Clearly further investigations are needed to support our tentative interpretation of the data. In particular analysis of bottom waters, surface sediments and their pore waters, and comparison of the results with the present results on surface waters and their suspended matter would indicate whether redox conditions played a role in determining the seasonal behaviour of Mn and Co, as occurs in the North sea (Dehairs *et al*; 1989).
Based on the available information which indicates no significant gain or loss in the mixing area, the gross riverine load of Ni, estimated to be 514 t/y (Camusso *et al.*, 1993) should represent the net efflux of this metal to the Northern Adriatic. For Co and Mn their net fluxes to the Northern Adriatic should be higher than the gross riverine inputs of 40 and 4332 t/y previously published (Camusso *et al.*, 1993) due to the observed release processes.

Acknowledgements

This study was part of the PRISMA multi-disciplinary project, supported by the Italian Ministry for Universities and Research, to investigate eutrophication in the Adriatic sea. We gratefully acknowledge the support of the region of Emilia-Romagna for making the oceanographic vessel Daphne II available for collecting marine samples. We also thank the crew of the vessel for their assistance and R. Balestrini, S. Valsecchi, A. De Paolis (IRSA) and A. Moretti (ENEL-CRAM) for their help in the treatment of samples. We also thank D.Ward for help with the English.

References

Boniforti, R., Ferraroli, P., Frigieri, P., Heltai, D. and Queirazza, G.: 1984, *An. Chim. Acta* **162**, 33-46.

Bourg, A.C.M.: 1983, *Trace metals in sea water,* Plenum Press, New York, Wong C.S, Boyle E., Bruland K.W., Burton J.D., Goldberg E.D. (eds), 195-208.

Boyle, E.A., Edmond, J.M. and Sholkovitz, E.R.: 1977, *Geochim. Cosmochim. Acta* **41**, 1313-1324.

Boughriet, A., Oudane, B., Fischer, J.C., Wartel, M. and Leman, G.: 1992, *Wat. Res.* **26**, 1359-1378

Camusso, M., Martinotti, W. and Pettine, M.: 1993, *Int. Conf. Heavy Metals in the Environment*, Toronto, Sept. 1993. **1**, 309-311.

Camusso, M., Balestrini, R., Muriano, F. and Mariani, M.: 1994, *Chemosphere* **29**, 729-745.

Chiffoleau, J-F, Cossa, D., Auger, D. and Truquet, I.: 1994, *Mar. Chem.* **47**, 145-158.

Dai, M., Martin, J-M. and Cauwet, G.: 1995, *Mar. Chem.* **51**, 159-175

Dehairs, F., Baeyens, W. and Van Gansbeke D.: 1989, *Estuar., Coast. and Shelf Sci.* **29**, 457-471

Degobbis, D. and Gilmartim, M.: 1990, *Oceanol. Acta* **13**, 31-45.

Dorten, W.S., Elbaz-Poulichet, F., Mart, L.R. and Martin, J.M.: 1991, *Ambio* **20**, 2-6.

Duinker, J.C., Nolting, R.F. and Michel, D.: 1982, *Thalassia Jugol.* **18**, 191-202

Elbaz-Poulichet, F., Guan, D.M. and Martin, J.M.: 1991, *Mar. Chem.* **32**, 211-224

Ferrari, G.M. and Ferrario, P.:1989, *Water, Air and Soil Pollut.* **43**, 323-343.

Malo, B.A.: 1977, *Environ. Sci. Technol.* **11**, 277-282

Martin, J.M and Whitfield, M.: 1983, *Trace metals in sea water,* Plenum Press, New York, Wong C.S, Boyle E., Bruland K.W., Burton J.D., Goldberg E.D. (eds), 265-296.

Martin, J.M., Huang W.W. and Yoon Y.Y.: 1994, *Mar. Chem.*, **46**, 371-386.

Millward, G.E. and Moore, R.M.: 1982, *Water Res.* **16**, 981-985.

Millward G.E. and Turner A.:1995, *Trace Elements in Natural Waters*, CRC Press, Boca Raton, B. Salbu, E. Steinnes (eds), 223-245

Morris, A.W., Bale, A.J. and Howland, R.J.M.: 1982, *Estuar., Coast. Shelf Sci.* **13**, 175-192.

Pettine, M., Camusso, M., Martinotti, W., Marchetti, R., Passino, R. and Queirazza, G.: 1994, *Sci. Total Environ.* **145**, 243-265.

Salomons, W. and Förstner, V.: 1984, *Metals in the Hydrocycle,* Springer-Verlag Berlin 212-257.

Yeats, P.A. and Bewers, J.M.: 1982, *Can. J. Earth Sci.* **19**, 982-992.

BENTHIC FLUXES OF DISSOLVED INORGANIC CARBON, NUTRIENTS AND OXYGEN IN THE GULF OF TRIESTE (NORTHERN ADRIATIC)

A. BERTUZZI[1], J. FAGANELI[2], C. WELKER[3] and A. BRAMBATI[1]

[1] *Department of Geological, Environmental and Marine Sciences, University of Trieste, Trieste, Italy,*
[2] *Marine Biological Station, Piran, Slovenia,* [3] *Laboratory of Marine Biology, Trieste, Italy*

Abstract. Benthic fluxes of dissolved inorganic N, Si and P nutrients, alkalinity, dissolved inorganic C (DIC), and O_2 from sediments in the Gulf of Trieste (northern Adriatic, Italy) were measured monthly in the period September 1995 - August 1996 using in situ incubated light benthic chambers. The highest effluxes of DIC, NH_4^+, PO_4^{3-}, $Si(OH)_4$, and NO_3^- influxes encountered in late summer - early autumn were the consequence of degradation of benthic microalgae, and in autumn mostly of sedimented phytoplankton. High NO_3^- efflux was observed in spring. Only NH_4^+ and $Si(OH)_4$ fluxes were significantly correlated with temperature. This correlation suggests that the rate of downward input and the quality and quantity of sedimentary organic matter (autochthonous and allochthonous) were superimposed on the temperature fluctuations. High DIC, NH_4^+ and $Si(OH)_4$ effluxes observed in July 1996 were due to the late spring - early summer degradation of sedimentary organic matter produced by benthic microalgae, while the autumn phytoplankton bloom was quickly reflected in enhanced benthic fluxes due to higher temperature. Significant correlations between NH_4^+, PO_4^{3-} and $Si(OH)_4$ fluxes suggested their parallel regeneration and utilization at the sediment-water interface. The nutrient fluxes were linked to O_2 consumption, suggesting that aerobic oxidation processes were important at the sediment-water interface in the Gulf. The N, P and Si nutrients released from sediment pore waters are probably utilized in benthic microalgal and bottom-water primary production. This indicates that pelagic and benthic communities in the central part of the Gulf of Trieste function relatively independently of each other.

Key words: Adriatic Sea, nutrients, benthic fluxes, carbon, nitrogen, silicon, phosphorus, budgets

1. Introduction

Early diagenetic reactions occurring in surficial sediments influence the marine biogeochemistry of nutrients (Froelich et al., 1979; Jorgensen, 1983). In shallow coastal waters, where the euphotic zone is present down to the sediment-water interface, benthic fluxes are also affected by the macro- and microphytobenthic primary production (Rizzo, 1990). Processes are further complicated by the fact that bioturbating infauna destroy the "normal" concentration gradients in pore waters by irrigation and particle reworking, introducing oxygenated overlying water (Aller, 1982). On the other hand, the temporal variations in the quantity and quality of sedimented organic matter are also reflected in pore water chemistry and temporal benthic flux variations. This is also the case of the Gulf of Trieste (northern Adriatic) where the bulk of pelagic organic matter is of marine origin, and more than half of the particulate C and N are decomposed in the water column or transported away (Faganeli, 1989; Faganeli et al., 1995). Although the photic zone reaches

Water, Air and Soil Pollution **99**: 305-314, 1997.

the benthos throughout most of the gulf, the benthic O_2 consumption exceeds benthic primary production (Herndl et al., 1989). Anaerobic degradation of sedimentary organic matter is important in these sediments (Faganeli and Herndl, 1991; Bertuzzi et al., 1996), and SO_4^{2-} and metal reduction rates appeared to be similar in magnitude to O_2 consumption rates (Hines et al., submitted). Previous studies in the Gulf of Trieste also showed that nutrients released from sediments are mostly utilized in benthic primary production (Herndl et al., 1986).

The aim of the present study was (1) to follow the temporal variation of in situ benthic fluxes of inorganic C, N, Si, and P, (2) determine the causes of those variations, and (3) assess their impact on the environmental condition in the water column of the shallow Gulf of Trieste (northern Adriatic).

2. Materials and methods

2.1. BENTHIC FLUX EXPERIMENTS

Benthic fluxes were determined on sampling site AA1 in the central part of the Gulf of Trieste (see Figure 1 in Cermelj et al., this issue) at the depth of 20 m using in situ deployed light benthic chambers between September 1995 and August 1996. The chamber consists of a 30x50x30 cm box, constructed in Plexiglass, which is open on the bottom, with valves on the lid for sampling and a flexible membrane consisting of polyethylene bag that served to compensate the volume removed during sampling. Attached to the inside of the chamber lid is a stirring mechanism coupled to a motor driven at a speed of 50 rpm. The chamber was carefully placed by SCUBA divers about 10 cm into the sediment. Samples were periodically, after 2, 8 and 24 hours, collected by divers with polypropylene syringes. The samples, except for oxygen, were filtered through 0.45 μm Millipore HA membrane filters. Fluxes of solutes across the sediment-water interface were evaluated from linear regressions of solute concentration against time. Some parallel measurements were performed in winter and late summer using two benthic chambers and no significant differences were observed.

2.2. ANALYSES

Nitrate, ammonium, phosphate, and silicate in chamber water were measured photometrically using standard procedures (Grasshoff et al., 1983). Alkalinity was measured by titration with 0.01 M HCl and then calculating DIC using pH and alkalinity data (Grasshoff, 1983). Oxygen concentrations were determined using a Winkler method (Grasshoff et al., 1983).

Microalgae in surficial sediment (down to the depth of 7 cm), sampled with corers (three replicates, 5 cm of internal diameter) by SCUBA divers, were fixed with hexamethylene tetramine-buffered formaldehyde solution. After stirring, subsamples of 20 μl were drawn off using a pipette, placed in a counting chamber, and then living cells, with plasmatic content, were countered under a Leitz inverted microscope and expressed as number of cells cm^{-3} (Welker and Nichetto, 1996).

3. Results and discussion

3.1. BENTHIC NUTRIENT FLUXES

The highest DIC and nutrient, except of NO_3^-, effluxes were observed in the late summer-early autumn period (Figures 1 and 2). High DIC, NH_4^+, PO_4^{3-} and $Si(OH)_4$ effluxes were observed in July 1996. The highest NO_3^- effluxes were observed in early autumn (October 1995) and spring (April 1996) while the highest influx in September 1995 in parallel with high O_2 consumption (Figure 2). O_2 consumption decreased from the summer - early autumn towards winter (Figure 2).

The highest benthic effluxes of all nutrients and DIC observed in the late summer - autumn period, as well as the smallest DIC, NH_4^+, $Si(OH)_4$ and PO_4^{3-} effluxes in winter, suggest that the temporal variations of these fluxes are temperature dependent (Figure 2). However, only NH_4^+ ($r=0.65$; $n=11$) and $Si(OH)_4$ ($r=0.60$; $n=11$) effluxes were significantly temperature dependent. This observation suggests that other processes, such as the rate of downward input of particulate matter and the quality of sedimented inorganic and organic matter (autochthonous and allochthonous), as well as the production of benthic organic matter, were probably superimposed on the above-described temperature fluctuations (Bertuzzi et al., 1996). In fact, high DIC, NH_4^+, PO_4^{3-} and $Si(OH)_4$ benthic effluxes observed in July 1996 were due to the degradation of sedimentary organic matter produced by a late spring - summer bloom of benthic microalgae. High summer nutrient effluxes supported intensive growth of benthic microalgae and phytoplankton or resuspended benthic microalgae in the bottom water below the pycnocline in July and August 1996. This utilization reduced the DIC and NH_4^+ effluxes and caused an influx of PO_4^{3-} and $Si(OH)_4$ (Figures 1 and 2). The intensive growth of benthic microalgae was evident from the temporal variation in the density of benthic microalgae (Figure 3) and phytoplankton biomass (from chlorophyll *a* concentrations; Figure 4). In that period no phytoplankton blooms in the upper layer of the water column were evident. Moreover, the analysis of benthic microalgae showed the absence of sedimented phytoplanktonic cells. The enhanced fluxes in early autumn were markedly influenced by phytoplanktonic bloom since the analysis of benthic microalgae showed 2-fold higher contents of phytoplanktonic than the microphytobenthic cells. The important role of benthic microalgae in the benthic nutrient biogeochemistry in the Gulf of Trieste in general is also supported by the depth distribution of the cells decreasing along the core depth (data not presented) in all periods except in winter.

Most the highest NO_3^- fluxes were oriented out of sediments, indicating that nitrification prevailed in surficial sediments. The NO_3^- produced also diffused downward and was reduced (Jahnke et al., 1982; Berelson et al., 1987; Devol and Christensen, 1993). The greatest influx of NO_3^- occurred in late summer (September 1995) during the bottom water

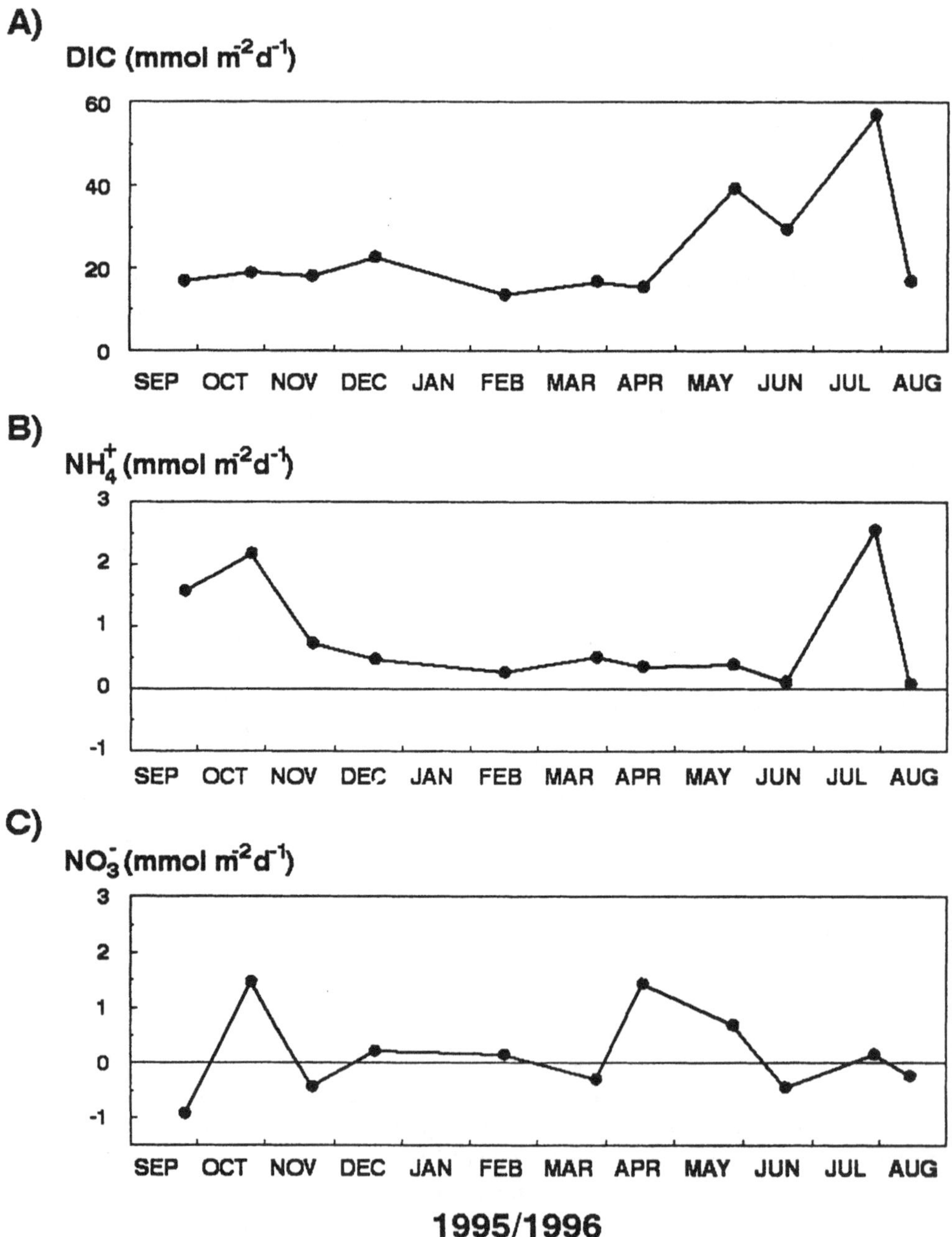

Fig.1. Variations of DIC (A), NH_4^+ (B) and NO_3^- (C) benthic fluxes at the sampling station AA1 in the period September 1995 - August 1996

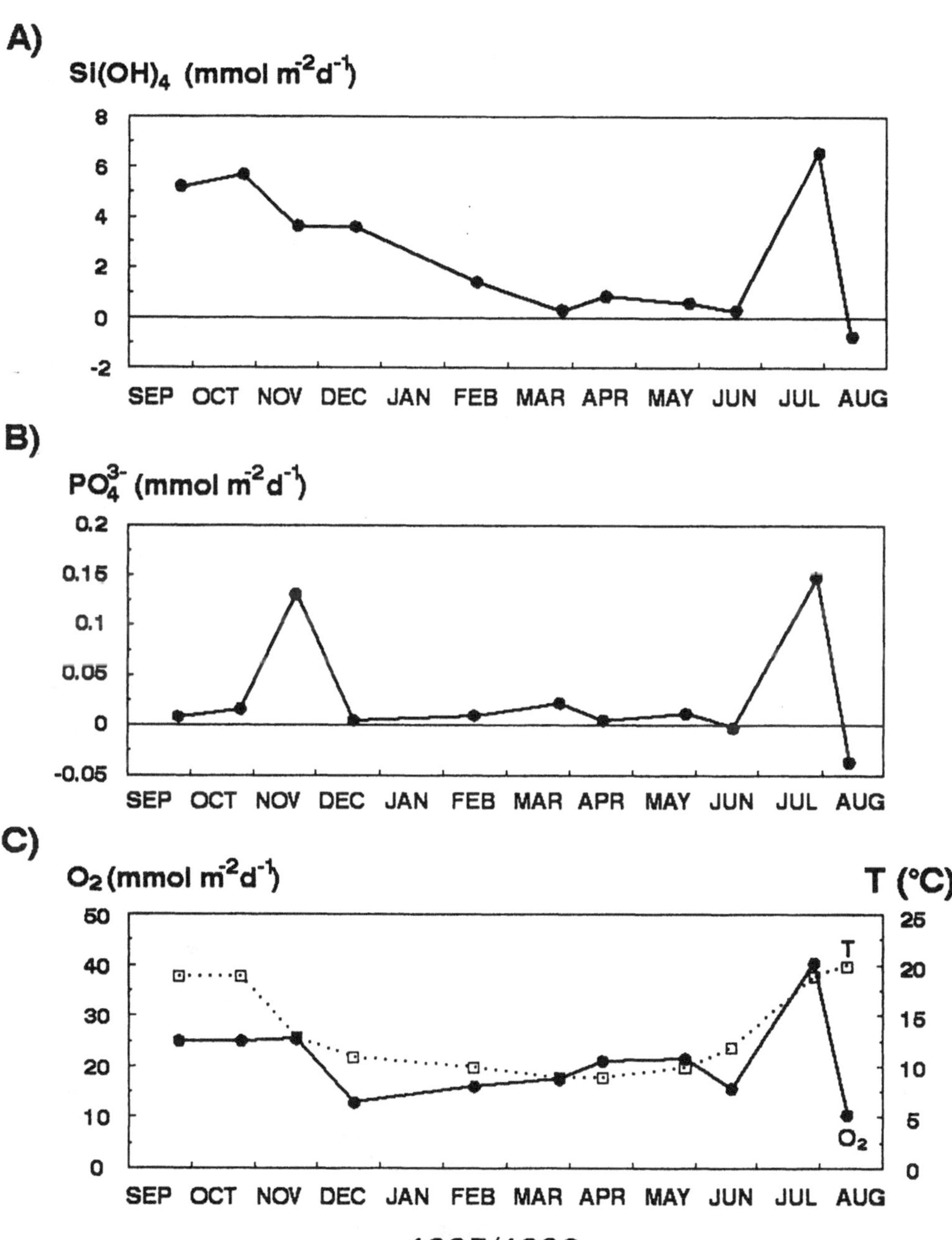

Fig.2. Variations of $Si(OH)_4$ (A) and PO_4^{3-} (B) benthic fluxes, O_2 benthic consumption (C) and bottom water temperature (C) at the sampling station AA1 in the period September 1995 - August 1996

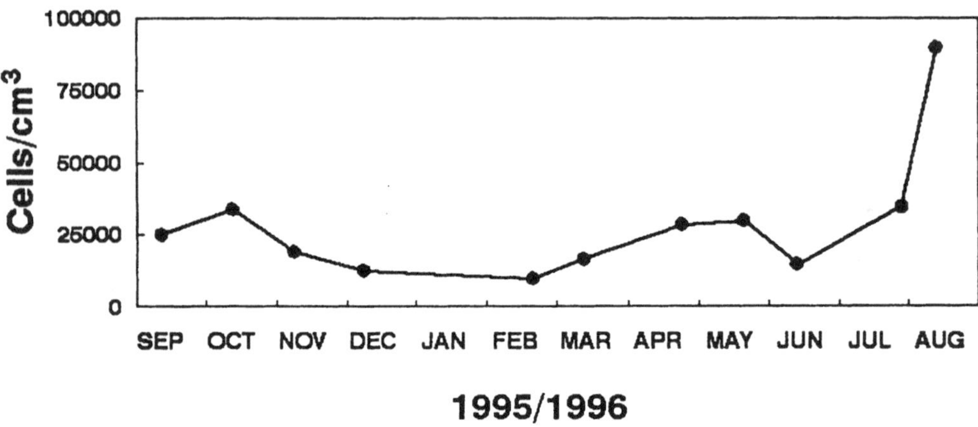

Fig.3. Variations of cell density of benthic microalgae in the surficial 7 cm layer at the sampling station AA1 in the period September 1995 - August 1996

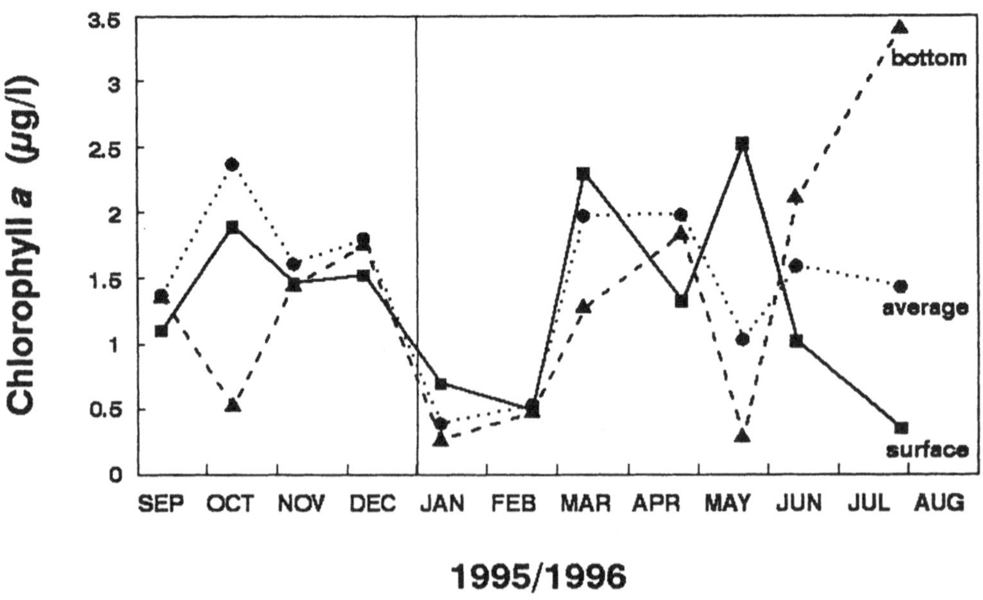

Fig.4. Variations of phytoplankton biomass (expressed as chlorophyll *a* concentrations) in surface (0.5 m) and bottom layers, with concentrations averaged over the water column depth, in the central part of the Gulf of Trieste in the period September 1995 - July 1996 (data are from the Marine Biological Station Piran)

O_2 depletion (O_2 concentration of 2.7 ml/l; data are from the Laboratory of Marine Biology, Trieste), indicating the important role of denitrification at that time superimposed to assimilation of benthic microalgae (Rysgaard et al., 1995).

Low PO_4^{3-} fluxes observed could be the consequence of assimilation of benthic microalgae (Rysgaard et al., 1995), and of the redox-dependent retention and accumulation of PO_4^{3-} by sorption on ferric oxyhydroxides (Buffle et al., 1989; Ingall and Jahnke, 1993). During the oxidative precipitation of Fe(III) oxyhydroxides under oxic conditions, the dissolved PO_4^{3-} is immobilized in the mineral phase (Krom and Berner, 1981). Fe(II) diffusing upward in pore waters precipitates as ferric oxyhydroxides in the aerobic sediment-water interface, acting as a sink of pore water PO_4^{3-}, which is later released during diagenesis (Froelich et al., 1988). Other possibilities include the formation of insoluble Fe-Mn phosphate complexes (Krom and Berner, 1981) and adsorption of PO_4^{3-} onto carbonate surfaces (Kitano et al., 1978; Hines and Lyons, 1982).

Comparisons with nutrient and DIC fluxes and sediment O_2 consumption rates previously obtained in the same station in the Gulf of Trieste using the laboratory incubation technique showed that the DIC, NH_4^+, NO_3^-, PO_4^{3-} and $Si(OH)_4$ fluxes measured in benthic chambers are about 2-fold higher while the O_2 sediment consumption rates are similar (Bertuzzi et al., 1996). This is probably due more to the interannual variation of benthic fluxes than to the methodological differences since Chanton et al. (1987) demonstrated that the laboratory sediment incubations at in situ temperatures give inorganic fluxes that are comparable to those measured in situ using benthic chambers. However, our summer benthic fluxes, except for PO_4^{3-}, are similar to those measured in the NW Adriatic off the Emilia-Romagna coast (Giordani et al., 1992), and other coastal zones, i.e., San Francisco Bay (Hammond et al., 1985), Narragansett Bay (McCaffrey et al., 1980), Tomales Bay (Dollar et al., 1991), and nearshore Georgia Bight (Hopkinson, 1987).

The DIC (r= 0.7; n=11), NH_4^+ (r=0.67; n=11), PO_4^{3-} (r=0.83; n=11) and $Si(OH)_4$ (r=0.61; n=11) benthic fluxes were linked to O_2 consumption suggesting that aerobic degradation of sedimentary organic matter is important in surficial sediments in the Gulf. On the other hand, recent measurements of anaerobic degradation of sedimentary organic matter, using Mn(IV), Fe(III), and SO_4^{2-} as electron acceptors (Hines et al., submitted), showed that the anaerobic processes are similar in magnitude to the benthic O_2 consumption in these sediments. Significant correlation between NH_4^+ and $Si(OH)_4$ (r= 0.91; n=11), and NH_4^+ and PO_4^{3-} (r=0.56; n=11) suggest their parallel benthic regeneration and assimilation.

TABLE I Mean (±standard deviation, n=11) annual (September 1995 - August 1996) benthic fluxes of DIC, N, P and Si nutrients, and O_2 consumption at the sampling station AA1 in the Gulf of Trieste (mmol m^{-2} d^{-1})

DIC	NH_4^+	NO_3^-	PO_4^{3-}	$Si(OH)_4$	O_2
24.1±14.4	0.8±0.7	0.17±0.73	0.029±0.05	2.59±2.3	-20.4±8.9

TABLE II Tentative budgets for C, N, P and Si at the sediment-water interface, benthic (Herndl et al., 1989) and phytoplankton (Faganeli et al., 1982) primary production in the Gulf of Trieste (mol m^{-2} y^{-1})

	C	N	P	Si
benthic fluxes	8.8	0.4	0.01	0.9
burial	0.5	0.1	0.01	1.9
benthic primary production	3.0	0.4	0.03	2.7
phytoplankton primary production	4.2	0.6	0.04	2.1

3.2. BENTHIC NUTRIENT BUDGETS

The fate of nutrients at the sediment-water interface can be decoded by constructing the nutrient budget (Hammond et al., 1985). A tentative annual benthic nutrient budget for the station AA1 in the central part of the Gulf of Trieste can be constructed from mean annual fluxes, obtained from daily fluxes (Table I), and assuming steady state conditions. The burial rate was estimated from the lowest organic C, total N and P, and biogenic Si contents in the deepest measured sediment layers at the station AA1 (12-13 cm; Bertuzzi et al., 1996), since they are nearly constant and less affected by biological activity operative mostly in the upper centimetres (Brizzi et al., 1994), and ^{210}Pb sediment accumulation rate (R) of 1.8 mm y^{-1} (Frignani, pers. comm.). The burial rate (S) of the element was calculated using the relation S = dcR(1-P) (Suess and Djafari, 1977) where d is the dry density of sediment (2.5 g cm^{-3}; Ranke, 1976), c the concentration of element and P porosity (0.75; Cermelj et al., this issue). The utilization of released nutrients by benthic microalgae and phytoplankton can be estimated using previously obtained data of ^{14}C phytoplankton production (Faganeli et al., 1982) and gross microphytobenthic production (Herndl et al., 1989) in the gulf, respectively. In transforming C values into production rates of N and P, the Redfield ratios were used. Si utilization in phytoplankton production was calculated using an Si/C molar ratio of 0.4 (Smith and Nelson, 1986) because on an annual basis about 40% of the phytoplankton in the Gulf of Trieste is composed of siliceous organisms (diatoms, silico-flagellates; Malej et al., 1995). On the other hand, benthic microalgae are nearly all composed of diatoms (Welker and Nichetto, 1996), and their Si utilization was computed using an Si/C molar ratio of 1 (Smith and Nelson, 1986).

A tentative nutrient budget at the sediment-water interface in the Gulf of Trieste (Table II) showed that about 95% of C and 80% of N are recycled before burial since the percentage of the nutrient burial is low. On the other hand, about 50% of P and the majority (70%) of biogenic Si are retained in sediments. The released N, Si, and P during benthic fluxes are probably consumed by benthic microalgal and bottom-water primary production (Herndl et al., 1986), while the majority of the released DIC could be supplied to phytoplankton. This tentative budget indicates that the nutrient supply for phytoplankton growth must principally originate from riverine inflows (Malej et al., 1995) and that during even the highest benthic effluxes in summertime the released nutrients remain in the bottom-water layer below the pycnocline (Faganeli et al., 1991). Thus, it seems that pelagic and benthic communities in the central part of the Gulf of Trieste function relatively independently of each other.

Conclusions

1. Benthic fluxes of DIC, NH_4^+, NO_3^-, $Si(OH)_4$, PO_4^{3-} and sediment O_2 consumption in the Gulf of Trieste (northern Adriatic) showed wide temporal variations. Only NH_4^+ and $Si(OH)_4$ effluxes were directly temperature dependent, suggesting that the rates of benthic microalgal production and particulate matter downward transport, and organic matter degradation, were superimposed on the temperature fluctuations. High effluxes of DIC, NH_4^+, PO_4^{3-} and $Si(OH)_4$ observed in summer (July 1996) were most probably due to the degradation of summer benthic microalgal bloom while the high autumn (October 1995) benthic effluxes were principally the consequence of the fast degradation of an autumn phytoplankton bloom. The high effluxes observed in July 1996 were followed by a significant peak of the density of benthic microalgae and phytoplankton biomass in the bottom water layer.
2. PO_4^{3-} benthic fluxes were low throughout this study because of the PO_4^{3-} removal at the oxic sediment-water interface, probably on the surfaces of Fe-Mn hydroxides and carbonates, superimposed to assimilation by benthic microalgae. Significant correlations observed between measured benthic fluxes and O_2 consumption rates suggest that the aerobic pathway of the degradation of sedimentary organic matter is important in the surficial sediments in the Gulf of Trieste.
3. Tentative nutrient budgets constructed for the sediment-water interface in the Gulf of Trieste indicate that about 95% of the input of C and 80% of N are recycled at the sediment-water interface while about 50% of P and 70% of biogenic Si are retained in the sediments. The released N, Si and P nutrients are probably utilized in the benthic production of microalgae, while part of the released DIC can be supplied to phytoplankton.

References

Aller, R. C.: 1982, In: McCall, D. L. and Tevezs, M. J. T. (Eds.), *Animal-sediment relations*. Plenum, New York, pp. 53-102.

Berelson, W. M. Hammond, D. E. and Johnson, K. S.: 1987, *Geochim. Cosmochim. Acta* **51**, 1345-1363.

Bertuzzi, A. Faganeli, J. and Brambati, A.: 1996, *P.S.Z.N. I, Mar. Ecol.* **17**, 261-278.

Brizzi, G. Aleffi, F. Landri, P. and Orel, G.: 1994, *29th Eur. Mar. Biol. Symp.* Vienna, Austria.

Buffle, J. DeVitre, R. R. Perret, D. and Leppard, G. G.: 1989, *Geochim. Cosmochim. Acta* **53**, 399-408.

Cermelj, B. Bertuzzi, A. and Faganeli, J.: 1997, *Water, Air, and Soil Pollut.* this issue.

Chanton, J. P. Martens, C. S. and Goldhaber, M. B.: 1987, *Geochim. Cosmochim. Acta* **51**, 1187-1199.

Devol, A. H. and Christensen, J. P.: 1993, *J. Mar. Res.* **51**, 345-372.

Dollar, S. J. Smith, S. V. Vink, S. M. Obelski, S. and Hollibaugh, J. T.: 1991, *Mar. Ecol. Prog. Ser.* **79**, 115-125.

Faganeli, J.: 1989, *Mar. Chem.* **26**, 67-80.

Faganeli, J. and Herndl, G. J.: 1991, In: Berthelin, J. (Ed.), *Diversity of environmental biogeochemistry*. Elsevier, Amsterdam, pp. 157-170.

Faganeli, J. Fanuko, N. Stegnar, P. and Vukovic, A.: 1982, *Acta Adriat.* **23**, 53-60.

Faganeli, J. Pezdic, J. Ogorelec, B. Herndl, G.J. and Dolenec, T.:1991, In Tyson, R. and Pearson, T.H. (Eds.), *Modern and ancient continental shelf anoxia*. Geological Society, London, pp. 107-117.

Faganeli, J. Kovac, N. Leskovsek, H. and Pezdic, J.: 1995, *Biogeochemistry* **29**, 71-88.

Froelich, P.N. Klinkhammer, G.P. Bender, M.L. Luedtke, G.R. Heath, D. Cullen, D. Dauphin, P. Hammond, D. Hartman, B and Maynard, V.: 1979, *Geochim. Cosmochim. Acta* **43**, 1075-1090.

Froelich, P. N. Arthur, M. A. Burnett, W. C. Deakin, M. Hensley, V. Jahnke, R. Kaul, L. Kim, K.-H. Roe, K. Soutar, A. and Vathakon, C.: 1988, *Mar. Geol.* **80**, 309-343.

Giordani, P. Hammond, D. E. Berelson, W. M. Montanari, G. Poletti, R. Milandri, A. Frignani, M. Lagone, L. Ravaioli, G. Rovatti, G. and Rabbi, E.: 1992, In: Vollenweider, R. A. et al. (Eds.), *Marine coastal eutrophication*. Elsevier, Amsterdam, pp. 251-275.

Grasshoff, K. Ehrhardt, M. and Kremling, K., 1983: *Methods of seawater analysis*. Verlag Chemie, Weinheim, 419 pp.

Hammond, D. E. Fuller, C. Harmon, D. Hartman, B. Korosec, M. Miller, L. G. Rea, R. Warren, S. Berelson, W. and Hager, S.: 1985, *Hydrobiologia* **129**, 69-90.

Herndl, G. J. Faganeli, J. Fanuko, N. Peduzzi, P. and Turk, V.: 1986, In Klekowski, R.Z. et al. (Eds.) *Proc. 21st Eur. Mar. Biol. Symp.* Ossolineim, Gdansk, pp. 297-308.

Herndl, G. J. Peduzzi, P. and Fanuko, N.: 1989, *Mar. Ecol. Prog. Ser.* **53**, 169-178.

Hines, M.E. and Lyons, W.B.: 1982, *Mar.Ecol.Prog.Ser.* **8**, 87-94.

Hines, M.E. Faganeli, J. and Planinc, R.: *Biogeochemistry*, submitted.

Hopkinson, C. S.: 1987, *Mar. Biol.* **34**, 127-142.

Ingall, E. and Jahnke, R.: 1993, *Geochim. Cosmochim. Acta* **58**, 2571-2575.

Jahnke, R.A.: 1990., *Jour. Mar. Res.* **48**, 413-436.

Jahnke, R. A. Heggie, D. Emerson, S. and Grundmanis, V.: 1982, *Earth Planet. Sci. Lett.* **61**, 233-256.

Jorgensen, B. B.: 1983, In Bolin, B and Cook, R.B. (Eds.) *The major biogeochemical cycles and their interactions*, SCOPE 21. Wiley, Chichester, pp. 477-509.

Kitano, V. Okomura, M. and Idogak, M.: 1978, *Geochem. Jour.* **12**, 29-37.

Krom, M. and Berner, R. A.: 1981, *Geochim. Cosmochim. Acta* **45**, 207-216.

Malej, A. Mozetic, P. Malacic, V. Terzic, S. and Ahel, M.: 1995, *Mar. Ecol. Prog. Ser.* **120**, 111-121.

McCaffrey, R. J. A. Myers, A. C. Davey, E. Morrison, G., Bender, M. Luedtke, N. Cullen, D. Froelich, P. and Klinkhamer, G.: 1980, *Limnol. Oceanogr.* **25**, 31-44.

Ranke, U.: 1976, *Senckenbergiana Marit.* **8**, 23-60.

Rizzo, W. W.: 1990, *Estuaries* **13**, 219-226.

Rysgaard, S. Christensen, P.B. and Nielsen, L.P.: 1995, *Mar. Ecol. Prog. Ser.* **126**, 111-121.

Smith, W. O. Jr and Nelson, D. M.: 1986, *Bioscience* **36**, 251-257.

Suess, E. and Djafari, D.: 1977, *Eart Planet. Sci. Lett.* **35**, 49-54.

Welker, C. and Nichetto, P.: 1996, *P.S.Z.N.I, Mar. Ecol.* **17**, 473-489.

CARBON AND NITROGEN STABLE ISOTOPE FRACTIONATION IN THE SEDIMENT OF LAKE BLED (SLOVENIA)

S. LOJEN[1], N. OGRINC[1] and T. DOLENEC[1,2]
[1] *"J. Stefan" Institute, Department of Environmental Sciences, Jamova 39, 1000 Ljubljana, Slovenia*
[2] *Faculty of natural sciences and technology, Dept. of Geology, Aškerčeva 12, 1000 Ljubljana, Slovenia*

Abstract: Stable isotope composition of carbon and nitrogen in the sediment and pore water of a eutrophic freshwater lake was studied. Based on changes in the $\delta^{13}C$ and $\delta^{15}N$ values of dissolved components and sediment fraction, possible processes involved in the decomposition of sedimentary organic matter are outlined. The relative importance of acetate fermentation and CO_2 reduction was estimated using known mathematical models, and ammonia assimilation by methanogenic bacteria is hypothesised to be the main process governing the isotope fractionation of dissolved nitrogen in pore water.

Key words: lake, sediment, carbon, nitrogen, stable isotope, fractionation

1. Introduction

Natural concentrations of biogenic elements such as carbon (C), nitrogen (N), sulphur (S), and phosphorus (P) in the atmosphere and hydrosphere have been significantly changed by human activity. The response of affected biogeochemical cycles is reflected in environmental changes. Artificially induced or promoted natural eutrophication of lakes by anthropogenic pollution is only one of the related processes which, in turn, can have many ecological and socio-economic effects on a local scale. In order to better understand the eutrophication process of freshwater lakes, an understanding of the cycling of biogenic elements in the sediment-water system is necessary. The stable isotope ratios of these elements in the inorganic component of the sediment, in the sedimentary organic matter (SOM), and in pore water are useful tools for explaining reaction pathways and the importance of some specific degradation processes of SOM.

The purpose of this research was to investigate the C- and N-biogeochemical cycles in the sediment of Lake Bled in more detail and to estimate the importance of particular processes using stable isotope techniques. Special attention was paid to methanogenesis and to evaluation of the importance of specific methane formation pathways (acetate fermentation or CO_2 reduction).

2. Methods

2.1. Sampling

Lake Bled is a highly eutrophic dimictic postglacial subalpine lake located in the Eastern Julian Alps at an altitude of 467 m a.s.l. with an area of 1.44 km^2 and mean depth of 17.9 m. The lake has all of the characteristics of eumictic freshwater basins with an anoxic hypolimnetic zone. Nevertheless, in some shallower bays, oxygen excess down to the bottom is typical for the water column due to biological activity. Following intensive restoration measures (Vrhovšek *et al.,* 1984), the water quality is

Water, Air and Soil Pollution **99**: 315-323, 1997.

improving, although a large accumulation of organic debris at the bottom of the lake still represents an important source of nutrients (Čermelj *et al.*, 1996).

Sediment samples were taken using a gravity core sampler with a plexi-glass tube (Meischner and Rumohr, 1974) in October 1994, and in April, August and November 1995 in the shallow (10-12m deep) Zaka bay in the western part of the lake, where the water column is supersaturated with oxygen down to the water/sediment interface. In August 1995, one sample was taken in the anoxic hypolimnion in the deepest part (30 m), also in Western basin (stations D and Z, see Ogrinc *et al.*, this issue). Bottom water temperatures measured in Zaka bay were 5°, 18°, 15° and 13°C in April, August, October and November, respectively. In the Western basin, the bottom temperature is more or less stable (4-6°C).

2.2. ANALYSES

Sediment cores (approx. 50 cm long) were cut into 1-4 cm sections and pore water was extracted under pressure of 0.4 MPa N_2 using a nylon squeezer and filtered through a 0.45μm membrane filter in a glove box with an inert atmosphere. pH, total alkalinity and the carbon isotope ($\delta^{13}C$) composition of dissolved inorganic C (DIC) were measured immediately. Samples for analyses of SO_4^{2-}, NO_3^-, NO_2^- NH_4^+ and acetate were refrigerated. Total alkalinity was determined using the Gran titration method (Gran, 1953). Sulphate was determined turbidimetrically, whereas other anions were measured using standard colorimetrical methods (Grasshof *et al.*, 1983). Acetate was analysed by high performance ion chromatography (Dionex 4000I). Sedimentary carbonates for $\delta^{13}C$ analysis were prepared by digestion in 100% H_3PO_4 at 50°C for 3h following the method of McCrea (1950). CO_2 released during acid treatment was analysed using a Varian MAT 250 mass spectrometer. Results are reported as deviations in per mil (‰) from the Chicago PDB standard with an analytical error of ±0.05‰ calculated as average standard deviation of at least three measurements. A Europa 20-20 Stable Isotope Analyser (Europa Scientific Ltd.) with Anca SL (solid-liquid) and Anca TG (trace gas) preparation module was used to determine the $\delta^{13}C$ of DIC, sedimentary organic carbon (SOC), $\delta^{15}N$ of total sedimentary nitrogen and total dissolved nitrogen in the pore water. $\delta^{13}C$ of DIC was measured following the modified procedure developed by Mook *et al.* (1974). The isotopic composition of SOC was determined on samples, which were pre-treated with 3N HCl and washed with distilled water. $\delta^{15}N$ of total sedimentary nitrogen was determined in bulk samples, whereas the total dissolved nitrogen was precipitated from the pore water as ammonium sulphate. The isotopic standard used for nitrogen is air. The analytical error for the measurements is ±0.1‰ for $\delta^{13}C$ and ±0.2‰ for $\delta^{15}N$.

2.1. CALCULATION PROCEDURES

LaZerte (1981) presented a model calculation for the estimation of a limiting $\delta^{13}C$-DIC value in pore water of lake sediments where methanogenic processes dominate. The reaction pathway of methane formation (acetate fermentation vs. CO_2 reduction) is taken into account. It was shown that DIC increases, whereas $\delta^{13}C$-DIC approaches a limiting value ($\delta^{13}C_f$), as methanogenesis proceeds. The limiting value depends on the ratios between the fluxes of methane formed by the two reaction pathways as follows:

$$\delta^{13}C' = \delta^{13}C_f + B' \times (DIC)^{((C/A)(1+1/\alpha c)-B/A-1)/(B/A-C/A+1)} \quad \text{where}$$

$$B' = (\delta^{13}C_i - \delta^{13}C_f) \times (\Sigma CO_2)^{(1+B/A-C/A(1-1/\alpha c))/(B/A-C/A+1)} \quad \text{and}$$

$$\delta^{13}C_f = ((1+(B/A)/\alpha_b) \times (\delta^{13}C_{org}+1000))/(1+B/A-(C/A)(1-1/\alpha_c)).$$

DIC is the concentration of total dissolved inorganic C with stable isotope ratio $\delta^{13}C'$, $\delta^{13}C_i$ is the initial $\delta^{13}C$-DIC value, $\delta^{13}C_{org}$ is $\delta^{13}C$ value of SOC, α_c is the C isotope fractionation factor between CO_2 and CH_4, α_b is the C isotope fractionation factor between SOC and CO_2 from the carboxyl group of acetate, and A, B and C are the fluxes of SOC to CO_2, of acetate to CO_2 and CH_4 and of CO_2 to CH_4, respectively.
The flux of NH_4^+ at the sediment/water interface was calculated using the approach of Čermelj *et al.* (1996) from the general equation for the mass balance of a dissolved nutrient in pore water using the solution by Van Cappellen and Berner (1988) and assuming that removal of ammonia by precipitation is negligible:

$$J_{(z=0)} = (-\Phi DR/DK_g + U(1 + K_{ads})) + \Phi U(K_{ads} + 1)C_{(z=0)}$$

where z is depth below the sediment/water interface, C = concentration, Φ = porosity, U = sedimentation rate (cm/yr), K_g = first order rate constant for organic N release (yr^{-1}), K_{ads} = coefficient of NH_4^+ adsorption, D = whole sediment diffusion coefficient of NH_4^+ (cm^2/yr), and R = NH_4^+ production rate from organic N ($mole/cm^3/yr$). Diffusion constants and their temperature dependence were calculated from data collected by Lerman (1979). Isotopic composition of NH_4^+ flux at the selected depth was estimated from the slope of the line $[NH_4^+]/[NH_4^+{}_{(z=0)}] - 1$ vs. $\delta^{15}N\text{-}NH_4^+ \times ([NH_4^+]/[NH_4^+{}_{(z=0)}])$ (Sayles and Curry, 1988).

3. Results and discussion

3.1. CARBON

Mesozoic carbonates from the hinterland of the lake show $\delta^{13}C$ values between -1.5‰ and +4.56‰ (Dolenec *et al.*, 1984), whereas sedimentary lacustrine carbonates are more depleted in ^{13}C. The $\delta^{13}C$ in the carbonate fraction of the bulk sediment in shallower nearshore parts of the lake is less negative (-3.3 to -2.5‰) than in the central part of the basin (-4.2 to -3‰). This suggests a bigger fraction of terrigenous carbonates in the Zaka bay than in the closed Western basin where in-situ produced carbonate is more abundant.

Isotopic composition of C in some plants, phyto- and zooplankton, as well as in some fishes was measured by Čermelj *et al.* (1996). The mean $\delta^{13}C$ values of phytoplankton and net zooplankton were -27.7 and -30.1‰, respectively. On the other hand, $\delta^{13}C$ of periphyton was rather high, averaging -11.4‰. Submerged plants show a much broader

$\delta^{13}C$ range: from -11.8‰ for *Myriophyllum spicatum* to -29.1‰ for *Nuphar lutetum*. $\delta^{13}C$ of different tissues of fishes varies between -29.74‰ (*Perca fluviatilis*, gill) and -32.42‰ (*Rutilus rutilus*, stomach). $\delta^{13}C$ of SOC in surficial sediment from different parts of the lake was measured to be between -28.5‰ and -37.87‰, indicating a significant depletion in ^{13}C compared to its source material. Èermelj *et al.* (1996) attributed this depletion to the presence of organic compounds synthesised by methanogenic bacteria.

Similar to carbonate, the isotopic composition of SOC in the well oxygenated Zaka bay shows less negative $\delta^{13}C$ values compared to the anoxic Western basin (-31.5 to -29‰ and -34.6 to -31.5‰, respectively). The good relationship between the $\delta^{13}C$ of carbonate and SOC (Fig. 1) suggests that decaying organic matter can be a major source of carbon for carbonate precipitation during the summer and the autumn, when the conditions for precipitation are favourable. Confirmation for this assumption is indicated by the calculation of the CO_2 fluxes (Ogrinc *et al.*, this issue).

Assuming that the dissolution of carbonate and the decay of SOC could be the most important sources of DIC in pore water, $\delta^{13}C$ value of DIC should be between those of carbonate and SOC. However, this is true only in the uppermost part of the sediment column (Fig. 2). Deeper below the sediment/water interface methanogenesis occurs, producing isotopicaly lighter methane and residual fraction of DIC, which is significantly enriched in ^{13}C (e.g. Nissenbaum *et al.*, 1972; LaZerte, 1981; Quay *et al.*, 1986; Kuivila *et al.*, 1988). The increase in $\delta^{13}C$-DIC values vs. depth is seasonally variable and is related to temperature changes, which are known to influence the rate of methanogenesis (Zeikus and Winfrey, 1976). The most common methane formation pathways in freshwater lacustrine sediments are reported to be (1) acetate fermentation and (2) reduction of CO_2, where acetate fermentation is more common (LaZerte, 1981; Oremland, 1988; Jędrysek. 1995). The ratio between methane produced by the two pathways was calculated as C/B value by fitting $\delta^{13}C$-DIC vs. DIC concentration profiles in pore water to the function developed in the model by LaZerte (1981). The $\delta^{13}C_f$ was assumed to be equal to the limiting $\delta^{13}C$-DIC values measured for each month. The α_c was calculated to be 1.05 from the difference between the mean $\delta^{13}C$ for CO_2 and $\delta^{13}C$ for CH_4 obtained in a laboratory incubation experiment with sediment taken from the Zaka bay in November 1995 (Lojen, 1996). The C/B ratio between methane produced by CO_2 reduction and that derived from acetate fermentation in November in the Zaka bay was calculated to be 24.7, which means that over 95% of methane was produced by CO_2 reduction. Acetate concentration in pore water in the November samples was below detection limit of 0.5 ppm, which supported our results since extended methane production from acetate can not proceed if its concentration does not reach some higher limiting value (Alperin *et al.*, 1992; Sugimoto and Wada, 1993). Assuming the same value for α_c, C/B ratios were calculated also for August in the Western basin and for April and August 1995 in Zaka bay. CO_2 reduction and acetate fermentation are of approximately the same importance in August in the Western basin (C/B = 0.99), whereas in this period of the year in the Zaka bay CO_2 reduction contributes about 80% to the total methane production. Acetate fermentation is more important only in the spring (April 1995) when only about 35% of produced methane is derived from the reduction of CO_2. Based on these results, the isotopic composition of methane produced during the year was calculated. For acetate-derived methane, a value of -55% was assumed (Woltemate *et al.*, 1984; Alperin *et al.*, 1992). It was estimated that the $\delta^{13}C$ of methane produced in the Zaka bay should be about -67‰ in the spring, -81‰ in summer and -87‰ in the autumn, and about -72‰ in summer in the Western basin. As reported in the literature (Sandbeck and Ward, 1981), the relatively high fraction of methane, produced by CO_2 reduction, can be explained by the low sulphate concentrations measured in pore waters, which are generally far below 20 μM at both sampling points.

3.2. NITROGEN

As shown in Fig. 3a, concentration of dissolved ammonia in pore water increases with depth, reaching up to 1.6 mM in the summer months. Nitrate is present in very low concentrations (up to 20 μM) only in the uppermost 3 cm. The diffusive flux of ammonia in the Zaka bay was calculated to be between 0.5 mM/m²/day in April and 1 mM/m²/day in August 1995 which is more than 3 times less than in the Western basin (Čermelj *et al.*, 1996).

The $\delta^{15}N$ of bulk sediment (Fig. 3b) in the oxygenated Zaka bay is almost independent of the depth (values range between +5.5 and +6.5‰), whereas in the anoxic Western basin, it reaches from +3.5‰ in upper 10 cm (similar as reported by Wada *et al.* (1991) for the shallow eutrophic lake Suwa in Japan) to up to +8.5‰ below 15 cm. One maximum of +6‰ at the depth of about 5 cm coincides with the maximal total C and N

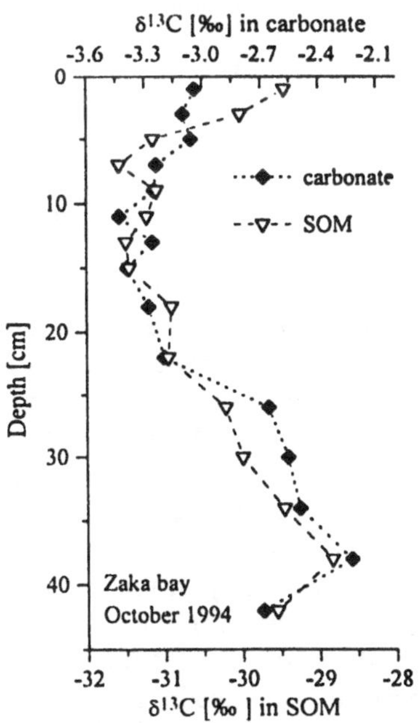

Fig. 1: $\delta^{13}C$ of carbonate and $\delta^{13}C$ of sedimentary organic matter (SOM) vs. depth (October 1994, Zaka bay)

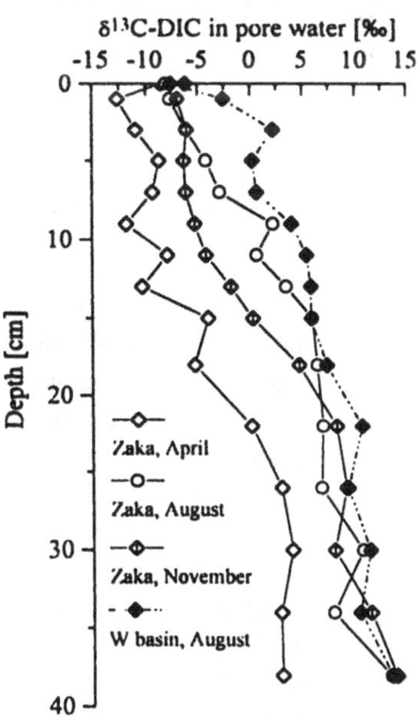

Fig. 2: Seasonal changes of $\delta^{13}C$ of dissolved inorganic carbon (DIC) in pore water vs. depth

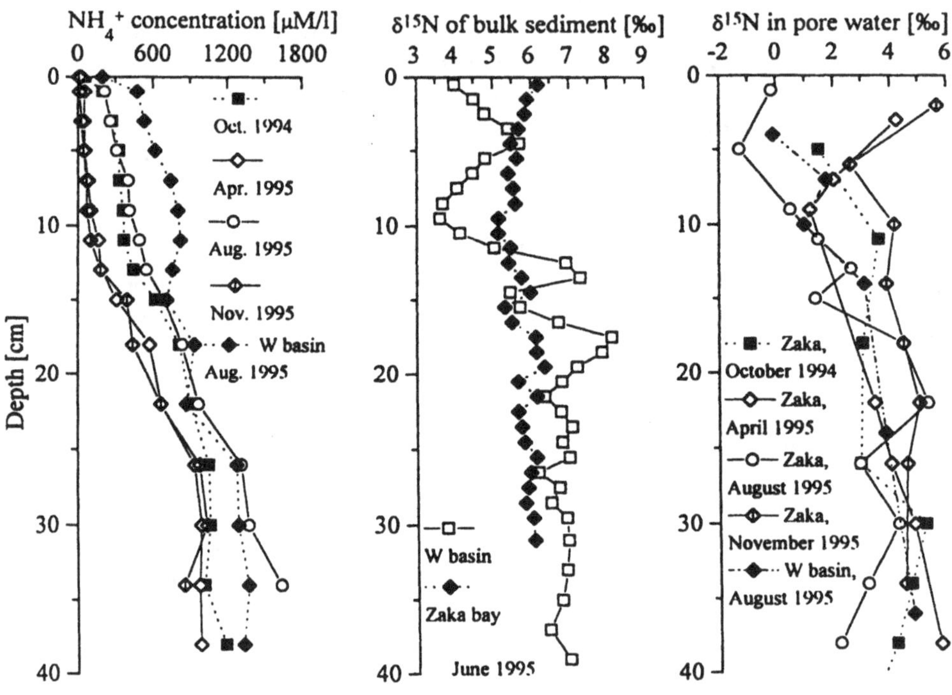

Fig. 3a: NH_4^+ concentration vs. depth

Fig. 3b: $\delta^{15}N$ of bulk sediment vs. depth

Fig. 3c: $\delta^{15}N$ of total dissolved N vs. depth

concentration in the sediment of Western basin as measured by Čermelj *et al.* (1996), thus confirming their observation that the isotopic composition of sedimentary organic matter is related to its concentration. The stability of $\delta^{15}N$ values of sediment in the Zaka bay can be attributed to excess oxygen concentration at the sediment/water interface and perceptible seasonal temperature changes, both of which enhance the decay of organic matter already in the water column before it settles at the lake's bottom. The increase in $\delta^{15}N$ with depth in Western basin is due to the degradation of organic matter and related enrichment of the residual fraction with heavy isotope. The $\delta^{15}N$ of total dissolved N shows a broad range of values especially in the uppermost part of the sediment (Fig. 3c) where it can be even higher than $\delta^{15}N$ of sedimentary nitrogen. The $\delta^{15}N_{dissolved}$ values are more uniform in the lower part of the sediment column (below 10 cm) where they generally increase with depth. However, no Rayleigh-type function for $\delta^{15}N$ is the sediment and pore water could be found, indicating that the isotopic composition of dissolved N is altered by some secondary processes. The assimilation of ammonia, which is related to big N-isotope fractionation (Hoch et al., 1992), as well as denitrification, which occurs in anaerobic sediments are supposed to be responsible for the enrichment of residual dissolved N with the heavy isotope.

The comparison of isotopic composition of dissolved and sedimentary nitrogen shows a linear dependence (Fig. 4a). The linear relationship was significant at the level of p = 99.9% in a whole sediment sample taken in October 1994 from the Zaka bay. Results from other samplings revealed different relationships for upper (above 15 cm) and lower sediment segments (below 15 cm). However, the relationship remains linear at significance levels higher than 80% in all cases. A linear relationship was found between the $\delta^{15}N_{dissolved}$ and $\delta^{13}C$-DIC in pore water (Fig. 4b), suggesting that secondary changes that affect isotopic composition of DIC involve dissolved N, too. Since the obtained pore water samples were too small to measure the $\delta^{15}N$ of dissolved ammonia and nitrate in the uppermost sediment segment separately, the isotope fractionation between nitrate and ammonia was calculated from the $\delta^{15}N$ of diffusive flux of ammonia (Sayles and Curry, 1988; Fig. 5) at a selected depth, where nitrate concentration dropped below the detection limit, and the isotope composition of total dissolved

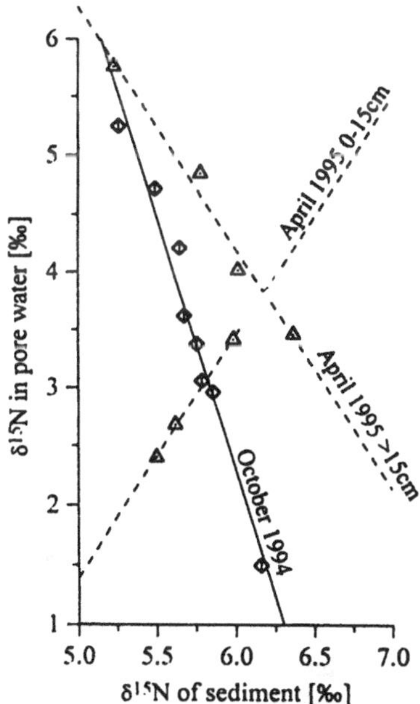

Fig. 4a: Relation between $\delta^{15}N$ of bulk sediment and $\delta^{15}N$ of total dissolved N

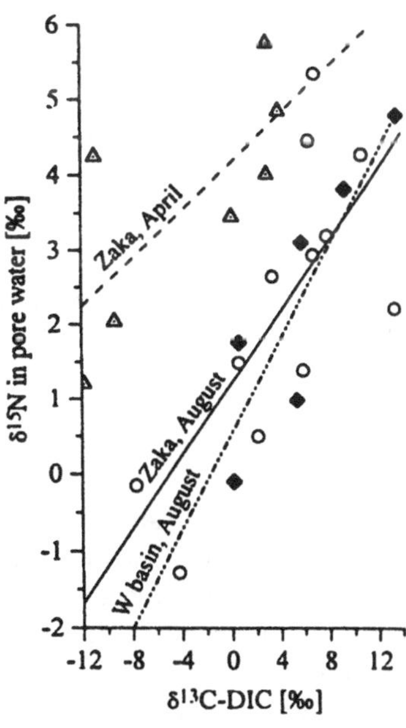

Fig. 4b: $\delta^{15}N$ of total dissolved N in pore water vs. $\delta^{13}C$ of DIC

nitrogen in a segment 1 cm above. Such a calculation was done on sediment from the Zaka bay in August when the concentration ratio of $[NH_4^+]$ to $[NO_3^-]$ at 1 cm below the sediment/water surface was 12, and in November, when it dropped to 1.5. It was found that in both cases, nitrate was significantly depleted in ^{15}N. In November, the $\delta^{15}N$-NO_3^- was calculated to be -7.4‰, whereas in August it should be even more depleted (-39.3‰), thus indicating a very high isotope separation. However, this are more or less approximate values. Nevertheless, such depletion seems reasonable for bacterially mediated processes (Delwiche and Stein, 1970), especially considering the low nitrate concentration compared to the initial ammonia pool. Further investigation is needed particularly in the areas of N speciation and related isotope fractionation effects in pore water.

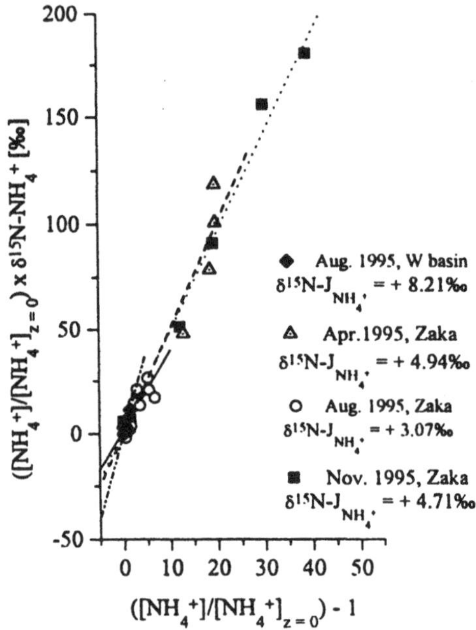

Fig. 5: $\delta^{15}N$ of diffus.ve fluxes of ammonia 3 cm below sediment/water interface, where nitrate concentration dropped below detection limit

4. Conclusions

Stable isotope analyses of DIC and total dissolved N in the pore water of an anoxic lacustrine sediment showed that both elements are affected by secondary processes, most probably governed by micro-organisms, which induce inter-related isotopic changes. Using model calculations from the literature (LaZerte, 1981), it was found that reduction of CO_2 is the dominant methane formation pathway, although, acetate fermentation can be equally important during the colder period of the year. Enrichment of dissolved ammonia with ^{15}N is attributed to its assimilation by micro-organisms (Hoch *et al.,* 1992). Methanogenic bacteria are suggested as the most probable assimilators.

Acknowledgements

This study was financially supported by the Ministry of Science and Technology of Slovenia. The authors would like to thank T. Donnelly and P. Ford for their helpful comments.

References

Alperin, M. J., Blair, N. E., Albert, D. B., Hoehler, T. M. and Martens, C. S.: 1992: *Global Biogeochem. Cycles* **6** (3), 271-291.
Delwiche, C. C. and Steyn, P. L.:1970, *Environ, Sci. Technol.* **4**, 929-935, in: Fritz P. and Fontes, J. Ch. (Eds.): 1980, Handbook of Environmental Isotope Geochemistry Vol.1, The Terrestrial Environment, A. Elsevier, pp.413-417.
Čermelj, B., Faganeli, J., Ogorelec, B., Dolenec, T., Pezdič, J., and Smodiš, B.: 1996, *Biogeochemistry* **32**, 69-91.
Dolenec, T., Pezdič, J., Ogorelec, M. and Mišič, M.: 1984, *Geologija* **27**, 161-170.
Gran, G.: 1952, *Analyst* **77**, 661-671.
Grasshof, K., Ehrhardt, M., and Kremling, K.: 1983, *Methods of Seawater Analysis.* Verlag Chemie, Weinheim, 419 pp.
Hoch, M. P., Fogel, M L. and Kirchman, D. L.:1992, *Limnol. Oceanogr.* **37** (7), 1447-1459.
Jędrysek, M. O.: 1995, *Geochim. Cosmochim. Acta* **59** (3), 557-561.
Kuivila, K. M., Murray, J. W., Devol, A. H.. Lidstrom, M. E. and Reimers, C. E.: 1988, *Limnol. Oceanogr.* **33** (4 part 1), 571-581.
LaZerte, B. D.: 1981, *Geochim. Cosmochim. Acta* **45**, 647-656.
Lerman, A.: 1979, Geochemical Processes Water and Sediment Environments. Wiley, New York, pp. 80-82.
Lojen, S.: 1996, Ph.D. Thesis, University of Ljubljana, Slovenia, 151 pp.
McCrea, J. M.: 1950, *J. Chem. Phys.* **18**, 849-857.
Meischner, D. and Rumohr, J.: 1974, *Senchenbergiana Maritima* **6**, 105-117.
Molnar, F. M., Rothe, P., Foerstner, U., Štern, J., Ogorelec, B., Šercelj, A. and Culiberg, M.: 1978, *Geologija* **21**, 93-164.
Mook, W. G., Bommerson, J. C. and Staverman, W. H.: 1974, *Earth Planet. Sci. Lett.* **22**, 169-176.
Oremland, R. S.: 1988, Biology of Anaerobic Microorganisms, ed. A. J. B. Zehnder, J. Wiley, New York, 641-705.
Nissenbaum, A,. Presley, B. J. and Kaplan, I. R.: 1972, *Geochim. Cosmochim. Acta* **36**, 1007-1027.
Quay, P. D., Emerson, S. R., Quay, B. M. and Devol, A. H.: 1986, *Limnol. Oceanogr.* **31** (3), 596-611.
Sandbeck, K. A. and Ward, D. M.: 1981, *Appl. Environ. Microbiol.* **41**, 775-782.
Sayles, F. L. and Curry, W. B.: 1988, *Geochim. Cosmochim. Acta* **52**, 2963-2978.
Sugimoto, A. and Wada, E.: 1993, *Geochim. Cosmochim. Acta* **57**, 4015-4027.
Van Cappellen, P. and Berner, R. A.: 1988, *Am. J. Sci.* **288**, 289-333.
Vrhovšek, D., Zupan, M. and Blejec, A.: 1984, *Biol. Vestn.* **32**, 45-56.
Wada, E., Lee, J. A., Kimura, M., Koike, I., Reebugh, W.S., Tundisi, J. G., Yoshinari, T., Yoshioka, T. and van Vuuren, M. M. I.: 1991, *Jpn. J. Limnol.* **52** (4), 263-281.
Woltemate, I., Whiticar, M. J. and Schoell, M.: 1984, *Limnol. Oceanogr.* **29** (5), 985-992.
Zeikus, J. G. and Winfrey, M.: 1976, *Appl. Environ. Microbiol.* **31**, 99-107.

C, N AND THEIR STABLE ISOTOPES IN SUSPENDED AND SEDIMENTED MATTER FROM THE PO ESTUARY (ITALY)

WALTER MARTINOTTI[1], MARINA CAMUSSO[2], LUIGI GUZZI[1], LUISA PATROLECCO[3], MAURIZIO PETTINE[3]

[1] *ENEL SpA-Material and Environment Research Centre, via Rubattino 54, 20134 Milan, Italy*

[2] *Water Research Institute-CNR, via della Mornera 25, 20047 Brugherio, Italy*

[3] *Water Research Institute-CNR, via Reno 1, 00198 Rome, Italy*

Abstract. C and N content, C/N (atomic) ratio, and C and N isotopic composition ($\delta^{13}C$ and $\delta^{15}N$) were determined on suspended particulate matter and sediment samples obtained from riverine, estuarine and marine environments in two cruises (September 1995 and March 1996) in the Po estuary (Italy). Isotopic tracers of C and N, reported for the first time for this environment, gave information on sources of organic matter and their distributions. An end-member mixing model based on $\delta^{13}C$ values was applied to estimate the relative importance of riverine and marine sources of organic matter in suspended particulate matter and sediments.

Key words: Carbon, Nitrogen, Isotopic composition, Estuary, Suspended matter, Sediments.

1. Introduction

Estuarine environments are efficient traps for river-borne particles as well as sinks for marine biogenic matter (Tan and Strain, 1979; Gearing *et al.*, 1984; Showers and Angle, 1986). The composition of suspended particulate matter (SPM) in these areas is governed by chemical and biological transformations and the mixing of riverine and marine material (Sholkovitz and Price, 1980; Morris *et al.*, 1982). Determination of the contributions of terrigenous and marine sources to suspended and sedimented matter in estuaries is important for understanding the distribution of surface-reactive pollutants and the fate of organic matter exported by rivers, as well as for distinguishing the fraction of organic matter produced *in situ* from other components. Carbon/nitrogen atomic ratios (C/N) and C and N isotope compositions have been used extensively to identify the origin of suspended and sedimentary organic matter in estuarine and marine environments (Fry and Sherr, 1984; Hedges and Parker, 1976; Shultz and Calder, 1976; Tan and Strain, 1979; Fontugne and Jouanneau, 1987; Wada *et al.*, 1987). These tracers have also been used to elucidate the sources and sinks of particulate material in several other environments (Saino and Hattori, 1980; Tan and Strain, 1983; Faganeli *et al.*, 1988).

The Po is the largest river in Italy and extensive environmental studies have been conducted along its course (Provini and Pacchetti, 1982; Grego and Mioni, 1985; Fossato, 1991; Pettine *et al.*, 1994). However, no stable isotope data exist for the Po estuarine system and C/N ratio data are limited.

This paper presents data on the C and N content and isotopic composition ($\delta^{13}C$ and $\delta^{15}N$) of the total and organic fraction of suspended and sedimentary matter collected from the Po estuary in surveys carried out in two seasons. These new data have revealed the seasonal variability of particulate organic matter in suspension and have permitted identification of the preferential settling areas of riverine terrigenous matter.

Water, Air and Soil Pollution **99**: 325-332, 1997.

2. Methods

2.1. STUDY AREA

The Po River is 646 km long, has a drainage area of about 70,000 km^2 and a mean annual water flow, averaged from long-term data, of 1,470 m^3/s. The maximum recorded flow has exceeded 10,000 m^3/s. After passing through the most heavily industrialized areas of Italy, the Po flows into the shallow and confined northern Adriatic basin, where it gives rise to a typical extended plume. The Po is the main freshwater source for the Adriatic. The suspended particulate matter transported by the river annually is estimated as 13.69 10^6 t/y (Dal Cin, 1983). This solid load settles differentially in the nearshore region (to a depth of 25 m). The estuary, which is non-tidal for most of the year, but has a salt wedge during the summer, has several channels, the most important of which, from the hydrological point of view, is Po di Pila.

2.2. SAMPLING

Cruises were carried out in the Po estuarine area over September 11-13, 1995 and March 4-6, 1996 when water discharge was medium, and tidal conditions were high and low respectively. Samples were collected at Pontelagoscuro (station R1), about 90 km from the mouth and representative of the closing-section of the Po basin, in Po di Pila (station R2 upstream of the saline intrusion zone and stations E1 and E2 in the tidal area), and in the Po diffusion plume in the coastal area (M1-9). A marine end-member sample was also collected at about 14 miles from the coast in an area rarely affected by the Po. Samples were collected from surface waters at the riverine and tidal stations and from both surface and bottom waters at marine stations. Suspended particulate matter (SPM) was obtained by filtering 0.5-2 L water samples through precombusted (2 hours at 500 °C) Whatman GF/F glass-fibre filters. Sediments were collected from each sampling site by means of box or cylindrical corers, care always being taken to avoid sediment resuspension. After removal of any large organic or mineral particles the cores were sectioned into surface layer samples about 1 cm thick.

2.3. ANALYSES

The C and N content and isotopic composition were simultaneously determined on aliquots of SPM or sediment sample using a continuous flow isotope ratio mass spectrometer (CF-IRMS) (TracerMass ANCA-MS, Europa Scientific Ltd, U.K.) connected to a biological sample converter (RoboPrep-CN, Europa Scientific Ltd, U.K.). Filters with SPM were dried at 60 °C and weighed; approximately half of each filter was then used to determine the C and N content and $\delta^{15}N$ composition in the total matter; the other half of the filter was used to determine the C, N and isotope composition in the organic fraction after treatment with 6 N HCl to remove carbonates. Blank filter corrections were made, although their contributions were generally insignificant. Sediment samples were dried at 60 °C, ground in a ball-mill and sieved through a 200 μm mesh prior to analysis. Elements in the organic fraction were determined after treatment with 6 N HCl. Sample sizes were approx. 15 and 50 mg (dry weight) for total sediment samples and their organic fractions respectively, and 15 mg (dry weight) for the total and organic fractions in SPM. Isotope data are expressed in the δ notation indicating depletion (–) or enrichment (+) of the heavy isotope (^{13}C, ^{15}N) compared to the lighter isotope (^{12}C, ^{14}N) relative to standard materials (Pee Dee Belemnite for C, atmospheric N_2 for N):

$$\delta X(‰) = (R_{sample}/R_{std} - 1) \cdot 10^3 \quad \text{(i)}$$

where X is ^{13}C or ^{15}N, R_{sample} is the isotope ratio ($^{13}C/^{12}C$ or $^{15}N/^{14}N$) of the sample and R_{std} is the isotope ratio of the standard. Analytical precision (±1SD), based on replicate analyses (n = 5) of $\delta^{13}C$ and $\delta^{15}N$ on a single sample was ± 0.3 and ± 0.4 ‰ respectively, for both SPM and sediments. Water quality parameters were measured by means of a multiparameter probe (Idronaut Srl, Italy).

3. Results and discussion

3.1. SUSPENDED PARTICULATE MATTER

The mean values of total (PC) and organic (POC) carbon were respectively 5.5±3.6 and 2.9±1.7 % in riverine (R) plus tidal (E) sites, 5.9±0.8 and 2.8±0.4 % in marine surface (Ms) sites and 3.2±0.7 and 1.3±0.4 in marine bottom (Mb) sites (Table 1). High PC and POC levels were found at station E1. Riverine and tidal samples had PC and POC concentrations in the ranges 2.1 to 10.8 % and 1.1 to 5.7% respectively, while the corresponding concentrations in coastal stations ranged from 2.1 to 7.2 % and 0.9 to 3.5 %. For samples taken in the coastal area, concentrations of PC and POC in the SPM of surface samples were significantly higher than those in bottom samples.

No significant differences between N content before and after acidification were observed, therefore the N values (PN) reported in Table 1 refer to the means of the total and organic fraction. Mean PN values were 0.6±0.5 % at both riverine and tidal sites, 0.6±0.1 % at marine surface sites, and 0.3±0.1 % at marine bottom sites. The distribution of PN values mirrored that of PC and POC values quite closely. Highest PN concentrations were found at station E1; concentrations of N in suspended particulate matter varied from 0.2 to 1.2 % at riverine and tidal sites, with a narrower range (0.2 to 0.7%) at coastal sites. PN levels in marine surface samples were higher than those in bottom samples.

C/N atomic ratios (POC/PN) were close to the Redfield ratio of about 6.6, with slightly higher values at riverine and tidal stations than coastal waters. If C/N values greater than 12 are assumed as representative of terrestrial organic C and values less than 8 representative of marine organic C (Milliman *et al.*, 1984), then our values indicate a dominant contribution of autochthonous (planktonic) organic C. High values of chlorophyll(a) found at the riverine and tidal sites in the September survey supported this view.

The mean values for POC $\delta^{13}C$ were –26.6±1.4 ‰ at riverine plus tidal sites, –21.9±1.0 ‰ at marine surface sites, and –21.2±1.1 ‰ at marine bottom sites. POC $\delta^{13}C$ values (Fig. 1) in riverine and tidal sites increased with salinity, i.e. going from estuarine to (surface) coastal sites (from –28.9 to –20.7‰). At coastal stations, POC $\delta^{13}C$ values tended to increase from surface to bottom waters (exceptions were sites M4 and M8). In September, ^{13}C enrichment was greater in the southern part of the marine area than in northern area. This finding indicates north-east advection of Po waters during this cruise (−23.8 and −23.1 ‰ at M1 and M5 sites, respectively) and a greater contribution of marine matter at other sites.

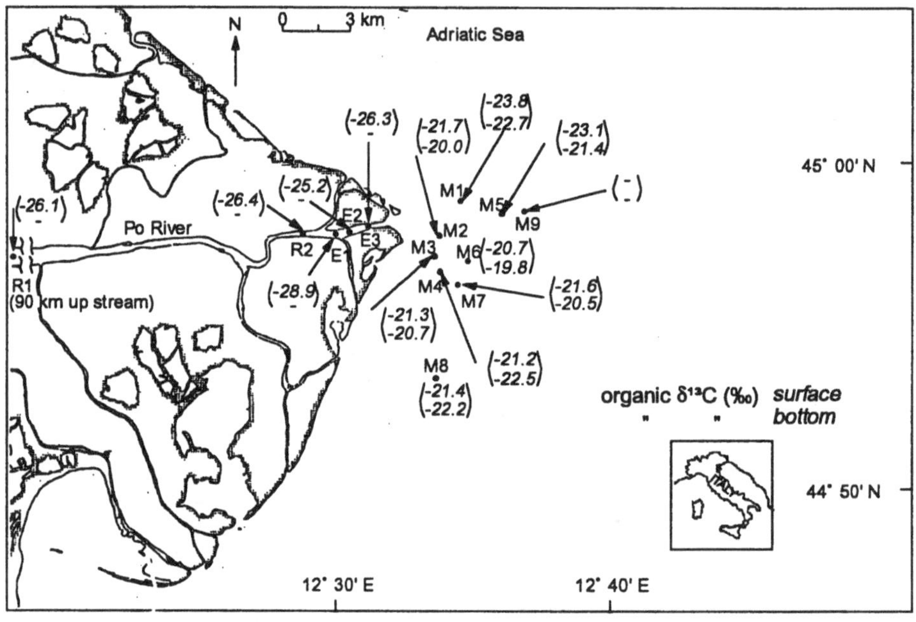

Fig. 1 - $\delta^{13}C$ values for particulate organic carbon in suspended particulate matter of water samples collected September 11-13, 1995.

The $\delta^{15}N$ values are reported in Table 1 for particulate N. Mean values were 5.8±0.7 ‰ at riverine plus tidal sites, 8.0±2.0 ‰ at marine surface sites, and 3.8±1.5 ‰ at marine bottom sites. The range of variation in riverine and tidal environments was narrower than at coastal sites. There was no obvious relationship between $\delta^{15}N$ of particulate N and salinity at riverine and tidal sites, while surface enrichment was observed in the coastal area. Furthermore, the distribution of C and N stable isotopes differed in the coastal area: the heavy isotope of C being enriched and the heavy isotope of N being depleted going from surface to bottom waters.

The results of the second survey, carried out in March 1996, are presented in Table 2. Comparison of data from the two surveys showed that PN at riverine and marine stations was lower in winter (March) than in summer (September). Similarly, $\delta^{15}N$ values on a site per site basis were lower in the March survey. Additionally, ^{13}C was significantly enriched in particulate organic matter sampled from marine sites in summer compared to winter. This finding reflects a larger contribution of marine material to POC in September than March, probably as a consequence of the phytoplankton bloom that occurred in September. The eastward influence of the Po plume was more marked in

March with $\delta^{13}C$ values at M5 of –23.8 and –23.0 ‰ for surface and bottom suspended organic matter.

Table 1- C and N content, C/N ratio, and $\delta^{13}C$ and $\delta^{15}N$ values in suspended particulate matter (total and organic fraction) of water samples collected along a longitudinal and salinity gradient in the riverine (R), tidal (E) and coastal (M) areas of the Po estuary, September 11-13, 1995. (s=surface sample; b=bottom sample).

Station	Salinity (‰)	C (‰) total	C (‰) organic	N(%)	C/N organic	$\delta^{13}C$ (‰) organic	$\delta^{15}N$ (‰)
R1 s	0.2	2.1	2.1	0.3	8.2	-26.1	5.6
R2 s	0.2	7.1	2.3	0.4	6.7	-26.4	7.0
E1 s	1.5	10.8	5.7	1.2	5.5	-28.9	5.3
E2 s	12.7	5.3	3.1	1.0	3.6	-25.2	5.7
E3 s	21.7	2.3	1.1	0.2	6.4	-26.3	5.3
M1 s	31.6	5.9	3.0	0.6	5.8	-23.8	12.7
M1 b	36.6	3.5	1.1	0.2	6.4	-22.7	2.9
M2 s	33.5	7.2	3.5	0.7	5.8	-21.7	7.3
M2 b	35.5	3.5	2.0	0.4	5.8	-20.0	7.0
M3 s	34.3	5.9	3.0	0.7	5.0	-21.3	7.5
M3 b	36.3	3.3	1.0	0.2	5.8	-20.7	3.9
M4 s	34.2	5.0	2.4	0.5	5.6	-21.2	6.6
M4 b	36.6	2.8	1.0	0.2	5.8	-22.5	4.3
M5 s	30.9	5.2	2.6	0.6	5.1	-23.1	7.0
M5 b	38.0	2.1	0.9	0.2	5.3	-21.4	1.9
M6 s	33.5	4.8	2.5	0.5	5.8	-20.7	7.1
M6 b	36.3	3.7	1.4	0.2	8.2	-19.8	3.9
M7 s	34.0	6.5	2.5	0.5	5.8	-21.6	7.1
M7 b	36.6	4.3	1.3	0.5	3.0	-20.5	3.2
M8 s	35.1	6.5	2.6	0.5	6.1	-21.4	8.6
M8 b	36.9	2.5	1.3	0.3	5.1	-22.2	3.3

Table 2- C and N content, C/N ratio, and $\delta^{13}C$ and $\delta^{15}N$ values in suspended particulate matter (total and organic fraction) of water samples collected along a longitudinal and salinity gradient in the riverine (R), tidal (E) and coastal (M) areas of the Po estuary, March 4-6, 1996. (s=surface sample; b=bottom sample).

Station	Salinity (‰)	C (‰) total	C (‰) organic	N(%)	C/N organic	$\delta^{13}C$ (‰) organic	$\delta^{15}N$ (‰)
R1 s	0.2	4.9	2.8	0.4	8.2	-26.7	3.1
R2 s	0.2	5.0	2.8	0.4	8.2	-25.4	0.2
E1 s	1.5	3.6	3.0	<0.5	n.d.	-27.3	n.d.
E2 s	21.7	3.1	2.0	0.3	7.8	-27.0	4.2
M1 s	14.0	3.8	3.0	0.6	5.8	-25.8	4.9
M1 b	37.5	3.2	1.4	<0.3	n.d.	-21.9	n.d.
M2 s	28.0	3.8	2.7	0.5	6.3	-23.4	6.5
M2 b	37.0	3.0	1.9	<0.3	n.d.	-24.0	n.d.
M3 s	18.0	3.4	2.3	<0.6	n.d.	-26.5	n.d.
M3 b	37.7	2.1	1.3	<0.4	n.d.	-25.9	n.d.
M4 s	32.0	3.3	2.9	0.5	6.8	-22.6	7.4
M4 b	37.4	3.0	1.5	<0.2	n.d.	-22.8	n.d.
M5 s	36.1	3.0	2.7	0.5	6.3	-23.8	7.1
M5 b	37.9	1.7	1.2	<0.4	n.d.	-23.0	n.d.

n.d.= not detectable

3.2. SEDIMENTS

The mean values of total and organic C in the September survey were respectively 2.1±1.1 and 0.5±0.6 % in the riverine plus tidal sites, and 4.2±0.6 and 1.1±0.3 % in marine sites (Table 3). Elemental C concentrations increased with salinity from riverine through tidal to marine sites. The N content at riverine and tidal sites increased from 0.01 to 0.2 % as a function of salinity, while at marine sites all values of N were in the range 0.1-0.2 % and showed no systematic variation. C/N ratios were in the range 4.7 to 11.7 at riverine plus tidal sites, and 7.0 to 15.2 at marine sites. The mean values for organic $\delta^{13}C$ were −26.0±0.9 ‰ at riverine plus tidal sites, and −23.5±0.7 ‰ at marine sites. The homogeneity of organic $\delta^{13}C$ values in sediments at coastal sites suggests a mixed riverine-marine source for organic matter in the area in front of the river mouth.

Table 3- C and N content, C/N ratio, and $\delta^{13}C$ and $\delta^{15}N$ values in sediments (total and organic fraction) collected along a longitudinal and salinity gradient in the riverine (R), tidal (E) and coastal (M) areas of the Po estuary, September 11-13, 1995.

Station	Salinity (‰)	C (‰)		N(%)	$\delta^{13}C$ (‰)	$\delta^{15}N$ (‰)
		total	organic		organic	
R1	0.2	1.3	0.1	0.01	n.d.	11.7
R2	0.2	1.2	0.1	0.02	n.d.	5.8
E1	1.5	2.1	0.3	0.04	n.d.	8.8
E2	12.7	2.0	0.4	0.1	-26.6	4.7
E3	25.0	4.0	1.5	0.2	-25.3	8.8
M1	36.6	4.9	1.5	0.2	-23.7	8.8
M2	35.5	4.3	1.2	0.2	-23.8	7.0
M3	36.3	3.4	0.6	0.1	-24.3	7.0
M4	36.6	3.7	0.9	0.1	-22.3	10.5
M5	38.0	3.6	0.7	0.1	-22.6	8.2
M6	36.3	4.2	1.3	0.1	-24.0	15.2
M7	36.6	5.1	1.0	0.1	-23.9	11.7
M8	36.9	4.5	1.4	0.2	-23.4	8.2
M9	36.0	4.0	1.4	0.2	-23.0	8.2

n.d.= not detectable

3.3. MIXING MODEL

In order to evaluate the importance of riverine and marine sources of organic matter in SPM and sediment samples, an end-member mixing model was applied to $\delta^{13}C$ values for organic C (Mulholland and Olsen, 1992). The marine fractional contribution (F) was calculated as:

$$F = 100\ (\delta^{13}C_R - \delta^{13}C_X)/(\delta^{13}C_R - \delta^{13}C_M) \qquad \text{(ii)}$$

where: $\delta^{13}C_R$ is the riverine end-member value; $\delta^{13}C_M$ is the marine end-member value; and $\delta^{13}C_X$ is the intermediate $\delta^{13}C$ value of surface samples of organic SPM or sediments. Taking mean values of −26.6 ‰ and −20.7 ‰ as riverine and marine end-members in the summer survey, the calculated marine contributions to SPM indicate that the river water flowed to the north-east after leaving the mouth. For the sediments, taking a mean value of −26.0 ‰ as riverine end-member and the value of the M4 station

(–22.3 ‰) as marine end-member, the pattern of calculated contributions of marine-derived organic C showed an influence of the river localized just in front of the mouth (Fig. 2).
Marine contributions to SPM calculated in the second survey, taking a mean value of –26.6 ‰ as riverine end-member and the value of the M4 station (–22.6 ‰) as marine end-member, indicate lateral advection of the Po outflow to the north-east (at site M5 the riverine contribution was still about 30%) and partially to the south-west very close to the coast (site M3). Marine contributions of organic C to sediments calculated assuming a mean value of –26.1 ‰ as riverine end-member and the value of the M5 station (–21.9 ‰) as marine end-member, indicate that terrigenous matter settles immediately in front of the mouth and in the southern coastal zone (45.2 and 40.5 % at M3 and M4 respectively), confirming the results obtained in the summer survey.

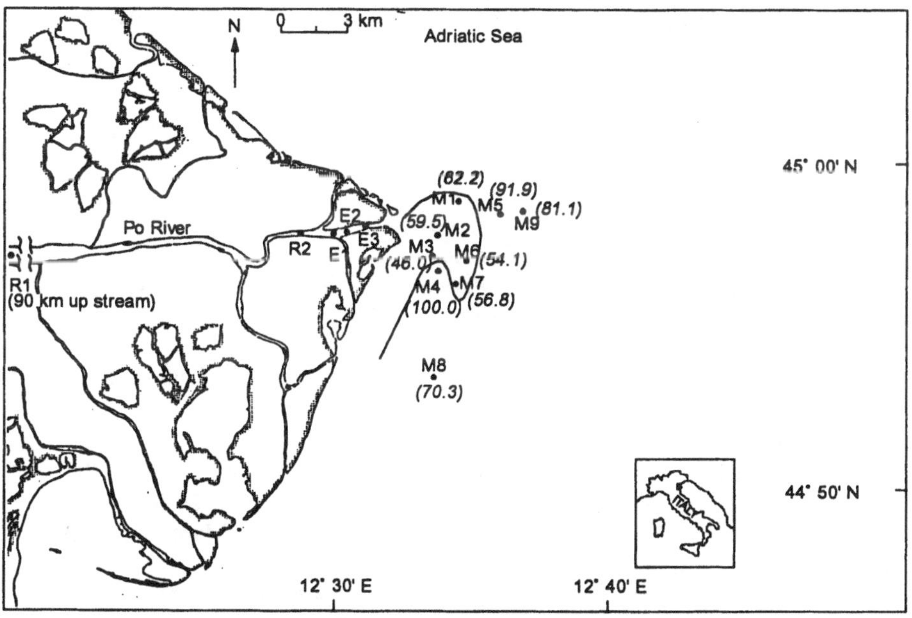

Fig. 2 - Percentage of marine-derived organic matter in sediments collected September 11-13, 1995.

4. Conclusions

The Po river discharges large quantities of organic matter and nutrients into the northern Adriatic sea which significantly affect the productivity and the trophic dynamics of the system. C and N content data and C and N isotope composition of SPM and sediments have been used for elucidating the sources of organic matter in this estuarine system. An end-member mixing model was applied to $\delta^{13}C$ values which gave the percentage contributions of marine and fluvial organic matter to both the suspended particulate and sedimentary matter in the coastal area under the influence of the Po. In addition to improving our knowledge of the extent of the coastal area affected by the Po plume in different seasons, this approach has also revealed the intensity of mixing processes which affect sedimentary matter, as shown by the large variation of C isotope composition.

Acknowledgements

We thank Francesco Bacciu and Domenico Mastroianni (Water Research Institute, Rome) for their help in the treatment of samples, Albino Moretti and Marco Arpini (ENEL SpA-CRAM) for their help, respectively, in field work/sample processing, and analyses. We gratefully acknowledge the support of Emilia-Romagna region for making the oceanographic vessel Daphne II available for collecting marine samples and we thank the crew of the vessel for their assistance. We also thank Don Ward for help with the English. This work was part of a program by the Water Research Institute (IRSA) of the National Research Council (CNR) (Italy), on estuarine processes in the Po-Adriatic mixing area.

References

Dal Cin, R.: 1983, *Boll. Soc. Geol. It.*, **102**, 9-56.
Faganeli, J., Malej, A., Pezdic, J. and Malacic, V.: 1988, *Oceanol. Acta* **11**, 4, 377-382.
Fontugne, M.R. and Jouanneau, J.: 1987, *Estuar., and Coast. Mar. Sci.* **24**, 377-387.
Fossato, V.U.: 1991, *Arch. Oceanogr. Limnol.* **17**, 125-139.
Fry, B. and Sherr, E.B.,: 1984, *Contrib. Mar. Sci.* **27**, 13-47.
Gearing, J.N., Gearing, P.J., Rudnick, D.T., Requejo, A.D. and Hutchins, M.J.:1984, *Geochim. Cosmochim. Acta* **48**, 1089-1098.
Grego, G and Mioni, F.: 1985, *Nova Thalassia* 7, 2, 27-87.
Hedges, J.I. and Parker, P.L.: 1976, *Geochim. Cosmochim. Acta* **40**, 1019-1029.
Milliman, J.D., Xie, Q.C. and Yang, Z.: 1984, *Am. J. of Sci.*, **284**,824-834.
Morris, A.W., Loring, D.H., Bale, A.J., Howland, R.J.M., Mantoura, R.F.C. and Woodward, E.M.S.: 1982, *Oceanol. Acta* **5**, 349-353.
Mulholland, P.J. and Olsen,C.R.: 1992, *Estuar. Coast. and Shelf Sci.* **34**, 95-107.
Pettine, M., Camusso, M., Martinotti, W., Marchetti, R., Passino, R. and Queirazza, G.: 1994, *Sci. Total Environ.* **145**, 243-265.
Provini, A. and Pacchetti, G.: 1982, *Ing. Ambientale* **11**, 3, 173-183.
Redfield, A.C.: 1934, in *James Johnston Memorial Volume*, Lverpool University Press, 177-192.
Saino, T. and Hattori, A.: 1980, *Nature* **283**, 752-754.
Sholkovitz, E.R. and Price, N.B.: 1980, *Geochim. Cosmochim. Acta* **44**, 163-171.
Showers, W.J. and Angle, D.G.: 1986, *Continental Shelf Res.* **6**, 227-244.
Shultz, D. J. and Calder, J.A.: 1976, *Geochim. Cosmochim. Acta* **40**, 481-485.
Tan, F. C. and Strain, P. M.: 1979, *Estuar., and Coast. Mar. Sci.* **8**, 213-225.
Tan, F. C. and Strain, P. M.: 1983, *Geochim. Cosmochim. Acta* **47**, 125-132.
Wada, E., Minagawa, M., Mizutani, H., Tsuji, T., Imaizumi, R. and Karasawa, K.: 1987, *Estuar., and Coast. Mar. Sci.* **25**, 321-336.

THE SOURCES OF DISSOLVED INORGANIC CARBON IN PORE WATERS OF LACUSTRINE SEDIMENT

N. OGRINC AND S. LOJEN
J. Stefan Institute, Jamova 39, 1000 Ljubljana, Slovenia

AND

J. FAGANELI
Marine Biological Station, Fornače 41, 6330 Piran, Slovenia

Abstract. The processes regulating the concentration and isotopic composition of dissolved inorganic carbon (DIC) in pore water were investigated in three different parts of Lake Bled. It was found that the isotopic composition of dissolved inorganic carbon (DIC) is strongly influenced by methanogenesis. Simple diagenetic model reproduces the observed carbon-isotope profiles and the DIC concentration data resonably well in investigated parts. This findings lead us to conclude that the majority of processes affecting DIC are taken into account in the model.

Key words: recent sediment, carbon stable isotope, methanogenesis, diffusion-reaction model

1. Introduction

Exchange between sediments and the water column resulting from the diagenetic reaction can be an important or even dominant component of the carbon cycle. Many reactions in the water column and sediments are related to the fixation of CO_2 to organic carbon and degradation of this organic matter. In high carbonate sediments the carbon cycle is also influenced by dissolution and precipitation of carbonate. Measurement of the carbon isotopic composition of pore water of surficial sediments provides an effective way to study these processes, since organic material is depleted in ^{13}C ($\delta^{13}C$ is approx. -25$^0/_{00}$) relative to inorganic carbonate ($0^0/_{00}$). The gradients in $\delta^{13}C$ of dissolved inorganic carbon (DIC) are controlled mainly by the oxidation of organic carbon and the dissolution of carbonate. These

Water, Air and Soil Pollution **99**: 333-341, 1997.

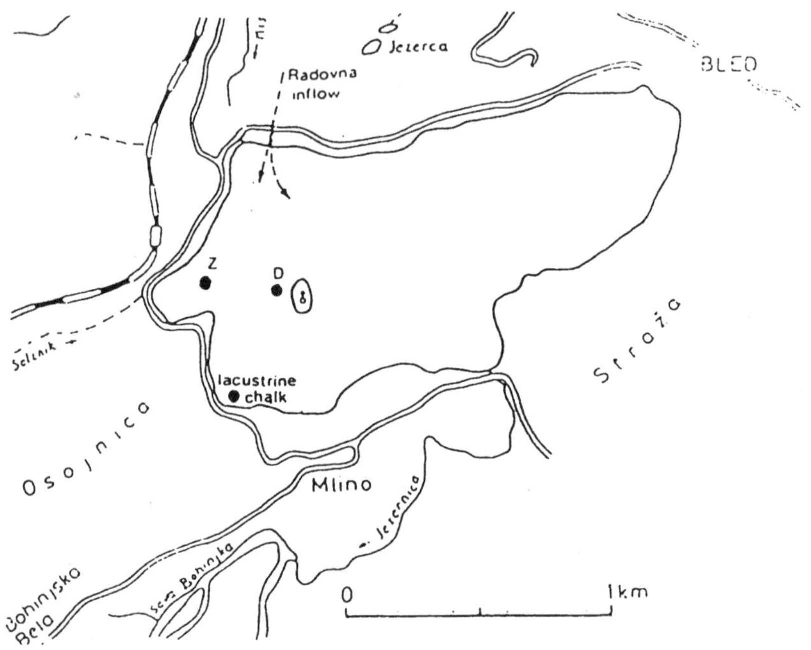

Figure 1. Sampling locations in Lake Bled

processes make the pore waters isotopically lighter (McCorkle *et al.*, 1985; Sayles and Curry, 1988; McNichol *et al.*, 1991). It was found that in strongly reducing environments pore water DIC becomes isotopically heavier and increases with depth. This enrichment has been related to the production of highly ^{13}C-depleted methane in these sediments (Quay *et al.*, 1988; Herczeg , 1988).

In this paper we coupled an investigation of the $\delta^{13}C$ of DIC in pore water with a study of pore water chemistry and sediment environments to define processes affecting the production of dissolved inorganic carbon, and to understeand observed variations in $\delta^{13}C$ in three different parts of Lake Bled.

2. Methods and site description

2.1. SITE DESCRIPTION

Lake Bled is an dimictic subalpine glacial lake located in the NW part of Slovenia. It has a surface area of 1.44 km^2 and a mean depth of 17.8 m. The lake is stratified most of the year, except during the early spring and autumn overturn. The hypolimnium (below approximately 15m) is anoxic, but in some parts oxidizing conditions prevail during most of the year. The lake has some small natural inflows and outflows. In the past two amelioration projects were undertaken: a fresh water inflow from the Radovna river and syphonic pumping of anoxic water from the eastern basin. The sediment is composed of clayey silts with carbonate contents of 50-80%. The highest carbonate contents of the sediment are localized in the SW part of the lake where calcite (so-called chalk) is precipitated (Dolenec *et al.*, 1984). The organic C contents range between 4.5 at the surface to 2.0% at a depth of 35 cm.

2.2. METHODS

Sediment samples were collected at three different parts of Lake Bled, at the station Z, so-called Zaka Bay, where oxidizing conditions prevail, at the station D, the deepest part of the lake where the environment is anoxic and at the part where calcite (lacustrine chalk) is precipitated (Figure 1). Undisturbed sediment cores to a depth of 30 cm were taken in May 1996, at all three stations using a gravity core sampler equipped with a plexiglass liner (Meischner and Rumohr, 1974). The cores were transported immediately to the laboratory and sectioned into 2 or 4 cm intervals in a N_2-filled glove bag. Pore waters were extracted from sediments using a nylon squeezer under nitrogen pressure through a 0.45 μm membrane filter.

Immediately pH, alkalinity and $\delta^{13}C$ of DIC were measured. Alkalinity was determined by the Gran titration method (Edmond, 1970). DIC was calculated from pH and alkalinity using the apparent dissociation constants of Hansson and Jagner (1970) (Millero, 1995). Phosphate, nitrate, ammonia and sulphide were analyzed using standard colourimetric methods (Grasshoff *et al.*, 1983). Ca^{2+} and Mg^{2+} were determined using flame AAS. Sulphate was analyzed turbidimetrically (Tabatabi, 1974).

Samples for $\delta^{13}C$-DIC were introduced to evacuated septum vial containing 100% phosphoric acid. The isotopic ratio of collected CO_2 was measured on a Europa 20-20 ANCA-TG mass spectrometer. The sediment in which $\delta^{13}C$ of organic carbon was determined was first dried and acidified with 3M HCl to remove carbonates. The redried sample was combusted and the isotopic composition of the resulting CO_2 was determined on a Europa 20-20 ANCA-SL mass spectrometer. Carbon and oxygen isotopes of bulk carbonates were measured in the CO_2 gas obtained by reacting them with 100% phosphoric acid for two hours at 50^0C. The isotopic analysis were performed on a Varian MAT-250 mass spectrometer. All stable isotopic results are reported as deviations in $^0/_{00}$ from the Chicago PDB standard. The precision of the analysis for $\delta^{13}C$ was $\pm$ $0.05^0/_{00}$ for carbonates, and $0.2^0/_{00}$ for $\delta^{13}C$-DIC and sedimentary organic carbon.

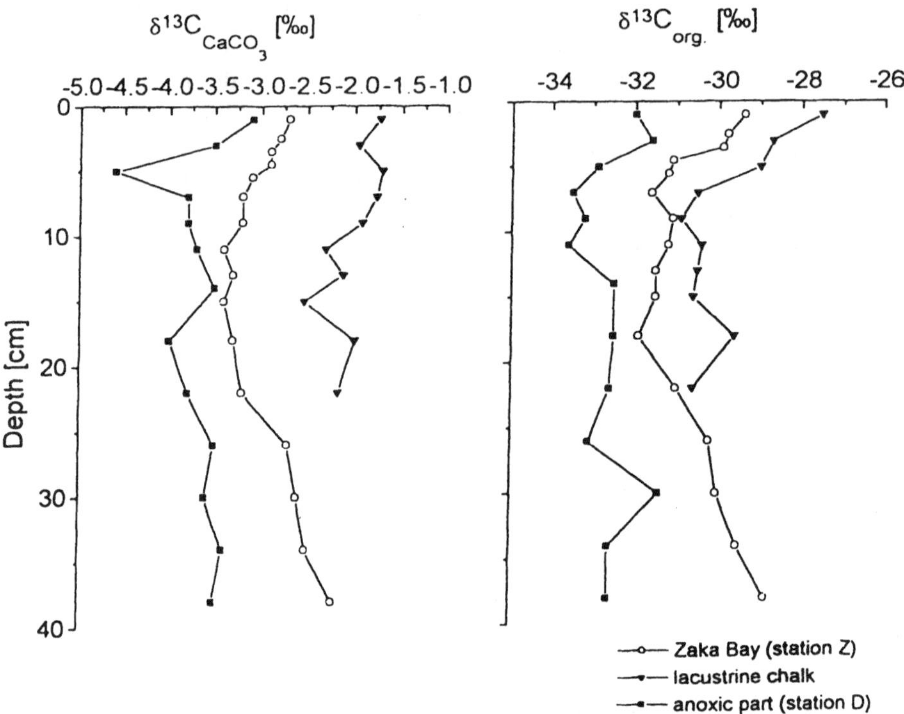

Figure 2. $\delta^{13}C$ of sedimentary organic carbon and carbonate at three stations of Lake Bled

3. Results and discussion

Depth profiles of $\delta^{13}C$ values of sedimentary organic carbon and bulk carbonates at all three stations are shown in Figure 2. The sedimentary organic carbon and carbonates in Zaka Bay exhibited higher $\delta^{13}C$ values in comparison to those from the anoxic part (point D). The isotopic composition of carbonate ranges between -2.6 to -1.8$^0/_{00}$ in the area where lacustrine chalk is precipitated where the main sources of organic carbon are water plants and terrigenous organic detritus from the surroundings ($\delta^{13}C$ about -28$^0/_{00}$) (Čermelj *et al.*, 1996).

The concentration of DIC and its isotopic composition increased along the depth profile (Figure 3). In the anoxic part and in the part where lacustrine chalk is present the values started to increase immediately below the sediment-water interface. In the upper part of profiles In Zaka Bay the values are similar to that of the overlying water, but below 15 cm the values started to increase. The positive gradient in $\delta^{13}C$-DIC must be the consequence of methane production within the sediments. It was found from an incubation experiment that the isotopic composition of methane ranged between -76 to -85$^0/_{00}$, which indicates that most of the methane is formed via CO_2 reduction (Lojen , 1996).

A simple diffusion-production model was constructed to describe the processes affecting DIC and its isotopic composition. The general diagenetic equation given by Berner (1980) was solved assuming steady state conditions. The sediment pore water profiles in Zaka Bay and in the lacustrine chalk sediment were divided into two zones, mixed and unmixed and the following equations were used, for Zone 1 (mixed, $0 \leq z \leq L_1$):

$$\frac{\partial C_1}{\partial t} = 0 = D_1(\frac{\partial^2 C_1}{\partial z^2}) + R_c(z) \quad (1)$$

and, for Zone 2 (unmixed, $L_1 \leq z \leq L_2$):

$$\frac{\partial C_2}{\partial t} = 0 = D_2(\frac{\partial^2 C_2}{\partial z^2}) + R_c(z) \quad (2)$$

where C is the concentration of the DIC, z [cm] is the depth, t is time, D_2 [cm^2 s^{-1}] is the diffusion coefficient of DIC corrected for tortuosity and temperature from the diffusion coefficients of Li and Gregory (1974), D_1 is the apparent diffusion coefficient and was estimated relative to the molecular diffusion coefficient. Changes in the concentration of DIC due to oxidation of organic matter or equilibration with $CaCO_3$ were described by a DIC-production rate, $R_c(z)$. The function $R_c(z)$ is defined as $R_c(z) = R_0 exp(-\beta z)$, where the subscript 0 refers to a reference depth (water-sediment interface), β is empirical constant and is often called the depth attenuation constant. Values of the parameters R_0 and β are obtained by fitting model to pore water profiles. For the appropriate boundary conditions the solution of Equations (1) and (2) is given by Aller (1982).

The best-fit model curves are presented in Figure 3. It can be seen that the model reproduces both data sets, the isotopic data and DIC concentration data, reasonably well in all three sites. The best-fit values for parameters are listed in Table 1.

The isotopic data were used to calculate the contribution of degradation of sedimentary organic matter, dissolution of calcite and methanogenesis to DIC using the mass balance equation:

$$\delta^{13}C_{DIC} * J_{DIC} = \delta^{13}C_{org} * J_{org} + \delta^{13}C_{Ca} * J_{Ca} + \delta^{13}C_{CH_4} * J_{CH_4} \quad (3)$$

$$J_{DIC} = J_{Ca} + J_{org} + J_{CH_4} \quad (4)$$

where J_{org}, J_{Ca}, J_{CH_4} are the fluxes derived from organic C degradation, carbonate dissolution and methanogenesis, respectively, and $\delta^{13}C_{org}$ and $\delta^{13}C_{Ca}$ are the mean $\delta^{13}C$ values of sedimentary organic C and calcite,

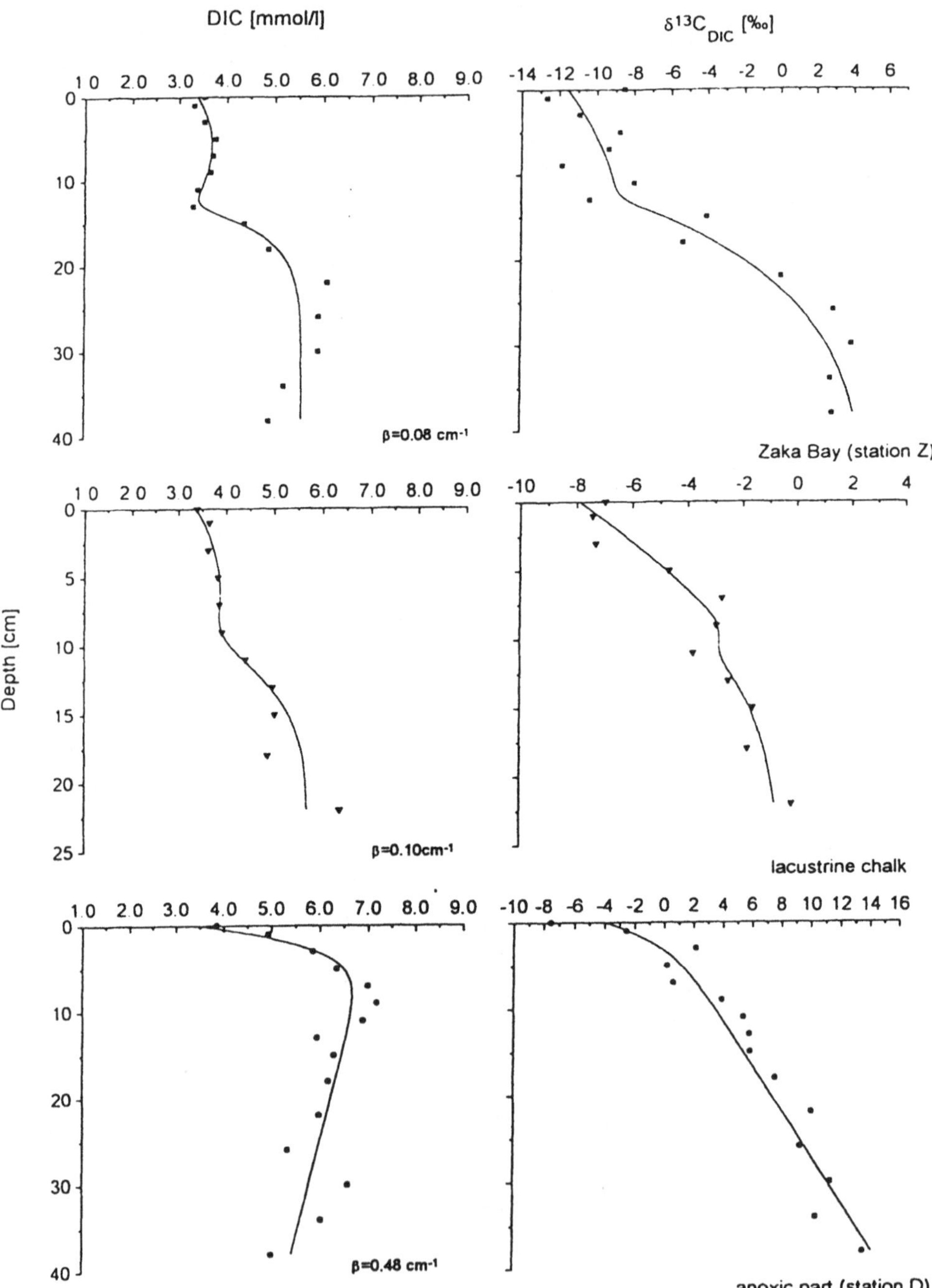

Figure 3. DIC and $\delta^{13}C$ of pore waters in three stations of Lake Bled. The model curves represent the estimates of β cm^{-1} (see text).

TABLE 1. Parameters used to fit the observed DIC isotopic and concentration data. The model fits are shown in Figure 3.

station	R_0 (x10^6) [mmol dm^{-3} s^{-1}]	β [cm^{-1}]	D_1
Zaka Bay	1.1	0.08	$8D_2$
lac. chalk	1.0	0.10	$4D_2$
station D	4.0	0.48	D_2

respectively. The isotopic composition of CO_2 derived from methanogenesis, $\delta^{13}C_{CH_4}$, was determined from the isotopic composition of sedimentary organic matter and the methane flux (-82$^0/_{00}$) and was estimated to be +26$^0/_{00}$ ± 7$^0/_{00}$ (Ogrinc *et al.*,, submit. 1996). $\delta^{13}C_{DIC}$ was determined from the regression line of (DIC_x/DIC_0)-1 vs $\delta^{13}C_{DIC}$*(DIC_x/DIC_0) (Sayles and Curry, 1988) (Figure 4). The C isotopic composition of the DIC flux at station Z (Zaka Bay) was +13.3$^0/_{00}$, +7.8$^0/_{00}$ in the area of lacustrine chalk and +15.7$^0/_{00}$ in the anoxic part of Lake Bled.

The calculated indices for carbonate minerals show that in Zaka Bay conditions for dissolution were favorable, while in the other two parts precipitation prevail. These findings lead us to conclude that the change in concentration and $\delta^{13}C$ of DIC at these two parts of the lake is not influenced by calcite. The contribution of calcite dissolution to DIC in Zaka Bay was estimated to be about 11% (Ogrinc *et al.*,, submit. 1996). Considering all the calculations in the Equations (3) and (4), we estimated that in Zaka Bay the contribution of organic matter degradation to DIC is about 17% and 72% of DIC is derived from methanogenesis. At sampling station D the influence of methanogenesis is more pronounced (about 81%), since the $\delta^{13}C$ DIC values started to increase just below the interface. This result is in reasonable agreement with the results found in other studies of lacustrine anoxic hypolimnia (Kelly *et al.*, 1982). Because of the high $\delta^{13}C$ of the DIC flux, we calculated that the isotopic composition of methane at part D should be the same as found in Zaka Bay. It is not easy to interpret the data in the area of lacustrine chalk from $\delta^{13}C$ of the DIC flux only. If we assume that the isotopic composition of CO_2 derived from methanogenesis is the same as found in Zaka Bay, then 67% of DIC is derived from methanogenesis, while 33% is derived from the degradation of sedimentary organic matter. Further studies are necessary to better understand the processes affecting DIC and its isotopic composition, since this environment is different from the others.

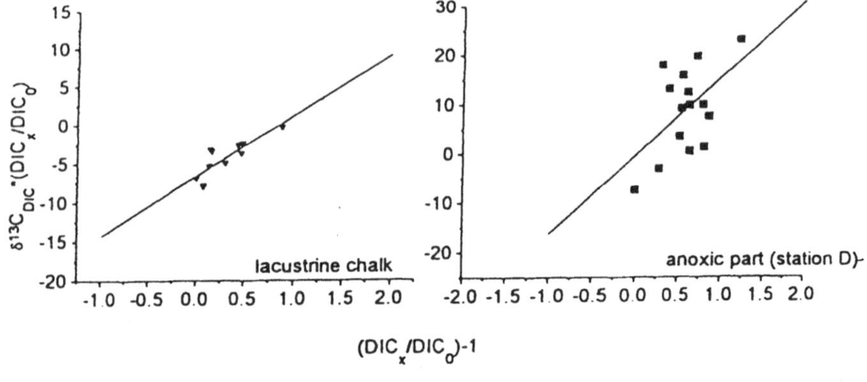

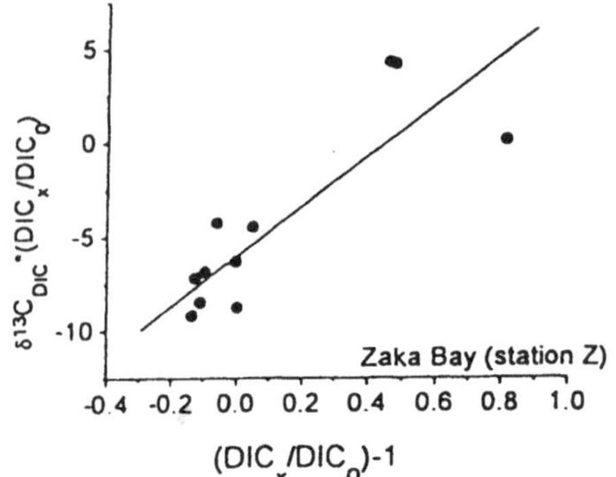

Figure 4. Parameterization of the data used to estimate $\delta^{13}C$ of the DIC flux as discussed in the text.

4. Conclusions

Modelling the DIC and δ^{13}C-DIC profiles shows that the ratio of the rate of organic matter degradation to sedimentation rate, (β), was 0.08, 0.10 and 0.48 cm^{-1} in Zaka Bay, in the area of lacustrine chalk, and at station D, respectively. The modelled profiles of δ^{13}C-DIC agree with the concentration data in all three stations.

The positive correlation between δ^{13}C-DIC and the concentration of DIC in the pore waters of the different sediment environments of Lake Bled is the consequence of methanogenesis. The major contribution to the DIC flux derives from methanogenesis (67-81%), less from sedimentary organic matter (17-33%), while the contribution of calcite is important only in Zaka Bay where it is about 11%.

References

Aller, R.C.: 1982, *Animal-Sediment Relatons: Chap. 2: The Effects of Macrobentos on Chemical Properties of Marine Sediment and Overlaying Water*, Ed. McCall, P.L.

and Tevesz, M.S.J., Plenum Publishing Corporation.
Berner, R.A.: 1980, *Early diagenesis: A Theoretical Approach.*, Princeton University Press.
Čermelj, B. Faganeli, J. Ogorelec, B. Dolenec, D., Pezdič, J. and Smodiš, B.: 1996, *Biogeochemistry* **32**, 69-91.
Dolenec, T., Pezdič, J., Ogorelec, B. and Mišč, M.: 1984, *Geologija* **32**, 161-170.
Edmond, J.M.: 1970, *Deep-Sea Res.* **17**, 737-750.
Grasshoff, K., Ehrhardt, M., and Kremeling, K.: 1983, *Methods of Seawater Analysis*, Verlag Chemie, Weinheim.
Herczeg, A.L.: 1988, *Chemical Geology* **72**, 199-209.
Kelly, C.A., Rudd, J.W.M., Cook, R.B., and Schindler, D.W.: 1982, *Limnol. Oceanogr.* **27**, 868-882.
Li, Y.-H., and Gregory, S.: 1974, *Geochim. Cosmochim. Acta*, **38**, 703-714.
Lojen S.: 1996, *Phase transition and isotope fractionation of light elements in recent sediment of Lake Bled*, PhD Thesis, University of Ljubljana, Slovenia.
McCorkle, D.M., Emerson, S.R. and Quay, P.D.: 1985, *Earth Planet. Sci.Lett.* **74**, 13-26.
McNichol, A.P., Druffel, E.R.M. and Lee, C.: 1991, *Carbon Cycle in Coastal Sediments: 2. An investigation of the sources of ΣCO_2 to pore water using carbon isotopes* in *Organic Substances and Sediment in Water*, **Vol. 2**. ed. R.A. Baker, Lewis Pub., Inc., Chelsea, MI, pp. 249-272.
Meischner, D. and Rumohr, J.: 1974, *Senckenbergiana Marit.* **6**, 105-117.
Millero, F.J.: 1995, *Geochim. Cosmochim. Acta* **59**, 661-677.
Ogrinc, N., Faganeli, J., Lojen, S. and Pezdič, J.: submitted in publication
Quay, P.D., Emerson, S.R., Quay, B.M. and Devol, A.H.: 1988, *Limnol. Oceanogr.* **31**, 596-611.
Sayles, F.L. and Curry, W.B.: 1988, *Geochim. Cosmochim. Acta* **52**, 2963-2978.
Tabatabi, M.A.: 1974, *Environ. Lett.* **7**, 237-243.

RELATIONSHIPS BETWEEN SUSPENDED AND SEDIMENT ORGANIC MATTER IN A SEMI-ENCLOSED MARINE SYSTEM: THE STAGNONE DI MARSALA SOUND (WESTERN SICILY)

A. PUSCEDDU[1], G. SARÀ[2], A. MAZZOLA[3] and M. FABIANO[3]

[1] *Istituto Scienze Ambientali Marine, Università di Genova, C.P. 79, 16138 S. Margherita L. (Genova, Italy),* [2] *Dipartimento di Biologia Animale, Università di Palermo, Via Archirafi, 18, 90123, Palermo (Italy),* [3] *Cattedra di Ecologia, Università di Ancona, Via Brecce Bianche 60131, Ancona (Italy)*

Abstract. To gather information on the interactions between the sediment and suspended organic matter pools in the Stagnone di Marsala, water and sediment samples were collected, on a monthly basis, at 11 stations. Water temperature and salinity showed a clear seasonality whilst particulate and sediment organic matter did not show any clear seasonal pattern. Relative abundances of suspended and sediment organic matter, on the other hand, appeared to be site-dependent and controlled mainly by the dynamic balance between resuspension and sedimentation. High quantities of both suspended and sediment total organic matter were present, while very low algal biomasses (in terms of chlorophyll-a concentrations) were observed both in the suspended and sediment pools thus showing the oligotrophy of the site. The contribution of phytoplankton and microphytobenthos to the total organic content of suspended and sediment matter was negligible. The low food availability of organic matter in the Stagnone di Marsala Sound may explain the low abundance of suspension-feeding molluscs, which are substituted by limnovore and detritivore species.

Keywords: suspended matter, sediment organic matter, chloroplastic pigments, mediterranean sound, trophic relationships.

1. Introduction

It is known that in shallow coastal environments, the interactions between continental and marine energy inputs are major factors affecting primary and secondary production. Problems arising from the ecological interpretation of the functioning of a shallow water system may be partially solved through the study of its spatial homogeneity and temporal stability (Carrada and Fresi, 1988). Nevertheless, it has been demonstrated that, in such environments, the occurrence of unpredictable changes in salinity, oxygen and turbidity may foul the "classic curves" of production and consumption of the primary energy resources (Alpine and Cloern, 1992; Millet and Cecchi, 1992; Pusceddu and Fabiano, 1996; Serra *et al.*, 1995; Pusceddu *et al.* in press). The suspended and sediment pools of particulate organic matter, given their importance in the energy transfer from the primary production level to consumers, are usefull tools in understanding the quality and the quantity of trophic interactions. The functioning of an aquatic system cannot be completely assessed without the analysis of the particulate organic matter fluxes (Parsons, 1977). In addition, sediments, as a result of pelagic-benthic coupling processes, may be considered a record of the biological phenomena occurring in the overlying waters (Graf, 1992).

Shallow waters are characterized by the lack of coupling between primary and secondary production: i.e., a large part of the phytoplankton and/or macrophyte production may be trapped in the microbial-detritus chain rather becoming directly available for secondary production (Newell, 1982). Therefore it is necessary to assess the relative importance of detrital and living fractions of suspended and sediment organic matter, their spatial distributions and their relative abundances in lagoons, ponds and sounds.

In the present paper spatial and temporal changes in the suspended and sedimental organic matter in the Stagnone di Marsala, a semi-enclosed marine system (Western Sicily, Mediterranean Sea), are studied. The main aims are to understand which spatial and temporal patterns of variables may better describe the study site and to evaluate the potential role of resuspension processes on the distribution and food availability of suspended organic matter.

2. Material and Methods

The study was carried out in the sound Stagnone di Marsala (Figure 1) which is a shallow depression with a 7 km N-S axis. A low calcarenitic platform (Grande Island) separates the basin from the open sea. The sound is very shallow (average 1.5 m) with depth ranging from 0.2 m along eastern shore of the Isola Grande to 0.50 m in the western area and increases gradually to about 2.5 m in the southernmost area close to the open sea. The northern mouth (Bocca S. Teodoro) and the southern mouth (Bocca Grande) are 450 and 1450 m wide respectively. The latter is open to a sea water inflow with evident internal tides and the prevalence of southern winds. The northern mouth is only occasionally influenced by turbulent inputs of marine waters. The two major islands (S. Maria and S. Pantaleo), mechanical obstacles to the water flow in the middle of the basin generate turbulence. The ribbon-leaved seagrasses, particularly luxuriant in the southern basin greatly affect currents and silting. No land inputs were present during the year of investigation.

Superficial water samples (0.5 m) were collected monthly from January to December 1994, using 10 l Niskin bottles, at 11 stations located along a North-South transect (Figure 1). At the same time duplicate samples of superficial sediment were collected using manual corers. Water temperature, salinity and dissolved oxygen were measured *in situ* using a Hydrolab Inc. multiprobe. Water samples were screened through a 200 μm mesh net in order to remove larger zooplankton and debris, then aliquots (100 to 500 ml) were filtered onto prewashed, precombusted (450°, 4 h) and preweighed Whatman GF/F filters (0.45 μm nominal pore size) for the analyses of total suspended matter (TSM) and photosynthetic pigments.

For the determination of TSM, Whatman GF/F filters (0.45 μm nominal pore size) were weighed after desiccation (60°C, 24h) using a Mettler M3 balance (accuracy ± 1μg). Organic suspended matter (OSM) content was determined by loss on ignition (450°C, 4h; Strickland and Parsons, 1968). Sediment organic matter relative to the top (0-2 cm) was determined by loss on ignition (550°C, 4h; Parker, 1983).

Chlorophyll-a and phaeopigments concentrations (90% acetone extraction) were calculated according to Lorenzen and Jeffrey (1980) for suspended material and according to Danovaro and Fabiano (1992) for sedimentary material.

3. Results

3.1 WATER TEMPERATURE, SALINITY AND DISSOLVED OXYGEN

Temperature patterns were quite similar at all the stations showing a clear seasonal trend (Figure 2). The minimum temperature was observed in December at station 6 (11.8° C), while the maximum was measured in August at station 7 (28.6°C). Salinity also showed a clear seasonal trend (Figure 3), the minimum being recorded at station 5 in February (33.1‰), the maximum measured at station 1 in September (45.5‰). Dissolved oxygen reached highest values (up to 13.6 ppm) at station 10 in January and was relatively low only in July (minimum 4.4 ppm at station 4).

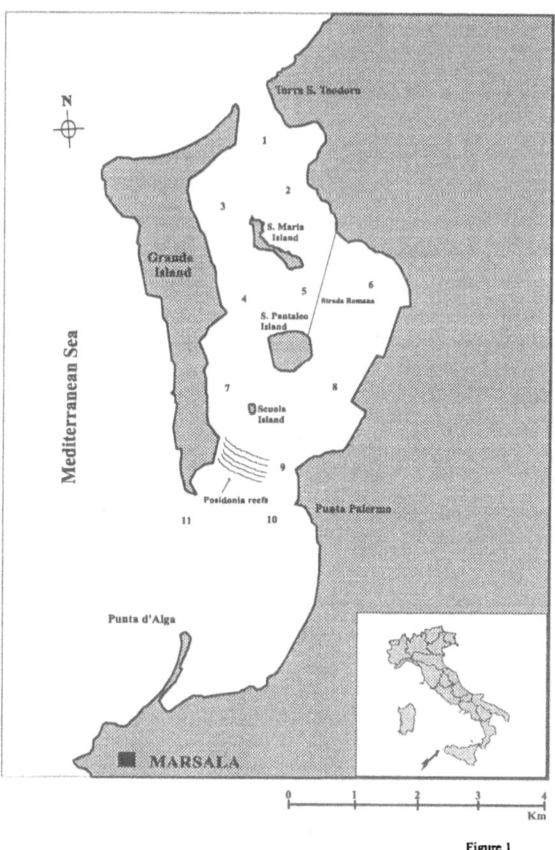

Figure 1. The Stagnone di Marsala Sound (Western Sicily) showing the 11 sampling stations, representing a transect from N to S.

The data were analysed by Principal Component Analysis (PCA) (Flury, 1988) using monthly environmental rank (Spearman rank transformation) transformed matrixes. For each set of variables it was possible to obtain two results; PCA organized by variable (Q-type analysis) and PCA organized by station (R-type analysis).

3.2 SUSPENDED MATTER

The annual mean concentration of TSM was quite high in 1994; varying between 2.4 ± 0.4 mg l^{-1} (August) and 12.0 ± 10 mg l^{-1} (February). The highest concentration was observed in January (49.0 mg l^{-1}) at Sta. 1, whilst the lowest was measured in April (1.6 mg l^{-1}) at Sta. 8. Suspended organic matter concentrations appeared quite constant at all sampled stations throughout the year (Figure 4). The maximum concentration (13.0 mg l^{-1}) occurred in February at Sta. 3, the minimum (0.20 mg l^{-1}) in October at Sta. 9. The Chlorophyll-a concentrations (Figure 5) ranged between 0.02 μg l^{-1} (Sta. 1, January) and 2.0 μg l^{-1} (Sta. 1, July). The highest values were measured in January and February (on average 0.55 ± 0.38 μg l^{-1}) whilst the lowest were recorded in June (0.19 ± 0.11 μg l^{-1}). Chlorophyll-a concentrations in the northern stations were on average significantly higher than those measured in the southern one. Phaeopigments were significantly correlated with chlorophyll-a ($r=0.98$; $p<0.001$) and showed the same temporal patterns (Figure 5). Highest concentrations were recorded in January (0.20 ± 0.19 μg l^{-1}) and March (0.20 ± 0.12 μg l^{-1}).

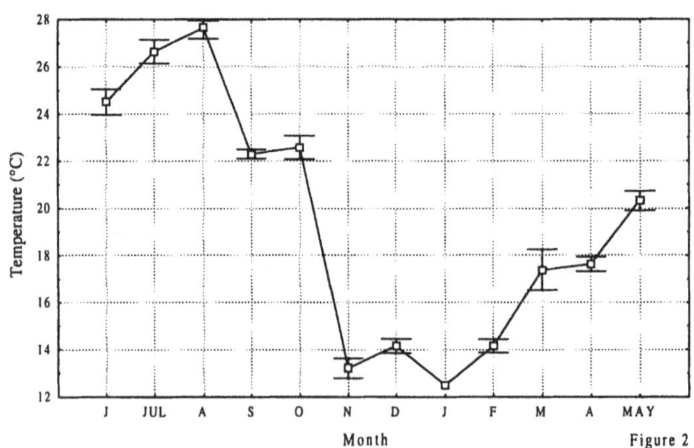

Figure 2. Seasonal pattern of water temperature (°C) in the Stagnone di Marsala Sound during 1994. Standard deviations from mean values of all 11 stations are reported.

3.3 SEDIMENTARY PARAMETERS

Total organic matter content of the sediments did not show evident seasonal changes (Figure 6a). Only in autumn, especially in November were sediments extremely rich in organic matter (305.5 ± 255.5 mg g^{-1}). The maximum organic matter content was recorded at Sta. 2 in November (972.2 mg g-1), the minimum at Sta. 1 in January (17.0 mg g^{-1}). Chlorophyll-a content of the sediments varied from 1.30 μg g^{-1} (January) to 5.28 μg g^{-1} (April). Concentrations of sedimental chloroplastic pigments (i.e. the sum of chlorophyll-a and phaeopigments, CPE, Figure 6b) showed an evident N-S gradient (Figure 7). The contribution of phytodetritus (as CPE concentrations), to the total sediment organic matter content, was very low (less than 0.2 ‰) but higher, however, in the mid stations (4-6) rather than in proximity to the open sea.

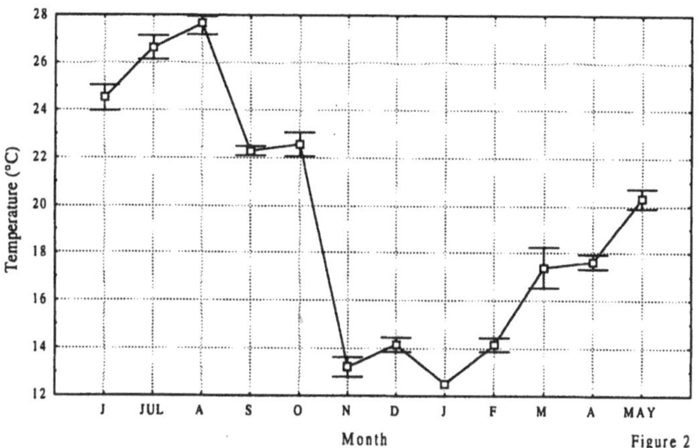

Figure 3. Seasonal pattern of salinity (‰) in the Stagnone di Marsala Sound. Standard deviations from mean values of all 11 stations are reported.

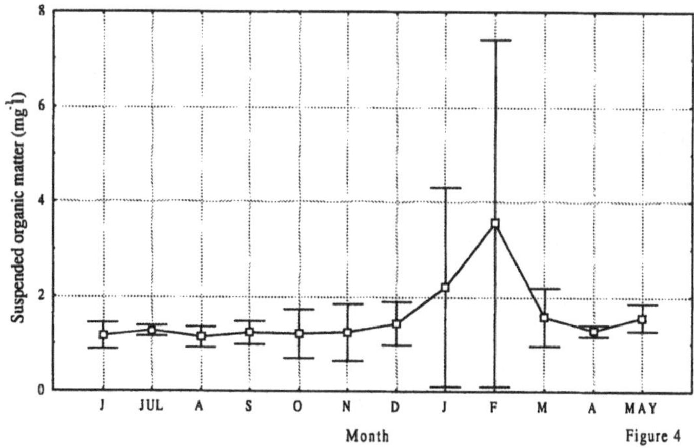

Figure 4. Concentrations of suspended organic matter (mg l^{-1}) in the Stagnone di Marsala Sound. Standard deviations from mean values of all 11 stations are reported.

4. Discussion

Sarà *et al.* (1995) highlighted the role of water exchange between the open sea and the interior sites of the Stagnone di Marsala Sound in determining short-term quantity and biochemical composition of particulate organic matter. They showed that the southern area of the Stagnone Sound is generally characterized generally by sea vivification and by allochtonous (marine) particulate organic matter. In contrast the northern area is

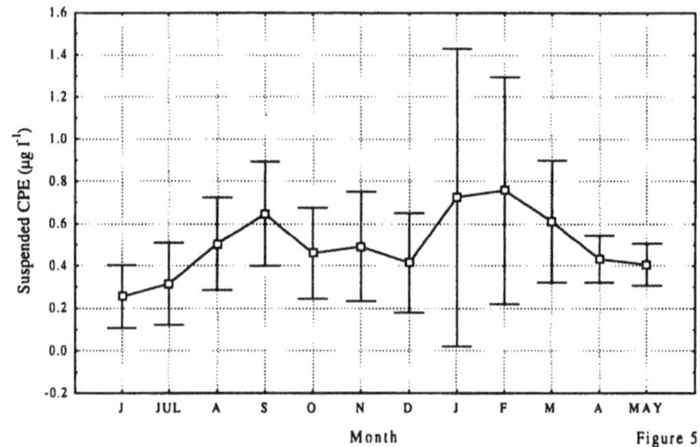

Figure 5. Concentrations of suspended chloroplastic pigments equivalents (μg l^{-1}) in the Stagnone di Marsala Sound. Standard deviations from mean values of all 11 stations are reported.

scarcely influenced by the open sea and is characterized by particulate organic matter deriving mostly from resuspension.

Results from the present study suggest that processes of resuspension and sedimentation, influence to different degrees both the suspended and sediment pools of organic matter in the whole study area. Total suspended matter and the concentrations of its autotrophic component (i.e. chlorophyll-a) decrease from the northern to the southern sampling sites (Figure 8). Although no measure of water currents or downward fluxes of particulate matter are available for the study area, we hypothesize that this pattern results from the attenuation of resuspension processes along a N-S transect according to the depth profile of the sound. The sediment organic matter (in terms of total organic matter and its autotrophic component expressed as CPE) reached the highest concentrations in the central stations (Figure 7), suggesting that sedimentation processes increase along the same N-S transect, becoming still greater as the *Posidonia oceanica* bed (Figure 1) is approached. This large seagrass bed acts as a mechanical barrier to the lateral drifting of suspended matter, giving rise to an increase in the sinking velocities of suspended particles.

Results of the principal component analysis (Table 1; Figure 9) only partially confirm our hypotheses about the role of resuspension and sedimentation processes in affecting the spatial distribution of organic matter in the Stagnone di Marsala Sound. Sta.1 (northern area) and Sta. 5 (mid-lagoon) clearly represented the spatial segregation of, respectively, resuspension and sedimentation processes, the latter being influenced by the *Posidonia oceanica* bed (Figure 1). All other stations showed no clear segregation or pattern.

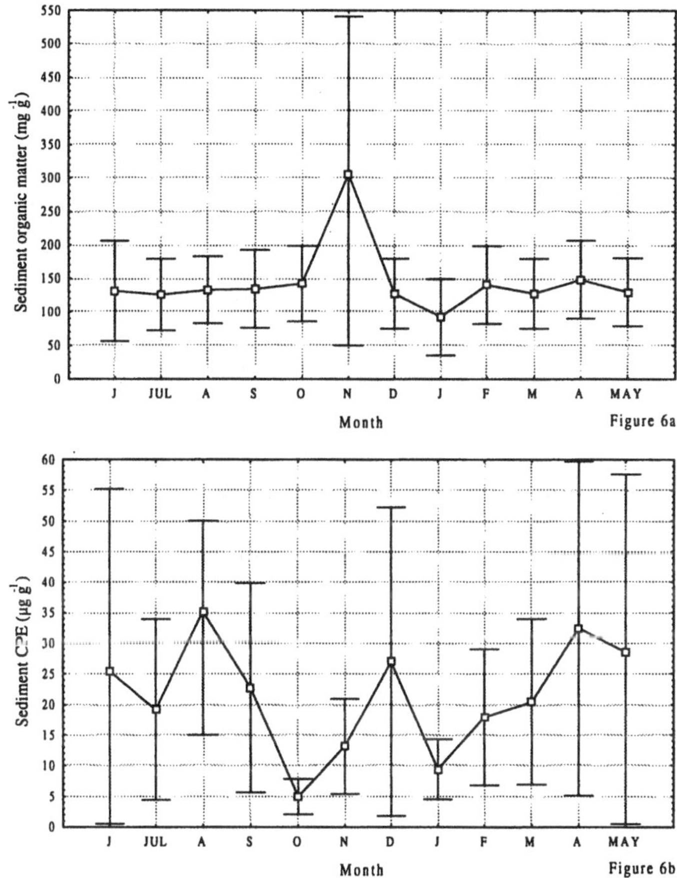

Figure 6. Total organic matter (a, mg g^{-1}) and chloroplastic pigments equivalents (b, µg g^{-1}) in the sediments (top 0-2 cm) of the Stagnone di Marsala Sound. Standard deviations from mean values of all 11 stations are reported.

Although further investigation of the sedimentation rates and the biochemical composition of organic matter should strengthen our hypothesis, we suggest that resuspension may be utilized profitably to explain some trophic interactions. Indeed, Hopkinson, (1985) suggested that resuspension determines the relative amounts of organic carbon as well as the sites and rates of organic matter degradation in the benthos and water column. Moreover, Wainright (1987; 1990) performing feeding experiments by means of heterotrophic microorganisms as consumers of resuspended organic matter, demonstrated that microbial growth was stimulated by resuspension of sediment. Thus, resuspension could significantly enhance microbial degradation of suspended organic matter and appear to influence POM cycling, persisting much longer than resuspension even itself.

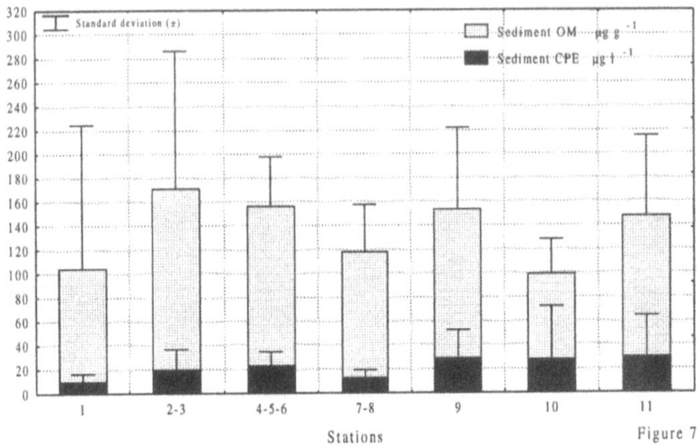

Fig. 7 - Spatial distribution (annual average) of chloroplastic pigments equivalents and total organic matter in the sediments (top 0-2 cm) of the Stagnone di Marsala Sound. Stations 2-3, 4-5-6 and 7-8 were averaged to obtain a N-S transect representation.

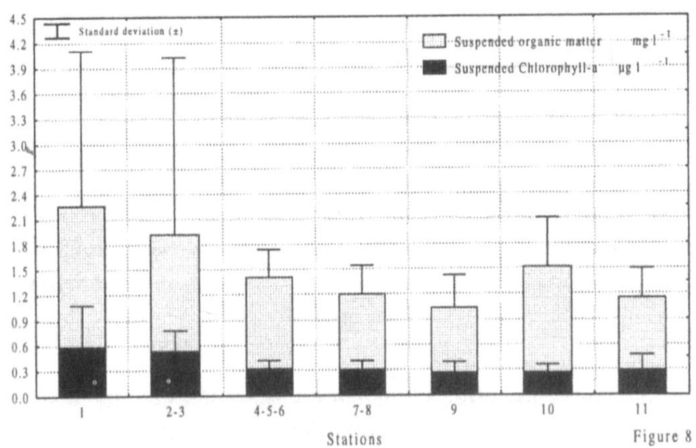

Figure 8. Spatial distribution (annual average) of suspended chlorophyll-a and total suspended matter in the Stagnone di Marsala Sound. Stations 2-3, 4-5-6 and 7-8 were averaged to obtain a N-S transect representation.

This appears even more probable in shallow water systems, where unpredictable changes in the major environmental variables affect sediment resuspension (Pusceddu and Fabiano, 1996; Pusceddu *et al.*, in press) and, consequently, plankton ecology (Alpine and Cloern, 1992; Millet and Cecchi, 1992). In oligotrophic sites, such as the Ligurian Sea (Albertelli and Fabiano, 1990) resuspended particles are principally composed of refractory matter and only contribute in a limited way to the nutritional needs of the suspension feeder community. In these areas benthic filter feeders meet their nutritional needs mainly from bacteria bound to suspended particles (Danovaro and Fabiano, 1995). In contrast, in eutrophic areas, such as Mediterranean coastal lagoons (Pusceddu and Fabiano, 1996) suspension feeders reach their food requirements by filtering particles which mainly originate from the resuspension of nitrogen enriched phyto-detritus rather than directly from living phytoplankton.

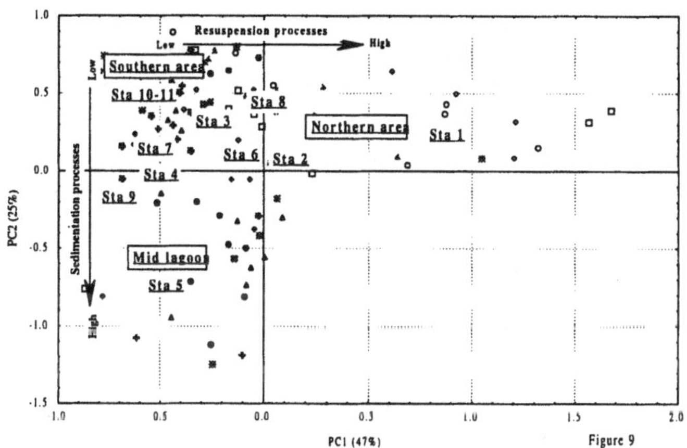

Figure 9. Results of the Principal Component Analysis organized by station.

Table I

Principal Components Analysis: variance explained by the system and relative contribution of each variable on the two factorial axis (PC1, PC2)

Variable	PC1	PC2
Suspended Organic Matter	0.80	-0.11
Suspended Inorganic Matter	0.87	-0.08
Suspended Chlorophyll-a	0.94	-0.08
Suspended Phaeopigments	0.95	-0.08
Sedimental Chlorophyll-a	-0.20	-0.88
Sedimental Phaeopigments	-0.15	-0.90
Sedimental Organic Matter	-0.03	-0.04
Total variance (%)	**47**	**25**

In this study phytoplankton and microphytobenthos biomasses, in terms of chlorophyll-a concentrations, were very low (on average less than 0.3 $\mu g\ l^{-1}$ and 2.9 $\mu g\ g^{-1}$ respectively), confirming the oligotrophy of the site, and contributing little to the suspended and sediment organic matter pools (on average less than 0.1 and 0.2 ‰ respectively). Thus, we may assume that most of the organic matter in the Stagnone di Marsala Sound is of detrital (non-living) and heterotrophic origin. Moreover, owing to the presence of the large *Posidonia oceanica* bed in the southern area of the sound, it is expected that a large fraction of the organic matter pool is refractory and, thus, unappetizing to consumers.

This hypothesis has been confirmed by the analysis of the biochemical composition of particulate organic matter performed in the study by Sarà *et al.* (1995). They found that the contribution of the labile fraction to the bulk particulate organic matter, used as a food index (Navarro *et al.*, 1993), was quite low (on average about 20%; range 2-30 %) when compared with values found in Mediterranean shallow coastal lagoons (on average 50%; range 30-100%, Pusceddu *et al.*, in press). The low food availability of organic particles may explain partially the low abundance of suspension-feeding molluscs, which are substituted by limnovore and detritivore species (Chemello, pers. comm.).

Acknowledgements

The authors would like to thank students from the Department of Animal Biology of the University of Palermo, for help during sampling. This work was funded by the Ministero Università Ricerca Scientifica e Tecnologica of the Italian Goverment. We thanks Dr. R. Chemello (University of Palermo) and Dr. R. Danovaro (University of Ancona) for useful suggestions and constructive criticisms. We are also grateful to two anonymous referees for their fundamental help in improving the manuscript.

References

Albertelli, G. and Fabiano, M.: 1990, *Atti VIII Congresso A.I.O.L.*, 59-66

Alpine, A.E. and Cloern, J.E.: 1992, *Limnol. Oceanogr.*, **37**(5), 946-955

Carrada, G.C. and Fresi, E.: 1988, Carrada G.C., Cicogna F. and Fresi E. (eds.), *Coastal lagoons: research and managment.* CLEM, 35-56

Danovaro, R. and Fabiano, M.: 1992, *Ist. Sci. Amb. Mar., Univ.Genova, Rapp. Tecn.*, **18**, 1-28

Danovaro, R. and Fabiano, M.: 1995, *Aquat. Microb. Ecol.*, **9**, 17-26

Flury, B.: 1988, John Wiley & Sons, New York, 258 pp.

Graf, G.: 1992, *Oceanogr. Mar. Biol. Annu. Rev.*, **30**, 149-190

Hopkinson, C.S. Jr.: 1985, *Mar. Biol.*, **87**, 19-32

Lorenzen, C.J. and Jeffrey, S.W.: 1980, *Unesco Technical Papers in Marine Science*, **35**, 1-20

Millet, B. and Cecchi, P.: 1992, *Limnol. Oceanogr.*, **37**, 140-146

Navarro, J.M. Clasing, E. Urrutia, G. Asencio, G. Stead, R. and Herrera, C.: 1993 *Est. Coast. Shelf Sci.*, **37**, 59-73

Newell, R.C.: 1982, *Oceanol. Acta, Proceedings Int. Symp. on coastal lagoons*, 347-355

Parker, J. G.: 1983, *Chemistry and Ecology*, **1**, 201-210

Parsons, T.R.: 1977, *Chem. Oceanogr.*, **3** , 365-383

Pusceddu, A. and Fabiano, M.: 1996, *Atti XI Congresso A.I.O.L*, 571-581

Pusceddu, A. Serra, E. Sanna, O. and Fabiano, M.: 1996, *Chemistry and Ecology*, **13**, 21-37

Sarà, G. Pusceddu, A. Mazzola, A. and Fabiano, M.: 1995, *Biol. Mar. Medit.*, **2**(2):127-129

Serra, E. Pusceddu, A. and Serra, A.: 1995, *Biol. Mar. Medit.*, **2**(2): 557-559

Strickland, J.D.H. and Parsons, T.R.: 1968, *Bull. Fish. Res. Board Can.*, 167, 310 pp.

Wainright, S. C.: 1987, *Science*, **238**, 1710-1712

Wainright, S. C.: 1990, *Mar. Ecol. Prog. Ser.*, **62**, 271-281

SOURCES AND PATHWAYS OF PARTICULATE ORGANIC CARBON IN A SUBMARINE CAVE WITH SULPHUR WATER SPRINGS

L. AIROLDI[1], A.J. SOUTHWARD[2], I. NICCOLAI [3] and F. CINELLI [1]

[1]*Dipartimento di Scienze dell'Uomo e dell'Ambiente, Università di Pisa, Via A. Volta 6, I-56100 Pisa, Italy, e-mail: laroldi@discat.unipi.it*
[2]*Marine Biological Association, Citadel Hill, Plymouth, PL1 2PB, U.K.*
[3]*ENEA-CREA S. Teresa, C.P. 316, I-19100 La Spezia, Italy.*

Abstract. The origin, quality and pathways of particulate organic carbon (POC) were studied from May 1991 to May 1995 in a submarine cave (Grotta Azzurra, Tyrrhenian Sea, Italy) with warm sulphur springs that support dense mats of sulphur-oxidizing bacteria. Multifactorial sampling designs were used to specifically address: (1) differences in the quantity and quality of suspended and sedimenting particles in two distinct regions of Grotta Azzurra, a weakly-illuminated outer region (Central Hall) and an innermost dark region (Snow Hall) characterized by the presence of sulphur waters; (2) the composition and fluxes of particulate material above and below the sulphurous boundary. The water and sediment trap samples were analysed for total particles, POC, particulate organic nitrogen (PON), chlorophyll *a* (Chl *a*) and phaeopigments (Phaeo). The microbial mats were assayed for carbon fixation and RubisCo activity. Stable carbon isotope ratios of the bacteria, benthic fauna and sediments were measured. The overall amount and flux of POC did not differ between the two regions of the cave. By contrast photosynthetic pigments decreased significantly from the outer to the inner region. The average POC:PON ratios of suspended and sedimenting material were lower than 10.2, and there was no apparent ageing or degradation from the Central Hall to the Snow Hall. Fluxes of POC above the boundary were on average 23.8% of fluxes measured below the boundary, and no Chl *a* was detected above the boundary. The mats of sulphur-oxidizing bacteria had a ^{13}C depletion of -30 to -31‰, providing a marker for tracing POC from this source. The majority of the benthic animals showed $\partial^{13}C$ from -20 to -24‰. Some polychaete worms and echinoderms were more depleted (-25‰), indicating greater consumption of bacterial carbon, while sponges were not depleted compared with the benthos of the control cave ($\partial^{13}C$ -18 to -19‰). It is estimated that bacteria supply about 31% of the suspended POC in the Snow Hall. A larger input may be available to animals living close to the bacterial mats and to those in the more sulphurous Grotta Sulfurea cave.

Key words: POC; sources; composition; sedimentation; sulphur bacteria; chemosynthesis; submarine caves.

1. Introduction

Supply of particulate food is considered a major limiting factor controlling the distribution and growth of animals in submarine caves (Harmelin *et al.*, 1985; Fichez, 1990a, 1991a). In dark caves photoautotrophic production is negligible, so that the heterotrophes rely on organic material carried in from outside (Fichez, 1990b, 1991b). Such caves generally show a decline of biomass and diversity of the benthic organisms from the entrance to the deeper parts (Pérès and Picard, 1949; Laborel and Vacelet, 1959; Gili *et al.*, 1986; Balduzzi *et al.*, 1989), which is thought to be related to progressive impoverishment of trophic resources due to filtering off, sedimentation and degradation of organic particles (Fichez, 1990b; Palau *et al.*, 1991; Garrabou and Flos, 1995). By contrast, some submarine caves with warm sulphur springs show a rich and abundant benthic fauna dominated by filter-feeding organisms (Cinelli *et al.*, 1994). These assemblages might be potentially dependent upon autochthonous sources of food produced by sulphur-oxidizing chemoautolithotrophic bacteria (Abbiati *et al.*, 1994).

In this paper we present an overview of the results of a multidisciplinary research devoted to investigate the origin, cycling and trophic pathways of particulate organic carbon (POC) in a submarine cave (Grotta Azzurra) with warm sulphur springs that support dense mats of sulphur-oxidizing bacteria. Ancillary studies were also made in a

Water, Air and Soil Pollution **99**: 353-362, 1997.

nearby cave with greater sulphur input (Grotta Sulfurea) and in a control cave without sulphur (Grotta Trombetta). An integrated approach was used, based on measurements of the concentration, biochemical composition and vertical fluxes of POC as well as analysis of autotrophic capability of the microbial mats and analysis of the stable carbon isotope ratios of bacteria, dominant benthic fauna and sediments. The study was carried out from May 1991 to May 1995, and part of the data have been published already (Mattison and Dando, 1994; Southward *et al.* 1996; Airoldi and Cinelli, 1996, 1997). The present paper will concentrate on large-scale patterns affecting the overall trophism of the cave, while causes and ecological significance of small-scale variability of trophic resources are discussed elsewhere (Airoldi and Cinelli, 1996).

2. Materials and Methods

2.1. STUDY AREA

The caves studied open in the limestone massif of Capo Palinuro (south-western coast of Italy). Grotta Azzurra, where most of the work was carried out, may be separated into two topographically distinct regions (Figure 1A): a weakly illuminated, outer region (Central Hall) and an innermost, dark region (Snow Hall), the latter characterized by the presence of hot sulphur springs that arise from fissures in the cave floor and flanks. The warm sulphur water stratifies above the ambient cooler sea water just below the vault, forming a sharp boundary at about 9 m depth (Figure 2B). Dense mats of sulphur-oxidizing bacteria (including attached *Beggiatoa*-like filamentous colonies up to 100 µm in diameter) line the rocky walls and the vault above the chemocline and act as primary producers, fixing CO_2 by means of the autotrophic enzyme ribulosebisphosphate carboxylase (RubisCo). Below the chemocline, rich faunal assemblages occur on both the walls (sponges and cnidarians) and the sediment floor (bivalves, ophiurids, tubicolous polychaetes). Further information on the characteristics of the habitat and the composition and structure of benthic assemblages may be found in Cinelli *et al.* (1994), Dando (1996) and Benedetti-Cecchi *et al.* (1997).

2.2. SUSPENDED AND SEDIMENTING PARTICULATE MATERIAL

Multifactorial sampling designs were used to specifically address whether there were differences in the quantity and quality of suspended and sedimenting particles in different regions of Grotta Azzurra. Two to 3 replicated seawater samples for suspended particulate material were collected at 4 stations along an outside-inside transect (ST1 outside the cave, ST2 at the entrance, ST3 in the Central Hall and ST4 inside the Snow Hall). Sampling was done at a depth of 12 m by SCUBA using 2.5 to 6-litre PVC bottles completely opaque to light. Sedimentation rates of different compounds were measured using 6 trap systems each consisting of 4 polypropylene cylindrical vessels (diameter 51 mm, aspect ratio: 3.9) positioned with their mouths at 1.5 m above the sea floor (the trap systems are described in Airoldi and Cinelli, 1996). The sediment traps were deployed at 6 sites selected at random, 3 in the Central Hall and 3 in the Snow Hall. Therefore, for each region (Central Hall vs Snow Hall) there were 3 replicated sampling sites, and for each site 4 replicated collection vessels. Measurements of both concentration and vertical flux of particles were repeated in February, May and October 1994 and in February 1995, in order to verify if patterns observed were consistent through time. An additional sediment trap deployment was run in May 1995 to address whether the composition and

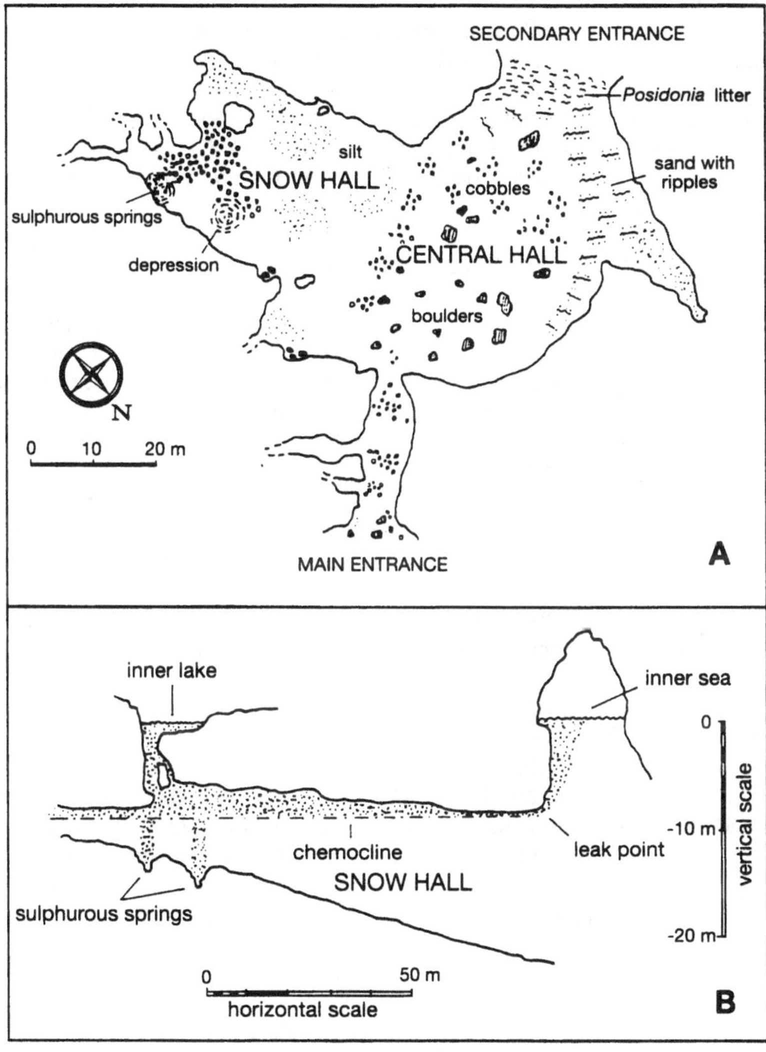

Fig. 1. (A) Sketch of the underwater portion of Grotta Azzurra cave and (B) cross section of Snow Hall (modified from Alvisi et al. 1994).

fluxes of particulate material differed above and below the sulphurous boundary. To this aim, sediment traps were deployed in the Snow Hall with their mouths 0.5 m above the boundary at 2 sites and 4 m below the boundary at another 2 sites. Therefore, for each position with respect to the boundary (above vs below) there were 2 replicated sampling sites, and for each site 4 replicated collection vessels. All samples were immediately transported to the laboratory and stored in a cooler until filtration, which was completed within 20 hours from the collection. Further information about sampling techniques are reported in Airoldi and Cinelli (1996, 1997).

Water and sediment trap samples were analysed to quantify the contents of coarse material (CM, dimensions >200 μm), total particulate material (TPM, dimensions <200 μm), particulate inorganic material (PIM), particulate organic carbon (POC) and nitrogen (PON), chlorophyll *a* (Chl *a*) and phaeopigments (Phaeo). Subsampling, filtration and chemical analysis were carried out as described by Airoldi and Cinelli (1996, 1997). Data were processed using techniques of analysis of variance (ANOVA) (Underwood, 1981). Most of the filtered samples were too low in carbon for analysis of stable isotopes by conventional methods.

2.3. AUTOTROPHIC ACTIVITY AND CARBON ISOTOPIC COMPOSITION OF THE BIOTA

The bacterial mats were sampled by scraping from a fixed area. Subsamples of the mats were weighed after removal of surplus water and added to aliquots of sea-water from above the chemocline enriched with ^{14}C-labelled bicarbonate. They were incubated in closed 50 ml vials with a head-space containing ambient air from outside the cave (oxygen is necessary for the oxidation of sulphide). Incubations were stopped after one hour by addition of acetic acid and the mats homogenised. The fixed carbon was assayed by established methods (Dando *et al.*, 1985, 1986). The approximate biomass of the mats was assessed by scraping circular patches into jars of 37 mm diameter and weighing the sample after removal of excess water. Assays of RubisCo activity in subsamples of the mats were made according to the techniques of Dando *et al.* (1985). It is possible that the homogenization techniques appropriate for bacterial symbionts in metazoans did not extract all the activity from the sulphur bacteria, so our figures may underestimate the autotrophic potential.

Samples of the dominant benthic fauna were taken from May 1991 to October 1994 for measurements of their stable carbon isotopes, in order to determine the relative contribution of isotopically distinct food sources to their nutrition (Conway *et al.*, 1989). Specimens were cleaned of epizoic organisms, acid-washed, dried and powdered. Powdered samples were combusted in sealed quartz tubes at 750-800°C. The resulting CO_2 was cryogenically purified at -78°C and analysed in a Finnigan MAT 251 isotope ratio mass spectrometer. The results are reported in the usual notation relative to the PDB standard as part per thousands. Further details on the methods are reported in Southward *et al.* (1996).

3. Results and discussion

3.1. DISTRIBUTION AND QUALITY OF PARTICLES

The quantity and quality of suspended and sedimenting particles in distinct regions of Grotta Azzurra were less variable than indicated for other submarine caves, where progressive impoverishment and degradation of TPM and its constituents were observed with increasing distance from the entrance (Fichez 1990b, 1991b). Mean concentrations and vertical fluxes of CM, TPM, PIM, POC and PON were similar among different regions of the cave (Tables 1 and 2), and no consistent spatial patterns were discernible over time. A clear spatial trend emerged only for chloropigments, whose concentrations and vertical fluxes significantly decreased from the Central Hall to the Snow Hall during most of the research (ANOVA, $F_{3,16}=4.259$ $p<0.05$ and $F_{1,4}=128$ $p<0.001$ for suspended and sedimenting Chl *a* respectively), suggesting lower inputs of photosynthetic material advected from the open sea in the form of both plankton and plant detritus.

TABLE 1

Mean concentrations and standard deviations (SD, n =12 for Chl *a* and Phaeo and 8 for the remaining components) of the different compounds of suspended particulate material calculated over 4 sampling periods at 4 stations (ST1, ST2, ST3 and ST4) in Grotta Azzurra. Symbols as in text

Component	ST1 mean (SD)	ST2 mean (SD)	ST3 mean (SD)	ST4 mean (SD)
TPM(μg/l)	490 (376)	662 (587)	515 (284)	798 (1127)
PIM (μg/l)	266 (240)	440 (489)	302 (197)	580 (924)
POC (μg/l)	52.9 (11.3)	46.7 (8.5)	62.2 (23.3)	49.2 (26.6)
PON (μg/l)	6.2 (2.07)	5.4 (1.6)	6.5 (3.1)	5.4 (2.5)
Chl *a* (ng/l)	223 (121)	211 (127)	213 (99)	142 (77)
Phaeo (ng/l)	95.3 (80.8)	78.3 (123.8)	62.7 (90.2)	41.8 (34.8)

TABLE 2

Mean vertical fluxes and standard deviations (SD, n = 36 for CM and 48 for all the other cases) of the different compounds of suspended particulate material pooled from 3 sites and 4 sampling periods in Central Hall and Snow Hall. Symbols as in text

	CM mean (SD)	TPM mean (SD)	PIM mean (SD)	POC mean (SD)	PON mean (SD)	Chl a mean (SD)	Phaeo mean (SD)
Central Hall	2219 (1851)	9099 (4191)	8283 (4017)	221 (138)	20.5 (8)	258 (160)	1111 (727)
Snow Hall	1847 (909)	7877 (2581)	7135 (2457)	171 (82)	16.4 (5)	131 (95)	460 (328)

Also the quality of particles was constant among different regions of the cave. A large fraction (on average 63 %) of suspended TPM was inorganic, and the inorganic component increased up to 88.9 % for sedimenting TPM (Tables 3 and 4). The relative content of POC was, consequently, low, but this was represented by fresh detritus of good nutritional quality. POC:PON ratios of both suspended and sedimenting organic material were meanly lower than 10.2, and no apparent ageing and degradation of organic material was observed going from the Central Hall to the innermost Snow Hall. Sedimenting POC was generally impoverished of Chl *a* with respect to suspended POC (Tables 3 and 4), but the analysis of the biochemical composition and energetic content of particles indicated in both cases a large importance of highly nutritive and easily degradable food material (Airoldi and Cinelli, 1996, 1997). Overall, the relative content of POC was constant among regions and did not match changes in the concentration and vertical flux of Chl *a*. These patterns clearly affected POC:Chl *a* ratios, which increased from the Central Hall to the Snow Hall at all sampling periods (Tables 3 and 4). This ratio estimates the relative content in plant pigments of organic detritus, and values higher than 100 are considered to indicate that some carbon originates from sources other than phytoplankton (Zeitzschel, 1970). The POC:Chl *a* ratio of suspended material in the Snow Hall was meanly 346 and increased up to 2582 for sedimenting material (Tables 3 and 4). These results suggest that a great proportion of particulate food potentially available to benthic fauna in the Snow Hall derives from sources different from photoautotrophic production. POM from chemolithoautotrophic production shows POC:PON ratios and biochemical composition comparable to that from photoautotrophic origin (Khripounoff and Alberic, 1991). Since POC:PON values were quite low and comparable to those found for natural living

TABLE 3

Mean ratios characterising the origin and composition of suspended particulate material at 4 stations in Grotta Azzurra. The ratios are computed from the data in Table 1. Pigm = total chloropigments (Chl *a* + Phaeo)

Ratio	ST1	ST2	ST3	ST4
PIM:TPM (%)	54.3	66.4	58.6	72.7
POC:TPM (%)	10.8	7.1	12.1	6.2
POC:PON (w/w)	8.5	8.6	9.6	9.1
POC:Chl *a* (w/w)	237	221	292	346
Chl *a*:Pigm (%)	70.1	73.4	77	77.3

TABLE 4

Ratios characterising the origin and composition of sedimented particulate material in Central Hall and Snow Hall. Data are means and standard deviations pooled from 3 sites and 4 sampling periods (n = 48)

	PIM:TPM mean (SD)	POC:TPM mean (SD)	POC:PON mean (SD)	POC:Chl *a* mean (SD)	Chl *a*:Pigm mean (SD)
Central Hall	88.9 (5.9)	2.49 (0.74)	10.2 (2.93)	1108 (1130)	25.6 (18.03)
Snow Hall	89.7 (3.6)	2.2 (0.71)	10.2 (2.86)	2582 (3319)	32 (24.69)

phytoplankton and microbial colonized detritus (Fukami et al., 1981), we may infer that a significative fraction of POC in the Snow Hall derives from bacterial autochtonous production. As a confirm filaments detached from the bacterial mats living on the vault of the cave were frequently recognized in CM collected with sediment traps in both Snow Hall and Central Hall (Airoldi and Cinelli, 1996).

3.2. VERTICAL FLUXES ACROSS THE BOUNDARY

With the exception of CM, mean vertical fluxes of all compounds of sedimenting particulate material were significantly higher below the sulphurous boundary than above (Figure 2 and Table 5). This pattern was especially evident for chloropigments, which were not detected in traps above the boundary, although data could not be analysed as the preponderance of zero values violated the assumption of homoscedasticity required for ANOVA (Underwood, 1981). Overall, POC collected by sediment traps below the boundary was more diluted by the inorganic fraction and showed higher POC:PON ratios than that collected above the boundary (Figure 2), although differences were not statistically significant.

Chl *a* is commonly considered a good descriptor of phytoplanktonic biomass (Parsons *et al.*, 1984) and in submarine caves may be used to quantify advection of organic particles from the open sea and to estimate rates of renewal of the water masses (Fichez, 1990a). Absence of Chl *a* in traps above the boundary suggests negligible upward transport of photosynthetic material advected from the open sea and a stable stratification of the water column. Fluxes measured above the boundary, therefore, may be considered a good estimation of primary inputs of fresh POC from authochthonous bacterial production. Such inputs represented, on average, 23.8 % of fluxes of POC measured

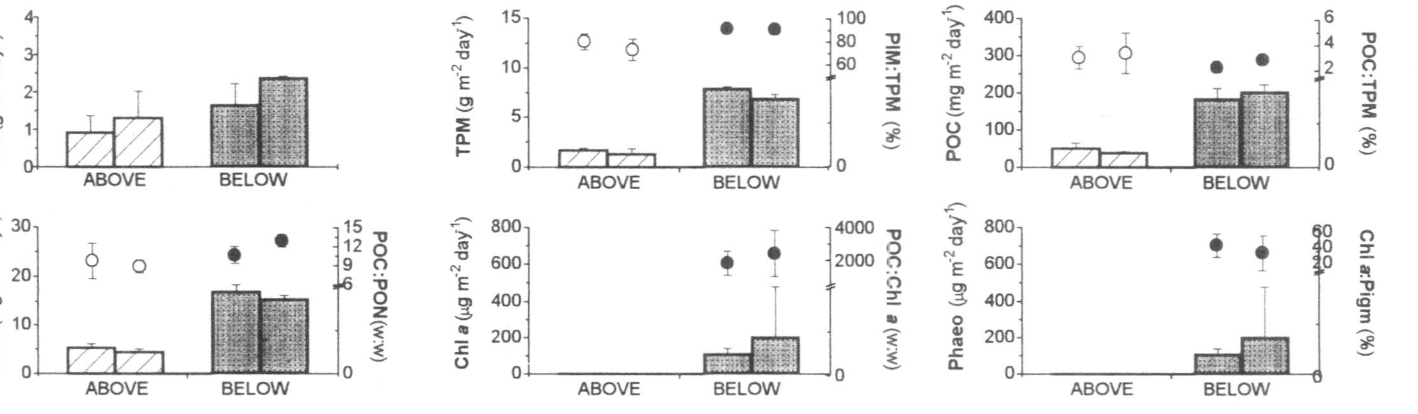

Fig. 2. Sedimentation rates of (left axes, columns) coarse material (CM), total particulate material (TPM), particulate inorganic material (PIM), particulate organic carbon (POC), particulate organic nitrogen (PON), chlorophyll *a* (Chl *a*), phaeopigments (Phaeo) and (right axes, scattered) PIM:TPM ratio (%), POC:TPM ratio (%), POC:PON ratio (w:w) and Chl *a*:Pigm (= Chl *a* + Phaeo) ratio (%) of sinking particles measured in the Snow Hall during May 1995 at different positions with respect to the boundary (above vs below). Two random sites were replicated for each position. Data are means + 1SD (n = 3 for CM and 4 for all the other cases).

TABLE 5

ANOVA table showing the effects of position with respect to the boundary (above vs below) and site on sedimentation rates of coarse material (CM), total particulate material (TPM), particulate inorganic material (PIM), particulate organic carbon (POC) and particulate organic nitrogen (PON) in the Snow Hall in May 1995. Position is a fixed factor, while site is a random variable nested in position. Significant p values ($p<0.05$) are shown in bold type

Source of variation	df	CM MS	CM *F*	CM p	TPM MS	TPM *F*	TPM p	PIM MS	PIM *F*	PIM p	POC MS	POC *F*	POC p	PON MS	PON *F*	PON p
Position	1	$2.4x10^6$	4.878	0.158	$138x10^6$	118.7	**0.008**	$124x10^6$	113.9	**0.008**	86471	178.7	**0.006**	514.2	154.7	**0.006**
Site (Position)	2	$0.5x10^6$	1.929	0.207	$1.2x10^6$	7.557	**0.007**	$1.1x10^6$	10.03	**0.003**	484	1.28	**0.313**	3.323	3.579	**0.06**
Residual	12*	$0.3x10^6$			$0.2x10^6$			$0.1x10^6$			378.1			0.928		
Transformation		none			none			none			none			none		
Cochran's C-test		C=0.44 p>0.05			C=0.46 p>0.05			C=0.45 p>0.05			C=0.55 p>0.05			C=0.57 p>0.05		

* df = 8 for CM

below the boundary, but this percentage is probably underestimated if we consider that fluxes below the boundary are likely affected by local resuspension of bottom sediments.

3.3. BACTERIAL AUTOTROPHY AND CARBON PATHWAYS

RubisCo activity of the bacterial mats (0.01 to 0.05 µmoles CO_2 g^{-1} min^{-1}) was an order of magnitude less than that shown by symbiotic sulphur bacteria. This may be due to difficulties in extraction of the enzyme (Southward *et al.*, 1996). CO_2 fixation by intact mats showed similar rates.

The carbon isotope ratio of the bacterial mats ranged from -30.7 to -30.1‰ (Table 7), providing a good marker for tracing POC from this source against less depleted photoautotrophic inputs. Results from these analyses confirm that POC produced by bacteria contributes to the bulk of suspended and sedimenting particulate food potentially available to benthic fauna in the Snow Hall and accumulates in bottom sediments. Using stable isotope ratio as an index of autochthonous chemosynthetic input and POC:Chl *a* ratios as an index of the relative amount of carbon of phytoplanctonic origin, the fraction of chemosynthetically derived carbon over the total amount of suspended POC in the Snow Hall has been estimated as about 31% (Airoldi and Cinelli, 1997). This estimate is consistent with that obtained by fluxes measured across the boundary.

Many animals (both filter and detritus feeders) living in the cave appeared to be able to feed on food from bacteria, the contribution from this source ranging from 0 to virtually 100 % depending on species and variation in local habitat. The majority of species showed ^{13}C depletion ranging from -24 to -20‰ (Table 7), suggesting a diet of mixed particles from allochthonous and autochthonous sources. Some polychaete worms (e.g. *Phyllochaetopterus socialis*) and echinoderms (e.g. *Cucumaria* sp.) were more depleted ($\partial^{13}C$ -25‰), indicating greater consumption of bacterial carbon. Contribution from bacteria was potentially larger for animals living close to the bacterial mats and to those in the more sulphurous Grotta Sulfurea ($\partial^{13}C$ -25 to -31‰). Conversely, sponges were not depleted compared with the benthos of control non sulphurous Grotta Trombetta ($\partial^{13}C$ -19.4 to -17.4‰), suggesting that these species rely mostly on photoautotrophic organic material carried in from outside the cave.

TABLE 6

ANOVA table showing the effects of position with respect to the boundary (above vs below) and site on PIM:TPM ratio (%), POC:TPM ratio (%) and POC:PON ratio (w:w) characterizing sedimented particulate material in the Snow Hall in May 1995. Position is a fixed factor, while site is a random variable nested in position

Source of variation	df	PIM:TPM MS	PIM:TPM *F*	PIM:TPM p	POC:TPM MS	POC:TPM *F*	POC:TPM p	POC:PON MS	POC:PON *F*	POC:PON p
Position	1	0.1213	14.11	0.064	0.111	1.818	0.31	0.2465	6.014	0.134
Site (Position)	2	0.0086	1.454	0.272	0.061	0.898	0.433	0.041	1.431	0.277
Residual	12	0.0059			0.068			0.0287		
Transformation		logarithmic			logarithmic			logarithmic		
Cochran's C-test		C=0.626 p>0.05			C=0.636 p>0.05			C=0.77 0.01<p<0.05		

TABLE 7

$^{13}C/^{12}C$ ratios of dominant benthic fauna, suspended particles and sediments of Grotta Azzurra (data from Southward *et al.*, 1996).

Species	$^{13}C/^{12}C$ ratio (‰)
Bacterial mat and filaments	-30.7 to -30.1
Sponges	-19.4 to -17.4
Cnidarians	-21.3 to -19.4
Polychaetes	
filter-feeders	-28.9 to -19.2
detritus-feeders	-25.0 to -20.8
carnivores	-23.5 to -19.2
Crustacea	-21.9 to -17.1
Bivalves	-21.2 to -19.2
Gastropods	-20.9 to -18
Echinoderms	
filter-feeders	-20.2 to -19.7
detritus-feeders	-25.0 to -19.1
grazers	-23.4 to -19.1
Suspended particles	-25.3 to -25.1
Sedimenting particles	-25.8 to -24.6
Sediments	-22.7 to -21.2

4. Conclusions

The results indicate an important effect of sulphur springs on the biology of Grotta Azzurra through enhanced inputs of fresh POC from autochthonous bacterial production. This suggests that submarine caves with sulphur springs may show some ecological similarity to deep-sea hydrothermal vent sites, where chemosynthetic production locally enriches invertebrate biomass through supply of large amounts of organic matter relative to the pelagic flux from surface primary production (Jannasch, 1985; Tunnicliffe, 1991).

In contrast to the general paradigm of caves as simplified oligotrophic ecosystems dependent upon outside sources of energy, Grotta Azzurra appears to be a complex environment, influenced by local processes acting within the cave. These processes potentially affect the modes of distribution and abundance of benthic organisms, which are more complex than generally reported from other mediterranean caves (Benedetti-Cecchi *et al.*, 1997). Further research is needed to explore in detail other potential influences of the sulphurous springs on the ecology of this unique environments, possibly involving non-trophic factors such as increased inputs of rare elements and potentially toxic compounds and patterns of water circulation.

Acknowledgements

We wish to thank all the numerous collegues who helped during this investigation. In particular, we are grateful to E.C. Southward, C.N. Bianchi, C. Morri, L. Benedetti-Cecchi and M. Abbiati for stimulating discussions during the work and for help in the field and laboratory activities. M.C. Kennicutt II and J. Alcalà-Herrera provided the carbon isotope analyses. This study was funded in part by the EC under MAST programme contract MAS2-CT93-0058.

References

Abbiati, M., Airoldi, L., Castelli, A., Cinelli, F. and Southward, A.J.: 1994, *Mém. Mus. Natn. Hist. Nat.* **162**: 323-329.

Airoldi, L. and Cinelli, F.: 1996, *Mar. Ecol. Prog. Ser.* **139**: 205-217.

Airoldi, L. and Cinelli, F.: 1997, *Mar. Biol.*: in press.

Alvisi, M., Barbieri, F., Bruni, R., Cinelli, F., Colantoni, P., Grandi, G.F. and Maltoni, P.: 1994, *Mem. Ist. It. Spel.* 2nd Ser. **6**, 51-56.

Balduzzi, A., Bianchi, C.N., Boero, F., Cattaneo-Vietti, R., Pansini, M. and Sarà, M.: 1989, *Scient. Mar.* **53**, 387-395.

Benedetti-Cecchi, L., Airoldi, L., Abbiati, M. and Cinelli, F.: 1997a, *J. Mar. Biol. Ass. U. K.*: in press.

Cinelli, F., Colantoni, P., Morri, C., Bianchi, C.N., Alvisi, M., Airoldi, L. and Abbiati, M.: 1994, *Mem. Ist. It. Spel.* 2nd Ser. **6**, 95-97.

Conway, N., Capuzzo, J.M. and Fry, B.: 1989, *Limnol. Oceanogr.* **34**: 249-255.

Dando, P.R. (ed.): 1996, *Oxic-anoxic intefraces as productive sites,* MAST Contract CT93-0058, Final Report, Volumes I and II.

Dando, P.R., Southward, A.J. and Southward, E.C.: 1986, *Proc. Royal. Society* B **227**: 227-247.

Dando, P.R., Southward, A.J., Southward, E.C., Terwilliger, N.B. and Terwilliger, R.C.: 1985, *Mar. Ecol. Prog. Ser.* 23: 85-98.

Fichez, R.: 1990a, *C. R. Acad. Sci. Paris* (Sér III) **310**: 155-161.

Fichez, R.: 1990b, *Hydrobiologia* **207**: 61-69.

Fichez, R.: 1991a, *Mar. Biol.* **110**: 137-143.

Fichez, R.: 1991b, *Mar. Biol.* **108**: 167-174.

Fukami, K., Simidu, U. and Taga, N.: 1981, *Bull. Plankt. Soc. Jap.* **1**: 33-41.

Garrabou, J. and Flos, J.: 1995, *Mar. Ecol. Prog. Ser.* **123**: 273-280.

Gili, J.M., Riera, T., Zabala, T.: 1986, *Mar. Biol.* **90**: 291-297.

Harmelin, J.G., Vacelet, J. and Vasseur, P.: 1985, *Téthys* **11**: 214-229.

Jannasch, H.W.: 1985, *Proc. Royal Soc. London* **225**: 277-297.

Khripounoff, A. and Alberic, P.: 1991, *Deep-Sea Res.* **38**: 729-744.

Laborel, J. and Vacelet, J.: 1959, *Cah. Océanogr.* **13**: 73-107.

Mattison, R.G. and Dando, P.R.:1994, *BRIDGE Newsletter* **7**: 10-11.

Palau, M., Cornet, C., Riera, T. and Zabala, M.: 1991, *Oecologia Aquatica* **10**: 299-316.

Parsons, T.R., Takahashi, M. and Hargrave, B.: 1984, *Biological oceanographic processes* 3rd edition, Pergamon Press, Oxford.

Pérès, J.M. and Picard, J.: 1949, *C. R. Soc. Biogéogr.* **227**: 42-45.

Southward, A.J., Kennicutt, M.C., Alcalà-Herrera, J., Abbiati, M., Airoldi, L., Cinelli, F., Bianchi, C.N., Morri, C. and Southward, E.C.: 1996, *J. Mar. Biol. Ass. U.K.* **76**: 265-285.

Tunnicliff, V.:1991, *Oceanogr. Mar. Biol. A. Rev.* **29**: 319-407.

Underwood, A.J.: 1981, *Oceanogr. Mar. Biol. A. Rev.* **19**: 513-605.

Zeitzschel, B.: 1970, *Mar. Biol.* **7**: 305-318.

SEASONAL VARIATIONS OF SULPHATE REDUCTION RATES, SULPHUR POOLS AND IRON AVAILABILITY IN THE SEDIMENT OF A DYSTROPHIC LAGOON (SACCA DI GORO, ITALY)

Giordani Gianmarco, Roberta Azzoni, Marco Bartoli, Pierluigi Viaroli.
Dipartimento di Scienze Ambientali, Università di Parma, Viale delle Scienze
43100 Parma, Italy

Abstract. The aim of this work was to analyse factors which regulate sulphide mobility in the sediments of a dystrophic lagoon (Sacca di Goro, the southern lagoon of the Po river delta, Italy). In 1995-96, sediment oxygen demand and variations in organic matter content, redox potential, iron availability, inorganic sulphur concentrations and sulphate reduction rates were measured in sediment profiles at three stations (G, 4 and 17) representative of the main areas of the lagoon. Stations differed mainly in their salinity range and primary producer communities. High concentrations of reactive iron (110-275 μmol ml^{-1}) and low sulphate reduction rates (0.8-16.1 mmol m^{-2} d^{-1}) were measured in the sediment. Moreover, high concentrations of reactive ferric iron were detected in winter and spring at the stations closest to the freshwater inputs. Nevertheless, in summer, high concentrations of free sulphides were detected in the porewater, although most of the reactive ferrous iron was not sulphide-bound, indicating that not all of the reactive iron pool was available to buffer against sulphide release and thus measures of reactive iron pools may not be a good measure of the true buffering capacity of the sediment. Furthermore, a considerable production of sulphide may occur in the decaying *Ulva* biomass in the water column, where its concentration will be independent of the potential buffering capacity of iron in the sediment. Therefore when assessing the vulnerability of coastal lagoons to dystrophic events, both the size and availability of the reactive iron pool as well as the site of sulphide production must be taken into consideration.

Key words: Dystrophy, sulphide, iron monosulphide, buffer capacities, iron.

1. Introduction

Despite its low energy yield, dissimilatory sulphate reduction is the dominant decomposition pathway in the anoxic sediments of coastal lagoons (Howarth, 1984). More than 50% of the organic matter available in the sediment is oxidised by sulphate reducing bacteria, particularly in summer when oxygen penetration is low (Jorgensen and Fenchel, 1974). The sulphides produced can be removed from the porewater by precipitation with metallic ions, e.g. ferrous iron (Jorgensen, 1977). In marine sediments, ferric iron occurs mostly as stable crystalline oxides, even in reduced sediments (Lovley and Phillips, 1987). The reactive forms of Fe(III) are amorphous and can be reduced by specialised microrganisms as well as by chemical reactions with reducing compounds such as sulphide. Ferrous iron is then released to the porewater where it can trap dissolved sulphide in amorphous insoluble compounds, the so-called "iron monosulphides" with the approximate composition FeS (Fenchel and Jorgensen, 1977). Furthermore, pyrite (FeS_2) can be formed by the reaction of FeS with elemental sulphur (Berner, 1984) and by the rapid reaction of ferrous iron with polysulphides (Luther, 1991). Therefore, the sediment capacity to buffer against free sulphide production is assumed to be controlled by ferrous iron availability (Giordani *et al.*, 1996). When dissolved sulphide (DS) production exceeds reactive iron availability, DS diffuses towards the overlaying water. In sheltered lagoons or embayments dominated by floating macroalgae, sulphate reduction may also occur in the water mass when macroalgal biomass starts to decompose (Viaroli *et al.*, 1996). The massive diffusion of DS into the water mass can result in an incomplete oxidation of DS leading to the

Water, Air and Soil Pollution **99**: 363-371, 1997.

formation of colloidal elemental sulphur, "white water" episodes (Valiela, 1984) or DS can be used directly by anaerobic phototrophic bacteria causing "red water" events (Caumette, 1986).

In this study we have investigated sulphide mobility in relation to sulphate reduction rates (SRR), and iron availability in a eutrophic lagoon affected by blooms of floating macroalgae and dystrophic crises with consequent mortality of the commercially important shellfish stocks (Viaroli *et al.*, 1992).

2. Study Area

The Sacca di Goro (44°47'-44°50'N and 12°15'-12°20'E) is the southernmost bay of the Po River Delta (Fig. 1). The total area is 26 km^2, tidal range is within 100 cm. The water temperature ranges between 5°C in January and 30°C in July. The lagoon can be divided into three different zones: 1) the western part (4 km^2) is affected by freshwater inputs from the Po di Volano River, 2) the central part is connected directly to the Adriatic Sea via a 1 km wide channel and 3) the eastern part (10 km^2), called the Valle di Gorino, is separated from the sea by a sand barrier and receives freshwater inputs from the Po di Goro. In the latter two areas blooms of floating macroalgae and dystrophic crises have been observed in recent years (Viaroli *et al.*, 1992).

Samples were collected at three stations representative of the above mentioned areas of the lagoon. Station G (mean depth 50 cm) is located in the western zone where water salinity undergoes wide changes (0-22 ‰).The granulometry of the sediment is silty-clay. Station 4 (mean depth: 150 cm, salinity range: 15-35 ‰) is located in the central part of the lagoon where salinity is on average higher than at station G and the sediment is mostly silt. Station 17 (mean depth: 50 cm salinity range: 15-30 ‰) is in the Valle di Gorino where the sediment is silty-clay.

Fig. 1. Sacca di Goro map. Sampling sites are indicated.

3. Materials and Methods

From February 1995 to January 1996, sediment cores (5 cm, internal diameter) were collected using hand corers at the three stations (Table I). Cores were maintained upright in the dark, in an aerated and temperature controlled water bath until analysed. All the analyses were carried out within 2 days of sample collection.

Redox potential (Eh) and dissolved sulphide (DS) profiles were measured with 15 cm needle Pt and Ag_2S electrodes (Microscale, Groningen, NL) driven by a Tesa Hite 300 micromanipulator. The reference electrode (Ag/AgCl REF 251, Radiometer) was positioned in the overlaying water. pH profiles were performed with a 3 mm-wide-tip pH combined electrode (Ingold LoT406-M3-S7/25). DS concentrations were calculated following Kuhl and Jorgensen (1992). Dissolved oxygen (DO) profiles were measured by means of needle electrodes (Diamond 768-22) and a reference Ag/AgCl electrode (REF 251 Radiometer). A Crison micro-pH2002 millivolt-meter was used for Eh and DS concentrations and a Keithley 485 picoampere-meter for oxygen determinations.

TABLE. I.
Variables and replicates considered for each sampling date.

Data	Eh	DS	DO	O.M.	SOD	AVS-CRS	SRR	Fe
09/02/95				1		1		3
06/06/95	3	3	3	1	3	1	3	3
13/07/95	3	3	3	1	3	1	3	3
12/09/95	3	3	3	1	3	1	3	3
14/11/95	3	3		1	3	1		3
23/01/96	3	3	3	1	3	1	3	3

Organic matter (O.M.) was determined as weight loss at 450°C for 24 hours. Sediment oxygen demand (SOD) was measured by incubating sealed cores in the dark in a temperature controlled water bath from 1 to 10 hours. Subsamples for oxygen determinations (A.P.H.A., 1975) were withdrawn at the beginning and at the end of each incubation period. SOD was then calculated as the oxygen consumed during the incubation, which was always less than 30% of the initial value to avoid hypoxia within the cores. The inorganic reduced sulphur in the sediments was separated into AVS (Acid Volatile Sulphide) and CRS (Chromium Reducible Sulphur) following the two step distillation method (Fossing and Jorgensen, 1989). Sulphate reduction rates (SRR) were determined on undisturbed sediment cores, 10 µl of carrier free $^{35}SO_4$ (expected activity 10 KBq cm^{-3} of sediment) were injected in the upper 10 cm of sediment at 1 cm depth intervals, via 3 lines of silicon sealed sampling ports. Cores were incubated in the dark at *in situ* temperature for 2 to 4 hours. At the end of the incubation the upper 10 cm of sediment was sliced into 2 cm sections which were fixed in 10 ml of 20% zinc acetate. Reduced ^{35}S was stripped off using the one-step distillation method (Fossing and Jorgensen, 1989) and trapped in 10 ml 20% zinc acetate traps. A 5 ml subsample was then added to 15 ml scintillation liquid (NBCS104, Amersham) and counted in a Minaxi TRI-CARB 4000 series (Camberra Packard) scintillation counter. In parallel, 0.2 ml of the supernatant, obtained by centrifugation of the slurry, was added to 8 ml of scintillation fluid to measure activity in $^{35}SO_4$.

SRR was estimated as nmol SO_4 ml^{-1} h^{-1}, according to the equation:

$$SRR=1.06\,(SO_4)(H_2{}^{35}S)/(((^{35}SO_4)+(H_2{}^{35}S))T) \quad (1)$$

where 1.06 is the defraction index (Sorokin, 1962), (SO_4) is sulphate concentration measured following Howarth (1978), $(H_2{}^{35}S)$ is the activity of reduced sulphur, $(^{35}SO_4)$ is the activity of labelled sulphate and T is the incubation time. The obtained rates were than transformed to mmol SO_4 m^{-2} d^{-1} by integration along the 10 cm sediment column.

Total iron, hydroxylammine-reducible-ferric iron and HCl-extractable ferrous iron were measured as Fe^{2+} using the ferrozine method, after extraction from the sediment (Lovley and Phillips, 1987). Iron reducible with hydroxylammine is considered as the more reactive form of ferric iron. The hydroxylammine-reducible-ferric iron plus the 0.5M HCl extractable-ferrous iron can be considered as total reactive iron, which can be involved in the redox processes (Lovley and Phillips, 1987). Total iron was measured following 37% HCl digestion of ash obtained from sediment samples after 24 hours at 550°C. After evaporation of the concentrated HCl, 20 ml of 0.5M HCl were added; 1 ml subsamples were added to 5 ml 10% ascorbic acid solution to reduce all the ferric iron. Reactive ferrous iron was measured after extraction of 1 hour with 0.5 M HCl. Ferric reactive iron was measured following extraction with 0.25M hydroxylammine-HCl (acidified to pH 0.5 with HCl 1M) for 1 hour.

4. Results and Discussion.

In 1995, *Ulva rigida* growth started in March in the eastern part of the lagoon. The maximum standing crop (98±11 g d.w. m^{-2}) was attained at the end of June, when macroalgal biomass underwent decomposition. In July, decay processes led to anoxia and "white water", which lasted for a few days at the end of the month. A minor bloom of *Ulva* and *Cladophora* spp. supported by high temperatures occurred from late September to early November and caused hypoxia (Bondavalli *et al.*, 1996).

The means and the SD of organic matter (O.M.) content in the surficial sediment at station 4 and station 17 was 11.4 ± 1.0 % d.w. and 6.4 ± 3.2 % d.w respectively. Significant seasonal differences in O.M. content (ANOVA, p=0.001) were found only at station G where O.M. content was 5.3 ± 1.0 % d.w. from February to June, 8.3± 1.0% from July to September and 13.7 ± 1.9 % from November to January.

A surface oxidised layer of about 5 cm was detected in winter and spring, whilst in summer, this became very thin and was restricted to the upper few mm. In September, strong reducing conditions were measured at stations 4 and 17 (Table 2). At station 4, Eh values lower than -270 mV coincided with high concentrations of DS in the sediment at depths below 5 cm from the sediment surface (Fig. 4). At all the stations, the porewater pH ranged between 7.57±0.16 in the superficial sediment and 7.34±0.11 at 10 cm depth. No significant differences were observed between sites.

TABLE II

Water temperature, redox potential (Eh) and reactive ferric iron measured in the 3 stations of the Sacca di Goro. Eh reported as means and SD of the values measured in the 5 to 10 cm sediment horizon. Reactive ferric iron is the average of the values determined in the first 10 cm of sediment.

	Water temperature	Eh (mV)			Reactive Ferric Iron (μmol ml^{-1})		
Data	°C	St. G	St. 4	St. 17	St. G	St. 4	St. 17
09/02/95	7	-	-	-	92±19	23±08	9±08
06/06/95	19	-156±61	-169±08	-188±80	64±20	0±00	74±26
13/07/95	27	-121±11	-128±41	-160±65	11±02	0±00	16±15
12/09/95	24	-174±07	-271±46	-203±32	5±05	0±00	0±00
14/11/95	17	-93±11	-46± 18	-64±12	4±01	16±15	9±08
23/01/96	6	-83±06	-70± 40	-78±18	48±05	4±03	5±02

Peaks of the sediment oxygen demand (SOD) were observed in July at station G (105 mmol m^{-2} d^{-1}) and station 4 (50 mmol m^{-2} d^{-1}); at station 17 a maximum SOD of 69 mmol $m^{-2}d^{-1}$ was measured in June, during the *Ulva* growth period (Fig. 2). Sulphate reduction rates (SRR) followed a similar seasonal trend. The highest SRR were observed in July at stations G (16.1 mmol SO_4 m^{-2} d^{-1}) and 17 (4.9 mmol SO_4 m^{-2} d^{-1}) and in September at station 4 (10.5 mmol SO_4 m^{-2} d^{-1}); minima were measured in January. Moreover, SOD and SRR followed patterns opposite to that shown by oxygen penetration into the surficial sediment. Oxygen penetration was estimated from profiles determined along the sediment horizon with a needle electrode measuring the distance between the water-sediment interface and the zero concentration depth. The minimum oxygen penetrations were observed in July, just before the dystrophic crises when SRR and SOD attained maximum values. By contrast, in January oxygen penetration reached 5 mm at all stations. The high oxygen penetration observed in June at station 17 may be due to the presence of polychaete burrows as demonstrated by the irregularity in the profile shapes. (Fig. 4). These burrows allow oxygenated water to penetrate deep into the sediment and stimulate SOD due to the increased surface area for O_2 exchange.

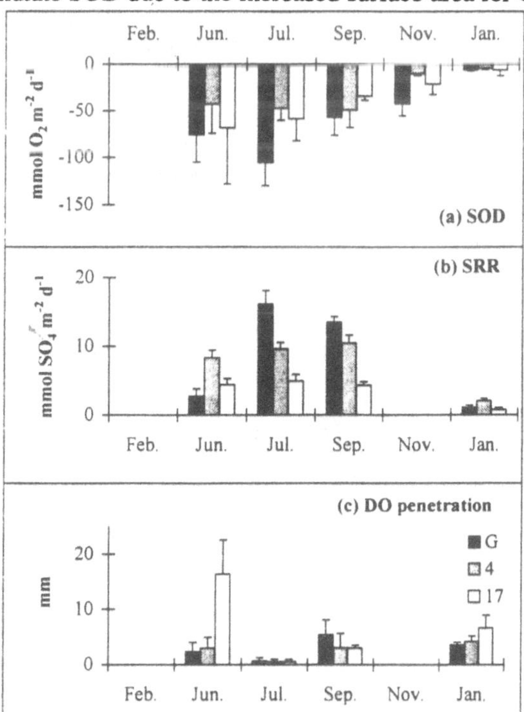

Fig. 2 a-b-c. Mean values of Sediment Oxygen Demand (SOD) (a), Sulphate reduction rates (SRR) (b) and Dissolved Oxygen (DO) penetration (c) measured in Sacca di Goro (SD are indicated as error bars).

Summer anoxia led to significant DS production (4.91 mM) in the sediment sub-surface at station 4 (Fig. 3). A smaller amount was also detected in the deeper sediment layers at station 17 (0.3 mM). The increased SOD and SRR conformed to variations of acid volatile sulphides (AVS) concentrations (Fig. 5a). In the upper sediment layer (10 cm) AVS concentrations increased from 5-10 μmol cm^{-3} in spring to 30-40 μmol cm^{-3} in autumn. In winter, the AVS pools decreased due to re-oxidation processes. CRS concentrations follow a similar pattern to those of AVS except at station G where the CRS maximum was observed in June (Fig. 5b). On average, AVS:CRS ratios increased during spring and summer reaching maximum values at the end of autumn (Table III) due to progressive trapping of sulphide in the AVS pool. The lower AVS:CRS ratios observed at station G, indicate that it remained more oxidised in spite of the higher summer SRR determined. Sediment at station 4 was the most reduced having AVS:CRS ratios higher than those observed at the other stations.

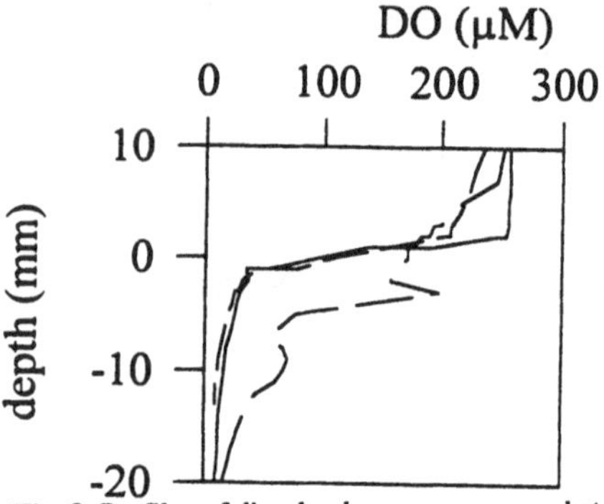

Fig. 3. Profiles of dissolved oxygen measured at station 17 in June 1995 (3 replicates are indicated).

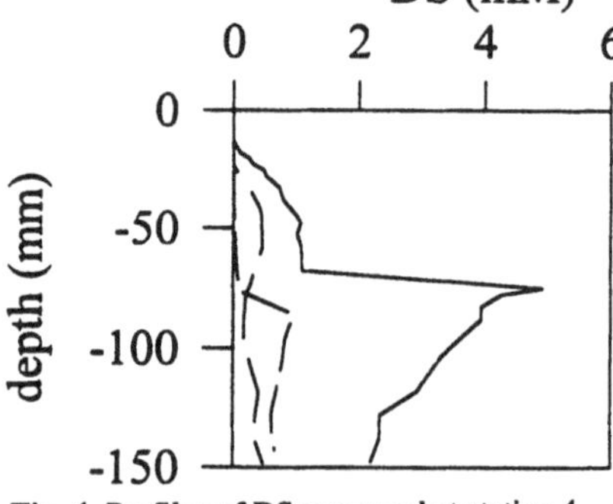

Fig. 4. Profiles of DS measured at station 4 in September 1995 (3 replicates are indicated).

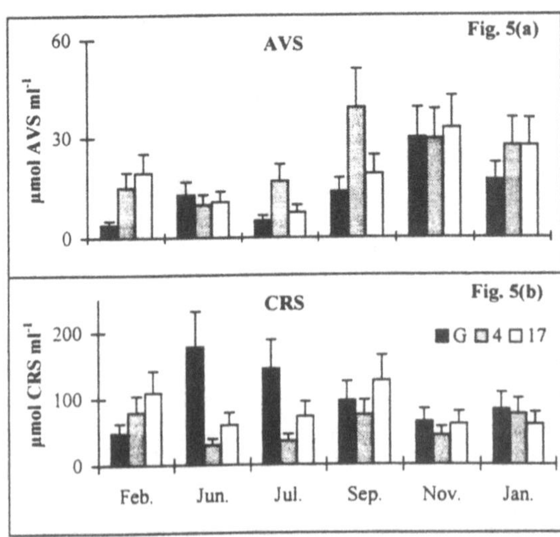

Fig. 5 a-b. Seasonal variations of Acid Volatile Sulphide (AVS) (a) and Chromium Reducible Sulphur (CRS) (b) in the Sacca di Goro. Mean values of the upper 10 cm are presented. Spatial variability was evaluated for each sites in 1 sampling data by means of 5 different replicates and it is indicated as error bar.

The total iron content of the upper 10 cm of sediment ranged between 200 and 500 μmol cm^{-3}. Seasonal variations in the reactive iron pools were not significant with

158±17 μmol ml^{-1} at station G, 138±24 μmol ml^{-1} at station 4 and 153±61 μmol ml^{-1}. At station G, a significant amount of reactive ferric iron was observed during winter and in spring (Tab. II), due either to re-oxidation processes or riverine discharge. A significant correlation was found between discharge and reactive iron concentration in the Po di Volano Canal ($r = 0.819$, $p < 0.001$). Riverine discharge can also explain the June peaks in reactive iron concentrations (275±43 μmol ml^{-1}) and the availability of reactive ferric iron (74±26 μmol ml^{-1}) observed at station 17, which followed a month of high freshwater input from the Po di Goro River (Bencivelli, 1995-96). However, only half of the total iron can be considered reactive and thus able to be involved in redox processes. We postulate that the reactive iron is almost completely reduced in summer. Presumably, ferric reactive iron can not persist due to the high organic matter availability, which supports strong decomposition processes leading to high SOD and SRR. The relationship between ferrous iron and sulphide is shown in Tab. III. The mean ratio of AVS to reactive ferrous iron increased from July (0.07±0.03) to November (0.25±0.04) due to AVS accumulation. The ratio started to decrease in January 96 to 0.18±0.02, a similar ratio to that observed in February '95 (0.14±0.06).

TABLE III
AVS:Fe(II), SOD:*SRR and AVS.CRS ratios observed in the 3 stations of Sacca di Goro.

	AVS:Fe(II)			SOD:*SRR			AVS:CRS		
Data	St. G	St. 4	St. 17	St. G	St. 4	St. 17	St. G	St. 4	St. 17
09/02/95	0.08	0.15	0.19	-	-	-	0.08	0.19	0.18
06/06/95	0.12	0.08	0.06	13.7	2.6	7.8	0.07	0.32	0.18
13/07/95	0.04	0.09	0.08	3.3	2.5	6.0	0.03	0.47	0.10
12/09/95	0.08	0.26	0.13	2.1	2.4	1.2	0.14	0.52	0.15
14/11/95	0.21	0.27	0.27	-	-	-	0.46	0.67	0.53
23/01/96	0.16	0.19	0.20	3.2	1.2	4.1	0.20	0.36	0.45

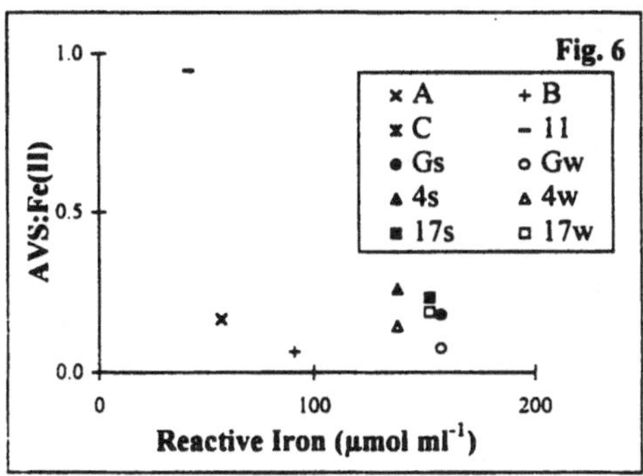

Fig. 6. AVS/Fe(II) ratio vs. AVS concentration in the sediment of 7 stations of three different lagoons. Values for stations A, B and C (Arcachon Bay, France) and 11 (Étang du Prévost) are from Giordani *et al.*, (1996). For the 3 stations of the Sacca di Goro, the winter (Gw, 4w, 17w) and the summer (Gs, 4s, 17s) values are obtained from the 22 February and 13 July 1995 sampling date of this work..

Considering that in coastal sediments about 90% of sulphide produced by sulphate reduction is re-oxidised by oxygen close to the sediment surface (Jorgensen, 1983), it is possible to evaluate the relative contribution of sulphate reduction to the SOD. For this purpose SRR were transformed into oxygen equivalents assuming O_2=$2S^{2-}$ (*SRR). At station 4, sulphate reduction accounted for approximately half of oxygen consumption, that is SOD:*SRR=2.15 ± 0.66 (Table III). At the other stations, the contribution of sulphate reduction to the total metabolism was several times lower as shown by the SOD:SRR ratios of 15.4±8.0 (station G) and 8.0±6.2 (station 17) measured in June.

5. Concluding remarks

In the sediment of Sacca di Goro, sulphate reduction was less intense than reported in other coastal lagoons affected by summer dystrophy (Finster *et al.*, 1994). High concentrations of easily reducible ferric iron were detected at the stations close to the freshwater inputs, probably due to the influence of riverine discharge. This ferric iron became reduced in summer when SOD and SRR attained their maxima. The capacity of the sediment to bind sulphide was exceeded in September '95, when high concentrations of DS were detected in the porewater of stations 4 and 17. At station G, the sediment was on average more oxidised since AVS:CRS and AVS:Fe(II) ratios were lower than those at the other stations, although it exhibited the highest SRR, due to the accumulation of reduced sulphur at this station in the CRS rather than the AVS pool. DS was never detected in the sediment of this station.

The vulnerability of lagoon ecosystems can be described by the seasonal changes in the sulphide-ferrous monosulphide-pyrite system (Golterman, 1995). Considering AVS:Fe(II) =1 as the saturation threshold of the sulphide-ferrous monosulphide buffer, we can argue that the sediment of Sacca di Goro should be well buffered against sulphide release since the ratios were considerably less than 1 throughout the year (Table III). In contrast in the highly eutrophic Étang du Prévost the iron buffering capacity was progressively depleted and the AVS:Fe(II) ratio approached 1 during summer (Fig. 6). However, despite the apparent excess of reactive iron in the sediments of the Sacca di Goro, free sulphides were still detected in the sediment porewater, indicating that a large proportion of the measured reactive iron pool was unavailable to buffer against sulphide release. These data demonstrate that simple measurements of sediment iron pools are not sufficient to determine the true sediment buffering capacity against sulphide release, as the availability of this iron may differ considerably between lagoons. Additionally, in highly eutrophic lagoons such as the Sacca di Goro and the Étang du Prévost a considerable production of free sulphide (up to 50 mmol $m^{-2}d^{-1}$; Viaroli *et al.*, 1996) may occur associated with the mineralization of primary producer biomass in anoxic bottom waters, where its concentration is independent of the potential iron buffering capacity of the sediment.

Acknowledgements

This research was partially supported by the EU project ROBUST: the role of buffering capacities in stabilising coastal lagoon ecosystems (contract No ENV4-CT96-0218). We gratefully acknowledge the support and facilities provided by the Amministrazione Provinciale di Ferrara, Battello Hydra (Mr S. Bencivelli and Mr. T.

Zappata). Many thanks are also due to Mr V. Gaiani (Dip. Biologia Evolutiva, Università di Ferrara) who contributed most of the field work and Dr. D.T. Welsh (Laboratoire d'Océanographie Biologique, Arcachon, France) who revised the manuscript. This paper is the ELOISE contribution No XX.

References

A.P.H.A.: 1975. Standard Methods for the Examination of Water and Wastewater. 14th Edition. APHA, Washington, 1193 pp.

Bencivelli S.: 1995-96. Monitoring activity in the Sacca di Goro. Monthly Reports 1995-1996. Amministrazione Provinciale di Ferrara, Servizio Ambiente.

Berner, R. A.: 1984. *Geochim. Cosmochim. Acta*, **48**: 605-615.

Bondavalli, C., Naldi M., Taje' L., Viaroli P.: 1996. Proceeding of 7th Congress of Italian Society of Ecology, 367-370 (in italian).

Caumette, P., 1986. *FEMS Microbiol. Ecol.* **38**: 113-124

Fenchel, T., Jorgensen B. B.: 1977. *Advances in Microbial Ecology*, **1**: 1-57.

Finster, K., Gammerlaad A. & Hansen L.S., 1994. Coastal Lagoon Eutrophication and Anaerobic processes. Progress Report, European Commission: pp 223-252. European Community Environment Programme, DO XII, Brussels.

Fossing, H., Jorgensen B.B.: 1989. *Biogeochemistry*, **8**: 205-222.

Giordani, G., Bartoli M., Cattadori M., Viaroli P.:1996. *Hydrobiologia* **329**:211-222.

Golterman, H. L.: 1995. *Hydrobiologia*, **297**: 43-54.

Howarth, R. W.: 1978. *Limnol. Oceanogr.*, **23(5)**: 1066-1069.

Howarth, R. W.: 1984. *Biogeochemistry*, **1**: 5-27.

Jorgensen, B. B., Fenchel T.: 1974. *Marine Biology*, **24**: 189-201.

Jorgensen, B. B.: 1977. *Limnol. Oceanogr.*, **22**: 814-832.

Jorgensen B.B.: 1983. SCOPE **23**, 477-509.

Kuhl, M., Jorgensen B. B.: 1992.. *Appl. Environ. Microbiol.*, **58(4)**: 1164-1174.

Lovley, D. R. Phillips E. J. P.: 1987. *Appl. Environ. Microbiol.*, **53**: 1536-1540.

Luther, G. W.: 1991. *Geochim. Cosmochim. Acta*, **55**, 2839-49.

Sorokin, Y. I.: 1962. *Microbiology* **31**: 329-335

Valiela, I.: 1984. Marine Ecological Processes. Springer-Verlag, New York, 546 pp.

Viaroli, P., Pugnetti A., Ferrari I.: 1992. Prec. Of 25th EMBS, Marine eutrophication and population dynamics, *Olsen and Olsen*, Fredensborg: 77-84.

Viaroli, P., Bartoli M., Bondavalli C., Christian R. B., Giordani G., Naldi M.: 1996. *Hydrobiologia* **329**: 105-119

HYPOLIMNETIC ALKALINITY GENERATION IN TWO DILUTE, OLIGOTROPHIC LAKES IN ONTARIO, CANADA

P.J. DILLON[1], H.E. EVANS[2] and R. GIRARD[1]

[1] *Ontario Ministry of Environment and Energy, P.O. Box 39, Bellwood Acres Road, Dorset, (Ontario) POA 1EO, Canada.* [2] *RODA Environmental Research Limited, P.O. Box 447, Lakefield (Ontario) KOL 2HO, Canada*

Abstract.

It has been observed that the alkalinity concentrations in the hypolimnia of many thermally stratified lakes increase over the duration of the summer. We have quantified the processes, both redox and exchange, that contributed to the alkalinity increases measured in two lakes that are situated on the Precambrian Shield of Ontario, Canada by measuring hypolimnetic mass balances for all substances involved in alkalinity-generating or -consuming reactions. These include nitrate, ammonium, base cations, iron, manganese, sulphate, organic anions, as well as alkalinity itself. In one lake, iron reduction was the dominant source of alkalinity; since this process is probably reversed at fall overturn when hypolimnetic waters mix with oxygenated surface waters, the alkalinity generated by this mechanism is likely temporary in nature. In the second lake, iron reduction and sulphate reduction were both important; the latter should provide more permanent alkalinity.

Keywords - alkalinity, mass balances, iron, sulphate, nitrate, base cations, manganese, ammonium

1. Introduction

It is well recognized that in-lake processes can generate more alkalinity in dilute, oligotrophic lakes (Schindler et al. 1986; Schindler 1986, 1988; Psenner 1988) than is supplied from the lakes' catchments. Many of these in-lake processes occur at the sediment-water interface, and in thermally stratified lakes, particularly in the hypolimnion where redox reactions prevail. However, some of these reactions, for example the dissimilatory reduction of Fe (III) to Fe (II), produce only "temporary" alkalinity, that is, the alkalinity is lost upon the breakdown of thermal stratification because Fe^{2+} is readily oxidized upon contact with the oxygen-rich upper water. Other processes are less easily reversed. For example, alkalinity generated by microbial denitrification (i.e., nitrate reduction to N_2O or N_2), is expected to generate a more "permanent" pool of alkalinity. Sulphate reduction generates permanent alkalinity if the reduced products are isolated from reoxidation by burial in the sediments, lost by outflow, or lost by gas exchange to the atmosphere after fall overturn.

In lakes on the Precambrian Shield, the hypolimnion is separated from the rest of the lake and its catchment during stratification, with diffusion and settling of particulates being the only inputs/outputs of substances. Thus, changes in the mass of a chemical species that occur in the hypolimnion throughout the period of stratification can be attributed either to net inputs/losses from diffusion and/or settling (which are relatively small), or to chemical reactions that occur in situ. Under the low oxygen or anoxic conditions frequently found in lakes' hypolimnia, exchange reactions and reduction processes prevail. Those reactions that produce or remove charged

Water, Air and Soil Pollution **99**: 373-380, 1997.

chemical species will affect alkalinity, as will some exchange reactions. Thus, those processes that can affect the acid-base status of the hypolimnion include: 1) exchange of base cations and H^+, 2) nitrate reduction (NO_3^- to NH_4^+), 3) denitrification (NO_3^- to N_2O to N_2), 4) manganese reduction, 5) iron reduction and 6) sulphate reduction. In addition, dissociated organic anions (A^-) contribute significantly to the charge balance in many dilute waters (LaZerte and Dillon 1984), especially those with substantial dissolved organic carbon (DOC). Therefore any process that results in the formation or loss of A^- also will affect alkalinity.

In this paper we examine the alkalinity generated in the hypolimnia of 2 lakes during summer stratification over periods of 7 and 4 years. The measured increases in hypolimnetic alkalinity observed during summer stratification are compared to those calculated from the sum of alkalinity-generating processes, and the relative importance of the various processes is evaluated.

2. Study Lakes

Chub Lake (lat. 45°13' long. 78°59') and Plastic Lake (lat. 45°11; long 78°50') are both situated in central Ontario in an area underlain by granitic bedrock of the Canadian Precambrian Shield. There are no measurable groundwater inputs to either lakes; measured hydrologic balances for 16 and 12 years for Chub and Plastic Lakes, respectively, demonstrate (Dillon, unpublished studies) that surface flow and precipitation account for all of the imputs of water. Chub Lake (A_o = 34.41 ha, z_{max} = 27 m, z_{mean} = 8.9 m) has a mean hypolimnetic volume of 8.69 x 10^5 m^3 and is usually anoxic below 19 m in the late summer; similarly, in Plastic Lake (A_o = 32.14 ha, z_{max} = 16.3 m, z_{mean} = 7.9 m) the hypolimnion (mean volume = 3.24 x 10^5 m^3) remains oxic until the latter part of the summer. The hypolimnia of both lakes begin at about 10 m depth and remain at that depth until just prior to fall overturn. The lakes and their catchments are described in detail in Dillon et al. (1987, 1994), Dillon and Molot (1990, 1996a,b) Dillon and Evans (1993), Dillon and LaZerte (1992) and Molot and Dillon (1993).

3. Methods

These studies were carried out at Chub Lake from 1986 - 1992, and in Plastic Lake from 1989 - 1992. Hypolimnetic samples were collected at least monthly during the period of thermal stratification at 2-m intervals (11-23 m in Chub Lake and 11-15 m in Plastic Lake) with a peristaltic pump. Concentrations of Ca^{2+}, Mg^{2+}, Na^+ and K^+ were summed (in equivalence units) and are called here total base cations (TBC). Total Fe, total Mn, NH_4^+, NO_3^-, SO_4^{2-}, DOC and the base cations were determined using standard analytical procedures (Ontario Ministry of the Environment 1983). Alkalinity was measured using the Gran titration procedure and [A^-] was calculated from DOC and pH according to Oliver et al. (1983). TBC concentrations in Plastic Lake were measured on only whole-hypolimnion volume-weighted samples rather than as profiles.

The measured alkalinity generated in the hypolimnia of the 2 lakes was determined from the increase in alkalinity during the period of summer stratification. The calculated alkalinity generated was determined as:

$$
\begin{aligned}
\text{Alkalinity generated} \quad = \quad & \text{Increase in [TBC]} + \text{Decrease in } [SO_4^{2-}] \\
+ \quad & \text{Increase in [Fe]} + \text{Increase in [Mn]} \\
+ \quad & \text{Decrease in } [NO_3^-] + \text{Decrease in } [A^-] \\
+ \quad & \text{Increase in } [NH_4^+] \qquad (1)
\end{aligned}
$$

where all terms in the equation are expressed in equivalence units.

Reduction of iron and manganese was estimated as the increase in their total dissolved concentrations in the hypolimnetic waters, since the reduced forms of these 2 metals are quite soluble whereas the oxidized forms are very insoluble (Stumm and Morgan 1981). Similarly, production of other base cations (NH_4^+ and TBC) was measured as an increase in their hypolimnetic concentrations during the summer. Alternately, reduction of NO_3^- and SO_4^{2-} was assumed to be equal to the total loss of these chemical species from the hypolimnetic water column during the period of summer stratification.

The changes in measured alkalinity, Fe, Mn, NH_4^+ and NO_3^- were determined as the difference in their hypolimnetic concentrations just prior to fall overturn versus the beginning of summer stratification. The resulting calculated changes in concentration of each parameter varied by 10-20% depending on the choice of initial and final dates. However, this effect is minimized by using the same 2 dates for each calculation in any year. Changes in A^-, TBC and SO_4^{2-} concentrations during summer stratification were very small relative to the mass of each in the hypolimnia, and, as their concentration changed (by visual inspection of plots), linearly over the course of the summer, the loss or gain of each was estimated from a linear regression of concentration versus time. The advantage of using the measured trends in these cases is that all available data are utilized, rather than only the initial and final concentrations, resulting in a more robust estimate. This method, of course, cannot be used if there is not a linear trend in concentration. These results were included only when the changes were significant. All volume-weighted hypolimnetic concentrations were converted to areal units by multiplying by the volume of the hypolimnion and then dividing by the hypolimnetic area (Chub Lake = 1.288×10^5 m^2; Plastic Lake = 1.123×10^5 m^2). Daily rates were calculated by dividing by the number of sampling days, which ranged between 84 and 168 depending on the lake and the year.

4. Results

In Table I the measured alkalinity generated (or lost) each year in the hypolimnia of the 2 study lakes is compared to the alkalinity generated as calculated from equation (1). In Chub Lake, the measured alkalinity generated ranged between 0.51 and 1.3 meq/m^2/d for 6 of the 7 study years. Unlike the other years, in 1992, there was a net loss of alkalinity (0.13 meq/m^2/d) in the hypolimnion during stratification. In Plastic Lake, the alkalinity generated during summer stratification was fairly consistent among the 4 study years ranging between 0.74 and 1.19 meq/m^2/d, with the lowest measured value (0.74) also occurring in 1992. This year was distinctive in that the summer was particularly cool, with temperatures being consistently below normal (Dillon et al. 1996). The onset of stratification in both lakes was delayed. In Chub Lake, there was oxygen at a depth of 22 m until mid-August. Therefore, it is not surprising that the alkalinity generating mechanisms dependent on redox potential were suppressed.

Table I

Alkalinity generated (or lost) in the hypolimnia of the 2 study lakes during each study year (meq/m²/d). Calculated alkalinity generation is based on equation (1).

	Chub Lake		Plastic Lake	
Year	**Measured**	**Calculated**	**Measured**	**Calculated**
1986	1.24	0.96		
1987	1.23	1.29		
1988	0.51	0.19		
1989	1.01	0.80	0.94	0.96
1990	1.30	1.41	0.90	1.02
1991	0.61	1.09	1.19	1.31
1992	-0.13	0.16	0.74	0.63
Mean	0.82	0.84	0.94	0.98

In almost all of the 11 cases, the calculated alkalinity generated (or lost) in the hypolimnia of the 2 lakes during summer stratification was consistent with the measured values (Table I). When the mean values for each lake are arranged for all years, the measured and calculated results are almost identical. However, in Chub Lake, the calculated value for 1991 (1.09 meq/m²/d) is almost twice that measured (0.61 meq/m²/d). Although we have no definitive explanation for this discrepancy, the difference between the calculated and measured values is due entirely to significant increases in TBC concentrations, specifically in Ca^{2+}, during the summer of 1991 (Table II). This is in contrast to all other years, when, because there was no significant change in TBC concentrations during the summer, TBC were not included in the alkalinity generation calculations for Chub Lake.

The relative importance of the various processes which contribute to summer alkalinity generation in the 2 study lakes is shown in Table II. Organic anions (A^-) are not included in this table because in Chub Lake there was no significant change in their concentration in any of the study years and in Plastic Lake, the alkalinity lost from increases in A^- during the summer represented <1% of the total calculated alkalinity generated. As mentioned earlier, an increase in TBC concentration was statistically significant in Chub Lake in only one year of 7; in that year (1991), TBC increased by 18 μeq/L or ca.9%. For the other 6 years, there were no significant changes in TBC concentrations in the hypolimnetic waters during summer stratification. Similarly in Plastic Lake, TBC changes contributed to alkalinity generation in only 1991 (Table II). Although it is possible that the alkalinity which was apparently generated from increases in TBC concentrations in 1991 is an artifact resulting from the methodological problem inherent in estimating small differences between 2 large numbers, the observation that this occurred in the same year in both lakes suggests otherwise.

Table II

The estimated rate (meq/m²/d) of the processes which contribute to hypolimnetic alkalinity generation in the 2 study lakes. The % contribution of each process to total hypolimnetic alkalinity production is given in brackets. (ns = not significant)

Year	Fe reduction	Mn reduction	NH_4^+ production	NO_3^- reduction	SO_4^{2-} reduction	TBC production
Chub Lake						
1986	0.78 (81)	0.03 (3)	0.13 (14)	0.02 (2)	ns	ns
1987	0.81 (63)	0.04 (3)	0.24 (19)	0.20 (15)	ns	ns
1988	0.55 (289)	0.03 (16)	-0.10 (-53)	-0.29 (-153)	ns	ns
1989	0.51 (64)	0.02 (2)	0.19 (24)	0.08 (10)	ns	ns
1990	0.63 (45)	0.03 (2)	0.21 (15)	0.14 (10)	0.40 (28)	ns
1991	0.57 (52)	0.03 (3)	0.05 (5)	-0.09 (-8)	ns	0.52 (48)
1992	0.23 (144)	0.01 (6)	-0.01 (-6)	-0.07 (-44)	ns	ns
Mean	0.58 (105)	0.03 (5)	0.10 (3)	0.00 (-24)	0.06 (4)	0.07 (7)
Plastic Lake						
1989	0.32 (33)	0.01 (1)	0.18 (19)	ns	0.45 (47)	ns
1990	0.65 (64)	0.02 (2)	0.01 (1)	0.06 (6)	0.28 (27)	ns
1991	0.35 (27)	0.04 (3)	0.04 (3)	0.03 (2)	0.24 (18)	0.60 (46)
1992	0.21 (33)	0.01 (2)	0.03 (5)	0.08 (13)	0.30 (48)	ns
Mean	0.38 (39)	0.02 (2)	0.07 (7)	0.04 (5)	0.32 (35)	0.15 (11)

From Table II it is evident that in both lakes, for all study years, reduction of Fe was the major contributor to hypolimnetic alkalinity generation. In Chub Lake, increases in Fe^{2+} during the summer ranged between 0.23 and 0.81 meq/m²/d and represent 45 to 289% of the net predicted alkalinity. Similarly, in Plastic Lake, alkalinity generated from Fe reduction ranged between 0.21 and 0.65 meq/m²/d representing 27 to 64% of the calculated alkalinity.

In contrast, the alkalinity produced from Mn reduction was of minor importance (generally <6%). Alkalinity generated or consumed as a result of the nitrogen cycle was variable both between the 2 lakes and also among the study years. In 1986, 1987, 1989 and 1990, increases in NH_4^+ in Chub Lake represented over 13% of the summer alkalinity generation, whereas in the other years, NH_4^+ production was of minor importance or even negative (i.e., NH_4^+ decreased in the hypolimnion). Similar results were observed for NO_3^-. In Plastic Lake, NH_4^+ production plus NO_3^- reduction together represented between 5 and 19% of the summer alkalinity production during the 4 study years. In both lakes, any increases in NH_4^+ in the hypolimnion during the summer appeared to be offset by corresponding decreases in NO_3^- and vice versa.

Reduction of SO_4^{2-} was not detectable (i.e., no significant decrease in SO_4^{2-} concentration during the summer) in Chub Lake for 6 of the 7 study years. However, in 1990, the sulphate reduction rate was 0.40 meq/m²/d representing 28% of the predicted alkalinity generated in the hypolimnion that summer. In Plastic Lake, SO_4^{2-} reduction (0.24 - 0.45 meq/m²/d) was an important source of alkalinity for all 4 study years, and represented 18 to 48% of the hypolimnetic alkalinity depending on the year. These results are in contrast to those of Kelly et al. (1987), who measured SO_4^{2-} reduction rates in the epilimnetic sediments of Plastic and Chub Lakes (in 1 month only, September 1982) as 0.091 and 0.29 meq/m^2/d, respectively.

5. Discussion

The measured and calculated rates of alkalinity generation in the hypolimnetic waters of these lakes, including the rates for those processes which contribute to alkalinity production/consumption (Table II) can be compared to rates which have been published in the literature for other lakes, although few data are available for dilute, oligotrophic lakes. In mesotrophic Mares Pond, USA, the predicted alkalinity (from the sum of the various processes) was 1.57 meq/m^2/d compared to a measured value of 1.38 meq/m^2/d (Kling et al. 1991). Similar to our results, Fe reduction represented about 38% (0.60 meq/m^2/d) of the total alkalinity generated whereas Mn reduction was negligible. Ammonium production contributed 13% (0.20 meq/m^2/d) to the total alkalinity. Kelly et al. (1982) reported rates of 0.60 - 1.87 and 0.04 - 0.13 meq/m^2/d for Fe and Mn reduction, respectively, in the hypolimnetic waters of 3 lakes in the Experimental Lakes Area (ELA) of Ontario, with NO_3^- reduction and NH^+ production being 0 - 0.43 and 0.65 - 1.67 meq/m^2/d, respectively. Similarly high NH_4^+ accumulation rates of 0.94 meq/m^2/d were measured also in the hypolimnion of mesotrophic Williams Bay (Jacks Lake, Ontario - Carignan and Lean 1991). With the exception of NH_4^+, all these rates are reasonably similar to those measured in the present study. The higher NH_4^+ production rates may be because Mares Pond and Williams Bay are mesotrophic and because 2 of the 3 ELA lakes had been fertilized (and eutrophied) previously with P and N (and for 1 lake, organic C). It has been suggested that increased productivity may increase internal alkalinity generation, possibly because of increased uptake of NO_3^- by algae (followed by burial), and/or increased denitrification in the sediments (Kelly et al. 1982). In contrast, Schafran and Driscoll (1987) reported that reduction of NO_3^- and production of NH_4^+ also contributed significantly (45 and 32%, respectively) to the production of hypolimnetic alkalinity in acidic Dart's Lake, NY, during the summer. In this lake the reduction of Fe and Mn was small (<0.01 meq/m^2/d). Dart's Lake differs from the other lakes discussed here in that its water replenishment rate is very rapid.

Many rates for SO_4^{2-} reduction that have been reported in literature are similar to those measured in Plastic Lake (0.24 - 0.45 meq/m^2/d) where SO_4^{2-} reduction is an important source of alkalinity (18 to 48% of the hypolimnetic alkalinity depending on the year). In Lake 302S, Kelly and Rudd (1984) measured SO_4^{2-} loss rates from incubated sediment cores of 0.2 - 0.72 meq/m^2/d depending on the temperature and initial SO_4^{2-}. Similarly, in Mares Pond, sulphate reduction rates were 0.59 meq/m^2/d (Kling et al. 1991). Kelly et al. (1982) found that SO_4^{2-} decreased by about 2 meq/m^2/d in the hypolimnetic waters of ELA lakes 226N, 227 and 223, the latter (Lake 223) of which had been previously acidified. In Lake 223, the reduction of SO_4^{2-} contributed as much as 60% to the total alkalinity produced in the hypolimnion during the summer (Cook et al. 1986). On the other hand, in acidic Dart's Lake, Schafran and Driscoll (1987) reported SO_4^{2-} reduction rates of only 0.18 meq/m^2/d. This represented 14% of the hypolimnetic alkalinity generated during the summer. Although there is some suggestion in the literature that SO_4^{2-} reduction may increase with higher levels of SO_4^{2-} (Giblin et al. 1990; Rudd et al. 1990), this is not firmly resolved (Urban 1994).

The non-detectable rates of SO_4^{2-} reduction in Chub Lake in most years may be a reflection of the relatively large volume of the hypolimnion, V_{hypo} relative to the total lake volume (V_{hypo}/V = 0.286); as a result, SO_4^{2-} reduction leads to very small changes in SO_4^{2-} concentration in the hypolimnion. On the other hand, in Plastic Lake the relative size of the hypolimnion is much smaller (V_{hypo}/V = 0.128) but because the hypolimnetic area is approximately the same proportion of the total lake area in both lakes (A_{hypo}/A = 0.37 and 0.35 in Chub and Plastic Lakes, respectively), the concentration change is more readily detected. This assumes that a significant portion of the SO_4^{2-} reduction takes place in the sediments or at the sediment-water interface. The morphometry of Plastic Lake appears to be more typical of other lakes where morphometric data have been published, and where hypolimnetic SO_4^{2-} reduction has been found to be significant. For example, in Mares Pond only 8% of lake volume and 23% of the lake area are in the hypolimnion. In Lake 223, V_{hypo}/A_{hypo} = 2.7 m, compared to Plastic Lake where V_{hypo}/A_{hypo} = 2.9 m. Thus, the large V_{hypo}/A_{hypo} of 6.75 m in Chub Lake results in a "dilution" in SO_4^{2-} changes in the hypolimnion which are not detectable by the mass balance approach.

6. Summary

The measured increases in the alkalinity in the hypolimnia of Chub and Plastic Lakes result largely from redox processes. Although there was substantial variability between years in the relative importance of these processes, Fe reduction (inferred from solublization) dominated in Chub Lake, while Fe reduction and SO_4^{2-} reduction predominated in Plastic Lake. Alkalinity generated by Fe reduction is likely lost when the lakes' thermal stratification breaks down in the fall and oxygenated waters mix with the hypolimnia.

Acknowledgements

The authors thank Ron Reid for assistance with the field programme and Marilyn Velenosi for typing the manuscript.

References

Carignan, R. and Lean, D.R.S.: 1991, *Limnol. Oceanogr.* **36**, 683-707.

Cook, R.B., Kelly, C.W., Schindler, D.W. and Turner, M.A.: 1986, *Limnol. Oceanogr.* **31**, 134-148.

Dillon, P.J. and Evans, H.E.: 1993, *Wat. Res.* **27**, 659-668.

Dillon, P.J. and LaZerte, B.D.: 1992, *Envir. Pollut.* **77**, 211-217.

Dillon, P.J. and Molot, L.A.: 1990, *Biogeochem.* **11**, 23-43.

Dillon, P.J. and Molot, L.A.: 1996a, *Biogeochem.* (in press).

Dillon, P.J. and Molot, L.A.: 1996b, *Wat. Res.* **30**, 2273-2280.

Dillon, P.J., Molot, L.A. and Futter, M.: 1996, *Int. J. Environ. Mon. Assess.* (in press).

Dillon, P.J., Reid, R.A. and de Grosbois, E.: 1987, *Nature* **329**, 45-48.

Dillon, P.J., Scheider, W.A., Reid, R.A. and Jeffries, D.S.: 1994, *Lake Reserv. Manage.* **8**, 121-129.

Giblin, A.E., Likens, G.E., White, D. and Howarth, R.W.: 1990, *Limnol. Oceanogr.* **35**, 852-869.

Kelly, C.A. and Rudd, J.W.M.: 1984, *Biogeochem.* **1**, 63-77.

Kelly, C.A., Rudd, J.W.M., Cook, R.B. and Schindler, D.W.: 1982, *Limnol. Oceanogr.* **27**, 868-882.

Kelly, C.A., Rudd, J.W.M., Hesslein, R.S, Schindler, D.W., Dillon, P.J., Driscoll, C.T., Gherini, S.A. and Hecky, R.E.: 1987, *Biogeochem.* **3**, 129-140.

Kling, G.W., Giblin, A.E., Fry, B. and Peterson, B.J.: 1991, *Limnol. Oceanogr.* **36**, 106-122.

LaZerte, B.D. and Dillon, P.J.: 1984, *Can. J. Fish. Aquat. Sci.* **41**, 1664-1677.

Molot, L.A. and Dillon, P.J.: 1993, *Biogeochem.* **20**, 195-212.

Oliver, B.G., Thurman, E.M. and Malcolm, R.L.: 1983, *Geochim. Cosmochim. Acta.* **47**, 2031-2035.

Ontario Ministry of the Environment.: 1983, *Handbook of Analytical Methods for Environmental Samples.*

Psenner, R.: 1988, *Limnol. Oceanogr.* **33**, 1463-1475.

Rudd, J.W.M., Kelly, C.W., Schindler, D.W. and Turner, M.A.: 1990, *Limnol. Oceanogr.* **35**, 663-679.

Schafran, G.C. and Driscoll, C.T.: 1987, *Environ. Sci. Technol.* **21**, 988-993.

Schindler, D.W.: 1986, *Wat. Air and Soil Pollut.* **30**, 931-944.

Schindler, D.W.: 1988, *Limnol. Oceanogr.* **33**, 1637-1640.

Schindler, D.W., Turner, M.A., Stainton, M.P. and Linsey, G.A.: 1986, *Science* **232**, 844-847.

Stumm, W. and Morgan, J.J.: 1981, *John Wiley & Sons, Inc., New York.* 780 p.

Urban, N.R.: 1994, *American Chemical Society, Washington.* 627 p.

GEOCHEMISTRY OF HCO_3^- AT THE SEDIMENT-WATER INTERFACE OF LAKES FROM THE SOUTHWESTERN CHINESE PLATEAU

F.C. WU[1], H.R. QING[2], G.J. WAN[1], D.G. TANG[1], R.G. HUANG[1] and Y.R. CAI[1]

[1] *State Key lab. of Environ. Geochem., Inst. of Geochem., Academia Sinica, P.R. China;*

[2] *Dept. of Geol., Royal Holloway, the University of London, UK*

Abstract. Sediments were cored, and the sediment-water interface and overlying waters were sampled in 5 lakes from the southwestern Chinese plateau during 1991-95. The geochemistry of HCO_3^- at the sediment-water interface was examined by studying detailed profiles of pH, HCO_3^- concentrations and $\delta^{13}C$ of dissolved inorganic carbon (DIC) in overlying lake water and porewater near the sediment-water interface. Dissolution-precipitation equilibrium of carbonates, diffusion flux, and the extent of the influence of diffusion on the whole lake were calculated. The results show that the HCO_3^- near the interface carried isotopic characteristics of decomposition of organic matter during early diagenesis, and that the porewater in surface sediments was unsaturated relative to calcite, and gradually saturated with depth. Furthermore, the interface is a source of HCO_3^- to the overlying water. Alkalinity (Alk) diffusion flux from sediments to the overlying water due to concentration gradients ranged from 0.51 to 24.33 x 10^{-4} mol cm^{-2} a^{-1}. The calculated contribution of the diffusion of Alk to the overlying water ranged from 0.46% to 49.42%. Diffusion is an important source of Alk in lakes with a long residence time and a relatively shallow depth.

Keywords: diffusion, alkalinity, sediment-water interface, southwestern Chinese plateau, carbonates, carbon isotopes

1. Introduction

HCO_3^- is a main component of alkalinity (Acid Neutralizing Capacity) in aquatic environments. With the seriousness of environmental problems, particularly acidification of lakes, many models have been built to predict evolution of acidity-Alk equilibrium (Dillon et al., 1984; Kelly et al., 1987). In most studies, concerns were focused on the influence of Alk from in-lake processes, e.g. the Alk from sediment-water interactions and sulfate reduction during early diagenesis (Carignan, 1985; Schindler et al., 1986; Kelly et al., 1987). Alk diffusion flux from sediments in various lakes also has been calculated (Carignan, 1985; Wu and Wan, 1996a,b). The geochemistry of HCO_3^- near the interface, however, needs to be further investigated, and the impact of the upward diffusion of dissolved substances to the overlying water needs to be quantified.

Carbon cycling is an important topic in the geochemistry of lake systems, particularly with increasing atmospheric CO_2 concentrations, which are believed to be responsible for the changes in global temperature and climate (Kelts and Hus, 1978; Oechel and Hastings, 1993). Therefore it is significant to investigate the response of natural geochemical processes at different interfaces to CO_2 increases, and their contribution to atmospheric CO_2.

The southwestern Chinese plateau is a large karst area, where the carbonate terrain is suffering severe weathering and erosion. Lake sediments record these processes and possibly influence the chemical composition of overlying water due to in-lake process and early diagensis. This study aims to evaluate possible impacts of interface diffusion on the acidity-Alk balance of overlying water, to characterize carbon isotopes and dissolution - precipitation equilibrium, and to investigate the geochemistry of HCO_3^- near the sediment-water interface.

2. Sampling and analysis

The selected lakes are located at the boundary between southeastern monsoon and southwestern monsoon on the eastern slope of the Qinghai-Xizang (or Tibet) Plateau of China. The physical and chemical characteristics of these lakes are presented in Table I. Lake Lugu (LG) (27°41′-25′N, 110°45′-50′E) and Lake Erhai (EH) (25°35′-58′N, 100°05′-17′E) are two deep lakes in Yunnan Province. Lake Aha (AH) and Lake Baihua (BH) are located in the suburb of Guiyang city, and are the main drinking-water reservoirs for Guiyang. Lake Hongfeng (HF) (26°25′-34′N, 106°20′-28′E) is the largest artificial reservoir in Guizhou province.

TABLE I

Physical and chemical features of 5 lakes from the southwestern Chinese plateau

	drainage area km^2	lake area km^2	elevation m	average depth m	water residence time(a)
Lake Lugu	171.4	50.5	2685	40.4	18.50
Lake Erhai	2470.0	249.8	1974	10.2	2.80
Lake Hongfeng	155.1	57.2	1240	9.3	0.26
Lake Aha	190.0	3.4	1108	13.2	0.44
Lake Baihua	183.2	14.5	1100	13.0	0.16

	Ca^{2+}	Mg^{2+}	K^+	Na^+ m mol L^{-1}	HCO_3^-	Cl^-	SO_4^{2-}	NO_3^-
Lake Lugu	0.81	0.28	0.014	0.39	2.22	0.42	0.005	0.012
Lake Erhai	0.90	0.54	0.049	0.30	2.58	0.36	0.15	0.019
Lake Hongfeng	1.23	0.49	0.041	0.11	2.65	0.49	0.21	0.032
Lake Aha	3.45	0.88	0.073	0.31	2.31	0.41	3.16	0.120
Lake Baihua	*	*	*	*	2.26	*	0.98	0.087

* = not determined

Samples of lake water were collected at about 5 m depth intervals. Sediments were cored at the center of each lake using a sediment-water interface sampler designed by ourselves (Yuang et al., 1993). The cores and porewater were not disturbed during the field coring process. Sediment cores were sliced *in-situ* into 0.5 to 2.0 cm sections, and porewater was extracted through 0.45 μm membrane filters following high-speed centrifugation. pH values were measured *in-situ* , SO_4^{2-} and NO_3^- ions were determined by chromatographic analysis, HCO_3^- by volumetric analysis, and Ca^{2+}, Mg^{2+}, Na^+, K^+, and dissolved Mn and Fe by AAS after *in-situ* acidification with superpure HNO_3. The $\delta^{13}C$ values of DIC were measured using mass spectroscopy.

3. Results and Discussion

3.1. DIFFUSION FLUX of ALK FROM POREWATER TO OVERLYING WATER

Profiles of pH and HCO_3^- concentrations in porewater and overlying waters of these lakes, shown in Figure 1, clearly demonstrate that HCO_3^- concentrations and pH rarely change in the overlying water. However, a major increase in HCO_3^- and a decrease in pH occurred at the interface in most profiles except for two. The HCO_3^- concentrations in the porewater were much higher than in the overlying water and increased gradually with depth. Other ions also showed obvious changes at the sediment-water interface. The SO_4^{2-} and NO_3^- concentrations of overlying water were higher than those of porewater, while the concentrations of dissolved Mn and Fe, Ca^{2+}, Mg^{2+}, K^+, Na^+ in porewater were greater than in overlying water (Wu,1995). Soluble substances could cross the sediment-water interface by molecular diffusion due to the concentration gradient (Glud et al.,1994; Wu and Wan,1996a,b). Therefore, SO_4^{2-} and NO_3^- may diffuse downward from overlying water into porewater, while dissolved Mn and Fe, Ca^{2+}, Mg^{2+}, K^+, Na^+ may diffuse upward, thus influencing the Alk near the interface.

Alk diffusion flux (J in mol cm^{-2} a^{-1}) can be calculated using the following equation if the coupling effect is ignored:

$$J = - \Phi Ds\, dc/dx \tag{1}$$

where Φ is the sediment porosity and equals 0.95 (Wu, 1995), dc/dx is the vertical concentration gradient at the interface (mol cm^{-4}), Ds is the sediment diffusion coefficient for sediments (which is a function of tortuosity,θ, and the molecular diffusion coefficient,D), and D (cm^2 s^{-1}) can be found in Li and Gregory (1974). In the following calculations we assume θ=1. Concentration gradients and the resulting fluxes of HCO_3^-, dissolved Fe, and Mn at the interface of these lakes were then calculated using equation 1 (Table II).

According to the balance of positive-negative ions and the diffusion flux near the interface, the contribution of various ions to Alk can be expressed as (Stumm and Morgan, 1981; Berner, 1980):

$$\begin{aligned} \text{Alk} &= 2mCa^{2+} + 2mMg^{2+} + mK^+ + mNa^+ - (2mSO_4^{2-} + mNO_3^- + mCl^-) \\ &= mHCO_3^- + mOH^- + 2mCO_3^{2-} - mH^+ \end{aligned} \tag{2}$$

Therefore, the flux of Alk across the interface can be approximated as:

$$J_{\text{Alk}} = J_{HCO_3^-} + J_{OH^-} + 2\,J_{CO_3^{2-}} - J_{H^+} \tag{3}$$

As OH^- and CO_3^{2-} concentrations were very low compared to that of HCO_3^-, they could be omitted from equation 3. In addition, oxidation of upwardly diffused reduced Fe and Mn can produce H^+ and reduce the flux. In this paper, we didn't differentiate between concentrations of different valencies of dissolved Fe and Mn in porewater. However, previous studies from Lake Hongfeng and other lakes from the southwestern Chinese plateau showed that concentrations of oxidized Fe and Mn in porewater were very low compared to those of reduced Fe and Mn (Chen, 1990; Wu, 1995). Thus we used diffusion fluxes of dissolved Fe and Mn to calculate the maximum influence of reduced Fe and Mn on Alk diffusion during early diagensis. Therefore, from equation 4, further estimates of their influence on Alk in overlying water can be derived:

$$J^0_{Alk} = J_{HCO3-} - 2\, J_{dissolved\ Fe} - 2\, J_{dissolved\ Mn} \qquad (4)$$

The flux of Alk calculated according to equation 4 (see Table II) ranged from 0.51 to 24.33 x 10^{-4} mol cm^{-2} a^{-1}.

3.2. CONTRIBUTION OF UPWARD DIFFUSION OF ALK TO THE OVERLYING WATER

If the amount(E) of element (i) accumulated in the overlying water via diffusion is: $Ei = Je \cdot Tw$ then the concentration (C_{in} in m mol L^{-1}) due to diffusion to the whole overlying water would be:

$$C_{in} = Ei / h = Je \cdot T_w / h \qquad (5)$$

The ratio between the concentration due to the diffusion (C_{in}) and total concentration (C m mol L^{-1}) in the overlying water is used here to indicate the contribution via diffusion for this element to the overlying water (Wu, 1995; Wu and Wan, 1996a):

$$C_{in} / C = Je \cdot T_w / h \cdot C = T_w / T \qquad (6)$$

where J_e is the diffusion flux of alkalinity (mol cm^{-2} a^{-1}), h is the effective water depth (cm), T_w is water residence time (a) and $h \cdot C/J_e$ can be defined as the residence time of the element (T_e).

Assuming that the waters were homogeneously mixed, the T_w /T_e ratio for HCO_3^- should represent the contribution of Alk diffusion to the overlying water which is associated mainly with diffusion flux, water residence time and effective depth. Based on J_e, T_w, h and C, the contribution of upward diffusion to the overlying water was calculated (see Table II). It can be seen from Table II that the contribution of upward diffusion of Alk varies in different lakes. In Lake Lugu, about one-third of Alk of waters came from diffusion, showing that interface diffusion played an important role in controlling acid-base equilibrium of lake water. In Lake Aha, the contribution of interface Alk diffusion to lake water is very small, only 0.46%. About one-tenth of Alk came from diffusion in the other three lakes.

TABLE II

Vertical Concentration gradients of dissolved Fe, Mn and HCO_3^-, their resulting diffusion fluxes, and the ratio of T_w to T_e at sediment-water interface of lakes from southwestern Chinese plateau

	LG9406-5	LG9405-1	LG9101	EH9402-2	EH9102	EH9101	AH9201	BH9402-3	BH9402-1	HF9101	HF9102
water depth (cm)	6000	5000	6500	1200	1600	1600	1900	1500	1500	2000	2000
lake alkalinity (m mol L^{-1})	2.52	2.49	2.27	2.39	2.54	2.54	2.05	2.31	2.31	2.74	2.74
residence time (a)	18.5	18.5	18.5	2.8	2.8	2.8	0.28	0.44	0.44	0.26	0.26
HCO_3^- gradient (10^{-6} mol cm^{-4})	1.01	0.56	0.96	0.15	0.54	0.57	0.23	3.15	1.55	6.89	4.23
HCO_3^- diffusion flux (10^{-4} mol cm^{-2} a^{-1})	3.54	1.96	3.36	0.53	1.04	2.00	0.81	11.03	5.43	24.33	14.83
dissol. Fe diffusion flux (10^{-7} mol cm^{-2} a^{-1})	1.45	8.31	2.70	5.07	14.02	8.47	21.05	22.77	1.78	0.195	0.195
dissol. Mn diffusion flux (10^{-7} mol cm^{-2} a^{-1})	15.34	103.10	38.68	6.39	13.89	2.81	65.43	296.50	206.5	0.089	0.08
J^o_{alk} (10^{-4} mol cm^{-2} a^{-1})	3.51	1.74	3.28	0.51	0.98	1.97	0.64	10.39	5.01	24.33	14.83
T_w/T_e	42.9	25.8	41.1	4.9	6.7	13.6	0.46	13.2	6.4	11.5	7.0

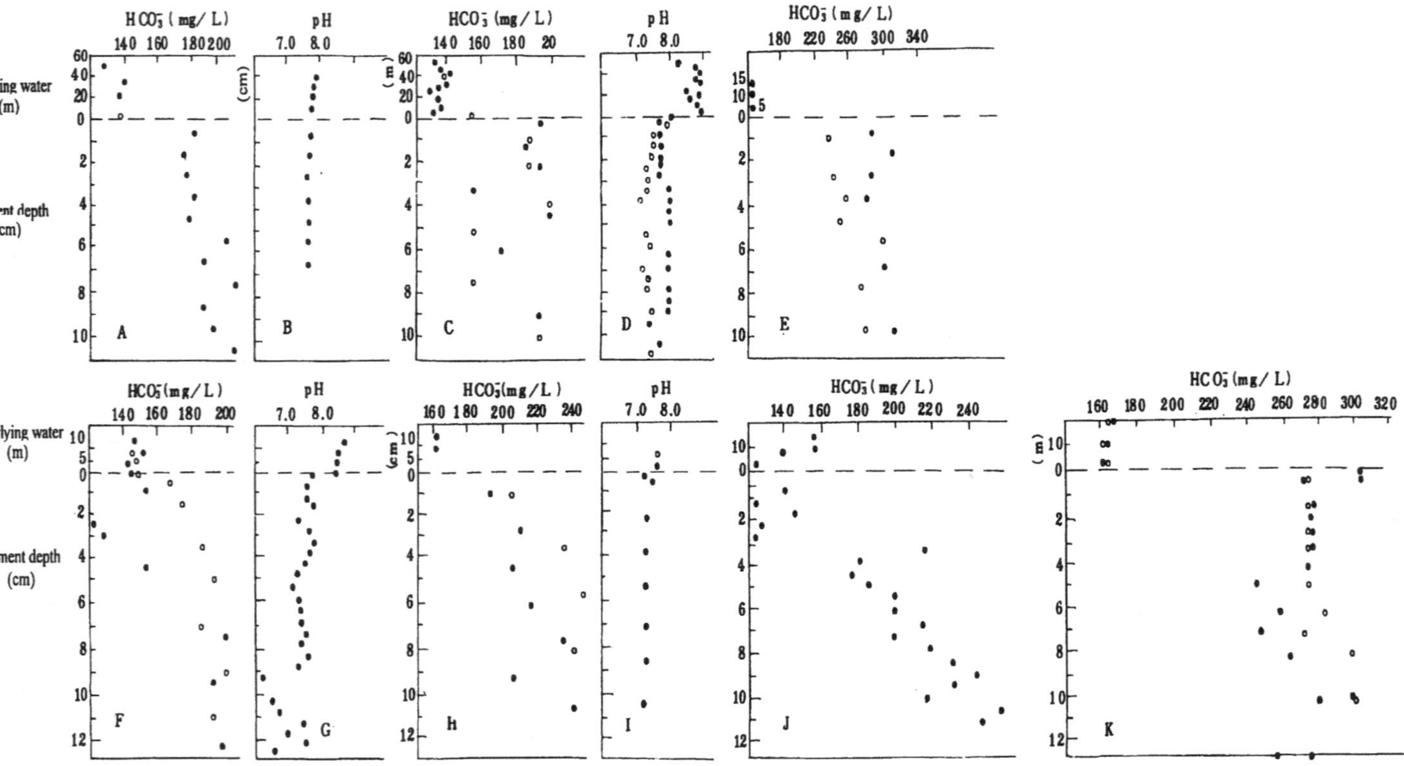

Figure 1 Profiles of pH and HCO_3^- concentrations in porewater and overlying waters of 5 lakes from southwestern Chinese plateau.

[A and B: LG9101; C and D: LG9405-1(●), LG9406-5(○); E: BH9402-3 (●), BH9402-1(○); F and G: EH9402-1(○), EH9401-2(●); H and I: EH9102(○), EH9101(●); J: AH9201; K: HF9101(○), HF9102(●)]

3.3. CARBONATE EQUILIBRIUM AT THE SEDIMENT - WATER INTERFACE

Based on the profiles of Ca^{2+} (Wu, 1995) and HCO_3^- concentrations in porewater and overlying water (Fig.1), we can characterize the solution geochemistry in a carbonate system using the saturated index (*SI*):

$$SI = IAP / K_{sp} \quad (7)$$

where K_{sp} is the solubility product of calcite and IAP is the ion activity product.

The dissolution and precipitation of calcites in sediments can be expressed by the following equations:

$$Ca^{2+} + CO_3^{2-} = CaCO3 \quad (8)$$
$$H_2CO_3 = H^+ + HCO_3^- \quad (9)$$
$$HCO_3^- = H^+ + CO_3^{2-} \quad (10)$$

Therefore, $IAP = aCa^{2+} \cdot aCO_3^{2-} = rCa^{2+} \cdot mCa^{2+} \cdot rHCO_3^- \cdot mHCO_3^- \cdot K_2 / m_{H+}$ (11)

$\therefore \log SI = \log mCa^{2+} + \log mHCO_3^- + \log rCa^{2+} + \log rHCO_3- + pH + P_{ksp} - P_{k2}$ (12)

Where rCa^{2+} and $rHCO_3^-$ are activity coefficients, $P_{ksp} = 8.37$, P_{k2} is the second-order dissolution constant of H_2CO_3 (equation 10) and equals 10.43 for a temperature of 15°C.

Based on mCa^{2+}, $mHCO_3^-$ and pH values, the profiles of *SI* in porewater and overlying waters of 5 lakes were constructed (see Figure 2) according to equation 12. From Figure 2, it can be seen that these lake waters were always saturated with respect to calcite, suggesting endogenic calcite precipitation from the overlying water. Near the sediment-water interface, the porewater was unsaturated relative to calcite, because decomposition of organic matter causes calcite dissolution. With increased burial, the porewater became anaerobic, the CO_2 from decomposition of organic matter in these fine sediments was greatly reduced and calcite precipitation resumed. From Figure 2, it can be seen that the porewater became saturated with increasing burial.

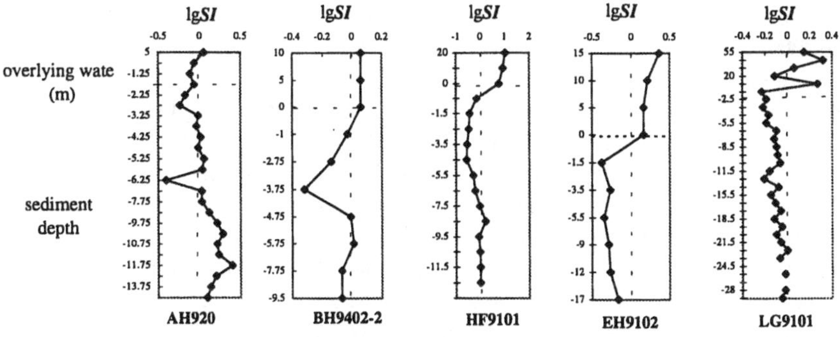

Figure 2 Profiles of saturated index (*SI*) of calcite in sediment porewater and overlying water.

3.4. CARBON ISOTOPES OF DIC DURING EARLY DIAGENSIS

Table III shows $\delta^{13}C$ characteristics of DIC in the porewater and overlying water in Lake Hongfeng. The $\delta^{13}C$ values of DIC in Lake Hongfeng range from -8.1 ‰ to -10.40 ‰. The DIC of surface waters has higher $\delta^{13}C$ values, whereas the deeper waters have lower $\delta^{13}C$ values. Because the influence of water atmospheric exchange on the carbon isotopic composition of DIC in lake water was very small (Wan et al., 1996), the lower $\delta^{13}C$ values at the sediment-water interface were caused probably by decomposition of organic matter, which releases ^{12}C-enriched carbon at the interface (Mckenzie, 1985; Wan et al., 1996).

The $\delta^{13}C$ values of DIC of sediment porewater are lower than that of overlying water (Table III), indicating the input of ^{12}C-enriched CO_2 from decomposition of organic matter. However, the $\delta^{13}C$ values of DIC in porewater increased with sediment depth. This indicates that decomposition of organic matter was greatly reduced with increasing burial due to gradual depletion of oxygen.

TABLE III

Carbon Isotopic Composition of DIC in porewater and overlying water of Lake Hongfeng

Water depth(m)	DIC, $\delta^{13}C$ (‰, PDB)
0	-8.10
10	-10.40
20	-10.10
Sediment depth(cm)	**DIC of porewater, $\delta^{13}C$ (‰, PDB)**
5	-12.10
8	-9.80
10	-8.70
20	-5.90

4. Conclusions

(1) The lower $\delta^{13}C$ values of DIC near the sediment-water interface characterize decomposition of organic matter during early diagenesis. The porewater in surface sediments was unsaturated relative to calcite, whereas authigenic carbonate precipitated from the porewater in deeper sediment.

(2) The sediment-water interface was a source of HCO_3^- to overlying water. The flux near the interface was calculated to be 0.51 to 24.33 x 10^{-4} mol cm^{-2} a^{-1}. The influence of diffusion on the whole lake depended chiefly on the flux, water depth and residence time of the lakes. This suggests that models on response of acid precipitation to acid-base equilibrium of lakes should take various Alk sources into account.

(3) Decomposition of organic matter is an important factor controlling geochemistry of HCO_3^- near the sediment-water interface, during which various acceptors (O_2, NO_3^-, Fe and Mn oxidation, SO_4^{2-}) were reduced, and CO_2 was released into the porewater. This process results in the increase in Alk and DIC, thus changing the saturated index of carbonate, and decreasing the carbon isotopic values.

(4) Carbonates in sediments contain much information about climate and environmental change. C and O isotopes were used to indicate past environmental change in oceans and lakes (Hodell et al., 1991; Talbot and Johannessen, 1992; Kelts and Hus, 1978; Wan et al., 1996). However, these indicators could be influenced by many geochemical processes (Chen, 1994; Talbot et al, 1990; Wu, 1995). Only when we understand these processes (e.g. early diagenesis), can we distinguish and extract environmental information from waters and sediments.

Acknowledgments

This study was supported by the Natural Science Foundation of China and the Chinese Academy of Sciences. Special thanks to Dr. Zhenlou Chen, Chaoyang Wei for their collaboration and discussions. The comments by two anonymous reviewers are also greatly appreciated.

References

Berelson, A.:1990, Geoochim.et.Cosmochim. Acta, 54, 3013-3020.

Berner, R. A.: 1980, Early Diagenesis: A Theoretical Approach. Princeton Univ. Princeton New Jersey Press.

Carignan, R.: 1985, Nature , 317(12), 158-160.

Chen, Z. L.: 1990 : M. S. thesis, Inst. of Geochem., Academia Sinica.

Dillon,P.J. and W.A.Scheider, In Modeling of Total Acid Precipitation Impacts(ed.Schoor L.), 121-154(Butterworth,london,1984)

Glud, R.N.,J.K.Gundersen,N.P.Revsech and B.B.Jorgensen: 1994, Limnol. Oceanogr.,39(2)462-467.

Hodell, D. A., J.H.Curtis, G.A.Jones,Higuera-Gundy,M.Brenner,M.W.Binford and K.T.Dorsey: 1991 Nature,352(6338), 790-793.

Kelts, K. and K. J. Hus : 1978, Lake: Chem.,Geo.and Phy.,Lerman(ed),Springer,Verlag, N.Y., 295-323.

Kelly, C. A. and others: 1987, Biogeochem.,3, 129-140.

Li, Y. H. and S. Gregory : 1974, Geochim.et.Cosmochim. Acta, 38, 703-714.

Mckenzie, J. A.: 1985, in W. Stumm (ed), Chemical Processes in Lake,Wiley, N.Y., p99-118.

Oechel, W. C., S. J. Hastings, G.Vourlitis, M.Jenkins,G.Riechers and N.Grulke: 1993, Nature, 361(11), 520-523.

Rudd, J. W. M. : 1990, Limnol. Ocearogr, 35(3), 663-679.

Schinder, D. W., A.Turner,M.P.Stainon, and G.A. Linsey: 1986, Science., 232, 844-847.

Stumm, W. and J. Morgan: 1981, Aquatic Chemistry, An Introduction Emphasizing Chemical Equlibria in Natural Waters(2nd), John Wiley.

Talbot, J. V.: 1990, Chem. Geo., 80, 261-279.

Talbot, M. R. and T. Johannessen: 1992, Earth Planet. Sci. Lett., 110(1), 23-37.

Tang, D. G.,G.J. Wan, and R.G.Huang: 1993, Progress in Physics,Chemistry and Mechanism, 3, 60-75.

Wu, F. C.: 1995, Ph.D. Dissertation, Inst. of Geochem., Academia Sinica.

Wu, F. C. and G. J. Wan :1996a, Chinese Journal. of Geochem.,15(3), 285-288.

Wu, F. C. and G. J. Wan: 1996b, Advance in Earth Sci., 16(2), 191-197 (in Chinese)

Wan, G. J., D. G. Tan, and F.C. Wu: 1996, Geo. Geochem., 2, 1-3 (in Chinese).

Yuang, Z. Q., D. S. Wuand R. G. Huang: 1993, Environ. Sci., 14(1) ,70-73 (in Chinese)

INTERACTIONS BETWEEN BENTHIC PHOSPHORUS RELEASE AND SULFUR CYCLING IN LAKE SCHARMÜTZELSEE (GERMANY)

ANDREAS KLEEBERG

Brandenburg Technical University of Cottbus, Faculty of Environmental Sciences,
Department of Water Protection, Seestraße 45, D-15526 Bad Saarow, Germany

Abstract. Sulfur (S) conversions were determined during summer stratification in 1995/96 to assess the extent to which benthic release of phosphorus (P) is influenced by the S cycling in eutrophic, dimictic, sulfate-rich (61.33 ± 10.41 mg SO_4^{2-} l^{-1}) freshwater Lake Scharmützelsee. Hypolimnetic SO_4^{2-} reduction (4.56 ± 0.73 g (S) m^{-2} d^{-1}) forming ΣH_2S (44.71 ± 17.57 mg ΣH_2S m^{-2} d^{-1}), leading to iron sulfide precipitation (5.62 ± 1.72 mg FeS m^{-2} d^{-1}) and dissolved iron depletion in the hypolimnion has a major influence on benthic P mobilization and release. The most important inorganic S pool is the CRS ($FeS_2 + S° + H_2S$; 15.1 % total S), being 1.3 to 6.6 times higher than the AVS ($FeS + H_2S$) in the uppermost 0 - 8 cm sediment. This diminishes the ability of the sediment to bind P (indicated by 14.6 % loosely bound P (NH_4Cl-P) and an exhaustion of the redox-sensitive P (BD-P)), leading to interstitial water P concentrations up to 10.8 mg l^{-1} and P release rates of 2.64 ± 0.56 mg P m^{-2} d^{-1}. As a consequence the P content of the lake increased fourfold within 58 days.

Key words: sulfate reduction, phosphorus release, sulfur speciation, phosphorus fractionation

1. Introduction

Phosphorus (P) has long been recognized as the most critical nutrient limiting phytoplankton growth in lakes (e.g., Golterman, 1984). The benthic release and pelagic cycling of P is in part regulated by the availability of and complexation with oxidized iron (e.g., Sulzberger et al., 1989). During thermal stratification in many lakes, hypolimnia become anoxic, and the reductive dissolution of ferric iron (both biotically and abiotically) releases ferrous iron (Murray, 1995) and sulfides from dissimilatory sulfate (SO_4^{2-}) reduction into the water column. The formation and precipitation of insoluble complexes of sulfide with ferrous iron can disrupt the iron-phosphate cycle which results in a lack of P binding partners and an excess of mobile P species. This in turn, can lead to secondary eutrophication processes (Caraco et al., 1993). The capacity of sediments to form iron sulfide depends on the supply of reducible iron (Cook & Kelly, 1992).
Based upon investigations of hypolimnetic and pore-water chemistry during summer stratification as well as on the distribution of the binding forms of sulfur (S) and P in recent sediments, the aim of this paper is to assess the extent to which benthic P release is influenced by S cycling in the 29.5 m deep, sulfate-rich Lake Scharmützelsee, where S and P cycling have not been studied previously.

2. Material and Methods

2.1. Study site

Dimictic, eutrophic soft water Lake Scharmützelsee is situated in eastern Germany near Berlin. Its morphometric and limnological characteristics are given in Table 1. Summer stratification usually lasts from April/May to October, and bottom waters become anoxic from May/June through October.

Water, Air and Soil Pollution **99**: 391-399, 1997.

Table 1. Morphometrical and limnological characteristics (epilimnion, 1995) of Lake Scharmützelsee, Germany (52° 15' N, 14° 03' E).

Morph. parameter (unit)		Limnol. parameter (unit)	min. ... max.
Surface area (km^2)	12.09	Total phosphorus ($\mu g\ l^{-1}$)	33.0 ... 58.0
Volume ($10^6\ m^3$)	108.23	Secchi depth (m)	0.7 ... 4.2
Mean depth (m)	8.8	Seston ($g\ dw\ l^{-1}$)	1.8 ... 9.1
Maximum depth (m)	29.5	Chlorophyll a ($\mu g\ l^{-1}$)	4.6 ... 46.9
Mean residence time (a)	16	Sulfate ($mg\ l^{-1}$)	29.5 ... 62.0
Drainage area (km^2)	112	Total iron ($mg\ l^{-1}$)	0.024 ... 0.028

2.2. Water sampling and analysis

Biweekly vertical profiles of water temperature, pH and dissolved oxygen (O_2) were taken from 07/20/1995 to 09/14/1995 (58 d) using a multiprobe (H20®, HYDROLAB) at the deepest location. Total sulfide concentration ($\sum H_2S$) was determined spectrophotometrically (methylene blue, 670 nm) (Rohde & Nehring, 1979). The concentration of the non dissociated, free hydrogen sulfide (H_2S), which is dependent on pH and water temperature, was calculated as:

$$H_2S\ [mg\ l^{-1}] = ((E_{670nm} \cdot 4.31) \cdot 10^{-pH}) / (1.14 \cdot 10^{-6} \cdot e^{(-0.10329 * \text{water temperature})} + 10^{-pH}). \quad (1)$$

After filtration (0.45 µm) the concentrations of ammonium (NH_4^+) and soluble reactive P (SRP) were measured spectrophotometrically according to standard methods (DIN 38406-E5-1) and (DIN 38405-D11), respectively. SO_4^{2-} concentration was determined by ion chromatography (DX-100, DIONEX). SO_4^{2-} reduction rates (SRR) were estimated from the net decrease in SO_4^{2-} in the anoxic portion of the hypolimnion (9 - 29.5 m) between spring and fall overturn (Cook & Schindler, 1983).

2.3. Sediment sampling and analysis

Intact sediment cores up to 25 cm long were taken in August 1996 with an UWITEC©-corer (flutter valve, acrylic glass tubes, i.d. = 6.0 cm) at 29.5 m depth. Sediment cores were kept no longer than 1 h at 4 °C until analysis. From each core the overlying lake water was siphoned off. After slicing the core (0 - 0.5, 0.5 - 1.0, 1 - 2, 2 - 4, ... 18 - 20 cm) aliquots were used for the determination of dry weight (d.w.; 48 h, 60°C), organic matter as loss on ignition (LOI; 3 h, 550 °C) and total inorganic carbon (TIC; 3 h, 900 °C). Another aliqout was homogenized and analysed for total S (TS; 1350 °C) and inorganic S at 520 °C (Holmer et al., 1994) (SC-432 model, LECO®) and total P (TP) after digestion ($K_2S_2O_8$, 2 h, 120 °C) (Gächter et al., 1992). Pore-water profiles were obtained by biweekly equilibration of an *in-situ* pore water sampler (peeper; 0.2 µm HT-Tuffryn 200®, GELMAN) (Hesslein, 1976). Samples were drawn into syringes (S-Monovettes®, SARSTEDT). The concentration differences at the sediment water boundary layer were used to calculate diffusive flux of SRP, given as P release rates PRR, towards the sediment surface according to Fick's first law of diffusion (Berner, 1980; Sinke et al., 1990).

Iron speciation

The diagenetically active iron (Fe_{HCl}) was defined as that removed by shaking ca. 100 mg fresh sediment with 5 ml 0.5 M HCl for 1 h at 22°C (Moeslund et al., 1994) and represented exchangeable Fe, most iron oxides, FeS and $FeCO_3$, and probably some Fe from silicates (Thamdrup et al., 1994).

After centrifugation dissolved Fe concentration was determined spectrophotometrically (Legler et al., 1986). Total iron (TFe) was determined by acid micro-wave digestion (MDS 2000; HNO_3:HF:HCl 2.5:2:0.5 2 h, 180 °C, + 0.05 g HBO_3) and inductively coupled plasma (ICP-AES, thermo jarrell ash).

Sulfur speciation

From wet sediment samples S species were determined. These included SO_4^{2-} of interstitial water (see above), ΣH_2S (N_2; centrifugation 5250 g, 3 min.), elemental S (S°) by methanol extraction (95 %, 5 min, 100 °C) (Canfield et al., 1986) and iodometric titration (Schmidt & Talsky, 1975), acid volatile S (AVS; FeS [+ H_2S]) by acid extraction (HCl_{conc}, 15 min, 100 °C) and chromium reducible S (CRS, FeS_2 + S° [+ H_2S]) by chromium reduction (Cr(II)/HCl/NH_3, 45 min, 100 °C) and indirect iodometric titration (Canfield et al., 1986; Fossing & Jorgensen, 1989). Organic S (S_{org}) was deduced from TS (see above) - ΣH_2S - SO42- - S° - FeS - FeS_2 = S_{org}, where CRS - S° = FeS_2.

Phosphorus speciation

Fresh sediment (0 - 0.5 to 18 - 20 cm) was successively extracted in duplicate with a) 1 M NH_4Cl for 0.5 h, b) 0.11 M $NaHCO_3/Na_2S_2O_4$ for 1 h (Bicarbonate-Dithionite, BD), c) 1 M NaOH for 16 h and d) 0.5 M HCl for 16 h (Psenner et al., 1984; Hupfer et al., 1995). In this way fractionation of sedimentary P into immediately available (loosely sorbed) P (NH_4Cl-P), reductant-soluble P (BD-P), Fe- and Al-bound P (NaOH-P) and Ca-bound P (HCl-P) was attempted. Non-reactive (organic) P (NRP) was deduced from the difference between TP - determined by digestion ($K_2S_2O_8$, 2 h, 120 °C) - and SRP of each extraction step. Residual P is the difference between TP of sediment (see above) and the sum of the fractions a) - d).

3. Results

Durin summer stratification in Lake Scharmützelsee a 58 day period from July to September 1995 is characterized in Figures 1 and 2. The thermal stratification is accompanied by O_2 consumption processes leading to anoxic hypolimnetic water and an increase in NH_4^+ and SRP concentration (Fig. 1).

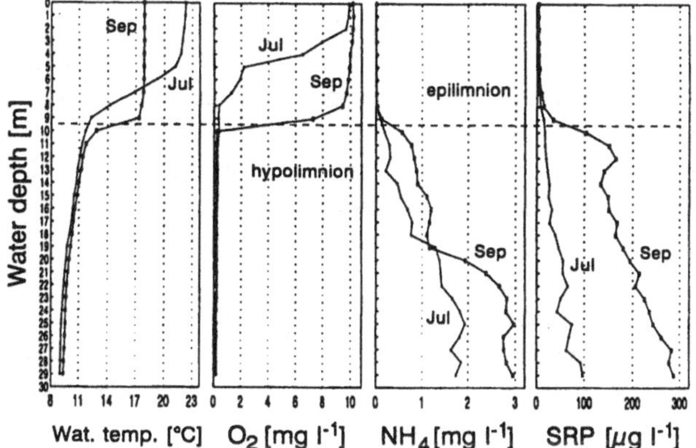

Figure 1. Vertical profiles of water temperature (wat. temp.), dissolved oxygen (O_2), ammonia (NH_4^+) and soluble reactive P (SRP) at the deepest location of Lake Scharmützelsee on July 20th and September 14th 1995 (58 d).

There is a strong linear relationship between both parameters in July: SRP (µg l^{-1}) = 0.036.NH_4^+ (µg l^{-1}) + 6.627; R^2 = 0.872, n = 30, ∝ = 0.05, as well as in September: SRP (µg l^{-1}) = 0.078.NH_4^+ (µg l^{-1}) + 35.165; R^2 = 0.889, n = 30, ∝ = 0.05. Due to benthic P release, the P content of the lake (6.347 t P) increased fourfold from 1.184 to 4.976 t P, where 60 % was derived from the sediment. As the hypolimnion became anoxic, SO_4^{2-} in the water column decreased (Fig. 2) due to SO_4^{2-} reduction processes at a rate of 142.61 ± 22.96 mg SO_4^{2-} m^{-2} d^{-1} leading simultaneously to an increase in ΣH_2S at a rate of 44.71 ± 17.57 mg ΣH_2S m^{-2} d^{-1} corresponding to 6.10 ± 3.42 mg H_2S m^{-2} d^{-1}.The ΣH_2S profiles are evidence for sulfides which have been removed from the lower strata. These cause a convex profile, the maximum of which (0.13 mg H_2S l^{-1}) moves up the water column with time until its progress is prevented by the presence of O_2 at water depths between 8 and 9 m (Fig. 2). About 19 % of the lake volume was filled with ΣH_2S. The ΣH_2S formation was accompanied by precipitation of iron sulfides at a rate of 5.62 ± 1.72 mg FeS m^{-2} d^{-1}. Hence, there was no DFe in the hypolimnion and the iron concentration in the unfiltered sample, given as TFe increased (Fig. 2).

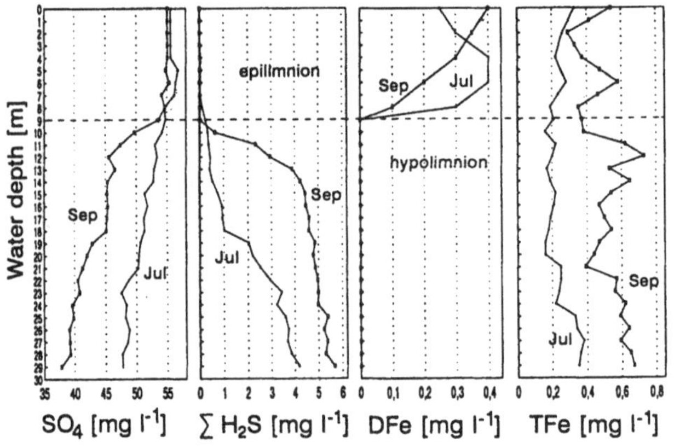

Figure 2. Vertical profiles of sulfate (SO_4^{2-}), total hydrogen sulfide (ΣH_2S), dissolved iron (DFe) and total iron (TFe) at the deepest location of Lake Scharmützelsee on July 20th and September 14th 1995.

The changes in the different S pools are summarized in a mass balance approach (Fig. 3) showing a distinct increase in the ΣH_2S content of the lake. About half of it was lost via iron sulfide precipitation. Besides the significant decrease in the SO_4^{2-} pool the remaining quantity of 6.95 t S, which was derived from the decomposition of organic matter corresponds to 7.9 % of the S content of the lake (Fig. 3). Distinct gradients are visible also at the sediment-water interface (Fig. 4). The annual mean SO_4^{2-} concentration (whole water column) amounted to 61.33 ± 10.41 mg SO_4^{2-}. This high concentration in the water leads to steep concentration gradients from 54 mg SO_4^{2-} at the sediment-water interface down to 3 mg SO_4^{2-} l^{-1} in the sediment. The NH_4^+ concentration starts to increase in the near bottom water and increases constantly with depth up to 48 mg N l^{-1}. Between spring and fall overturn the pore-water SRP concentration increased up to 10.8 mg l^{-1}. Its vertical distribution (0 - 30 cm) is relatively constant (10.24 ± 0.67 mg P l^{-1}) (Fig. 4). From the steep gradient at the sediment-water interface a mean PRR of 2.64 ± 0.56 mg P m^{-2} d^{-1} can be calculated over the period of 58 days with a maximum of 48.85 mg P m^{-2} d^{-1}. Besides a decreasing concentration of interstitial DFe with depth, a distinct subsurface minimum becomes visible (Fig. 4).

The sediment itself is poor in TIC, with a surface enrichment of organic matter, P and iron (Fig. 5). The mean TFe content (0 - 20 cm) amounts to 21.58 ± 3.73 mg Fe g d.w.$^{-1}$. The percentage of diagenetically derived iron (Fe_{HCl}) avarages 28.4 ± 15.7 % TFe and decreases with depth from 48.3 to 15.8 % TFe. The TS averages 5.7 ± 2.6 % LOI and decreases down core from 2.0 to 8.5 % LOI.

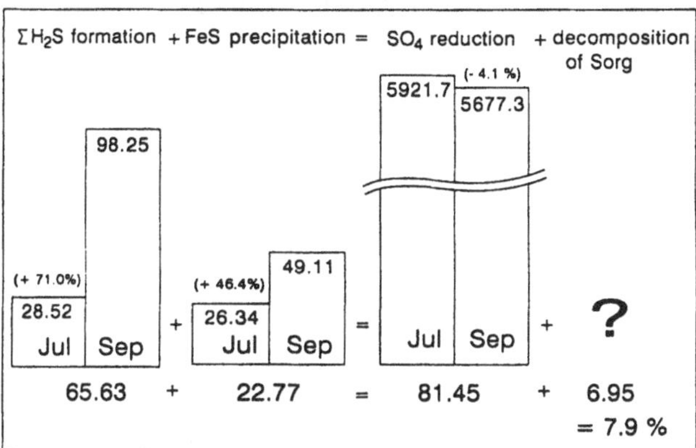

Figure 3. Changes in the main sulfur pools (given in tons) of Lake Scharmützelsee between July 20th and September 14th in 1995.

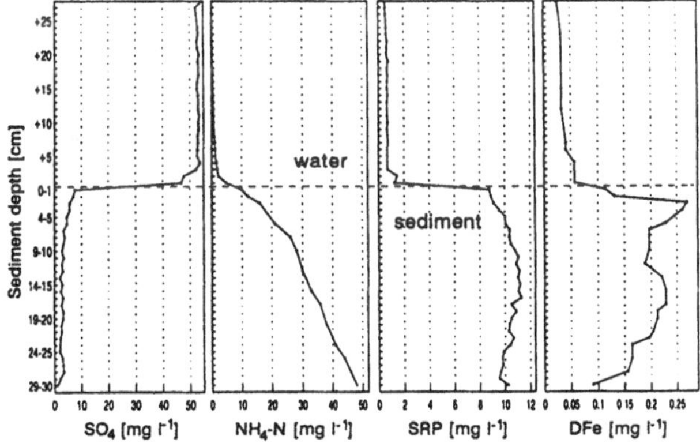

Figure 4. *In-situ* pore-water profiles of sulfate (SO_4^{2-}), ammonia (NH_4^+), soluble reactive P (SRP) and dissolved iron (DFe) at the deepest location of Lake Scharmützelsee in August 1996.

The atomic Fe:S ratio decreases with depth from 1.45 to 0.7. The most important inorganic S pool is the CRS (FeS_2 + S° + H_2S), being 1.3 to 6.6 times higher than the AVS (FeS + H_2S). There is higher percentage of FeS and FeS_2 in the 0 - 8 cm layers (Fig. 6). The SO_4^{2-} pool in the uppermost sediment layers is about two times higher than the pool of H_2S, which is only 0.2 to 0.4 % TS. There is no distinct trend in the S° distribution, and its percentage amounts to 1.3 ± 0.2 % TS (Fig. 6). The difference between TS and the sum of all inorganic S fractions (Sorg; not shown) increased from 49.3 % d.w. in the 0 - 0.5 cm layers to 81.6 % d.w. in the 18 - 20 cm layer (mean 64.3 ± 15.9 % d.w.). The difference between TS (1350 °C, Fig. 5) and the inorganic S portion (520 °C, Fig. 5) amounted, on average, to 20.9 ± 8.6 % d.w. giving a residual S portion of 14.8 ± 7.8 % d.w..

The fractional composition of P shows that the portion of interstitial water P, i.e., P loosely adsorbed to surfaces (NH_4Cl-P) is decreasing with depth (Fig. 7). Its percentage is relatively high, 14.6 ± 2.2 %

TP, and its vertical distribution is constant in agreement with SRP (Fig. 4). The fraction of redox sensitive P (BD-P) is minimal, 7.7 ± 1.8 % TP, and not highly correlated with Fe_{HCl} (BD-P (mg P g d.w.$^{-1}$) = 0.005.Fe_{HCl} (mg Fe g d.w.$^{-1}$) + 0.057; R^2 = 0.587, n = 24, α = 0.05). The NaOH fraction is the most dominant P fraction.

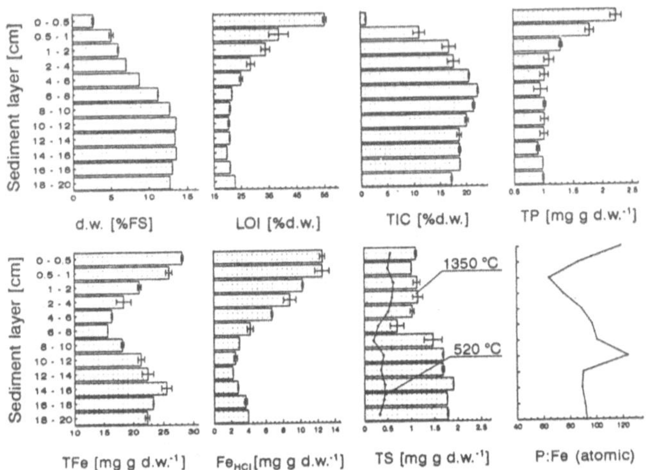

Figure 5. Vertical profile (mean values ± SD) of dry weight (d.w.), loss on ignition (LOI), total inorganic carbon (TIC), total P (TP), total iron (TFe), HCl-extractable iron (Fe_{HCl}), total S (TS, The difference between S cobusted at 1350 °C and 520 °C represents the organic S.) and the atomic P:Fe ratio in the interstitial water of two sediment cores from the deepest location of Lake Scharmützelsee in August 1996. Note the different scale of the uppermost sediment layers.

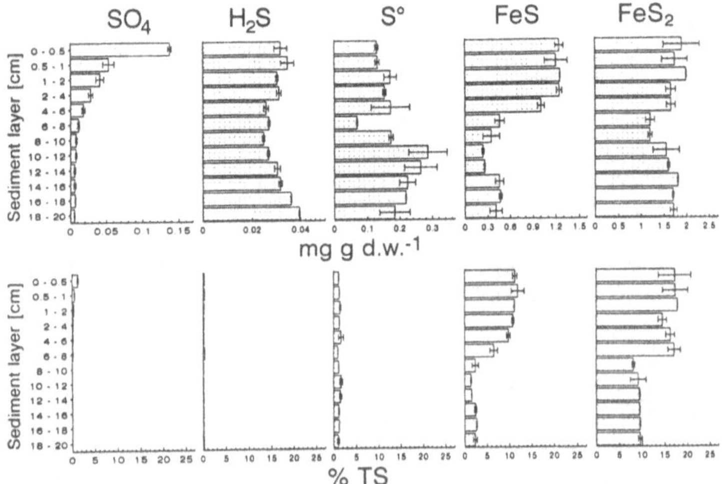

Figure 6. Vertical distribution of inorganic S species (mean values with SD): sulfate (SO_4^{2-}), hydrogen sulfide (H_2S), elemental S (S°), iron-monosulfide (FeS) and iron-disulfide (FeS_2) in mg g d.w.$^{-1}$ and in % of total S (TS), respectively, of two sediment cores from the deepest location of Lake Scharmützelsee in August 1996. Note the different scale of the uppermost sediment layers.

NaOH-SRP which represents P bound to metal oxides (mainly to Fe; NaOH-SRP (mg P g d.w.$^{-1}$) = 0.006.Fe_{HCl} (mg Fe g d.w.$^{-1}$) + 0.035; R^2 = 0.817, n = 24, α = 0.05), slightly decreases with depth and NaOH-NRP, which represents the organic bound P (NaOH-NRP (mg P g d.w.$^{-1}$) = 0.003.LOI (mg g d.w.$^{-1}$) - 0.291; $R^2 = 0.963$, n = 24, ∝ = 0.05), follows the vertical distribution of TP (Fig. 5): (NaOH-NRP (mg P g d.w.$^{-1}$) = 0.871.TP (mg P g d.w.$^{-1}$) - 0.416; $R^2 = 0.962$, n = 24, ∝ = 0.05). The TP itself is strongly correlated to sedimentary organic matter: TP (mg P g d.w.$^{-1}$) = 0.037.LOI (% d.w.) + 0.172; $R^2 = 0.948$, n = 24, ∝ = 0.05. The portion of HCl-P, which represents the carbonatic and apatitic bound P, is low in the uppermost layers and increases sharply at the 6 - 8 cm layer to 7.7 - 11.6 % TP. The refractory P (Pr) shows no distinct trend and amounts on average to 17.0 ± 8.0 % TP (Fig. 7).

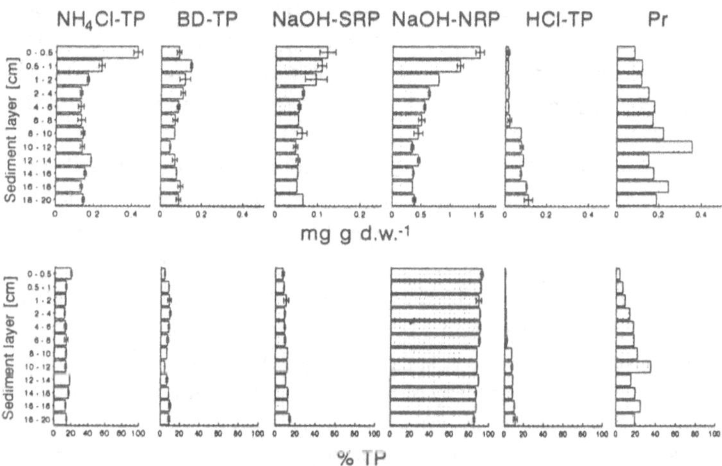

Figure 7. Vertical distribution of P fractions (mean values with SD): interstitial water and immediately available P (NH_4Cl-P), redox-sensitive P (BD-P), metal oxide bound P (NaOH-SRP) and organic P (NaOH-NRP), carbonatic and apatitic bound P (HCl-P) and residual P (Pr) of two sediment cores from the deepest location of Lake Scharmützelsee in August 1996. Note the different scale of the uppermost sediment layers.

4. Discussion

On a global scale freshwater lakes usually have SO_4^{2-} concentrations between 2.2 and 24.1 mg SO_4^{2-} l^{-1} (reviewed by Cook & Kelly (1992)) and have a different potential for SO_4^{2-} reduction to ΣH_2S by bacteria. For the majority of lakes studied, the rates of SO_4^{2-} reduction range from 0.6 to 6 mg (S) m^{-2} d^{-1}. Lake Scharmützelsee is situated in a region with lakes of remarkably high SO_4^{2-} concentrations ranging from 30 to 110 mg SO_4^{2-} l^{-1} (Nixdorf & Kleeberg, 1996). Both, the high productivity and the high SO_4^{2-} concentration in Lake Scharmützelsee (Fig. 2) result in a very high hypolimnetic SO_4^{2-} reduction rate of 4.56 ± 0.73 g (S) m^{-2} d^{-1} which exceeds that of most freshwater lakes. It is rather comparable to the highest rate reported by Cook & Kelly (1992), 14.4 g (S) m^{-2} d^{-1}, which was measured in an impoundment with high SO_4^{2-} concentrations (288 - 1920 mg SO_4^{2-} l^{-1}) derived from acid mine drainage. It is important to note that in Lake Scharmützelsee part of the ΣH_2S was derived from organic matter decomposition, obviously from decaying filamentous blue-geen algae, mainly *Oscillatoria limnetica* and *Limnothrix redekei* as described elsewhere (Nixdorf & Kleeberg, 1996). This is supported by the strong correlation between hypolimnetic NH_4^+ and SRP (Fig. 1).

The steep SO_4^{2-} gradient between sediment and overlying water (Fig. 4) by a factor of 6 indicates both, diffusion into the sediment of a few centimeters (usually < 5 cm; Cook & Schindler, 1983) and also a rapid consumption of SO_4^{2-} by bacterial reduction within the uppermost sediment layers. However, in accord with Lovley & Klug (1986) the interstitial SO_4^{2-} concentration of 7.48 mg SO_4^{2-} l^{-}

[1] (0 - 1 cm) down to 3.29 mg SO_4^{2-} l^{-1} (8 - 9 cm) is high enough that SO_4^{2-} reduction is not limited by SO_4^{2-} and predominates over methanogenesis at SO_4^{2-} concentrations > 5.76 mg SO_4^{2-} l^{-1}. This high hypolimnetic and benthic SO_4^{2-} reduction potential of Lake Scharmützelsee undoubtedly has consequences for its fertility in different ways:

First, during summer stratification the hypolimnetic chemistry is completely sulfide dominated, i.e., there is a continuation of oxidation of organic matter under anoxic conditions in which bacteria may use SO_4^{2-} as a terminal electron acceptor, and where breakdown of S-containing organic matter accounts for 8 % of the total reduced S production. Hence, a biologically mediated transformation from relatively inert SO_4^{2-} ions into labile sulfides (exhaustion of hypolimnetic DFe) is a key factor in the redox conditions in the lake and is an effective means of incorporating S into a more dormant pool of the S cycle in Lake Scharmützelsee.

Second, these S conversions are reflected in the vertical distribution of inorganic and organic S forms. The penetration depth of SO_4^{2-} (Fig. 4) as well as the „formation horizon" of ΣH_2S (Fig. 6) are in good agreement with the distribution of the AVS and the CRS (Fig. 6). Further, it explains the subsurface minimum of DFe (Fig. 4), indicating that the formation of FeS and FeS_2 seem to be Fe limited. Although, about 90 % of the reduced sulfides may reoxidize (Thamdrup et al., 1994), the precipitation of iron sulfides in various forms is an important process in terms of S burial in Lake Scharmützelsee. This is supported by a constant percentage of FeS and FeS_2 in the sediment depth from 8 to 20 cm (Fig. 6). The major pathway of S incorporation, however, is obviously the burial of Sorg, which increases with depth and is on average relatively high (5.7 ± 2.6 % LOI) in comparison to 1 % LOI according to Mackereth (1966). The residual S portion found may result from S losses due to drying of sediment (Amaral et al., 1989).

Third, the formation of iron sulfides diminishes the ability of the sediment to bind P. Hence, the Fe_{HCl} pool was generally low corresponding to the vertical distribution of AVS (Fig. 6) and a small portion of the BD-P (Fig. 7). This indicates the exhaustion of the redox-sensitive P and that other P forms are becoming more important for P retention and P release, respectively. The high P:Fe ratio (Fig. 5) indicates P mobility and is in agreement with the NH_4Cl-P portions (Fig. 7) indicating that mobilization of organic bound P by decomposition of organic matter, e.g., via SO_4^{2-} reduction, as described above, may contribute considerably to a redox-independent release of P. This is supported by a similar vertical distribution of TP and NaOH-NRP (Figs. 5 and 7). Assuming, that the constant TP content within 2 to 20 cm depth (0.99 ± 0.04 mg P g $d.w.^{-1}$) (Fig. 7) represents that amount of P which is permanently buried in the sediment, the difference of 0.79 mg P g $d.w.^{-1}$ relative to the TP content of the uppermost 2 cm layer (1.78 ± 0.47 mg P g $d.w.^{-1}$), i.e., about 80 % of the P reaching the sediment surface was obviously released on an annual basis. The PRR for 1995 (based on pore-water profiles) is in good agreement with the PRR calculated by mass balance for different years (1993: 2.11 ± 0.44 mg P m^{-2} d^{-1}, 1994: 1.53 ± 1.13 mg P m^{-2} d^{-1} and 1995: 2.74 ± 0.56 mg P m^{-2} d^{-1}) (Nixdorf & Kleeberg, 1996). Moreover, once the inventory of Fe_{HCl} is exhausted, the capacity for S and P retention by reaction with iron is limited by the rate of supply of iron.

Fourth, the high hypolimnetic and benthic SO_4^{2-} reduction potential of Lake Scharmützelsee enhances the P availability for the reuse by phytoplankton by controlling the Fe:P ratio in anoxic bottom waters. Tessenow's (1974) and Caraco's et al. (1993) results suggest that when the atomic Fe:P ratio in bottom waters is less than 2.2, bottom waters are far more likely to „leak" P to surface waters. This ratio in the water layers between 22 and 29.5 m decreased from 2.4 ± 0.4 in July to 0.7 ± 0.4 in September indicating a reduced potential for P precipitation during overturn.

5. Conclusions

The intensive SO_4^{2-} reduction in Lake Scharmützelsee controls P mobilization processes to a large extent. This leads to a permanent surplus in sedimentary mobilized P forms coupled to a high release of P, maintaining a high level of productivity, which in turn enables intensive SO_4^{2-} reduction. This phenomenon is typical also for other SO_4^{2-} rich lakes in the region. It means, that future restoration measures should be directed first to a reduction in the external nutrient loading to diminish the level of productivity and second to an increase of the portion of permanently bound P species.

Acknowledgements

I am grateful to the whole staff of the research station in Bad Saarow, especially to I. Henschke and W. Terlinden for field assistance, to G. Lippert, T. Hanke and D. Jendritzki for laboratory assistance. The support of the staff of the central analytical laboratory (BTU Cottbus) is gratefully acknowledged.

References

Amaral, J.A. Hesslein, R.H. and Rudd, J.W.M.: 1989, Limnol. Oceanogr. **34**, 1351-1358.
Berner, R.A.: 1980, Early diagenesis. A theoretical approach. Princeton Univ. Press, New York, 241 pp.
Canfield, D.E. Raiswell, R. Westrich, J.T. Reaves, C.M. and Berner, R.A.: 1986, Chemical Geology **54**, 149-155.
Caraco, N.F. Cole, J.J. and Likens, G.E.: 1993, Hydrobiol. **253**, 275-280.
Cook, R.B. and Schindler, D.W.: 1983, Ecol. Bull. (Stockholm) **35**, 115-127.
Cook, R.B. and Kelly, C.A.: 1992 in Howarth, R.W., Stewart, J.W.B. and Ivanov, M.V. (eds.), Sulphur cycling on the continents: Wetlands, terrestrial ecosystems and associated water bodies, John Wiley & Sons, Chichester • New York, 145-187.
Deutsche Einheitsverfahren zur Wasser-, Abwasser- und Schlammbehandlung (DIN 38406-E5-1 and DIN 38405-D11).
Fossing, H. and Jorgensen, B.B.: 1989, Biogeochem. **8**, 205-222.
Gächter, R. Tessier, A. Szabo, E. and Carignan, R.: 1992, Aquatic Sciences **54**, 1-9.
Golterman, H.L.: 1984, Verh. Intern. Verein. Limnol. **22**, 23-59.
Hesslein, R.H.: 1976, Limnol. Oceanogr. **21**, 912-914.
Hupfer, M. Gächter, R. and Giovanoli, R.: 1995, Aquat. Sci. **57**, 305-324.
Holmer, M. Kristensen, E. Banta, G. Hansen, K. Jensen, M.H. and Bussawarit, N.: 1994, Biogeochem. **26**, 145-161.
Legler, Ch. Breitig, G. Steppuhn, G. and Vobach, V.: 1986, Ausgewählte Methoden der Wasseruntersuchung. Gustav Fischer Verlag, Jena, 517 pp.
Lovley, D.R. and Klug, M.J.: 1986, Geochim. Cosmochim. Acta **50**, 11-18.
Mackereth, F.J.H.: 1966, Philos. Trans. R. S. London Ser. B250 **765**, 165-213.
Moeslund, L. Thamdrup, B. Jorgensen, B.B.: 1994, Biogeochem. **27**, 129-152.
Murray, T.E.: 1995, Can. J. Fish. Aquat. Sci. **52**, 1190-1194.
Nixdorf, B. and Kleeberg, A.: 1996, BTUC-UW **2**, 7-106.
Psenner, R. Pucsko, R. and Sager, M.: 1984, Arch. Hydrobiol. Suppl. **70**, 111-155.
Rohde, K.H. and Nehring, D.: 1979, Geod. Geoph. Veröff. IV **27**, 24-27.
Schmidt, M. and Talsky, G.: 1975, Z. anorg. allgem. Chem. **289**, 1683-1686.
Sinke, A.J.C. Cornelese, A.A. Keizer, P. Van Tongeren, O.F.R. and & Cappenberg, Th. E.: 1990, Freshw. Biol. **23**: 587-599.
Sulzberger, B. Suter, D. Siffert, C. Banwart, S. and Stumm, W.: 1989, EAWAG News **26**, 18-22.
Tessenow, V.U.: 1974, Arch. Hydrobiol. Suppl. **47**, 1-79.
Thamdrup, B. Fossing, H. Jorgensen, B.B.: 1994, Geochim. Cosmochim. Acta **58**, 5115-5129.

FACTORS REGULATING THE FLUX OF PHOSPHATE AT THE SEDIMENT - WATER INTERFACE OF A SUBTROPICAL CALCAREOUS LAKE: A SIMULATION STUDY WITH INTACT SEDIMENT CORES

W. ECKERT, A. NISHRI and R. PARPAROVA

Israel Oceanographic and Limnological Research, The Yigal Allon Kinneret Limnological Laboratory, P.O.Box 345, Tiberias 14-102, Israel

Abstract. Different factors which interactively control the flux of soluble reactive phosphorus (SRP) at the sediment-water interface (SWI) of Lake Kinneret were studied seasonally. The influence of pH, Eh and microbial activity on SRP flux at the SWI was investigated by manipulating the conditions in the overlying water of intact sediment cores. The calculated diffusive SRP flux out of the sediment was lower in cores sampled during winter and spring than during the period of amixis. Potential SRP release, as measured in the absence of microbial activity, was strongly enhanced upon the transition from oxic to anoxic conditions indicating P release from iron(III)-bound phosphorus. In spring and summer cores, an enhanced SRP flux from sediments at pH 7 in comparison to pH 8 indicated P release from carbonate-bound P which sedimented previously as result of high pH values during the algal spring bloom. Microbial uptake at the SWI was the most important sink for SRP and no net-flux occured under oxic conditions. The higher net-flux of P under anoxic conditions was linked to carbon limitation of the bacteria at the SWI.

Keywords: phosphorus, P flux, microbial activity, redox, simulation, Lake Kinneret, sediment, accumulative P release

1. Introduction

The regulatory effect of sediments for the phosphorus (P) cycle of freshwater lakes is well established (e.g. Li et al., 1972; Baccini, 1985; Boström et al., 1988). On one hand, sediments may act as a P sink by accumulating high concentrations of allogenic apatite minerals and organic and inorganic P complexes. On the other hand, they may act as a P source releasing phosphorus back into the water column. Sedimentary P release is triggered by the concentration gradient between dissolved phosphorus in the porewater and the overlying water (OW). Elevated P concentrations in porewater are the result of dissolution and desorption processes and mineralization of organic-bound P (Boström et al., 1982; Enell and Löfgren, 1988).

Following the classical works of Einsele (1936), Ohle (1938) and Mortimer (1941) sedimentary P cycling is considered to be linked mainly to the iron cycle. As the redox conditions drop below ca. 200 mV, iron(III) is reduced chemically causing the mobilization of the sorped phosphate (Boström et al., 1982). In sediments with an oxidized surface layer, upward diffusing Fe^{2+} is reoxidized forming a micro-layer with a high sorption capacity for phosphate at the sediment water interface (SWI) (Löfgren & Boström, 1989; Sinke, 1992). The stability of the iron(III)-P complex is strongly pH-dependent. An increase in pH favors P desorption due to substitution of phosphate by OH^-. In calcareous lakes this P may be re-precipitated as hydroxy apatite or adsorbed to $CaCO_3$ (Boström et al. 1988). In contrast to

Water, Air and Soil Pollution **99**: 401-409, 1997.

non-calcareous lakes, P immobilization may occur in calcareous lakes at higher pH levels due to co-precipitation or adsorption of P on calcite (Otsuki and Wetzel, 1972). This carbonate-bound P can be mobilized at lower pH values (Golterman, 1988).

Until the mid-eighties, it was considered that P cycling at the SWI of freshwater lakes was regulated mainly by abiotic processes (Baccini, 1985); involvement by bacteria was perceived to be indirectly via the regulation of redox conditions. This view changed with the discovery that bacteria can directly participate in P cycling by sinking large amounts of phosphorus under oxic conditions and releasing P under anoxic conditions (Fleischer, 1986; Gächter et al., 1988; Hupfer and Uhlmann, 1991; Sinke and Cappenberg, 1988). Gächter and Meyer (1993) claimed that in oligotrophic lakes the P flux across the SWI is regulated by benthic bacteria which may contribute to the terminal burial of phosphorus in the sediment by the production of refractory P compounds.

The aim of the present study was to evaluate the differential impact of pH, Eh and microbial activity on P release from the calcareous sediments of Lake Kinneret. For this purpose we exposed the overlying water (OW) of intact sediment cores to controlled (regulated) *in situ* pH and Eh changes in the presence and absence of microbial activity while following the accumulation of soluble reactive phosphorus (SRP) in the OW.

2. Materials and Methods

2.1. STUDY AREA

Lake Kinneret (12 x 22 km, 168 km^2) is a warm monomictic calcareous lake in the northern part of the Afro-Syrian rift valley (Serruya, 1978). Maximum and average depths are 42 and 24 m, respectively. The lake is stratified from April until December with surface water temperatures ranging between 15 °C in winter and 30 °C in summer. A prominent biological event during the annual lake cycle is the spring bloom of the dinoflagelate *Peridinium gatunense* (Berman and Pollingher, 1974). During the bloom, epilimnetic pH values may rise to 9.6 accompanied by a massive sedimentation of authigenic $CaCO_3$ (8 g m^{-2} day^{-1} were reported by Serruya et al., 1974).

The composition of the sediments in the central area of the lake is dominated by calcium carbonate (51% of dry weight) while iron and phosphorus portions constitute 2.5 and 0.1 %, respectively (Serruya, 1978). The portion of the sedimentary organic matter content varies seasonally within the uppermost 0.5 cm (12 - 30 %) dropping to 8 % below the depth of 2 cm (Koren, 1993).

2.2. SAMPLING AND EXPERIMENTAL

Sediment cores from the central lake station (maximum depth) were collected with perspex tubes (ϕ 50 mm) by use of a Tessenow sampler (Tessenow et al., 1977) during 1992-94. During each season 4 parallel cores denoted as A, B, C, and D were collected and transported in the dark to the laboratory. In core A the upper 10 cm were sectioned immediately to one

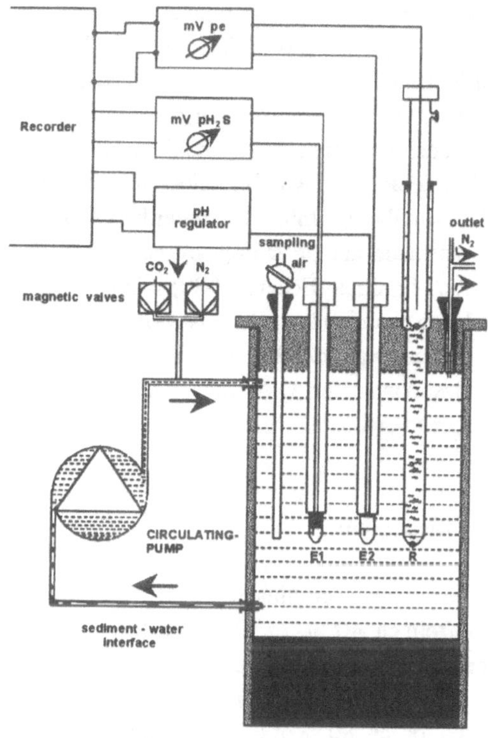

Fig. 1. Experimental setup of sediment-water system designed to manipulate the hydrochemical conditions in the overlying water (E1: pH_2S electrode; E2: combined redox-glass electrode; R; double-junction reference electrode)

0.5 cm slice and nine 1-cm slices under a nitrogen atmosphere and the porewater extracted by N_2 - pressure filtration through a 0.45 μm filter (Whatman). Sub-samples from each slice were used to determine different particulate-P fractions (Hieltjes and Lijkema, 1980) and water content. The porewater was analyzed for SRP and total dissolved phosphorus (TDP) using standard analytical procedures (APHA, 1985). The gradient between the SRP concentration in the OW and in the porewater of the uppermost cm of the sediment was used to calculate the expected diffusive flux according to Fick's second law with the following constants: porosity (ϕ = 0.86) and the square of tortuosity (θ^2 = 1.4) were determined previously for Kinneret sediments (Stiller,1974) while the molecular diffusion coefficient for $H_2PO_4^-$ at 20°C ($D_i = 7.49 \cdot 10^{-6}\ cm^2 \cdot s^{-1}$) was taken from Li and Gregory (1974).

Cores B, C and D were incubated at constant temperature of 20°C under the conditions shown in Table I while the concentration of SRP in the OW was measured daily (APHA , 1985). In cores B and C microbial activity was inhibited by the addition of sodium azide (150 mg·L^{-1}) and antibiotics (50 mg·L^{-1} each of penicillin and streptomycin) to

TABLE I

Treatment of sediment cores for studying accumulative P release.

core	microbial activity	pH	redox conditions during mixis (winter & spring)		redox conditions during amixis (summer & fall)
			initial	final	
B	-	7.0	oxic	anoxic	anoxic
C	-	8	oxic	anoxic	anoxic
D	+	variable	oxic	anoxic	anoxic

the OW. Core D served as control. The pH in the OW of core B was maintained at 7.0 via a feedback-controlled simulation unit (Figure 1) using regulator gasses (CO_2, N_2). In the OW of cores C and D, pH values were measured daily while in core C it was readjusted daily to pH 8. Redox conditions were manipulated either via bubbling of air or a sodium sulfide solution adjusted to the ambient pH value. The transition from oxic to anoxic conditions in the OW of the inhibited cores was accomplished by de-aeration with N_2 followed by addition of sulfide (final concentration = 10^{-4} mol $HS^{-} \cdot L^{-1}$) simulating ambient summer hypolimnetic sulfide concentrations. The OW of core D was made anoxic simply by cutting off the atmospheric gas exchange. According to previous findings (Eckert, 1989) this procedure results into sulfide formation after 3-5 h. In order to examine the role of carbon availability on microbial P uptake, in some experiments glucose was added to the OW of core D. In three experiments when stable SRP concentrations in the OW indicated steady state conditions, the anoxic OW was re-aerated in order to evaluate P resorption.

3. Results and Discussion

Particulate phosphorus (PP) in Lake Kinneret sediments is dominated by HCl -extractable (calcium-bound) P showed a vertical increase from ca. 40 % of PP in the top 0.5 cm to ca. 58 % in 10 cm (Table II). The opposite trend was observed for residual P (organically-bound P) which decreases from ca. 30 % to 23 %. Within the investigated 10 cm of sediment, the sum of residual and HCl-P accounted for ca. 80 % of PP. The remaining 20 %

TABLE II

Vertical distribution of PP, different PP fractions, porewater TDP and porewater SRP in Lake Kinneret sediments averaged over seasonal cores from a two year period

depth	% of PP				PP	porewater	
(cm)	NH_4Cl-P	NaOH-P	HCl-P	Residual-P	($mg \cdot g_{dw}^{-1}$)	TDP ($mg \cdot L^{-1}$)	SRP ($mg \cdot L^{-1}$)
0-0.5	7±2	22±9	40±4	31±22	1.11±0.17	0.75±0.30	0.65±0.36
0.5-1.5	2±2	16±4	47±11	35±14	1.36±0.22	1.11±0.21	0.96±0.43
1.5-2.5	5±1	17±10	52±9	25±13	1.21±0.26	1.35±0.41	1.16±0.46
2.5-3.5	8±2	11±5	48±5	33±10	0.97±0.07	1.56±0.63	1.26±0.33
3.5-4.5	10±2	12±5	55±3	22±5	0.95±0.07	1.63±0.12	1.44±0.44
4.5-5.5	10±2	9±4	50±3	31±8	1.03±0.09	1.84±0.69	1.48±0.50
5.5-6.5	12±2	8±3	55±2	25±10	0.96±0.09	1.97±0.78	1.42±0.49
6.5-7.5	13±1	8±3	51±5	28±18	1.01±0.13	1.73±0.34	1.41±0.58
7.5-8.5	12±1	8±2	54±7	27±11	1.08±0.06	1.80±0.33	1.55±0.64
8.5-9.5	10±1	8±1	58±2	23±10	1.11±0.10	1.71±0.63	1.42±0.61

comprised NH_4Cl (labile) P that slightly increased with sediment depth and NaOH (Fe- Al-bound) P. The vertical increase of HCl-P accompanied by the decrease in residual P and NaOH-P are an indication for diagenetic processes within the sediments. P released as the result of organic decomposition and of the reduction of iron (III) is immobilized with the formation of authigenic apatite minerals (Berner, 1974). The increase in labile P is most probably related to an intensified adsorption to calcium carbonate with increasing sediment depth (Pettersson et al. (1988). The high variability of organic P in the top layer reflects the seasonal change in organic matter sedimentation.

The verical decrease of the sedimentary organically-bound and metal-bound P indicate continuous P mobilization processes. As such, the porewater is strongly enriched in dissolved P. Average porewater concentrations of SRP which in the upper 3 cm accounted for ca. 87% of the total dissolved P ranged from 0.65 mg $P{\cdot}L^{-1}$ within the top layer to 1.5 mg$\cdot$ PL^{-1} at a depth of 9 cm (Table II). The SRP concentration in the OW of Lake Kinneret on the other hand ranges between 4 and 40 µg $P{\cdot}L^{-1}$ (Serruya, 1972). The gradient between SRP in the porewater and the OW showed a strong seasonality as shown by the calculated diffusive P-flux in Figures 2 - 4.

In the presence of microbial activity (control core D) we observed no P release to the OW of cores sampled during winter (Figure 2A, B) as long as the OW remained oxic. However, the induction of anoxic conditions caused SRP release at a rate similar to that of the expected diffusive flux. In the cores without microbial activity SRP accumulated continu-ously in the OW at a rate that was strongly enhanced under anoxic conditions. The redox effect on this potential SRP flux indicated P release from iron(III)-bound P at the SWI due to the reduction of iron by sulfide followed by the formation of FeS (Boström et al., 1982). The similar release patterns of SRP at pH 7 and pH 8 indicated that pH-dependent P exchange processes do not affect P release during winter. Microbial P uptake apparently prevented SRP release under oxic conditions but not under anoxic conditions. This

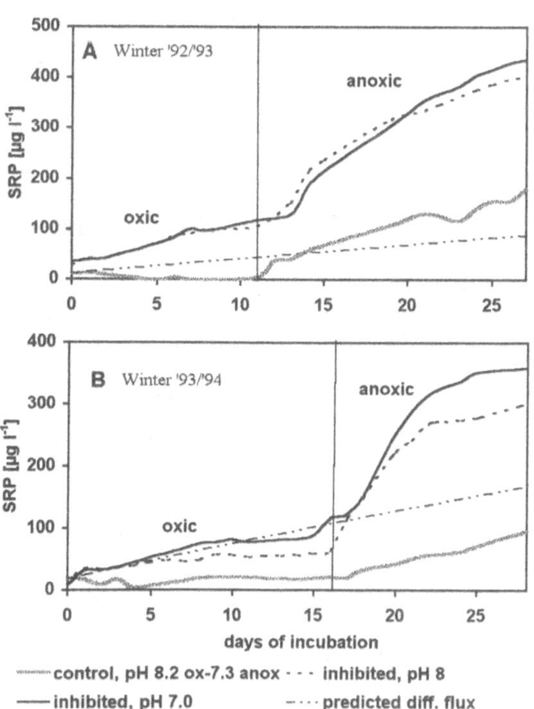

Fig. 2. Accumulation of SRP in the OW of manipulated and untreated intact sediment cores and predicted diffusive flux: A: winter '92/'93, B: winter '93/'94

observation may have two explanations:

1. Microbial uptake of upwards diffusing P under oxic conditions was followed by microbial P release upon decreasing Eh. This explanation corresponds to the findings of Hupfer and Uhlman (1991).
2. The efficiency of microbial P uptake is a function of the availability of suitable carbon sources (Sinke and Cappenberg, 1988).

According to the first explanation, one would expect a higher P flux in the control due to additional redox dependent microbial P release. However, we are inclined to accept the second explanation as P release was much higher in the inhibited cores (Figure 2). We obtained further evidence for this interpretation during our experiments carried out with cores sampled in spring '92 (Figure 3A) when no net flux of SRP was observed. Due to the sedimentation of organic matter during the spring algal bloom the bacteria at the sediment-water interface were P- rather than carbon-limited (Hadas and Pinkas, 1992). Accordingly, microbial P uptake prevailed under oxic as well as anoxic conditions. It must be stated, however, that in spring '92 we measured the lowest porewater concentrations of SRP throughout our experimental period. During spring '93 (not shown) and '94 (Figure 3B) when P net flux occurred under anoxic conditions, porewater concentrations of SRP were considerably higher. Similar to the experiments carried out during winter, the shift from oxic to anoxic conditions caused an increase of SRP release in the cores without microbial activity. Unlike in winter, we observed a pH effect with higher SRP release at pH 7 in comparison to pH 8. This pH effect might be attributed to desorption and/or dissolution of carbonate-bound P that was

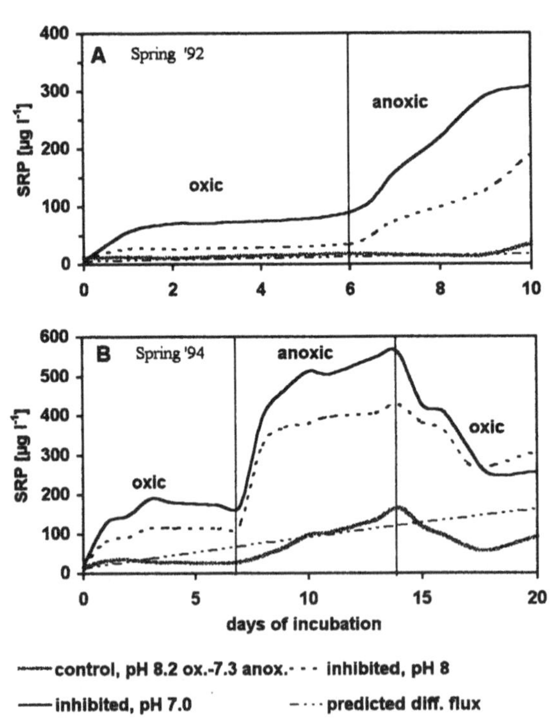

Fig. 3. Accumulation of SRP in the OW of manipulated and untreated intact sediment cores and predicted diffusive flux: A: spring '92, B: spring '94

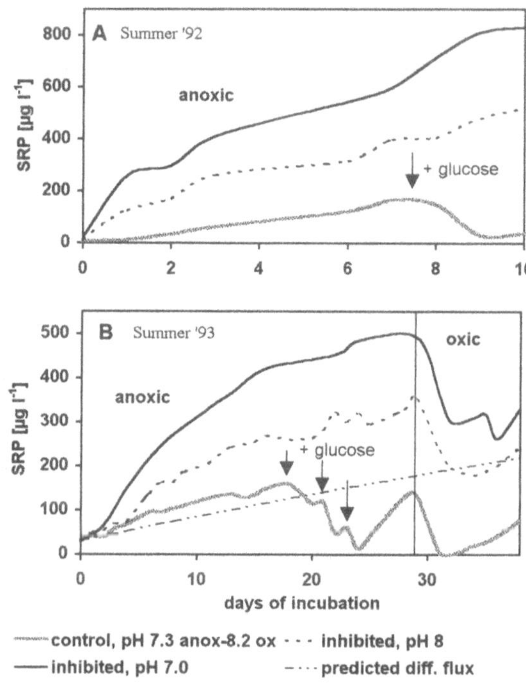

Fig. 4. Accumulation of SRP in the OW of manipulated and untreated intact sediment cores and predicted diffusive flux: A: summer '92, B: summer '93

precipitated out of the photic zone during the algal bloom (Otsuki and Wetzel, 1972; Serruya et al. 1974).

The same effect of pH on the release of SRP was observed during summer (Figure 4). As such, net release of SRP occurred in the control core at a rate similar to the predicted diffusion (Figure 4B). During this summer experiment, we also examined the role of carbon availability on the microbial P. Upon the repeated addition of glucose to the control core, SRP concentration in the OW dropped dramatically to concentrations below the predicted diffusive flux (Figure 4B). This observation indicates that during summer the bacteria at the SWI were carbon limited. The change in the OW conditions from anoxic to oxic (simulating fall turnover) caused the SRP concentrations in the control core to drop to zero while in the inhibited cores oxygenation yielded a drop in SRP by ca. 30 %. A similar effect of aeration was observed during experiments carried out during spring '94 (Figure 3B) and fall '94 (Figure 5B). The drop in the SRP concentration following the re-aeration of the OW of the inhibited cores can be attributed to P sorption onto iron(III) hydroxide formed during the oxygenation (Lijklema, 1980).

The SRP release patterns during fall '92 and '94 (Figure 5) were similar to those found during winter under simulated anoxic conditions (e.f. Figure 2). No pH effect was detectable in the inhibited cores. Note the similarity of potential, net and diffusive fluxes in Figure 5B. Obviously carbon limitation of bacteria assemblages at the SWI was most expressed during fall when diffusion of SRP seemed to become the dominating P release process. The relatively high net flux during fall reflects the situation in the hypolimnion of Lake Kinneret towards the end of amixis when under similar hydrochemical conditions (pH $\approx$ 7.2, total sulfide $\approx 10^{-3.5}$ mol$\cdot$ L^{-1}; Eckert and Trüper, 1993) SRP accumulation averages arround 50 μg$\cdot$L^{-1}.

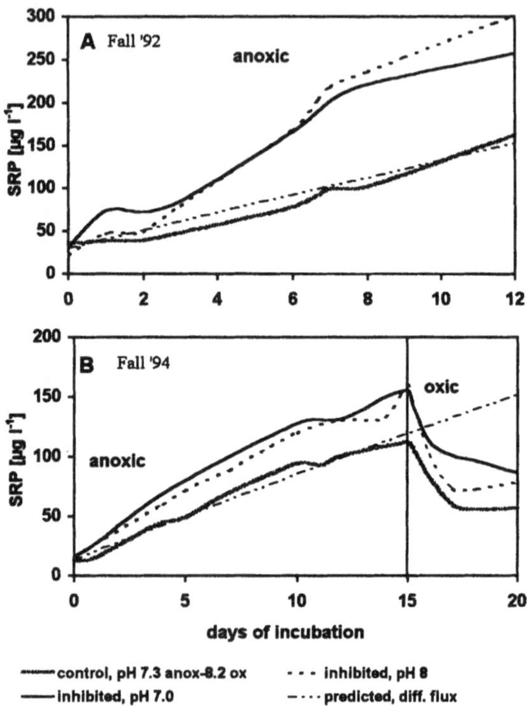

Fig. 5. Accumulation of SRP in the OW of manipulated and untreated intact sediment cores and predicted diffusive flux: A: fall '92, B: fall '94

4. Conclusions

- The predicted diffusive flux of SRP as the dominating fraction of TDP in the porewater varied seasonally with relatively lower rates during winter and spring and increasing rates following stratification.
- Maximum SRP release to the OW occurred under anoxic conditions. The observation that, in the absence of bacterial activity, P flux was strongly enhanced upon the transition from oxic to anoxic conditions is an indication for P release from iron(III) bound phosphorus.
- The effect of pH on the SRP flux varied seasonally with no measurable effect during fall and winter. During simulated spring and summer conditions we obtained an enhanced release of P from sediments at pH 7 in comparison with pH 8. We attribute this effect to P release from carbonate-bound P sedimenting during the algal spring bloom.
- Microbial uptake at the SWI was the most important sink for SRP in all seasonal simulations.. The relatively higher net-flux of P under anoxic conditions was shown to be linked to carbon limitation of the bacteria at the SWI.

Acknowledgements

We thank Dr. K. D. Hambright for his careful review of the manuscript.

References

APHA Standard Methods for Examination of Water and Wastewater.:1985, 16th edition, Am. Publ. Health Ass., Washington.

Baccini, P. : 1985, In: W. Stumm (ed.) *Chemical processes in lakes* Wiley, N.Y., 189-205.

Berman, T.and Pollingher, U.: 1974, *Limnol. Oceanogr.* **19**, 31-54.

Berner, R. A.: 1974, In: E. D. Goldberg (ed.) *The Sea*, Wiley, New York, pp. 427-450.

Boström, B., Jansson, M. and Forsberg C.: 1982, *Arch. Hydrobiol. Beih. Ergebn. Limnol.* **18**, 5-59.

Boström, B., Andersen, J. M., Fleischer, S. and Jansson, M.: 1988, *Hydrobiologia* **170**, 229-244.

Eckert, W.: 1989, Phd thesis, University of Bonn, Germany.

Eckert. W. and Trüper, H. G.: 1993, *Biogeochemistry* **21**, 1-19.

Einsele, W.: 1936, *Arch. Hydrobiol.* **29**, 664-686.

Ennel, M. and Löfgren, S.:1988, *Hydrobiologia* **170**, 103-132.

Fleischer, S.: 1986, *Arch.Hydrobiol.* **107**, 269-272.

Gächter, R. and Meyer, J. S.: 1993, *Hydrobiologia* **253** 103-121.

Gächter, R., Meyer, J. S. and Mares, A.:1988, *Limnol.Oceanogr.* **33**, 1542-1558.

Golterman, H. L.: 1988, *Hydrobiologia* **159**, 149-151.

Hadas, O. and Pinkas, R.: 1992, *Hydrobiologia* **235/236**, 295-301.

Hakanson, L. and Jansson, M.: 1983, *Principles of lake sedimentology* Springer. Berlin. 316p.

Hieltjes, A. H. and Lijklema, M. L.: 1980, *J.Environ.Qual.* **9**, 405-407.

Hupfer, M. and Uhlmann, D.: 1991, *Verh. Intern. Limnol.* **24**, 2999-3003.

Koren, N: 1993, MSc thesis, The Haifa Technion.

Li, W. C., Armstrong, D. E., Williams, J. D. ,. H., Harris, R. F. and Syers, J. K.: 1972, *Soil Sci.Soc.Am.Proc.* **36**, 279-285.

Li, Y-H. and Gregory, S.: 1974, *Geochim. Cosmochim. Acta.* **38**, 703-714.

Lijklema, L.: 1980, *Environ. Sci. Technol.* **14**, 537-541.

Löfgren, S. and Boström, B.: 1989, *Water Res.* **9**, 1115-1125.

Mortimer, C. H.: 1941, *J. Ecol.* **29**, 280-329.

Ohle, W.: 1938, *Vom Wasser* **13**, 87-97.

Otsuki, A. and Wetzel, R. G.: 1972, *Limnol Oceanogr.* **17**, 763-767.

Petterson, K., Boström, B. and Jacobsen, O.-S.: 1988, *Hydrobiologia* **170**, 91-101.

Serruya, C., Edelstein M. , Pollingher U. and Serruya S.: 1974, *Limnol Oceanogr.* **19**, 489-508.

Serruya, C. (ed.): 1978, *Lake Kinneret.* Dr.W.Junk Publisher, Amsterdam.

Sinke, A. J.C.: 1992, PhD thesis. University of Wageningen. The Netherlands: pp 121.

Sinke, A. J. and Cappenberg, T. E.: 1988, *Arch. Hydrobiol. Beih. Ergebn. Limnol.* **30**, 5-13.

Stiller, M.: 1974, PhD thesis, The Weizman Institute of Science, Rehovot, Israel, pp 241.

Tessenow, U., Frevert, T., Hofgartner, W. and Moser, A.: 1977, *Arch.Hydrobiol. Supl.* **48**, 438-452.

EXPERIMENTS ON WATER–SEDIMENT NUTRIENT PARTITIONING UNDER TURBULENT, SHEAR AND DIFFUSIVE CONDITIONS

PANAGIOTIS D. SCARLATOS
Department of Ocean Engineering, Florida Atlantic University, Boca Raton, Florida 33431 U.S.A.

Abstract. Cultural eutrophication from excessive input of nutrients is a major problem for many water bodies around the world. Phosphorus and to a lesser degree nitrogen constitute the limiting elements for growth of plankton cells. Mobility, speciation and partition of nutrients in aquatic ecosystems depend on a number of physicochemical parameters. Experiments have been conducted for quantification of nutrient partition between ambient water and cohesive sediments. The experiments included nitrate and soluble phosphorus (superphosphate – 46% P_2O_5 fertilizer) partitioned between tap water and sediment slurries. The slurries involved kaolinite and bentonite as well as natural organic mud from, Lake Okeechobee, Florida. The nutrient exchange was promoted by sediment resuspension. Resuspension was induced either under homogeneous turbulent conditions in an oscillating–grid tank or by shear flow in a lock–exchange flume. The effects of phosphorus or nitrogen concentration, sediment concentration, water temperature, pH and salinity on nutrient partition were quantified. The results obtained through this study appear to be in agreement with data from other similar laboratory or field studies.

KEY WORDS: Adsorption; Desorption; Diffusion; Nitrate; Nutrients; Phosphate; Sediments; Turbulence

1. Introduction

Intensive anthropogenic activities such as urbanization, industrialization and agriculture have placed a tremendous ecological stress on the receiving water bodies. Most of the lakes, rivers and estuaries around the world have increased levels of nutrients, trace metals, Cl–pesticides, PAHs, PCBs and other ecologically harmful chemical compounds. The problem is further complicated by the fact that most of these pollutants, dissolved or particulate, tend to adhere to suspended organic and inorganic particles, which eventually end up in bottom sediments. At the same time, contaminants are continuously released from the sediments by diffusion or during bottom sediment resuspension events. Thus, there is a continuous recycling of pollutants between the ambient water and the bottom sediments (Medina and McCutcheon, 1989).

The processes involved i.e. speciation, mobilization, precipitation, bioavailability, metabolism, etc. of the various chemical constituents in aquatic ecosystems are affected by the combined effects of temperature, dissolved oxygen, salinity, pH and the presence of other chemicals (Stumm and Morgan, 1981). Furthermore, chemical exchange/partition between the water and solid particulate phase of such systems depends not only on their physicochemical state but also on the hydrodynamic characteristics of the system (Bonner et al, 1994).

One of the most common problems of aquatic contamination is cultural eutrophication (Berner and Berner, 1996). Abundant supply of nutrients from animal waste, fertilizers, detergents, domestic and industrial sewage waters can lead to to high biological productivity. An eutrophic water body can support as much as 700 g C/m^2/yr. Algal blooms, are aesthetically unpleasant and can emit disagreeable odors. More important, they can have devastating effects on fish and bottom fauna by reducing the levels of dissolved oxygen or producing toxins. It is well–known that the general stoichiometry of phytoplankton cells is $C_{106}H_{263}O_{110}N_{16}P_1$. From these five fundamental elements carbon (C), hydrogen (H)

Water, Air and Soil Pollution **99**: 411-425, 1997.

and oxygen (O) are abundant and readily available in nature. Therefore, availability of the two remaining elements, nitrogen (N) or phosphorus (P), constitute the limiting element for algal growth. In oxygen–rich waters settled phosphorus remains permanently buried in the bottom sediments. Under anaerobic conditions, phosphorus is released from the sediments into the anaerobic hypolimnion where hydrogen sulfide (H_2S) and ammonia (NH_4^+) accumulate while nitrate (NO_3^-) is denitrified (Gächter and Meyer, 1990). Thus, in most cases phosphorus enrichment is the main cause of eutrophication. Reduction of nutrient input could therefore improve the ecological state of an aquatic system.

In aquatic systems nitrogen can be found in various inorganic forms from the most reduced NH_4^+ to the most oxidized NO_3^-. Domestic wastewaters (urea) and agricultural fertilizers (ammonium salts) can create very high oxygen demands on receiving water bodies (Carberry, 1990). However, even without anthropogenic inputs nitrogen–consuming bacteria can uptake nitrogen directly from the atmosphere.

In contrast to nitrogen, phosphorus is generally found in its most oxidized state as phosphate ion (PO_4^{3-}). Also, since phosphorus reacts easily to form calcium, iron and aluminum phosphates which are relatively insoluble minerals, most of the phosphorus is transported in particulate form (Berner and Berner, 1996). Phosphorus inputs from animal waste, fertilizers and detergents exceed by far the amount of phosphorus generated from rock and mineral weathering. In most natural aquatic systems phosphorus constitute the limiting element.

To understand nutrient behavior in natural water–sediment systems a large number of controlling parameters should be collected and accounted for (Scarlatos, 1996). Generally, analysis of field data may provide information about the general trends of contaminant transport and fate, but cannot specify the special effect that a certain physicochemical parameter may have on a specific chemical species. Laboratory experiments on the other hand can provide more detailed information on a particular parameter or on an individual processes by using appropriate controls.

The purpose of this study was to quantify the nutrient water–sediment partition under different physicochemical conditions. For that purpose nitrate (NO_3^-) and soluble phosphorus (46% P_2O_5 fertilizer) were selected as the representative nutrients. The sediment media used included clay minerals and natural organic mud mixed with tap water. The experiments were focused on the nutrient partition resulting from diffusion and sediment resuspension events. Two modes of resuspension were investigated. One induced by the motion of an oscillating grid over the sediment bed and another caused by shear production on the surface of a sediment–slurry gravity current. The importance of ambient water temperature, salinity and pH on nutrient partition were also assessed.

2. Experimental Set–Up

2.1 EXPERIMENTAL APPARATUSES

The experiments were conducted using two different type of experimental apparatuses, an oscillating–grid tank and a lock–exchange flume. The specific characteristics of the these apparatuses are given in the proceeding.

2.1.1 OSCILLATING–GRID TANK

The oscillating–grid tank comprised of a plexiglass tank with a 50x50 cm base and a height of 67 cm (maximum holding capacity of 167.5 liters). At one side, the tank was equipped with sampling ports located at six different heights: 0.5, 10.5, 20.5, 30.5, 40.5 and 50.5 cm from the bottom.

Turbulence was induced by a metal grid oscillating vertically within the water column. The grid was made of steel bars forming sixteen 10x10 cm squares and was driven by a 1 HP motor with a rated speed of 1725 rpm (28.75 Hz). The oscillating frequency was controlled by a system of disks, and was reduced to 2.87 Hz (Figure 1). At equilibrium position the grid was 15 cm from the bottom of the tank. The amplitude of the oscillation was 6.1 cm. A 10 cm layer of sediment was placed on the bottom of the tank and was subject to resuspension. The oscillating motion of the grid generated conditions similar to that of homogeneous turbulence (Itsweire et al., 1986; Scarlatos, 1992). In nature, these conditions develop when turbulence in the upper mixing layer propagates downwards and reaches the sediment bed.

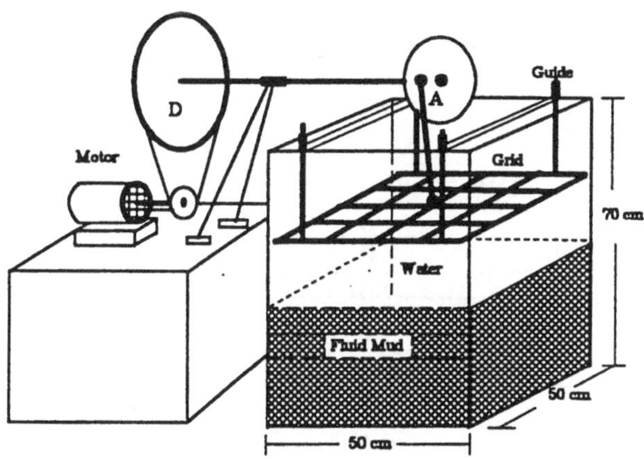

Fig. 1. Oscillating–grid tank (after Scarlatos, 1992).

2.1.2 LOCK–EXCHANGE FLUME

The lock exchange flume was comprised of a plexiglass channel 243 cm long having a cross section of 30x30 cm. The water depth was maintained at 20 cm. At a distance of 10 cm from the one end, a vertically–operating wooden partition was installed, forming a small compartment (Figure 2). Within this compartment sediment slurry was introduced while the rest of the flume was filled with water at the same level as the slurry. When the gate was lifted, the heavier sediment slurry propagated as a density current into the receiving water. The shear production at the sediment slurry–water interface caused sediment entrainment into the water column. Using a 50 ml Class A volumetric pipette with a suction bulb, samples were collected at three locations spaced at 60, 133 and 201 cm from the location of the removable partition. At each location samples were taken at a distance of 5 and 15 cm from the bottom. These conditions of sediment resuspension are found typically in many natural parallel shear flows (Scarlatos and Wilder, 1990).

2.1.2 NUTRIENTS AND SEDIMENTS USED

Diluted standard nitrate solutions and soluble phosphorus (superphosphate – 46% P_2O_5 fertilizer) were the chemicals used to simulate nutrient behavior. The sediments applied were kaolinite, bentonite, kaolinite/bentonite mixtures and natural organic mud from three different locations in Lake Okeechobee, Florida. The ambient water, as well as that used to form the sediment slurries was tap water. The background levels of nutrients in the tap water were found to be 0.01 to 0.03 mg/l of nitrates (NO_3–N) and 7 to 11 μg/l of phosphorus (PO_4–P). These levels did not affect the accuracy of the results since they are one order of magnitude less than the experimental data.

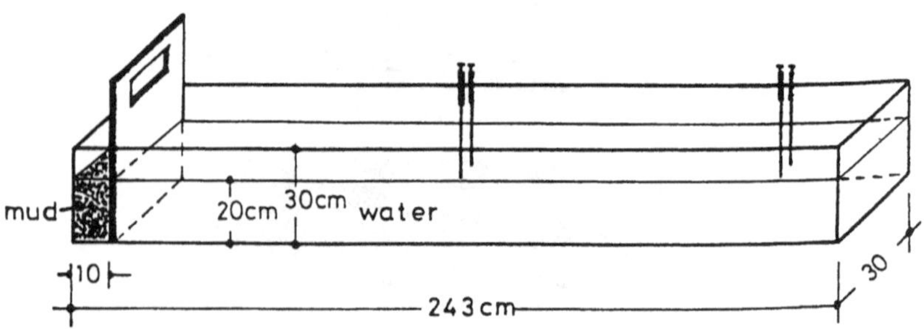

Fig. 2. Lock–exchange flume (after Scarlatos and Wilder, 1990).

2.2 METHODOLOGY

The various sediments were thoroughly mixed with tap water to form concentrations of 50, 66, 200 or 603 g/l dry weight. Then the slurry was "spiked" with either nitrate or soluble phosphorus. The nutrient concentrations varied from 0.5 mg/l to 10 mg/l for nitrates and from 2.5 to 14 mg/l for superphosphate. Once properly prepared, the sediment slurry was carefully placed in the experimental apparatus and the rates of nutrient partition between ambient water and sediments under diffusing (non–mixing) and turbulent (mixing) conditions were quantified. The duration of most of the experiments lasted from 24 to 34 hours . However, in order to investigate long–term effects, in some experiments the sampling period was extended up to five days. In the mixing experiments, the time of oscillatory motion was about fifteen seconds, which was enough to create a uniform mixing of the suspended particulates. Sampling started immediately after the mixing and continued under non–mixing conditions for the specified period of time. In the lock–exchange experiments the mixing/entrainment lasted only few seconds. After the opening of the gate, the high–concentration sediment slurry moved very fast as a near–the–bottom density current. However, a much lighter concentration of suspended sediments moved throughout the water column and remained in suspension for a considerably longer period of time. The water samples collected were filtered through a 1 μm Whatman filter paper or through a combination of a 11 μm Whatman filter and diatomaceous earth (Celite) and then were chemically analyzed. In some of the experiments, instead of the sediments the ambient water was "spiked" with nutrients. However, the rest of the experimental procedure remained the same. By measuring the time variation of nutrient concentration in the ambient water the adsorption/desorption of nitrates or phosphorus by the sediment particles were properly estimated.

2.3 CHEMICAL ANALYSES

The chemical analysis of the samples was conducted by using either a Spectronic–601 Spectrophotometer or an Alpkin Rapid Flow Analyzer (ARFA). The spectrophotometer had a wave length varying from 195 nm to 999 nm. Some control samples were analyzed by using both instruments for quality assurance of the data. For all of the analyses the "Standard Methods for the Examination of Water and Wastewater" were adopted (American Public Health Association, 1980).

2.3.1 NITROGEN (NO_3)

Nitrates were analyzed using the cadmium reduction standard method. The nitrate (NO_3^-) was reduced to nitrite (NO_2^-) in the presence of cadmium (Cd) granules treated with copper sulfate ($CuSO_4$) solution. The analytical limits of this method for nitrate concentrations range from 0.01 to 1.0 mg/l.

2.3.2 PHOSPHORUS (P_2O_5)

Phosphorus was analyzed by using two different procedures. For the analysis of samples taken from the oscillating–grid tank experiments the stannus chloride method was used. The analytical limits of this method for phosphorus concentrations range from 3 to 7,000 μg/l.

All samples taken from the lock–exchange flume were analyzed using a high–precision Alpkin Rapid Flow Analyzer. The chemical reactions involved were as follows. After perchloric acid digestion of the sample, phosphate was analyzed by the above procedure.

3. Results and Discussion

A total number of sixty five experiments were conducted. For an easier comparison and interpretation of the results, the experiments were grouped into three categories such as: oscillating–grid tank nitrate experiments (OGN), oscillating–grid tank phosphorus experiments (OGP), and lock–exchange flume phosphorus experiments (LEP).

3.1 OSCILLATING–GRID TANK EXPERIMENTS FOR NITROGEN

A total number of 32 experiments were conducted in this experimental group. All of the these experiments were focused on the water–sediment partition of nitrates. The specifics of these experiments are listed in Table I. Samples were collected at all six depths during 0, 1, 6, 12 and 24 hrs. Samples along the depth were numbered upwards: #1 being the one near the bottom and #6 the one near the water surface. After the analysis, the data were plotted in terms of nitrate concentration at various depths versus time. Typical plots for diffusive and turbulent exchange of nitrogen are given for experiments OGN13 and OGN14 (Figures 3 and 4 respectively). In Figure 3, the diffusive transfer of nitrates from the sediments to the ambient water is evident. In Figure 4, nitrates that have been released in the water column during turbulent mixing are adsorbed back to the sediments after the mixing ceases. It is clear however, that this exchange takes place mainly within the first hour of the experiment. Steady–state conditions are reached approximately after six hours from the initiation of the observations.

A critical review of the experimental data resulted in the following general results (Kari, 1993). For kaolinite the partition coefficient of nitrates was found to decrease with increasing sediment concentration. This result is in agreement with the conclusions of O'Connor and Connolly (1980). However, the opposite was found to be true for bentonite, while for the organic mud the results were rather inconclusive. A possible explanation may lie in the fact that bentonite is an expanding clay while kaolinite is not. The rates of nitrogen release from the organic mud was much higher than those from the inorganic minerals. This was attributed to the fact that the organic mud contained much larger particle sizes than either kaolinite or bentonite. Temperature decrease from 20.1°C to 15°C did not significantly affect the rate of nitrate exchange. However, when the temperature was raised from 20.1°C to 25°C nitrates were released from the sediments into the water column (Holdren and Armstrong, 1980).

3.1 OSCILLATING–GRID TANK EXPERIMENTS FOR PHOSPHORUS

A total number of 27 experiments were conducted in this experimental group. All

TABLE I
Oscillating–Grid Tank Experiments with Nitrogen – Nitrates.

Experiment Number	Mixing Conditions	Nitrate Conc. (mg/l)	Sediment Type	Sediment Conc. (g/l)	Filter Size (μm)	Special Effects
OGN1	Diffusive	0.5	Kaolinite	200	11	
OGN2	Turbulent	0.5	Kaolinite	200	11	
OGN3	Diffusive	1.0	Kaolinite	200	11	
OGN4	Turbulent	1.0	Kaolinite	200	11	
OGN5	Diffusive	1.0	Kaolinite	50	1	
OGN6	Turbulent	1.0	Kaolinite	50	1	
OGN7	Diffusive	5.0	Kaolinite	50	1	
OGN8	Turbulent	5.0	Kaolinite	50	1	
OGN9	Diffusive	10.0	Kaolinite	50	1	
OGN10	Turbulent	10.0	Kaolinite	50	1	Temp. 25°C
OGN11	Diffusive	0.5	Bentonite	50	11	
OGN12	Turbulent	0.5	Bentonite	50	11	
OGN13	Diffusive	1.0	Bentonite	50	11	
OGN14	Turbulent	1.0	Bentonite	50	11	
OGN15	Diffusive	0.5	Organic mud 2	50	11	
OGN16	Turbulent	0.5	Organic mud 2	50	11	
OGN17	Diffusive	1.0	Organic mud 2	50	11	
OGN18	Turbulent	1.0	Organic mud 2	50	11	
OGN19	Diffusive	0.5	Organic mud 3	50	11	
OGN20	Turbulent	0.5	Organic mud 3	50	11	
OGN21	Diffusive	0.75	Organic mud 3	50	11	Temp. 15°C
OGN22	Turbulent	0.75	Organic mud 3	50	11	Temp. 15°C
OGN23	Diffusive	1.0	Organic mud 3	50	11	
OGN24	Turbulent	1.0	Organic mud 3	50	11	
OGN25	Diffusive	1.0	Organic mud 3	50	1	
OGN26	Turbulent	1.0	Organic mud 3	50	1	
OGN27	Diffusive	5.0	Organic mud 3	50	1	
OGN28	Turbulent	5.0	Organic mud 3	50	1	
OGN29	Diffusive	0.5	Organic mud 4	50	11	
OGN30	Turbulent	0.5	Organic mud 4	50	11	
OGN31	Diffusive	1.0	Organic mud 4	50	11	
OGN32	Turbulent	1.0	Organic mud 4	50	11	

NOTES: 1) All samples were analyzed with a Spectronic–601 Spectrophotometer using the cadmium reduction method. 2) In all of the experiments nitrates were mixed in the sediment slurry. 3) In OGN10 the temperature was gradually raised within 24 hours from 20.1°C to 24.9°C. In OGN21 and OGN22 the temperature was decreased to 15°C immediately after mixing. 4) In all of the experiments the pH of the ambient water ranged from 6.2 to 6.7. 5) In all of the experiments the temperature of the ambient water ranged from 20.1°C to 20.9°C.

of the experiments were focused on phosphorus exchange between sediments and ambient water under diffusive and mixing conditions. The specifics of those experiments are listed in Table II. Again, data were collected at 0, 1, 6, 12 and 24 hrs intervals unless the experiment lasted longer than one day. As with the previous group of experiments, the data were plotted in terms of phosphorus concentration at various depths versus time elapsed. The experiments were conducted into two phases. During the first phase, phosphorus enriched sediment was kept undisturbed in the bottom for a 24 hr period during which the diffusion release rates were observed. Then the sediment was thoroughly mixed and the observations continued for at least another 24 hrs.

An overall analysis of the data revealed the following general conclusions (Khan, 1993). The phosphate–clay exchange consisted of two stages; a 24 hour rapid exchange

TABLE II

Oscillating–Grid Tank Experiments with Soluble Phosphorus – Superphosphate 46%.

Experiment Number	Mixing Conditions	Phosphorus Concentration (mg/l)	Sediment Type	P–Spiked Medium	Special Effects
OGP1	Diffusive	2.5	Kaolinite	Sediment slurry	
OGP2	Turbulent	2.5	Kaolinite	Sediment slurry	
OGP3	Diffusive	5.0	Kaolinite	Sediment slurry	
OGP4	Turbulent	5.0	Kaolinite	Sediment slurry	
OGP5	Diffusive	5.0	Kaolinite	Ambient water	pH≃4.85
OGP6	Turbulent	5.0	Kaolinite	Ambient water	pH≃4.85
OGP7	Diffusive	5.0	Kaolinite	Ambient water	
OGP8	Turbulent	5.0	Kaolinite	Ambient water	
OGP9	Diffusive	7.5	Kaolinite	Sediment slurry	
OGP10	Turbulent	7.5	Kaolinite	Sediment slurry	Temp. 25°C
OGP11	Diffusive	10.0	Kaolinite	Sediment slurry	
OGP12	Turbulent	10.0	Kaolinite	Sediment slurry	Temp. 27°C
OGP13	Diffusive	5.0	Bentonite	Sediment slurry	
OGP14	Turbulent	5.0	Bentonite	Sediment slurry	
OGP15	Diffusive	5.0	Bentonite	Sediment slurry	Salinity
OGP16	Turbulent	5.0	Bentonite	Sediment slurry	Salinity
OGP17	Diffusive	7.5	Bentonite	Sediment slurry	
OGP18	Turbulent	7.5	Bentonite	Sediment slurry	Temp. 25°C
OGP19	Diffusive	5.0	Organic mud 3	Sediment slurry	
OGP20	Turbulent	5.0	Organic mud 3	Sediment slurry	
OGP21	Turbulent	5.0	Organic mud 3	Sediment slurry	Salinity
OGP22	Diffusive	7.5	Organic mud 3	Sediment slurry	
OGP23	Turbulent	7.5	Organic mud 3	Sediment slurry	
OGP24	Diffusive	10.0	Organic mud 3	Sediment slurry	
OGP25	Turbulent	10.0	Organic mud 3	Sediment slurry	Temp. 25°C
OGP26	Diffusive	5.0	Organic mud 4	Ambient water	
OGP27	Turbulent	5.0	Organic mud 4	Ambient water	

NOTES: 1) All samples were analyzed with a Spectronic–601 Spectrophotometer using the stannous chloride method. 2) All samples were filtered through a 1 μm pore size filter. 3) The concentration for all sediment slurries was 50 g/l. 4) In all of the experiments the pH of the ambient water ranges from 6.2 to 6.8. 5) In all of the experiments the temperature of the ambient water ranged from 20.8°C to 21.2°C. Temperature increase was applied after: 1 hr in OGP10, 48 hrs in OGP12 and OGP25, and 24 hrs in OGP18. 6) OGP15, OGP16 and OGP21 were continued from OGP13, OGP14 and OGP20 by increasing the salinity of the ambient water to 35 ppt after a period of 24 hours.

followed by a much slower exchange that lasted for several days. The second stage may be significant under acidic conditions since dissolution of aluminum from the clay lattice results in formation of metal–phosphate crystals (Chen et al., 1973). Indeed, it was found that for acidic conditions phosphorus adsorption rates were increased (Figures 5 and 6) (Edzwald et al., 1976). For kaolinite, an increase in temperature promoted release of phosphorus while the process was reversed when the temperature was allowed to decrease (Figure 7). The same was true for the organic mud. The bentonite experiments appeared to have little effect from temperature changes. An increase in salinity, did not affect the organic mud experiments. However, an increase in salinity resulted in desorption of phosphorus from the bentonite (Figure 8). This can be explained on the basis that salinity promotes aggregation resulting in reduction of the specific area of the particles. In all of the experiments, as was expected, higher amounts of superphosphate concentrations resulted in higher rates of phosphorus disorption/diffusion.

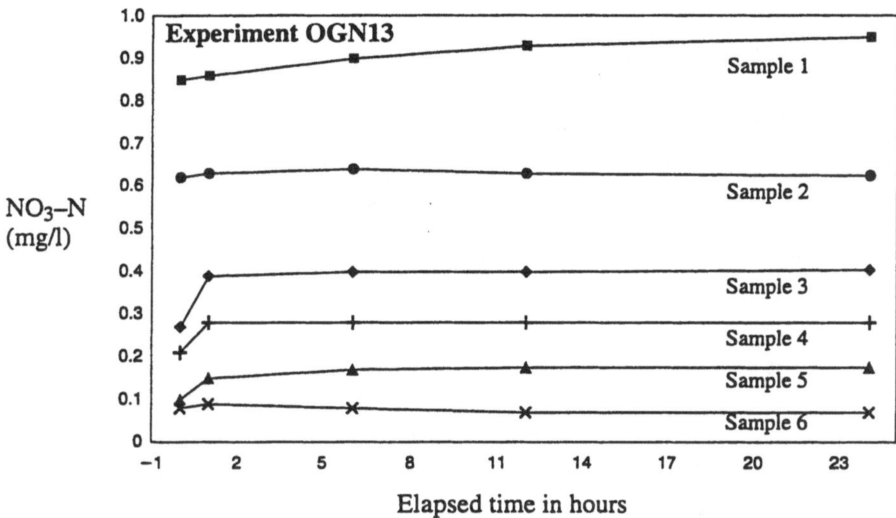

Fig. 3. Experiment OGN13 – Nitrate concentration versus time.

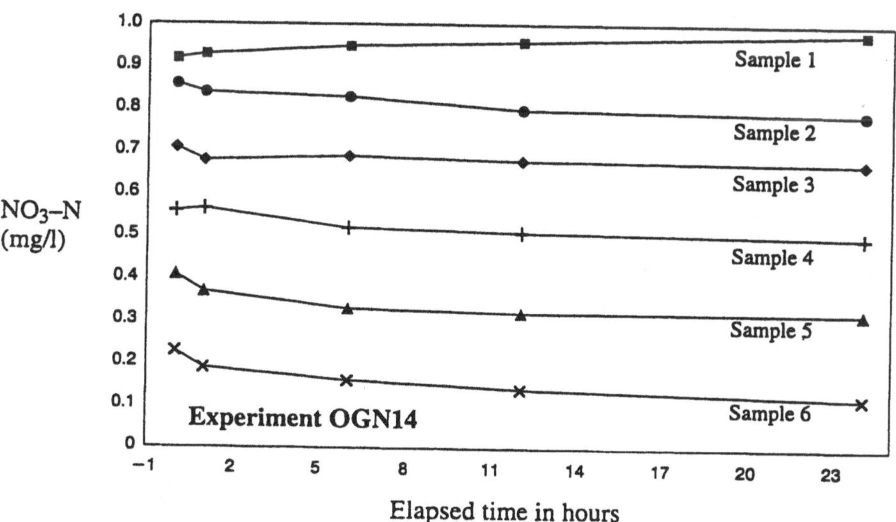

Fig. 4. Experiment OGN14 – Nitrate concentration versus time.

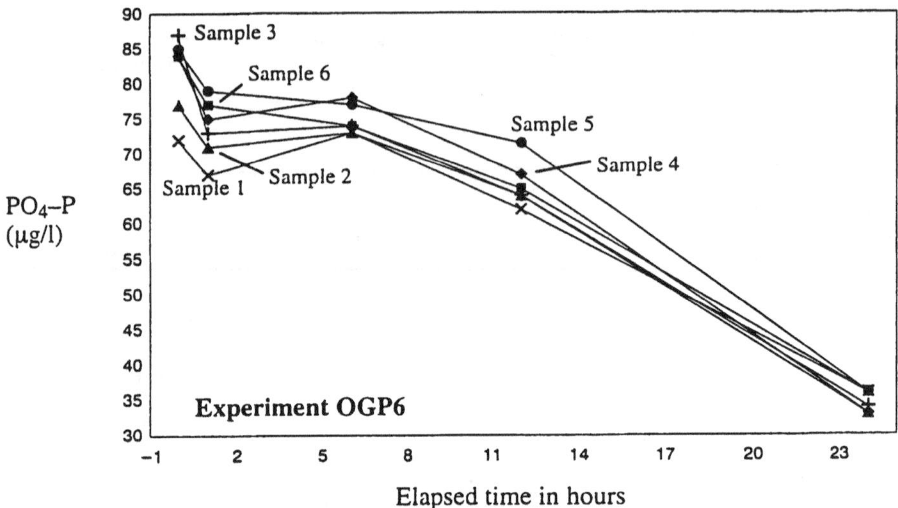

Fig. 5. Experiment OGP6 – Phosphate concentration versus time.

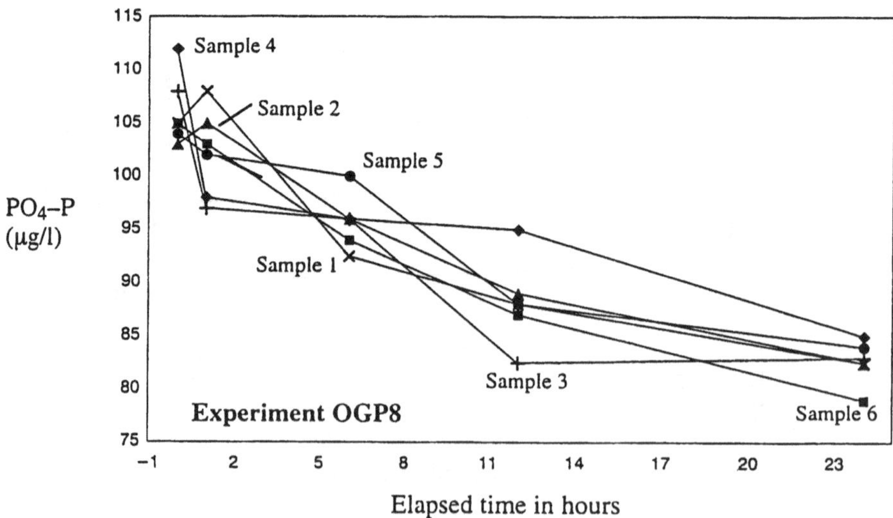

Fig. 6. Experiment OGP8 – Phosphate concentration versus time.

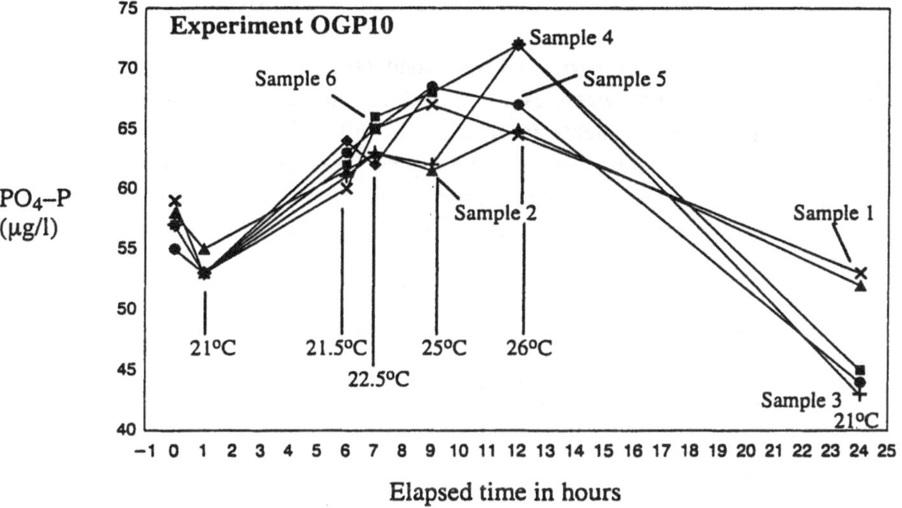

Fig. 7. Experiment OGP10 – Phosphate concentration versus time – Temperature effects.

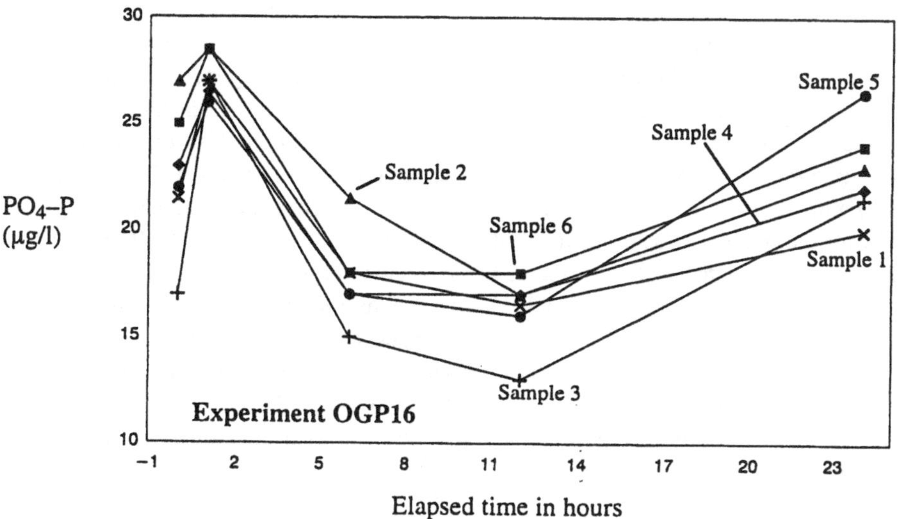

Fig. 8. Experiment OGP16 – Phosphate concentration versus time – Salinity effects.

3.2 LOCK–EXCHANGE FLUME EXPERIMENTS FOR PHOSPHORUS

A total number of 6 experiments were conducted in this experimental group. The purpose of these experiments was to estimate the diffusion coefficients for total phosphorus. The specifics of these experiments are listed in Table III. Six samples were collected at three sections along the flume at two different depths. The samples were numbered starting from the movable partition as 1, 2, 3 for the bottom and similarly 4, 5, 6 for the surface samples. Data were collected at 1, 3, 6 12, 24 and 36 hrs.

TABLE III

Lock–Exchange Flume Experiments with Soluble Phosphorus – Superphosphate 46%..

Experiment Number	Mixing Conditions	Phosphorus Concentration (mg/l)	Sediment Type	Sediment Concentration (g/l)
LEP1	Shear/Diffusion	4.6	Kaolinite	603
LEP2	Shear/Diffusion	9.3	Kaolinite	603
LEP3	Shear/Diffusion	13.9	Kaolinite	603
LEP4	Shear/Diffusion	4.6	Kaolinite/Bentonite	33/33
LEP5	Shear/Diffusion	9.1	Kaolinite/Bentonite	33/33
LEP6	Shear/Diffusion	13.7	Kaolinite/Bentonite	33/33

NOTES: 1) All samples were analyzed with an Alpkin Rapid Flow Analyzer using the ascorbic acid method. 2) All samples were filtered through a 1 μm pore size filter. 3) In all of the experiments nitrates were mixed in the sediment slurry.

Analysis of the data led to the following general conclusions (Donovan, 1993). Initially, phosphorus concentration near the bottom was higher than near the water surface. However, as time elapsed diffusion effects resulted in a more homogeneous system (Figure 9). The estimated diffusion coefficients for low and high phosphorus concentrations ranged from 0.84 to 0.06 cm^2/day for kaolinite and from 0.91 to 0.014 cm^2/day for kaolinite/bentonite mixtures (Donovan, 1993). These values are comparable to other diffusion coefficients reported in the literature (Van Rees et al., 1991) and (Krom and Berner, 1980). One reason for the higher diffusion coefficients found between experiments LEP1 to LEP3 in comparison to LEP4 to LEP6 is attributed to the fact that during resuspension, kaolinite particles entrained the water column and remained in suspension for long periods of time. Thus, diffusion occurred both from the bed and the suspended particulates. On the contrary, during the release of the kaolinite/bentonite mixture the slurry was combined primarily near the bottom without any substantial entrainment into the water. In addition, kaolinite is generally inactive in adsorbing water and any dissolved matter in its lattice compared to the very active bentonite.

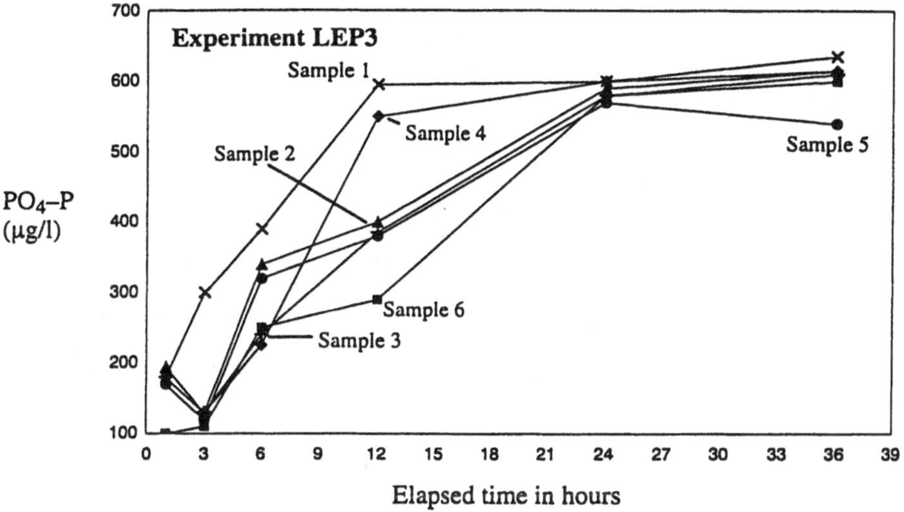

Fig. 9. Experiment LEP3 – Phosphate concentration versus time.

4. Summary and Conclusions

Experiments were conducted in order to assess the exchange of nitrate and soluble phosphorus between cohesive sediments and ambient water under different physicochemical conditions. The general conclusions derived from this study are:

- The partition coefficients of nitrates in water–kaolinite systems tend to decrease with increasing sediment concentration. The opposite is true for water–bentonite mixtures.
- During turbulent mixing, nitrates are released into the ambient water. Part of the released nitrates are adsorbed back into the sediments after ceasing of the mixing.
- The rates of molecular diffusion of nitrates from the sediments are much higher during the first hour. The rate is substantially decreased during the next few hours. Equilibrium is achieved after six to twelve hours.
- Larger particle diameters may lead to higher diffusion rates.
- Temperature variation from 20°C to 15°C did not affect the rates of nitrate desorption. However, the rates were increased when the temperature was raised from 20°C to 25°C.
- Phosphorus diffusion from the sediments has a much higher rate during the first 24 hours of the phenomenon. The rate is substantially reduced after that initial period of time.
- Acidic conditions lead to phosphorus desorption from kaolinite sediments.

- Temperature increase promotes phosphorus desorption from kaolinite and organic sediments. The desorbed phosphorus is adsorbed back into the sediments whenever the temperature is reduced to its original level. Diffusion from bentonite sediments appears to be insensitive to temperature variations.
- Increased salinity promotes phosphorus desorption from bentonite while organic sediments are not affected. The explanation given is that salinity promotes aggregation which subsequently reduces the specific surface of clay particles.
- Higher phosphorus concentrations in sediments lead to higher rates of diffusion.
- Under shear flow conditions kaolinite may result in higher rates of phosphorus diffusion than bentonite. This is due to the fact that kaolinite particles entrain very easy into the water column and remain in suspension for long periods of time (few days). In addition, kaolinite is relatively inactive in adsorbing water along and any dissolved materials into its lattice, while bentonite is very active.

Acknowledgements

The author would like to acknowledge of the financial support provided by the U.S. Geological Survey (Grant # 14–08–001–G2012) and the South Florida Water Management District (Contract # C91–2237). Also special thanks are extended to my graduate students Dr. W.C. Donovan, R. Kari and M.A.B. Khan that conducted the experiments and assisted in the interpretation and analyses of the results.

References

American Public Health Association: 1980, *Standard Methods for the Examination of Water and Wastewater*, 16th Ed., Port City, Maryland.

Berner, E.K. and Berner, R.A.: 1996, *Global Environment – Water, Air, and Geochemical Cycles*, Prentice Hall, Englewood Cliffs, New Jersey.

Bonner, J.S., Ernest, A.N., Autenrienth, R.L and Ducharme, S.L.: 1994, Parameterizing Models for Contaminated Sediment Transport, in: *Transport and Transformation of Contaminantsd Near the Sediment–Water Interface*, (DePinto, J.V., Lick W. and Paul, J.F., Eds) Lewis–CRC Press, Boca Raton, Florida.

Carberry, J.B.: 1990, *Environmental Systems and Engineering*, Saunders College Publishing, Philadelphia.

Chen, Y.S., Butler, J.N. and Stumm, W.: 1973, *Environ. Sci. and Tech.*, **7**, 327–332.

Donovan, W.C.: 1993, *The Laboratory Measurement of Soluble Phosphorus Diffusion Coefficients in Kaolinita and Bentonite Clay Sediments*, MS Thesis, Dept. of Ocean/Civil Engineering, Florida Atlantic University, Boca Raton, Florida.

Edzwald, J.K., Toensing, D.C. and Leung, M.C.Y.: 1976, *Environ. Sci. and Tech.*, **10**, 485–490.

Gächter, R. and Meyer, J.S.: 1990, Mechanisms Controlling Fluxes of Nutrients Across the Sediment/Water Interface in a Eutrophic Lake, in: *Sediments: Chemistry and Toxicity of In–*

Place Pollutants, (Baudo, R., Giesy, J. and Muntau, H., Eds.) Lewis Publishers, Chelsea, Michigan.
Holdren, G.C. and Armstrong, D.E.: 1980, *Environ. Sci. and Tech.*, **14**, 79–87.
Itsweire, E.C., Helland, K.N. and Van Atta, C.W.: 1986, *J. Fluid Mech.*, **162**, 299–338.
Kari, R.: 1993, *Nitrate partitioning Between Water and Sediments*, MS Thesis, Dept. of Ocean/Civil Engineering, Florida Atlantic University, Boca Raton, Florida.
Khan, M.A.B.: 1993, *Oscillating–Tank Experiments for Quantification of Water–Sediment Phosphorus Exchange*, MS Thesis, Dept. of Ocean/Civil Engineering, Florida Atlantic University, Boca Raton, Florida.
Krom, M.D. and Berner, R.A.: 1980, *Limnology and Oceanography*, **25**(2), 327–337.
Medina, A.J. and McCutcheon, S.C.: 1989, Fate and transport of sediment–associated contaminants, in: *Hazard Assessment of Chemicals*, (Saxena J., Ed.) Hemisphere Publishing Corp., New York, New York.
O'Connor, D.J. and Connolly, J.P.: 1980, *Water Research*, **14**, 1517–1523.
Scarlatos, P.D.: 1992, Erosion of a Thin Lutocline Under Homogeneous Turbulence, in: *Hydraulic Engineering* (Jennings, M. and Bhowmik, N.G., Eds.) ASCE, Baltimore, Maryland, 263–268.
Scarlatos, P.D.: 1996, Ecohydrodynamics, Chapter 10, In: *Environmental Hydraulics* (Singh, V.P. and Hager, W.H., Eds.) Kluwer Academic Publishers, Amsterdam, 349–397.
Scarlatos, P.D. and Wilder, B.J.: 1990, Experimental Investigation of Density–Driven Hyperconcentrated Flows, in: *Hydraulics/Hydrology of Arid Lands*, (French, R.H., Ed.) ASCE, San Diego, California, 633–638.
Stumm, W. and Morgan, J.J.: 1981, *Aquatic Chemistry*, 2nd Ed., John Wiley and Sons, New York, New York.
Van Rees, K.C.J., Sudlcky, E.A., Rao, S.L. and Reddy, K.R.: 1991, *Environ. Sci. and Tech.*, **25**(9), 1605–1611.

VARIATIONS OF PHOSPHORUS RELEASE FROM SEDIMENTS IN STRATIFIED LAKES

T. GONSIORCZYK, P. CASPER and R. KOSCHEL

Institute of Freshwater Ecology and Inland Fisheries, Department of Limnology of Stratified Lakes
Alte Fischerhuette 2, D-16775 Neuglobsow, Germany

Abstract. The aim of this investigation was to study the temporal variation in phosphorus release from the sediments and its influence on water quality of stratified lakes. The concentrations of soluble reactive phosphorus (SRP), calcium and sulfate in the interstitial water and the pH in the wet sediments of dimictic lakes were investigated during the spring circulation and at the end of summer stratification. Multiple regression analysis using the calculated diffusive fluxes of SRP out of the sediments and the morphometric characteristics of the lakes (reduced water depth), explained 73 % of the variance of the SRP-accumulation in the hypolimnia during summer stagnation. At the end of summer stratification diffusive fluxes of SRP out of the sediments increased and pH-values and sulfate-concentrations decreased at the sediment surface (0-2 cm) and in the hypolimnia. The maximum diffusive flux of SRP was calculated to be 5.8 $mg/m^2/d$ at the end of summer stagnation. Probable reasons for these higher diffusive fluxes of SRP at the end of summer stagnation are higher supply of labile organic matter and thereby higher mineralization rates, lower redox potential and thus higher dissolution of redox sensitive P-binding forms and/or dissolution of phosphorus being bound to Ca-phases at lower pH.
Key words: Lakes, sediments, interstitial water, phosphorus release, seasonal variations

1. Introduction

The concentrations of most dissolved components in the profundal of stratified lakes show drastic concentration gradients at the sediment-water interface (Weiler 1973, Emerson 1976). They are the result of diagenetic processes in the sediment and indicate the extent of exchange of dissolved compounds between the interstitial water and the water above the sediment (Kamp-Nielsen 1974). Microbial processes as well as desorption and dissolution processes of inorganic substances lead to a concentration increase of some dissolved substances in the interstitial water that may diffuse out of the sediments (Emerson 1976, Sinke *et al.*, 1990). On the other hand, there is a decrease in the concentrations of dissolved components that are consumed, precipitated or adsorbed within the sediments. In stratified lakes diffusion is the most important transport mechanism to exchange dissolved substances between interstitial water and the water above the sediment (Boström *et al.*, 1982). The release of phosphorus out of the sediments may have an important influence on the water quality of lakes. The mobilization of phosphorus is affected by a multitude of factors (e.g. mineralization, redox potential, pH) and in various lakes different factors can be dominant (Boström *et al.*, 1982, Caraco *et al.*, 1991).

Water, Air and Soil Pollution **99**: 427-434, 1997.

In this study we estimated the extent of P-release from the sediments and its effect on the SRP-accumulation in the hypolimnia of stratified lakes of north eastern Germany. During the summer stratification drastic changes in the chemical conditions of the hypolimnia usually occur. Their effects on the spatial and temporal variations of the concentrations in the interstitial water were investigated. The concentrations of soluble reactive phosphorus (SRP), calcium and sulfate in the interstitial water and the pH in the wet sediments were analyzed in 20 stratified lakes during spring circulation. Twelve of these lakes were investigated again at the end of summer stagnation when changes in the chemical conditions in the hypolimnia expected to reach an extreme.

2. Materials and methods

The lakes investigated are located in north eastern Germany (State of Brandenburg). All lakes with the exception of oligotrophic Lake Stechlin were characterized by an anoxic hypolimnion at the end of summer stagnation (Mietz *et al.*, 1994). Important trophic and morphometric characteristics of the lakes are listed in Table I. As a result of calcite precipitation the sediments of most of the studied lakes are rich in calcium carbonate (> 30 % of dry weight, 0-5 cm, except Pe, Wi, Ns, Wu, Nn) (Gonsiorczyk and Casper 1995).

The investigations were carried out during 1994. In general, the sediment cores (diameter: 5 cm, length: 12-40 cm) were taken at the deepest location of the lakes by use of a Jenkin-surface-mud sampler and subsequently were cut into slices of 2 cm (Casper 1992). The interstitial water was obtained by centrifugation followed by filtration through cellulose-nitrate-filters (0.45 µm). The interstitial water samples for determination of Ca^{2+} concentrations were fixed by adding 0.2 ml 0.1 N HCl to 10 ml of sample volume and were stored at 4°C. All others were frozen.

SRP and Ca^{2+} of the interstitial water were analyzed with a FIAstar 5010-Analyzer (Tecator) as well as the concentrations of NH_4^+ and total phosphorus (TP) of the lake water. TP was transformed into SRP by adding 0,5 g $K_2S_2O_8$ to 50 ml of lake water and by allowing it to react for 30 min at 134°C. Sulfate was analyzed with an ion chromatograph (Jasco) equipped with a conductivity detector (Shodex). Values for pH were determined in wet sediments with a micro processor pH-Meter pH 3000 (WTW).

The diffusive fluxes were calculated according to Fick´s first law of diffusion (Berner 1976):

$$J = \Phi D_s \, dc/dx \qquad (1)$$

with J = diffusion rate ($g/m^2/d$), Φ = porosity (m^3 interstitial water/m^3 wet sediment), D_S = diffusion coefficient in the sediment (m^2/d), dc = concentration difference (g/m^3), dx = diffusion distance (m). As summarized by Clavero *et al.*, (1992) the diffusion coefficients in the sediment can be expressed as:

$$D_s = D_i \, \Phi/F \qquad (2)$$

with $F = 1.28/\Phi^2$ (3)

and D_i = molar diffusion coefficients for various substances (m^2/d). The molar diffusion coefficient for H_2PO_4 was interpolated from data of Li and Gregory (1974) for a temperature of 5°C ($D = 4.08\ 10^{-5}\ m^2/d$). The water temperature above the sediments was on average 4.0 ± 1.5 °C during the spring circulation and 6.5 ± 1.0 °C at the end of summer stagnation. The concentration gradients between the upper layer of the sediments (0-2 cm) and the water above the sediment cores were used as the basis for the diffusion calculations (dx = 0.01 m). The porosity was calculated from the composition of dry material according to Röhrs (1986) and was on average 0.967±0.017.

TABLE I

Trophic and morphometric characteristics of investigated lakes and interstitial water composition
z = depth at the sampling point, chl a = chlorophyll a, TP = total phosphorus (chl a and TP: Mean for epilimnion, 1992-1994 (max. 4 samples per year), n.d. = not determined
[1]according to Mietz et al. (1994) [2]mean concentrations during spring circulation (0-10 cm), sulfate measured above the sediment cores [3]TP and chl a Data of 1995 (12 samples per year) [4]Lakes that were also investigated at the end of summer stagnation [5]calculated from the median H^+-concentration

Lake name		Trophic state[1]	Morphometry z (m)	A (km^2)	Lake water chl a (µg/l)	TP (µg/l)	Interstitial water[2] SRP (mg/l)	Ca^{2+} (mg/l)	SO_4^{2-} (mg/l)	pH (-)
Stechlinsee[3,4]	(St)	oligotrophic	32	4.25	1.8	17	0.37	54	31	7.65
Wummsee[4]	(Wu)	mesotrophic	35	1.52	1.5	16	0.24	66	32	7.37
Wittwesee	(Wi)		10	1.63	1.3	19	0.03	54	34	7.43
Nehmitzsee-South.[3,4]	(Ns)		18	0.60	2.1	20	0.03	80	20	7.12
Nehmitzsee-North.[3,4]	(Nn)		18	1.00	2.1	20	0.03	76	21	7.34
Peetschsee[3,4]	(Pe)	slightly	16	1.00	2.8	19	0.75	41	13	6.70
Roofensee[4]	(Ro)	eutrophic	17	0.59	3.2	24	1.30	104	27	7.07
Boberowsee	(Bo)		11	0.20	5.6	30	0.43	89	24	7.51
Zootzensee	(Zo)		20	1.67	6.0	29	3.15	103	23	7.05
Gr. Küstrinsee[4]	(Kü)		18	2.25	7.8	43	0.66	112	31	7.21
Haussee (Hardenbeck)[4]	(Hs)	eutrophic	35	1.72	5.3	50	0.93	100	70	7,20
Zechliner See	(Ze)		36	2.16	5.6	45	1.72	92	59	7.11
Wurlsee[4]	(Wl)		28	1.64	5.8	46	1.78	69	n.d.	7.05
Netzowsee	(Ne)		10	1.15	7.0	44	0.98	95	63	7.53
Zenssee[4]	(Zn)		24	1.12	9.7	43	1.44	92	21	7.16
Tornowsee	(To)		7	1.32	14.8	48	0.07	82	22	7.77
Rheinsberger See	(Rh)		27	2.70	25.8	52	2.00	95	72	7.01
Gr. Lychensee[4]	(Ly)	highly	18	2.87	19.8	73	3.97	100	14	6.91
Ruppiner See-South.[4]	(Rs)	eutrophic	20	6.00	36.0	80	4.27	101	92	6.97
Zermützelsee	(Zr)		7	1.33	54.3	97	0.24	98	56	7.53
Median value			**18**	**1.58**	**5.7**	**43**	**0.84**	**92**	**31**	**7.14**[5]
Standard deviation			9	5.64	13.3	22	1.25	19	22	0.30
Maximum			36	6.00	54.3	97	4.27	112	92	7.77
Minimum			7	0.20	1.3	16	0.03	41	13	6.70

A box model was used to compare the calculated diffusive fluxes of SRP out of the sediments with the accumulation rates of SRP in the hypolimnia (0.5 m above the sediment). It was assumed that SRP that is diffusing out of a sediment surface of 1 m² is accumulating in a volume of 1 m³ above the sediment. The reduced water depth (z_{red}) was calculated to describe the morphometric characteristics of the lakes (Mietz and Vietinghoff 1994):

$$z_{red} = z_{max} / \sqrt[4]{A} \quad (4)$$

with z_{max} = maximum water depth (m) and A = surface area (m^2).

Regression analyses were done using Spearman coefficients.

3. Results

During spring circulation the median SRP-concentrations in the interstitial water were 43 times (range = 9-332) higher than in the lake water (Table II). During the summer stratification SRP accumulated in the hypolimnia of all lakes. At the end of summer stagnation the median concentrations of SRP in the hypolimnia were 34 (range = 2-134) times higher than in the lake water during spring circulation. The concentrations of SRP in the upper 2 cm of the sediments and the concentration gradients at the sediment-water interface increased drastically at the end of summer stagnation. Accordingly, the calculated diffusive fluxes of SRP were higher at the end of summer stagnation (Median: 2.3 ± 1.7, 0.2-5.8 $mg/m^2/d$) than during the spring circulation (Median: 0.5 ± 1.3, 0.0-4.7 $mg/m^2/d$).

During the spring circulation only 22 % (range = 8-58 %) of the TP in the pelagial was present as SRP (Table II). However, at the end of summer stagnation 84 % (range = 32-100 %) of the TP was SRP.

Similarly, the concentrations of NH_4^+ are increasing drastically in the hypolimnia during the summer stratification (Table II). The concentrations of SRP and NH_4^+ in the hypolimnia at the end of summer stagnation were highly correlated:

$$SRP\ (mg/l) = 0.023 + 0.148\ NH_4^+\ (mg/l),\ (r^2 = 0.79,\ p < 0.001,\ n = 12) \quad (5)$$

TABLE II

Variability of dissolved substances in the interstitial water (0-2 cm) and lake water of 12 dimictic lakes (Median ± standard deviation, minimum and maximum)

[1]measured above the sediment cores [2]not determined in Wl [3]not determined in St

	SRP (µg/l)	TP (µg/l)	NH_4 (µg/l)	SO_4^{2-} (mg/l)	Ca^{2+} (mg/l)	pH
Spring circulation						
Pelagial (0 m)	5 ± 10	23 ± 23	50 ± 53	27 ± 24[1]	58 ± 15	8.39 ± 0.31
	(2-35)	(15-100)	(0-180)	(13-92)	(32-81)	(8.07-9.04)
interstitial	213 ± 507	n.d.	n.d.	14 ± 12[2]	81 ± 18[3]	7.44 ± 0.30
water	(18-1780)			(1-45)	(47-113)	(6.77-7.88)
End of summer stagnation						
0.5 m above the	172 ± 235	204 ± 257	1205 ± 1423	17 ± 13[1]	59 ± 15	7.16 ± 0.26
sediment	(11-750)	(22-830)	(14-4850)	(7-51)	(30-83)	(6.59-7.70)
interstitial	1083 ± 1009	n.d.	n.d.	4 ± 7[2]	81 ± 20[3]	7.31 ± 0.30
water	(114-3294)			(0-25)	(32-100)	(6.89-8.04)

In all lakes the sulfate concentrations decreased with increasing sediment depth. In relation to the concentration of sulfate above the sediment a decrease of already more than 90 % occurred in the sediments (0-2 cm) of some lakes during the spring circulation (Ly, Zn, Rh, Hs). The sulfate concentrations in the interstitial water and in the water above the sediments declined during summer stagnation (Table II).

During spring circulation the median calcium concentrations in the interstitial water were 1.4 (range = 1.1-2.1) times higher than in the lake water (Table II). In all sediments a slight increase in the calcium concentrations was observed with increasing sediment depth. The highest concentration gradients were found usually at the sediment-water interface.

During spring circulation the pH-values in the lake water were clearly higher than in the sediments (Table II). In all sediments the pH decreased with increasing sediment depth, whereby the highest gradients where found at the sediment-water interface. The pH-values in the hypolimnia and in the upper 4 cm of most of the sediments decreased during the summer stagnation.

4. Discussion

4.1. Effects of Diagenetic Processes

The highest concentration gradients of SRP were found usually at the sediment-water interface and may be a product of mineralization of recently sedimented organic substances (Eckerrot and Pettersson 1993, Sinke *et al.*, 1990). Steep gradients of sulfate concentrations at the sediment-water interface (see also Table II) indicate a high rate of microbial sulfate reduction in the upper sediment layers. However, the concentrations of SRP continue to rise with increasing sediment depth (especially during spring circulation) indicating that the maximum production zone of SRP is located in deeper sediment layers (Staudinger *et al.*, 1990).

The mineralization of organic substances can lead to a decrease in pH due to the production of CO_2 (Weiler 1973, Driscoll *et al.*, 1993). The rise of the calcium concentrations in the interstitial water (Table II) can be attributed to the dissolution of calcite crystals in the CO_2-rich interstitial waters (Weiler 1973, Emerson 1976). However, we did not observe an accumulation of calcium in the hypolimnia during the summer stagnation.

4.2. SRP-Diffusion out of the Sediments and Accumulation in the Hypolimnion

The increase of TP in the hypolimnia is first of all a result of the accumulation of SRP (Table II). However, the accumulation of SRP could be a result of P-release from the sediments or from sedimenting particles. With the box model 51 % of the variance in SRP-accumulation in the hypolimnia during the stratification period could be accounted for using the calculated mean diffusive fluxes of SRP (Figure 1 a). Forty-five percent of the variance in SRP-accumulation in the hypolimnia could be accounted for using the reduced water depth (Figure 1b). Multiple regression analysis was able to explain 73 % of the variance in SRP-accumulation in the hypolimnia during the summer stagnation:

$$ACC = 1.81 + (0.52\ DIFF - 1.89\ z_{red}),\ (r^2 = 0.73,\ p = 0.003,\ n = 12) \quad (6)$$

ACC = Mean accumulation rate of SRP above the sediment (April-October) in ($mg/m^3/d$)
DIFF = Mean diffusion rate of SRP out of the sediment (April and October) in ($mg/m^2/d$)
z_{red}: see (4)
This indicates that the release of phosphorus from the sediments is most relevant to the accumulation of SRP in the hypolimnia.

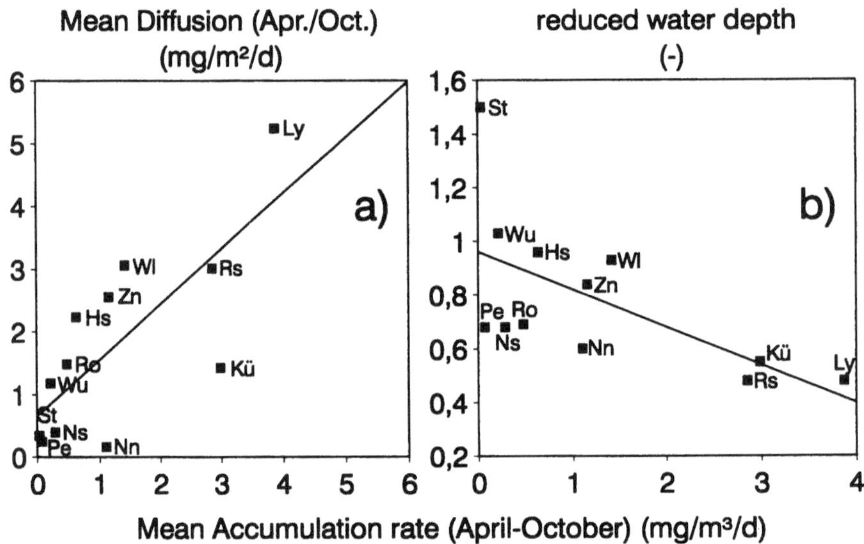

Fig. 1. SRP-accumulation 0.5 m above the sediment and
a) mean SRP-diffusion out of the sediments ($ACC = 0.12 + 0.64\ DIFF$, $r^2 = 0.51$, $p = 0.009$, $n = 12$)
b) reduced water depth ($ACC = 3.43 - 2.77\ z_{red}$, $r^2 = 0.45$, $p = 0.018$, $n = 12$)

4.3. Hypothesis for the Temporal Variation of Phosphorus Release

The ratio between accumulation and consumption processes during the summer stagnation led to an increase in the SRP-concentrations and a decrease in sulfate concentrations and pH values in the hypolimnia and at the sediment surface (Table II). However, the increase of the SRP-diffusion out of the sediments must be a result of changes in the physical/chemical and/or biological milieu. In polymictic lakes an increased P-release in the summer months often was found and attributed to higher temperatures, pH-values and/or a higher supply of labile organic matter (Boers and de Bles 1991, Eckerrot and Pettersson 1993). Also, in dimictic lakes an increase in P-release was observed during the summer (Wisniewski and Planter 1987, Driscoll *et al.*, 1993). In contrast to polymictic lakes the pH-values in the hypolimnia of the most stratified lakes decrease during the stagnation periods. The very small temperature differences in the hypolimnia could only have a minimal influence on the P-release. In stratified lakes the increased SRP-diffusion at the end of summer stagnation may be due to the following:

a) Higher supply of unstable organic matter leading to higher mineralization rates
b) Lower redox potential leading to higher dissolution of redox sensitive P-binding forms
c) Dissolution of phosphorus that is bound to Ca-phases at lower pH

a) By the degradation of organic matter, C-, N- and P-compounds are apparently released with a similar ratio (Boers and de Bles 1991, Sinke *et al.*, 1990). Accordingly, an enhanced mineralization rate of organic substances leads to a higher mobilization of these compounds. Many investigations have shown a positive relationship between the release of SRP and dissolved inorganic carbon (Driscoll *et al.*, 1993, Sinke *et al.*, 1990, Gonsiorczyk *et al.*, 1995). The N/P-ratio of SRP and NH_4^+ in the hypolimnia at the end of summer stagnation was 6.8 g N/ g P (equation 5) very close to the Redfield-ratio (Redfield *et al.*, 1963) of 7.2 g N/ g P. This indicates that the P-release is completely induced by the microbial degradation of organic matter.

b) The mineralization of organic substances also could have an indirect effect on P-release, due to lowering the redox potential (Boström *et al.*, 1982, Eckerrot and Pettersson 1993). The depletion of oxygen and the decrease in sulfate concentrations in the hypolimnia at the end of summer stagnation (Table II) are indicative of a lowering of the redox potential. This could lead to the dissolution of redox sensitive binding forms like Fe- and Mn-bound phosphorus.

c) Many investigations have shown an enhanced P-release with increasing pH (Pettersson and Boström 1990, Boers 1991). However, in calcareous sediments a decrease in pH may lead to a dissolution of phosphorus that is bound to Ca-phases (Driscoll *et al.*, 1993, Staudinger *et al.*, 1990). In the investigated lakes higher SRP-concentrations in the interstitial water coincided with lower pH values ($r^2 = 0.50$, $p < 0{,}001$, $n = 20$). Furthermore, in the sediments having a calcium carbonate content of more than 30 % of dry weight (0-5 cm), the diffusion rates of SRP and calcium were positively correlated to each other at the end of summer stagnation ($r^2 = 0.70$, $p < 0.001$, $n = 8$).

5. Conclusions

For mass balances of phosphorus in lakes, it is imperative to know the extent and the mechanisms of P-release from the sediments. Diffusion of SRP from the sediments is the most important cause of SRP-accumulation in the hypolimnia of stratified lakes. Furthermore, distinction between chemical and microbial induced P-release is important. The high correlation between the concentrations of SRP and NH_4^+ in the hypolimnia at the end of summer stagnation indicates that the mineralization of organic matter is most relevant to P-release. The increase in SRP-diffusion at the end of summer stagnation coincides with a decrease in pH and sulfate concentrations in the hypolimnia and at the sediment surface. At the end of summer stagnation the chemically induced P-release due to lower redox potential and/or lower pH may have a higher influence on P-release than in early summer.

Acknowledgments

This investigation was carried out in the framework of the project „Seenkataster Brandenburg" financed by the Ministerium für Umwelt, Naturschutz und Raumordnung des Landes Brandenburg. We thank Olaf Mietz for providing unpublished data on water quality, Peter Kasprzak, Dominik Hepperle, Karl Erich Lindenschmidt and Brook Lemma for their comments on the manuscript and Peter Dietrich, Karin Teske, Uta Mallok, Carola Kasprzak and Ute Beyer for their help in analyzing the samples.

References

Berner, R. A.:1976, Earth Planet Sci. Lett. **29**, 333-340.

Boers, P.: 1991, *Wat. Res.* **25**, 309-311.

Boers, P. and de Bles, F.: 1991, *Wat. Res.* **25**, 591-598.

Boström, B., Jansson, M. and Forsberg, C.: 1982, *Arch. Hydrobiol. Beih. Ergebn. Limnol.* **18**, 5-59.

Caraco, N. F., Cole, J. J. and Likens, G. E.: 1991, *Verh. Internat. Verein. Limnol.* **24**, 2985-2988.

Casper, P.: 1992, *Limnologica* **22**, 121-128.

Clavero, V., Fernandez, J. A. and Niell, F. X.: 1992, *Hydrobiologia* **235/236**, 387-392.

Driscoll, C. T., Effler, S. W., Auer, M. T., Doerr, S. M. and Penn, M. R.: 1993, *Hydrobiologia* **253**, 61-72.

Eckerrot, Å and Pettersson, K.: 1993, *Hydrobiologia* **253**, 165-177.

Emerson, S.: 1976, *Geochim. Cosmochim. Acta* **40**, 925-934.

Gonsiorczyk, T., Casper, P. and Koschel, R.: 1995, *Limnologica* **25**, 365-379.

Gonsiorczyk, T. and Casper, P.: 1995, DGL, Jahrestagung 1995, Berlin, *Erw. Zsfg.*, 108-113.

Kamp-Nielsen, L.: 1974, *Arch. Hydrobiol.* **73**, 218-237.

Li, Y. H. and Gregory, S.: 1974, *Geochim. Cosmochim. Acta* **38**, 703-714.

Mietz, O., Arp, W., Riemer, A., Vietinghoff, H., Henker, K., Wöbecke, K. and Gabrysch, I.: 1994, Bericht des Projektes „Seenkataster Brandenburg". Förderprojekt des Ministeriums für Umwelt, Naturschutz und Raumordnung des Landes Brandenburg.

Mietz, O. and Vietinghoff, H.: 1994, *Wasserwirtschaft* **84**, 662-667.

Pettersson, K. and Boström, B.: 1990, In: Tilzer, M. and Serruya, C. (ed.): Brock/Springer series in contemporary bioscience, Springer-Verlag Berlin, Heidelberg, 427-435.

Redfield, A. C., Ketchum, B. H. and Richards, F. A.: 1963, In: M. N. Hill (ed.): *The Sea*, vol. 2. Interscience, 26-77.

Röhrs, J.: 1986, *Documenta naturae* (München) **31**, 1-105.

Sinke, A. J. C., Cornelse, A. A., Kelzer, P. van Tongeren, O. F. R. and Cappenberg, T. E.: 1990, *Freshwater Biol.* **23**, 587-599.

Staudinger, B., Peiffer, S., Avnimelech, Y. and Berman, T.: 1990, *Hydrobiologia* **207**, 167-177.

Weiler, R. R.: 1973, *Limnol. and Oceanogr.* **18**, 918-931.

Wisniewski, R. J. and Planter, M.: 1987, *Ekol. Pol.* **35**, 219-236.

MODELLING OF PORE WATER NUTRIENT DISTRIBUTION AND BENTHIC FLUXES IN SHALLOW COASTAL WATERS (GULF OF TRIESTE, NORTHERN ADRIATIC)

B. CERMELJ [1], A. BERTUZZI [2] and J. FAGANELI [1]

[1] *Marine Biological Station,Piran, Slovenia,* [2] *Department of Geological, Environmental and Marine Sciences, University ofTrieste, Italy*

Abstract. Pore waters, extracted monthly from short cores at two sedimentologically and biologically different locations (AA1 and F) in the Gulf of Trieste (Northern Adriatic), were analyzed for NH_4^+, NO_3^-, PO_4^{3-}, $Si(OH)_4$, DIC and periodically for DOC, DON and DOP. Nutrient concentrations were used to model vertical profiles using a diffusion-reaction model which included the macrofaunal influence on sediment-water exchange rates. Winter nutrient profiles showed nearly an exponential increase, or decrease in the case of NO_3^-, in nutrient concentrations with depth while the profiles from other seasons exhibited concentration maximum at 3-6 cm, a minimum around 8-10 cm, and then, except for NO_3^-, a gradual increase or constant values. This vertical distribution is attributed to seasonal variations in the benthic infauna activity, mostly composed of polychaetes and bivalves and concentrated in the top 4-5 cm, being less active during the winter. The vertical profiles of DOC and DON showed the vertical distribution described above in all periods, while DOP was similar to that of PO_4^{3-}. The comparison of modelled fluxes of nutrients across the sediment-water interface at the location AA1 and those measured using *in situ* benthic chamber showed quite good agreement for NH_4^+ and PO_4^{3-} fluxes but not for NO_3^- and $Si(OH)_4$. Discrepancies could be caused by a topography effect and for $Si(OH)_4$ by an additional dissolution of the solid phase balancing the diffusive loss into burrows and lessening the effect of bioturbation.

Key words: Adriatic Sea, marine sediments, pore waters,bioturbation, nutrients, modelling

1. Introduction

In order to determine the nutrient recycling rates at the sediment-water interface, it is necessary to estimate their benthic fluxes. Benthic fluxes may be determined from diagenetic models based on pore water and solid phase profiles (Lerman, 1979; Berner, 1980) and compared to those obtained with benthic chambers deployed *in situ*. However, there are still uncertainties in modelling the pore water chemistry as a means of determining benthic fluxes especially where the sediments are bioturbated (Aller, 1982).

In addition to the investigations of *in situ* nutrient benthic fluxes in the Gulf of Trieste, Northern Adriatic (Bertuzzi et al., this issue), we have also modelled the benthic fluxes using the pore water nutrient profiles. The data obtained were compared to the benthic nutrient fluxes measured in a flux chamber deployed *in situ*.

Water, Air and Soil Pollution **99**: 435-444, 1997.

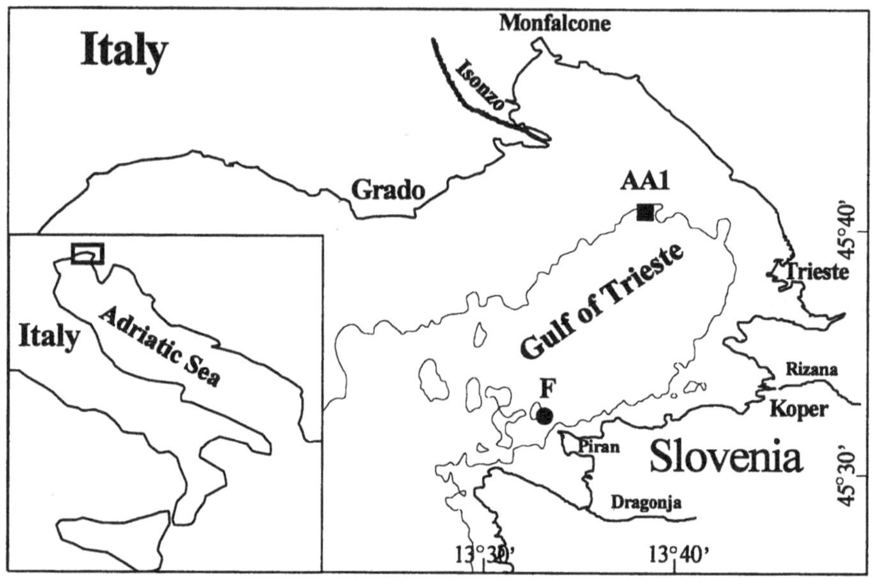

Fig. 1. Location of sampling sites AA1 and F in the Gulf of Trieste (Northern Adriatic)

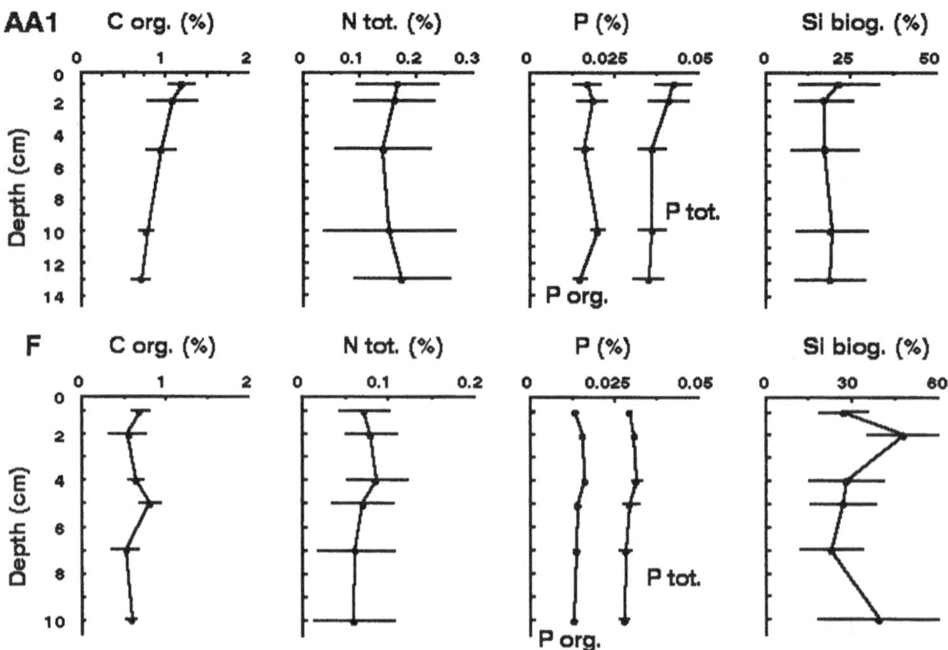

Fig. 2. Vertical distribution of mean values (with standard deviations, n = 15 for AA1 and n = 6 for F) of organic C, total N and P, organic P and biogenic Si contents at sampling sites AA1 and F

2. Materials and methods

2.1. SAMPLING LOCATIONS

The sampling sites AA1 and F were located in the central and SW part of the Gulf of Trieste (Northern Adriatic) at a depth of 20 m (Figure 1). The Gulf of Trieste is a 500 km^2, shallow (max. depth of 25 m) basin isolated from the rest of the Adriatic by a shoal. The salinity of the waters ranges between 33 and 35.5 psu and bottom water temperatures between 8 °C in winter and 22 °C in summer. A density gradient in late summer often results in bottom water hypoxia or even anoxia. The sedimentary detrital material to the southeast is derived mostly from the erosion of the flysch in the hinterland while the northern and western sediments originate primarily from the Isonzo River delivering carbonate material from the Julian Alps (Ogorelec et al., 1991). The sediments are composed primarily of clay and silt along the eastern shore (site AA1) which grade quickly into silty sand (site F) and sand towards the west and northwest (Ogorelec et al., 1991). The ^{210}Pb accumulation rates are 1.8 mm yr^{-1} at site AA1 (Frignani, pers. comm.) and 1.2 mm yr^{-1} at site F (Ogorelec et al., 1991). The carbonate content increases from <30% to >70% along this gradient. Mineralogically, this area is composed predominantly of calcite, aragonite and dolomite, followed by quartz and in lesser amounts of feldspars, heavy (magnetite and micas) and clay (illite, chlorite, kaolinite, montmorillonite) minerals (Ogorelec et al., 1991). The biogenic component is abundant and composed mostly of brittle stars, sponges and tunicates. The sediments are bioturbated further by polychaetes and bivalves. The surficial sediment is populated by a microalgal mat, composed mostly of diatoms (Bertuzzi et al., 1996).

2.2. SAMPLING

Site AA1 was sampled in the period June 1992 - October 1993 and September 1995 - June 1996, and site F in the period November 1991 - December 1992. Virtually undisturbed sediment cores (in duplicate) were collected by SCUBA divers who inserted a polyethylene tube (6.5 cm of internal diameter) into the sediment.

The cores were transported immediately to the laboratory, supernatant water was collected (about 5 cm thick layer above the sediment-water interface) and the sediment was extruded and sectioned into 1 or 2 cm intervals in a N_2-filled glove bag. Anoxic conditions were maintained throughout processing. The wet sediments from each interval were placed in 50 cm^3 tubes and centrifuged at 7000 rpm for 15 min, or placed in a nylon squeezer containing a 0.45 μm Millipore HA filter and driven by N_2 flow under pressure of 0.4 MPa (Reebourgh, 1967) at room temperature. Pore waters extracted by centrifugation were filtered through 0.45 μm Millipore HA filters in an inert atmosphere. No significant difference between the two methods was observed. The solid sediment was used for analyses of organic C, total N and P, organic P, and biogenic Si.

2.3. ANALYSES

Nitrate, ammonium, phosphate, and orthosilicate in pore waters were measured photometrically using standard procedures (Grasshoff et al., 1983). Alkalinity was measured by titration with 0.01 M HCl and then calculating DIC using pH and alkalinity data. The HCl acidified filtrates were analyzed for DOC using a high temperature catalytic oxidation (Shimadzu TOC-500 Total Carbon Analyzer). DON and DOP were determined by the wet oxidation method using potassium persulphate as an oxidant (Grasshoff et al., 1983).

Organic C and total N were determined on freeze-dried and homogenized samples, after acidification (Hedges and Stern, 1984), using a Carlo Erba and Perkin Elmer C-H-N elemental analyzers. Total P was analyzed by extraction with HCl after ignition, and organic P by the difference in P contents of HCl extracts before and after ignition of the samples (Aspila et al., 1976). Dissolved PO_4^{3-} in extracts were determined photometrically (Grasshoff et al., 1983). Biogenic ("amorphous") silica was extracted with Na_2CO_3 (DeMaster, 1981) and dissolved $Si(OH)_4$ determined photometrically (Grasshoff et al., 1983). Porosity (ϕ) was determined by measuring the weight loss of sediments dried overnight at 110 °C and calculating

$$\phi = (M_w/1.0) / [(M_s/2.5)+(M_w/1.0)], \quad (1)$$

where M_w is the weight of water lost on drying, M_s is the weight of dry sediment, and 1.0 and 2.5 are the densities (g/cm^3) of water and sediment, respectively (Johnson et al. 1982).

3. Results and discussion

3.1. SOLID PHASE

Sedimentary organic C, total N and P levels were higher at site AA1 compared to F (Figure 2), but the opposite was true for biogenic Si. At both sites the C, N and P contents generally decreased with depth, while Si showed irregular depth variation. Organic C:N ratios (atomic) at site AA1 ranged mostly between 5-20, while organic N:P (atomic) ratios were mostly in the range 10-30, thus being somewhat higher than the Redfield ratio (Romankevich, 1984), probably due to the preferential degradation of sedimentary organic N and P. On the other hand, the organic C:N and N:P ratios (atomic) at site F were 5-10 and about 16, respectively, being similar to the Redfield ratio and suggesting that the sedimentary organic material is mostly of marine origin (Faganeli et al., 1991). The porosities at site F and AA1 were 0.61 and 0.75, respectively.

3.2. PORE WATERS

A general increase in concentration and steep near-surface gradients were observed for pore water NH_4^+, PO_4^{3-}, $Si(OH)_4$ at site AA1 (Figure 3a) and NH_4^+, PO_4^{3-}, $Si(OH)_4$, DOC, DON and DOP at site F (Figure 3b). Winter nutrient profiles of DIC, NH_4^+, PO_4^{3-}, $Si(OH)_4$ showed a nearly exponential increase in concentration with depth The profiles in other seasons exhibited a concentration maximum at a depth between 2-6 cm followed by a gradual decrease and then a gradual increase or constant values. The behaviour of NO_3^- at site AA1 (Figure 3a) was opposite showing in winter a nearly exponential decrease. In other periods a subsurface minimum, and subsequently a maximum at a depth of about 6-8 cm and then a decrease was found. The described vertical distribution can be attributed more to the seasonal variations of activity, connected with temperature variations, than to the benthic infauna abundance since the temporal variation of abundance at both sites was minimal (Brizzi et al., 1994). The vertical profiles of DOC and DON at site F exhibited larger subsurface maxima in all periods which are due probably to the predominantly macromolecular (humic) nature of the pore water DOM resulting in lower mobility (Burdige et al., 1992). DOP profiles were similar to those of PO_4^{3-} suggesting that DOP is not associated with humic compounds.

3.3. PORE WATER AND FLUX MODELLING

To fit the data we used the cylinder microenvironment concept of Aller (1982) created for cases when populations of infauna of variable sizes and spacing are present in sediment and the dimensions of burrows represent a weighted average of those actually present in the deposits. The steady-state distribution of solute concentration within any cylinder is given by:

$$\frac{\partial C}{\partial t} = 0 = D_s\left(\frac{\partial^2 C}{\partial x^2}\right) + D_s\left\{\frac{\partial}{\partial r}[r(\partial C/\partial r)]\right\} + R \qquad (2)$$

where concentration C has the dimensions of mass per unit volume of pore water, x is the sediment depth (cm), D_s is diffusion coefficient ($cm^2\ sec^{-1}$) corrected for porosity (fraction) and temperature (Li and Gregory, 1974; Krom and Berner, 1980) and r is the radial distance (cm) from the tube axis. R is the rate of production of dissolved solute by organic matter decomposition (in mass per unit volume of pore water per unit time):

$$R = R_0 \exp(-\alpha x) + R_1 + k(C_{eq} - C) \qquad \text{(Aller, 1982)} \qquad (3)$$

where R_0, α, R_1, k are constants, C_{eq} is the equilibrium concentration at which solutes precipitate. In pore water modelling two end-members are considered. The first is the pore water solute controlled by pseudo-zero-order reaction kinetics so that $R = R_0\exp(-\alpha x) + R_1$ which approximates the NH_4^+, PO_4^{3-},DOC, DON and DOP production. The second is the case of $Si(OH)_4$ controlled by the first-order reaction, or $R = k(C_{eq} - C)$ and k is a rate constant (yr^{-1}). If the sediment solute distributions are controlled by one-dimensional vertical transport and reaction processes, the diffusive flux (J_x) of a constituent across the sediment-water interface could be calculated simply from Fick's first law:

$$J_x = -\phi D_s \left(\frac{\partial C}{\partial x}\right)_{x=0} \qquad (\text{mol cm}^{-2}\ \text{yr}^{-1}) \qquad (4)$$

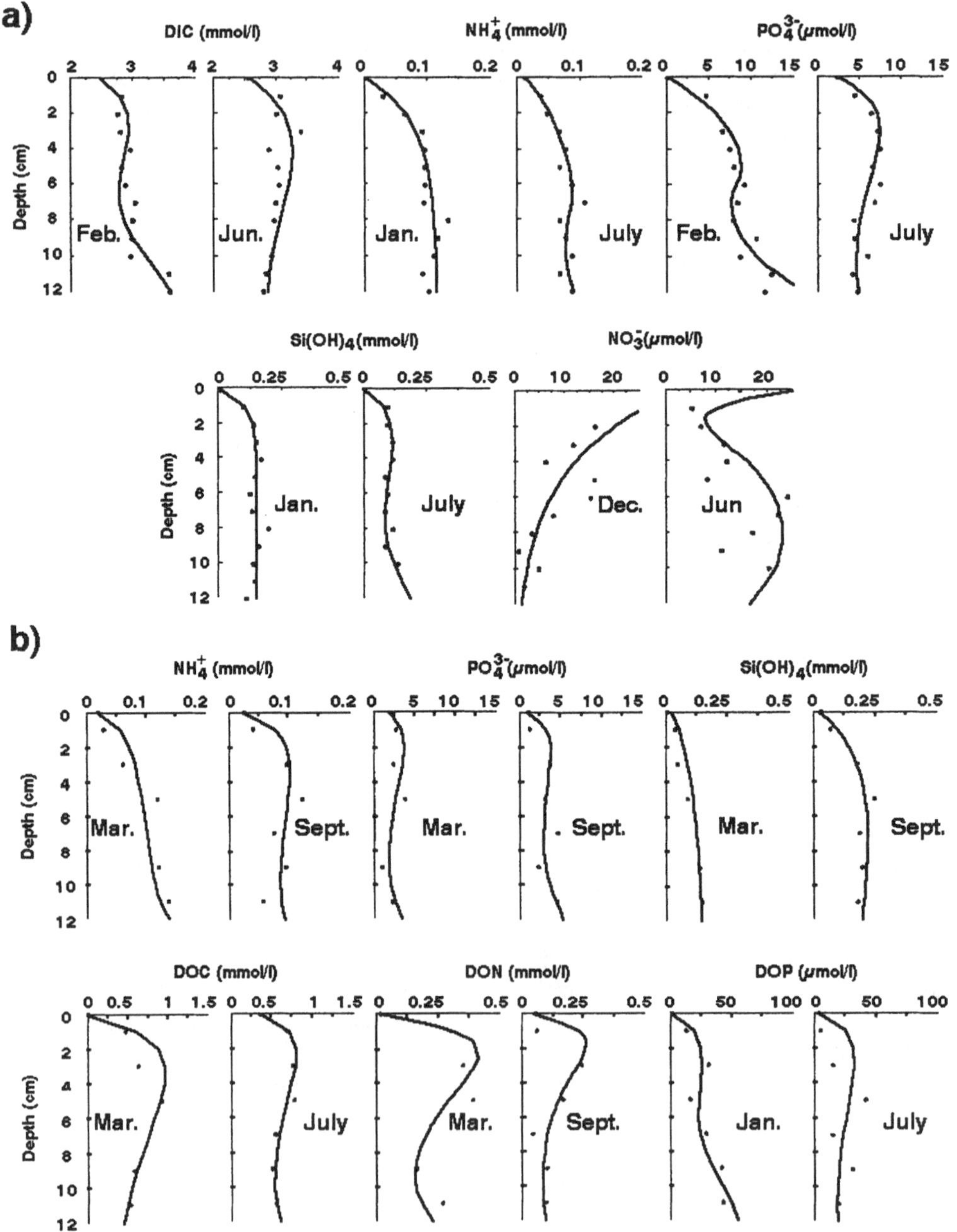

Fig. 3. **a)** Concentration profiles of pore water DIC, NH_4^+, PO_4^{3-}, $Si(OH)_4$ and NO_3^- concentrations at sampling site AA1 in the period 1992/1993 with best fit curves for determining concentration gradients at sediment - water interface.

b) Concentration profiles of pore water NH_4^+, PO_4^{3-}, $Si(OH)_4$, DOC, DON and DOP concentrations at sampling site F in the period 1991/1992 with best fit curves for determining concentration gradients at sediment-water interface.

At steady state, the diffusive flux of NH_4^+, PO_4^{3-}, DOC, DON and DOP is independent of transport (Aller, 1982). If $R = R_0\exp(-\alpha x) + R_1$, over the interval $0 \leq x \leq L$ (L is the depth of the bioturbated zone in cm), the flux at the sediment-water interface (J_T) is:

$$J_T = \frac{\phi R_0}{\alpha}[1 - \exp(-\alpha L)] - \phi R_1 L, \quad (5)$$

assuming no significant flux from the lower layer. For $Si(OH)_4$, subjected only to a first-order reaction rate $R = k(C_{eq} - C)$. If the concentration gradient across L is 0 and the value $(k/D_s)L$ is large the flux at the sediment-water interface is:

$$J_T = \phi\sqrt{k D_s}(C_{eq} - C). \quad (6)$$

In this case the flux is not independent of transport: the greater the effective transport rate (diffusion coefficient) the larger the flux.

In the case of PO_4^{3-} the concentration is dependent on pseudo-zero-order reaction kinetics so that $R = R_0\exp(-\alpha x) + R_1$ and presumably controlled by the first-order reaction as well, assuming that phosphate precipitation occurs from supersaturated solution. However, calculating the phosphate flux, the precipitation has been neglected, due to low concentrations of dissolved phosphate. In the nearby sediments of the River Po delta, pore waters were supersaturated with respect to carbonate fluorapatite only at PO_4^{3-} concentrations higher than 30 µmol/l (Barbanti et al., 1992).

NO_3^- reduction represents a reactive substrate depletion with depth. A reduction rate is then defined as $R = -R_0\exp(-\alpha x)$. The concentration gradient B at the base of the bioturbated zone (L) was set to $B = 0$.

Since the buildup of pore water solutes depends strongly on the size and abundance of infauna, the mean pore water concentration of both solutes NH_4^+ and $Si(OH)_4$ depends on the infauna size (r_1) and the half distance between burrows (r_2) (Aller, 1982). The data on the abundance of the infauna in both sites F and AA1 suggest different values of r_2. The concentration of benthic macrofaunal organisms at the site AA1, mostly composed of polychaetes and bivalves, reaches according to Brizzi et al. (1994) approximately 130 individuals/m^2 in the first cm of the sediment, decreasing rapidly with depth to a value of approximately 30 individuals/m^2 in the depth interval from 5 to 10 cm. This would imply a value of r_2 from 5 to 11 cm. On the other hand, the abundance and the number of species of macrobenthos at site F are much higher. In the depth interval from 7 to 15 cm there are still over 1000 individuals/m^2 in the spring period and approx. 1400 individuals/m^2 in summer implying lower values of r_2 (from 1 to 2 cm). The average concentration from 27 profiles of NH_4^+ at the site AA1 is 105 ± 161 µmol/l. Like the average concentration of $Si(OH)_4$ (116 ± 63 µmol/l) it would imply a much lower half distance between burrows, namely from 1.5 to 3 cm. It appears, hence, that the model of Aller (1982) is difficult to apply in sediments characterized by low density and variable activity of macrofaunal organisms because no activity is included in the model. The flux calculations and concentration profile fitting were therefore performed with lower values of r_2, using higher values (5 - 10 cm) only for the winter situation where lower bioturbation activity is present.

Once established the approximate depth of bioturbation according to the measured NH_4^+ and PO_4^{3-} concentration profiles, at the point of concentration minimum, the value for L was applied to all other solutes taken into consideration. The fitted profiles were generated by fixing r_1 and by varying the r_2 value until the best fit was obtained. The initial value for r_1 of 0.1 cm was chosen on the basis of the size of the polychaetes which were numerically dominant among the species found in the sediment at the site F.

The mean fluxes at the sediment-water interface at sites AA1 and F in the Gulf of Trieste are presented in Table I.

Table I. Mean values (± standard deviation) of modelled (J_T) and diffusive fluxes (J_x) at the sediment water interface of DIC, NO_3^-, NH_4^+, SI(OH)$_4$, PO_4^{3-}, DON, DOC, and DOP at sites AA1 and F in the Gulf of Trieste (mmol m^{-2} d^{-1}) (the number of samplings is in parentheses).

Site		DIC	NO_3^-	NH_4^+	Si(OH)$_4$	PO_4^{3-}	DON	DOC	DOP
modelled fluxes									
F[a]	**(12)**			2.6±3.3	1.5±0.6	0.013±0.01	3.8±3.0	4.4±0.7	0.021±0.017
AA1[b]	**(15)**	2.6±1.4	-0.10±0.08	1.1±0.9	0.6±0.3	0.005±0.005			
AA1[c]	**(12)**		-0.16±0.05	0.9±0.2	0.9±0.2	0.009±0.009			
diffusive fluxes									
F[a]	**(12)**			0.15±0.2	0.18±0.14	0.002±0.002			
AA1[b]	**(15)**			0.27±0.38	0.23±0.14	0.001±0.001			
AA1[c]	**(12)**			0.3±0.5	0.45±0.14	0.004±0.002			

[a] 1991/1992, [b] 1992/1993, [c] 1995/1996

3.4. COMPARISON BETWEEN DIFFUSIVE, MODELLED AND MEASURED FLUXES

The ratios of modelled fluxes (from burrow walls and the sediment-water interface, J_T), to the diffusive fluxes (J_x) across the sediment-water interface (J_T/J_x) at station F were higher confirming the observed higher abundance of benthic infauna and benthic activity. The average ratio J_T/J_x, for NH_4^+ was 15.7 ± 14.5 (n=14), for PO_4^{3-} was 13.8 ± 15.9 (n=14) and for Si(OH)$_4$ was 9.5 ± 2.5 (n=12). At the AA1 station the ratios were lower and more dependent on the different benthic activity within different seasons of the year. The average ratio J_T/J_x (n = 27), for NH_4^+ was 3.94 ± 4.19, for PO_4^{3-} was 1.8 ± 1.8 and for Si(OH)$_4$ was 3.2 ± 0.36. Hence, the flux ratios at both sites reflect differences between the bulk sediment molecular diffusion coefficient and effective diffusion coefficient (Aller, 1982).

The comparison between modelled and *in situ* measured (Bertuzzi et al., this issue) benthic fluxes at site AA1 in 1995/96 showed quite good agreement for NH_4^+ and PO_4^{3-} fluxes but not for NO_3^- and Si(OH)$_4$. This could be attributed to a topography effect caused by an underestimation of the actual sediment surface area when performing one-dimensional calculations of modelled fluxes. Another possible reason for the discrepancy in Si(OH)$_4$ fluxes is the additional dissolution of the solid phase, as already pointed out by Aller (1982), which tends to balance the diffusive loss into burrows and lessens the effect of bioturbation on the pore water concentration profile.

4. Conclusions

1. Pore water concentrations of DIC, NH_4^+, PO_4^{3-}, $Si(OH)_4$ at two sampling sites (F and AA1) in the Gulf of Trieste (Northern Adriatic) showed in wintertime nearly exponential increase, and for NO_3^- a decrease with depth. In other periods the profiles exhibit first a subsurface maximum then decrease, and successively, except for NO_3^-, increase or constant values. The described distribution is attributed to seasonal variations in the benthic infauna activity. DOC and DON showed the described vertical pattern in all periods probably because they are bonded to macromolecular compounds with lower mobility. The DOP profiles were similar to those of PO_4^{3-}.
2. The fluxes calculated from vertical profiles using the diffusion-reaction model which included the macrofaunal influence on sediment-water exchange rates, were 10-15 fold higher than diffusive fluxes at site F and 2-4 fold higher at site AA1. The differences between the two sites are suggested to reflect primarily sedimentological conditions, and the density and activity of benthic infauna.
3. The comparison between modelled and *in situ* measured fluxes at site AA1 showed quite good agreement for NH_4^+ and PO_4^{3-} but not for NO_3^- and $Si(OH)_4$ fluxes. The discrepancy observed could be caused by an underestimation of the actual sediment surface area when performing one-dimensional calculations of modelled fluxes (topography effect) and for $Si(OH)_4$ by an additional dissolution of the solid phase tending to balance the diffusive loss into burrows and lessening the effect of bioturabation on the pore water concentration profiles.

References

Aller, R.C.: 1982, In: McCall D.L. and Tevezs M.J.S. (Eds.), *Animal-sediment relations*, Plenum, New York, pp. 53-102.

Aspila, K.I. Agemian, H. and Chau, H.S.Y.: 1976, *Analyst* **109**, 189-197.

Barbanti, A. Ceccherelli V.U. Frascari, F. Reggiani G. and Rosso, G.: 1992. *Hydrobiol.* **228**, 1-21.

Berner, R. A.: 1980, *Early diagenesis: A theoretical approach*, Princeton Univ. Press, Princeton, NJ.

Bertuzzi, A. Faganeli, J and Brambati, A.: 1996, *P.S.Z.N. I Mar. Ecol.* **17**, 261-278

Bertuzzi, A. Faganeli, J. Welker, C. and Brambati, A.: this issue.

Brizzi, G. Aleffi, F. Landri, P. Goriup, F. and Orel, G.: 1994, *29th Eur. Mar. Biol. Symp.* Vienna, Austria.

Burdige, D.J. Alperin, M.J. Homstead, J. and Martens, C.S.: 1992, *Geophys. Res. Lett.* **16**, 687-690.

DeMaster, D.: 1981, *Geochim. Cosmochim. Acta* **45**, 1715-1732.

Faganeli, J. Planinc, R. Pezdic, J. Smodis, B. Stegnar, P. and Ogorelec, B.: 1991, *Mar. Geol.* **99**, 93-108.

Grasshoff, K. Ehrhard, M. and Kremling K.: 1983, *Methods of seawater analysis*. Verlag Chemie, Weiheim, 419 pp.

Hedges, J. and Stern, J.H.: 1984, *Limnol. Oceanogr.* **29**, 657-663.

Johnson, T. C. Ewans, J. E. and Eisenreich, S. J.: 1982, *Limnol. Oceanogr.* **27**, 481-491.

Krom, D.M. and Berner, R. A.: 1980, *Limnol. Oceanogr.* **25**, 327-337.

Lerman, A.: 1979, *Geochemical processes water and sediment environments*, Wiley, New York.

Li, Y.-H. and Gregory, S.: 1974, *Geochim. Cosmochim. Acta* **38**, 703-714.

Ogorelec, B. Misic, M. and Faganeli, J.: 1991, *Mar. Geol.* **99**, 79-92.

Reebourgh, W.S.: 1967, *Limnol. Oceanogr.* **12**, 163-165.

Romankevich, E.A.: 1984, *Geochemistry of organic matter in the ocean*, Springer, Berlin.

IMPACT OF NITRATE ADDITION ON PHOSPHORUS AVAILABILITY IN SEDIMENT AND WATER COLUMN AND ON PLANKTON BIOMASS - EXPERIMENTAL FIELD STUDY IN THE SHALLOW BRACKISH SCHLEI FJORD (WESTERN BALTIC, GERMANY)

M. Feibicke

Technical University Berlin, FG Limnology, Hellriegelstr. 6, D-14195 Berlin, Germany

Abstract. The Schlei is a hypertrophic brackish fjord of the Baltic. In order to assess the optimal method for restorative sediment treatment, an *in-situ*-experiment was carried out to oxidize the uppermost sediment strata by direct nitrate injection. An area of 3,5 ha, covered with sapropelic sediment, was treated with 50 t of commercial calcium-salpetre. Half of this area was treated once with a dose of 140 g NO_3-N m^{-2}, whereas the other half was treated twice receiving a dose of 280 g NO_3-N m^{-2}. Two enclosures were installed on these treated areas. The reaction in the interstitial water of the sediment (nitrate, phosphate, sulphate, iron) and in the water bulk (tot-N, tot-P) are documented and compared with an untreated reference area. Moreover, the impact of the sediment treatment on the phyto- and zooplankton biomass is discussed.

1 Introduction

In many shallow hypertrophic lake-like water systems a strong increase in phosphorus can be observed regularly in early summer, indicating an internal P-mobilization from sapropelic sediments (RIPL 1983, 1990) . After the diversion of external P-sources recovery can be retarded considerably in such cases (RIPL et al. 1993, SONDERGAARD et al. 1993, CHORUS 1996). One mechanism, promoting P-mobilization is based on sulphate reduction which occurs right below the sediment surface. This process is stimulated by the sedimentation of organic matter drawn from the actual algal bloom and the retreat of effective electron-acceptors like oxygen (O_2) and nitrate (NO_3^-).

Due to the formation of hydrogen sulphide (H_2S), a metabolic end product of the sulphate reduction, a strong impact on the iron bound phosphorus is triggered, following a two step process:

I. the reduction of trivalent to bivalent iron and the formation of molecular sulfur (S^o)

II. and the formation of relatively insolute iron sulphide (FeS).

Due to a sufficient supply of sulphate and easy degradable organic carbon the amount of H_2S formation directly depends on the process of desulfurication. The metabolic end product is turned over by means of FeS formation. A surplus of free H_2S limits further sulphate reduction. Based on these reactions (I, II) the iron and sulfur reach a maximum molecular ratio of 1:1,5 in the solid phase. As a further result the iron bound phosphate, replaced by sulphide, concentrates in the interstitial water and diffuses into the open water (RIPL 1978, 1994a, WOLTER 1985, KOPPELMEYER et al. in press). Based on this process there is an effective internal recycling of phosphorus, which stimulates primary production and sedimentation of algal matter. The latter supports those anaerobic putrefying degradation processes in the sediment, that were mentioned above.

To resolve the problem of internal P-mobilization a sufficient supply of an effective electron-acceptor such as nitrate is suggested: By increasing the supply of nitrate, denitrification is stimulated, organic matter is decomposed and the chemical energy content of the labile degradable organic matter is lowered. Furthermore the reduced FeS is oxidized

Water, Air and Soil Pollution **99**: 445-456, 1997.

to iron oxide / hydroxide (FeOOH), and absorbs phosphate concentrated in the interstitial water. Thus, if the organic bound energy content is lowered to a sufficient extent, the internal P-mobilization can be reduced (RIPL & FEIBICKE 1992). In order to keep phosphate largely within the sediment, the available iron has to reach a sufficient concentration. If these conditions are fulfilled and external P-inputs are diverted successfully, a significant reduction in primary production as well as in sedimentation of organic matter is brought about.

The introduction of nitrate can be realized in different ways:

- a permanent introduction of nitrified and phosphate removed sewage works effluent into the recipient as proposed by RIPL & LINDMARK (1978). This has been carried out in „Tegeler See“ (Berlin, Germany, since 1994) (CHORUS 1996).
- a single addition of nitrate to the water body or direct injection into the sapropelic sediment. Examples of restorated lakes are Lake Lillesjön (South Sweden, in 1975) (RIPL 1978) and the oxbow of the river Danube (Vienna, Austria, in 1994-96). In addition both water bodies were treated with an iron-(III) chloride solution (RIPL 1994b).

In the present case - the innermost Schlei, a Baltic lowland fjord (also called firth or fjard) - both measures, a permanent nitrate introduction by means of a sewage plant and a temporary addition by nitrate injection, are recommended in order to control the internal P-mobilization. During a restoration planning project the nitrate injection was tested in a half-scale field experiment (RIPL 1990).

The results of this field study are presented in this paper. Besides the reaction within the sediment, brought about by the nitrate injection, the effects on the nutrient regime and plankton biomass in the water body will be focused on. Therefore suitable enclosures were installed on the treated sediment areas.

2 Location

The Schlei is a narrow, shallow and brackish fjord of the Baltic Sea (42 km length, 2,9 m mean depth) situated in the North of Germany between the cities of Kiel and Flensburg (Figure 1A). The innermost part, the so called "Innere Schlei", forms a lake-like enlargement of about 20 km^2 (2,5 m mean depth) (Figure 1B).

Due to inflow of brackish water from the Baltic Sea from the east and inflow of freshwater from the western part of its catchment (670 km^2 catchment area) a gradient of salinity, representing about 3-6 ‰ salinity in the innermost part to 15-20 ‰ in the outer part of the Schlei has been built up. The average water exchange is 100-150 days in the innermost part during summer, shortened to about 30 days in winter.

The Schlei is one of the most nutrient enriched fjord systems of Germany especially the innermost part of it. There the gross primary production of the phytoplankton reaches 600 to 800 g C a^{-1} m^{-2} (SCHIEMANN 1974, FEIBICKE 1994b, c). As a result of incomplete mineralization processes dark sulphide enriched sapropelic layers have been formed which cover about 60 % of the bottom of the innermost part of the Schlei (20 km^2 by area). The annual sediment accumulation amounts to approximately 3,5 mm a^{-1} of wet sediment (RIPL 1990).

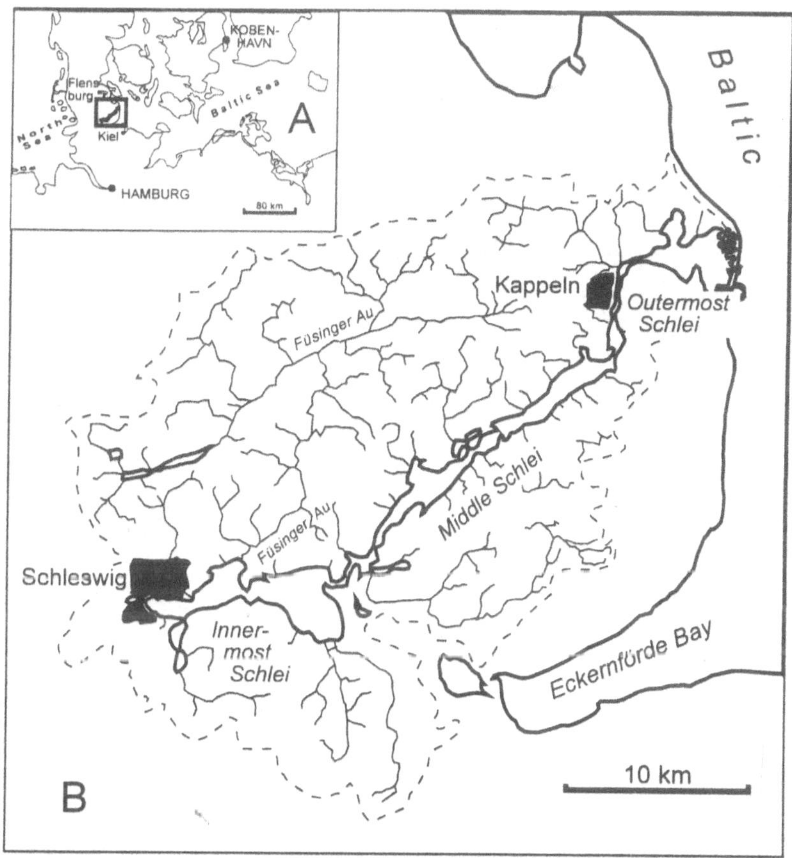

Fig. 1. Schlei fjord. A. Situation in the Baltic (inlet). B. Catchment.

The reasons for this situation have been sewage effluents from the town of Schleswig since the end of the last century as well as inputs from the surrounding farming area. The latter have sharply increased since the end of second world war (RIPL 1990, FEIBICKE 1994a, b).

Whereas the total nitrogen concentration in the innermost Schlei is high in winter, caused by a high input of nitrate, the total phosphorus concentration is relatively low (Figure 2). In summer, when the drainage is rather poor, the external nutrient loading as well as the internal nitrate concentration decrease. On the other hand the phosphorus concentration increases rapidly strongly promoting a high primary production, dominated by cyanobacterial phytoplankton communities. During this period about 1 g P m^{-2} is released from the sediment, while short term maximum rates of 30 mg P $m^{-2} d^{-1}$ appear (RIPL 1990).

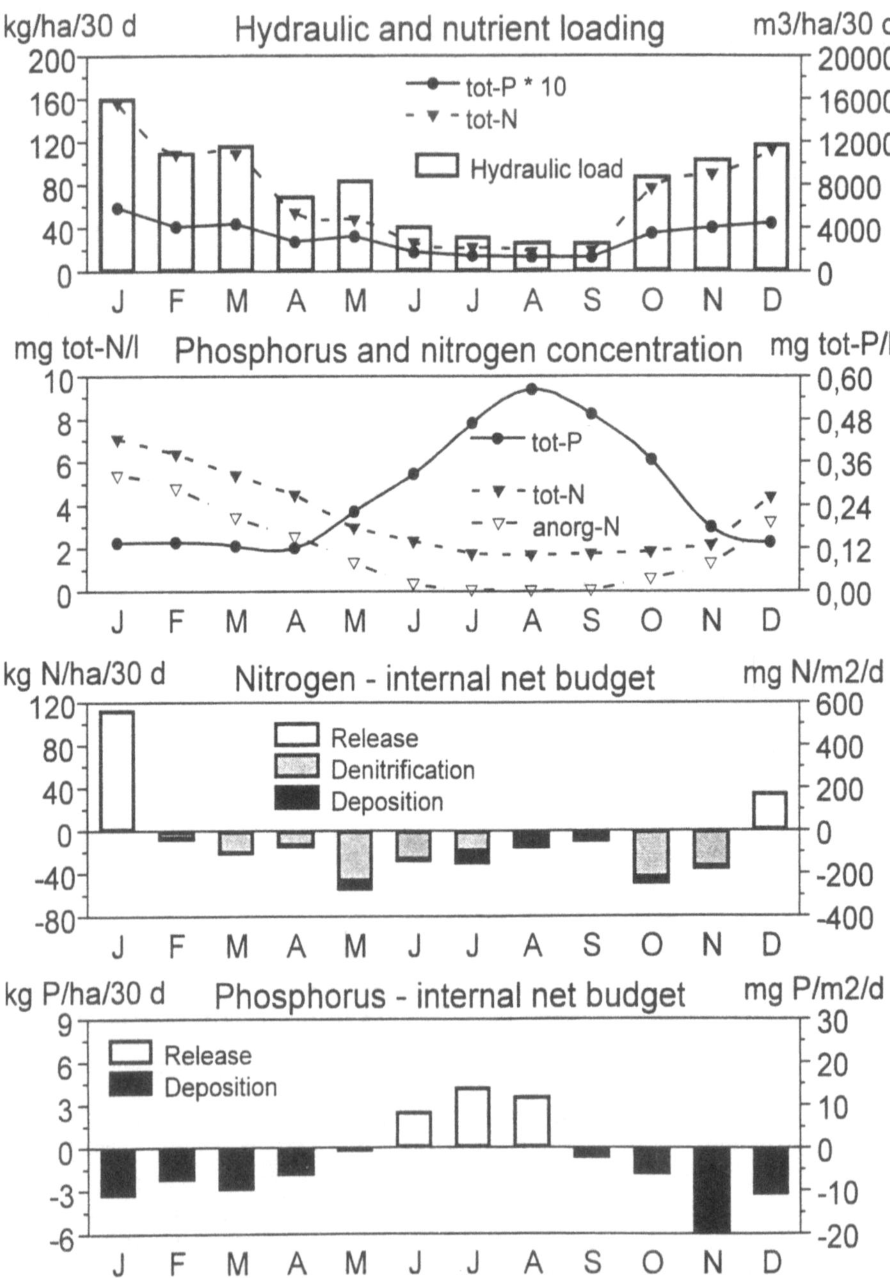

Fig. 2. Loadings, nutrient concentrations and budgets of the innermost Schlei (according to RIPL 1990, mean annual course, based on monitoring of 9 sampling stations in the Schlei and 16 stations at its tributary during 1981-1983).

3 Technical implementation and experimental design

The nitrate injection was performed using a method developed in cooperation with the Atlas Copco Comp. An area of 3,5 ha, situated in the Northeast part of the innermost Schlei („Kleine Breite") (Figure 3A, B), covered with sapropelic sediments (>50 cm thick) was treated with 50 t of commercial calcium-saltpetre. Sediment characteristics of the upper layer are summarized in Table 1.

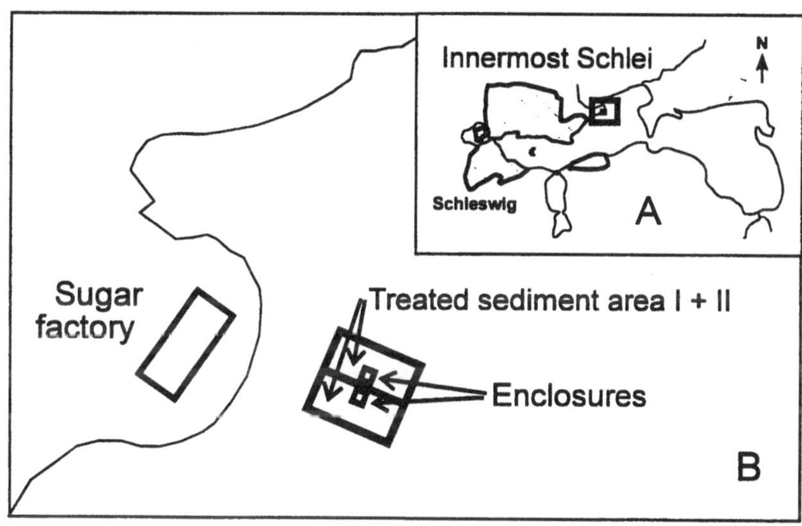

Fig. 3. Study area. A. Innermost Schlei (inlet). B. Situation of the nitrate treated sediment areas and enclosures.

Table I

Sediment characteristics of the upper 25 cm sediment layer (total mean and standard deviation) (RIPL 1990) (25 sediment cores form the innermost Schlei, each core sliced into 10 strata of 2,5 cm height, N=250).

Parameter	Mean	(SD)	Parameter	Mean	(SD)
Dry weight (% FW)	20,0	(1,8)	Iron (mg g^{-1} DW)	27,0	(0,8)
Loss of ignition (% DW)	19,9	(1,2)	Manganese (mg g^{-1} DW)	0,95	(0,04)
Kjehldahl-nitrogen (mg g^{-1} DW)	9,2	(0,5)	Calcium (mg g^{-1} DW)	19,3	(1,5)
Phosphorus (mg g^{-1} DW)	1,15	(0,05)	Magnesium (mg g^{-1} DW)	6,4	(0,5)

The calcium-saltpetre solution was injected by using a sediment-harrow dragged over the sediment surface (Figure 4). Half of the treated area received a nominal dosage of 140 g NO_3-N m^{-2} (single treatment), whereas the rest was treated twice receiving a double nominal dosage of 280 g NO_3-N m^{-2} (double treatment).

For logistical reasons the enclosures could be build up on the experimental area only two weeks after the treatment (Figure 3B). The enclosures were setup on the single treated as well as on the double treated test area. They consisted of a butyl-rubber collar and a heavy open iron frame, which sank into the sediment. The volume of water within each enclosure was 30 m^3 (mean water depth 3 m).

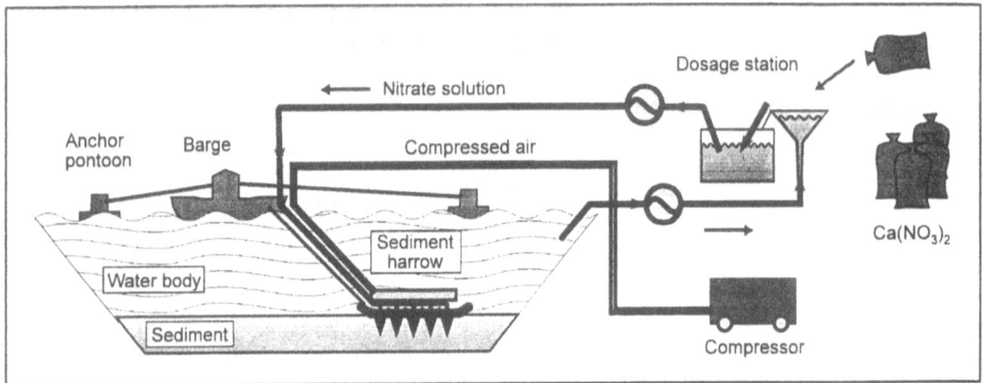

Fig. 4. Nitrate injection of calcium-salpetre solution into the sediment. Schematic illustration of the technical application.

The conditions in the open water and in the interstitial water inside the enclosure, installed on the treated sediment areas, as well as in the open water of the Schlei and on untreated reference areas (situated in the immediate vicinity to the nitrate injection experiment) were analyzed from April to October 1994 (11 sampling days).

4 Materials and methods

Transparency was measured with a Secchi disk (24 cm diameter). For chemical-physical analyses water samples were taken with a Ruttner sampler at 0,5 m water depth.

O-phosphate was determined using the molybdenum blue reaction of MURPHY & RILEY (1962). Nitrate was reduced to nitrite by coppered cadmium granulate (WOOD et al. 1967) and afterwards determined by means of the Gries-Ilosvay-Reaction (BENDSCHNEIDER & ROBINSON 1952). After a digestion with potassium persulphate the determination of total-phosphorus and total-nitrogen (KOROLEFF 1976a, b) followed the methods for o-phosphate and nitrate respectively (see above).

Plankton samples were obtained by means of a plugable tube made of Plexiglas (inner diameter 50 mm) which collected a water column about 1,5 m in depth. For the analysis of phytoplankton and protozoa unfiltered water samples were taken, fixed with an acetic acid Lugol-solution. For the analysis of larger zooplankton species (rotifers and crustaceans) a sample volume of 5 l was filtered through a meshsize of 45 µm and fixed with buffered formaldehyde solution. The phytoplankton and zooplankton were counted in sedimentation chambers with an inverted microscope according to UTERMÖHL (1958). Gelatinous cyanobacteria were first disintegrated by sonification. For further details see FEIBICKE (1994b).

For interstitial water sampling sediment cores were taken with a core sampler according to BERGGREN (1972). The pore water sediment sampling was based on a dialysis technique according to RIPL (1978). A chambered and a half sliced glass tube, filled with destilled water, and covered with a dialysis membrane (Visking Type 20/32) were introduced and incubated for 3 days immediately after core sampling. The determination of phosphate, nitrate and sulphate in the interstitial water followed the methods for flow injection analysis (FIA) according to RUZICKA & HANSEN (1975). Iron was determined by means of flame AAS-technique (Perkin-Elmer). For further details see RIPL (1990).

5 Results and discussion

5.1 Processes in the sediments

Based on the interstitial water analysis of 30 sediment cores taken immediately after the treatment from the test area only 57 g NO_3-N m^{-2} were found again in the area treated once, whereas 129 g NO_3-N m^{-2} were found in the area treated twice. During the treatment strong water movement as a result of stormy weather caused a considerable flushing of nitrate from the sediment. As expected there were strong differences in the values from core to core since the injection nozzles of the sediment harrow were attached at intervals of 20 cm. Therefore the nitrate distribution in the sediment had to be quite different immediately after the injection.

After the installation of the enclosures there was a significant release of gas. The released gas could not be analyzed under field conditions. But gas chromatographic analysis of nitrate incubated sediment cores carried out under laboratory conditions revealed that nitrogen (N_2) and carbon dioxide (CO_2) were the main components (98 % Vol.). Traces of methane and oxygen were also found (LINKE 1986).

The interstitial nitrate concentrations in the treated areas decreased rapidly, so that after 100 days only 1 % percent of the original start concentration remained.For the single and for the double treated area average nitrate consumption rates of 0,8 or 1,8 g N m^{-2} d^{-1} were determined. Maximum rates reached 1,12 or 2,24 g N m^{-2} d^{-1}. Laboratory experiments carried out on sediment cores incubated with 145 g $N0_3$-N m^{-2} over 90 days (LINKE 1986) revealed similar nitrate demand rates of 1,1 to 1,3 g N m^{-2} d^{-1} on average.

Additional field experiments using *in-situ*-through-flow-enclosure systems and a nitrate addition to the water column showed that the average nitrate demand rates ranged from 0,4 to 0,8 g N m^{-2} d^{-1} depending on the open water nitrate concentration (range 5-15 mg NO_3-N l^{-1}). On the basis of a nitrogen net budget for the innermost Schlei a nitrogen consumption of about 0,2-0,3 g N m^{-2} d^{-1} is calculated (in the presence of nitrate concentration of 3-5 mg NO_3-N l^{-1}) for winter and spring (RIPL 1990, FEIBICKE 1994c).

The mean vertical distribution of the nitrate concentration in the sediment pore water clearly showed a nitrate enrichment which reached its maximum at 10-20 cm sediment depth (Figure 5). The high standard deviation in these strata indicates a high spatial diversity but also a high temporal turnover in the treated sediment areas. A nitrate enrichment was observed down to a sediment depth of 30-40 cm. Unlike this, on the untreated reference area only a low amount of nitrate found its way into the upper sediment layer during spring, when there is a higher nitrate level in the water column.

Whereas the sulphate profile in the untreated area showed a remarkable decrease starting immediately under the sediment surface indicating a strong sulphate reduction, the profiles in the treated areas showed different patterns.

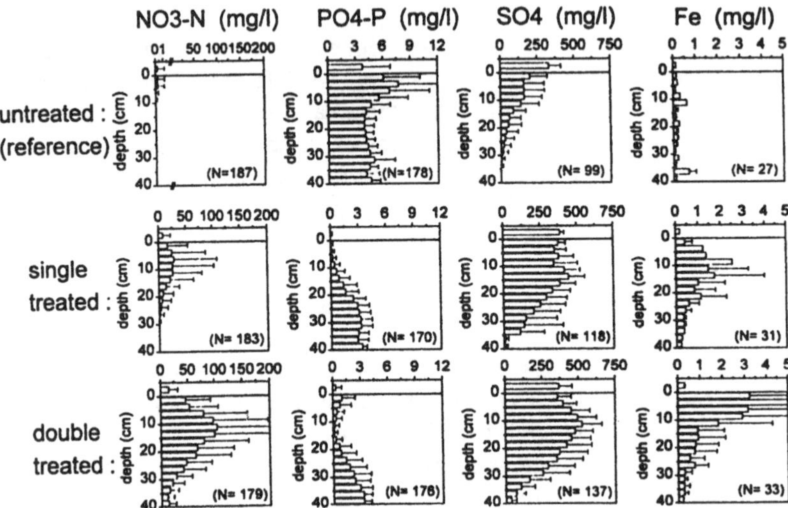

Fig. 5. Depth profiles of soluble components (nitrate, phosphate, sulphate, iron) in the interstitial water of nitrate treated (single and double injection) and untreated sediments (depth specific mean and standard deviation, based on 11 sampling days from April-October).

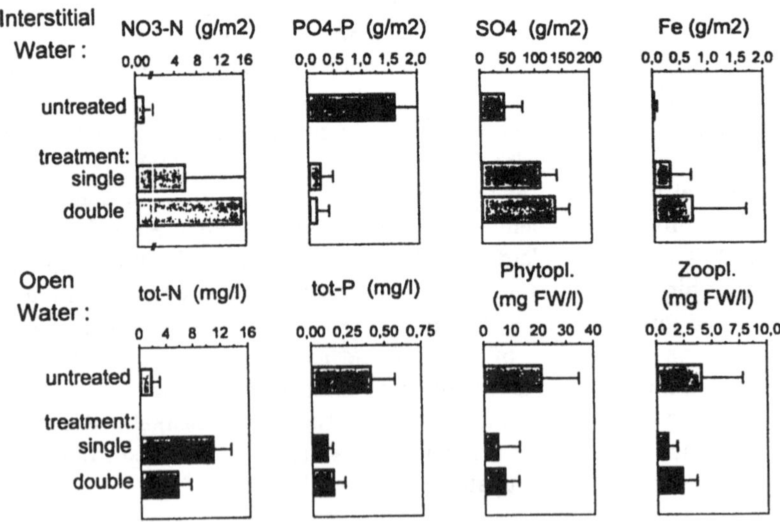

Fig. 6. Concentrations of soluble components (nitrate, phosphate, sulphate, iron) in the interstitial water (0-15 cm sediment depth) in comparison with nutrient concentration and plankton biomass (total nitrogen, phosphorus, phyto- and zooplankton biomass) in the water bulk of the enclosures and the neighbouring Schlei, used as a reference (mean and standard deviation, based on 11 sampling days from April-October).

There decreasing sulphate concentrations appeared first below the nitrate enriched strata, which indicated a suppression of desulfurication in the presence of nitrate anda displacement of that process into deeper sediment layers. Especially in the double treated area increasing sulphate concentrations occurred, exceeding sulphate concentration levels in the open water body. A nitrate induced bacterial oxidation of reduced sulphur compounds is to be expected in these strata.
The contents of soluble iron showed significantly higher values within the treated sediment strata than in the untreated regions, indicating the increasing availability of iron released by the oxidation of FeS.

In the treated areas the vertical distribution of interstitial o-phosphate corresponded inversely to the distribution of nitrate. Phosphate minima were correlated with nitrate maxima and vice versa. However in the untreated area immediately below the sediment surfaces phosphate maxima coincided with sulphate minima.

The vertical distribution of ammonium showed on average a slight increase in the double treated area in comparison to the untreated one. In contrast to the ammonium content of the commercial calcium saltpetre with a ammonium amount of 1.1 %, only residues were observed in the interstitial of the treated sediment areas.

5.2 Processes in the water body

Due to leaking of injected nitrate solution out of the sediment, the total nitrogen content in the water column increased reaching maximum values of 10 and 13 mg tot-N l^{-1} respectively (8,4 or 10,1 mg NO_3-N l^{-1}). During the course of the experiment the total nitrogen as well as nitrate decreased due to sedimentation and mainly to denitrification processes in the enclosures (for further details and plots of time series see RIPL 1990). At the end of the experiment the nitrate content was 1,1 in the single treated and 3,7 mg NO_3-N l^{-1} in the double treated enclosure. On the average the total nitrogen was three- to sixfold higher in the enclosure than outside in the innermost Schlei (Figure 6).

In the Schlei, the total phosphorus content increased from 0,14 mg tot-P l^{-1} in April to 0,74 mg tot-P l^{-1} in August due to P-mobilization from the sapropelic sediment (average 0,41 mg tot-P l^{-1}). On the other hand in the enclosures the total phosphorus concentrations reached 0,11 and 0,15 mg tot-P l^{-1}on average. Thereby a P-reduction of 66 to 77 % was reached in comparison to the reference area outside. In the enclosure, medium o-phosphate concentrations of 0,007 or 0,010 mg PO_4-P l^{-1} were found, whereas outside a remarkable higher PO_4-P-concentration of 0,150 mg PO_4-P l^{-1} was observed (Figure 6).

If there is a relatively low amount of phosphate (<0.01 mg PO_4-P l^{-1}), the growth of phytoplankton is controlled by the mineralization capacity of the zooplankton. As a result material cycling processes are more effective and losses of putrefactive organic substances to the sediment are reduced.

Due to the considerable reduction in the P-supply in the water column of the enclosures, the phytoplankton biomasses also were reduced remarkably. In the nitrate treated enclosures average phytoplankton biomasses of 5,2 to 7,6 mg FW l^{-1} were reached, whereas outside at the reference station a mean phytoplankton biomass of 21,2 mg FW l^{-1} occurred. Correspondingly the Secchi depths increased, reaching 1,1 to 1,4 m on the average inside the enclosure, unlike 0,5 m outside. Also the zooplankton biomasses in the enclosures were reduced reaching 1,0 and 2,3 mg FW l^{-1} respectively in contrast to outside with a mean biomass of 4,1 mg FW l^{-1} (Figure 6).

The succession and the species composition of phytoplankton showed remarkable differences between the enclosure systems and the reference area. Whereas in the innermost Schlei cyanobacteria (*Microcystis*) took the place of small coccal green algae (*Dactylosphaerium*, *Monoraphidium*) in summer and presented 83 % of the phytoplankton biomass in average, in the enclosures also diatoms (*Chaetoceros, Nitzschia*), euglenoids (*Euglena, Phacus*), dinoflagellates (*Peridinium*) and haptophytes (*Phaeocystis*-like stages) succeeded temporarily. Further detailed information on species composition and temporal variation will be published elsewhere.

External phosphorus loadings from outside occurred by water penetration during a storm event (only double treated enclosure) and by excrement inputs of sea birds in autumn at the end of the experiment (both enclosures). In both cases, primary production was stimulated and cyanobacteria or haptophytes were favored. The stimulation effect triggered by water penetration lasted about 10 weeks. Afterwards the phosphorus and phytoplankton biomass decreased reaching levels comparable to those in the unaffected enclosure. Without these unexpected loading events the reduction of the P-concentrations and plankton biomass triggered by nitrate treatment would be even more pronounced.

6 Conclusions

This half scale experiment clearly shows that by means of a specific and responsible use of nitrate, desulfurication and P-mobilization in the sediment surface of sapropels can be prevented. Due to the reduced phosphorus supply in the water column positive effects on the plankton community like the reduction of biomass, production and sedimentation can be triggered. When the nitrate treatment is carried out in winter or spring these effects can be induced within the current vegetation period.

By means of nitrate oxidation of sapropelic sediments further ecosystem responses are initiated. In the medium-term a re-invasion of the macrozoobenthos can take place as was observed on the test area some years later. By its feeding, burrowing and ventilating activity the macrozoobenthos contributes to a further degradation of the organic substances and stabilizes oxidized condition in the upper sediment. In the long-term conditions for a re-invasion of submerged macrophytes are improved due to an increased transparency. The colonization of the formerly deserted sediment by higher plants covered by peripython will stimulate further positive effects like a rising nutrient competition with the plankton community.

Although nitrate oxidation of sapropelic sediments can be used as a control for internal P mobilization, it must be coupled with a sustained diversion of external phosphorus sources. This was shown by the effects of both unexpected external P-loading events during this experimental course.

The successful and sustainable reduction of the internal P-loading by nitrate addition depends on a sufficient dosage of nitrate to oxidize the sapropelic layer and the presence of a sufficient amount of iron in the upper sediment. Furthermore, to prevent losses of nitrate e.g. to groundwater the specific hydrological regime has to be taken into account, too.

The method of nitrate injection tested here needs further improvement because losses of nitrate by flushing reached about 54-59 % of the nominal input on condition of a wind exposed coastal site.

It is proposed to use such improved technique for a single treatment of the thickest sapropelic strata (>50 cm thickness, 1 km^2 area) in the innermost Schlei to accelerate the oxidation of the sapropelic sediment (RIPL 1990).

A permanent input of nitrate in the open water of the innermost Schlei should be realized by the introduction of nitrified and phosphate removed sewage works effluent. In the innermost Schlei the nitrogen input by the sewage plant and other municipal effluents contribute only c. 10 % of the input by their tributaries. The introduction of nitrate enriched sewage works effluent instead of ammonium enriched does not increase the municipal nitrogen loading but guarantees a sufficient supply of effective electron-acceptor during summer, when the nitrate loadings from the surrounding farming area decrease and the internal P-mobilization appears.

In case of the Schlei the high recent external loadings of nitrogen and phosphorus, carried in by agricultural runoff during winter, as well as the non-seasonal loadings by municipal effluents have to be reduced effectively by different measures (for further detail see RIPL 1990).

If the proposed external and internal measures for the restoration of the innermost Schlei will be realized, the actual high loadings of nitrogen and phosphorus from the Schlei into the Baltic Sea (GERLACH 1990) will also decrease in the long-term.

Acknowledgment

The project "Restaurierungsplanung Schlei" was carried out under the supervision of the Department of Limnology (Technical University Berlin) and in cooperation with the „Landesamt für Wasserhaushalt und Küsten" (Schleswig-Holstein). The financing was provided by the „Bundesministerium für Forschung und Technologie" as well as the „Umweltbundesamt" and the „Bundesland Land Schleswig-Holstein". My special gratitude applies for Mr. L. Schmitt and Mrs. K. Wöbbecke for their examination of the manuscript and language corrections.

7 References

Bendschneider, K. & Robinson, R.J.: 1952, A new spectrometric determination of nitrite in sea water. *J. Marine Res.* **11**, 87-96.

Berggren, H.: 1972, Sedimentprovtagning med rörhämtare. (Sediment sampling with a core sampler). *Vatten* **28**, 374-377.

Chorus, I.: 1996, Seesanierung in Berlin. - *DGL-Tagung* 24.-29.9.1995 in Berlin, Germany. **II**, 177-181.

Feibicke, M.: 1994a, Innere Schlei, Stoffeinträge und Prozesse. *Beiträge zur angewandten Gewässerökologie Norddeutschlands* **1**, 66-73.

Feibicke, M.: 1994b, *Strukturelle und funktionelle Studien zum Phytoplankton und Zooplankton in einer hypertrophen Brackwasserförde (Innere Schlei, Westliche Ostsee)*. - Dissertationes Botanicae **225**, 144 pp. + appendix.

Feibicke, M.: 1994c, Wege und Verbleib des Kohlenstoffs in der Inneren Schlei. - 2. Wiss. Workshop "Ökosystem Boddengewässer - Organismen und Stoffhaushalt", 29.-30.9.94 Kloster/Hiddensee *Bodden* **2**, 205-217.

Gerlach, S.A.: 1990, Nitrogen, phosphorus, plankton and oxygen deficiency in the German Bight and in Kiel Bay. - *Kieler Meeresforschungen*, Sonderheft 7: 341 pp.

Koppelmeyer, B., Feibicke, M., Heller, S., Markwitz, M., Wöbbecke, K., Wolter, K.-D. & Ripl, W.: in press, A restoration concept for a polytrophic shallow lake (Schwielochsee/Ostbrandenburg, Germany). - IVL-Conference "Shallow Lakes '95" 21.8.-26.8.95 in Mikolaajki, Poland. *Hydrobiologia*: in press.

Koroleff, F.: 1976a, Determination of phosphorus. - In: Grasshoff, K. (ed..)*Methods of seawater analysis*, 117-125.

Koroleff, F.: 1976b, Total and organic nitrogen. - In: Grasshoff, K. (ed.): *Methods of seawater analysis*, 167-173.

Linke, W.: 1986, *Sedimentchemische Laboruntersuchung an Faulschlammsedimenten der „Inneren Schlei" und deren Veränderung durch Nitratbehandlung*. - Dissertation, submitted for diploma, FU Berlin: 55 pp.

Murphy, J. & Riley, J.P.: 1962, A modified single solution method of phosphate in natural waters.*Anal. Chim. Acta* **27**, 31-36.

Ripl, W.: 1978, *Oxidation of lake sediments with nitrate. A restoration method for former recipients*. Coden Lunbds / (NBLI-1001)/1-151/(1989) ISSN 0348-0798.

Ripl, W.: 1983, *Dümmersanierung*. - Limnological Report, TU Berlin, Inst. f. Ecology, Dep. Limnology: 154 pp.

Ripl, W.: 1990, Restaurierung der Schlei. Bericht über ein Forschungsvorhaben. *Ökosystemforschung und Gewässerbewirtschaftung* **1**, 86 pp. + appendix.

Ripl, W., Heller, S., Koppelmeyer, B., Markwitz, M. & Wolter, K.-D.: 1993, *Limnologische Begleitstudie zur Entlastung des Tegeler Sees*. - Final Limnological Report, GFG mbH & TU Berlin, Inst. f. Ecology, Dep. Limnology: 51 pp. + Appendix.

Ripl, W.: 1994a, Sediment treatment. - In: Eiseltová, M. (ed.) Restoration of lake ecosystems. A holistic approach. - *IWRB Publication* **32**, 75-81.

Ripl, W.: 1994b, *Sanierung Alte Donau*. Zwichenbericht. - Aquaterra-Consult-Gesellschaft mbH: 26 pp. + appendix.

Ripl, W. & Feibicke, M.: 1992, Nitrogen metabolism in ecosystems. A new approach. *Internat. Rev. ges. Hydrobiol.* **77**, 5-27.

Ripl, W. & Lindmark, G.: 1978, Ecosystem control by nitrogen sediment metabolism. - *Vatten* **34**, 135-144.

Ruzicka, J. & Hansen, E.H.: 1975, Flow injection analysis. I. A new concept of fast continuous flow analysis. - *Anal. Chim. Acta* **78**, 145-157.

Schiemann, S.: 1974, *Die Primärproduktion des Phytoplanktons der Schlei und des Windebyer Noors im Jahre 1972*. - Thesis, Christian-Albrechts-Universität, Kiel.

Sondergaard, M., Kristensen, P. & Jeppesen, E.: 1993, Eight years of internal phosphorus loading and changed in the sediment phosphorus profile of Lake Sobygaard, Denmark. - *Hydrobiologia* **253**: 345-356.

Utermöhl, H.: 1958, Zur Vervollkommnung der quantitativen Phytoplankton-Methodik. *Mit. Internat. Verein. Limnol.* **9**, 1-38.

Wolter, K.-D.: 1985, *Stoffwechselprozesse in den Faulschlammsedimenten der Schlei*. Dissertation, submitted for diploma, FU Berlin: 58 pp.

Wood, E.D., Armstrong, F.A.J. & Richards, F.A.: 1967, Determination of nitrate in sea water by cadmium-copper reduction to nitrite. - *J. mar. biol. Ass. U.K.* **47**, 23-31.

Bottom sediments in a humic lake with artificially increased calcium content: sink or source for phosphorus?

M. RZEPECKI

Department of Hydrobiology, Institute of Ecology, Polish Academy of Sciences, Dziekanów Leśny, 05-092 Łomianki, Poland

Abstract. Lake Flosek (north-eastern part of Poland) is a small shallow and without outflow lake which has been limed in 1970. The concentration of Ca was increased from 3-4 mg L^{-1} to 17 mg L^{-1} in the water and from 0.2-0.3% dry weight to 0.9-1.7% dry weight in sediments (5 cm upper layer) due to $CaCO_3$ addition to the lake.

In the spring-summer seasons of 1992 and 1993, an experimental study was conducted in Lake Flosek to assess the capacity of bottom sediments to uptake and release mineral phosphorus. The rate of phosphorus exchange between sediments and near-bottom water was experimentally measured under conditions of high (100%), and of reduced (10%) oxygen saturation in near-bottom water.

To determine the component of sediments responsible for the uptake of most phosphorus, the proportions of phosphorus forms in sediments were analysed.

Sediments of Lake Flosek showed a slight tendency to release phosphates. The rate of this process was similar under high (100%) and low (10%) oxygen saturations ranging from - 0.161 to + 0.200 mg P m^{-2} d^{-1}. This is much lower (by 1-2 orders of magnitude) than reported from other harmonic, non-humic lakes.

In the total phosphorus pool, the highest content of phosphorus was found in the organic and residual phosphorus fractions (over 70% of the total phosphorus in sediments). The largest part of the readily extractable phosphorus was found in the fraction bound to Al and humic substances (41%). Both these fractions determine a weak exchange of phosphorus between sediments and water. No difference in P-release related to P-fraction compound was found in the cores taken from three sites in the lake.

Keywords: sediment, phosphorus, fractionation, release, humic lake

1. Introduction

Dystrophic or humic lakes are typical components of the biome of coniferous and mixed forests of the boreal zone (mostly in Scandinavian countries) and also of the hilly lakeland landscape of north-eastern Poland (Kondracki, 1972). They increase the biodiversity of this area, though their contribution to the total number of lakes in Poland is small (Hillbricht-Ilkowska *et al.*, 1977). Their basic features include: the watershed covered by coniferous forest with peatland or marshland, occurrence of *Sphagnum* moss around the lake, high content of humic substances in the water and sediments which are of the "dy" type, low pH (4.5-6.0) and Ca content in water (< 5 mg L^{-1}) and sediments (< 0.2%) (Wetzel, 1983). The lakes usually have low productivity and relatively high transparency. They also have low capacity to neutralize acid precipitation. Sediments of type "dy" are decomposed to only a small degree, and do not release P and other nutrients. Such sediments are mostly P-sinks (Boström *et al.*, 1982).

Dystrophic lakes in Poland are as a rule small, shallow, and without outflow, but of great importance to the hydrologic cycle. Because of their small size, and limnological type, these lakes are particularly sensitive to environmental changes due to human influences such as acidification and eutrophication (Hillbricht-Ilkowska *et al.*, 1977). Liming is a common treatment applied in such lakes to counteract a decrease in pH due to acid rain (Eriksson *et al.*, 1983; Hörnström and Ekström, 1986; Broberg, 1988; Calderoni, *et al.* 1991).

Water, Air and Soil Pollution **99**: 457-464, 1997.

TABLE I.

Past (before liming) and actual (i.e. 20 years after liming) sediment characteristics of Lake Flosek in comparison to the sediments from humic and non-humic eutrophic lakes in Great Masurian Lakes District (in % of dry weight).

	Flosek lake		Non-humic lakes[2/]	Humic lakes[2/]
	past	actual[1/]		
water content	-	95.5 (94 - 97)	78 - 97*	-
organic matter	83	84	12 - 56**	76.73 - 83.41**
Ca	0.2	1.3 (0.9 - 1.7)	0.8 - 23**	0.30 - 3.90**
Fe	-	0.24 (0.19 - 0.28)	0.11 - 0.37**	-
Al	-	0.16 (0.11 - 0.21)	0.07 - 0.32*	-
Mg	-	0.063 (0.049 -0.078)	0.25 - 0.56	-
Mn	-	0.032 (0.002 - 0.06)	0.006- 0.02	-
$P_{tot.}$	0.10	0.10 (0.09 - 0.11)	0.12 - 0.28	-

[1/] average and range for 3 sites
[2/] range for 4-5 lakes

* Wiśniewski and Planter, 1985
** Rybak, 1969

Data concerning the influence of liming on lake chemistry are ambiguous. Some authors found that phosphorus concentration in the water column more or less increased after liming whereas some others did not. (Yan *et al.*, 1977; Hultberg and Andersson, 1982; Scheider and Brydges, 1984; Broberg, 1987).

Uptake/release properties of bottom sediments under various physico-chemical conditions such as pH, redox potential, organic matter decomposition rate, temperature, etc., depend on the type and stability of chemical compounds (Boström and Pettersson, 1982; Boström, 1984; Golterman, 1984).

To identify the effects of liming in a naturally weakly acid ecosystem (pH 5.6-6.0) and the possibility of counteracting the effects of acid precipitation, a study was conducted in one of the typical dystrophic lakes of the Great Masurian Lake District, in north-eastern Poland. The lake was limed 20 years earlier. As a result of this treatment, many chemical parameters of the lake were changed more or less permanently, e.g., the concentration of Ca increased in the water (up to 14-17 mg L^{-1}) and in bottom sediments (up to 0.9-1.7% dry weight), pH increased to 7-8, and the decomposition rate of organic matter increased (Table I) (Hillbricht-Ilkowska *et al.*, 1977).

To determine the role of bottom sediments in the lake changed by liming, (the humic one, but with increased Ca content in sediments and water, and with neutral pH), the direction and rate of phosphate exchange between sediments and near-bottom water were analyzed under different oxygen conditions. Also, the content of P in the fractions of bottom sediments was determined to identify the fraction which plays the most important part in uptake/release of phosphorus. It can be expected that, under conditions of neutral pH and a low redox potential, phosphorus will be released from sediments at a higher rate due to the acceleration of decomposition processes, similar to eutrophic lakes.

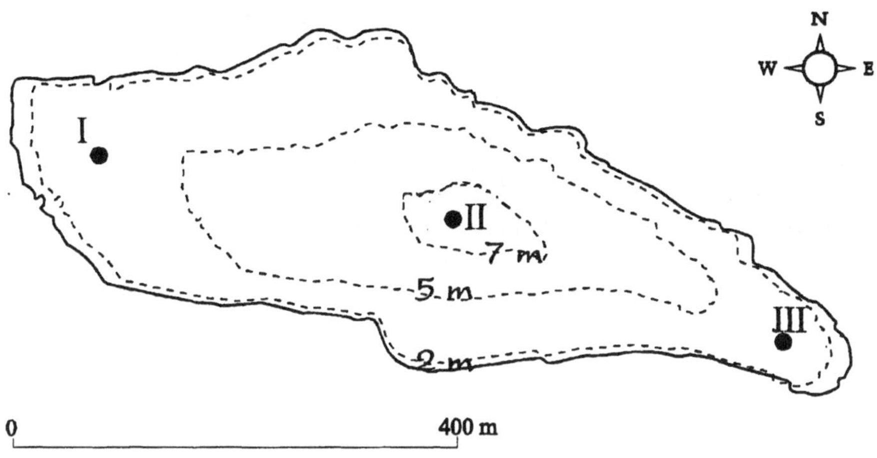

Fig. 1. Bathymetric map and the location of sampling sites (I, II, III) in Flosek Lake.

At high Ca concentrations, the capacity of phosphorus sorption by humic-metal-complexes declines due to the substitution of Fe and Al by Ca, which has lower sorption capacity especially in these circumstances. It can be expected that this would increase the pool of phosphorus in ferric-phosphate compounds, more sensitive to changes in redox potential (Ohle, 1964; Boström *et al,* 1982; Pettersson and Olsson, 1986) and thus enhance P release.

2. Study area, and methods

The study was carried out in a small (4 ha), shallow (maximum depth of 7 m), humic lake situated in a coniferous and mixed forests area with *Sphagnum* in the Great Masurian Lake District (north-eastern part of Poland).

The sediments samples were taken at three sites located along the longer axis of the lake.To encompass different structures and stages of decomposition of organic matter in sediments the samples were taken at different depths. The samples ranged from sediments with the lowest degree of decomposition of organic matter (in part near the shore of the lake) to sediments with relatively uniform structure and semi-fluid consistency (in the central part of the lake). Site I was located near the shore of the lake at a depth of 3 m, site II in the central part of the lake at a depth of 7 m (maximum depth of the lake) and site III was located near the peat bogs at a depth of 3 m (Figure 1).

Undisturbed sediment cores 15-20 cm long including the overlying water were collected with a Kajak bottom sediment sampler (Kajak *et al.* 1965) using plexiglas tubes, 10 cm in diameter and 0.5 m long. The samples were transported to the laboratory and kept in darkness under thermal conditions similar to those in the lake during sampling (10-15°C).

Experiments on the exchange of phosphates between sediments and near-bottom water were carried out in a series of four replicates at 2 oxygen levels: 100% and 10% oxygen.

For a period of 8 days at two-day intervals, 50-ml samples of near-bottom water were taken with a syringe. The samples were filtered through a Whatman GF/C glass filter. Phosphate release from sediments was calculated as the difference in P concentration between successive water samples.

The content of phosphorus forms was analyzed in sediment samples taken from the same sites as the experimental sediment cores. Phosphorus fractions were analyzed in the upper sediment layer (0-5 cm) using the method of selective extraction according to Psenner *et al.* (1991). With this method, the following phosphorus fractions could be identified: 1) labile P, extractable with NH_4Cl; 2) P bound to Fe and Mn, mostly to their oxides/hydroxides, extractable with buffered dithionite, denoted by Fe-P; 3) P bound to Al, adsorbed to oxidized metals and in humic complexes, extractable with NaOH, denoted by Al-P; 4) P bound to Ca - total apatite phosphate, except for stable minerals, extractable with HCl, denoted by Ca-P.

The proportion of residual P and organic P (P-res.) was calculated as the difference between the total phosphorus and the sum of the above fractions. The content of minerals in the upper layer of sediments (0-5 cm) was analyzed by atomic adsorption spectrometry after a mild digestion in HCl. The content of water in sediments was calculated as the difference between the wet mass of sediments and its mass after drying at 105^oC for 24 hours. The content of phosphates in water samples and in solutions obtained during fractionation of sediments was determined by the molybdate blue method (A.P.H.A., 1971). Total phosphorus was determined by the molybdate blue method after combustion (ibid.).

3. Results and discussion

Chemical composition of the bottom sediments in Lake Flosek is shown in Table I. In comparison to non-humic lakes of the Great Masurian Lake District, it is characterized by a high content of water and organic matter. The content of organic matter lies within the range known for other dystrophic lakes. The sediment-samples show a permanently increased calcium content compared to sediment samples that were taken before the liming (Hillbricht-Ilkowska *et al.*, 1977). However, it is still 10-20 times lower, than that found in mesotrophic and eutrophic lakes (Wiśniewski and Planter, 1985).

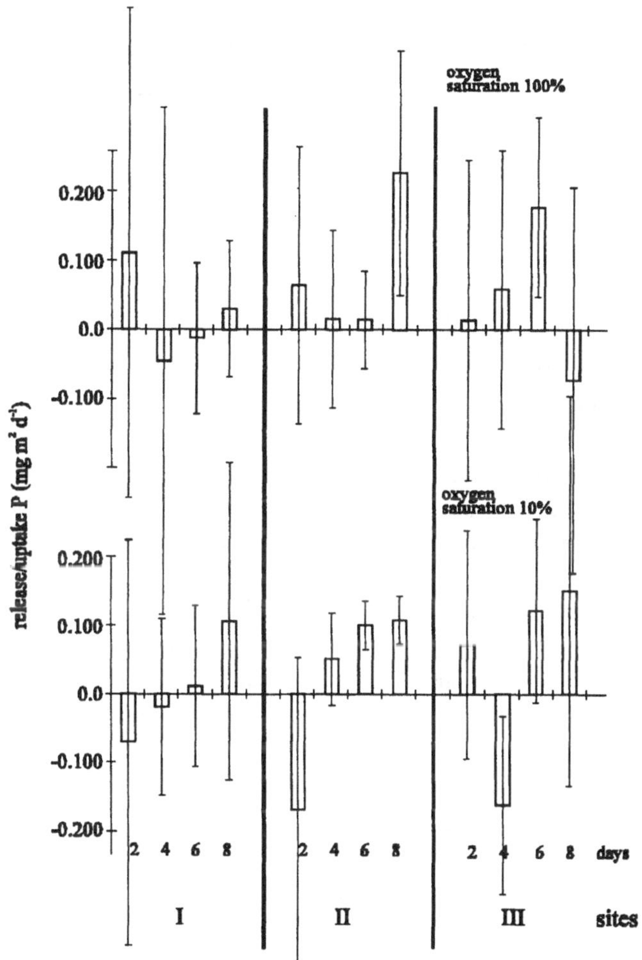

Fig. 2. Phosphorus uptake/release from undisturbed sediment cores of humic lake Flosek (columns - average of 4 cores ± SD).

The content of magnesium also was lower (by an order of magnitude) relative to that in non-humic lakes. However, the contents of iron and aluminum fell within the range known for lakes with higher trophy (Table I). No data are available on the content of minerals (Al, Mn, Fe, Mg) in Lake Flosek before liming.

No significant differences were found in the uptake/release of phosphates at high and low oxygen concentrations. Mean values for 4 cores (with standard deviations) are shown in Figure 2. The range of standard deviations obtained for 4 cores at each site shows, that the uptake and release of phosphates were balanced, with a slight net tendency for release. However, the mean values, from -0.161 to +0.20 mg P m^{-2} day^{-1}, appeared to be at least 10 to 100 times lower, than those in non-humic, meso-, eu-, and hypertrophic lakes. According to Wiśniewski

TABLE II.

Phosphorus fractions in the upper layer (0-5 cm) of bottom sediments in Lake Flosek (in μg P g^{-1} dry weight).

Station (depth)	labile–P	Fe–P	Al–P	Ca–P	res.–P
I (3 m)	17	11	140	9	934
II (7 m)	70	77	317	25	300
III (3m)	28	30	194	11	823

and Planter (1985) 5 to 15 mg P m^{-2} day^{-1} were released in these types of lakes. This means that after an increase in Ca content in the sediments (as a result of liming), the rate of P release remained very low and independent of oxygen saturation. The variation in P release was similar at both oxygen concentrations.

Using the method of selective extraction, five fractions of phosphorus bound in sediments were distinguished (Table II, Figure 3). The organic and residual phosphorus fractions (P-res.) were dominant, over 75% of the total phosphorus at sites I and III, and only 38% at site II (Figure 3).

The largest part of the readily extractable phosphorus was represented by Al-P. It amounted to 140 μg g^{-1} dry weight of the sediments (13%) at site I, 194 μg g^{-1} (18%) at site III, and 330 μg g^{-1} (40%) at site II (Figure 3, Table II). Sedimentary P is bound to humic complexes indirectly through metals (eg. iron, aluminium) (Boström, 1984). These compounds are not sensitive to changes in the redox potential, thus relatively stable. This implies that most of the phosphorus retained in the sediments of Lake Flosek are resis-

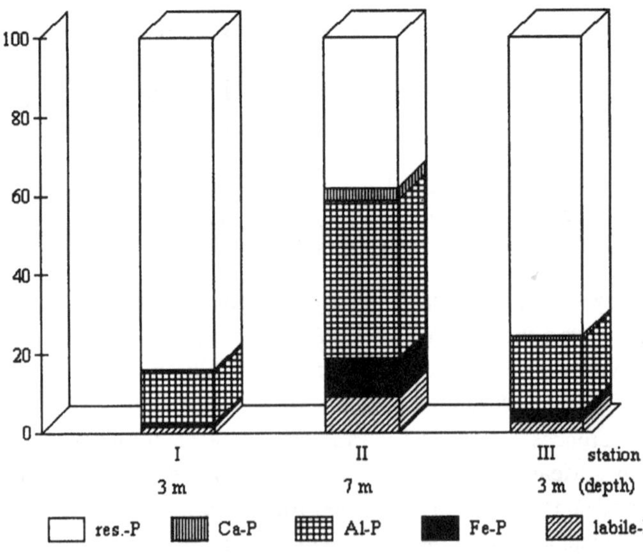

Fig. 3. Phosphorus fractions in the sediment of lake Flosek in % of P_{tot}.

tant to release and recycling.

The contribution of the remaining fractions represented a small part of the total phosphorus (3% at site I, 6% at site III, and 22% at site II). Phosphorus bound to oxidized metals such as Fe and Mn, as well as adsorbed on the surface of their hydroxides (Fe-P) accounted for 10% of the total P in sediment at site II, and for 1 and 3% at sites I and II, respectively. Notable is a low proportion of labile P (2-9%). It was highest in the central part of the lake at site I (9%), compared to the remaining two sites. It is also interesting that the content of P bound to Ca (Ca-P), was not increased as a result of the liming. This phosphorus fraction accounted for merely 1 to 6% of the phosphorus content in sediments (Figure 3).

There are no significant differences between the P-release rates at differences sites, times sequences and oxygen saturation (Figure 2). The standard deviations are overlapping. The maximum average values of P-release were obtained with the sediments cores from sites II i.e. of maximum Al-P, Fe-P and Ca (sites II) and maximum content of labile-P (~8%) at 100% oxygen saturation but not in reduced saturation. However, these differences are not significant (3-WAY Repeated Measures ANOVA for II sites vs. I and III - $F_{2,3} = 1.40$, $p = 0.37$; for oxygen saturation - $F_{1,4} = 1.27$, $p = 0.32$)

These results are consistent with data for humic sediments of acid lakes with a high sorption capacity of phosphorus. The largest fraction of sedimentary phosphorus is represented by organic phosphorus (ca. 75%), a lower fraction by phosphorus bound to Al and Fe (ca. 23%), whereas the labile fraction and the fraction bound to Ca are very small, even after liming (Boström, 1984; Pettersson and Olsson, 1986).

The liming intensified the decomposition processes of organic matter by increasing the pH from 5.6-6.0 to 7-8 and the increase in Ca content in sediments from 0.2 to ~1% dry mass. However, it seems that these changes had no significant effect on the rate of exchange of phosphorus compounds between sediments and water in the lake. Thus, the liming of the study lake in 1970 did not change the sediments from functioning as a sink for phosphorus. This situation determines the low trophic status of this lake. The present results confirm the data of other authors (Yan *et al.*, 1977; Hultberg and Andersson, 1982; Scheider and Brydges, 1984) who also did not find significant changes in the lake phosphorus pool after liming.

4. Conclusions

Fractions of organic, residual, and humic phosphorus in Lake Flosek represent the bulk of sedimentary phosphorus resistant to decomposition. Consequently, the exchange of phosphorus between sediments and water is almost nonexistent and it does not change under conditions of increased pH and Ca. Such situation seems to determine the permanently low trophy of this lake.

The results show that sediments of the shallow, humic Lake Flosek are a sink and not a source of phosphorus to the lake. The release of phosphorus is very low and independent of the redox potential. This is opposite to the situation found in eutrophic lakes.

References

American Public Health Association (A.P.H.A.): 1971, *Standard methods for the examination of water and wast water. 13 th ed., New York,* 874 pages

Boström, B.: 1984, *Int. Revue ges. Hydrobiol.* **69,** 454-474.

Boström, B., Jansson, M., Forsberg, C.: 1982, *Arch. Hydrobiol.* **18,** 5-59.

Boström, B. and Pettersson K.: 1982, *Hydrobiol.* **92,** 415-429.

Broberg, O.: 1987, *Hydrobiol.* **150,** 11-24.

Broberg, O.: 1988, *Ambio* **1,** 22-27.

Calderoni, A., Mosello, R. and Quirci, A.: 1991, *Arch. Hydrobiol.* **122,** 4: 421-439.

Eriksson F., Hörnström E., Mossberg, P. and Nyberg, P.: 1983, *Hydrobiol.* **101,** 145-164.

Golterman, H. L.: 1984, *Verh. inter. Ver. Limnol.* **22,** 23-59.

Hillbricht-Ilkowska, A., Rybak, J. I., Kajak, Z., Dusoge, Z., Ejsmont-Karabin, J., Spodniewska, A., Wêgleñska, T. and Godlewska-Lipowa, W. A.: 1977, *Ekol. pol.* **25,** 3: 379-420.

Hörnström, E., and Ekström, C.: 1986, *Acidification and liming Effects on Phyto- and Zooplankton in some Swedish West Coast Lakes.* Swedish Environment Protection Board, Report 1864, 108 pages

Hultberg, H., and Andersson, I.B.: 1982, *Wat. Air, Soil Pollut.* **18,** 311-331.

Kajak, Z., Kacprzak, K. and Polkowski, R: 1965, *Ekol. pol. B,* **11,** 2: 159-165

Kondracki, J.: 1972, *Geografia fizyczna Polski, PWN, Warszwa.* 271 pages

Ohle, W.: 1964, *Helgol. Wiss. Meeresunters.* **10,** 411-429.

Pettersson, K., and Olsson, H.: 1986, *Mobility and fractional composition of phosphorus in the sediments of oligotrophic non-acidified, acidified and limed lakes.* 12th Nordic Symp. on Sediments, Skallingen, Denmark. K. Henriksen (ed.) 13-35.

Psenner, R., Boström, B., Dinka, M., Pettersson, K., Pucsko, R. and Sager, M.: 1991. *Arch. Hydrobiol.* **30,** 83-112.

Rybak, J.I.: 1969, *Ekol. pol.* **35,** 611-662

Scheider, W. A. and Brydges, T. G.: 1984, *Fisheries* **9,** 17-18.

Starmach, K., Wróbel, S. and Pasternak, K.: 1976, *Hydrobiologia. Warszawa, PWN,* 620 pages

Yan, N. D., Jones, J., Cave, B., Scott, L., and Powell, M: 1977, *The effects of experimental elevation of lake pH on chemistry and biology of Nelson Lake near Sudbury, Ontario,* Ontario Ministry of the Environment Technical Rep., Toronto, 27 pages

Wetzel, R. G.: 1983, *Limnology,* Saunders College Publishing (Second edition), 767 pages

Wiśniewski, R. J., and Planter, A.: 1985, *Verh. int. Ver. Limnol.* **22,** 3345-3349.

SEDIMENT AND PHOSPHORUS EXPORT FROM A LOWLAND CATCHMENT: QUANTIFICATION OF SOURCES

B. KRONVANG, R. GRANT and A.L. LAUBEL

National Environmental Research Institute, Department of Streams and Riparian Areas, Vejlsøvej 25, DK-8600 Silkeborg, Denmark.

Abstract. Storm event and annual export of suspended sediment (SS) and particulate phosphorus (PP) was measured during three hydrological years (June 1993 to May 1996) in Gelbæk stream, a Danish lowland stream draining a 11.6 km^2 arable catchment area. The contribution of subsurface drainage water, surface runoff and stream bank and bed erosion to catchment SS and PP losses was estimated using three different strategies: 1) Simultaneous and comparative monitoring of subsurface water. 2) A mass-balance and budget approach dividing the Gelbæk catchment into two sub-catchments. 3) Application of the fingerprinting technique to single storm events. Subsurface drainage water proved to be a significant SS and PP source. Subsurface drainage water from half of the catchment area accounted for 9.8-15% of the total annual SS loss from the Gelbæk catchment and 9.6-18.2% of the annual PP loss. The mass-balance and budget approach showed stream bank and bed erosion to be the major source of SS and PP in this channelized and highly managed lowland stream. These findings were consistent with the fact that the annual loss of SS and PP from an upper culverted stream sub-catchment was significantly lower than that estimated from a mass-balance for a lower sub-catchment with an open stream channel. Comparison of the tracer content (e.g. ^{137}Cs) of SS collected during four storm events with that of topsoil and subsoil using a simple mixing model revealed subsoil to be a major source of SS.

Key words: Suspended sediment, particulate phosphorus, catchment budget, fingerprinting technique.

1. Introduction

In many countries, diffuse pollution is presently the most important environmental problem, contributing to the eutrophication of aquatic water and to contamination of aquatic sediments and the drinking water supply (e.g. Kronvang *et al.*, 1993; European Environmental Agency, 1995a; European Environmental Agency, 1995b). As a consequence, there has been a growing need to develop monitoring strategies and methods able to identify critical source areas and quantify the importance of different delivery pathways for diffuse pollution at the catchment level (e.g. Reinelt *et al.*, 1988; Kronvang, 1996).

The important role of suspended sediment (SS) in the transport of pollutants such as phosphorus, heavy metals, pesticides, etc., has attracted increasing attention during recent years (e.g. Walling, 1988). For example, sediment-associated particulate phosphorus (PP) has been shown to account for major parts of the total phosphorus loss from agricultural catchments (Johnson *et al.*, 1978; Kronvang, 1990; Dorioz and Ferhi, 1994; Svendsen *et al.*, 1995).

Major sources of SS and P loss from arable fields are soil erosion (e.g. Sharpley and Smith, 1989; Boardman, 1990) and, as more recently shown, transport through the soil column via macropores in clayey structured soils (Grant *et al.*, 1996; Kronvang *et al.*, 1996). In small catchments other important SS sources are stream bank and bed erosion (Prestegaard, 1988; Foster *et al.*, 1990; Peart and Walling, 1988).

Water, Air and Soil Pollution **99**: 465-476, 1997.

The large spatial extent and temporal variation in the delivery, transport and fate of SS and PP in catchments presents a problem in the identification of source areas and quantification of the importance of different delivery pathways (surface runoff, subsurface drainage, bank erosion, etc.). One important method of assessing SS sources and yields at the catchment level is the establishment of sediment budgets (e.g. Trimble, 1981; Prestegaard, 1988; Foster *et al.*, 1990). The importance of specific SS and PP delivery pathways can be investigated at the field level and then linked to the total SS and PP export from the catchment as a whole or from sub-catchments differing in potential SS source areas. Another, albeit indirect method, is the "fingerprinting" of sediment sources using differences in the physical and chemical properties of SS to trace their sources (e.g. Peart and Walling, 1988; Walling and Woodward, 1995).

The present paper focuses on the identification and quantification of source areas and delivery pathways for SS and PP within an intensively farmed catchment area drained by a small lowland stream. Three different strategies for tracing SS and PP source areas and delivery pathways are compared and their general applicability discussed.

2. Study site and methods

The study was undertaken in Gelbæk stream catchment area located in eastern Jutland, Denmark. The stream drains a 11.6 km^2 area of morainic deposits from the latest glaciation period (Weichel). The relief is gentle with elevation ranging from 20-135 m a.s.l. Gelbæk catchment is classified as a low-risk area for soil erosion based on mapping of the potential erodible agricultural areas in the catchment using a modified USLE model (Hasholt *et al.*, 1990).

In the late 1940's the upper part of the stream was culverted, thereby dividing the catchment into two sub-catchments differing in potential sediment delivery pathways: An upper sub-catchment of 4.4 km^2 and a lower sub-catchment of 7.2 km^2. Both sub-catchments are intensively farmed >90% being agricultural land and the remaining part being small woods and paved areas. The proportion of land devoted to agriculture in Denmark is higher (63%) than the average for the whole of Europe (about 42%) (European Environment Agency, 1995b). About half of the Gelbæk catchment is artificially tile-drained.

Measurements were conducted at two stream stations (S1 and S2) at the boundary dividing the two sub-catchments. In addition, measurements were conducted at two drain stations (D1 and D2) draining arable fields of 4.4 and 13.3 ha, respectively (Figure 1).

2.1 SAMPLING TECHNIQUES

Comparative measurements of suspended sediment (SS) and particulate phosphorus (PP) at stream station S2 and drain stations D1 and D2 were conducted for three hydrological years (June 1993 to May 1996). Measurements at station S1 were only conducted during the two last hydrological years. Sampling at all four monitoring stations was conducted with automatic samplers collecting hourly water samples.

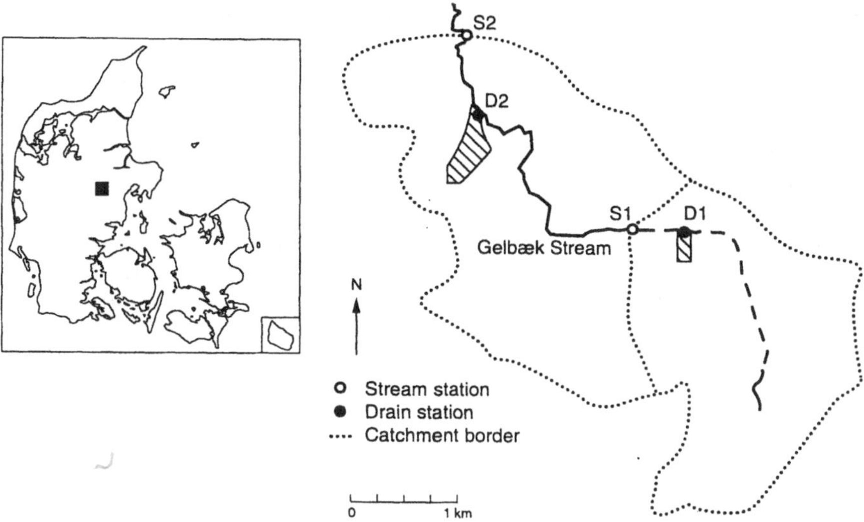

Fig. 1: Map of Gelbæk catchment indicating the two stream stations (S1 and S2) and the two drain stations (D1 and D2).

The samples were pooled in groups of eight in one-litre polyethylene bottles to give 8-hour composite samples, the latter being collected from the field once weekly. Normally, 21 bottles were pooled to give a weekly composite sample but in case of a rise in water level, 3 bottles were pooled to give a daily composite water sample covering the storm period. During single storm events, a separate automatic sampler at each station was triggered by a predetermined rise in water level to collect 24 samples within a 24 hours period.

The water level in the stream and in a drop manhole with a V-notch shaped weir at the drain stations was recorded every 10 minute and stored on a data logger. Discharge was measured every fortnight in the stream and weekly in drainage water by means of standard methods.

2.2 ANALYSIS

The water samples were analysed for suspended sediment (SS), total P and total dissolved P. The concentration of SS was measured by filtering 100-500 ml water samples through pre-weighed 1.2 μm Whatmann GF/C glass microfibre filters. The concentration of total P and total dissolved P was measured by converting to dissolved reactive P using persulphate digestion in an autoclave (Koroleff, 1983). Dissolved reactive P was determined colorimetrically according to the method of Murphy and Riley (1962). Particulate P (PP) was determined by subtracting total dissolved P from total P.

Soil samples were collected from topsoil and subsoil within the Gelbæk catchment. The samples were wet-sieved in the laboratory and the fractions <63 μm dried and stored for further analysis. Sediment samples from drainage water and stream water during selected storm events were obtained by sedimentation of SS from 24 one litre water samples. The stored soil and sediment samples were subjected to gammaspectrometric analysis and analysis of organic carbon (C) and organic nitrogen (N) using an elemental analyzer (RoboPrep-C/N) in line with a mass spectrometer (Tracer mass, Europe Scientific, U.K.). Grain size analysis was conducted on oxidized and dispersed soil and sediment samples using a laser diffraction method.

2.3 CALCULATIONS

Daily mean discharge at stream station S2 was calculated from stage-discharge relationships. Daily mean discharge from stream station S1 was obtained from a relationship between measured discharge at station S1 and the instantaneous discharge one hour later at station S2. Daily mean discharge at the two drain stations was calculated using a basic head-discharge calibration for each of the two weirs (Bos, 1976). Transport of SS and PP was calculated by summing the product of discharge and concentration for the period in question (Kronvang and Bruhn, 1996).

3. Results

3.1 CATCHMENT HYDROLOGY AND LOSSES OF SS AND PP

Annual precipitation and runoff from the Gelbæk catchment varied considerably between the three hydrological years studied (Table I), being higher than the long-term average in the hydrological years 1993/94 and 1994/95, but lower in the hydrological year 1995/96.

Annual loss of SS and PP from Gelbæk catchment to some extent followed the variations in catchment hydrology (Table I), being lowest in the driest hydrological year (1995/96) but highest in the hydrological year 1994/95, during which runoff was near to the long-term average for the catchment (Table I).

Seasonal loss of SS and PP varied considerably during the three hydrological years studied (Figure 2), monthly losses being highest in the winter period (e.g. March 1994 and February 1995). The highest monthly loss of SS for each of the hydrological years 1993/94, 1994/95 and 1995/96 accounted for 36%, 51% and 63%, respectively, of the annual loss. The corresponding figures for PP were 33%, 46% and 49%, respectively.

3.2 SOURCE AREAS AND DELIVERY PATHWAYS FOR SS AND PP

Subsurface drainage water losses of SS and PP proved to be an important but highly variable source of catchment SS and PP losses (Figure 3). On average, drainage water from artificially drained arable areas in the Gelbæk catchment accounted for 50% of annual runoff (range: 30-65%), 12.3% of annual SS loss (range: 9.8-15%) and 12.5% of annual PP loss (range: 9.6-18.2%).

Annual loss of SS and PP from the two sub-catchments differed considerably. Thus annual losses from the culverted upper sub-catchment amounted to 34.1 kg SS ha^{-1} and 0.146 kg PP ha^{-1} y^{-1} during the wetter of the two hydrological years (1994/95) as opposed to 114.7 kg SS ha^{-1} y^{-1} and 0.468 kg PP ha^{-1} y^{-1} from the lower sub-catchment where the stream is open. The corresponding figures for the dryer of the two hydrological years (1995/96) were 7.78 kg SS ha^{-1} y^{-1} and 0.051 kg PP ha^{-1} y^{-1}, as opposed to 18.4 kg SS ha^{-1} y^{-1} and 0.132 kg PP ha^{-1} y^{-1}.

TABLE I

Annual precipitation, runoff and loss of suspended sediment and particulate phosphorus from Gelbæk catchment during three hydrological years (June-May). For comparison long-term figures for precipitation and runoff is included.

	1993/94	1994/95	1995/96	Reference period[1]
Precipitation (mm y^{-1})	927	956	433	712
Runoff (mm y^{-1})	369	227	68.8	204
Suspended sediment (kg ha^{-1} y^{-1})	70.8	84.1	14.4	-
Particulate phosphorus (kg P ha^{-1} y^{-1})	0.315	0.346	0.101	-

[1]: Reference period for precipitation: 1961-90 and for runoff: 1974-94.

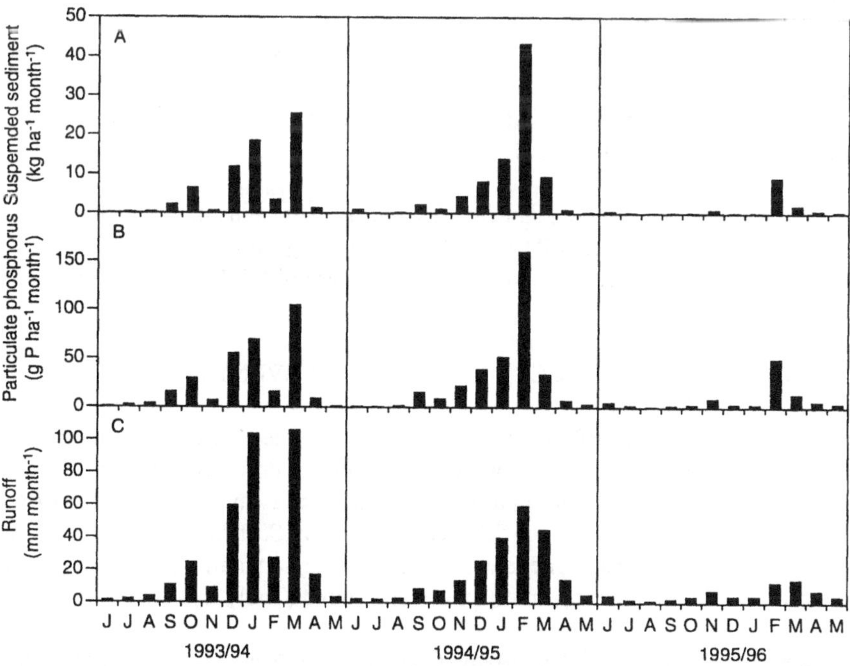

Fig. 2: Monthly suspended sediment loss (A), particulate phosphorus loss (B) and runoff (C) from the Gelbæk catchment during the three hydrological years studied (June 1993 to May 1996).

The difference between annual losses of SS and PP in the two sub-catchments during the two hydrological years (1994/95 and 1995/96) is only partly attributable to a difference in runoff since the open:culverted sub-catchment ratio in the two years was 1.2 and 1.8, respectively, for runoff but 3.4 and 2.4 for SS and 3.2 and 2.6 for PP. The open stream sub-catchment must therefore include important source areas and delivery pathways that do not deliver to the culverted stream sub-catchment.

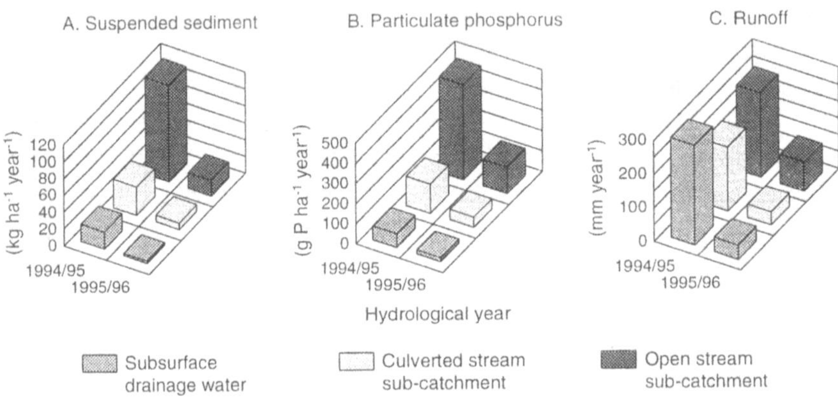

Fig. 3. Annual suspended sediment loss (A), particulate phosphorus loss (B) and runoff (C) from the open stream sub-catchment, the culverted stream sub-catchment and the two drain stations (average value) during two hydrological years (June 1994 to May 1996).

Mobilization and delivery of SS and PP from soil erosion and surface runoff is a potential important SS and PP source. However, early spring inspection of potential erodible areas within the open sub-catchment revealed no sign of rill erosion and surface runoff following the winters of 1994/95 and 1995/96.

A mass-balance for the open sub-catchment thus revealed stream bank and bed erosion to be the major source of SS (92%) and PP (93%) in both hydrological years. Sub-surface drainage water was the only other source, accounting for 8% in case of SS delivery to the stream and 7% of PP delivery.

3.3 ESTIMATION OF SS SOURCES DURING STORM EVENTS

The concentration of SS and PP in stream and drainage water varied considerably during single storm events (Figure 4). The importance of subsurface drainage water for SS delivery to Gelbæk stream was estimated during four individual storm events in different seasons. The percentage of the catchment losses of SS accounted for by SS from drainage water varied considerably during the four storm events, being highest in an autumn storm and lowest in a spring storm event with thawing snow on frozen ground (Table II).

The trace element (^{137}Cs and ^{210}Pb), organic C and organic N content of SS collected from stream station S2 and drain station D2 during the four storm events was compared to that of arable subsoil from the Gelbæk catchment. The ^{137}Cs content of SS from stream and subsurface drainage water was significantly ($p=0.003$) correlated to SS median grain size. During the storm event on 9 January 1995, median grain size of SS was 2.6 μm in sub-surface drainage water as compared to 8.2 μm in stream water. SS median grain size in stream water during the other three storm events was 4.0-5.0 μm. Differences in grain size distribution between SS in sub-surface drainage water and SS in stream water thus indicate that other important sediment sources deliver SS to the stream during storm events. Nevertheless, the trace element content of SS in drainage water was chosen as the fingerprint for topsoil source areas as selective mobilization and transport of sediment must occur from all potential topsoil source areas.

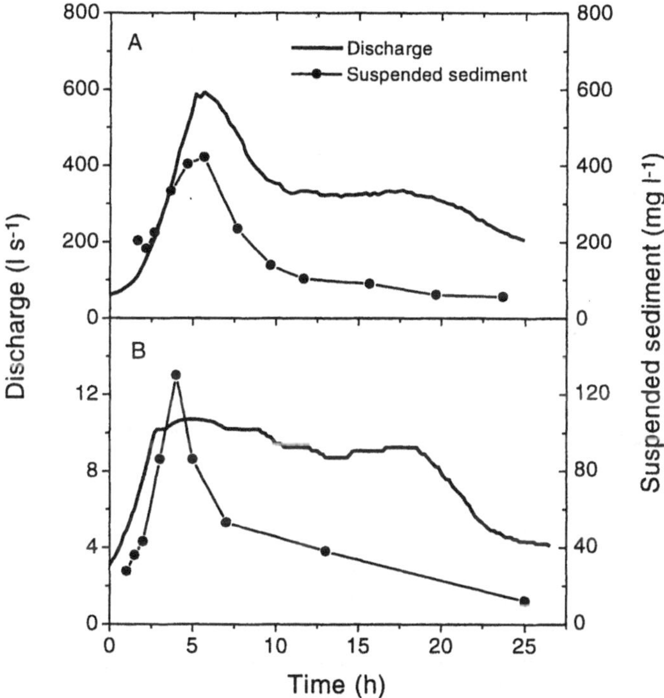

Fig. 4. Concentration of suspended sediment and discharge in Gelbæk stream at station S2 (A) and in subsurface drainage water (D2) (B) during the storm event on 9 January 1995.

TABLE II

Contribution of subsurface drainage water to catchment losses of SS at station S2 in Gelbæk stream during four storm events each lasting 24 hours.

Date of storm event	% of catchment SS losses
7 March 1994	3.9 %
3 October 1994	45 %
9 January 1995	20 %
17 February 1995	16 %

Utilizing the four SS tracers in a simple mixing model revealed that topsoil compartments account for 41-76% of SS during the three storm events analysed (Table III). In comparison, utilizing ^{137}Cs as the sole and possibly most reliable tracer revealed that topsoil compartments accounted for 48-71% of the SS exported from Gelbæk catchment during the four storm events (Table III).

TABLE III

Contribution of topsoil to catchment losses of SS in Gelbæk catchment estimated by fingerprinting using a simple mixing model with either four tracers (^{137}Cs, ^{210}Pb, organic C and organic N content in subsoil and of SS from stream water (S2) and drainage water (D2)), or ^{137}Cs alone.

Fingerprinting elements	7 March 1994	3 October 1994	9 January 1995	17 February 1995
Four tracers	Not analysed	75% (37-118%)	41% (4-66%)	76% (67-88%)
^{137}Cs as tracer	48%	51%	51%	71%

4. Discussion

The range found for annual losses of SS (14.4-84.1 kg ha^{-1}) and PP (0.101-0.346 kg P ha^{-1}) from this small, arable lowland catchment covers the catchment response to extreme hydrological conditions, both wet and dry. The annual loss of PP was highly significant from the viewpoint of eutrophication of shallow lakes (e.g. Kronvang *et al.*, 1993).

The apparent lack of a relationship between annual catchment hydrology and SS and PP losses during the hydrological year 1994/95 possibly reflects the impact of stream management on sediment and phosphorus export. Thus, extremely high precipitation in August and September 1994 following a preceding wet year caused flooding of arable areas in many parts of Denmark. Many streams were therefore channelized the following autumn to enhance their drainage capacity. A 200-400 m reach of the upper Gelbæk just downstream from station S1 was thoroughly channelized by excavation of both the banks and the bed. This left the stream banks unprotected by vegetation during the following winter period (1994/95), thereby facilitating bank erosion.

Simultaneous and comparative measurement of the export of SS and PP from two small tile-drained arable fields and a stream station (S2) draining the whole Gelbæk catchment was used in a first attempt to quantify the annual importance of subsurface drainage water as a source of SS and PP. Subsurface drainage water proved to be a significant delivery pathway for SS and PP to the stream water. Studies from other countries indicate the same and that PP may account for great proportions of total P (Sharpley and Syers, 1979; Skaggs *et al.*, 1994). The source of SS and PP in subsurface drainage water has previously been identified as the topsoil of structured clayey soils, transport taking place through macropore flow (Grant et al., 1996). A hydrological short-cut thus seems able to transport appreciable amounts of the nutrient-enriched and possibly also pollutant-enriched topsoil directly to surface waters.

The second method used in the present investigation to detect important SS and PP source areas and delivery pathways in Gelbæk catchment was the establishment of a mass-balance and budget for two sub-catchments: An upper sub-catchment in which the stream was culverted and therefore devoid of erosional sources (soil, bed and bank erosion), and a lower sub-catchment where the stream was open and input from erosional sources therefore possible. The mass-balance approach revealed large differences in SS and PP export from the two sub-catchments during the two study years, annual export from the open stream sub-catchment being 2.4-3.4 times greater than from the culverted catchment in the case of SS and 2.6-3.2 times greater in the case of PP.

This difference between the two hydrological years was only partly attributable to differences in annual runoff between the two sub-catchments, thus implying that erosional sources comprise an important delivery pathway in the lower Gelbæk sub-catchment. As rill erosion and surface runoff was not seen in this sub-catchment during either hydrological year, stream bank and bed erosion is the major source of SS and PP annual losses (92-93%). Such a major role for stream bank and bed erosion in annual SS catchment budgets is rare (Kronvang, 1996), although Foster *et al.* (1990) reported channel erosion to be a major sediment source (94%) in catchments where cattle trampled the stream banks.

The third method applied to the Gelbæk catchment to assess important SS source areas was "fingerprinting" such as used by e.g. Walling *et al.* (1993) and Walling and Woodward (1995). For comparison, the importance of subsurface drainage water as a SS source was also determined by direct quantification. Fingerprinting of SS topsoil and subsoil source areas undertaken during three single storm events using a suite of four tracers (^{137}Cs, ^{210}Pb, organic C and organic N) in a simple mixing model revealed that topsoil compartments on average accounted for 41-76% of the SS losses. However, the range for the four tracers used was very high, especially in an early autumn storm event (3 October 1994) in which the content of organic N content of stream water SS exceeded that of either the subsoil or topsoil. A possible explanation is resuspension of SS retained on the stream bed during the previous summer, as previously described for this stream by Svendsen et al. (1995 and 1996).

Caesium-137 seems to be an important tracer element for distinguishing between topsoil and subsoil compartments in complex geological catchments such as Gelbæk. Thus several surveys of soil profiles within the catchment have documented that the ^{137}Cs content of topsoil (< 63 μm fraction) is much higher (average of 8 samples: 22.1 Bq kg^{-1} DW) than in the subsoil (average of 8 samples: < 1.01 Bq kg^{-1} DW) (Vægter and Iversen, 1994). Utilizing ^{137}Cs to fingerprint the topsoil as a SS source in the Gelbæk catchment during four storm events revealed that the topsoil accounted for 48-71% of total losses. For each of the four storm events, fingerprinting of topsoil SS sources by means of ^{137}Cs indicated a higher contribution to total SS losses than that determined by quantitative assessment of subsurface drainage water.

Subtracting the amount of SS accounted for by subsurface drainage water in each of the four storm events revealed that other topsoil compartments contributed 44% (7 March 1994), 6% (3 October 1994), 31% (9 January 1995) and 55% (17 February 1995). The apparent increase in the importance of topsoil compartments other than with subsurface drainage water with increasing discharge from autumn to winter could be attributable to an increase in bank collapse during high flow conditions. During lower flow conditions, in contrast, subsoil compartments within the stream channel should be the dominant source. In the Gelbæk stream, however, subsoil compartments seemed to be a very important source of SS during all four storm events (29-52%). These findings are in accordance with those obtained using the mass-balance approach in that the latter also proves that subsoil compartments (channel banks and bed) are extremely important SS sources.

5. Conclusion

Catchment management aimed at combatting diffuse pollution necessitates a thorough knowledge of the important source areas and delivery pathways for sediment and sediment-associated compounds. Such knowledge is seldom attainable through normal monitoring programmes, and alternative methods have therefore to be developed.

The study showed (1) that a continuous and comparative samling strategy is necessary to achieve reliable SS and PP transport estimates and (2) that such a strategy can successfully quantify the importance of different SS and PP source areas and delivery pathways at the catchment or sub-catchment level during single storm events and on an annual basis.

The study also indicates that using ^{137}Cs to indirectly discriminate between topsoil and subsoil compartments as source areas for SS loss during single storm events is a valuable first screening tool for determining the relative importance of two major sources (channel erosion, soil erosion). However, fingerprinting methods cannot be used alone in a catchment management context, but should be followed by direct quantitative measurements and mapping of potential source areas.

Acknowledgements

We gratefully acknowledge the financial support of the Centre for Root Zone Processes under the Danish Environmental Research Programme. We are also grateful to David I. Barry for revising the English.

References

Boardman, J. 1990: Soil erosion on the South Downs: A review. In: Boardman, J., Foster, I.D.L. and Dearing, J.A. (Eds), *Soil Erosion on Agricultural Land*, John Wiley & Sons, 87-106.

Bos, M.G. 1976: *Discharge measurements structures*. International Institute for Land Reclamation and Improvement/ILRI, Wageningen, The Netherlands.

Dorioz, J.M. and Ferhi, A. 1994: Non-point pollution and management of agricultural areas: Phosphorus and nitrogen transfer in an agricultural watershed. *Wat. Res.* **28(2)**, 395-410.

European Environment Agency 1995a: European Rivers and Lakes. *EEA Environmental Monographs* **1**, 122 p.

European Environment Agency 1995b: *European Environment The Dobris Assessment*, Stanners, D. and Bourdeau, P. (Eds). EEA, Copenhagen, Denmark, 676 p.

Foster, I.D.L., Grew, R. and Dearing, J.A. 1990: Magnitude and frequency of sediment transport in agricultural catchments: A paired lake-catchment study in Midlands England. In: Boardman, J., Foster, I.D.L. and Dearing, J.A. (Eds), *Soil Erosion on Agricultural Land*, John Wiley & Sons, 153-172.

Grant, R., Laubel, A., Kronvang, B., Andersen, H.E., Svendsen, L.M. and Fuglsang, A. 1996: Loss of dissolved and particulate phosphorus forms in drainage water from four arable catchments on structured soils in Denmark. *Wat. Res.* (In press).

Hasholt, B., Madsen, H.B., Hansen, A.C. and Platou, S.W. 1990: Erosion og transport af fosfor til vandløb og søer. [Erosion and transport of phosphorus to streams and lakes]. *NPo research from the Danish Environmental Protection Agency*, No. **C12**, 120 p. (In Danish).

Johnson, A.H., Bouldin, D.R., Goyette, E.A. and Hedges, A.H. 1978: Phosphorus loss by stream transport from a rural watershed: quantities, processes and sources, *Environ. Qual.* **5**, 148-157.

Koroleff, F. 1983: Determination of phosphorus. In: Grasshoff, K., Ehrhardt, M and Kremling, K. (Eds), *Methods of Seawater Analysis, 2nd Edition*, Chemie, Weinheim, New York, USA, 125-131.

Kronvang, B. 1990: Sediment-associated phosphorus transport from two intensively farmed catchment areas. In: Boardman, J., Foster, I.D.L. and Dearing, J.A. (Eds), *Soil Erosion on Agricultural Land*, John Wiley & Sons, 313-330.

Kronvang, B. 1996: *Suspended sediment and nutrient loads: Diffuse sources. Delivery pathways, fate and management. PhD Thesis*. National Environmental Research Institute, Department of Stream and Riparian Areas and University of Aarhus, Department of Soil Science (In press).

Kronvang, B., Ærtebjerg, G., Grant, R., Kristensen, P., Hovmand, M. and Kirkegaard, J. 1993: Nationwide monitoring of nutrients and their ecological effects: State of the Danish aquatic environment. *AMBIO* **22** (4), 176-187.

Kronvang, B., Grant, R., Larsen, S.E., Svendsen, L.M. and Kristensen, P. 1995: Non point nutrient losses to the aquatic environment in Denmark: Impact of agriculture. *Mar. Freshwater Res*, **46**, 167-177.

Kronvang, B., Laubel, A. and Grant, R. 1996: Suspended sediment and particulate phosphorus transport and delivery pathways in an arable catchment, Gelbæk, Denmark. *Hydrol. Process.*, (In press).

Kronvang, B. And Bruhn, A.J. 1996: Choice of sampling strategy and estimation method for calculating nitrogen and phosphorus transport in small lowland streams. *Hydrol. Process.* **10**, (In press).

Meals, D.W. 1985: Detecting changes in water quality in the Laplatte river watershed following implementation of BMPS. *Lake and Reservoir Management*, **3**, 185-193.

Murphy, J. and Riley, J.P. 1962: A modified single solution method for determination of phosphate in natural waters. *Analyt. Chim. Acta* **27**, 31-36.

Peart, M.R. and Walling, D.E. 1988: Techniques for establishing suspended sediment sources in two drainage basins in Devon, UK: A comparative assessment. In: Bordas, M. and Walling, D.E. (Eds.), *Sediment Budgets, Proc. Porto Alegre Symp., Int. Assoc. Hydrol. Sci.*, Publ. **174**, 269-280.

Prestegaard, K.L. 1988: Morphological controls on sediment delivery pathways. In: Bordas, M. And Walling, D.E. (Eds), *Sediment Budgets, Prc. Porto Alegre Symp., Int. Assoc. Hydrol. Sci.*, Publ. **174**, 533-540.

Reinelt, L.E., Horner, R.R.and Mar, B.W. 1988: Nonpoint source pollution monitoring program designs. *Journal of Water Resources Planning and Management* **114(3)**, 335-352.

Ritter, W.F., Chirnside, A.E.M. and Lake, R.W. 1989: Influence of best management practices on water quality in the Appoquinimunk watershed. *J. Environ. Sci. Health* **A24(8)**, 897-924.

Skaggs, R.W., Brevé, M.A. and Gilliam, J.W. 1994: Hydrologic and water quality impacts of agricultural drainage. *Crit. Rev. Environ. Sci. Technol*, **24**, 1-32.

Sharpley, A.N. and Syers, J.K. 1979: Loss of nitrogen and phosphorus in tile drainage as influenced by urea application and grazing animals. *N. Z. J. Agric. Res*, **22**, 127-131.

Sharpley, A.N. and Smith, S.J. 1989: Prediction of soluble phosphorus transport in agricultural runoff. *J. Environ. Qual*, **18**, 313-316.

Svendsen, L.M., Kronvang, B., Kristensen, P. and Græsbøll, P. 1995: Dynamics of phosphorus compounds in a lowland river system: Importance of retention and non-point sources. *Hydrol.l Process.* **9**, 119-142.

Svendsen, L.M., Kronvang, B., Laubel, A.R., Larsen, S.E. and Andersen, B. 1996: Phosphorus retention in a Danish lowland river system. *Verh. Internat. Verein. Limnol.* (In press).

Trimble, S.W. 1981: Changes in sediment storage in the Coon Creek Basin, Driftless Area, Wisconsin, *Science* **214**, 181-183.

Vægter, B. and Iversen, H.L. 1994: *Mobilisering og transport af partikulært stof - kildeopsporing af suspenderet stof i et mindre østjysk vandløb. [Mobilisation and transport of particulate matter and source apportionment of suspended matter in a small stream in eastern Jutland]. MSc Thesis*, University of Aarhus, Department of Soil Science, 247 p. (In Danish).

Walling, D.E. 1988: Erosion and sediment yield research - some recent perspectives. *J. Hydrol.* **100**, 113-141.

Walling, D.E., Woodward, J.C. and Nicholas, A.P. 1993: A multi parameter approach to fingerprint suspended sediment sources. *IAHS* Publ. **215**, 329-338.

Walling, D.E. and Woodward, J.C., 1995. Tracing sources of suspended sediment in river basins: A case study of the River Culm, Devon, UK. *Mar. Freshwater Res.* **46**, 327-336.

SEDIMENT ACCUMULATION OF NUTRIENTS (N, P) IN THE EASTERN GULF OF FINLAND (BALTIC SEA)

J. LEHTORANTA[1], H. PITKÄNEN[1] and O. SANDMAN[2]

[1]*Finnish Environment Institute, Box. 140, FIN-00251 Helsinki, Finland*

[2]*South Savo Regional Environment Center, Jääkärinkatu 14,FIN-50100 Mikkeli Finland*

Abstract. Dry weight (DW), ignition loss (IL) and concentrations of total nitrogen (TN) and total phosphorus (TP) of the sediment surface layer (0 to 10 cm, 1 cm slices) were analyzed from 20 sites in the eastern Gulf of Finland. The distance of the sampling sites from the mouth of the River Neva explained the nutrient concentrations of the sediments well, while the effect of water depth was negligible. The increase of TN and the decrease of TP along the transect from the river mouth towards the open Gulf were caused by the diminishing share of allochthonous material supplied from the River Neva. The mean TN concentration of the different accumulation areas was about 40 % higher in the sediment surface than in the deeper layer (9 to 10 cm). The corresponding difference for TP varied from 53 to 56 %. The results suggest considerable netflux of nutrients from sediment to water. The net sediment accumulation of nutrients were estimated as 6.0 g m^{-2} a^{-1} of N and 1.7 g m^{-2} a^{-1} of P corresponding 22 000 t a^{-1} of N and 6 100 t a^{-1} of P for the whole eastern Gulf.

Key words: sediment, nitrogen, phosphorus, organic matter, cluster analysis, Gulf of Finland, estuaries

1. Introduction

The general trophic status (chlorophyll-*a* concentration and biomasses of auto- and heterotrophic organisms) is far higher in the eastern Gulf of Finland than in the western Gulf or in the other parts of the northern Baltic. This is principally due to the heavy nutrient loading from the St. Petersburg area and the River Neva, coupled to the hydrodynamics of the area (Pitkänen et al., 1993; Kauppila et al., 1994).

According to mass balance calculations the eastern Gulf of Finland retains amounts corresponding 69 % of the land-derived and atmospheric N and 100 % of the corresponding P inputs (Pitkänen, 1994). The suspended particulate matter (SPM) is sedimented primarily in the Neva Estuary. Biological processes can explain a part of the losses of P from the euphotic surface layer (Pitkänen and Tamminen, 1995). However, it is also possible that a part of the P co-precipitates with iron (Fe) and manganese (Mn) oxides already in the water column (Golterman, 1980). The general features of Fe and P removal, involving rapid aggregation of riverborne colloids, have been reproduced in laboratory simulations of estuarine mixing (Bale and Morris, 1981).

The objective of this study is to classify different types of sedimentation areas of the eastern Gulf of Finland and to find out the factors causing the observed differences in concentrations of organic matter and nutrients between various areas. With a multivariate method (cluster analysis) it is possible to classify the sampling sites to different groups according to the measured variables. This information is simplified and used for new hypotheses.

Water, Air and Soil Pollution **99**: 477-486, 1997.

2. Materials and methods

2.1. STUDY AREA

The eastern Gulf of Finland (13 000 km^2) is the area east from the Gogland island (Figure 1). The mean depth of the area is ca. 30 m and the average residence time of water ca. one year. The River Neva discharging to the easternmost corner of the Gulf is the largest river flowing in to the Baltic Sea (mean flow ca. 2 600 m^3 s^{-1}). It corresponds to 75 % of the total inflow into the Gulf of Finland and to 20 % of the total riverine inflow into the whole Baltic Sea (Ehlin, 1981). The surface salinity of the study area increases from 0 psu in the Neva Bay to 4 to 5 psu in the open eastern Gulf, whereas in bottom-near waters salinities vary from 4 to 8 psu (Pitkänen et al., 1993). According to echographs the areas of erosion, transportation and accumulation vary strongly spatially due to variations in morphometrical and hydrodynamical conditions.

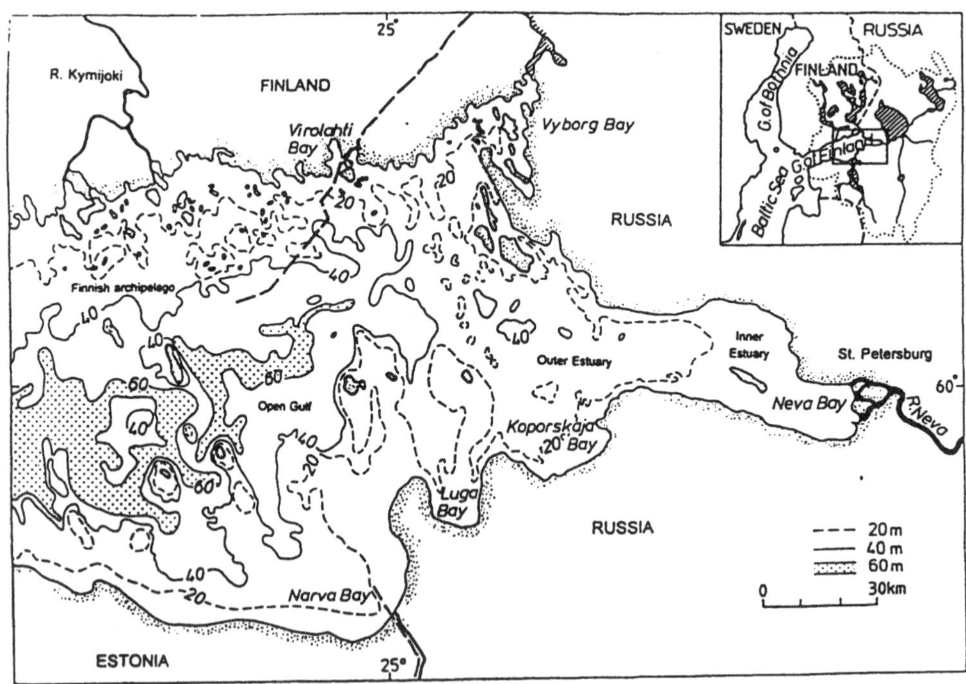

Figure 1. The eastern Gulf of Finland.

2.2. SEDIMENT SAMPLING AND CHEMICAL ANALYSES

On the basis of the echographs (Atlas Deso 10 echosounder, 30 kHz), sediment samples and bathymetric maps, the bottoms of active accumulation were distinguished from erosion and transportation bottoms. The sediment samples were taken with a gravity corer at 20 stations in July-August 1992 to 93. The cores were sliced into 10 mm slices from the surface to the depth of 10 cm. Each slice was frozen in an air-sealed plastic bag. Dry weight

(DW) and ignition loss (IL) were analyzed according to the Finnish standard SFS 3008 (1981). The concentration of organic carbon was calculated from the values of IL according to the ratio 1:2 (Håkanson and Jansson, 1983). Total nitrogen (TN) and total phosphorus (TP) were analysed by co-digestion using the method of Zink-Nielsen (1975). Organic matter was decomposed by strong sulphuric acid. Nitrate and nitrite were reduced to ammonia with Devarda solution. The formed ammonium sulphate was distilled and ammonium was titrated (Starck and Haapala, 1984). Inorganic phosphate complexes and organic P were converted to orthophosphate by sulphuric acid and analysed after SFS 3025 (1986).

2.3. STATISTICAL ANALYSES

With cluster analysis it is possible to establish groups, which are cognate according to measured variables. SAS (PROC CLUSTER; SAS Institute, Inc., 1985) was used for statistics. The data were clustered with an average distance method (Sokal and Michener, 1958) and it tends to join clusters with small variances and is slightly biased towards producing clusters with same variance.

The cluster analysis has been applied e.g. in studies where several metal concentrations have been measured from sediment samples (Hodkinson et al., 1986; Albani et al., 1989). The clusters have constituted regional groups for which causal effects have been sought. In this study the cluster analysis was done by sampling sites (individual directed analysis) using each measurement as a discrete variable.

Because the correlations between DW, IL and TN were high compared with correlations to TP (Figures 2a and b), two separate analyses were performed to get more detailed information concerning organic matter content and TN, and TP concentrations in various areas. In the first cluster analysis concentration of DW, IL and TN for each of the 1 cm sub-layers were used as discrete variables (30 variables). In the second cluster analysis (10 variables) every measurement of the TP concentration of each site was used as a discrete variable.

3. Results

3.1. GENERAL CHARACTER OF THE SEDIMENT SAMPLES

According to echographs the largest accumulation areas are located in the deepest parts of the open Gulf (ca. 60 m) and in the more shallow (30 to 40 m) Neva Estuary. In all sampled cores (mostly clayey gyttja) a light brown layer (max. 4 cm thick) was found in the upper sediment. Beneath this layer the sediment was reduced and coloured by black sulfides. In the black layer, most of the cores were colonised by white filamentous bacteria, probably sulphide-oxidising *Beggiatoa*. The core nearest to the River Neva mouth contained sand sized particles.

3.2. NUTRIENT CONCENTRATIONS VERSUS WATER DEPTH AND DISTANCE FROM THE RIVER NEVA MOUTH

The water and organic matter content (IL) of the sediment surface (0 to 1 cm) were mainly high, ranging from 76.9 to 96.7 % and from 6.8 to 23 % DW in the surface layer, respectively. The mean TN content decreased from 8.0 mg g^{-1} DW (2.7 to 12.6 mg g^{-1} DW) at the surface to 4.6 mg g^{-1} DW (1.7 to 7.3 mg g^{-1} DW) at 9 to 10 cm depth. The corresponding figures for TP were from 2.9 (1.1 to 4.4 mg g^{-1} DW) to 1.3 mg g^{-1} DW (0.7 to 2.1 mg g^{-1} DW).

The correlations between variables (DW, IL, TN and TP) were similar for the surface sediment (0 to 1 cm) and for the whole sediment column (0 to 10 cm). The correlation coefficient between IL and TN in the whole column was as high as 0.98 (Figure 2a), while the corresponding correlation between IL and TP was clearly poorer (r=0.53, Figure 2b).

The water depth in areas of accumulation did not explain the mean TN content (r^2=0.15) nor the TP content (r^2=0.006). The mean values of TN increased while TP decreased with distance from the river mouth (Figures 3a and 3b). The C:N ratio of terrestrial organic matter is higher than that of marine organic matter (respectively >12 and 6-9, Bordovskiy, 1965; Prahl et al. 1980). The C:N ratio of the surface layer decreased from 12.5 in the inner estuary to 8.9 in the Finnish archipelago possibly indicating an increase of autochtonous matter in the sediment. However C:N as a provenance indicator is not very reliable, because particular organic matter source signature may be lost by biochemical alteration after deposition (Thornton and McManus, 1994). Both TN and TP concentrations were low at two sites (Figures 3a and b), which were classified as transportation bottoms.

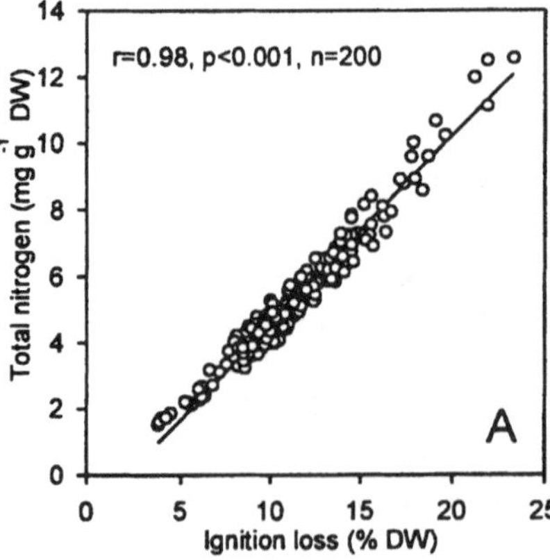

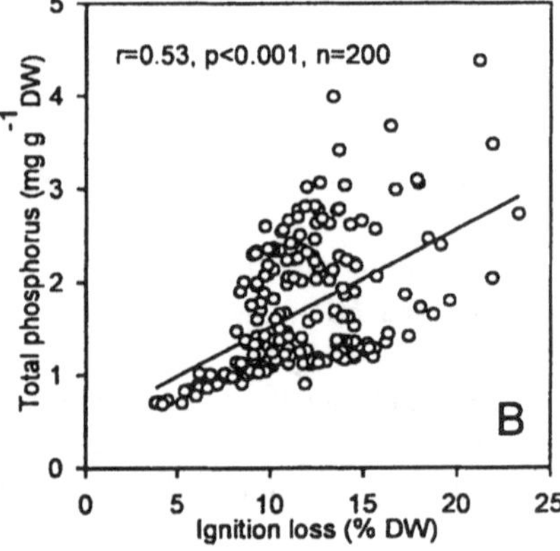

Figures 2a and b. The relationships between organic matter and nutrients (N, P).

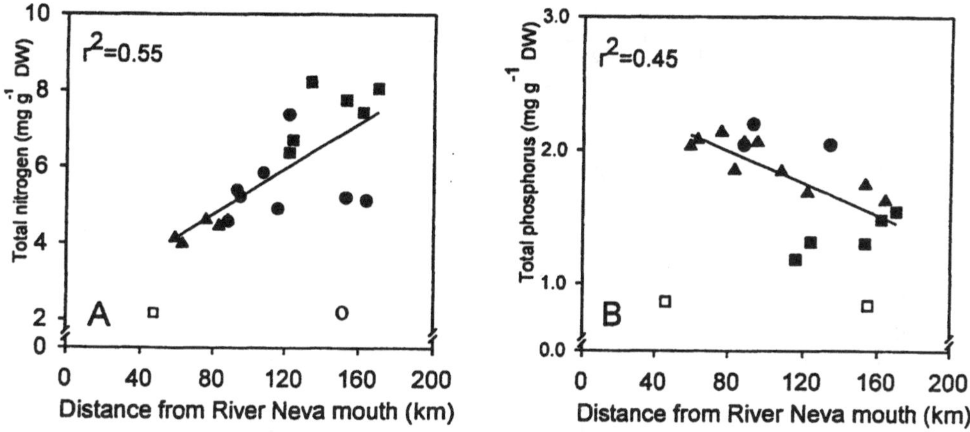

Figures 3a and b. The mean values of a) TN and b) TP concentrations of each site in 0 to 10 cm sediment layer vs. distance from river mouth. The transportation areas (empty symbols) were not included in calculations. The symbols are same used for the clusters in Figure 4.

3.4. GROUPING OF THE SAMPLING SITES

On the basis of the first cluster analysis (including DW, IL and TN) the sites were divided into five clusters (Figure 4). In the inner estuary the sediment contained sand and the area was classified as a transportation bottom (cluster N1, Figure 4).

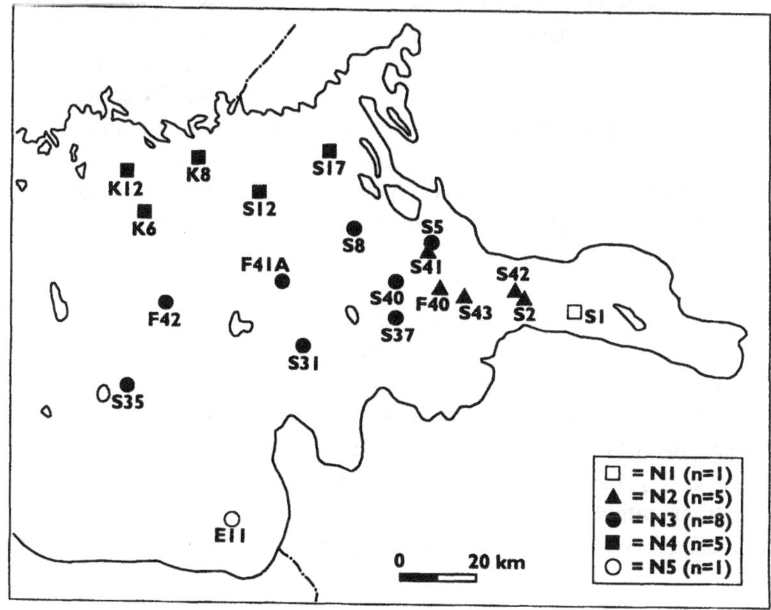

Figure 4. The locations of the clusters N1-N5.

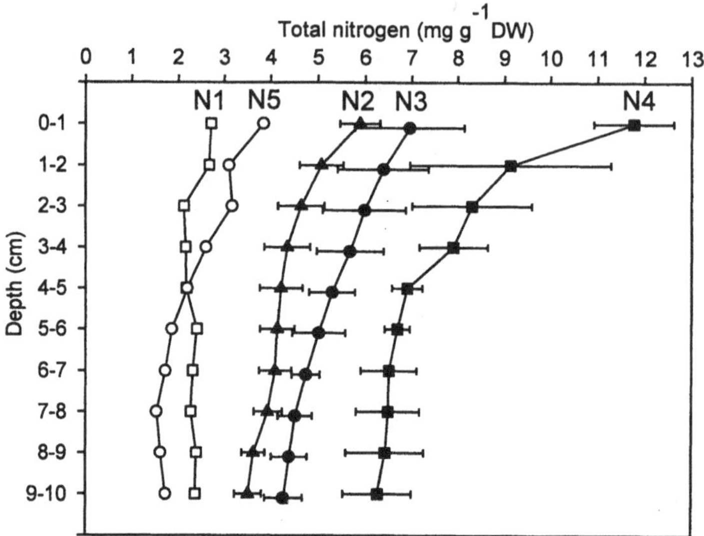

Figure 5. The vertical mean distributions with standard deviation bars for the clusters N1-N5.

The Narva Bay (Figure 1) was characterized as a transportation bottom as well (cluster N5). The openness and relatively shallow water depth (max. 36 m) expose the bottom to waves and currents. The cluster N2 corresponded well to the basin of the outer Neva Estuary, while the sampling sites of the cluster N3 situated in the transition zone between the estuary and the open Gulf (Figure 1). In the Finnish archipelago and in the open area east from the archipelago the accumulation bottoms located mainly in small, steeply sloped bottoms surrounded by large, shallow transportation and erosion areas (cluster N4).

TN concentrations of the Narva Bay and the inner estuary (clusters N1 and N5) were low (<4 mg g^{-1} DW, Figure 5) in the whole sediment column. The mean TN content of clusters N1, N2, N3 and N4 increased (in this order) from the River Neva mouth towards the west. Thus the mean TN concentrations of the Finnish archipelago (cluster N4) were clearly higher (11.8 mg g^{-1} DW) in the surface layer than those of the other accumulation areas (6.0 to 7.0 mg g^{-1} DW). The average TN concentrations of the accumulation areas (clusters 2, 3 and 4) were 39 to 41 % higher in the surface (0 to 1 cm) than in the deeper layer (9 to 10 cm).

On the basis of the second cluster analysis the sites were divided into four clusters (Figure 6). In the transportation areas (cluster P1) the mean surface TP concentration was low (1.1 mg g^{-1} DW, Figure 7). In the Finnish archipelago the TP decreased sharply from 2.5 mg g^{-1} DW at the surface to 1.5 mg g^{-1} DW at the depth of 1 to 2 cm (cluster P4). At most of the estuarine and open Gulf sites (cluster P2) TP decreased gradually from the sediment surface (2.7 mg g^{-1} DW) to the depth of 8 cm (1.2 mg g^{-1} DW). However, at some sites the decrease was strong within the top few centimeters (cluster P3). The mean TP of the accumulation areas (P2, P3 and P4) was 53 to 56 % higher in the sediment surface than in the deeper sediment.

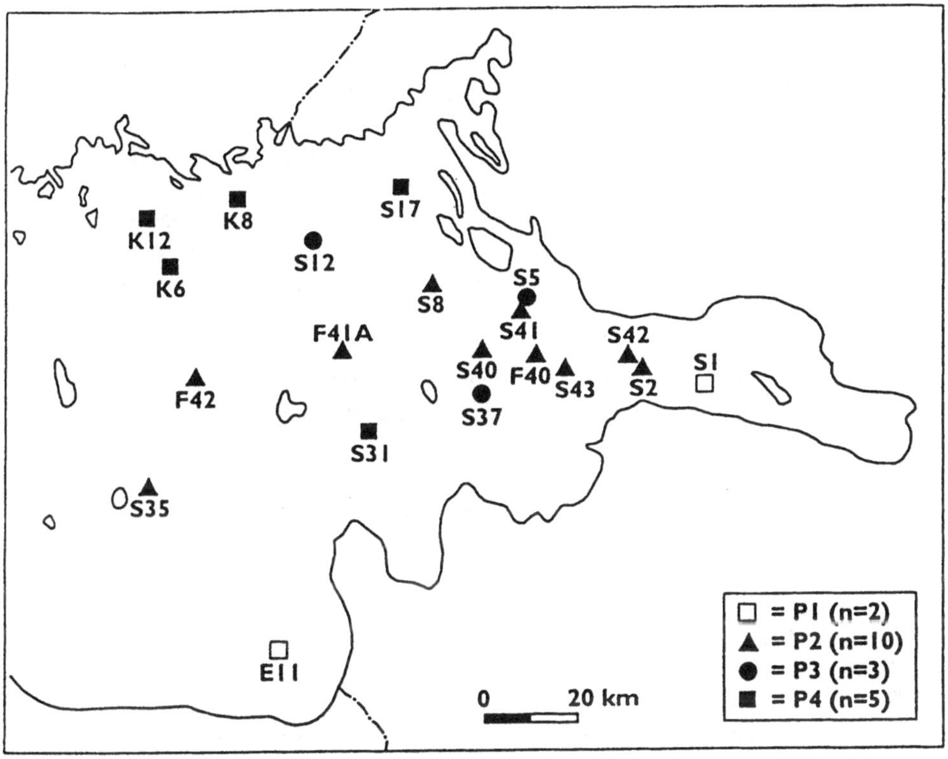

K12
K8
S17
S12
K6
S8
S5
S41
F41A
S42
S40
F40
S43
S2
S1
F42
S37
S31
S35
E11
0
20 km
□ = P1 (n=2)
▲ = P2 (n=10)
● = P3 (n=3)
■ = P4 (n=5)

Figure 6. The locations of the clusters P1-P4.

Total phosphorus (mg g-1 DW)
0
1
2
3
4
5
P1
P4
P2
P3
Depth (cm)
0-1
1-2
2-3
3-4
4-5
5-6
6-7
7-8
8-9
9-10

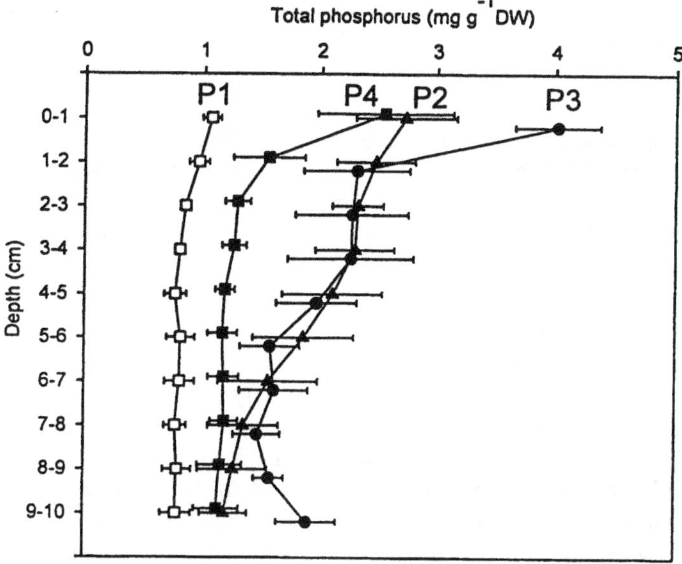

Figure 7. The vertical mean distributions with standard deviation bars for the clusters P1-P4

4. Discussion

4.1. EFFECTS OF WATER DEPTH AND THE DISTANCE FROM THE MOUTH OF THE RIVER NEVA ON NUTRIENT CONCENTRATIONS

The mean TN concentration in the upper 2 cm of sediment in the eastern Gulf (7.4 mg g^{-1} DW) was similar to that obtained for the central Baltic Sea by Koop et al. (1990), while the mean surface TP (2.6 mg g^{-1} DW) was almost twice the concentration measured from the same area (1.3 to 1.6 mg g^{-1} DW, Carman and Wulff, 1989). Also in the anoxic layer (4 to 6 cm) the mean TP concentration was higher (1.7 mg g^{-1} DW) than in the central Baltic Sea (1.0 to 1.2 mg g^{-1} DW). In addition the mean sedimentation rate of the eastern Gulf of Finland (11 mm a^{-1}, Pitkänen, 1994) is higher than that of the Baltic Proper (1.5 to 4.2 mm a^{-1}, Jonsson et al., 1990). Thus the net sediment accumulation of P appears to be almost one order of magnitude higher in the eastern Gulf of Finland than in the main basin of the Baltic Sea.

In the central Baltic TP concentrations decrease with increasing water depth (Carman and Wulff, 1989; Koop et al., 1990). This is explained by the low near-bottom oxygen concentrations in the deep Baltic Proper. A positive correlation has been found, on the other hand, in the shallow Kymijoki Estuary in southeast Finland (Pitkänen, 1994). The poor depth dependence of TP in the eastern Gulf of Finland in this study is a result of good oxygen conditions in the deep water (50 to 60 % in late summer, Pitkänen et al.,1993). and the effective sedimentation of P in the Neva Estuary.

The decreasing TP along the estuarine gradient towards the open Gulf suggests that most of the P is bound to inorganic matter in the sediment. Also in the Delaware Estuary sediment TP concentration decreases with increasing salinity and this variation was found to be correlated with decreasing iron oxyhydroxide content in the sediment (Strom and Biggs, 1982). In the Neva Estuary the accumulation of nutrients is most efficient in the deep part (30 to 45 m), 60 km from the river mouth. Currents prevent the permanent accumulation of small particles in the shallow (< 20 m) innermost estuary.

In the Chesapeake Bay (USA) as much as 99 % of the sedimented N is organic (Keefe, 1994). Because TN of the surface sediment was strongly dependent on the concentration of organic matter also, the increase of TN concentration along the estuarine gradient was more due to autochthonous material induced by the heavy nutrient loading than to allochthonous (partly inorganic) material supplied by the River Neva. This hypothesis is supported also by the decreasing C:N ratio along the estuarine gradient.

4.2. NET SEDIMENTATION OF NUTRIENTS

The decomposition rate of organic matter decreases exponentially with sediment depth (Aller, 1980). Also in the present study both organic matter and TN decreased most strongly in the surface sediment (0 to 2 cm) showing that microbial decomposition occurred primarily in this layer. The TN concentration decreased by 39 to 41 % between the sediment surface and the sediment depth of 10 cm. Though the mean organic matter content varied from 13 to 22 % DW and the sedimentation rate varied from 3 to 20 mm a^{-1} in different accumulation areas (Pitkänen, 1994), the results suggest that nearly a constant proportion of the buried TN has been mineralized in the 0 to 10 cm layer at accumulation areas.

The concentration of TP decreases with sediment depth, because of the upward migration of phosphate produced by the mineralization of organic P and the reduction of iron oxides to which phosphate is adsorbed (Krom and Berner, 1981; Balzer, 1986). It has been suggested that most of the phosphate is released in the deep sediment from iron oxides undergoing reduction (Yamada and Kayama, 1987; Sundby et al., 1992). TP concentrations in the different accumulation areas were 53 to 56 % higher in the sediment surface than in the anoxic deeper sediment (9 to 10 cm). In the surface layer the fraction of labile organic P and Fe and Mn bound P was probably large, because TP decreased strongly in the black reduced layer. Thus it seems that about half of the sediment surface P is in refractory organic and inert inorganic form in the eastern Gulf of Finland. In the western Baltic Sea (Århus Bay) 40 to 50 % of TP is Fe and Mn bound P (Mortensen et al., 1993). In the Gulf of St. Lawrence approximately half of the sedimentation flux of particulate P is mobilized within the sediment and returned to the water column (Sundby et al., 1992).

^{137}Cs datings based on 10 cores suggest that the mean accumulation rate of wet SPM of the eastern Gulf is 11 mm a^{-1} (Pitkänen, 1994). The mean dry weight of the sediment (11.1 % of wet weight) and the mean density of 1.1 g cm^{-3} (Niemistö et al., 1978) gave the mean accumulation rate of 1 300 g m^{-2} a^{-1} of dry particulate matter for the accumulation areas of the eastern Gulf. The mean sediment concentrations of 4.6 mg g^{-1} for N and 1.3 mg g^{-1} for P (the anoxic layer from 9 to 10 cm) of the present study correspond to mean accumulation rates of 6.0 g m^{-2} a^{-1} of N and 1.7 g m^{-2} a^{-1} of P. In the whole eastern Gulf, 3 600 km^2 of the bottoms has been classified as accumulation areas (Pitkänen, 1994). Accordingly this 21 600 t a^{-1} of N and 6 100 t a^{-1} of P have sedimented permanently in the area. These estimates are ca. 20 % lower than an earlier estimates based on nutrient concentrations from the anoxic 5 to 10 cm sediment depth (Pitkänen, 1994).

5. Conclusions

The results support the earlier conclusion that the Neva Estuary retains a large proportion of the high nutrient load inflowing from the River Neva and the St. Petersburg area. The increase of the TN concentration from the inner estuary towards the open Gulf is due to increased share of autochthonous material induced by the heavy nutrient loading and diminishing share of allochthonous material (partly inorganic) supplied from the River Neva. The decreasing TP suggests that most of the P is bound to sedimented inorganic matter in the estuarine waters. The results also suggest that there is a large pool of mobilisable P in the sediments of the Neva Estuary, while the corresponding pool of P is smaller in the Finnish archipelago and in the parts of open Gulf.

Acknowledgements

We are grateful to Hannu Rita and Heikki Peltonen for practical advice regarding statistical analyses and to Päivi Tahvanainen and Yki Laine for drawing the figures.

References

Albani, A.D., Rickwood, P.C., Favero, V.M. and Serandrei Barbero, R.: 1989, *Mar. Pollut. Bull.* **20**(9), 438.

Aller, R.C.: 1980, *Adv. Geophysics* **22**, 237.

Bale, A.J. & Morris, A.W.: 1981, *Estuar. Coast. Shelf Sci.* **13**, 1.

Balzer, W.: 1986, *Ophelia* **26**, 19.

Bordovskiy, O.K.:1965, *Marine Geology* **3**, 3.

Carman, R. and Wulff, F.: 1989, *Estuar. Coast. and Shelf Sci.* **29**, 447.

Ehlin, U.: 1981, in A. Voipio (ed.) Baltic Sea. Elsevier, Amsterdam., pp.123-134.

Golterman, H.L.: 1980, *Hydrobiologia* **72**(1-2), 61.

Hodkinson, R., Cronan, D.S., Glasby, G.P. and Moorby, S.A.: 1986, *N Z J.Geol.* **29**, 335.

Håkanson, L. and Jansson, M.:1983, Principles of lake sedimentology, Springer-Verlag, Berlin. p. 316.

Jonsson, P., Carman, R. and Wulff, F.: 1990. *Ambio* **19**(3), 152.

Kauppila, P., Hällfors, G., Kangas, P., Kokkonen, P. and Basova, S.:1994, *Ophelia* **42**, 179.

Keefe, C. W.: 1994, *Estuaries* **17**(1B),122.

Koop, K., Boynton, W.R., Wulff, F. & Carman, R.: 1990, *Mar. Ecol. Prog. Ser.* **63**, 65.

Krom, M.D. & Berner, R.A.: 1981, *Geochim.Cosmochim. Acta* **45**, 207.

Mortensen, P.B., Jensen, H.S., Rasmussen, E.K. & Thamdrup, B.: 1993, in P.C.M. Boers, T.E. Cappenberg and W. van Raaphorst (eds.), Proceedings of the third international workshop on phosphorus in sediments, *Hydrobiologia* **253**, 101.

Niemistö, L., Tervo, V. and Voipio, A.: 1978, *Finnish Marine Research*, **244**, 36.

Pitkänen, H.: 1994, *Eutrophication of the Finnish Coastal Waters: Origin, fate and effects of riverine nutrient fluxes*, Publications of Water Research Institute, National Board of Waters and Environment No.18, Helsinki, Finland.

Pitkänen, H. and Tamminen, T.: 1995, *Mar. Ecol. Prog. Ser.* **129**(1-3), 283.

Pitkänen, H., Tamminen, T., Kangas, P. Huttula, T., Kivi, K., Kuosa, H., Sarkkula, J., Eloheimo, K., Kauppila, P. and Skakalsky, B.: 1993, *Estuar. Coast. and Shelf Sci.* **37**, 453.

Prahl,F.G.,Bennett, J.T. and Carpenter,R.:1980, *Geochim.Cosmochim.Acta* **44**, 1967.

SAS Institute, Inc.:1985. *User's Guide: Statistics*, 5th edn. Cary, North Carolina, 956 p.

SFS 3008: 1981, Suomen Standardisoimisliitto. (in Finnish). p.4.

SFS 3025: 1986, Suomen Standardisoimisliitto. (in Finnish). p.10.

Sokal, R.R. & Michener, C.D.: 1958, *University of Kansas Science bulletin* **38**, 1409.

Starck, B. and Haapala, K.: 1984, *The analysing of nitrogen in waste water, extension study*. National Board of Waters and the Environment, Finland. Vesihallituksen monistesarja No. 257, Helsinki, Finland. (mimeographed, in Finnish).

Strom, R.N. and Biggs, R.B.: 1982, *Estuaries* **5**(2), 95.

Sundby, B., Cobeil, C., Silverberg, N. & Mucci, A.: 1992, *Limnol. Oceanogr.* **37**(6), 1129.

Thornton, S.F. and McManus, J.:1994, *Estuar.Coast.and Shelf Sci.* **38**, 219.

Yamada, H. & Kayama, M.: 1987, *Oceanol. Acta* **10**, 311.

Zink-Nielssen, I.: 1975, Intercalibration of chemical sediment analyses. Nordforsk Miljövårdssekretariatet. Publication no. 6, 1.

ALGAL BLOOMS IN THE DARLING-BARWON RIVER, AUSTRALIA

T.H. Donnelly[1], M. R. Grace[2] and B. T. Hart[2]
[1] CSIRO Division of Water Resources, Canberra, ACT, 2601
[2] Monash University, Water Studies Centre, Melbourne, Victoria, 3162

Abstract. Australian waterbodies have long water residence times, stratification is common, and eutrophication is driven mainly by the internal loads. The 1991 blue-green algal bloom on the Darling-Barwon River was at a time of low river flow (~100's ML/day) and hot/still conditions. The sustained low flow allowed significant influx of a sulfate-rich saline groundwater and this caused clay flocculation, water clarification and increased photosynthesis in the surface water, and increased sulfate reduction, pyrite formation and Fe mobilisation in anoxic bottom sediments. Since this time similar optimum bloom periods have not produced blooms, or Fe mobilisation, despite high soluble P concentrations. Algal growth during optimum bloom periods is known to be N-limited and it is possible that, in this case, essential trace elements are limiting N-fixation - clay flocculation is a very efficient process of removing trace elements from the water column. Optimum bloom periods can potentially set up a feed-back involving the bottom sediments and increasing sulfate concentration, to resupply surface water with P and trace metals. It is suggested that this operated in late 1991, but not during later optimum bloom periods as no further Fe mobilisation is recorded. The management implication is to maintain sufficient river flow to prevent any significant groundwater influx.

Key words: Cyanobacterial blooms, Australian rivers, flow, stratification, clay/Fe/P association, nutrients, N fixation.

1. Introduction

In late 1991 the world's largest riverene cyanobacterial (blue-green algal) bloom occurred along 1000km of Australia's Darling-Barwon River (DBR) (Fig.1). This event caused Federal and State Governments to react to what was actually a widespread Australian problem. Because of the scale and urgency of the problem, water quality managers were often forced to act on the basis of a 'best guess' approach. However, it is important that as new information comes forward from research, old concepts of how to deal with the problem are reassessed. Considering the large amounts of money being used for remedial programs, it is important that these programs are accurately targeted. In this paper we present some background to a number of these 'best guess' approaches, often based on the North American experience, and indicate some of the reasons why managing Australian rivers and catchments needs a distinctly Australian approach.

The Darling Basin occupies some 640,270km^2 of two Australian states (Fig. 1) and is an important water resource to agriculture, urban and industrial users within the basin. The Bourke Weir Pool (BWP) is downstream of the major subcatchments to this river, the rest of this very long low gradient river is essentially channel and fine sediment transported by this river is suggested to have a common origin (Woodyer, 1978). The first occurrence of the 1991 algal bloom was in the channel section of the river. The BWP therefore can be studied as an analogue of much of the river system. The DBR has high concentrations of soluble phosphorus (P) and there is a need to understand why a major bloom occurred in late 1991, but not at other times of

Water, Air and Soil Pollution **99**: 487-496, 1997.

apparently identical environmental conditions. In this study we examine sediment/water interactions in the fine sediments deposited in the BWP to try to determine the processes which caused a bloom to occur at this time and the management options for controlling the health of the river.

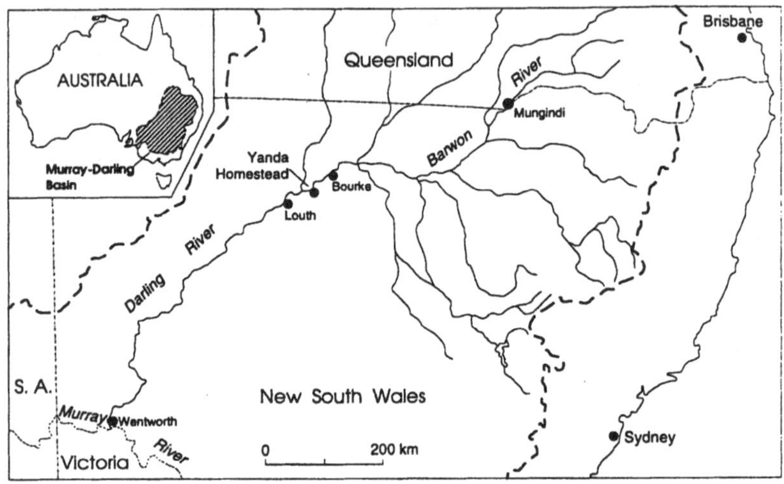

Fig. 1. The Darling-Barwon River showing the Bourke Weir Pool and other locations discussed in this study.

2. Background

Discussion of the sources of P in Australian aquatic environments have tended to accept as axiomatic that the predominant sources are the result of human impact. Certainly in the more heavily populated parts of the world (e.g. Europe and North America), where the problems of eutrophication have been studied for decades, there is considerable evidence for increases in P concentrations as a result of human activities (e.g. Morse et al., 1993).

There are few examples of direct human impact in Australian waterways (e.g. Birch, 1982; Hall, 1992) and it is becoming increasingly evident that the overseas experience is not directly relevant to the Australian environment. The processes/mechanisms are the same, but the balance and relative importance is not necessarily so, particularly because of the greater levels of turbidity in many of our inland waterways. High levels of turbidity reduce light penetration and the strong particle-P association means less of the total P (TP) is immediately available for algal growth (e.g. Oliver et al., 1993; Douglas, 1993). High TP concentrations can be tolerated in many of our waterways without causing excessive algal growth (Hart et al., 1993). While in general, this strong particle P association limits soluble P concentrations in Australian waterways to low levels, the DBR is an exception with a high TP concentration supporting a high soluble P concentration. The DBR is an example of a river system in Australia that is normally not P limited.

In Australia even relatively shallow waterbodies will stratify. Because of the hot/dry climate almost all rivers are regulated to some degree and many water storages have been constructed to overcome the drought periods. The control of flow has resulted in the creation of many standing water bodies that undergo stratification and thus have the potential for resolubilisation of sediment-bound P. Harris (1995) noted the relatively long residence time for water in many Australian waterbodies and commented that in this country eutrophication is mainly the result of in situ P-release processes from bottom sediments.

2.1. RESOLUBILISATION OF SEDIMENT-BOUND P

Fleischer (1978) found that a biochemical process regulated the release of P from anoxic sediments rather than the absence of oxygen. This finding supported the earlier work by Hasler and Einsele (1948) that suggested that sulfate could affect the availability of P in bottom waters, because the sulfide produced by the sulfate-reducing bacteria disrupts the interconnections between the iron and P cycles. Recently, Caraco et al. (1989 and 1991) used data from 23 different aquatic systems to indicate that the control of P-release from sediments by oxygen is not substantiated. These authors state that: (1) increased sulfate content of the waters was the major factor critical to the control of P release from sediments (through the activity of the sulfate-reducing bacteria); and (2) the difference between P-release under oxic and anoxic conditions is actually quite small, when sulfate concentrations are low. A recent study by Chiswell et al. (1995) confirmed the role of sulfate in P release from freshwater sediments, and a study of natural environments by Barbanti et al. (1995) confirmed the relationship between relatively large pore water P concentrations in anoxic bottom sediments and the activity of the sulfate-reducing bacteria.

To be active these bacteria need: (1) an anoxic environment in the bottom sediments; (2) soluble sulfate concentrations >10 mg/L, which in continental environments generally comes from influx of sulfate-rich saline groundwater, or from use of sulfate-rich fertilisers in the catchment; and (3) metabolisable organic matter. There are a number of mechanisms by which the bottom water can move into the surface water and they are by: (1) natural seasonal overturn; (2) increase in wind strength leading to currents forcing some degree of water circulation; and (3) relatively small diurnal thermal changes in the water body leading to some leakage of bottom water to the surface. Given the potential of bottom sediments to be relatively a very large source of soluble P (e.g. Lawacz, 1985; Nurnberg, 1987), any leakage to surface waters would be significant.

3. Methods

All sediments analysed come from the BWP (Figure 1), an approximately 13km long weir pool with a maximum water depth of around 5m near the weir. The sediments analysed were suspended river sediments collected using a continuous flow centrifuge, bottom grab sediments from along the weir pool and a 66cm sediment core from a depression site in the BWP; the banding in the core represents a sequence of sediments deposited over, at least, some tens of years. All sediment samples were analysed by standard X-ray fluorescence methods (Norrish, 1968) and the minerals in 11 sediment samples were analysed by standard X-ray diffraction methods. As part of a larger study of the DBR, water quality monitoring was also carried out in the BWP and in a deep hole

(~4.5m depth) in the river off Yanda Homestead (Figure 1). Turbidity and conductivity were routinely measured, Fe was determined by AAS, and soluble reactive P (SRP) was P in the <0.2μm fraction of the water sample analysed by the standard colorimetric method. Iron sulfide concentrations were determined on a number of samples by sulfide evolution using HCl to determine iron monosulfide (FeS) and the chromium reduction method of Howarth and Merkel (1984) for pyrite (FeS_2) concentration.

4. Results and Discussion

The development of an algal bloom requires the appropriate conditions in a waterbody which include: a hot/still environment with low (or no) flow and good light penetration; an abundance of the major nutrients C/N/P and minor nutrients such as Fe and Mo; and low zooplankton grazing. The dominant blue-green algal species forming the major bloom in the DBR in late 1991 was the potentially toxic Anabaena circinalis. Using the SRP concentrations shown in Table 1 for the DBR, and Redfield stoichiomentry, the calculated population of algal cells which could be formed, assuming no resupply of SRP, is around 4-8 times the Australian alert level (15,000 cell/ml) for algal blooms; this river is not P limited for excess algal growth.

In the DBR however, excess algal growth is generally light limited because of high turbidity (Oliver, 1996). Bek (1985) reported median turbidities in the BWP of 250 NTU (n = 91), but Grace et al. (in press) observed that turbidity levels markedly decrease during periods of sustained low river flow (~100's ML/day). Turbidity is dominated by the suspended clays and these authors showed that rapid clay flocculation during low river flow was caused by influx of saline groundwater. Clay flocculation under these conditions can be a very efficient process of increasing water clarification (Liss, 1976; Boyle et al., 1977; Aston, 1978). Williams (1991) noted that an influx of saline groundwater can potentially occur along the length of the DBR during low river flow, and Donnelly et al. (1992) noted that this saline groundwater was sulfate-rich and indicated its potential to increase the intensity of sediment/water interactions in anoxic bottom sediments.

4.1 THE DARLING RIVER OFF YANDA HOMESTEAD

An example of stratification and sediment/water interactions occurring in the DBR during a period of low river flow and hot/still conditions, is shown in Table I. These water conditions were monitored in a relatively deep hole in the DBR off Yanda Homestead (Figure. 1). Typically high SRP concentrations were observed and because of low flow this waterbody eventually became stratified; temperature differences between surface and stratified bottom water were as high as 10°C in both October and December 1994. Dissolved oxygen (DO) levels in the surface water were 12.6 mg/L (i.e. above the theoretical saturation value of 8.6 mg/L at 22.5°C) reflecting strong algal productivity, while in the bottom water values of <1 mg/L reflect the strong respiration occurring in this deep hole. The anoxic bottom water became euxinic (i.e. H_2S-bearing) with high soluble P concentrations (Table I).

The sequence of events were: (1) low river flow, influx of a sulfate-rich saline groundwater, and stratification; (2) flocculation, clarification and increased photosynthesis; (3) sulfate reduction, and P-release from bottom sediments; (4) decreasing surface water SRP concentration due to algal growth. Resupply of SRP is

TABLE I

Water quality data from a deep hole in the DBR off Yanda Homestead.

Date	Surface Water		Bottom Water	
	SRP (μg/L)	DO (mg/L)	SRP (μg/L)	DO (mg/L)
16/10/94	44±1	9.0	201±18	0
14/12/94	22±3	12.6	276±2	0.2

limited by lack of flow, absence of particle-bound P and thermal stratification. Harris (1995) noted that because of the Australian climate in situ sources of P in Australian waterways will be a more important part of the process of algal bloom formation than found in most overseas studies.

4.2 THE BOURKE WEIR POOL

Australian sediments are highly weathered and the calculated weathering values (McLennan, 1993) of five BWP samples (84 ± 2) indicate the almost complete removal of the alkali and alkaline earth elements and dominance of clay minerals. Mineral identification using XRD confirmed the weathering calculations indicating that the fine sediments in the BWP are mixtures of quartz and clays with some trace feldspar.

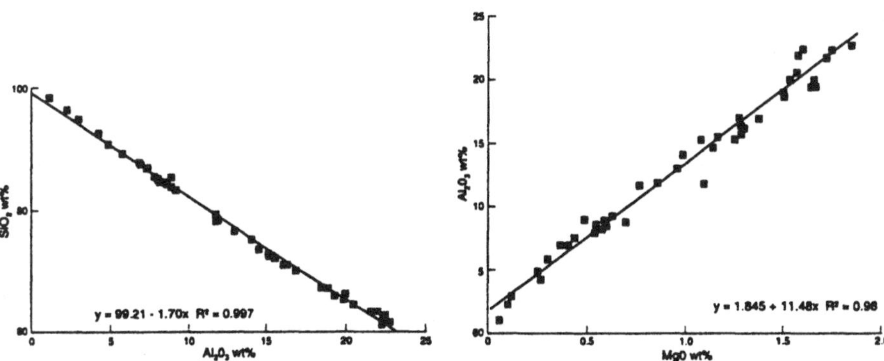

Fig. 2. The plot of SiO_2 versus Al_2O_3 shows the well sorted nature of fine and coarse minerals in the Darling River.

Fig. 3. The plot of Al_2O_3 versus MgO shows the consistency of the clay mineral composition in space and time.

Clastic sediments transported over long distances in a low relief arid landscape, such as the Darling Basin, would be expected to show clear chemical distinctions (Argast and Donnelly, 1987). The strong negative correlation (r^2= 0.997; n = 50) between SiO_2 and

Al_2O_3 for the BWP data (Figure. 2) is an example of these clear chemical distinctions indicating an essentially two component mixing system between the quartz and clay minerals (Argast and Donnelly, 1987). Similarly, the good positive correlation ($r^2 = 0.98$; n = 63) between Al_2O_3 and MgO (Figure. 3) reflects the consistency of the clay mineral composition in the BWP through space and time. The significant positive correlation ($r^2 = 0.94$; n = 63) between P and Fe in all sediment samples (Figure. 4) supports the widespread observation of the clay/Fe/P relationship found Australian soils (e.g. Norrish and Rosser, 1983).

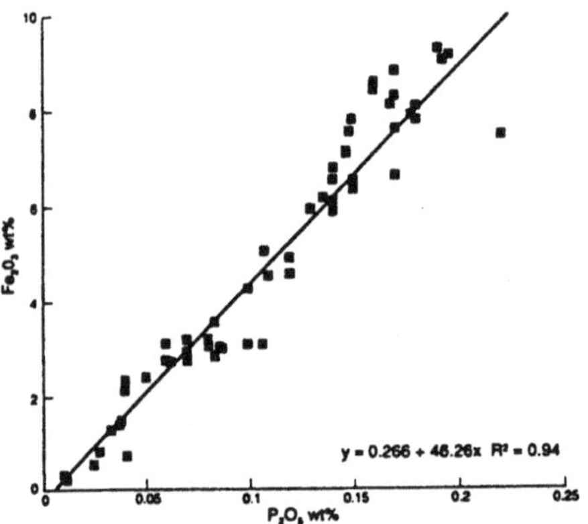

Fig. 4. The plot of Fe_2O_3 versus P_2O_5 showing typical strong clay/Fe/P association found in Australian soils and sediments.

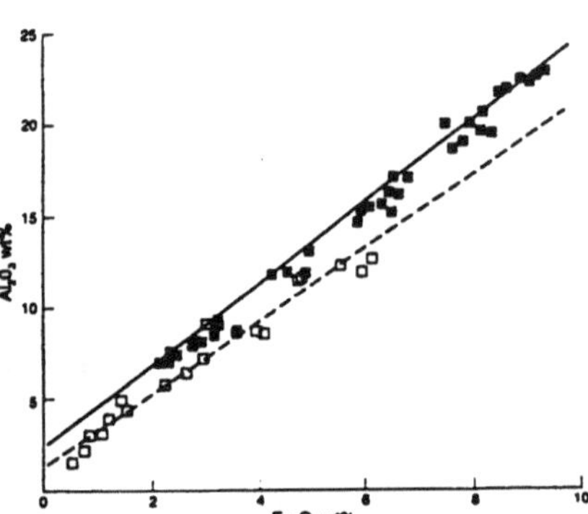

Fig. 5. The plots of Al_2O_3 versus Fe_2O_3 show that each data set produces a good relationship, but with different slopes. Open squares are data from the late 1991 algal bloom.

The relationship between Al and Fe in these well sorted minerals in the DBR also shows a very significant positive correlation and can be used to examine BWP samples for possible changes in the Fe systematics relative to an element (Al) representing the unchanging clay mineral composition in the weir pool. In Figure 5 two plots of Al_2O_3 versus Fe_2O_3 are shown. Sediment samples from the BWP representing before and after the late 1991 period ($r^2 = 0.99$; n = 42) are compared to samples collected during the late 1991 algal bloom ($r^2 = 0.96$; n =18). With such strong relationships between clay minerals and their iron oxyhydroxide coatings, it should be possible to examine whether the period of the late 1991 bloom caused changes to the Fe systematics in deposited sediments, compared to long-term averages.

The two data sets produce fitted lines with statistically different slopes at the 95% confidence limit. The calculated 95% confidence intervals of 2.21-2.34 for the non-bloom period and 1.74-2.15 for the bloom period were not significantly altered by changes in assumptions about the relative magnitudes of the errors. Iron mobilisation occurred in the BWP bottom sediments during the late 1991 bloom, but apparently identical environmental conditions since this time have not caused similar algal bloom problems, or significant Fe mobilisation.

4.3 BACTERIAL SULFATE REDUCTION

It was suggested previously that the addition of sulfate to freshwater systems leads to disruption of the Fe/P cycle. Iron sulfide is precipitated in anoxic sediments and pore waters and anoxic bottom waters become enriched in P. Under these conditions the activity of the sulfate-reducing bacteria can be an indicator of the intensity of sediment/water interactions.

During the low flow periods of 1991 a significant influx of a sulfate-rich saline groundwater occurred in parts of the DBR. Soluble sulfate concentrations in the BWP during the late 1991 algal bloom were measured at 40 mg/L (Donnelly et al., 1992) and at these levels should not have limited the activity of the sulfate-reducing bacteria. Anoxic sediment samples, taken from along the BWP at the time of the bloom, contained abundant FeS_2 with some minor amounts of FeS. As indicated by Berner (1970) the transformation of FeS to FeS_2 occurs relatively quickly in modern sediments. The calculated degree of pyritisation (DOP = pyrite Fe/reactive Fe; Raiswell et al., 1988) in these anoxic samples have a mean and standard deviation of 0.52 ± 0.06 ($n = 12$), indicating similar activity by the sulfate reducers in the weir pool. The values are high and, as Raiswell et al. (1988) have shown, indicate, at least, very low bottom water oxygen concentrations at the time of the bloom. This DOP calculation however, has not taken into account the FeS proportion and the values will be underestimated. Raiswell et al. (1988) has recorded DOP values for the Black Sea, a modern euxinic environment, of up to 0.71, and it is likely that the BWP was at times also euxinic (i.e. similar to the conditions found in the deep hole off Yanda Homestead).

In marine sediments it has been shown that the amount of organic matter that can be metabolised is the most important factor limiting pyrite formation (Berner, 1984). The envelop shown in Figure 6 encloses a large data base of S/C ratios from modern marine anoxic sediments with overlying oxic water. These data have a significant correlation, with a mean S/C ratio of 0.36, and the line of best fit extends through zero (Berner,

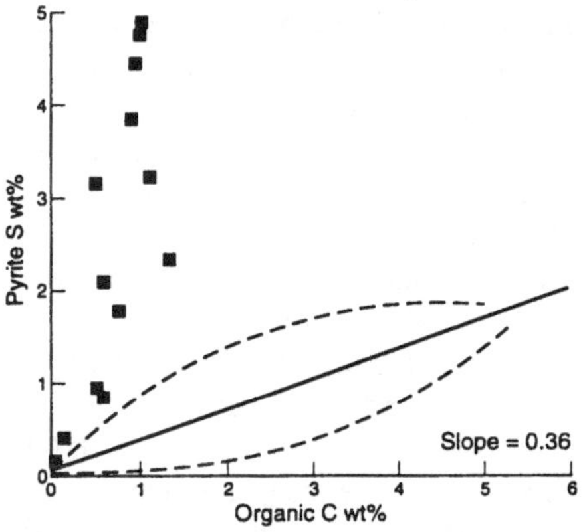

Fig. 6. Pyrite S/organic C ratios in twelve BWP anoxic bottom sediments collected during the 1991 bloom. The envelope shown encloses a large data base of S/C ratios from modern marine anoxic sediments with overlying oxic water.

1984). The S/C ratios from the BWP anoxic sediments, collected at the time of the late 1991 bloom, lie outside the envelop and extend to high ratios with little increase in organic C concentration. The S/C ratios from BWP anoxic sediments are similar to the ratios found in bottom sediments of euxinic environments (Berner and Raiswell, 1983), where FeS_2 forms both in the water column and in the anoxic sediments (i.e. suggesting that the BWP bottom water was euxinic, at least, at times).

Raiswell et al. (1988) have shown that in environments where sedimentary organic matter is deposited, Fe and detrital organic matter are good surrogates for clay minerals. There is no correlation between Fe and organic C for the late 1991 samples from the BWP. This coupled with high activity by the sulfate reducers suggests an organic C source which was predominantly dead algae - a not unexpected easily metabolisable source in this river at this time. Vascular plant-derived organic matter is less readily decomposed and metabolised by the sulfate-reducing bacteria than algae (Lyons and Gaudette, 1979). In the BWP therefore, in late 1991, these bacteria had the best of conditions to significantly increase their activity, produce a high degree of pyrite formation and cause significant Fe mobilisation in anoxic bottom sediments.

5. Conclusions

The DBR normally carries relatively high concentrations of clay minerals. While this river also has high TP and SRP concentrations it is these relatively high turbidities which limit excess algal growth when river flows are ~1000's ML/day. However, when drought, or excess irrigation abstraction, causes river flows to drop to ~100's ML/day or less, the river is susceptible to influx of a sulfate-rich saline groundwater. It has been shown that this influx causes the clays to flocculate and the resulting water clarification with high SRP concentrations leads to significant increases in photosynthesis. It was a period of low flow and hot/still conditions in late 1991 which lead to a major algal bloom at this time. In this study the processes leading the formation of algal blooms in the DBR are examined.

The BWP, which can be considered an analogue of much of this river system, was selected to examine these processes. Sediments deposited in the BWP over time were used to examine the Fe systematics in the weir pool at a point in time (i.e. the late 1991 bloom) compared to long term averages. Mineral sorting which occurs as a result of sediment transport over long distances in Australia produces very good element relationships. Because of these good relationships it was shown that Fe in the weir bottom sediments was mobilised during the period of the late 1991 bloom, but not later during other times for optimum algal growth. The increase in the intensity of sediment/water interactions at this time was shown to be related to the increase in sulfate concentration to the river during this low flow period. During this time the sulfate-reducing bacteria had the best of conditions, an easily metabolisable source of organic matter (i.e. dead algal material) and abundant sulfate, and their high activity was reflected in a high degree of FeS_2 formation and a significant mobilisation of reactive Fe. The generally high S/C ratios in the BWP anoxic sediments are also in agreement with an anoxic, at times euxinic, stratified bottom water (Berner and Raiswell, 1983) at this time. That is, conditions similar to what was more recently found in our study of the DBR off Yanda Homestead (Table 1). Further confirmation of the importance of these in situ processes, to bloom formation, comes from a study of the Western Australian Swan Estuary. In this estuary depression sites in the base of the river, where fine sediment and organic ooze accumulate, are sites of anoxic bottom water, very active bacterial sulfate reduction and annual phytoplankton blooms (Deeley et al., 1993).

While Fe mobilisation occurred in the BWP anoxic sediments in late 1991, similar environmental conditions since this time have not caused bloom problems, or significant Fe mobilisation. Such a period of intense sediment/water interactions in the stratified BWP would have resulted in the bottom water becoming a significant source of SRP for resupply of any depletion of surface water concentrations. Addition of SRP to the surface water may have been necessary to generate the biomass observed during the late 1991 bloom, but not for the initiation of the bloom. However, an hypothesis is proposed from this study that such intense sediment/water interactions, in stratified sections of the DBR, were necessary in late 1991 for algal blooms formation in this river.

Algal growth experiments, at times of optimum algal growth in the DBR post 1991, have shown a nitrogen (N) limitation. This is not unusual in Australian waters and has led to a competitive advantage for the blue-green algae, some of which can use atmospheric N. However, it is possible that the N limitation is actually due to the fact that N-fixation, during times of optimum algal growth in the DBR, is trace element limited (Howarth et al., 1988). Clay flocculation is a very efficient method of removing Fe (dissolved and colloidal) from the water column, even at low salinity's (Liss, 1976; Boyle et al., 1977; Aston, 1978). Howarth et al. (1988) noted that Fe and Mo may interact to control N-fixation and that the availability of Fe is normally highly correlated with the availability of Mo. During periods of low flow in this low gradient river clay flocculation and stratification occurs which both clarifies the water column and separates the bottom water from the surface water. It is possible that during periods of optimum algal growth that N-fixation in the surface water becomes trace element limited. If resupply of essential trace elements for N-fixation in the surface water is controlled by a feed-back from the anoxic bottom sediments, then the reason why there are not always blooms in this P-rich river could be a function of the intensity of sediment/water interactions. The implication for management therefore, is to maintain sufficient river flow to prevent a significant influx of the sulfate-rich saline groundwater.

Acknowledgements

The assistance of Gary Caitcheon, Jon Olley and Peter Wallbrink for river sample collections is acknowledged, as is the support funding from the Murray-Darling Basin Commission through their National Resource Management Strategy. The authors are grateful to Dr Jeff Woods for his statistical analysis proving iron remobilisation had occurred and to Drs Rod Oliver and Graham Skyring for critically reviewing a final draft of the paper.

References

Argast, S. and Donnelly, T. W.: 1987, *J. Sed. Petrol.* **57**, 813-823.

Aston, S. R.: 1978, Estuarine chemistry. In: *Chemical Oceanography* 2nd ed., De. J. P. Riley and R. Chester, New York: Acad. Press, pp. 361-440.

Barbanti, A., Bergamini, M. C., Frascari, F., Miserocchi, S., Ratta, M. and Rosso, G.: 1995, *Mar. Freshwater Res.* **46**, 55-67.

Bek, P.: 1985, Darling River and tributaries - quality of waters, *NSW Dept. of Water Resources Technical Report No 5.*

Berner, R. A.: 1970, *American J. Sci.*, **268**, 1-23.

Berner, R. A.: 1984, *Geochem. Cosmochem. Acta* **48**, 605-615.

Berner, R. A. and Raiswell, R.: 1983, *Geochem. Cosmochem. Acta* **47**, 855-862.

Birch, P. B.: 1982, *Aust. J. Mar. Freshwater Res.* **33**, 23-32.

Boyle, E. A., Edmond, J. M. and Sholkovitz, E. R.: 1977, *Geochim. Cosmochim. Acta* **41**, 1313-1324.

Caraco, N. F., Cole, J. J. and Likens, G. E.: 1989, *Nature* **341**, 316-318.

Caraco, N. F., Cole, J. J. and Likens, G. E.: 1991, A cross-system study of phosphorus release from lake sediments. In: *Comparative Analyses of Ecosystems - Patterns, Mechanisms and Theories* (ed. Cole, J., Lovett, G. and Findlay, S.) Springer-Verlag, New York Inc., pp. 241-258.

Chiswell, B., O'Halloran, K. and Wall, J.: 1995, The role of sulfur in phosphorus release from sediments. RACI Proceedings 4th Environmental Chemistry Conference, Darwin, July 1995, EO16-1 - 3.

Deeley, D. M., Donohue, R. and Parsons, G.: 1993, Nutrient loads in surface drainage from Ellen Brook from 1987 to 1992. In: *The Effects of Land Use on the Quality of Water Discharged to Ellen Brook.* CSIRO Division of Water Resources, Annual Report 1992-93.

Donnelly, T. H., Olley, J. M., Murray, A. S. and Wasson, R. J.: 1992, Algal blooms in the Darling River: Run of River' study Wentworth to Collarenabri December 1991. CSIRO Cons. Report to the MDBC, pp.37.

Douglas, G. B.: 1993, Characterization of suspended particulate matter in the Murray-Darling River system. PhD Thesis, Monash University.

Fleischer, S.: 1978, *Naturwissenschaften* **65**, 109-110.

Grace, M. R., Hart, B. T., Hislop, T. M. and Beckett, R.: 1996, *Colloids Surfaces A.*, (in press)

Hall, D. N.: 1992, Port Phillip Bay Environmental Study: Status Review. (ed. Douglas N. Hall) Tech. Rep. No 9, Pub. from the CSIRO INRE Project Office, Canberra - Port Phillip Bay Environmental Study ISSN No 1039-3218.

Hasler, A. C. and Einsele, W. G.: 1948, *Trans. North Am. Wildlife Confer.* **13**, 527-555.

Harris, G.: 1995, *Water J.* **22**, 9-12.

Hart, B. T., Campbell, I. C., Angehrn-Bettinazzi, C. and Jones, M. J.: 1993, *J. Aquatic Ecosystem Health* **2**, 151-163.

Howarth, R. W. and Merkel, S.: 1984, *Limnology and Oceanography* **29**, 598-608.

Howarth, R. W., Marino, R. and Cole, J. J.: 1988, *Limnology and Oceanography* **33**, 688-701.

Liss, P. S.: 1976, Conservative and nonconservative behavior of dissolved constituents during estuarine mixing. In: *Estuarine Chemistry,* Ed. J. D. Burton and P. S. Liss, New York: Acad. Press., pp. 93-130.

Lawacz, W.: 1985, Factors affecting nutrient budgets in lakes of the Jorka River watershed Masurian Lakeland, Poland. XI. Nutrient budget with special consideration to phosphorus retention. *Ekol. Pol.* 33, 357-382.

Lyons, W. B. and Gaudette, M. E.: 1979, *Org. Geochem.* **1**, 151-155.

McLennan, S. M..: 1993, *J. Geology* **101**, 295-303.

Morse, G. K., Lester, J. N. and Perry, R.: 1993, The economic and environmental impact of phosphorus removal from wastewater in the European community. Pub. Div. Selper Ltd., Chiswick. UK, pp.90.

Norrish, K.: 1968, Some phosphate minerals in soils. Trans. 9th Conf. International Soil and Sci. Soc., Adelaide, Vol. **2**, (ed. J. W. Homes), Pub. Angus and Robertson, Sydney, Australia.

Norrish, K. and Rosser, H.: 1983, Mineral Phosphate. In: *Soils: An Australian Viewpoint*, Division of Soils CSIRO, Ch. 24, pp. 335-361.

Nurnberg, G. K.: 1987, *Limnol. Oceanogr.* **32**, 1160-1164.

Oliver, R. L., Hart, B. T., Douglas, G. B. and Beckett, R.: 1993, *AWWA Water J.* **20 (4)**, 24-26 and 29.

Oliver, R. L.: 1996 Australian Soc. for Limnology. 35th Cong., Program and Abs., pp. 112.

Raiswell, R., Buckley, F., Berner, R. A. and Anderson, T. F.: 1988, *J. Sed. Petrol.* **58**, 812-819.

Williams, R. M.: 1991, Groundwater inflows to the Darling River Mungindi to Wentworth: Run of river study 1990/91. NSW DWR Tech. Services Div., Aug. 1991, pp. 107.

Woodyer, K. D.: 1978, Proc. Royal Society of Victoria, Australia, 90, 139-147.

SPATIAL DISTRIBUTIONS OF BIOGEOCHEMICAL PARAMETERS IN SURFACE SEDIMENTS

OSTROVSKY, I., D. WYNNE, T. BERGSTEIN-BEN DAN, A. NISHRI, H. LI, Y. Z. YACOBI, N. KOREN & R. PARPAROVA

Israel Oceanographic & Limnological Research Company, The Yigal Allon Kinneret Limnological Laboratory, POB 345, Tiberias, 14102, Israel.

Abstract: The spatial variability of several sedimentological, chemical and biological parameters in the uppermost layer of bed sediment (ULBS) in Lake Kinneret, was studied during the development of anoxic conditions in the hypolimnion (May, 1995). ULBS samples were taken along a transect from the littoral to the pelagic zones, during the crash of the *Peridinium gatunense* bloom, about 2 months after the onset of stratification and when oxygen in the hypolimnion was almost completely depleted. The 2-3mm of theULBS, collected by SCUBA diver, contained relatively fresh material (as shown by high Chl c content) but differed from that of intact *Peridinium* cells. In the ULBS, the C:N atomic ratio averaged 8 and was similar all over the lake bottom and the average Chlorophyll a:c ratio was 4 (reaching a value of ~7 at 5m). These ratios in *Peridinium* cells were about 14.3 and 2, respectively. In addition, $\delta^{13}C$ in the organic matter from the ULBS was lighter than that of *Peridinium*. This data suggests that substantial degredation of the organic matter already occurs in the water column. Grain size distribution suggests the occurance of intense focussing processes in Lake Kinneret. Chemical and biological parameters in the ULBS exhibited clear depth dependant patterns, suggesting changes in physical and chemical processes occur. Three different zones can be distinguished in Lake Kinneret bottom sediments. a. The littoral oxic photic zone (down to 5-7m), which is influenced by intensive biodegredation and high turbulence. b. Transition zone (7-20m), with an oxic-photic gradient and developed focussing processes. c. Profundal, anoxic zone (>20m), where all fine, settled, organic rich material concentrates and undergoes anaerobic decomposition.

Keywords: surface sediments; spatial distribution; Chl a & c; C, N & P; bacterial activity; Lake Kinneret

1. Introduction

The spatial distribution of various parameters of settled material in lakes have been extensively studied over the past decades. The distribution pattern may depend on various chemical, physical, sedimentological and biological processes (Hakanson and Jansson, 1983: Lebo and Reuter, 1995). Of special interest is the upper-most layer of the bottom sediment (ULBS), because it may contain freshly deposited material. This layer is often a site of intensive biological and chemical activity and is charaterised by high bacterial populations and meiobenthic organisms (Wetzel, 1983). In many cases, processes occurring in this layer, reflect the transition between oxic and anoxic conditions.

The aim of this study was to investigate the biological and chemical characteristics of the 2-3mm ULBS, and their depth dependant distribution patterns, in a subtropical, stratified lake.

Water, Air and Soil Pollution **99**: 497-505, 1997.

The results are interpreted in terms of sediment focussing, near-bottom water conditions affecting the decomposition and biological activity of the freshly-settled sediments.

2. Materials and methods.

2. 1. STUDY SITE.

Lake Kinneret, a warm, subtropical, monomictic lake, situated in the north of Israel, has been extensively studied over the past 25 years (see Serruya, 1978; Berman et al., 1995). The lake has a surface area of 166 km^2 and a maximum depth of about 43 m. Homothermy usually occurs between December and March. During stratification (March-December) the thermocline deepens from 15 m to 30 m. The winter-spring (March-May) bloom of *Peridinium gaunense* accounts for about 60% of annual lake primary production. The epilimnion is characterized by relatively high pH (8.2-9.6) and dissolved oxygen, whereas the hypolimnion is anoxic, H_2S enriched and has lower pH values (7.3-8.0).

The degradation products of organic matter, both dissolved nitrogen, as NH_4^+, and dissolved phosphorus, accumulate in the hypolimnion. Bulk sedimentation rates vary from relatively high values in the littoral zone (10-20kg m^{-2} per year) to lower levels (of about 1kg m^{-2} per year) in the pelagic region of the lake (Nishri & Koren, 1993). In the pelagic zone, the lake sediment is mainly clay or silt-clay. Depending on location, the upper 0.5-2cm sediment layer is composed of jelly-like material, while the underlying sediments are more compact. Bottom sediment is formed mainly from allochthonous material (Koren, 1993; Singer et al., 1972).

2. 2. SAMPLING AREA AND ANALYSIS.

Water samples were taken by Jenkin sampler and analysed for DO, pH and nitrate (APHA, 1992). Bottom sediment samples were collected, in duplicate, by SCUBA diver, along a transect from the littoral zone (5m water depth) on the North-Western side of the Kinneret, to the lake center (39m). The sampling was carried out during May 1995, two months after the onset of stratification, and at the beginning of anoxic conditions in the hypolimnion. This transect crossed the thermocline at about 15m. At each site, the upper 2-3 mm sediment layer was taken cautiously, to prevent resuspension, from a large bottom area (0.5 - 1 m^2) to diminish the effect of micropatchiness. This method permits the collection of freshly settled material unlike grabs and corers which disturb the ULBS (Ostrovsky, submitted)

Samples were kept in ice until they were brought to the laboratory, and then subdivided for the various analyses. Subsamples for chemical analyses were dried at 80°C for 24h, then ground in a mortar. From each sample, an aliquot (about 0.1-0.3 g) was taken, weighed with a precision of 10 μg, ignited in a muffle furnace at 500 °C for 4 h, then reweighed. The losses on ignition, equivalent to the organic matter content (OMC) were calculated from the dry weight of the sample. Grain size composition was estimated using a standard hydrometer method (eg Hakanson and Jansson, 1983).

Concentrations of Chlorophyll *a* and *c* were measured in ULBS subsamples, filtered onto GF/C filters, frozen, extracted with 90% acetone and then kept in the dark at 4°C, overnight. Subsequently, the extract was cleared by centrifugation and measured spectrophotometrically, using the trichromatic equations (Jeffrey and Humphrey, 1975). The remaining suspended matter was oven dried, weighed and pigment concentrations normalized to sediment dry weight.

Analyses of organic carbon, nitrogen and $\delta^{13}C$ were carried out at the Geology Department, The Hebrew University of Jerusalem and were measured, as follows (Zohary et al., 1994): Calcium carbonate minerals were removed by adding 10% HCl to the weighed, dried (105°C for 2h) sediment sample until bubbling had ceased. The sample was then dried again, in a hood, with a drying lamp for 2h and placed in an Automated Nitrogen & Carbon Analyzer (ANCA; Europa Scientific) connected to a VG-602 mass spectrometer. Results for organic C and N are presented as % of bulk sediment dry weight; carbon isotope ratios ($\delta^{13}C$) are given as ‰.

Phosphorus in the sediment samples was analysed by sequential extraction with 1M NH_4Cl and 0.5M HCl, which reflect two different phosphorus fractions in the sediment: labile P (LP) and Ca bound P (CP), respectively (Hieltjes and Lijklema, 1980; Pettersson et al., 1988). Alkaline phosphatase activity (APA), in sediment, was measured spectrophotometrically, using p-nitrophenylphosphate (PNPP) as substrate (Wynne, 1977, 1981; Wynne and Bergstein-Ben Dan, 1995), as follows: 0.5M Tris-HCl, pH8.9 buffer (0.27ml), 3mM PNPP (0.1ml) and 0.1M $MgCl_2$ (0.027μl) were added to duplicate samples of wet sediment (0.5 ml) and the volume adjusted to 2.7ml with double distilled water. After 2h incubation at 37°C, 1M NaOH (0.3ml) was added to each sample, which was then clarified by centrifugation. The absorbance at 410nm was measured and AP activity calculated per g dry wt. of sediment.

Bacterial numbers were measured at the sediment-water interface, as follows: Sediment ($30g_{wet\ wt}$) was shaken for 3 min with 300 ml of bottom water and the sediment particles allowed to settle for 10 min. Subsamples were taken from the overlying water for bacterial counts. Photosynthetic bacteria in sediment was determined using a serial dilution technique, with agar shake cultures (Bergstein et al., 1979). Pfennig medium (Pfennig, 1965) was used for counting *Chlorobium* spp. To estimate heterotrophic bacterial numbers, water samples were plated on Plate Count Agar and incubated at 37°C for 24 hrs (APHA, 1992).

Bacterial activity (^3H-thymidine incorporation) in the same samples was measured under standard conditions (20°C, in the dark) as described by Berman et al. (1994). For the photosynthetic bacteria, potential photosynthesis was also measured. Water samples (20 ml) were incubated for 2 hrs, under a light intensity of 50μEin m^{-2} s^{-1}, at 30°C, as described by Bergstein and Cavari (1983). Dichloromethylurea (DCMU), at a concentration of 5.10^{-7} M, was used to inhibit algal photosynthesis in these samples. Potential algal photosynthesis was measured in situ by a modified ^{14}C uptake technique (Steemann-Nilsen, 1952). All results are presented as the mean of duplicate measurements. Individual measurements did not differ by more than 10%.

3. Results

3.1 Water column characteristics, Spring, 1995

In March, 1995, the entire water column was fully oxygenated (data not shown). During May (Fig. 1), the epilimnion was characterised by oxygenated water with a high pH (about 9.1) and high nitrate concentrations (about 0.15mg N L^{-1}). At intermediate depths (>20m), oxygen becomes depleted and the pH drops. In the deepest layer (40m), the water was devoid of oxygen and nitrate (due to denitrification) and pH had declined to 7.6-7.7. April-May, 1995, was characterised by extremely high Chlorophyll and primary production

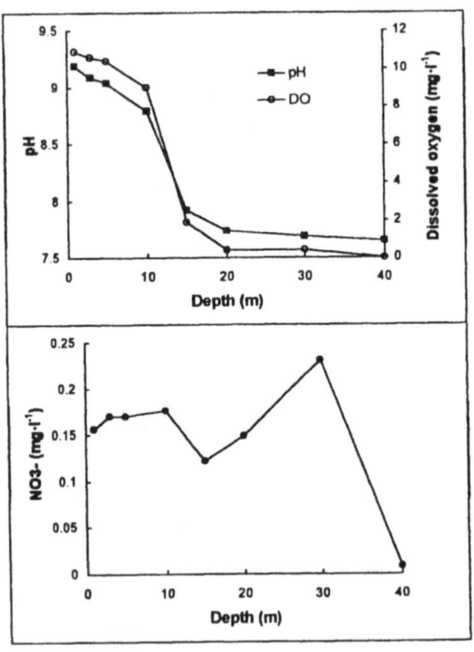

Fig. 1. Depth profile of pH, dissolved oxygen and nitrate concentrations in the water column at the pelagic station of Lake Kinneret, May 1995.

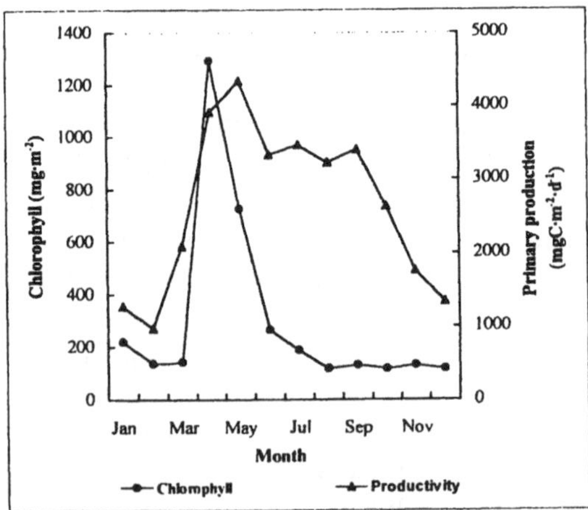

Fig. 2. Chlorophyll a content and primary production of phytoplankton from Lake Kinneret, 1995.

of the freshwater dinoflagellate *Peridinium gatunense* (Fig. 2).

3.2 Bottom sediment charateristics.

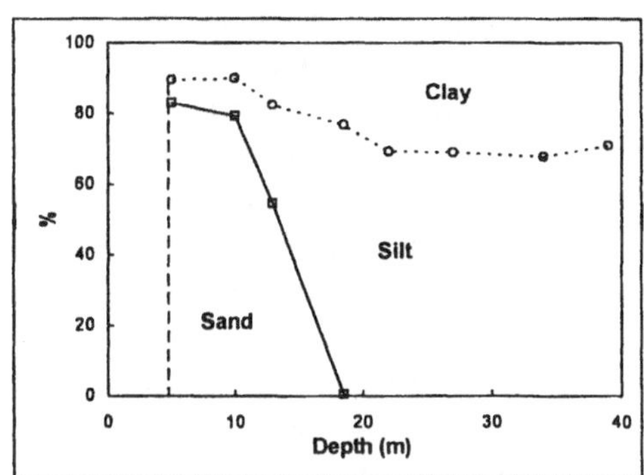

Fig. 3. Granulometric characteristics of the ULBS, along a transect in Lake Kinneret, May 1995.

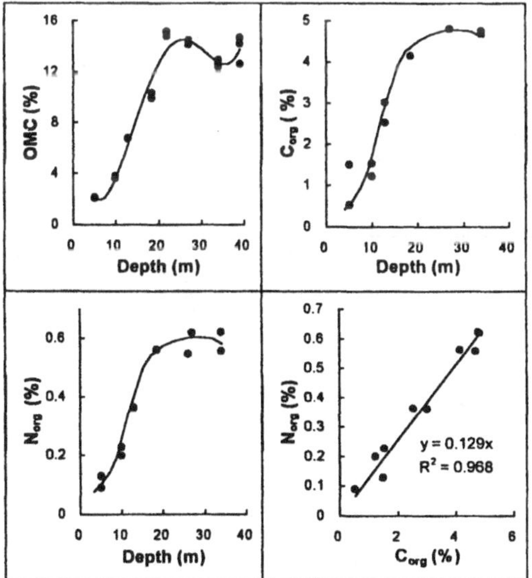

Fig. 4. Organic matter characteristics of the ULBS, along a transect in Lake Kinneret, May 1995.

Grain-size distribution. Sand comprised about 80% of the sample at 5 and 10m, then drastically decreased to zero at 18m. Along the cross-section the sediment character gradually changes from coarse sand at 5 m to fine sediment (silt + clay) at 18 m water depth and deeper. Below 18m, the clay and silt composition was constant with a silt:clay ratio of 4. (Fig. 3).

Organic matter (Fig. 4), showed a gradual increase with depth (5-20m); at the deeper levels, organic matter was approximately constant (12-15%). Both organic N and C showed similar patterns and were highly correlated. The C:N atomic ratio was 8 and was independent of depth. C:Org matter was approximately 2.8. Carbon isotope ratios ($\delta^{13}C$) of the organic matter ranged between -23.04 and -27.38 with an average of -26.14‰ (Table 1).

Table 1. Carbon isotopic ratios ($\delta^{13}C$) in Kinneret bed sediments.

Depth (m)	$\delta^{13}C$ (‰)
5.1	-24.63
10.0	-26.75
13.0	-26.85
18.5	-26.65
26.0	-27.30
34.1	-26.89

Using a sequential extraction procedure, we separated sedimentary P into two different phosphorus fractions (Fig. 5): labile P (LP) and Ca bound P (CP). LP increased with depth, whereas CP, which represents the bulk

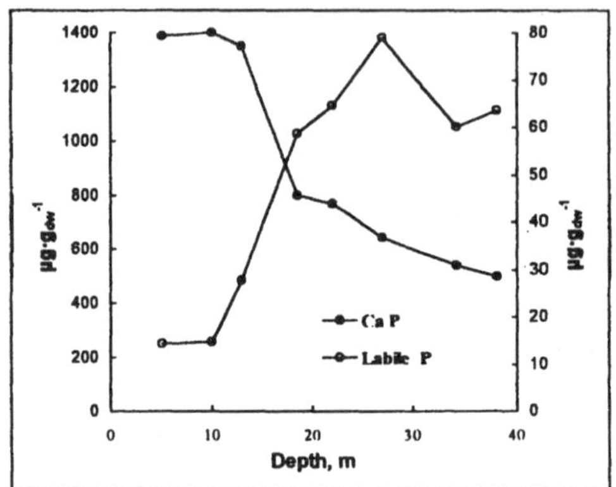

Fig. 5. Labile and calcium bound phosphorus fractions in the ULBS, along a transect in Lake Kinneret, May 1995.

of total phosphorus, showed a quite different pattern and was high in the epilimnetic sediments, dropped sharply between 13 and 18m, then gradually decreased with depth in the hypolimnetic samples.

Chlorophylls *a* and *c* in the ULBS tended to increase with depth (Fig. 6). Between 10 and 39m the Chl *a:c* ratio remained constant (averaging about 4). At 5m, however, this ratio was about 7.

In shallow (5-10m) bed sediments Thymidine uptake, estimated under standard conditions, was high, ranging from 7-16 .10^{-5} nmoles mL^{-1} h^{-1}(Fig. 6). At 13m and deeper, Thymidine uptake was significantly lower and varied between 0.87-1.31.10^{-5} nmoles

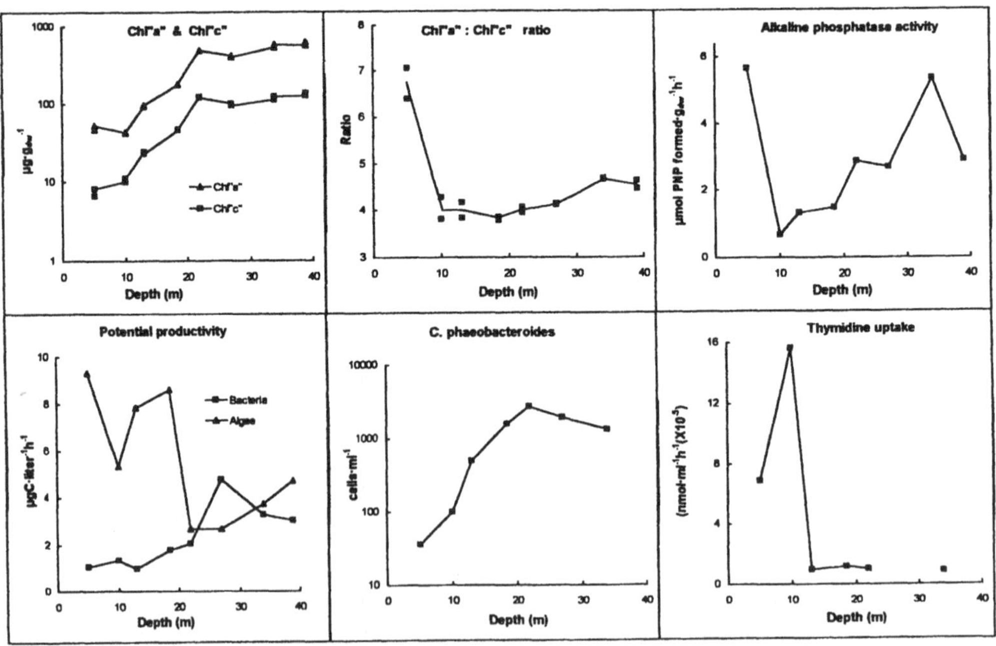

Fig. 6. Biological characteristics of the ULBS, along a transect in Lake Kinneret, May 1995.

$mL^{-1} h^{-1}$. Total heterotrophic bacterial numbers were constant and independent of depth, varying between 10^3-10^4 cells mL^{-1}. Numbers of the photosynthetic bacterium *Chlorobium phaeobacteroides* (Fig. 6) increased with depth until about 20m, then remained constant (at ~3000 cells mL^{-1}). Potential bacterial photosynthesis showed a similar pattern to cell numbers (Fig. 6) whereas potential algal photosynthesis showed the opposite pattern; C uptake activity was much higher in the shallower zones than in samples from more pelagic regions of the lake (Fig. 6). Alkaline phosphatase activity was high in the upper photic zone (5m) but dropped dramatically at 10m. Activity then gradually increased to a maximum at the deepest sampling sites (Fig. 6).

4. Discussion

The fate of newly settled organic matter in the aquatic environment is important since its decomposition can cause a high internal nutrient load, resulting in possible eutrophication. In Lake Kinneret, the large bloom of the freshwater dinoflagellate *Peridinium gatunense*, may supply significant amounts of organic matter to the bottom sediment. Consequently, it is important to know whether this material reaches the bottom sediments or is decomposed in the water column, as suggested by Hertzig et al. (1981).

Several of our results suggest that sediment samples collected during this study represent freshly deposited material. a. They contained relatively high levels of Chl c, originating from *Peridinium*. Chlorophyll *c* rapidly decomposes once the cells harboring it disintegrate, such that no Chl *c* was found at the bottom, 3-4 months after the crash of the bloom (Yacobi et al., 1991). Furthermore, the Chl *a*: Chl *c* ratio in the ULBS averaged 4, whereas in *Peridinium* cells this ratio averaged 2 (Wynne et. al., 1982). Since aged material contains high amounts of only Chl *a* (Yacobi et al., 1991), mixtures of fresh and aged organic material should result in much higher Chl *a*: Chl *c* ratios than found in this study. b. The hypolimnetic sedimentation rate in May, 1995, was $9.1 g_{dwt}\, m^{-2}\, d^{-1}$ (Koren, unpublished). Using a porosity of 94.6% (Ostrovsky, unpublished) this would result in a deposit of 0.16mm d^{-1} of upper fluffy sediment layer i.e. a 3mm layer would be collected over about 20 days in the deeper pelagic zone.

Since the *Peridinium* bloom is the major source of organic matter at this time (>95% biomass; Berman et al., 1995), it would be expected that the organic matter composition of these samples would resemble that of intact *Peridinium* cells, if little decomposition had occurred. However, average $\delta^{13}C$ values of the organic matter in the ULBS averaged -26.51‰, much lighter in comparison to values for intact *Peridinium* cells (-18‰; Zohary et al., 1994). It is, therefore, suggested that despite the predominance of Pyrrophyta in spring , the ULBS contained a considerable portion of organic matter derived from other sources. Moreover, the C: N atomic ratio of the ULBS was constant over the entire lake bottom and averaged about 8, much lower than that found for *Peridinium* (14.3) taken from a bloom in the lake (Wynne et al., 1982). These results suggest that a considerable portion of organic matter, initially originating from *Peridinium*, was significantly transformed during sedimentation.

The transect followed in our investigations, crossed the thermocline (about 15m). Most of the changes observed in the samples occurred in the more shallower sites (up to 20-25m). Organic matter (% dry wt) increased considerably with depth together with a parallel increase in organic C, N and labile P as well as Chl *a* and Chl *c*.

The high Chl *a* : Chl *c* ratio (~7) at 5m depth, compared to that at other depths, suggests the local fast cycling of organic matter (Yacobi et al., 1991). This is also supported by the high potential algal photosynthesis and by high thymidine uptake activity (and possibly also APA) at shallower depths. Consequently, the low concentration of organic matter at 5m could result from its fast degredation, as well as from redeposition to the deeper part of the lake. Such processes are also suggested from the grain size distribution. The coarser grains and the presence of DO in the overlying waters of the littoral zone, permits active, aerobic bacteria to be present in this area and results in very low numbers and activity of the obligate anaerobic photosynthetic bacterium *Chlorobium phaëobacteroides*. Since numbers of these photosynthetic bacteria increase with depth (until a maximum below 22m), we suggest that the size of the anaerobic micro-environment within the ULBS also increases.

At 5m depth, the sedimentary "easily exchangable" phosphorus fraction is relatively low. If this fraction is related to bioavalable phosphorus, it could also partially explain the high APA at this depth. Our results show that depth dependant changes in calcium bound P (but not labile P) appeared to be completely different from that of C and N. We suggest that the low concentration of organic matter in the littoral ULBS results from both intensive biodegredation and focussing.

In summary, we could distinguish between different zones in Lake Kinneret. a. The littoral oxic photic zone (down to 5-7m), which is characterised by intensive cycling of organic matter. b. Transition zone (7-20m), with an oxic-photic gradient and developed focussing processes. c. Profundal, anoxic zone (>20m), where all fine, settled material concentrates and undergoes aerobic decomposition. A theoretical basis for such zones has been shown previously by Rowan et al., (1995)

5. References

APHA.: 1992 *Standard Methods for the Examination of Water and Wastewater.* 18th ed. Greenberg, A. E., Clasceri, L. S. and Eaton, A. D. (eds.), American Public Health Association pp. 9-32 - 9-34.

Bergstein, T. and Cavari, B. Z.: 1983. *Hydrobiologia.* 106, 241-246.

Bergstein, T., Henis, Y. and Cavari, B. Z.: 1979. *Can. J. Microbiol.* 25, 999-1007.

Berman, T., Hoppe, H-G. and Gocke, K.: 1994. *Mar. Ecol. Prog. Ser.* 104, 173-184.

Berman, T., Stone, L., Yacobi, Y. Z., Kaplan, B., Schlichter, M. , Nishri, A. and Pollingher , U.: 1995. *Limnol. Oceanogr.* 40, 1064-1076.

Hakanson, L. and Jansson, M.: 1983. *Principles of Lake Sedimentology*. Springer-Verlag, NY pp316.
Hertzig, R., Dubinsky, Z. and Berman, T.: 1981. *Developments in arid zone ecology and environmental qualitiy*. Shuval, H. (ed.) Balaban, Philadelphia pp179-185.
Hieltjes, A. H. M. and Lijkema, L.: 1980. *J. Environ. Qual.* 9, 405-407.
Jeffrey, S.W., and Humprey, G. F.: 1975. *Biochem. Physiol. Pflanzen*, 167, 191- 194.
Koren, N.: 1993. *Sedimentation rates and particle dynamics in the northern and central part of Lake Kinneret.* MSc. thesis, Haifa University. pp157.
Lebo, M. E. and Reuter, J. E.: 1995. *Mar. Freshwater Res.* 46, 321-326.
Nishri, A. and Koren, N. 1993.: *Verh. Tnternat. Verein. Limnol.* 25, 290-292.
Pettersson, K., Bostrom, B. and Jacobsen, O-S.: 1988. *Hydrobiologia* 170, 91-101.
Pfennig, N.: 1965. *Hyg. Abt. I.* 1, 179-189.
Rowan, D. J., Cornett, R. J., King, K. and Risto, B.: 1995. *J. Paleolimnol.* 13, 107-118.
Serruya, C. (ed.).: 1978. *Lake Kinneret.* Dr. W. Junk Publ., The Hague.
Singer A., Gal, M. and Banin, A.: 1972. *Sediment. Geol.*, 8, 289-308.
Steeman-Nielsen, E.: 1952. *J. Cons. Cons. Int. Explor. Mer.* 18, 117-140.
Wetzel, R. G.: 1983. *Limnology* CBS College
Wynne, D.: 1977. *Physiol. Plant.* 40, 219-224.
Wynne, D.: 1981. *Hydrobiologia* 83, 93-99.
Wynne, D., Patni, N. J., Aaronson, S. and Berman, T.: 1982. *J. Plankton Res.* 4, 125-136.
Wynne, D. and Bergstein-Ben Dan, T.: 1995. *Can. J. Microbiol.* 41, 278-283.
Yacobi, Y. Z., Mantoura, R. F. C. and Llewellyn, C. A.: 1991. *Freshwater Biol.* 26, 1-10.
Zohary, T., Erez, J., Gophen, M., Berman-Frank, I. and Stiller, M.:1994. *Limnol. Oceanogr.* 39, 1030-1043.

ELEMENTAL DISTRIBUTION IN A SEDIMENTARY DEPOSIT ON THE SHELF OFF THE TAGUS ESTUARY (PORTUGAL)

P. Paiva[1], J.-M. Jouanneau[2], F. Araújo[1], O. Weber[2], A. Rodrigues[3] and J.M.A. Dias[4]

[1]Dep. Química, ITN, Estrada Nacional 10, 2685 Sacavém, Portugal, [2]Dép. Géologie et Océanographie, URA, CNRS, Univ. Bordeaux I, Avenue des Facultés, 33405 Talence Cedex, France, [3]I. H., DO, Rua das Trinas 49, 1200 Lisboa, Portugal, [4]UCTRA, Campus de Gambelas, Univ. Algarve, 8000 Faro, Portugal

Abstract. The Tagus estuary, located at the western Iberian coast in front of Lisbon is the largest one in Portugal, where it drains highly populated and industrialised regions. The amount of sediment transported by this river is so great that the submarine delta is one of the largest in the Iberian margin, very well defined and reaching the 70m isobath. The geochemical pattern of the muddy deposit off the Tagus river has been determined by elemental analyses of bulk surficial sediments, collected at the Portuguese margin adjacent to the Tagus estuary. Measurements were carried out by energy-dispersive X-ray fluorescence spectrometry (EDXRF). Sediments were usually composed of fine particles (silts and clays) and were high in organic matter (measured as loss on ignition). Elemental concentrations for Zn and Pb clearly indicate an estuarine contamination probably associated with discharges from urban centers or due to the influence of the industries located downstream. Apparently the depletion on the Cu contents is caused by its release (soluble complexes) into the marine environment. Variations observed in the elemental distribution of the sediments off the Tagus with distance from the estuary seem to be dependent on the grain size distribution related to the dynamics of the sediment transport.

key words: elemental composition, fine sediments, Tagus, Atlantic shelf, sediment transport, pollution.

1. Introduction

The Tagus river is the largest of the Iberian Peninsula, draining about 81000 km^2. Its estuary is one of the largest estuaries on the west coast of Europe, covering an area of 320 km^2. It is composed by a deep, straight and narrow inlet channel, and a broad shallow inner bay.

In the Tagus basin pollutants are discharged particularly into the lower estuary, a densely populated region with several urban centers (Lisbon and others) and where many industries (e.g. petrochemicals, chemicals, metallurgy, etc.) are located.

The discharge of freshwater normally shows a pronounced dry season/wet season as well as large inter-annual variation. The residence time of freshwater in the estuary is highly variable and may range seasonally from 65 to 6 days. The tides are semi-diurnal, with amplitudes at Lisbon that range from 1 m at neap tide to about 4 m at spring tide (Vale and Sundby, 1987).

The river discharge is the main source of fine-grained sediments to the continental shelf. In the oceanic domain, sediments are transported by nepheloid layers, turbidity currents, flows and mass-movements. When the supply exceeds the dispersal

Water, Air and Soil Pollution **99**: 507-514, 1997.

rate, the particulate matter settles and covers the continental shelf off the river mouth (Gorsline, 1984).

The inner continental shelf just in front of Tagus estuary is predominantly sandy. In this region, bottom currents are about 45 cm/s (IH, 1988, 1993), which is strong enough to prevent the deposition of fine sediments. These particles remain in suspension until they reach the middle and outer shelf (Garcia *et al.*, 1995). As a result of this, a large muddy deposit is located off the Tagus estuary, covering the continental shelf from the submarine delta front (~50 m) until about 130 m depth.

The portuguese margin in front of Tagus is characterized by a narrow shelf cut by several canyons (Mougenot, 1988) which extend into the continental shelf. The Tagus estuary is quite well studied in terms of geochemistry and sedimentary budget (Vale and Sundby 1987; Vale 1990). However, information pertaining to the adjacent continental shelf is still lacking.

In the present study, the elemental distribution of the sediments covering the Portuguese inner and middle shelf off the Tagus estuary was determined by energy-dispersive X-ray fluorescence spectrometry (EDXRF) and compared to previously published values measured in sediments collected at the estuary. Besides the chemical characterization of the fine-grained sedimentary deposit, the transport of metals with an anthropogenic origin from the estuary into the shelf was investigated.

2. Methods

Twenty-six surficial sediment samples were collected at the Portuguese margin in front of Tagus during the cruise PLUTUR I/93, from 26 June - 3 July 1993 (Fig. 1), which was promoted by the Portuguese Hydrographic Institute.

Sediment samples were collected with a Smith-McIntyre grab. Subsamples were taken from the central part of the grab to avoid metal contamination, stored in plastic boxes and deep-frozen prior to oven-drying. About 50 g of each dried sediment sample were ground to obtain a grain size smaller than 63 μm. The homogenised ground material was dried at 110°C for 24 h. Then the ground sediments were processed into pressed pellets, by adding an organic binder to two grams of the material and then pressing it under 15 ton of pressure.

The sediment pellets were analysed for aluminium (Al), silicon (Si), potassium (K), calcium (Ca), titanium (Ti), chromium (Cr), manganese (Mn), iron (Fe), nickel (Ni), copper (Cu), zinc (Zn), rubidium (Rb), strontium (Sr), zirconium (Zr) and lead (Pb) by a commercially available energy-dispersive X-ray spectrometer, a Kevex Delta XRF Analyst System. Spectral data of sediment samples and standards were acquired using three different excitation modes: direct excitation (at 4kV and 0.18 mA) and two secondary targets, Ge (at 15 kV and 2.0 mA) and Ag (at 35 kV and 1.2 mA).

The accuracy and the precision of the analytical technique have been checked by analysing several geological international standards (SRM2704, MAG1 and SGR1) and found to be better than 10 %. The detection limits are around 10 mg/kg for most of the elements, while for the lower atomic number elements ($Z < 20$) they are about 100 mg/kg. A detailed description of the sample preparation and quantitative analysis has been given recently (Araújo *et al.*, 1996).

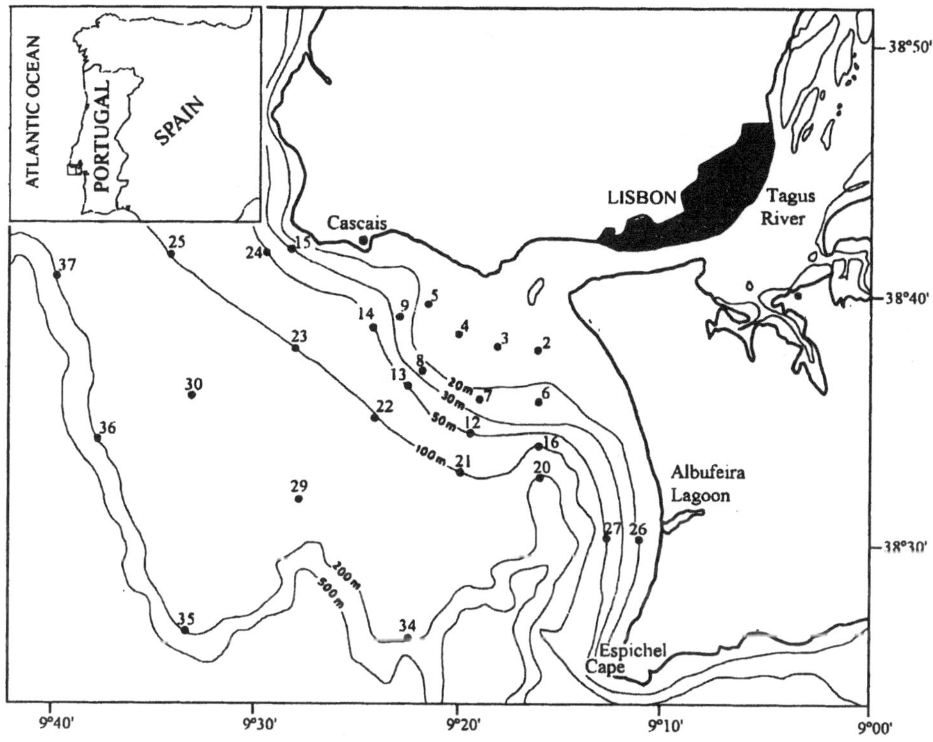

Fig. 1: Map of the Portuguese shelf adjacent to Tagus estuary, with sampling locations.

The organic matter contents of the sediment samples were determined by loss of ignition at 550°C (LOI).

3. Results and discussion

The elemental concentrations determined in the analyses of the sediments by EDXRF and the LOI determinations are listed in Table 1.

The Al and Si concentrations of the analysed sediments were found to be highly variable (1.1 - 8.6 % and 12.0 - 43.1 %, respectively), reflecting the occurrence of various granulometric fractions (from clays to sands), which is the result of different susceptibilities to weathering displayed by the several source lithologies (e.g., schists and granites). Also, Ca distribution varied appreciatively ranging between 2.86 and 25.2 %. The highest concentrations (around 20%) were found at the outer shelf at depths below 100 m, where benthic activities are prominent (samples 30, 35, 36 and 37). Undoubtedly, this is a consequence of the presence of biogenic coarse remains of shells (mostly calcium

TABLE 1
Elemental composition of the sediments samples collected during PLUTUR I/93 cruise (values in mg/kg, unless otherwise indicated).

Station	Al (%)	Si (%)	K (%)	Ca (%)	Ti (%)	Cr	Mn	Fe (%)	Ni	Cu	Zn	Rb	Sr	Zr	Pb	LOI (%)
2	2.36	35.3	2.04	8.23	0.09	<10	102	0.46	<10	16	15	96	433	54	23	9.1
3	2.55	34.4	2.02	8.61	0.13	12	142	0.60	<10	17	30	91	436	93	23	7.6
4	1.08	43.2	0.94	4.27	0.02	22	87	0.17	<10	17	<10	41	224	33	<10	5.3
5	2.31	38.7	1.81	6.01	0.17	<10	147	0.58	<10	21	29	85	313	165	21	7.1
6	2.52	38.7	2.09	3.85	0.16	11	113	0.57	<10	17	19	107	217	133	23	4.8
7	4.29	35.3	2.31	5.11	0.34	19	243	1.56	13	28	115	125	266	310	63	8.5
8	5.17	32.8	2.33	4.96	0.53	27	309	1.93	17	31	142	124	219	652	69	8.2
9	3.98	31.1	2.09	6.61	0.38	24	295	1.49	10	37	101	104	313	373	50	11.7
12	5.62	31.2	2.47	4.16	0.43	44	200	2.06	15	27	141	136	186	344	63	8.4
13	7.75	31.0	2.69	3.02	0.49	65	325	3.37	22	44	276	155	150	260	111	12.9
14	8.50	30.3	2.72	2.86	0.50	72	345	3.72	25	50	306	159	140	272	117	11.9
15	2.58	31.7	2.12	8.94	0.16	11	205	0.71	<10	14	15	100	470	178	24	9.7
16	7.09	31.2	2.60	3.55	0.49	54	290	2.94	22	39	197	149	150	315	75	11.4
20	7.85	28.8	2.61	3.89	0.47	59	260	3.35	25	39	209	151	170	257	84	17.1
21	8.44	29.8	2.71	3.11	0.52	74	272	3.49	30	38	224	161	139	257	86	13.2
22	8.44	28.3	2.70	3.15	0.48	92	291	3.76	22	46	289	158	148	205	119	14.0
23	8.55	28.1	2.71	3.71	0.49	64	298	3.72	27	35	209	159	166	183	87	17.5
24	3.64	30.0	2.16	9.32	0.28	39	225	1.11	<10	17	34	106	441	275	33	12.4
25	6.70	24.6	2.47	6.14	0.44	67	250	2.89	24	26	118	136	259	190	51	17.8
26	3.41	36.9	3.04	5.10	0.16	11	62	0.49	<10	14	11	143	211	161	24	6.0
27	3.03	40.5	2.65	4.09	0.18	22	48	0.48	<10	14	13	126	179	83	19	5.1
29	7.52	26.2	2.56	5.78	0.43	72	232	3.29	24	28	94	143	219	183	45	15.7
30	3.57	13.8	1.38	22.1	0.18	37	141	2.11	12	15	47	67	831	77	30	26.4
34	4.44	29.7	2.64	8.19	0.25	48	166	3.27	18	17	54	112	267	167	26	14.1
35	1.07	12.0	1.24	22.6	0.04	67	164	4.30	12	10	31	31	805	26	13	17.1
36	1.52	16.4	1.82	18.8	0.07	66	98	4.18	<10	12	26	56	724	47	12	22.6
37	1.35	13.1	0.87	25.2	0.06	41	79	1.67	13	11	19	30	656	30	16	20.3

carbonate). Other, "anomalously" high concentrations of Ca (up to 8%), measured for other surficial sediments collected at the shelf, around the fine sedimentary deposit, result from a mixture of terrigeneous and marine particles. Therefore, the low concentration values of Al measured for some of the sediments ($1 < \%\ Al < 3.5 - 4$) seem to be a good indicator of the grain size of the sedimentary material. Also, for these samples, the transition element contents were close to our detection limits. As a consequence, based upon the lower percentages of Al which always correspond either to elevated contents of Si (coarse quartz particles) or of Ca (coarse biogenic particles), we could define the samples collected at the sedimentary formation and our following discussion will be focused mainly on these finer grain sized sediments (7, 8, 9, 12, 13, 14, 16, 20, 21, 22, 23, 24, 25, 29 and 34).

The major, minor and trace elements determined in the fine sediments are plotted in Fig. 2. Stations are grouped approximately according to their isobaths, following a northwestern direction.

The main differences observed are for samples 7, 8, 9, 24, 25, 29 and 34 that are located on the border (Fig. 1) of the sedimentary formation, in a mixing area. Even

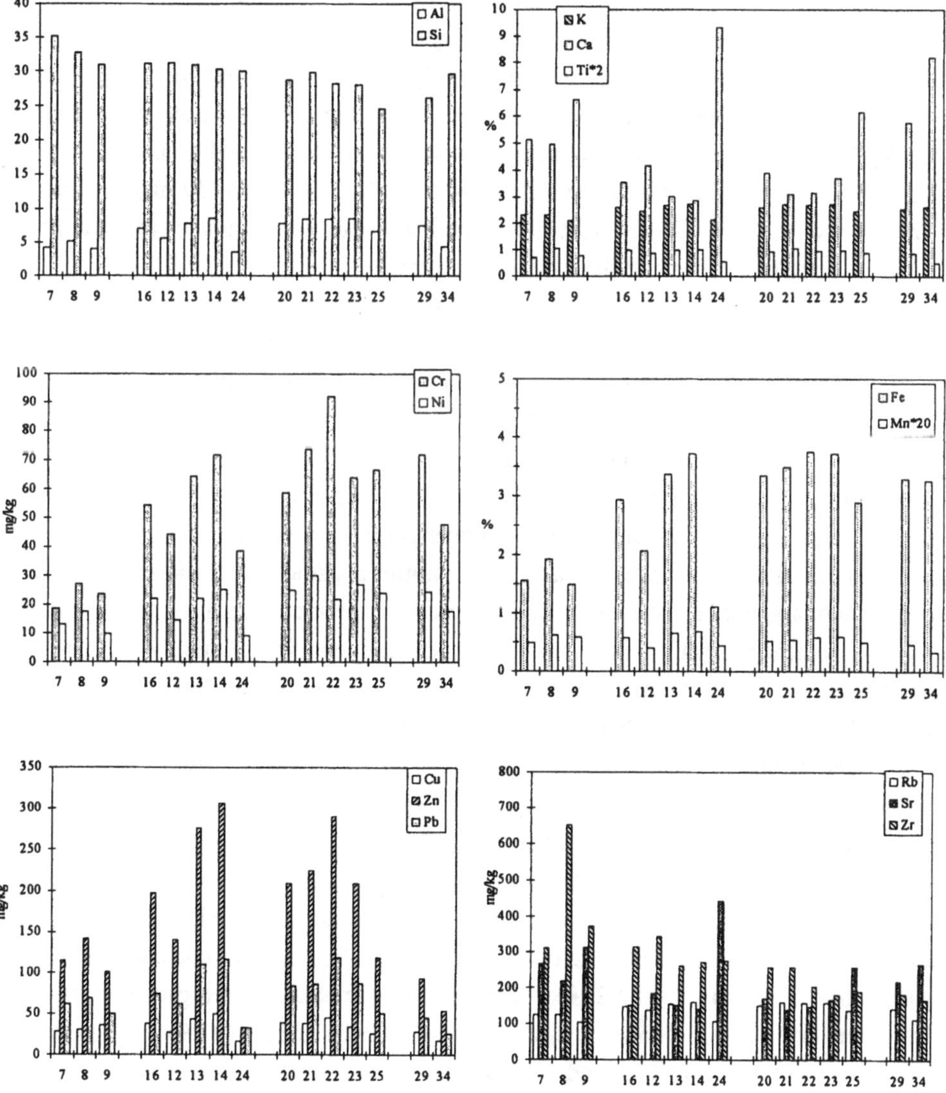

Fig. 2: Elemental distribution in fine sediments from the continental shelf off the Tagus estuary (PLUTUR I/93 cruise)

though Al values are appreciable (4-6%), relatively higher Ca values (5-8%) together with low content in heavy metals indicate somehow a mixture of estuarine and authigenic/biogenic materials. The Zr contents in sediments 7, 8 and 9 are higher, probably due to a faster deposition of the heavy minerals.

Elemental distribution for the heavy metals (which may have an anthropogenic origin dependent on local influences) are within the values usually referred as background (Salomons and Förstner, 1984; Taylor and Lennan, 1985). The exceptions are the Zn and Pb concentrations, that are certainly due to an anthropogenic contribution. In a previous work (Vale, 1990), a gradual decrease in the concentration of these metals with depth has been reported in a gravity core sampled at the muddy area outside the inlet channel. The higher concentrations (Zn ~ 300 -400; Pb ~ 100-130 mg/kg) obtained in the surface layers (~ 10 cm) were rather similar to our values and were attributed to the human activities in the middle and lower estuary.

Although in Fig. 2 only the fine grained sediments are plotted, significant differences in elemental distribution are found for Al, Si and heavy metals. This may be due to different particle sorting. Many authors have used different geochemical approaches where the natural trace metal variability is normalized by the concentration of other elements (Salomons and Förstner, 1984). The most common approach utilizes Al as a normalizer since it has a high natural abundance and it is not commonly associated with an anthropogenic origin (Goldberg *et al.*, 1979). Therefore, we calculated the metal/Al ratios to investigate a possible anthropogenic enrichment. Distribution profiles show a generally common trend for most of the metals, e.g., a rather constant behaviour, indicating that the distributions of these elements in the muddy sediments are regulated mainly by grain size. The most noticeable differences are found for the Zn/Al and Pb/Al ratios, confirming their nondetrital origin.

The results for percentage of organic matter (LOI) ranged from 4,8 to 26,4 %. The highest values were found in samples 30, 35, 36 and 37 at depths below 100 m, directly correlated with the calcium carbonate content.

Correlation coefficients (r) were calculated among sixteen variables (chemical elements and LOI) for the above discussed 15 sediments collected at the mud field. Results showed that Al presents highly significant ($p<0.05$) positive correlations with all the transition metals (0.72 - 0.92) indicating a similar mobility and a probable transport as clay minerals. A strong positive correlation between Zn and Pb (0.99) was found, also between each of these metals and Cu (0.93 and 0.92, respectively) which means a common source and/or a similar enrichment mechanism. Although the Cu content can be considered as being within background values or even depleted, the correlation factor with the pollutants might indicate a release of Cu as has been reported for some other polluted marine sediments from different regions (Araújo *et al.*,1989; Ravichandran *et al.*,1995; Windom *et al.*, 1989). The release of Cu from the sediments can be explained by the increasing capability of divalent ions of the 1st elements transition series with the atomic number (up to Cu and decreasing for Zn), to form metal ion soluble complexes largely independent of ligands (Shriver *et al.*, 1990).

The correlation coefficients between organic matter and metals showed that Cr, Fe and Ni are positively correlated with LOI, indicating that adsorbed organic binding sites on the sediment particles are an important phase carrier for these metals.

4. Conclusions

The geochemical pattern of the muddy sediment deposit off the Tagus (more than 50 m depth) clearly record an estuarine influence. Sediment cover at the inner shelf is dominated by terrigeneous materials, mostly coarse quartz particles. At the outer shelf (deeper than ~130m), the sandy deposits are composed of elevated levels of carbonates (biogenic).

An anthropogenic contamination, particularly for Pb and Zn, typically associated with discharges from urban centers (Windom *et al*, 1989), was measured at the sedimentary deposit at the shelf at about the same concentrations as the surficial layer of sediments collected at the inner channel. This points to a rapid fine grained deposition directly from the river discharge. The high positive correlations among Cu, Zn and Pb seem to indicate a common (anthropogenic) source of these elements. However, while Zn and Pb are enriched in sediments, Cu is depleted, probably because of its ability to form soluble complexes in the marine environment.

Significant correlations among some transition elements (iron, chromium, nickel) and organic matter suggest their probable presence in organic coatings.

Acknowledgements

The authors would like to thank the captain, the crew and the scientific team of the cruise PLUTUR I, on board the R. V. Auriga, during which sediments were collected. P. Paiva acknowledges the PRAXIS XXI for the master grant.

References

Araújo, M.F., Bernard, P.C. and Van Grieken, R.E.: 1989, *Mar.Pollut. Bull.* **19**, 269-273.
Araújo, M.F., Valério, P. and Jouanneau, J.-M.: 1996, *X-Ray Spectrometry* (to be published).
Bruland, K., Bertine, K., Koide, M. and Goldberg, E.:1974, *Environ. Sci. Technol.*, **8**, 425-432.

Garcia, C., Oliveira, A., Jouanneau, J.-M., Dias, J.A., Rodrigues, A. and Weber, O.: 1995, *Memórias nº 4 - 4ºCongresso Nacional de Geologia*, Universidade do Porto, 941-945.

Goldberg, E.D., Griffin, J.J., Hogdge, V., Koide, M., Windom, H.:1979, *Environ. Sci. Technol.*, **13**, 588-594.

Gorsline, D.S.: 1984 in *Fine-grained sediments: deep-water processes and facies*, Geological Society, Special Publication, nº15.

IH: 1988, REL.FT.MC.1/88 (II Parte). *Recolha e processamento de dados de marés e correntes no estuário do Tejo.* Divisão de marés e correntes, p. 220.

IH: 1993, REL.FT.MC.11/93. *Medição de correntes, temperaturas e salinidades nas imediações da doca dos Olivais e foz do Trancão, vol 1.* Divisão de ondas e marés, p. 368.

Mougenot, D.:1988, *Géologie de la marge portugaise.* Thèse Doct. d'État, Paris VI, p. 153.

Ravichandran, M., Baskaran, M., Santschi, P.H. and Bianchi, T.S.: 1995, *Environ. Sci. Technol.* **29**, 1495-1503.

Salomons, W. and Förstner, U.:1984, *Metals in the Hydrocycle*, Springer-Verlag, Berlin.

Shriver, D.F., Atkins, P.W. and Langford, C.H.: 1990, *Inorganic chemistry.* Oxford University Press, Oxford.

Taylor, S.R. and Lennan, S.M.: 1985, *The Continental Crust: its Composition and Evolution*, Blackwell Scientific Publications, Oxford.

Vale, C.: 1990, *The Science of the Total Environment*, **97/98**, 137-154.

Vale, C. and Sundby, B.: 1987, *Estuarine, Coastal and Shelf Science*, **25**, 495-508.

Windom, H.L., Schropp, S.J., Caldet, F.D., Ryan, J.D., Smith, R.G. Jr., Burney, L.C., Lewis, F.G. and Rawlinson, C.H.:1989, *Environ. Sci. Technol.*, **23**, 314-320.

Remobilization of trace elements from polluted anoxic sediments after resuspension in oxic water

W. Petersen, E. Willer, C. Willamowski
GKSS Research Centre, Institute of Hydrophysics, D-21502 Geesthacht, Germany

Abstract. Polluted sediments are periodically subjected to resuspension processes resulting from natural events (e.g. storms, strong waves) as well as from anthropogenically induced activities (e.g. dredging). The main part of the resuspended material is initially in an anoxic state and will be reoxidized more or less quickly in the oxic water column.

In laboratory experiments reflecting, as far as possible, natural conditions (e.g. constant pH) the release of Cd, Cu and Zn during this reoxidation phase was investigated. Up to 2% of the particulate bound heavy metals were remobilized from the sediments. In addition the evolution of the concentrations of the anions PO_4, SO_4, NO_3 and NH_4 were measured to examine the influence of microbial processes on the release of trace elements. Cell counts and microbial activity of certain micro-organisms during the release processes were also investigated.

The investigations illustrated that biological activity has a significant effect on release. In all sediment samples the release of cadmium was delayed in comparison with the other elements even in sediments from different river systems. The influence of different microbial processes on this divergent behavior was examined. The significance of dredging activities to the remobilization processes during reoxidation of anoxic sediments in the Elbe River is discussed.

Key words: sediment, dredging, trace metal, nutrients, bacteria, reoxidation, remobilization, microbial processes

1. Introduction

In industrialized countries most of the rivers are highly polluted with contaminants. A significant part of these pollutants are associated with suspended matter (Förstner et al., 1985; Müller & Sigg, 1990; Jenne, 1995). As a consequence the bottom sediments are severely contaminated with trace elements in the fine grain fraction of the sediments (Ackermann et al., 1983). Toxic organic pollutants can be partly degradated by micro-organisms (Wotzka et al., 1993), the toxicity of heavy metals can be changed only by their physico-chemical forms, i.e. their speciation. In many cases the sediments may act as a sink for the pollutants due to the adsorption of contaminants to particulate matter. In the upper oxic layer of the sediments trace elements may be released to the aqueous phase by mineralization of the freshly deposited organic material (Petersen et al., 1995). Some heavy metals will be immobilized very efficiently by the formation of sulfide minerals in deeper layers of the sediments where anoxic conditions prevail (Engler et al. 1975; Giblin et al. 1986).

The sediments of the Elbe and in particular the sediments of Hamburg harbor are highly contaminated (Lichtfuss & Bruemmer, 1977; Förstner et al, 1990). They become anoxic within a few millimeters due to high contents of organic matter (mean value of organic carbon 5%) and strong microbial activity. As long as the sediments remain undisturbed most of the trace elements are strongly fixed to the sediment (Petersen et al., 1995). However, if these anoxic sediments are exposed to an oxic environment the sulfidic minerals may be reoxidized resulting in the release of trace elements to the aqueous phase or the transformation at the solid phase to more bioavailable forms (Khalid et al. 1978).

This resuspension may occur due to natural processes such as high current velocities caused by tidal action or storm events as well as by human activities such as shipping or dredging. In the lower part of the Elbe river dredging is necessary to maintain shipping activities. At Hamburg harbor alone dredging activities produce 2.5 million m^3 dredged spoil per year consisting of about 40% sand and 60% contaminated mud (Tent, 1987; Netzband, 1992).

In earlier studies Förstner et al. (1986) pointed out four main factors (pH, redox conditions, organic complexation, salinity) which influence the mobility and bioavailability of

Water, Air and Soil Pollution **99**: 515-522, 1997.

sediment-bound metals. The effect of oxidation on the release of trace metals during dumping of dredged material has been discussed by Kersten et al. (1985). The main effects were caused by acid production (decreasing pH) during the oxidation process in the spoil area. Berghahn et al. (1986) and Prause et al. (1985) investigated the remobilization of trace elements after dumping of dredged material in the marine environment. Prause et al. observed that Cd was released in long-term experiments due to the formation of complexes of chlorine. They also found out that the release was interrupted after the addition of antibiotics. Remobilization of Zn and Cd by periodic alternation between oxic and anoxic conditions in surface layers of an intertidal sediment have been described by Kerner & Wallmann (Kerner & Wallmann, 1992). Other publications deal with the different characteristics of anoxic sediments related to their potential mobility (Kersten & Förstner, 1991; Calmano et al., 1992)

In natural rivers during resuspension or dumping of anoxic sediments in oxic water the pH does not decrease noticeably due to the buffering capacity of the water. Thus the release of trace elements during the resuspension in oxic water under natural conditions has to be carried out under stable pH. The aims of this work were (1) to investigate the influence of reoxidation of anoxic sediment on the release of trace elements and in particular under stable pH conditions, (2) to study the effect microbial processes have on this release and (3) to estimate the significance of dredging activities in the Elbe river on these processes.

2. Materials and Methods

Wet sediments equivalent to ≈18 g dry weight were obtained from the river Elbe (experiments poisoned with $HgCl_2$: location Tesperhude, km 579, other experiments: Schnakenbek, km 571) and subsequently suspended under controlled conditions in 1800 ml of artificial river water (5 mM NaCl, 2 mM $CaCl_2$) or natural water from the sampling location for up to 600 hours. The suspensions with artificial water were buffered with 10 ml buffer solution (40g HEPES (2-[4-(2-hydroxyethyl)-1-piperazyl]-ethansulfoniumacid in 0.5 M NaOH). During the experiment the pH was adjusted daily to pH=7.5 with small volumes of 1 M NaOH. The suspensions were stirred with a magnetic stirrer and the air was filtered using a wash bottle through which the air was continuously bubbled into the suspension. All suspension vessels were kept in the dark to inhibit algae growth. If not other mentioned all experiments were carried out at 20°C. In order to inhibit biological activity 10 ml of 5 % $HgCl_2$ was added to each one of the "20°C-vessels". For studying the influence of nitrifying bacteria, in further experiments to each one of the 20°C vessels 2 mg/l allyl-thiourea (ATU) was added in order to inhibit the nitrification process. In this series the suspensions with artificial water were no longer buffered with HEPES.

The vessels were sealed with a silicon lid having two holes. Through one of the holes the suspensions were aerated and through the other samples were taken. The sediments were incubated in nitrogen for 3.5 hours before being aerated.

Samples were taken before the experiment began, one hour after start of the aeration and then on a daily basis. Before each sampling the pH was adjusted and the samples were then filtered through a 0.2 μm membrane filter (Nalgene). An aliquot of 10 ml was analysed using anodic stripping voltammetry (ASV), a second aliquot (3 ml) was immediately acidified to pH=2 for analysis with atomic absorption spectrometry (AAS) and a third aliquot (5 ml) was frozen at -15°C for nutrient analysis. 18 ml of new artificial river water were then added through the filters into the suspensions to equalize the volume losses. Nitrate and sulfate were determined by ion chromatography (Dionex) and phosphate by a FIA-system with photometric detection (Tecator). After digestion with concentrated HNO_3 at 90°C the solids were analysed for heavy metals by atomic absorption spectrometry.

The bacterial population was characterized by the number of total aerobic heterotrophic bacteria, by the number of lithoautotrophic nitrifying bacteria and the concentration of the enzyme phosphatase. Bacteria numbers were estimated by the most probable number (MPN) technique (Koops et al., 1991). The heterotrophic bacteria were incubated on a standard medium (Merck 7882). For the nitrifying bacteria a medium of 0.05g NH_4Cl, 0.6g NaCl, 0.05g $MgSO_4 7 \cdot H_2O$, 2.5g $CaCO_3$, 12g Agar in 1l was used. Four samples were collected from the suspensions after 42, 186, 354 and 545h. Phosphatase was measured according to Obst & Holzapfel-Pschorn (1988) at 400 nm using a spectral photometer. The incubation time was 24h at 20°C. The measured changes in concentration were computed in activities (nmol/l/h) with regard to the incubation time.

3. Results and Discussion

3.1. IMPACT OF POISONING ON THE RELEASE OF TRACE ELEMENTS

The impact of microbial activity on the release of trace elements was investigated by parallel incubation of sediment suspensions under different conditions. The first suspension was kept at 20°C and the second at 5°C in order to reduce microbial activity. In the third suspension, which was also kept at 20°C, the microbial activity was totally inhibited by addition of $HgCl_2$. The evolution of sulfate and nitrate as indicators of microbial activity and the corresponding release of the trace elements cadmium, zinc and copper are shown in Figure 1.

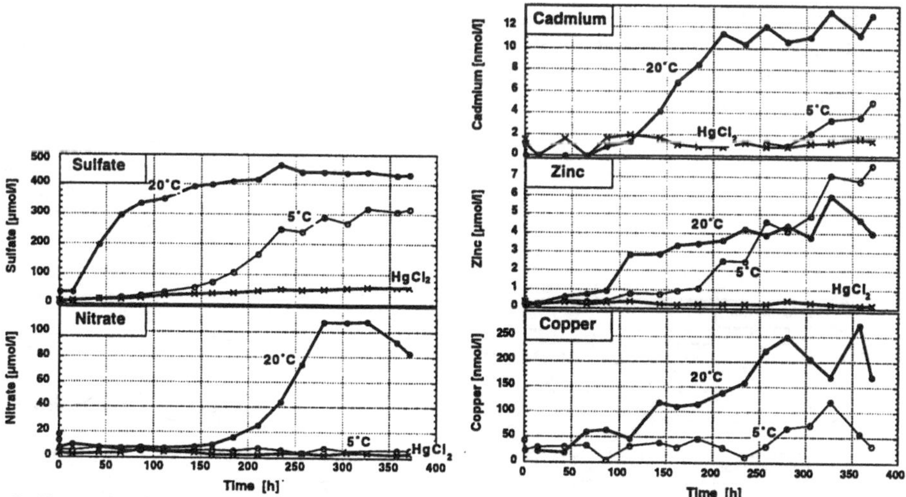

Fig. 1: Changes in of sulfate, nitrate, cadmium, zinc and copper in oxic suspensions of anoxic sediments

After the start of the experiment sulfate immediately increased in the solution kept at 20°C whereas in the 5°C suspension the onset of sulfate appeared much later, after ≈150 h. In the solution poisoned with $HgCl_2$ only a very small increase could be observed indicating that the reoxidation of sulfide to sulfate is strongly connected to microbial activity and that chemical oxidation plays no significant role in this process. The evolution of nitrate was observed only in the suspension kept at 20°C after 170h. This was probably the length of time the population of nitrifying bacteria required for growth up. In the two other solutions nitrification did not occur because the solution was too cold and poisoned by $HgCl_2$, respectively.

The different microbial activity is also reflected in the release of the trace elements cadmium, copper and zinc. For all three elements no release was observed in the poisoned suspensions. Cadmium shows a similar behavior as that of nitrate but was also partially released at 5°C after approximately 10 days. What in particularly interesting is the delay in cadmium evolution in comparison to the elements zinc and copper in the 20°C suspension. Cadmium mobilization begins at nearly the same time as the onset of the nitrification process, indicating that cadmium release is possibly connected to nitrification processes. After the start of the experiment zinc continuously increases in the water phase. However, at 5°C this release occurred much later, after ≈150h, the same time as when the rise of sulfate indicates the oxidation of sulfides. The behavior of copper was similar to that of zinc. In the 20°C solution a continuous increase was observed and at 5°C only a small increase was detectable after 10 days.

As already shown in earlier investigations this behavior of trace elements was very similar even for sediments of different river systems (Petersen et al., 1996). Furthermore the comparison among sediments of the river Main, Neckar and Elbe showed that the release of cadmium was delayed in all three cases.

3.2. INFLUENCE OF DIFFERENT MICROBIAL PROCESSES ON THE RELEASE

In further experiment the influence of microbial processes on the release of trace elements was investigated. Experiments were carried out using artificial as well as natural river water, the density of total bacteria and ammonia oxidizing bacteria was determined and in particular the nitrification process was inhibited by addition of allyl-thiourea (ATU).

A comparison of the evolution of ammonia and phosphate as well as of the trace elements cadmium, copper and zinc between resuspension in natural and artificial river water respectively is shown in Figure 2.

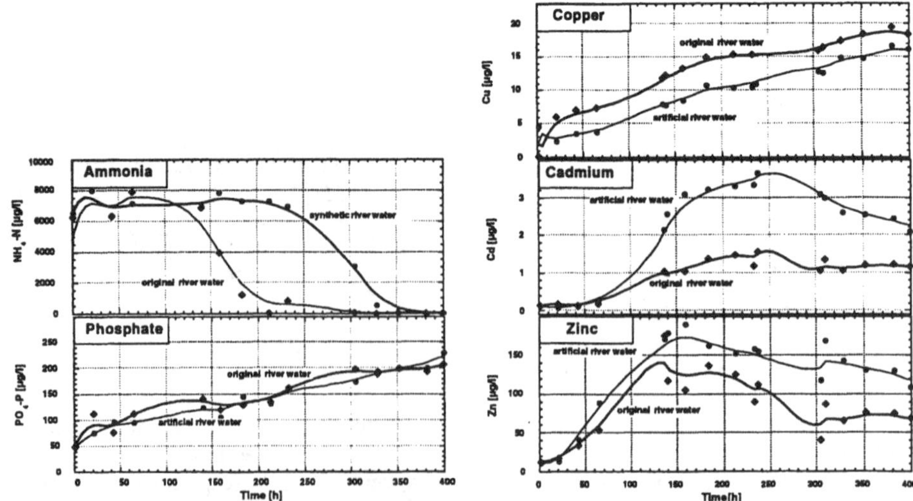

Fig. 2: Comparison of the evolution of ammonia, phosphate and the trace elements zinc, copper and cadmium in suspensions of a sediment from the Elbe river in natural and artificial river water

The curve for ammonia depict a delayed nitrification process in the artificial river water. The ammonia concentration decreased after 110h in the suspension with natural river water whereas in the case of artificial river water ammonia decreases but only after 250h. Higher population densities of certain bacteria in natural river water are probably the reason for this decrease although the numbers of total heterotrophic bacteria cells showed no changes. A change in bacterial activity is not reflected in the rate of release of phosphate either. Both curves show a similar pattern. Phosphate is released continuously after the reoxidation process starts. Both copper curves are comparable and showed no significant difference between the suspension in artificial and natural river water. In contrast zinc and in particular cadmium show a different behavior in the divergent suspensions. Surprisingly both elements are released to a higher degree in artificial water although the bacterial activity obviously is lower. This may be the result of higher amounts of organic matter in the natural water with more or stronger adsorption sites where cadmium in particular can be readsorbed. Partially the bacteria themselves may act as readsorption sites. Remarkably, there is no time shift in the release of cadmium although the nitrification process starts much later in the artificial suspension. Both curves show a delay of cadmium release of approximately 100h.

In further experiments the influence of the microbial processes and the influence of nitrification in particular was investigated in more detail. For this purpose the bacterial population was analysed (total and lithotrophic nitrifying bacteria) and phosphatase as an indicator of microbial activity was determined in parallel incubations. Additionally in some suspensions the nitrification process was inhibited by the addition of ATU.

Figure 3 depicts the determined density of lithotrophic ammonia oxidizing bacteria in suspensions in natural river water (Figure 3a) and in artificial water (Figure 3b).

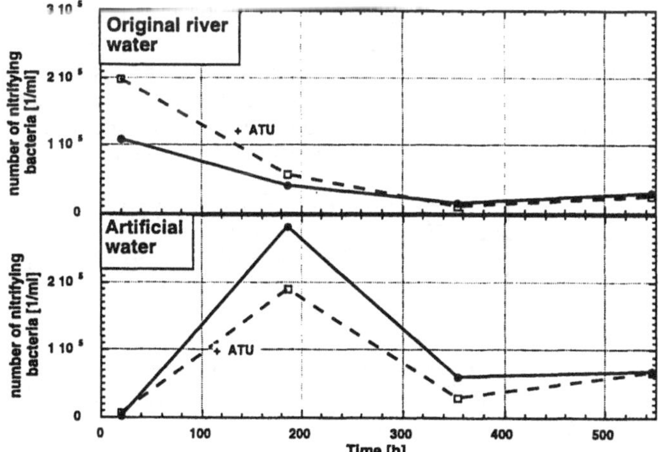

Fig. 3: Density of lithotrophic ammonia oxidizing bacteria in suspensions (sediment from the Elbe river) of natural and artificial river water with and without added allyl-thiourea (ATU)

Each second curve shows the behavior of the suspension with added ATU. The results indicate that in the suspension of natural river water the number of nitrifying bacteria is high at the beginning of the experiment and decreases continuously as the experiment ends. The reason is that in this case the nitrifying process starts much earlier. Figure 3b reveals that in artificial water the population of nitrifying bacteria was approximately zero at the start of the experiment and required a growing period. The number of nitrifying bacteria in the suspension with added ATU was significantly and consistently lower.

In order to measure the bacterial activity the phosphatase enzyme activity was determined. Phosphatase is a general parameter of bacterial activity because it is used for the reaction of adenosion-triphosphate (ATP) to adenosin-diphosphate (ADP). Figure 4 depicts the phosphatase activity in both the original and artificial river water.

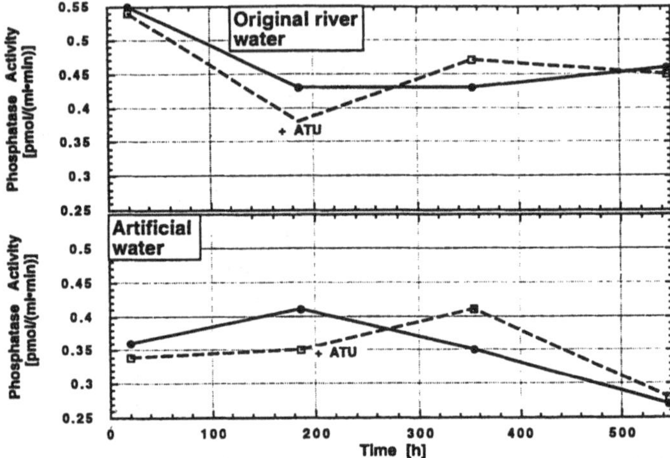

Fig. 4: Phosphatase activity in suspensions (sediment from the Elbe river) of natural and artificial river water with and without added allyl-thiourea (ATU)

As was the case for nitrifying bacteria density, bacterial activity was also significantly higher in the original river water. In both cases the activity decreased following the addition of ATU. However, the effects of inhibition of the nitrification process only lasted a certain time; after that the concentration of phosphatase and the resultant microbial activity reaches similar values to those found in the suspension without the added ATU.

Figure 5 shows the comparison of the evolution of ammonia and nitrate as well the released cadmium in order to verify the assumed influence on the cadmium release delay.

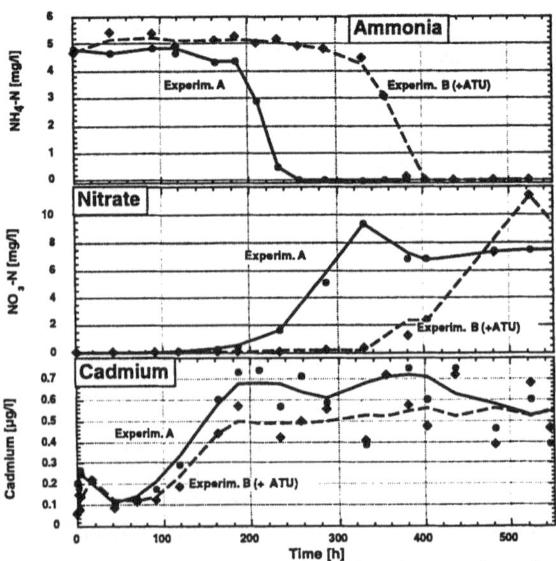

Fig. 5: Comparison of the evolution of ammonia, nitrate and cadmium in suspensions of artificial river water with and without added allyl-thiourea (ATU)

The nitrification process was not completely inhibited but significantly delayed by the addition of ATU which agrees well with the measured phosphatase activity shown in Figure 4. However, there was no significant postponement of the mobilization of cadmium. Both results, the comparison of artificial and original river water as well as the comparison of suspension inhibited with or without ATU, reveal that during the resuspension of anoxic sediments in oxic water the cadmium release was not obviously influenced by the main microbial process of nitrification as first assumed. First of all, during the reoxidation process the more easily degradable compounds will be mineralized and because cadmium is bound more strongly to sulfidic binding sites it is possible that it will to a greater degree be associated with more stable compounds such as humic substances.

3.3. Significance in Relation to Dredging Activities

The significance of trace element release during reoxidation processes was examined for the dredging activities at Hamburg harbor. At Hamburg harbor approximately $2.5 \cdot 10^6$ m^3/a are dredged in order to maintain acceptable water depths in the harbor area. This dredged material contains about 300 kg of sediment per m^3 (Netzband, 1992). The mean concentrations of the trace elements in this sediment are 10 mg/kg Cd, 500 mg/kg Cu and 2000 mg/kg Zn (Netzband, 1992). Assuming the total quantity of dredged material ($2.5 \cdot 10^6$ m^3) would be deposited back into the river (Table I), the amount of remobilized trace elements are estimated from the percentage of released metals measured in the experiments carried out with sediments suspended in natural river water. Additionally, the theoretical remobilized fraction is compared with the total material flow of the trace elements.

TABLE I
Significance of the release of trace elements during dredging activities in Hamburg harbor

	mean sediment concentration [mg/kg]	measured release [%]	remobilized fraction [kg/a]	total material flow in the Elbe river (Arge, 1993) [t/a]	percentage related to total materials flow [%]
Cd	10	1.1	83	10	0.8
Cu	500	0.7	2600	126	2.1
Zn	2000	0.8	12000	650	1.9

The table reveals that, if only those trace elements which are released into the water phase are examined, only a low percentage of the total material flow in the Elbe river would result from the resuspension of dredged material in the river. However, for a realistic estimation, changes of the speciation of the particulate bound part of trace elements have to be taken into account, and these changes strongly influence their bioavailability as has been shown by different authors (Förstner et al., 1986). For an integral risk assessment of the disposal of dredged material in the water, ecotoxicological investigations of toxicity are required (Burton, 1991; Krebs, 1992).

4. Conclusions

The results demonstrate that a certain part of the particulate bound trace elements can be released during the reoxidation of anoxic sediments and that this release is strongly affected by microbial processes. Thus parameters such as available substrates, temperature and bacteria population will influence this release.

Although sediments from different rivers generally behaves in a similar manner, different elements are released at different stages of the reoxidation process. For example the release of cadmium was significantly delayed in all experiments. The assumption that the release of cadmium was connected to the nitrification process as assumed in an earlier publication (Petersen et al. 1996), could not be verified. The activity of nitrifying bacteria seems not to influence the trace element release. Suspensions in natural and artificial river with different bacterial activities as determined by the phosphatase enzyme had no significant influence on the evolution of trace elements at certain temperatures. Even the elements cadmium and zinc seem to be released to a greater extent in artificial water with lower bacterial activity. Presumably readsorption to the particulate phase play an important role. The microbial investigations did not elucidate which microbial processes control metal release. Parameters such as phosphate or sulfate which are much easier to measure are just as good indicators of total microbial activity and show a strong connection with the trace element release.

To understand the specific influence of different mineralization processes on trace element release, further investigations must be carried out which should be able to identify the remaining microbial processes, an example of which is the use of gene probes to identify sulfur oxidation.

The significance of the disposal of dredged material is low because, as yet, it is just the release of this material into the water body which has been examined. However, in an integrated assessment the dramatic change in the speciation and therefore in the bioavailability in the particulate phase must be taken into account.

Acknowledgments

We wish to thank Dr. D. Klages for carrying out the microbial investigations and K. Wirth for nutrient analysis.

References

Ackermannn, F.; Bergmann, H.; Schleichert, U.: 1983, *Env. Techn. Lett.* **4**, 317-328
Arbeitsgemeinschaft für die Reinhaltung der Elbe (ARGE): 1993, *Wassergütedaten der Elbe, Zahlentafel 1992*. Hamburg, Germany
Berghahn, R.; Karbe, L.; Seidel, U.; Burchert, S.; Zeitner, R.: 1986, *Vom Wasser*, **66**, 211-224
Burton Jr., G. A.: 1991, *Environmental Toxicology and Chemistry* **10**, 1585-1627
Calmano, W.; Hong, J.; Förstner, U.: 1992, *Vom Wasser* **78**, 245-257
Engler, R. P.; Patrick, W. H. Jr.: 1975, *Soil Sci.* **119**, 217-221
Förstner, U.; Ahlf, W.; Calmano, W.; Kersten, M.; Salomons, W.: 1986, In: Sly, P. G.: *Sediments and Water interactions*. Springer, New York
Förstner, U.; Calmano, W.; Schoer, J.: 1985, *Vom Wasser* **64**, 1-16
Förstner, U.; Schoer, J.; Knauth, H. -D: 1990, *Sci. of Total Environm.* **97/98**, 347-368
Giblin, A. E.; Luther III, G. W.; Valiela, I.: 1986, *Coastal and Shelf Science* **23**, 477-498
Jenne, E. A.: 1995, *Mar. Freshwater Res.* **46**, 1-18
Kersten, M.; Förstner, U.; Calmano, W.; Ahlf, W.: 1985, *Vom Wasser*, **65**, 21-35
Kersten, M.; Förstner, U.: 1991, *Geo-Marine Letters* **11**, 184-187
Kerner, M.; Wallmann, K.: 1992, *Estuarine, Coastal and Shelf Science* **35**, 371-393
Khalid, R. A.; Patrick, W. H.; Gambrell, R. P.: 1978, *Estuarine Coastal Mar. Sci.* **6**, 21-35
Koops, H. -P; Böttcher, B.; Möller, U. C.; Pommering-Röser, A.; Stehr, G.: 1991, *J. Gen. Microbiol.* **137**, 1689-1699
Krebs, F.: 1992, *Deutsch. Gewaesserk. Mitt.*, **36**, 165-169
Lichtfuss, R.; Bruemmer, G.: 1977, *Naturwissenschaften* **64**, 122-125
Müller, B.; Sigg, L.: 1990, *Aquatic Science* **52**, 75-92
Netzband, A.: 1992, *4. Magdeburger Gewässerschutzseminar*, Spindleruv Mlyn, CSFR, 22.9.-27.9.1992
Obst, U.; Holzapfel-Pschorn, A.: 1988, *Enzymatische Tests für die Wasseranalytik*. Oldenbourg, München,Wien
Petersen, W.; Wallmann, K.; Li, P.; Schroeder, F.; Knauth, H.-D.: 1995, *Marine and Freshwater Research* **46**, 19-26
Petersen, W.; Hong, J.; Willamowski, C.; Wallmann, K.: 1996, *Arch. Hydrobiol. Spec. Issues Advanc. Limnol.* 47, 295-305
Prause, B.; Rehm, E.; Schulz-Baldes, M.: 1985, *Environmental Technology Letters* **6**, 261-266
Tent, L.: 1987, *J. Hydrobiologia* **149**, 189-199
Wotzka, J.; Pfitzner, S.; Giest, B.: 1993, *Deutsche Gewaesserk. Mitt.*, **37**, 106-113

TROPHIC STATUS AND LAKE SEDIMENTATION FLUXES

G. TARTARI, G. BIASCI
CNR Istituto di Ricerca Sulle Acque, Località Occhiate, I-20047 Brugherio, Milan, Italy.

Abstract. In limnological studies the measure of sedimentation fluxes of seston is neglected, in spite of the importance it can have in determining water quality, studying biogeochemical cycles, evaluating the distribution of chemical species, etc. Often sedimentation is obtained only from mass balance models, not taking into account the fact that the uncertainty of determining inputs and outputs makes this evaluation from their difference rather unreliable; other factors of the balance, such as exchanges with the atmosphere, between water and sediments, are equally difficult to define. Though the direct measurement of sedimentation also presents some methodological and logistic difficulties, such as resuspension of material from the bottom, grazing, etc., this does not justify the very scarce attention paid to this kind of determination.
This paper reports the sedimentation fluxes of 39 lakes, in different parts of the world, having different limnological and trophic characteristics ($0.4 < TP < 369$ µg L^{-1}; $0.5 <$ chlorophyll $a < 50$ mg m^{-3}). The fluxes of PM, C, N and P show a log-log relationship ($r \approx 0.6$, $p \leq 0.05$) with the common trophic variables (Secchi disk, total phosphorus, chlorophyll and primary production), independently of the morphometric characteristic of lakes. Hence sedimentation seems not to be an intrinsic property of the environment but is related to the trophic state of the system. The results achieved tend to confirm that the nature of sedimenting seston is generally autochthonous, even though the poor correlation between PM and the same trophic variables suggests that PM is more influenced by allochthonous material.

Key words. Sedimentation, sediment traps, nutrients, elemental ratio, lakes, trophic state.

1. Introduction

Sedimentation is the phenomenon which transfers and distributes the chemical species from the source (euphotic or coastal zones) to the bottom of lakes (Lastein, 1976; Bloesch *et al.*, 1977). The sedimenting material sinks as particles which together make up seston. The structure and the chemical composition of seston are often complex, since it is comprised of living organisms, biological debris (tripton), organic macromolecules, clays, minerals and different oxides (Stabel, 1985). In lakes these particles come from the watershed (allochthonous matter) or they originate directly in the lake waters (autochthonous matter). The former consists of inorganic minerals or the organic product of the decomposition processes acting on terrestrial biomass. The autochthonous matter is prevalently organic in nature (produced by planktonic algae or coastal macrophytes), though considerable amounts of $CaCO_3$ can originate from supersaturation (Bloesch and Wehrli, 1995).

Generally, most of the suspended particles are being created inside the lake or put back into the water through resuspension. It follows that a relationship between trophic variables and sedimentation fluxes is to be expected (Bloesch, 1994). In some lakes a direct relationship has been found (Stabel, 1985) between the standing crop of biomass in the euphotic zone and the flux to the bottom of particulate organic matter (POM), a relationship confirmed in Galvez *et al.* (1989) who compared lakes from different trophic categories.

Water, Air and Soil Pollution **99**: 523-531, 1997.

The predominantly autochthonous origin of seston in lakes is generally confirmed by the good agreement of the carbon (C), nitrogen (N) and phosphorus (P) elemental composition of particles with the values indicated in Redfield's formulation (1934) concerning phytoplankton (Biasci *et al.*, 1992).

Sedimentation fluxes are measured directly (sedimentation traps, rate of sedimentation through sediment cores) or indirectly (mass-balance estimations). There are some problems with trap measurements (resuspension, mineralization, grazing, periphyton, fouling etc.) such that they are not net flux estimations (Kozerski, 1994), but are gross sedimentation measurements which are, however, of great importance in studying most of the environmental problems of lakes (Håkanson, 1994). Though a good knowledge of the phenomenon has been acquired in recent years, sedimentation measurements are still not widely employed. There are also few attempts to predict sedimentation fluxes from other parameters, and models using chemical variables or topographic and morphometric parameters are often based on data from a limited number of lakes (Evans and Håkanson, 1992).

Using literature data, we searched for general relationships between fluxes of the principal nutrients and the major morphometric and trophic parameters of lakes, with the aim of providing models which describe the sedimentation phenomenon.

2. Source of data

This study includes data from 36 papers for lakes spanning a broad range of trophic conditions (Table I). The reference dates demonstrate a lack of studies of sedimentation fluxes during recent years, with most of them concentrated in the late seventies and early eighties. The broad time span and methodological differences between these studies made it imperative to introduce some criteria to ensure consistency. We preferred annual or pluriannual campaigns, except for some particular cases in which environmental conditions make such a long period of trap exposure impossible (i.e. lake surface ice-covered for most of the year). The paper considers only those studies meeting the technical specifications in Bloesch and Burns (1980). We preferred to use daily fluxes (mg m^{-2} d^{-1}) instead of annual rates because of the heterogeneity in time-exposure of the traps. For the trophic variables total phosphorus (TP) and chlorophyll *a* (Chl *a*), we chose their mean concentrations during thermal overturn, because they are more representative of the whole trophic state of the environments.

Where they were indicated by authors, we used organic particulate carbon (POC) instead of total particulate carbon fluxes to avoid the overestimation by $CaCO_3$ precipitation. The distinction among sub-basins and whole lake basins or from pelagic and coastal waters for large lakes was kept where indicated.

Altogether we assembled sedimentation values for particulate matter (PM), C, N and P from 39 lakes for a total of 44 flux values including those obtained in sub-basins of Lakes Lugano, Biel, Costanza and Windermere, as distinct from the whole lake. For the specific methods of sampling and chemical analysis of the variables see the description given in each study. All the elaborations were developed using log-transformed values of flux (mg m^{-2} d^{-1}), due to the large ranges of magnitude of the data.

TABLE I
Lake sedimentation fluxes.

LAKES	COUNTRY	Morfometric features				Trophic features				Sedimentation fluxes					REFERENCES
		AREA km²	DEPTH (m) Mean	Max	τ_w yr	TP µg L^{-1}	Chl a µg L^{-1}	PPR mgCm^{-2}d^{-1}	S.d. m	SAMPLING PERIOD	PM d.w.	C	N	P (mg m^{-2}d^{-1})	
Fayettville Green	USA	0,3	28	53	3	0,4		795	16	III-VI 67	820	10	6	0,4	Brunskill, 1969
Ennerdale Water	GB	3,0	18	42		2	0,5			IX 70-V 72	560	84	11		Pennington, 1974
Wastwater	GB	2,9	40	76		2	1,0			IX 70-V 72	450	76	10		Pennington, 1974
Orta	I	18	71	144	9	5	5,3	614	5	1986-1990	940	142	58	5,1	Tartari *et al.*, 1996
Chuzenji	J	12	95	163	5	6	1,6	202	8	VII 81-VIII 83	2960	137	15	3,2	Fukushima *et al.*, 1989
Koenigsee	D	5,2	98	190	2	6	1,6	211	10	1980	510	70			Stabel, 1985
Michigan (offshore)	USA	58	81	281	99	7	1,3	380		V-XI 87	700	63	6	0,8	Eadie *et al.*, 1984
Ontario (offshore)	CDN-USA	19009	86	244	8	7	5,8	1080	4	III-XI 81	1600	179	27	3,0	Rosa, 1985
Mergozzo	I	1,2	45	73	6	8	5,0	725	4	I-V 83	600	134		3,7	Bertoni *et al.*, 1986
Little rock	USA	18	4	10	9	9	2,4			VII 85-VII 86	540				Baker *et al.*, 1988
Castle lake	USA	0,2	11	35	0,4			1680	13	VI-XI 75	460	41	5		Kimmel and Goldman, 1977
Thingvalla	ISL	84	34	114	0,8	12	0,5	247		VIII-XI 75	1280	73	10	2,9	Lastein, 1983
Kinneret	IL	170	25	42		12		1960		X 72-IX 73	1810	128	11	4,1	Serruya, 1976
Lawrence	USA	0,1	6	13			1,5			I 72-I 74	760	83		1,6	White and Wetzel, 1975
Mezzola	I	4,9	40	69	0,2	16	1,8		4	III 94-III 95	26200	275	166	25,6	Tartari *et al.*, 1996
Erie	USA-CDN	25821	18	64	3	16	4,3	250	3	VII-IX 78	3550	273	29	3,4	Rathke *et al.*, 1981
Washington	USA	88	33	165	2	18	5,0		7	1971-78			23	5,2	Edmonson and Lehman, 1981
Windermere (whole lake)	GB	15	25	64		20	10,0			X 40-XI 41	500				Pennington, 1974
Windermere (south basin)	GB	6,7	18	42		20				IX 80-IX 81	2800	342	50		Hamilton-Taylor and Wills, 1984
Lucerna (Horw bay)	CH	1,7	43	72	4	20		1137		I 69-II 70	3490	172	34	3,3	Bloesch *et al.*, 1977
Hallwill	CH	10	29	48		21		1318		V 82-V 83	2430	310		7,1	Bloesch and Uehlinger, 1986
Rimov (reservoir)	CZ	2,1	17	43	0,3	36	12,8	1700	3	IV 86-XII 86	1100			7,0	Porcalova, 1989
Costanza	D-CH-A	539	100	252	4		2,5	904		1982				1,4	Stabel, 1984
Zuerich	CH	67	50	136	1,4	65	2,8	783	5	VII 83-XII 84	3800	274		5,4	Sigg *et al.*, 1987
Norvikken	S	2,6	5	12	0,8	66	17,0	1055	1	1973-74-75		1104	190	19,7	Ulen, 1978
Lemano	CH	584	152	310	12	68	5,2	819	11	I 86-I 87	2830	243	16	3,4	Gandais et Vernet, 1988
Costanza (Ueberlinger See)	CH			147		91		795		1981-82	2200	375		1,5	Stabel 1984; Stabel, 1988
Pusiano	I	4,9	14	24	0,8	94	1,9		9	IV 94-III 95	3300	369	42	6,4	Tartari *et al.*, 1996
Biel (Tuescherz basin)	CH			74		96		795		IV 76-IV 78	3100	800			Wright *et al.*, 1980
Biel (Neuenstadt basin)	CH			35				795		IV 76-IV 78	2200	800			Wright *et al.*, 1980
Biel (Luescherz basin)	CH			55				410		IV 76-IV 78	1600	400			Wright *et al.*, 1980
Blelham Tarn	GB	0,1	7	15			13,1	500		V-X 72	950	191	20	4,4	Jones, 1976
Onondaga	USA	12	12	21	0,3	100	50,0		1	V-X 80				29,4	Wodka, 1985
Lugano (Agno basin)	CH	2,1	57	87	2	132	17,5	1900	3	VII 75-VII 76	1300			2,8	Premazzi and Rossi, 1984
Zug (north basin)	CH	22	52	64		157				III 82-III 83	2310	271		6,7	Bloesch and Sturm, 1986
Sempach	CH	15	44	87	17	165		1096		1984	2640	179			Weilenmann *et al.*, 1989
Lugano (Ponte Tresa basin)	CH-I	1,0	33	50	0,6	184	13,4	1315	3	X 79-XI 80	2460	320	41	4,9	Premazzi and Marengo, 1982
Zug (south basin)	CH	16	127	197		205				III 82-III 83	1910	242		6,1	Bloesch and Sturm, 1986
Esroem	DK	17	12	22	9	238	9,5	507	4	VIII 71-VIII 72	1730				Lastein, 1976
Rotsee	CH	0,5	9	16	0,3	369		1400		I 69-II 70	2710	405	44	5,9	Bloesch *et al.*, 1977
Itasca	USA			14						VII-X 88	2670	599	76		Hicks *et al.*, 1994
Canadarago	USA	7,6	7	13	0,6					V-XII 69	3000			8,5	Fuhs, 1973
Piecek	PL	0,1		23				662		II-XI 71	5700	700	125		Kajak and Lawacz, 1976
Dgal Maly	PL	0,2		14				931		II-XI 71	7700	601	123		Kajak and Lawacz, 1976

3. Results

Table I lists the selected morphometric variables for each lake: renewal time (τ_w, yr), surface area (km^2), average (z_{med}) and maximum (z_{max}) depth (m). As a trophic index of water quality the table also reports: the total phosphorus concentration (TP, μg L^{-1}) and the concentration of chlorophyll *a* (Chl *a*, μg L^{-1}) during the thermal overturn, the rate of primary production (PPR, mgC m^{-2} d^{-1}) and the transparency (m) as Secchi disk (S.d.). The sampling period over which the mean settling fluxes were obtained are given along with the flux estimates.

TABLE II

Linear correlation coefficients between logarithmic values of variables. For units see Table III.

	Area	z_{med}	z_{max}	τ_w	TP	Chl *a*	PPR	S.d.	PM	C	N	P
Area	1											
z_{med}	0.433	1										
z_{max}	0.596	0.950	1									
τ_w		0.462	0.478	1								
TP					1							
Chl *a*					0.646	1						
PPR						0.719	1					
S.d.						-0.703		1				
PM					0.607				1			
C					0.816	0.676		-0.656	0.749	1		
N			-0.428		0.638	0.666		-0.736	0.777	0.877	1	
P				-0.563	0.608				0.490	0.792	0.791	1

Correlations significant at $p < 0.05$

3.1. DATA CONSISTENCY

The data shown in Table I represent all trophic categories and the lakes were classified according to TP concentrations. Other variables (Chl *a*, PPR and S.d.) were considered to evaluate their agreement with relationships well known in literature (Reckhow and Chapra, 1983).

For example, Table II shows an acceptable correlation coefficient between logarithmic values of transparency and chlorophyll *a* (r=-0.703, n=14, p<0.005). A similar linear relationship was found between the logarithmic values of TP and chlorophyll *a* (r=0.646, n=20, p<0.002) and between daily primary production and chlorophyll *a* (r=0.719, n=16, p<0.002).

3.2. SEDIMENTATION FLUXES

A summary of settling flux values is reported in Table III. The data represent lakes with varying morphometric characteristics (0.1<area<25 10^3 km^2, 10<z_{max}<310 m, 0.2<τ_w<99 yr) and with a wide range from ultra-oligotrophy to hypertrophy. 80% of TP concentrations fall within a range of 5 to 178 μg L^{-1}, relating to chlorophyll *a* concentrations from 1.2 to 15 μg L^{-1} and consistent with the transparency (2.3 and 11 m). The number of lakes for each variable ranges from a minimum of 24 for nitrogen (N) to a maximum of 40 for PM. The sedimentation fluxes were generally distributed over a wide range, but with a significant reduction of scattering if the10th to 90th percentile (%ile) interval is considered (Table III). A striking example is the high value of the PM flux in lake Mezzola (26.2 g m^{-2} d^{-1}), determined by the predominantly allochthonous origin of the seston (Tartari *et al.*, 1996). From the analysis of other variables (C, N and P), two other lakes clearly emerged as completely different from the others, Norvikken and Onondaga. As in Mezzola, these two lakes have very short water renewal times (0.3 and 0.8 yr) and small surface areas (2.6 and 12 km^2), but are much shallower (12 and 21 m) than Mezzola. TP and the concentrations of other trophic variables in these three lakes were anomalous and together with the high values of settling fluxes leads to the conclusion that there are unknown factors involved in these waterbodies. For this reason these lakes were not included in the following analyses.

TABLE III
Main statistical characteristics of lake and sedimentation fluxes.

	Area	z_{max}	τ_w	TP	Chl *a*	PPR	S.d.	PM d.w.	C	N	P
	km^2	m	yr	μg L^{-1}	μg L^{-1}	$mgCm^{-2}d^{-1}$	m	mg $m^{-2}d^{-1}$			
N	39	44	27	34	26	30	19	40	35	24	29
Min	0.1	10.0	0.2	0.4	0.5	202	1.0	450	10	4.8	0.4
Max	25 10	310	99	369	50	1960	16	26200	1104	190	29
Mean	1195	84	7.4	67	7.4	899	6.0	2704	299	47	6.3
Median	7.6	60	2.4	20	4.6	795	4.1	2055	243	28	4.4
10%ile	0.2	14	0.3	5	1.2	250	2.3	537	71	7.5	1.5
90%ile	244	195	10.1	178	15	1682	11	3575	660	124	11

3.3. ELEMENTAL RELATIONSHIPS IN SEDIMENTATION FLUXES

According to Vollenweider (1985) it is possible to compare the elemental ratios (C:N:P) of suspended particulate in fresh waters with those characteristic of the biota (106:16:1; Redfield, 1934). Other authors (Gachter and Bloesch, 1985, Hecky *et al.*, 1993) showed that the stoichiometric ratio among these elements can vary widely from that indicated.

Starting from these considerations we analysed molar ratios of fluxes which, in spite of the wide range ($28<C:P<645$), $6<N:P<31$), present median values in good agreement with Redfield's formulation. The agreement is particularly good for the N:P ratio, which has an average value of 16 and a median of 17, while the C:P ratio has an average of 147 and a median of 131, values which are however within the interval ($159<C:P<75$) indicated by Goldman *et al.* (1979) as consistent with a fixed composition of particulate matter. To examine more closely the variability among nutrient fluxes we calculated the linear correlations between the logarithmic values of PM and nutrients. As Table II shows all the correlations are significant ($p<0.05$), with similar values of r (0.66 - 0.88) except for the relationship between PM and phosphorus which, though significant, is lower (0.49). This behaviour may be associated with different factors, all of them due to the chemical characteristics of phosphorus, which in oligotrophic lakes undergoes a strong recycling, whilst in eutrophic environments it is absorbed directly on the surface of sedimenting particles (Stumm, 1992).

3.4. SETTLING FLUXES AND TROPHIC STATUS

Table II reports the linear correlation coefficients among the logarithmic values of PM, C, N and P fluxes and TP obtained during the winter thermal overturn. All the relationships are significant at least to 95% with r ranging from 0.607 to 0.816. This indicates the key role of phosphorus in regulating trophic activity in these lakes and suggests the predominantly autochthonous origin of the nutrients measured in sedimentation fluxes.

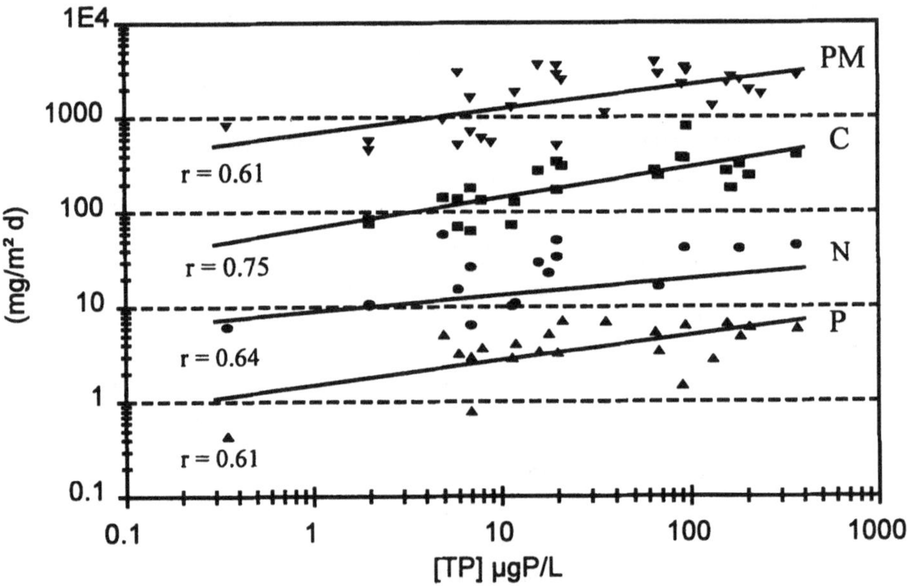

Figure 1. Linear relationships between sedimentation fluxes and in lake total phosphorus concentrations during maximum overturn.

Figure 1 shows a clear tendency for the regression lines to run parallel. If we consider all the data, the slope of the regression of C on TP was 0.41. This result, however, is the consequence of a surprisingly low organic carbon flux of Fayettville Green, the most oligotrophic lake in Table I. Taking into account the likely increased error in estimation of such low values of POC, plus the short exposure period of the traps (Table I), we decided to recalculate the regression excluding this lake. The value so obtained (0.32) is similar to that of the other variables (see below), though the regression coefficient is lower (0.752), than the previous one (0.820). This might be a consequence of the large number of values for eutrophic lakes, resulting in a bias of the relationships toward greater flux values. Therefore, the final equations obtained among the PM, C, N, P fluxes and TP concentration in lakes are:

$$\log PM = 2.84 + 0.253 \log TP \qquad r = 0.607$$
$$\log C = 1.83 + 0.321 \log TP \qquad r = 0.752$$
$$\log N = 1.00 + 0.270 \log TP \qquad r = 0.638$$
$$\log P = 0.177 + 0.261 \log TP \qquad r = 0.608$$

Finally, a test of comparison of regression lines (Sokal and Rohlf, 1981) has shown that the slopes of the relationships are not significantly different ($F=0.193$; $p>0.900$).

4. Discussion

In all the lakes we examined, the low POC values show that the particles are mainly composed of tripton, partly mineralised; the ratios among the nutrients, however, remain constant. Besides, we can consider the intercepts of the regressions indicated in Figure 1 as the extrapolated values of the individual variables under conditions of marked oligotrophy (TP=1 µg L^{-1}). Since markedly oligotrophic lakes do not normally have a high input of allochthonous matter, it may be assumed that these intercepts could be used as an estimation of the lower values of the tripton fluxes. These calculations give a molar ratio of $C_{115}N_{15}P$, surprisingly better than the previous ones and in good agreement with the expected values (Stumm and Morgan, 1982). This result strengthens the relationship between fluxes and TP concentration in lake waters, and introduces the possibility of forecasting the magnitude of fluxes from TP concentrations in lakes. As TP is definitely easier to measure than settling fluxes, it will be desirable to further validate the described relationships with more data obtained by similar methods, which are now available and well consolidated (Bloesch and Burns, 1980). For example, in some cases the availability of only the average values calculated on the entire column, and the different exposure periods of traps could have strongly influenced the variability of the correlations. Therefore, if in spite of these influencing factors, we obtained such good relationships between fluxes and TP, it is reasonable to hypothesise that a functional relationship exists between them.

Another important aspect which emerges is that there is no significant functional relationship with the major morphometric variables, such as lake area, depth and theoretical water renewal time (correlations significant at $p<0.05$), surely due to the wide range of environments considered, but also to the importance of in-lake tripton production.

5. Conclusion

This paper indicates the existence of a general relationship between nutrient fluxes and trophic variables, in particular for TP concentration measured at thermal overturn

The possibility of estimating settling fluxes from TP concentration measurements is attractive, being less expensive and involving fewer methodological problems than measurement by sedimentation traps. It also has a potentially wide application in limnology, not only for evaluating the accuracy of mass balance estimations in studies of nutrient circulation in waters, but also in more complex studies such as those regarding removal mechanisms of metals.

Though these results could be used for some simple extrapolations, examination of additional data is essential before they can be used for wider applications. We would therefore hope to see more gross-sedimentation measurements being performed by traps in the future, as well as greater collaboration among researchers, with the aim of improving and making more uniform the procedures used to treat seston during and after sampling, and also to furnish sedimentation models with a wider range of data.

References

Baker, L.A., Tacconi, J.E. and Brezonik, P.L.: 1988, *Verh. Internat. Verein Limnol.* **23**: 346-350.
Bertoni, R., Calleri. C., De Marco, e Contesini. M.: 1986, In: Atti del 7° Congresso AIOL, Trieste. 145-154.
Biasci, G., Renoldi, M., Tartari, G., Colombo, D. and Muntau, H.: 1992, In: Atti del V° Congresso Nazionale della Società Italiana di Ecologia, Milano. 535 - 540.
Bloesch J.: 1994, *Hydrobiologia* **284**: 1-3.
Bloesch, J. and Burns, N. M.: 1980, *Schweiz. Z. Hydrol.* **42**: 15-55.
Bloesch, J. and Sturm, M.: 1986, In: P. G. Sly (Ed.), *Sediment and water interactions*, Springer-Verlag, New York. 481-490.
Bloesch, J. and Uehlinger, U.: 1986, *Limnol. Oceanogr.* **31** 1094-1109.
Bloesch, J. and Wehrli, B.: 1995, *Eawag News* **38**: 10-12.
Bloesch, J., Stadelmann, T and Buehrer, H.: 1977, *Limnol. Oceanogr.* **22**: 551-526.
Brunskill, G.J.: 1969, *Limnol. Oceanogr.* **14**: 830-847.
Eadie, B.J., Chambers, R.L., Gardner, W.S. and Bell, G.L.: 1984, *J. Great Lakes Res.* **10**: 307-321.
Edmonson, W.T. and Lehman, T.: 1981, *Limnol. Oceanogr.* **26**: 1-29.
Evans, R. D. and Håkanson, L.: 1992, *Hydrobiologia* **235/236**: 143-152.
Fuhs, G.W.: 1973, *Limnol. Oceanogr.* **18**: 989-993.
Fukushima, T., Aizaki, M. and Muraoka, K., 1989, *Hydrobiologia* **176/177**:279-295.
Gachter, R. and Bloesch, J.: 1985, *Hydrobiologia* **128**: 193-200.
Galvez, J. A., Niell, F. X. and Lucena, J.: 1989, *Arch. Hydrobiol. Beih.* **33**: 9-18.
Gandais, V. and Vernet, J-P.: 1988, *Rapp. Comm. int. prot. eaux Léman contre pollut.*, Campagne 1987, 1988: 97-118.
Goldman, J.C., McCarthy, J.J. and Peavey, D.G.: 1979, *Nature* **279**: 210-215.
Håkanson, L. 1994, *Hydrobiologia* **284**: 19-43.
Hamilton-Taylor, J. and Wills, M.: 1984, *Limnol. Oceanogr.* **29**: 695-710.
Hecky, R. E., Campbell, P. and Hendzel, L. L.: 1993, *Limnol. Oceanogr.* **38**: 709-724.
Hicks, R.E., Owen, C.J. and Aas, P.: 1994, *Hydrobiologia* **284**: 79-91.
Kajak, Z. and Lawacz W., 1976, In: H. L. Golterman (ed.), *Interaction between sediments and freshwater.* Dr W. Junk Publ. 72-75.
Kimmel, B. L. and Goldman, C. R.: 1977, In: H. L. Golterman (ed.), *Interaction between sediments and freshwater.* Dr W. Junk Publ. 148-155.
Kozerski, H. P.: 1994, *Hydrobiologia* **284**: 93-101.
Jones, J.B.: 1976,*J. Ecol.* **64**: 241-278.
Lastein, E.: 1976, *Oikos* **27**: 44-49.
Lastein, E.: 1983, *Oikos* **40**: 103-112.
Pennington, W.. 1974, *J. Ecol.* **62**:215-251.
Porcalova, P.: 1989, *Arch. Hydrobiol. Beih.* **33**: 349-353.
Premazzi, G. and G. Marengo.: 1982, *Hydrobiologia* **92**: 603-610.
Premazzi, G. and Rossi, G.: 1984, *Phosphorus cycle in a eutrophic subalpine lake.* Final Rept. Eur 9116, JRC Ispra Establishment.
Rathke, D.E., Bloesch, J., Burns, N.M. and Rosa, F.: 1981, *Verh. Internat. Verein Limnol.* **21**: 383-388.
Reckhow, K.H and Chapra, S.C.: 1983, *Engineering approaches for lake management.* Vol. 1: Data analysis and empirical modeling. Butterworth Publ., Boston. 340 pp.

Redfield, A. C.: 1934, On the proportions of organic derivates in sea water and their relation to the composition of plankton. In: *James Johnston Memorial Volume*, Liverpool University Press: 176-192.
Rosa, F.: 1985, *J. Great Lakes Res.* 1985:13-25.
Serruya, C.: 1976, In: H. L. Golterman (ed.), *Interaction between sediments and fresh water*. Dr W. Junk Publ. 48-56.
Sigg, L., Sturm, M. and Kistler, S.: 1987, *Limnol. Oceanogr.* **32**: 112-130.
Sokal, R.R. and Rohlf, F.J.: 1981, *Biometry*, 859 pp.
Stabel, H.H.: 1984, *Verh. Internat. Verein Limnol.* **22**: 964-969.
Stabel, H.H.: 1985, In: W. Stumm (Ed.), *Chemical processes in lakes*, 143-167.
Stabel, H.H.: 1988, *Verh. Internat. Verein Limnol.* **23**: 700-706.
Stumm, W.: 1992, *Chemistry of the Solid-Water interface*, J. Wiley & Sons, New York, 428 pp.
Stumm, W. and Morgan, J.J.: 1982, *Aquatic chemistry*. J. Wiley & Sons, New York, 780 pp.
Tartari, G., Patrolecco, L., Prina, M., Quattrin, B. and Biasci, G.: 1996, Proc. 7th Int. Symp. "The interactions between sediments and water". *Water, Air, Soil Pollut.* submitted.
Ulen, B.: 1978: *Schweiz, Z. Hydrol.* **40**: 262-286.
Vollenweider, R.A.: 1985, *Arch. Hydrobiol.* **105**: 11-29.
White, W.S. and Wetzel, R.G.: 1975, *Verh. Internat. Verein Limnol.* **19**: 330-339.
Weilenmann, U., O'Melia, C.R. and Stumm, W.: 1989, *Limnol. Oceanogr.* **34**: 1-18.
Wodka, M. C., Effler, S.W. and Driscoll, C.T.: 1985, *Limnol. Oceanogr.* 30: 833-843.
Wright, R.F., Matter, A, Schweingruber, M. and Siegenthaler, U.: 1980, *Schweiz, Z. Hydrol.* **42**: 101-126.

RELATIONSHIP BETWEEN BENTHIC FLUXES AND MACROPHYTE COVER IN A SHALLOW BRACKISH LAGOON

P. VIAROLI[1], M. BARTOLI[1], I. FUMAGALLI[2] and G. GIORDANI[1]

[1]*Dipartimento di Scienze Ambientali, Università, viale delle Scienze, 43100 Parma, Italy,* [2]*ENEL S. p. A. - CRAM, via Rubattino 54, Milano, Italy*

Abstract. The relationship between macrophyte cover and benthic fluxes of oxygen, nutrients and sulphide has been examined in a shallow fishpond with a nearly homogeneous meadow of *Ruppia cirrhosa* (Petagna) Grande (Bassin d'Arcachon, western France). In 1993 and 1994, benthic fluxes were measured in early and late summer. These periods were selected to represent the production and decay phases of *Ruppia* in order to determine the effect on benthic processes. Benthic fluxes of elements were measured by means of multiple dark and light benthic chambers in the presence or absence of community components. In summer 1994, at the end of the incubation period, profiles of acid volatile sulphide (AVS) and chromium reducible sulphur (CRS) were measured also in the 0-5 cm sediment horizon in cores withdrawn from the dark benthic chambers and from the sediment outside the chambers. Oxygen production and consumption were closely related to macrophyte cover, whilst the contributions of plankton and microphytobenthic communities were less significant. In the water column, dissolved inorganic nutrients were almost totally depleted, while dissolved organic nitrogen attained concentrations up to 200 μM. In late summer, *Ruppia* biomass underwent a significant decay due to the build up of a thick epiphyte layer, mostly around floating leaves. The epiphyte slime was rich in labile organic matter, the decomposition of which led to a significant oxygen uptake as well as to sulphide production. Therefore, we postulate that epiphyte growth can cause disturbance in the aquatic system keeping dissolved sulphide at very high levels. Biogeochemical reactions, such as precipitation of iron sulphide, can exert a control lowering the amplitude of such disturbances.

Key words. *Ruppia cirrhosa*, epiphyte slime, oxygen, sulphide, disturbance

1. Introduction

Submerged aquatic macrophytes (SAV) support high primary productivity and maintain high stability and habitat values in shallow aquatic environments (Stevenson *et al.*, 1988). The structure or form of plant communities may mitigate against external forcing factors ensuring substrate stabilisation and providing other beneficial effects. For example, SAV contribute to sediment oxygenation through oxygen release from roots reducing levels of sulphide and other toxic ions (Sand-Jensen *et al.*, 1982; Thursby, 1984). Exploitation of coastal areas, water pollution, and eutrophication are known to have a negative impact on SAV, mostly due to decreased water transparency, build up of epiphytes and the occurrence of reducing conditions in the bottom water and surficial sediments (Twilley, *et al.* 1985, Stevenson *et al.*, 1988, Sand-Jensen and Borum, 1991, Nienhuis, 1992). The competition of SAV with or the replacement of SAV by other primary producers can affect also benthic processes leading to changes in ecosystem function, which in turn lead to a further deterioration in water quality (Stevenson *et al.*, 1988, Sand-Jensen and Borum, 1991). For example, the substitution of SAV by fast growing seaweeds has been considered to be one of the most important factors in determining dystrophy in several lagoons (Sfriso, *et al.* 1992, Viaroli *et al.*, 1995).

The submerged angiosperm *Ruppia cirrhosa* (Petagna) Grande (Potamogetonaceae) is found commonly in Europe in shallow and sheltered brackish water, ranging from coastal environments to interior saline lakes (Verhoeven, 1980). In the last few decades, substantial losses of *Ruppia* meadows have occurred in coastal lagoons receiving pulsed freshwater inputs from ricefields (Comin *et al.*, 1991) as well as in large impoundments that have been exploited heavily for aquaculture (Sorokin *et al.*, 1996).

This work aims to analyse benthic fluxes of oxygen, sulphide and nutrients in relation to the macrophyte cover in a *Ruppia* meadow located in the brackish lagoon system of Certes in Arcachon Bay, Western France (44°40' N, 1°10' W). Research was carried out in a abandoned fishpond with a mean depth of ca. 30 cm and surface area of ca. 30,000

Water, Air and Soil Pollution **99**: 533-540, 1997.

m^2. This lagoon is man-regulated and the water residence time ranges from two weeks to one month (Escaravage, 1990; Thimel and Labourg, 1992). Therefore, the aquatic environment is almost totally isolated with prolonged water stagnation. As a consequence, we can postulate that benthic exchanges are controlled by biological processes, first of all by primary producers, with a minor contribution of hydrodynamics. In 1993 and 1994, benthic fluxes were measured twice a year in early and late summer. These periods were selected to represent the production and decay phases of *Ruppia*. Benthic fluxes were measured using multiple dark and light benthic chambers in the presence or absence of community components to determine their role in benthic exchanges. A preliminary discussion of results obtained in May and August 1994 has been reported by Viaroli *et al.* (1996). In this paper, we discuss the complete data set with emphasis on oxygen and sulphide fluxes and redox conditions.

2. Materials and Methods

Fluxes of nutrients, oxygen and dissolved sulphide were determined using dark and light cylindrical benthic chambers with a volume of 14 litres and cross-sectional area of 700 cm^2. Benthic chambers were incubated without stirring to avoid any disturbance of the macrophyte bed. Experiments were carried out from 3 to 4 June and 30 August to 2 September 1993, 23 to 26 May and 23 to 25 August 1994 (Table I). Each experiment began in the early afternoon and lasted approximately 48 hours. Water samples were collected at different times from the benthic chambers with a syringe connected to an internal pipe (length=18 cm, inner diameter = 3 mm) with side inlets every 2 cm. At the same time additional water samples were taken from a fixed station outside the chambers. Approximately 30 mL of water was fixed immediately for oxygen determination, while 30 mL was used for pH, Eh and sulphide analyses. One hundred mL was filtered through pre-rinsed filters (GF/C Whatman) for nutrient determinations.

Oxygen was determined using the iodometric titration of Winkler (APHA 1975). Sulphides were analyzed by inflection-point titration with silver nitrate (TIM 90, F1212S and K711 electrodes, Radiometer). In parallel, pH (GK 2401 C electrode, Radiometer) and redox (platinum P101 and reference K711 electrodes, Radiometer) were measured with the same apparatus. Salinity was determined by means of conductometry (CDM 83, CDC 304 cell, and T 801 temperature compensation, Radiometer). Nitrite and nitrate (reduction by cadmium columns) were determined by spectrophotometry after diazotation (APHA, 1975); ammonium was measured by the indophenol-blue method (Koroleff, 1970), soluble reactive phosphorus (SRP) by the ascorbic acid method (Valderrama, 1977), and dissolved reactive silica by the molibdate stannous chloride method (APHA, 1975). Total dissolved phosphorus and nitrogen were determined as SRP and nitrate after acid-alkaline digestion (Valderrama, 1981).

The above-ground biomass of *Ruppia* was sampled across a transect by direct harvesting inside circular cores (30 cm i.d.). The harvested plants were sorted, washed with both lagoon and tap water, weighed and oven dried to constant dry weight (DW) at 70°C. The epiphyte slime covering *Ruppia* was removed carefully from the host plants by means of forceps and oven dried at 70°C. Dried subsamples in the range 100-200 mg were used to determine total carbon (TC) content of the harvested biomasses (LECO 600 CHN elemental analyzer). Water extractable carbon (DOC) was also determined by means of wet oxidation with $K_2Cr_2O_7$ and water soluble carbohydrates (DCH) were determined with the phenol-sulphuric acid method.

On 25 August 1994, sediment cores were withdrawn from the dark benthic chambers as well as from the sediment outside. Profiles of acid volatile sulphide (AVS) and chromium reducible sulphur (CRS) were then recorded along the surficial sediment horizon (0-5 cm) of individual cores. AVS and CRS were determined following the two step distillation procedure proposed by Fossing and Jørgensen (1989).

3. Results and discussion

Ruppia distribution in the lagoon followed a similar pattern both in 1993 and 1994. Maximum above-ground biomass was measured in May, with highest values in the central part of the lagoon. In late summer, the emerging fronds were covered by a thick layer of epiphytes with the above-ground biomass inversely related to the amount of epiphyte slime (Table I, Fig. 1).

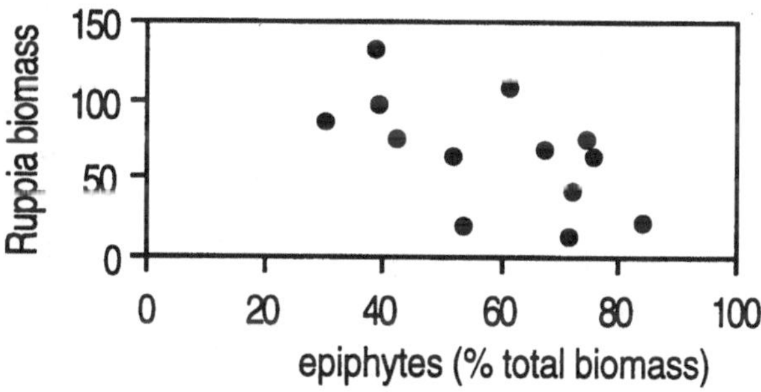

Fig. 1. Relationship between above ground biomass of *Ruppia* (g DW m^{-2}) and the relative abundance of epiphytes (26 August 1994).

Oxygen concentrations followed a clear diurnal cycle (Fig. 2A). Maximum oxygen production and consumption estimated from daily trends were closely related to the standing stock of *Ruppia*, while the role played by the plankton and the benthic microphytes was less significant (Table I, Fig. 2A). Dissolved reactive silica (DRSi) concentrations showed that the contribution of diatoms to the phytobenthos was negligible in spring. DRSi increased during the incubation in all benthic chambers and differences between dark and light conditions were not significant (Fig. 2B). In late summer 1993 and 1994, *Ruppia* underwent decay due to the extensive growth of a dense epiphyte slime around the leaves, mostly in the uppermost water layer. Epiphyte aggregates consisted of diatoms cells enveloped in a mucillagineous substratum. The presence of large populations of diatoms reduced DRSi to undetectable levels. During each sampling, under all the experimental conditions tested nitrate, nitrite, and soluble reactive phosphorus in the water remained below the level of detection. In all the benthic chambers, most of the dissolved nitrogen remained in the organic pool, whilst ammonium was the only inorganic form detected (Table II).

The epiphyte slime was rich in readily soluble carbohydrates (TC=20.94±0.90% DW, DOC=4.40±0.59% DW, DCH=1.42±0.29% DW), which fed microbial activities. Slime decomposition resulted in a significant oxygen demand and high free sulphide (DS) concentrations were measured in the water within the dark benthic chambers with epiphytes in excess of *Ruppia* (Table I).

TABLE I

Net maximum oxygen production (NP_{max}, μmol m^{-2} h^{-1}) and dark oxygen consumption (DR, μmol m^{-2} h^{-1}) measured from benthic chambers with different macrophyte biomass. Maximum sulphide concentration in the water within the dark benthic chambers is also reported (DS, μM) B: biomass: g dry weight m^{-2}, R/E: ratio of *Ruppia* to epiphyte biomass (by dry weight), #: epiphytes not detectable.

No	Date	B	R/E	NP_{max}	DR	DS
1	03.06.93	80	#	13.93	-7.53	0
2	03.06.93	7	#	3.86	-3.77	0
3	30.08.93	87	0.2	18.57	-13.40	239
4	30.08.93	59	0.2	14.23	-10.86	68
5	30.08.93	0	#	3.83	-8.54	56
6	23.05.94	110	#	12.41	-24.47	134
7	23.05.94	70	#	3.79	-6.00	300
8	23.05.94	0	#	1.95	-3.05	172
9	23.08.94	92	0.3	12.50	-13.15	591
10	23.08.94	55	0.8	17.20	-12.97	430
11	23.08.94	0	#	5.50	-11.00	291

TABLE II

Ammonium and dissolved organic nitrogen (DON) concentrations (mmol m^{-2}) in the water at the beginning (t_0) and after 48 hour incubation in the light (L_{48}) and dark (D_{48}) benthic chambers. No, date and B as in TABLE I.

			Ammonium			DON		
No	Date	B	t_0	L_{48}	D_{48}	t_0	L_{48}	D_{48}
3	30.08.93	87	0.00	0.14	2.30	45.7	69.9	79.7
4	30.08.93	59	0.00	0.29	5.41	47.1	43.9	78.6
5	30.08.93	0	0.00	0.39	0.47	44.0	39.7	49.3
6	23.05.94	110	1.32	1.90	4.23	25.4	28.0	40.6
7	23.05.94	70	1.32	1.10	0.91	23.6	42.9	46.1
8	23.05.94	0	1.32	1.51	6.16	31.6	33.9	41.4

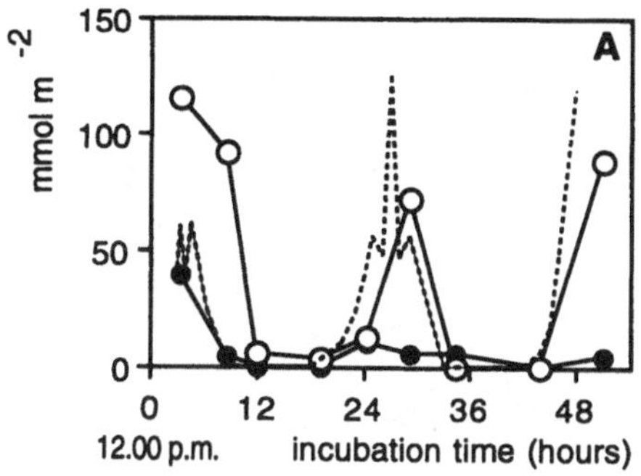

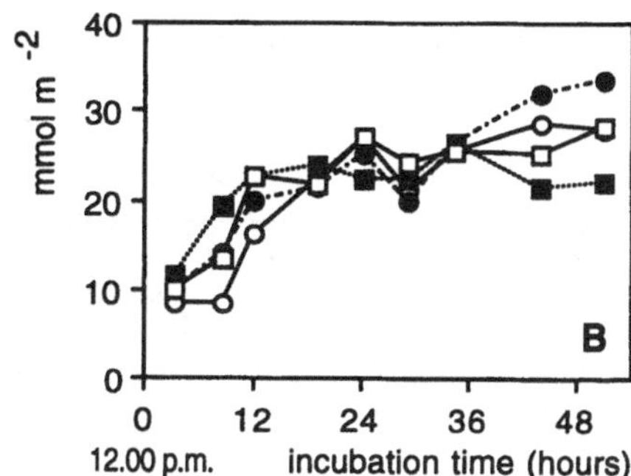

Fig. 2. A: Daily changes of oxygen in light benthic chambers without (dark circle) and with *Ruppia* (open circle). The dotted line represents illuminance (klux, scale as for oxygen). B: Daily changes of dissolved reactive silica concentrations in the water inside light (open symbols) and dark (dark symbols) benthic chambers without (square) and with *Ruppia* (circle). The incubation started on 23 May 1994, at 3.00 p.m.

The experiments, which lasted about two days, were also designed to analyse the resistance of the different plant communities against stresses induced by the prolonged incubation. Here we report on the potential oxygen availability (or deficiency) which can be considered as an indicator of system disturbance. For this purpose, sulphide concentrations were transformed into oxygen equivalents ($O_2 = 2S^{2-}$) according to Jørgensen (1983) and then subtracted from the measured oxygen concentrations. Moreover, changes in oxygen equivalents correspond to redox potential trends (Fig. 3). In the light benthic chambers with macrophytes and their epiphytes, potential oxygen availability showed a daily trend with a pronounced night deficiency corresponding to free-sulphide production. In the light benthic chamber with *Ruppia* in excess of epiphytes, oxidising conditions were recovered during the next light periods, while in the benthic chamber with a large excess of epiphytes reoxidation did not take place after the second dark period. Abrupt daily changes in oxygen availability were observed also within *Ruppia* stands in the lagoon, showing that this aquatic ecosystem was subject to strong disturbances (Fig. 4). The prolonged dark incubation which simulated the effect of early decomposition within the dense macrophyte mats led to highly reducing conditions. After 48 hours, dissolved sulphide concentration in the water column was 291 μM in the benthic chamber with unvegetated sediment, 430 μM with *Ruppia* in excess of epihpytes and 591 μM with epiphytes in excess of macrophytes. We postulate that the large amount of sulphides originated from respiratory processes which were fuelled by the labile organic carbon coming from epiphyte slime. Since sulphate reduction was occurring in the water column, biogeochemical reactions buffering against sulphide accumulation occurred mostly at the water-sediment interface, where we found a significant accumulation of AVS (Fig. 5). The maximum AVS concentration (81 μmol cm^{-3}) was observed in the uppermost sediment layer (0-1 cm) in the benthic chamber with epiphytes in excess of macrophytes, while in the unvegetated sediment outside the benthic chambers AVS was approximately three times lower (26 μmol cm^{-3}). AVS content decreased with depth in the vegetated sediments, while in the bare sediments AVS concentrations were highest in the deeper layers, probably as a result of the thickness of the microphytobenthic mat that was built up by settled epiphytes. In the uppermost sediment layer, CRS followed the same trend shown for AVS. In the deeper sections of vegetated sediment, CRS accumulation (400-800 μmol cm^{-3}) might depend on processes which occurred before the beginning of incubations, when oxygen transport into the rhizosphere was active (Sand-Jensen *et al.*, 1982, Giordani *et al.*, 1996).

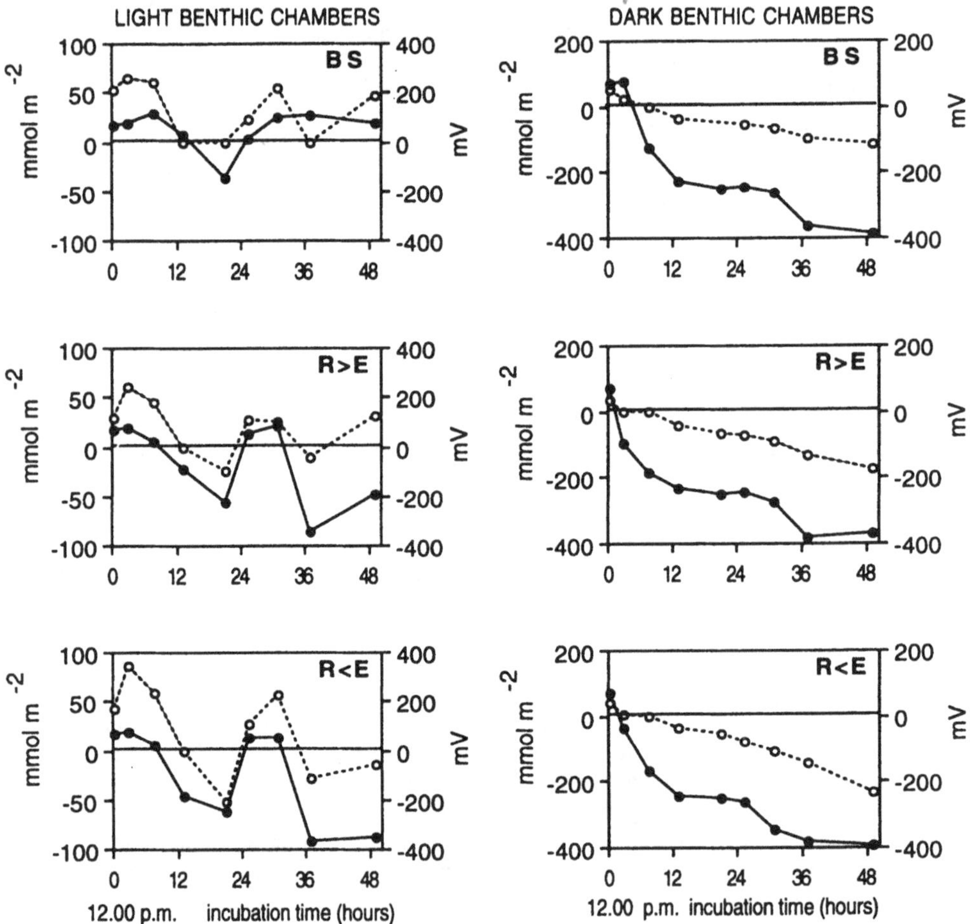

Fig. 3. Oxygen availability as oxygen equivalent units (dotted line, mmol m^{-2}) and redox potential (continuous line, mV) in benthic chambers with different macrophyte cover. BS: bare sediment, R>E: *Ruppia* in excess of epiphytes, R<E: epiphytes in excess of *Ruppia*. The incubation started on 23 August 1994, at noon.

4. Conclusions

Seasonal changes in the structure of macrophyte communities can affect organic matter decomposition and determine disturbance. Epiphytes which extend over host macrophytes lead to light attenuation causing oxygen deficit and accumulation of toxic ions, e.g. ammonia and sulphide (Comin *et al.*, 1991; Turpin, 1991; Smolders et al. 1995). Biogeochemical reactions can buffer against dissolved sulphide lowering the amplitude of such disturbances (Jørgensen, 1983; Smolders and Roelofs, 1993). The significant fluxes of sulphide observed in the dark and light benthic chambers can be explained by the production of free-sulphide by the epiphytes growing on the leaves of the *Ruppia* or by the epiphytes covering the sediment surface. These results demonstrate the importance of epiphytes in establishing strong microbial activities at the sediment-water interface and around the *Ruppia* leaves. In the short-term, the epiphytic pathway

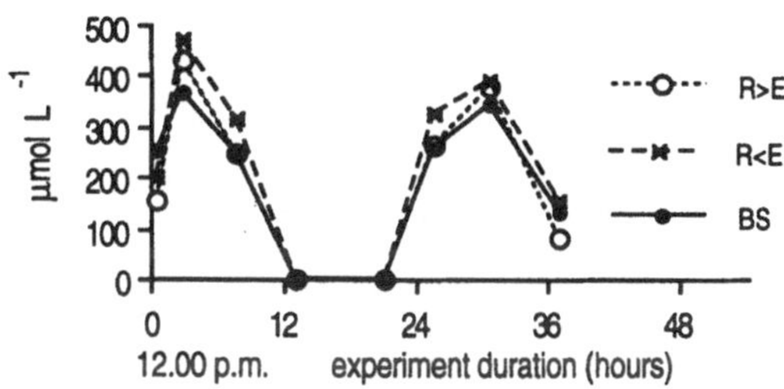

Fig. 4. Daily variations of oxygen concentrations in the water (depth=10 cm) outside benthic chambers at sites characterised by different macrophyte biomass from 23 to 25 August 1994. Symbols as in Fig. 3.

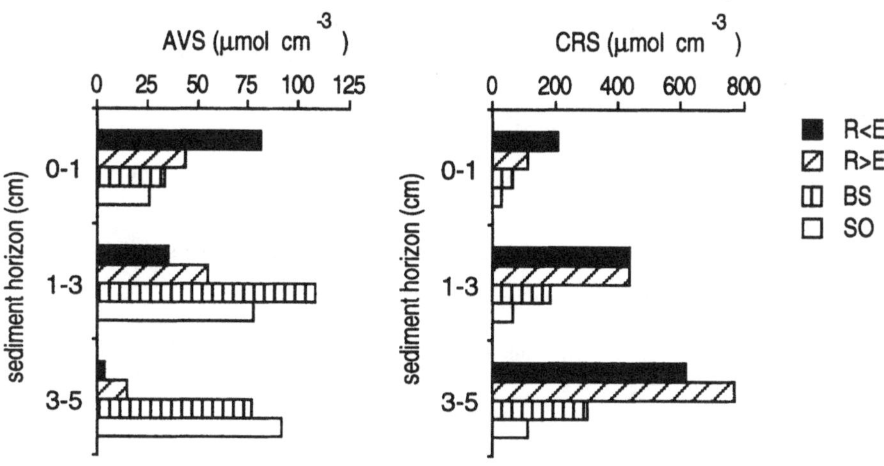

Fig. 5. Depth profiles of acid volatile sulphide (AVS) and chromium reducible sulphur (CRS) in the dark benthic chambers shown in Fig. 3 and in the sediment outside the chambers (SO). Symbols as in Fig. 3.

may operate almost independently from that in the sediment and may thus lead to significant fluxes of sulphide to the water column. When sulphide production is located in the water column, sedimentary biogeochemical control becomes ineffective causing dystrophy to increase. Therefore, ecosystem stability is weakened and the occurrence of a positive feedback may lead to a progressive deterioration in the resistence and/or resilience of the macrophyte community (Burkholder *et al.*, 1994).

Acknowledgements

This research was supported by the EU projects CLEAN (Contract No EV5V-CT92-0080) and ROBUST (Contract No ENV4-CT96-0218). Authors wish to thank Dr M. Naldi (Dipartimento di Scienze Ambientali, Università di Parma) who contributed data and helpful discussion on macrophytes, and Dr D. T. Welsh (Laboratoire d'Océanographie Biologique, Arcachon, France) who revised the manuscript.
This paper is the ELOISE contribution No *XX*.

References

A.P.H.A., A.W.W.A., W.P.C.F.: 1975, *Standard methods for the examination of water and wastewaters,* 14th ed. A.P.H.A, Washington, 1193 pages.
Burkholder, J. M., Glasgow, H. B. jr, Cooke, J. E.: 1994, *Mar. Ecol. Prog. Ser.* **105**, 121-138.
Comin, F. A., Menendez, M., Martin, M.: 1991, *Mem. Ist. ital. Idrobiol.* **48**, 9-22.
Escaravage, V.: 1990, *Hydrobiologia* **207**, 131-136.
Giordani, G., Bartoli, M., Cattadori, M., Viaroli, P.: 1996, *Hydrobiologia* **329**, 211-222.
Fossing, H., Jørgensen, B. B.: 1989, *Biogeochemistry* **8**, 205-222.
Jørgensen, B. B.: 1983, *The major biogeochemical cycles and their interactions, SCOPE 21*, J. Wiley, New York, 477-509.
Koroleff, F.: 1970, *Information on techniques and methods for seawater analysis.* I.C.E.S. Interlaboratory Rep. No. **3**: 19-22.
Nienhuis, P. H.: 1992, *Estuaries* **15**, 538-548.
Sand-Jensen, K., Borum, J.: 1991, *Aquat. Bot.* **41**, 137-175.
Sand-Jensen, K., Prahl, C., Stokholm, H.: 1982, *Oikos* **38**, 349-354.
Sfriso, A., Pavoni, B., Marcomini, A., Orio, A. A.: 1992, *Estuaries* **15**, 517-528.
Smolders, A., Roelofs, J. G. M.: 1993, *Archiv fur Hydrobiologie* **133**, 349-365.
Smolders, A., Nijboer, R. C., Roelofs, J. G. M.: 1995, *Freshwater Biology* **34**, 559-568.
Sorokin, Yu. I., Sorokin, Yu. P., Gnes A.: 1996, *Mar. Ecol. Prog. Ser.* **133**, 57-71.
Stevenson, J. C., Staver, L. W., Staver, K. W.: 1988, *Estuaries* **16**, 346-361.
Thimel, A., Labourg, P. J.: 1992, *Vie Milieu* **42**, 185-192.
Thursby, G. B.: 1984, *Mar. Ecol. Prog. Ser.* **16**, 303-305.
Twilley, R. R., Kemp, W. M., Staver, K. W., Stevenson, J. C., Boynton, W. R.: 1985, *Mar. Ecol. Prog. Ser.* **23**, 179-191.
Turpin, D. H.: 1991, *J. Phycol.* **27**, 14-20.
Valderrama, J. C.: 1977, *Rep. Baltic Intercalibration Workshop*, Interim Commission for the Protection of the Environment of the Baltic Sea, 14-34.
Valderrama, J. C.: 1981, *Mar. Chem.* **10**, 109-122.
Verhoeven, J. T. A.: 1980, *Aquat. Bot.* **8**, 1-85.
Viaroli, P., Bartoli, M., C. Bondavalli, C., Naldi, M.: 1995, *Fresenius Environmental Bulletin* **8**, 381-386.
Viaroli, P., Bartoli, M., C. Bondavalli, C., Christian, R. R., Giordani, G., Naldi, M.: 1996, *Hydrobiologia* **329**, 105-119.

WATER-SEDIMENT EXCHANGE OF NUTRIENTS DURING EARLY DIAGENESIS AND RESUSPENSION OF ANOXIC SEDIMENTS FROM THE NORTHERN ADRIATIC SEA SHELF.

Federico Spagnoli and Maria Cristina Bergamini.
Istituto di Geologia Marina, Consiglio Nazionale delle Ricerche, Bologna, Italy

Abstract. This paper presents the results of a study on nutrient exchange at the sediment-water interface which is caused by early diagenesis and resuspension of bottom sediments. The research was carried out on anoxic silty-clay sediment cores collected south of the Po river delta (Northern Adriatic Sea, Italy) in late summer.

The early diagenetic processes were investigated by means of the integrated study of pore-water chemistry and solid phase composition. Exchange at the sediment-water interface was studied by comparing the fluxes measured in incubated cores with the fluxes calculated by modelling pore-water profiles. Nutrient exchange during resuspension was analysed by simulating a storm event in the laboratory.

The high production of nutrients near the sediment-water interface is mainly caused by the anoxic degradation of organic matter and the successive reductions of Mn and Fe-oxyhydroxides and, to a lesser extent, of sulphate. The oxic degradation of organic matter occurs only at the sediment-water interface.

In the incubation experiment the increases of phosphate, ammonia, nitrate, silica, and Fe in bottom waters were measured. The comparison between calculated and measured fluxes showed that: a) the fluxes are mainly controlled by molecular diffusion; b) phosphate and Fe sink because of the Fe-oxyhydroxide precipitation and c) nitrification process influences the ammonia and nitrate fluxes.

Resuspension caused the release of: a) phosphate through surficial desorption and authigenic apatite dissolution; b) ammonia by means of the oxic degradation of organic matter; and c) dissolved silica generated by biogenic silica dissolution. Resuspension also caused a weak removal of Fe. The more oxic conditions following resuspension favoured the formation of a Fe-oxyhydroxide film at the sediment-water interface which inhibited the phosphate fluxes from sediments to the water column.

Keywords: Diagenesis, resuspension, sediment, pore-water, marine

1. Introduction

In coastal marine environments the chemical transformations that take place at the sediment-water interface determine the cycling of nutrients between sediments and waters. The former can constitute a source or a sink for nutrients, so that in areas with shallow waters sediments can be one of the major factors which control the trophic level of the aquatic system. Many of these chemical reactions are biologically mediated; their relative importance depends on several factors, such as sediment composition, sedimentation rate, hydrodynamics, bioturbation and irrigation, as well as the physical and chemical characteristics of bottom waters.

Early diagenesis can produce solute concentration differences between pore waters and sea waters, mainly because of organic matter degradation and variations of pH and redox potential. Such differences cause fluxes through diffusion, bioturbation and irrigation (Hammond *et al.*, 1985) at the sediment-water interface, where also adsorption/desorption processes can take place (Santschi *et al.*, 1990).

Many authors investigated the release of nutrients from sediments by means of *in situ* measurements (Hammond *et al.*, 1985; Westerlund *et al.*, 1986; Bender *et al.*, 1989; Berelson *et al.*, 1987b, 1990, 1994), micro and mesocosms (Sundby *et al.*, 1986; Petersen *et al.*, 1995), water dynamics modelling (Berelson *et al.*, 1987a) or pore water profile

Water, Air and Soil Pollution **99**: 541-556, 1997.

modelling (Aller, 1980; Berner, 1980; Klump and Martens, 1981; Berelson *et al.*, 1987a; Hammond *et al.*, in press)

Resuspension may influence the chemical composition of the water column because of pore water and bottom water mixing, sorption-desorption on solid surfaces, precipitation-dissolution processes, and an enhanced organic matter degradation. Solid-water exchanges during resuspension have been relatively neglected because of sampling difficulties during storms (Oviatt *et al.*, 1981; Fanning *et al.*, 1982)

The Northwestern Adriatic Sea is an economically important region undergoing heavy eutrophication and pollution pressure (Marchetti, 1992) and where nutrient release from sediments plays a major role (Giordani *et al.*, 1992). Fluxes at the sediment-water interface have been measured *in situ* (Giordani and Hammond, 1985; Giordani *et al.*, 1992), by means of incubation or pore water modelling (Barbanti *et al.*, 1995). The chemical effects of resuspension were investigated only by Frignani and Turci (1981) and, in lagoonal areas, by Sfriso *et al.* (1990) and Calvo *et al.* (1991).

The aim of this work was to investigate the role played by anoxic sediments as nutrient sinks or sources, as well as the changes in speciation on the particulate phase of some elements caused by early diagenesis and resuspension processes. Fluxes at the sediment-water interface in low hydrodynamic conditions, and the water-solid matter interactions during resuspension were studied by means of a laboratory experiment on intact cores.

2. Sampling area

The cores were collected from the 22.5 m deep seafloor in the southern prodelta area of the Po River, North Adriatic Sea (Fig. 1). This area has a sedimentation rate as high as 0.88 cm/y (Frignani and Langone, 1991) and silty clay sediments which are mainly supplied by the Po River, the largest Italian river (mean water discharge 1500 $m^3 s^{-1}$, mean solid discharge 14 million tonnes $year^{-1}$; Dal Cin, 1983) which drains very industrialised and intensively cultivated areas. The sediments have high concentrations of continental and marine organic matter and host the typical populations of muddy seabeds. Polychaetes, such as *Aricidea assimilis, A. claudiae, Prionospio malmgreni, P. cirrifera, Maldane sarsi,* and *Nephtys incisa* dominate the macrofauna (Crema *et al.*, 1991).

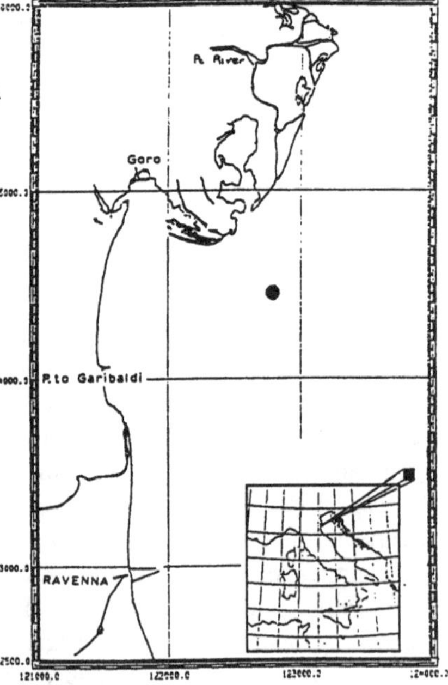

Figure 1 Study Area

The water circulation of the area follows the general circulation of the Northern Adriatic Sea, with seasonal dynamics which are strongly conditioned by thermohaline and barocline factors and, to a lesser extent, by winds (Franco *et al.*, 1982; Orlich *et al.*, 1992; Artegiani *et al.*, in press). In summer the circulation of waters is rather slow, with horizontal stratification in the water column mainly caused by the outflow of the Po River (Nelson, 1970; Orlich *et al.*, 1992). In winter a cyclonic water circulation prevails, with a southward fast-moving current and a more homogenous water column which mostly consists of marine waters, even if the Po River waters may be occasionally discernable.

This area has frequent resuspension events which are caused chiefly by the bora and scirocco winds and secondarily by tidal currents (Drake *et al.*, 1992). The dissolved oxygen in bottom waters varies considerably: it is maximum in winter and very low in summer when waters often reach an anoxic state.

3. Materials and methods

3.1. Sampling operations

Field operations were performed on September 1st, 1992. Continuous temperature, salinity, dissolved oxygen, chlorophyll *a* and pH profiles were performed by CTD probe. Dissolved oxygen in the bottom waters indicate anoxic conditions. The water column was sampled by means of a Niskin bottle, while bottom water was collected by a diver. Using Plexiglas tubes of 10.4 cm in diameter, divers also collected the cores close to one another, to minimize disparity in sediment composition. The cores were immediately stored at 4°C and kept refrigerated for about five hours, the time required to take them to the laboratory.

One core which was about 30 cm long, was used to characterise the sediments. Core extrusion and pore water extraction were performed in a nitrogen atmosphere within ±2°C of the *in situ* temperature. The core was cut into 0.5 to 3 cm thick slices, using the higher resolution near the sediment-water interface. Each slice was placed in a polypropylene tube and centrifuged at 5000 rpm at the *in situ* temperature for 15 mins. The supernatant was then filtered through a 0.45 μm Durapore membrane and split into two aliquots: one was left unacidified for nutrient, alkalinity, sulphate, sulphide and salinity determinations, while the other was acidified to pH 1-2 with Suprapure HNO_3 for Ca, Mg, Fe, Mn and trace metal analysis (Spagnoli, 1993). The pH and redox potential were determined through precalibrated punch-in electrodes. Visual descriptions of the core were recorded. The entire procedure was completed in 6 hours. After centrifugation, sediments were freeze-dried and ground by means of an agate pestle and mortar for subsequent chemical analyses: organic and inorganic C, total content of Ca, Mg, Fe, Mn, Ti, Al, Cu, Zn, Cr, Cd, Pb and P, as well as Fe, Mn, Cu, Zn, Cr, Cd, Pb and P sequential extraction (Spagnoli, 1993). Grain sizes were determined with the wet samples.

3.2. Laboratory experiment procedures

Seven cores were used for laboratory experiments. All preparatory operations were performed at 20±2°C in a nitrogen atmosphere. Cores A and B were used for simulating resuspension events and core C for measuring the fluxes at the sediment-water interface in low hydrodynamic conditions (stirring experiment). The remaining four cores were used for slice "shaking experiments".

The supernatants of the A, B and C cores were removed carefully and replaced with 4 l of natural, oligotrophic and filtered seawater. The seawater was added very slowly through Teflon capillary tubing so that it trickled gently down the tube wall above the sediment; the replacement took 3 to 4 h. Subsequently, a 8.5 cm wide Teflon paddle was placed about 18.5 cm above the sediment-water interface, in order to create resuspension (high angular speed) or to keep the diffusive boundary layer thickness at the sediment-water interface unchanged during incubation (low angular speed: 2-3 rpm). After setting up these apparatuses, all cores were kept at 20±2°C with their tops exposed. Supernatant samples were collected by means of a syringe. At the end of the stirring experiment, the surficial sediments (0.5 cm) were collected.

Fig. 2 shows the different steps of the stirring experiment for the A and B cores. The paddle rotated for the first 45 mins at low speed in order to bring the added seawater and sediments to a balance. During this time the suspended solids ranged from 1 to 23 mg/l. In the second and third steps the speed of the paddle was increased further for 3.5 h and 8.25 h, respectively. The maximum stirring speed for the A core was faster than that for the B core. The resuspended solids of the A core increased at the beginning and at the end of the maximum stirring step, while those of the B core increased continuously throughout the same step. The maximum resuspended solid value for A was 0.89 g/l, while it was 0.11 g/l for B. The distance between paddle and sediment interface, the fastest angular speed and the maximum stirring time were set to simulate the duration (about 7-8 h) and magnitude (about 600 mg/l) of an average resuspension event as recorded by the Geoprobe deployed in the sampling area (Drake *et al.*, 1992).

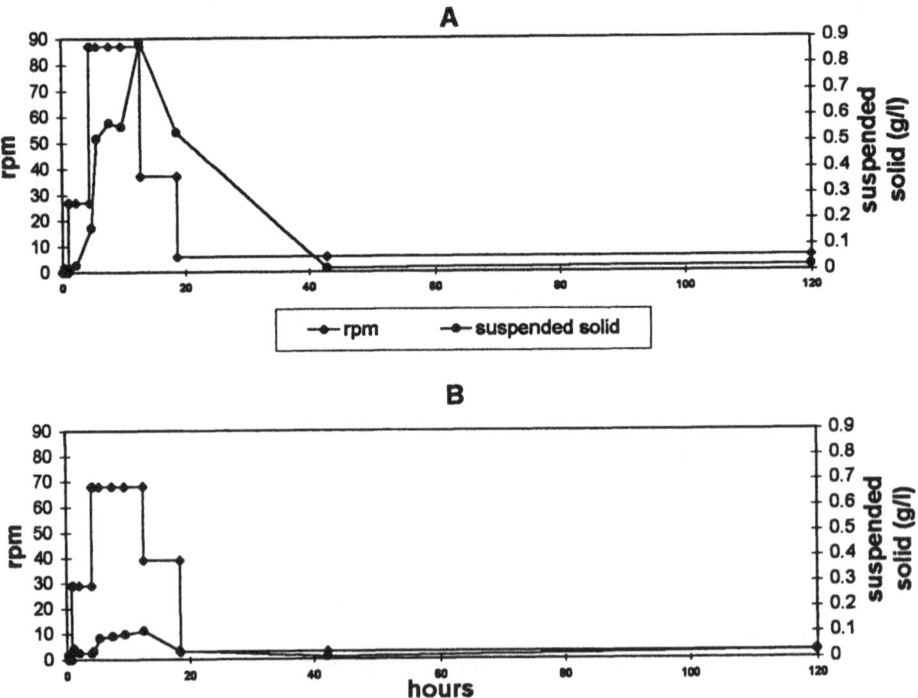

Figure 2. The RPM of the stirring paddles and suspended solids in cores A and B.

In contrast with the literature, at the end of the maximum stirring step no resuspension steady state was reached (it took about 10 minutes according to Tsai and Lick, 1986). We presume that this was caused by a degradation process of the highly reactive organic matter (Westrich and Berner, 1984) at the interface following the more oxic conditions created by our resuspension. The measurement of maximum solid resuspension made it possible to determine the maximum thickness of sediment erosion. With a dry density of 2.7 and a porosity of 0.83, the erosion was 1 mm in core A and 0.12 mm in core B.

After resuspension the paddle was slowed down to intermediate and low speeds in two steps. The last slow-speed step (incubation) lasted four days. The same slow speed was used throughout the experiment with the C core (low hydrodynamic conditions). During such incubation the suspended solids ranged from 5 to 35 mg/l.

In the shaking experiment, one slice per core (the first 0.1, 0.5, 1 and 2 cm, respectively) was collected and split into four subsamples. The subsamples then were put in to sealed vessels with 100 ml of natural, oligotrophic, filtered seawater and shaken for 1, 2, 10 and 21 h, respectively. In order to assess the solid-water fluxes for each chemical, the pore water content of each species was subtracted from its final amount in the 100 ml after shaking. The final content/solid matter ratios were also determined.

After pH, Eh, salinity, O_2 and chlorophyll *a* determinations, samples of the stirring and shaking experiments were filtered through pre-weighed, 0.45 μm Nucleopore filters. Then filters were washed with Milli-Q water, freeze-dried, and weighed again. The resulting weights were used to determine the resuspended particle concentrations. The filtered solutions and solid samples were analyzed for the same sediment parameters as above. No algal growth was found during either the stirring or shaking experiments.

3.3. CHEMICAL ANALYSIS

3.3.1 Water analysis

Dissolved oxygen and salinity were measured by suitable electrodes; chlorophyll *a* was determined by means of a fluorimeter. Dissolved phosphate, ammonia, nitrite and nitrate were determined using the colorimetric technique of Strickland and Parsons (1972). Alkalinity was determined by titration with HCl (Gieskes and Rogers, 1973); Total carbonate was calculated from alkalinity, pH, salinity and temperature. Sulphides were removed by adding $ZnCl_2$ first to prevent oxidation to sulphates and then determined with the method of Cline (1969). Sulphates were measured by the indirect titration of precipitated barium sulphates (Howarth, 1978). Fe, Mn, Ca, Mg and other trace metals (Spagnoli, 1993) were determined on the acidified aliquot by means of ICP-AES.

3.3.2 Solid phase analysis

Grain size analysis was performed after the complete destruction of aggregates by hydrogen peroxide and sonication. The > 63 μm fraction was separated from the mud by wet sieving, and the silt-clay fractions were subsequently analysed by means of an X-ray Sedigraph. Porosity was calculated from the sediment weight loss after drying at 60°C (Berner, 1971). Mineralogical composition was determined using X-ray diffraction analysis (Cook *et al.*, 1975). Organic carbon and total nitrogen were determined using a CHN elemental analyser after removal of the inorganic carbon with HCl (Froelich, 1980).

Iron speciation was determined using the procedure of Forstner and Salomons (1980) which was integrated with the procedure of Tessier *et al.* (1979). This method (F&T) (Barbanti and Sighinolfi; 1988) separates and quantifies the exchangeable fraction (exc), the carbonate fraction (carb), the easily and moderately reducible fraction (red), the organic and sulphide fraction (oxid) and the residual mineral fraction (res) of metals associated with the solid phase. The total iron content was determined by total digestion of the sediment samples with concentrated mineral acid mixtures (Guerzoni *et al.*, 1984). Fe concentrations were determined in extracts by means of ICP-AES.

P speciation was obtained using two different selective, sequential chemical extraction techniques: the Golterman and Booman method (1988) as modified by Barbanti *et al.* (1994) and the F&T procedure satisfactorily tested by Barbanti and Sighinolfi (1988) for P. The modified Golterman and Booman sequence (G&B) identifies exchangeable P (P-exc), P bound to Fe (P-Fe) and P bound to Ca (P-Ca). Total P was determined with the same method used for total iron. Organic P (P-org) in G&B was determined as the difference between total P and inorganic P. Reactive phosphates in all extracts were determined using the colorimetric technique of Strickland and Parsons (1972). The total P extracts were pre-treated by digestion with fuming HNO_3-$HClO_4$.

4. Results and discussion

4.1. DIAGENETIC PROCESSES

The distribution of organic matter degradation products in pore waters (total carbonate, ammonia and phosphate) (Fig. 3) shows a near-surface peak and a regular increase at deeper sediments levels. The anoxic degradation of organic matter by the successive reduction of Mn, Fe-oxyhydroxides and sulphates takes place just below the sediment-water interface; this is supported by the negative Eh values and by the blackish grey colour of the sediments, the typical shade of anoxic conditions. The slight and gradual increase of total carbonate, ammonia and phosphate at deeper levels can be accounted for only by a small sulphate depletion, indicating that the reactivity of the buried organic carbon is probably too low to drive significant sulphate reduction. The oxic degradation of organic matter is confined to the sediment-water interface.

The diagenesis of phosphorus deserves special consideration, since its concentration in pore waters is controlled not only by the organic matter degradation, but also by the dissolution-precipitation of Fe-oxyhydroxides with which phosphate tends to associate under oxic conditions (Berner, 1980; Krom and Berner, 1981). The influence of the reduction process of Fe-oxyhydroxides on phosphate regeneration in the uppermost sediments is confirmed by the profile of the P-Fe fraction (Fig. 4): the P-Fe concentration decreases rapidly in the solid phase and a rapid rise in pore water phosphate concentrations occurs at the same time. Below the 4 cm level phosphate the pore water phosphate increases gradually, while the P-Fe remains constant; this suggests that the contribution from the organic matter degradation process prevails. The precipitation of authigenic apatite is another important process controlling the phosphate concentration in pore waters. The slight increase of P-Ca fraction with depth (Fig. 4) indicates a diagenetic removal of P for precipitation with the authigenic solid phase (Ruttenberg and Berner,

1993). Furthermore, calculations of saturation degrees (Spagnoli, 1993) confirm the possibility of authigenic fluoroapatite precipitation in sediments.

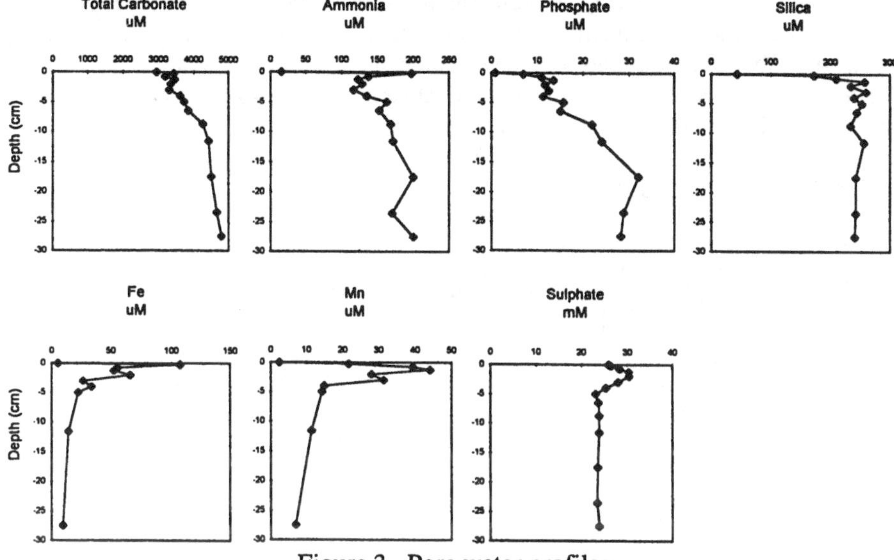

Figure 3. Pore water profiles

We calculated the saturation index of the other minerals, sulphides excepted, which are more common in anoxic marine sediments: calcite, siderite, struvite, vivianite and rhodochrosite. Pore waters were oversaturated with calcite a few centimeters below the interface. Silica concentrations increase in the uppermost sediments because of the dissolution of biogenic silica, until reaching an equilibrium with the solid phase (Sholkovitz, 1973).

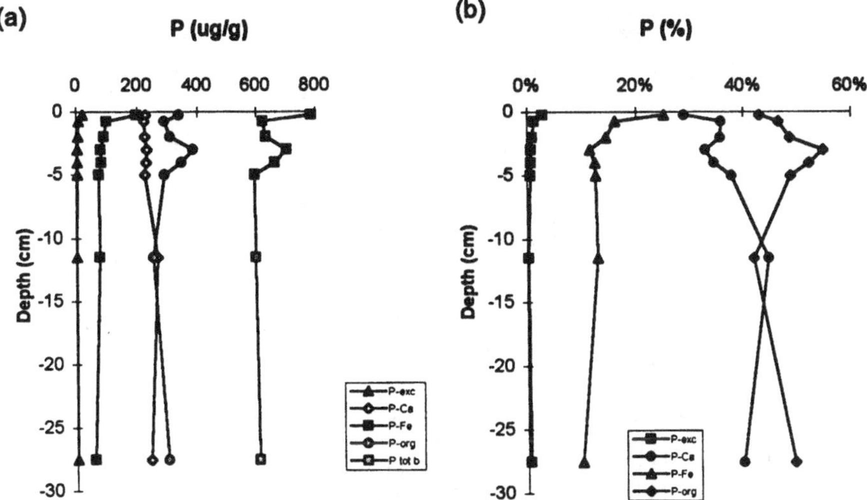

Figure 4. P speciation in sediment core following modified Golterman and Booman sequence: a) P contents in different fractions; b) percentage values with respect to total P content.

4.2. LABORATORY EXPERIMENT

At the beginning of the stirring experiment the concentrations of all chemicals were irregular, because the replacement of the supernatant with fresh seawater caused a slight resuspension of bottom sediments and the appearance of more oxic conditions.

During resuspension (Fig. 5) the phosphate release was proportional to the resuspended solid matter. In the few days following resuspension the phosphate concentration increases in the supernatant from the C core (low hydrodynamic conditions) exceeding that for the A and B cores.

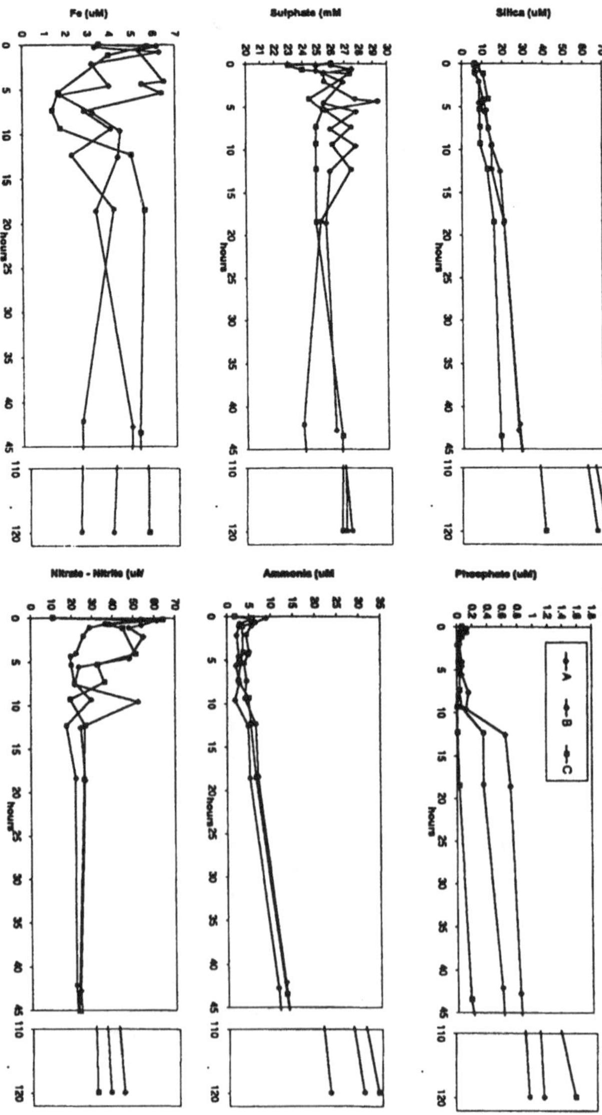

Figure 5. Dissolved chemical concentrations versus time in A,B, and C cores of the stirring experiment.

In the shaking experiment, the net phosphate concentration increased (Fig. 6). This indicates that phosphate was released from the solid phase: such release was greater in the three thinner slices. The relative phosphate release per gram of shaken sediment was much higher in the thinnest slice and proportional to the shaking time in the 0.1 and 0.5 cm slices.

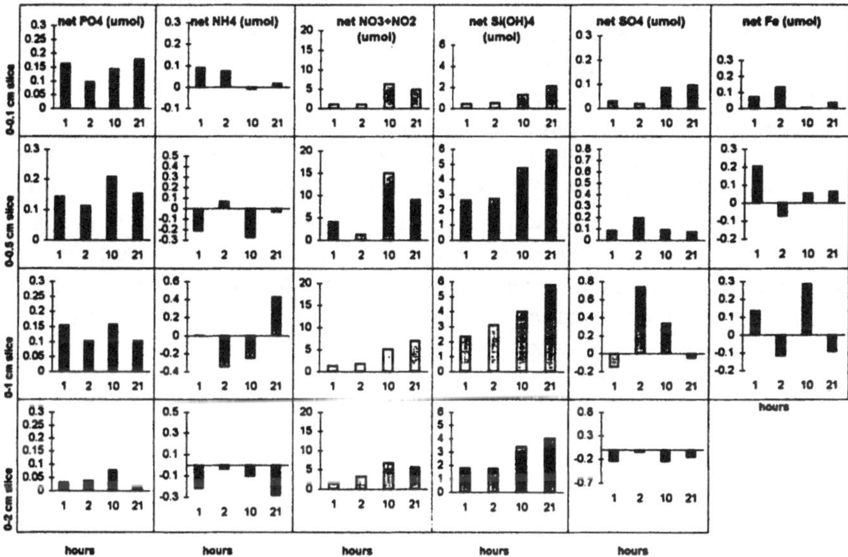

Figure 6. Concentrations of dissolved chemicals in the supernatants versus shaking time in the shaking experiment. Net amounts of chemicals are obtained by subtracting the pore water and initial contents from final quantities in the supernatant.

The results of the two sequential extraction procedures (Fig. 7) used in the shaking experiment show that the P-exc fraction decreased in time in the 0.5 cm slice, suggesting that the phosphate release during resuspension was caused partly by desorption of the P weakly bound to the solid surfaces. Furthermore, the P-Ca fraction decreased in the solid phase from the 0.5 and 1 cm slices and in the surface sediments taken at the end of the stirring experiment. This indicates that P was released because of the dissolution of the fresh authigenic apatite which precipitated near the interface. At the same time, the P-Fe fraction increase in the solid phase of the shaken samples, indicates that P precipitated together with Fe-oxyhydroxides. The high concentrations of the P-red (associated with the Fe-oxyhydroxide) fraction in the suspended solids of the F&T sequences indicates that P probably precipitated with this mineral phase. However, the extent of this P sinking process was lower than the release processes of the other P fractions. Undoubtedly, the high P-red concentrations were caused partly by the diagenetic enrichment process near the interface. The existence of such a process is confirmed also by the fact that the high P-total concentrations found in the resuspended solids and in the samples from the thinnest slice are in contrast to the general phosphate releases recorded at the end of the shaking of the 0.1 cm slice samples. The P-exc concentrations increased in the shaken samples of the thicker slices and were higher than in the samples of the thinnest slice. This was due to the stronger adsorption processes determined by the presence of partially dissolved Fe-oxyhydroxides at deeper levels.

This observation together with the P-Fe fraction increase and the reduced phosphate release in the samples of the thickest slice suggest that the phosphate release is low when resuspension involves anoxic sediments and is greater when resuspension involves oxic sediments. The oxidizing conditions in the water column determine a considerable removal of phosphate, because there are larger amounts of dissolved Fe which precipitates in oxyhydroxide forms, as well as numerous free sites for adsorption on the surface of partially dissolved oxyhydroxides.

The increase of P-Fe and P-red fractions in the surface sediments of the A and B core indicate that after resuspension the oxidizing conditions at the sediment-water interface of both cores lead to the formation of a Fe-oxyhydroxide film acting as a chemical trap for the P which diffuses into the water. The same indication comes from the after-resuspension phosphate fluxes detected for the A and B cores which were lower than the sediment-water interface fluxes of the incubated C core. The decrease of the P-Fe, P-red, and P-exc fractions in the surface sediments from the C core indicates reducing conditions at the sediment-water interface which cause the dissolution of Fe oxyhydroxides. These in turn release the P bound to them.

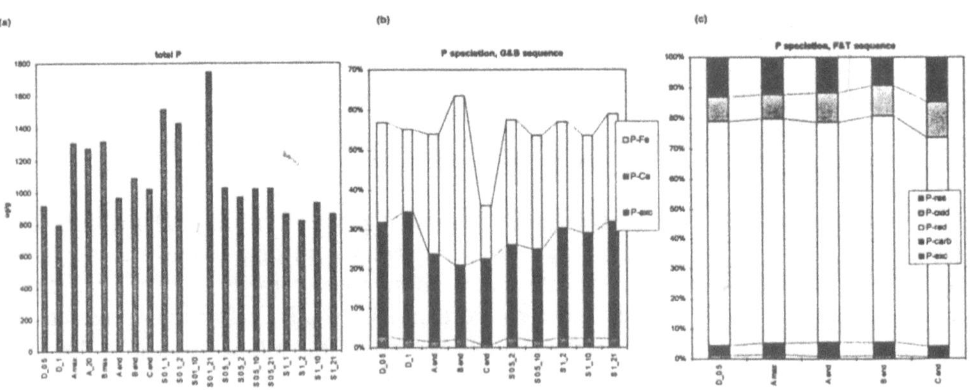

Figure 7. Total phosphorus concentrations (a) and percentages of (a) obtained with the G&B (b) and F&T (c) sequences. D_0.5 and D_1 indicate solid phase samples before the stirring and shaking experiment (number is sample thickness). A,B, and C = stirring experiment samples, max = suspended solid sample at end of maximum stirring, end = surficial sediment sample at end of stirring, S= solid phase samples from shaking experiment where first number is thickness and second number is shaking time.

During resuspension the ammonia concentrations (Fig. 5) increased slightly while in the shaking experiment the ammonia release (Fig. 6) was relevant in the thinnest slice. The lower ammonia increase in the cores A and B with respect to the shaken experiment was due to a partial nitrification of ammonia. This was confirmed by the nitrite+nitrate increase in the stirring and in the shaking experiments (Fig. 5 and 6). The ammonia release in the both experiments is caused by organic matter degradation which is stronger near the interface where the content of reactive organic matter is greater.

After resuspension in cores A and B, the ammonia concentration gradient was less than in the incubated C core. In fact, the nitrification process was stronger for the A and B cores, because more oxidizing conditions were created by stirring the supernatant. Also in this situation the nitrification process is confirmed by the trends of nitrite+nitrate: their gradients are proportional to the degree of stirring (greater for A, medium for B and lower for C) and inverse to the ammonia gradient (greater for C, medium for B and lower A). The resuspension in the water column causes the release of nitrogen as ammonia which then tends to be transformed successively into nitrite and nitrate.

During resuspension the dissolved silica concentration (Fig. 5) shows a slight increase which continues also during the final incubation step. Indeed, for the C core the dissolved silica gradient increased less than for the A and B cores.

In the shaking experiment samples (Fig. 6) the dissolved silica concentration increased with shaking time. The release of silica was caused by its desorption from the solid phase (Sholkovitz, 1973) and by the dissolution of fresh biogenic silica, the content of which was greater in the first millimeters of sediments. In fact, the released silica amount was lower for the thicker slices. The silica increase observed for the C core and, after the end of resuspension, also for the A and B cores was mainly caused by diffusive fluxes from pore waters.

During resuspension, the sulphate concentrations (Fig. 5) increased slightly and then decreased to steady values similar to those determined in the initial stirring step and for the C core. In the two thinner slices of the shaking experiment (Fig. 6) a slight sulphate release was noticed. This was caused by the dissolution of the unstable sulphides which were poorly crystallised and present in larger amounts near the interface (Canfield, 1989) because of their recent precipitation.

With respect to alkalinity and calculated total carbonate, it must be said that the stirring experiment was set up as an open system; therefore, the dissolved oxygen and CO_2 reached an equilibrium with the atmosphere and it was impossible to observe the processes occurring at the sediment-water interface. In fact, these two parameters had only minor variations. However, during resuspension a slight increase in alkalinity concentrations could be identified, which was attributable mainly to organic matter oxidation and ammonium nitrification. In the shaking experiment alkalinity concentrations decreased in all sample slices, perhaps because the Fe-oxyhydroxide reduction processes prevailed over the oxidation of sulphides and ammonium (Berelson *et al.*, 1987a).

During resuspension iron concentrations (Fig. 5) were variable. The correlation between sulphate and iron concentrations for the A and B cores was nearly perfect. This suggests that any sudden Fe concentration increase produced by the dissolution of particularly unstable Fe-sulphides was compensated by the successive re-precipitation of Fe as oxyhydroxides.

After resuspension the Fe concentration decreased in the A and B cores. In the incubated C core the Fe concentration had a rapid, initial increase because of the fast dissolution of the fresh Fe-oxyhydroxides formed immediately after the bottom water replacement. In the following days the dissolution of more crystalline Fe-oxyhydroxides and the diffusive fluxes from the sediment-water interface caused a weak increase in the Fe concentration.

In the shaking experiment, Fe concentrations (Fig. 6) increased slightly in the two thinner slices. The reason for this is that the oxidation of newly formed, poorly cristallized sulphides (present in larger amounts in the first millimeters of sediment) dominated over the Fe-oxyhydroxide precipitation. On the contrary, this precipitation is prevalent at deeper levels because of the larger quantities of dissolved Fe and adsorption sites on partially dissolved Fe-oxyhydroxides.

The total Fe content in the solid phase (Fig. 8) indicates that considerable amounts of this element was present in both resuspended solids of A and B cores and the 0.1 cm slice samples because of Fe reprecipitation and recycling at the sediment-water interface. The results of sequential extractions do not help in understanding these Fe-related processes, since resuspension created Fe concentration variations in the solid phase which were lower than analytical error. Only the Fe-carb and Fe-oxid fractions have recognizable trends. The Fe-carb fraction decreased in the surficial sediments collected from the cores after the stirring experiment and in all shaked subsamples. This decrease indicates that Fe is desorbed from this mineral phase. The Fe-oxid fraction decreased in the 0.5 cm slice, whereas it was almost constant in the remaining slices; this confirms that the oxidation processes of sulphides and organic matter prevail at surficial levels. The Fe-res fractions and the Fe-red fractions from the thicker slices were higher and lower, respectively, than the corresponding fractions detected from the thinnest slice. The reason for this is the ageing process of Fe-oxyhydroxides which, with increasing depth and time, acquire a more stable crystal form.

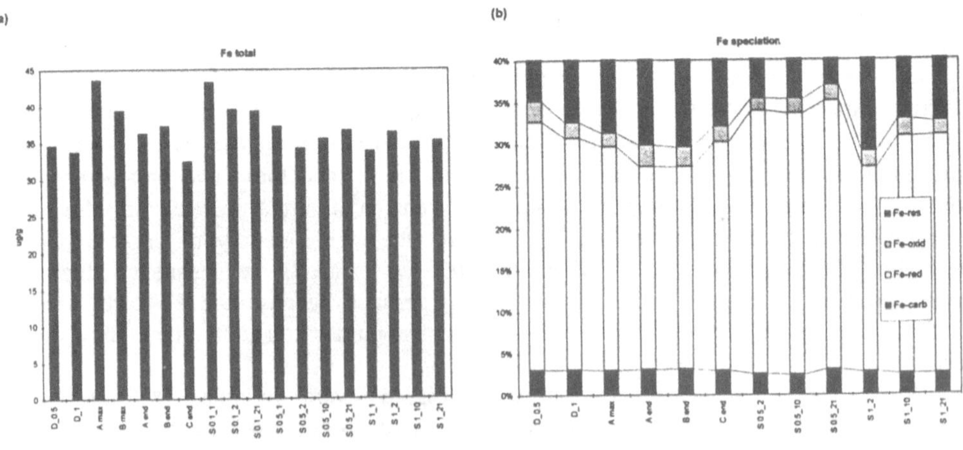

Figure 8. Total iron concentrations (a) and percentages of (a) obtained with the F&T sequence (b). A,B,C, max, end and S are the same as indicated in Figure 7.

In general, Fe tended to precipitate as oxyhydroxides during resuspension in oxidizing environments. Such phenomena would reverse following Fe-sulphide oxidation and, to a lesser extent, Fe desorption from the carbonate fraction. Fe precipitation was fast but limited, and this perhaps was caused by the complexation of Fe with organic and inorganic colloids which were released during resuspension and prevented Fe precipitation (Hunter and Liss, 1982). After resuspension (cores A and B), the oxidizing conditions at the water-sediment interface promoted the precipitation of Fe-oxyhydroxides. On the contrary, a reducing environment determined by stable hydrodynamic conditions (core C) promoted the dissolution of Fe-oxyhydroxides.

4.3. CALCULATED FLUXES

The fluxes at the sediment-water interface were calculated applying Fick's First Law (Berner, 1980): $J_i = -\phi D_{i(sed)} (\frac{\delta C_i}{\delta z})_{z=0}$ where J_i is the diffusive flux of the solute *i* (mmol m^{-2} day^{-1}), ϕ is the porosity, $D_{i(sed)}$ is the bulk sediment diffusion coefficient of the solute *i* (cm^2 s^{-1}) and $\delta C_i / \delta z$ is the concentration gradient of the solute *i* at depth *z*=0. The bulk sediment diffusion coefficient was calculated as proposed by Ullman and Aller (1982) for nearshore sediment when porosity is > 0.7. The free solution diffusion coefficient of solute *i* (D_i) (Wollast and Garrel, 1971; Li and Gregory, 1974) was corrected for temperature and salinity with the function $D_i = D_0(1+aT)(1-bS)$, where D_0 is the free solution diffusion coefficient at temperature 0°, *T* is the temperature, *S* is the salinity, *a* and *b* are two interpolation constants. To calculate the concentration gradients at the sediment-water interface we applied an exponential function describing the pore water profile of dissolved silica and a quadratic function for phosphate, nitrate and Fe (Berelson *et al.*, 1987a). Then we derived the appropriate function for depth *z*=0. For dissolved silica we have: $C = C_\infty(1-e^{-bz}) + C_0(e^{-bz})$, where C is the silica concentration at depth z, C_∞ is the silica concentration at the solid-liquid phase equilibrium and b is a regression constant. The concentration gradient for depth z=0 is: $b(C_\infty - C_0)$, where C_0 is the silica concentration at z=0. For phosphate, nitrate and Fe we have: $C_i = C_0 + a_1 z + a_2 z$ where C_i is the concentration of solute *i*, and a_1 and a_2 are two regression constants. The concentration gradient for depth z=0 is a_2. If only one useful sample is available near the water-sediment interface (such as for ammonia), we calculate a linear concentration gradient (Klump and Martens, 1981). The fluxes are then corrected for the benthic boundary layer effect by successive approximations. The concentration gradients were calculated using the *in situ* bottom water concentrations.

Table 1. Calculated and measured fluxes at the sediment-water interface (positive values indicate upward fluxes).

Fluxes (mmol m^{-2} d^{-1})	phosphate	ammonia	nitrate	silica	Fe
Calculated	0.14	5.66	0.09	3.16	0.27
Measured	0.19	2.69	0.71	2.40	0.03
Measured after resuspension	0.05	1.27	1.27	3.07	A -0.14 B -0.03

The calculated phosphate fluxes are comparable to the measured ones. This means that diffusion is the major transport mechanism. However, there is a flux component due to the degradation of organic matter near the sediment-water interface. This component can balance phosphate precipitation with Fe-oxyhydroxides. The phosphate fluxes measured after resuspension are smaller than the calculated ones by one order of magnitude, because of the precipitation of Fe-oxyhydroxides which act as sinks for phosphate.

The difference between calculated and measured ammonia fluxes is attributable to the nitrification which takes place in the supernatant where oxidizing conditions occurs In the A and B cores the nitrification process was stronger and fluxes were lower accordingly. The nitrification process is evidenced by the nitrate fluxes determined during incubation and after resuspension which are higher than the calculated nitrate fluxes. In low hydrodynamic conditions the ammonia produced in the sediment by organic matter degradation flows out of sediments mainly by molecular diffusion.

Silica fluxes support the prevalence of diffusive process. Calculated iron fluxes are higher than the measured ones by one order of magnitude. This difference is attributed to a process which took place near the sediment-water interface but was not recorded in the core sampled, i.e. the precipitating Fe-oxyhydroxide induced sinking of the iron flowing out of sediments. This same process does not influence the measured and calculated phosphate fluxes, because it was balanced by the production of organic matter degradation. The more oxic conditions following resuspension favour a considerable Fe-oxyhydroxide precipitation which determines the Fe negative flux.

5. Conclusions

During resuspension, phosphate release occurs mainly by desorption from solid surfaces and dissolution of authigenic apatite. If the resuspension event involves the oxic layer of sediments, the phosphate release is proportional to the resuspended matter. If the resuspension event involves the anoxic layer too, the phosphate release is more limited. Following resuspension, the oxic environment favours the precipitation of Fe-oxyhydroxides at the sediment-water interface, which act as sinks for phosphate. This process creates lower phosphate fluxes than those measured in low hydrodynamic conditions. The phosphate fluxes are mainly diffusive in low hydrodynamic conditions, although the degradation of organic matter at the sediment-water interface gives some contribution.

During resuspension, ammonia is released because of the degradation of organic matter, but the more oxic conditions favour the fast transformation of ammonia into nitrate. After resuspension, ammonia diffusive fluxes prevail, as in a low hydrodynamic environment, but the oxic conditions of waters cause a strong nitrification process.

During resuspension, silica is released mainly because of biogenic silica dissolution. In low hydrodynamic conditions the silica fluxes are mainly diffusive. During resuspension, iron sulphides with low cristallinity dissolve and iron associated with the carbonate fraction desorbs. Only a portion of the free iron precipitates as Fe-oxyhydroxides, because some iron becomes bound with organic colloids. After resuspension, the removal of Fe from the water column prevails because of the precipitation of Fe-oxyhydroxides at the sediment-water interface. In low hydrodynamic conditions low diffusive Fe fluxes prevail over the precipitation of Fe-oxyhydroxides at the sediment-water interface.

Acknowledgements

We wish to thank Dott.ssa Franca Frascari of the Istituto per la Geologia Marina-CNR for her technical, spiritual and financial support during this research; Prof. Antonio Brambati of the Trieste University for funding the grant which made this work possible; Dott. Stefano Miserocchi, Mr. Angelo Magagnoli, Mr. Maurizio Mengoli and Ms. Gabriella Rovatti of the CNR Institute of Marine Geology, as well as Dott.ssa Carla Ferrari of the U.O. Daphne of the Emilia Romagna Regional Administration for their assistance with the laboratory analyses and technical support. This is contribution n. 1070 of Istituto di Geologia Marina-CNR, Bologna, Italy.

References

Aller, R.C.: 1980, *Advances in Geophysics* **22**, 237-350.

Artegiani, A., Bregant, D., Paschini, E., Pinardi, N., Raichic, F. and Russo, A., *J.Physical Oceanogr.*, in press.

Barbanti, A. and Sighinolfi, G.P.: 1988, *Environ. Technol. Lett.* **9**, 127-134.

Barbanti, A., Bergamini, M.C., Frascari, F., Miserocchi, S. and Rosso, G.: 1994, *J. Environ. Qual.* **23**, 1093-1102.

Barbanti, A., Bergamini, M.C., Frascari, F., Miserocchi, S., Ratta, M. and Rosso, G.: 1995, *Mar. Freshwater Res.* **46**, 55-67.

Bender, M., Jahnke, R., Weiss, R., Martin, W., Heggie, D.T., Orchardo, J. and Sowers, T.: 1989, *Geochim. Cosmochim. Acta* **53**, 685-697.

Berelson, W.M., Hammond, D.E. and Johnson, K.S.: 1987a, *Geochim. Cosmochim. Acta* **51**, 1345-1363.

Berelson, W.M., Hammond, D.E., Smith, K.L., Jahnke, R.A., Devol, A.H., Hinga, K.R., Rowe, G.T. and Sayles, F.: 1987b, *Mar. Technol. Soc. J.* **21**, 26-32.

Berelson, W.M., Hammond, D.E., O'Neill, D., Zu, .M., Chin, C. and Zukin, J.: 1990, *Geochim. Cosmochim. Acta* **54**, 3001-3012.

Berelson, W.M., Hammond, D.E., McManus, J. and Kilgore, T.E.: 1994, *Global Biogeochem. Cycles* **8**, 219-235.

Berner, R.A.: 1971, *Priciples of chemical sedimentology*, McGraw-Hill (eds.), New York, p. 240.

Berner, R.A.: 1980, *Early diagenesis: a theoretical approach*, Princeton Univ. Press (eds.), N.J., p. 241.

Calvo, C., Donazzolo, R., Guidi, F. and Orio, A.: 1991, *Water Res.* **25**, 1295-1302.

Canfield, D.E.: 1989, *Geochim. Cosmochim. Acta* **53**, 619-632.

Cline, J.D.: 1969, *Limnol. Oceanogr.* **14**, 454-458.

Cook, H.E., Johnson, P.D., Matti, C. and Zemmels, I.: 1975, *Methods of sample preparation and X ray diffraction data analysis. X ray mineralogy laboratory, deep-sea drilling project*, University of California, Riverside, Init. Rep. DSDP **28**, 999-1007.

Crema, R., Castelli, A. and Prevedelli, D.: 1991, *Mar. Pollut. Bull.* **22**, 503-508.

Dal Cin, R.: 1983, *Boll. Soc. Geol.* **102**, 9-56.

Drake, E.D., Cacchione, D.A., Frascari, F., Bortoluzzzi, G., Tate, G., Miserocchi, S., Thede, J. and Viall, R.: 1992, *Rapp. Comm. Int. Mer Medit.* **33**.

Fanning, K.A., Carder, K.L. and Betzer, P.R.: 1982, *Deep-Sea Res.* **29**, 953-996.

Forstner, U. and Salomons, W.: 1980, *Environ. Technol. Lett.* **11**, 494-517.

Franco, P., Jeftic, L., Malanotte Rizzoli, P., Michelato, A. and Orlic, M.: 1982, *Oceanologica Acta* **5**, 379-389.

Frignani, M. and Turci, C.: 1981, *CNR, AQ/2/12* **101**.

Frignani, M. and Langone, L.: 1991, *Continental Shelf Res.* **11**, 525-542.

Froelich, P.N.: 1980, *Limnol. Oceanogr.* **25**, 242-248.

Gieskes, J.M. and Rogers, W.C.: 1973, *J. Sedim. Petr.* **43**, 272-277.

Giordani, P. and Hammond, D.E.: 1985, *Technical Report No. 20*, Consiglio Nazionale delle Ricerche, Istituto per la Geologia Marina (Bologna), p. 33.
Giordani, P., Hammond, D.E., Berelson, W.M., Montanari, G., Poletti, R., Milandri, A., Frignani, M., Langone, L., Ravaioli, M., Rovatti, G. and Rabbi, E.: 1992, *Sci. Total Environ.* **Supplement**, 251-275.
Golterman, H.L. and Booman, A.: 1988, *Verh. Int. Ver. Limnol.* **23**, 904-909.
Guerzoni, S., Frignani, M., Giordani, P. and Frascari, F.: 1984, *Environ. Geol.* **6**, 111-119.
Hammond, D.E., Fuller, C., Harmon, D., Hartman, B., Korosec, M., Miller, L.G., Rea, R., Warren, S., Berelson, W. and Hager, S.W.: 1985, *Hydrobiologia* **129**, 69-90.
Hammond, D.E., McManus, J., Berelson, W., Kilgore, T. and Pope, R., *Deep-Sea Res.*, in press.
Howarth, R.W.: 1978, *Limnol. Oceanogr.* **23**, 1066-1069.
Hunter, K.A. and Liss, P.S.: 1982, *Limnol. Oceanogr.* **27**, 322-335.
Klump, V.J. and Martens, C.S.: 1981, *Geochim. Cosmochim. Acta* **45**, 101-121.
Krom, M. and Berner, R.A.: 1981, *Geochim. Cosmochim. Acta* **45**, 207-216.
Li, Y. and Gregory, S.: 1974, *Geochim. Cosmochim. Acta* **38**, 703-714.
Marchetti, R.: 1992, *Sci. Total Environ.* **Supplement**, 21-33.
Nelson, B.W.: 1970, *Deltaic sedimentation - modern and ancient*, Morgan T. Ed., Soc. Ec. Paleon. Sp. Pubbl. **15**, 152-184.
Orlich, M., Gacic, M. and La Violette, P.E.: 1992, *Oceanologica Acta* **15**, 109-124.
Oviatt, C.A., Hunt, C.D., Vargo, G.A. and Kopchynski, K.W.: 1981, *J. Mar. Res.* **34**, 605-626.
Petersen, W., Wallmann, K., Pinglin, Li, Schroeder, F. and Knauth, H.D.: 1995, *Mar. Freshwater Res.* **46**, 19-26.
Ruttenberg, K.C. and Berner, R.A.: 1993, *Geochim. Cosmochim. Acta* **57**, 991-1007.
Santschi, P., Hohener, P., Benoit, G. and Bucholtz-ten Brink, M.: 1990, *Mar. Chem.* **30**, 269-315.
Sfriso, A., Donazzolo, R., Calvo, C. and Orio, A.A.: 1990, *Environ. Technol.* **12**, 371-379.
Sholkovitz, E.: 1973, *Geochim. Cosmochim. Acta* **37**, 2043-2073.
Spagnoli, F.: 1993, *Diagenesi precoce e processi di scambio di sostanze nutrienti e metalli in tracce tra acqua e sedimento in condizioni di quiete e di risospensione di un'area marina a sud del delta del Po*, PhD Dissertation, Univ. Trieste, Italy.
Strickland, J.D.H. and Parsons, T.R.: 1972, A practical Handbook of seawater anakysis, Fish. Res. Board Canada, Ottawa, p. 310.
Sundby, B., Anderson, L.G., Hall, P.O.J., Iverfeldt, A., Rutgers Van der Loeff, M.M. and Westerlund, S.F.G.: 1986, *Geochim. Cosmochim. Acta* **50**, 1281-1288.
Tessier, A., Campbell, P.C.G. and Bisson, M.: 1979, *Anal. Chem.* **51**, 844-850.
Tsai, C.H. and Lick, W.: 1986, *J. Great Lakes Res.* **12**, 314-321.
Ullman, W.J. and Aller, R.C.: 1982, *Limnol. Oceanogr.* **27**, 552-556.
Westerlund, S.F.G., Anderson, L.G., Hall, P.O.J., Iverfeldt, A., Rutgers Van der Loff, M.M. and Sundby, B.: 1986, *Geochim. Cosmochim. Acta* **50**, 1289-1296.
Westrich, J.T. and Berner, A.: 1984, *Limnol. Oceanogr.* **29**, 236-249.
Wollast, R. and Garrels, R.: 1971, *Nature London Phys. Sci.* **229**, 94.

FLUXES OF SUSPENDED MATERIALS IN THE NORTH ADRIATIC SEA (PO PRODELTA AREA)

Gabriele Matteucci and Franca Frascari
Istituto di Geologia Marina, Consiglio Nazionale delle Ricerche, Bologna, Italy

Abstract

Two time-series sediment traps were deployed south-east of the Po River delta (North Adriatic Sea) in 23m of water at 12 meters and 21 meters depths from 12/20/89 to 06/26/90. Temperature, conductivity, salinity, pH, Eh, O_2, transmittance profiles, wave height, speed and direction of currents were measured. Bottom sediments and river suspended matter were periodically collected during the same period. Organic carbon, Al, Fe and Mn concentrations and partitioning of Phosphorous on the solid phase in sediment samples were determined. This made it possible to calculate vertical fluxes at two water levels, and then to quantify and separate vertical fluxes of sinking particles into fluxes of lateral transports (river input), primary production and resuspension. To do so, the approach was based on the identification, comparison and integration of two different label substances (POC and Mn). This led to the finding that sediment fluxes and biogeochemical characters of suspended matter are in good agreement with seasonal differences and short-term events such as river floods, algal blooms and marine storms. A decrease in organic carbon along the water column (5.13% mean value at 12 m, 2.84% mean value at 21 m) with minimal values in the bottom sediment (1.49% mean value) was detected. Resuspension and re-oxygenation increase the mineralization of organic matter (occurring already in water column). Differences in adsorption and desorption in the fine particulates of fundamental elements for primary production such as phosphorous were detected at different levels in the water column and in the bottom sediments.

Keywords: suspended sediment, transport, resuspension, Mn, POC

1. Introduction

The Po river is the largest Italian river (mean liquid discharge: 1500 m^3/sec; mean solid discharge: 14 million tonnes/yr, Dal Cin, 1983). It drains the most industrialized Italian regions and influences the particle, nutrient and pollutant budgets of the entire Adriatic Sea. Only a few documented studies exist concerning transport, deposition, resuspension and transformation of the organic and inorganic matter in the marine regions influenced by the Po river (Pigorini, 1968; Nelson, 1970; Frascari *et al.*, 1988; Puškaric *et al.*, 1992). In the period from winter 1989 to spring 1990 a systematic sampling programme on the vertical fluxes of suspended particles was set up in a site located SE of the Po river delta. Sampling operations involved the deployment of two suspended sediment traps at two depth levels.

The objectives of the research were: a) the quantitative evaluation of particle fluxes along the water column; b) the differentiation of direct river input, primary production supply and resuspension contributions; c) the assessment of particle composition changes with respect to both bottom sediments and seasonal variations; and d) the evaluation of how some elements, in particular P, are bound to particles at different water levels.

Water, Air and Soil Pollution **99**: 557-572, 1997.

2. Materials and methods

2.1. STUDY AREA

The water circulation in the Northern Adriatic Sea is influenced strongly by thermohaline factors which determine a markedly cyclonic circulation in winter. In summer the water column is stratified and the water circulation becomes more complex because local oceanographic conditions may prevail (Franco *et al.*, 1982; Orlich *et al.*, 1992; Artegiani *et al.*, in press). The input from the Po river characterizes the sedimentation throughout the area in front of the delta (Boldrin *et al.*, 1988). Only in some prodelta zones and at particular times are there contributions from northern areas because of the cyclonic circulation (data in press).

The selected study site (Figure 1) lies in a prodelta area which is influenced by the Po river throughout the year. The local grain size of the 23 m deep seabed is characterized as silty clay. The position of the site was marked by buoys and instruments were deployed during the period from December 1989 to June 1990 (Figure 2).

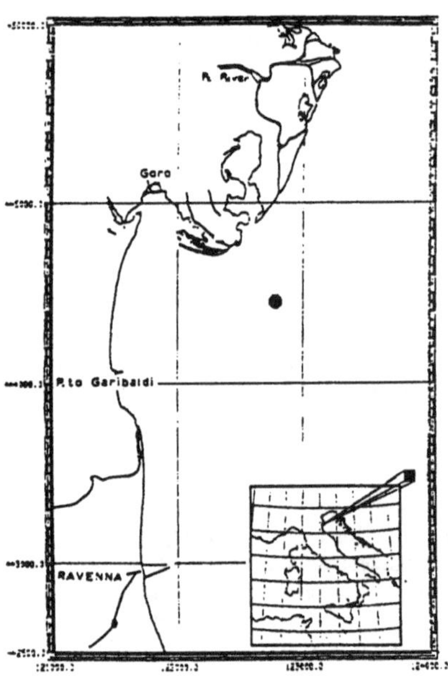

Figure 1. Study area and site.

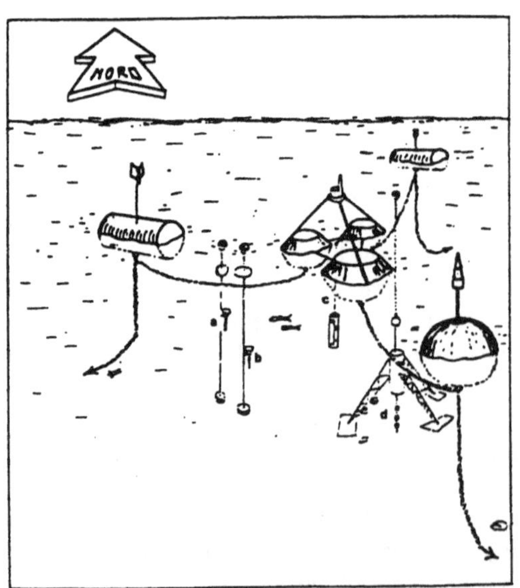

Figure 2. Instrumental set-up. a) Sediment trap at 12 m depth; b) sediment trap at 21 m depth; c) CTD profiler; d) tripod with the instruments for measuring oceanographic physical parameters at the sediment-water interface.

2.2. WATER COLUMN MEASUREMENTS

Salinity, temperature, dissolved oxygen and pH were measured by means of an Ocean Seven 501 probe. This was associated with a Sea Tech Inc., Side view 25 cm transmissometer used for turbidity determinations. From December 25, 1989 to May 18, 1990 the above parameters were recorded along the entire water column with 2-hour intervals through a fully automatic CTD profiler which radio-transmitted data to a mainland station. From January 13, 1990 to May 21, 1990 a GEOPROBE tripod (Cacchione & Drake, 1979) was deployed at the sediment-water interface. This instrument produced continuous records of wave height, current speed and direction (at 18.5, 50, 87.5, and 122 cm from the sea floor), as well as total suspended matter (at 30, 100, and 220 cm from the sea floor).

2.3. SAMPLE COLLECTION

The two suspended sediment traps deployed are of the same type as described by Anderson (1977). The trap placed at 12 m from the seabed has a 25 cm diameter and a 2.10 m length; the trap placed at 21 m from the seabed has the same diameter and a 1.40 m length. The uppermost trap was programmed with 10 day sampling intervals, whereas initially the lowermost trap had sampling intervals of 7 days which were then reduced to 2 days in order to obtain a better characterisation of events with a short time scale. Each trap was provided with a dispenser of poison, sodium azide (NaN_3), which was diffused into the sea waters in the form of hypersaline solution. The NaN_3 hypersaline solution was used to keep samples as integral as possible.

Problems with the time-marking device made it impossible to separate the time intervals in the uppermost trap effectively on the first and last deployments which, therefore, produced samples of 57 days and 46 days, respectively. In the trap at 21 m depth the particles sampled in the periods March 26-30, April 09-12 and April 22-26 were underestimated, since the maximum capacity of the sampling tube was exceeded.

Field studies included also the collection of bottom sediment samples by means of box-corers.

2.4. SAMPLE PREPARATION AND CHEMICAL ANALYSIS

Trap samples were sieved through a 150 μm screen with filtered sea water taken from the site; the oversize particles were then washed with distilled H_2O to eliminate the residues of the NaN_3 hypersaline solution. After this the material was centrifuged, freeze-dried and weighed. Particulate organic carbon (POC) and particulate nitrogen (PN) were determined by means of a CHN Elemental Analyser (Carlo Erba mod. 1106) after removing inorganic carbon with 2N HCl and then drying at 60°C (Froelich, 1980; Hedges and Stern, 1984). Hereinafter the PN content data will be considered as the expression of a tendency, not as absolute values, since the NaN_3 used as poison might have contaminated the results, although it should have been removed by the sample treatments since it does not bind to particles (US GOFS, 1989). Total phosphorus (P), aluminium (Al), iron (Fe) and manganese (Mn) concentrations were determined after the samples had been completely digested by strong acids under high temperature and pressure conditions (Guerzoni *et al.*, 1984). The metal concentrations of the extracts were determined by means of atomic absorption spectroscopy. Moreover, a sequential selective extraction of P was made to determine the easily exchangeable P fraction adsorbed by particle surfaces (exchangeable P), the P fraction bound to iron minerals (Fe-bound P) and the P fraction bound to calcium minerals (Ca-bound P) following the procedure of Golterman and Booman (1988) as modified by Barbanti *et al.* (1994). Phosphorus and its fractions were determined by colorimetric technique (Strikland and Parsons, 1972). The same analytical procedure was followed for the bottom sediment samples. Table I shows the concentration values of the elements analysed.

3 Results

3.1. Po River Discharge

In 1990 the lowest average discharge of the last 73 years was recorded ($Q_{l(1917\text{-}1990)}$ 1500 $m^3\ sec^{-1}$, $Q_{l(1990)}$ 930 $m^3\ sec^{-1}$). During the sampling period (Figure 3e) it was possible to distinguish a winter season characterized by a low water regime with constant discharges ($Q_l \approx 750\ m^3\ sec^{-1}$) and a spring season with markedly higher discharges, and peaks of 2000 $m^3\ sec^{-1}$ in April and 2300 $m^3\ sec^{-1}$ in late May-early June (Regione Emilia Romagna, 1989; Regione Emilia Romagna, 1991).

3.2. Hydrology

At the beginning of the sampling period the water column (Figures 3a-b) was stratified with cold, less saline waters above and warm, more saline waters below, but in mid-February a homogenization process started and continued until the end of April. This process was caused by colder surface waters from the river which compensated for the density differences in the deeper waters, and was triggered by a minor storm (Figure 3). In springtime the increases in temperature and river discharge determined a new stratification where water temperatures were opposite to the winter ones. In the course of the study period we determined a surficial pycnocline within the first metres and a deeper pycnocline within the 8-15 m interval which involved the uppermost trap placed at 12 m depth. During the periods of marked water stratification the deeper pycnocline had strong salinity and temperature gradients (2 °C; 2 PSU) which decreased during water homogenization (0.5 °C; 0.5 PSU) and became zero only temporarily following the late March and mid-April storms.

During the period considered the dissolved oxygen also showed seasonal trends, similar to temperature and salinity. However, no dissolved oxygen deficiency was ever detected along the water column or close to the seafloor. The minimum average values of 5-6 ppm were recorded near the seabed because of the normal oxidation processes involving the organic matter at the sediment-water interface. In early and late February, as well as in early May, the greatest values for dissolved oxygen (>13 ppm) and pH (as high as 8.5 in early February) found in the upper water levels reflected three episodes of considerable primary production (algal blooms).

Table I

Fluxes and element concentrations of settling particulate matter and bottom sediments. The mean, median, maximum (max) and minimum (min) values and the number of samples (n) are given.

		gross sed. rate	POC	PN	Exchangeable P	Fe-bound P	Ca-bound P	Total P	Al	Fe	Mn
		$g\ m^{-2}\ d^{-1}$	%		$mg\ kg^{-1}$				%		$mg\ kg^{-1}$
12/20/89 - 06/26/91											
trap at 12 m $n = 9$	mean	1.45	5.13	0.77	46.9	364.2	138.8	1194	6.65	3.42	581
	median	1.88	4.75	0.66	40.2	402.5	179.1	1151	7.03	3.89	510
	max	3.08	7.29	1.40	98.4	631.4	276.4	1448	8.43	4.18	758
	min	0.24	2.21	0.32	20.2	131.3	60.6	904	4.80	2.32	499
12/20/89 - 05/21/89											
trap at 21 m $n = 64$	mean	8.27	2.84	0.35	39.1	255.4	197.9	1237	6.64	3.36	750
	median	5.48	2.46	0.32	34.1	253.7	198.7	1137	6.63	3.49	733
	max	88.73	6.52	0.86	92.1	389.8	258.7	2006	8.29	4.67	1000
	min	0.79	1.63	0.20	11.6	177.0	124.2	638	4.47	1.76	584
09/08/89 - 06/26/91											
bottom $n = 9$	mean		1.49	0.20	11.3	151.4	226.9	858	7.20	4.52	635
	median		1.49	0.19	12.1	146.3	222.5	840	7.15	4.32	611
	max		1.71	0.25	18.8	209.0	243.0	1100	8.94	6.44	979
	min		1.18	0.17	1.6	86.6	215.3	657	5.73	3.45	463

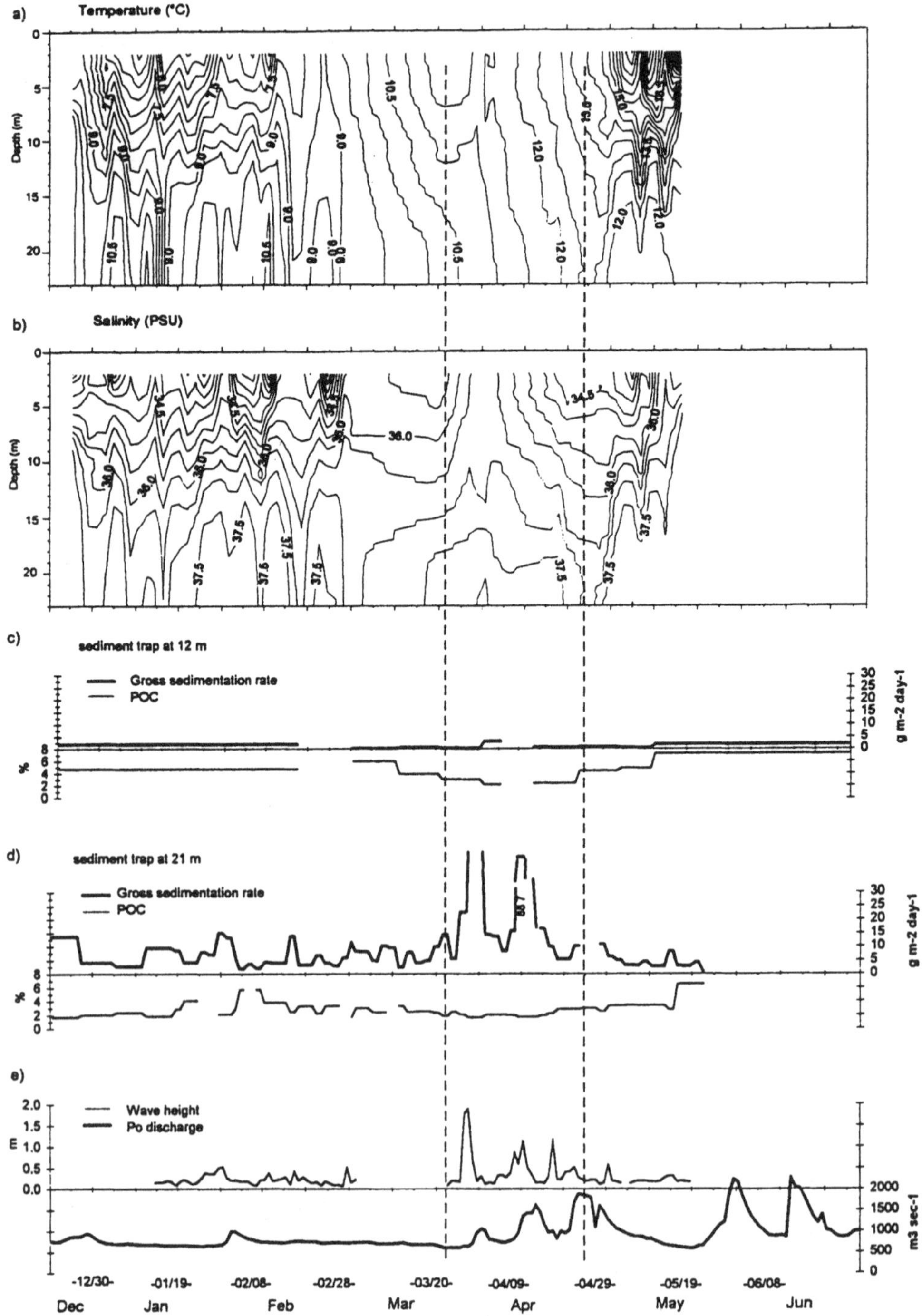

Figure 3. a) Water temperature; b) water salinity; c) gross sedimentation rate and POC of the trap at 12 metres; d) gross sedimentation rate and POC of the trap at 21 metres; e) wave height and Po river discharge. The vertical dashed lines include the period from March 22 to April 22, when meteomarine dynamics were stronger.

3.3. WAVE HEIGHT, CURRENT AND TOTAL SUSPENDED MATTER

In the periods March 23-25 and April 11-12 two N-NE storms hit the area. During these events, wave height peaks exceeded 1 m (Figure 3e) and the maximum current speeds (> 30 cm sec-1) and suspended particle amounts were measured (Figure 4).

The turbidity profiles recorded along the water column showed larger quantities of suspended particles at the two pycnocline levels (river input). At lower water levels turbidity decreased and showed a slight increase close to the seabed. The calculated Total Suspended Particulate Matter (TSM) within 2 m of the sea floor ranged from 1 to 5 mg l^{-1} with calm sea conditions. This indicates that tidal effects play a negligible role at the level of the lowermost trap. During the two storm events the TSM values were maximum (more than 40 mg l^{-1}), but they decreased rapidly with distance from the sea floor (Figures 4b-c-d). In this respect, Drake *et al.* (1992) showed that bottom particles behave like fine quartz sand (d = 0.01 cm) even if they are made of silty clay. Considering that these sediments are cohesive, it seems reasonable to assume that bottom particles are mostly resuspended in the form of large aggregates which, therefore, tend to remain close to the sea floor.

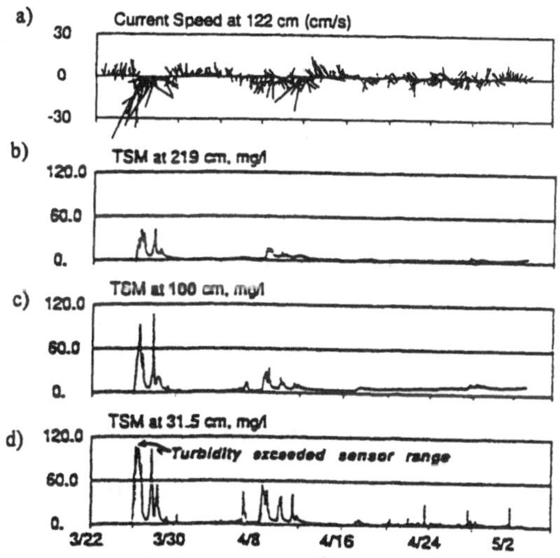

Figure 4. Current speeds and directions at 122 cm from the sea floor, and Total Suspended Matter (TSM) measured at three levels within 2 m from the sea floor, 03/22/90-05/06/90 (Reprinted from Drake *et al.*, 1992).

3.4. FLUX OF PARTICULATE MATTER

3.4.1 Gross sedimentation rate

Figures 3c-d show the fluxes which were expressed as g m^{-2} day^{-1} for both traps. Throughout the study period the lowermost trap collected more material than the uppermost trap. In winter the average particle flux for this trap was 1.49 g m^{-2} day^{-1}, while it was 6.64 for the lowermost trap. From late March to early April the former recorded a lesser flux (1.00 g m^{-2} day^{-1}), while the latter showed a three-fold increase in its flux (17.94 g m^{-2} day^{-1}) with a peak of 88.73 g m^{-2} day^{-1} during the storm of April 11-12. From late April to late June the trap deployed at 12 m depth showed its highest average flux (1.58 g m^{-2} day^{-1}); in contrast to this, the trap placed at 21 m depth (which operated until late May) registered its minimum flux (4.70 g m^{-2} day^{-1}). Throughout the period considered the gross sedimentation rate determined by the lowermost trap was closely linked with wave height: even moderate increases in wave height caused corresponding increases in the gross sedimentation rate and this shows the strong influence of resuspension on bottom sediments (Figures 3d-e).

3.4.2 Particulate organic carbon and particulate nitrogen

The POC and PN concentrations recorded by the uppermost trap (5.13% and 0.77% of dry weight, respectively) were higher than those of the lowermost trap (2.84% and 0.35%, respectively) and the sea floor, where the minimum POC and PN values were found (1.49% and 0.20%, respectively). This is in line with a photic zone limited to the upper water levels (Faganeli *et al.*, 1989; data in press). In winter the average POC and PN concentrations in the uppermost trap (4.81% and 0.68%, respectively) are twice those of the trap deployed at 21 m depth (2.74% and 0.34%, respectively). However, in early February the lowermost trap showed POC and PN peaks (5.70% and 0.86%, respectively) which exceeded average concentrations and coincided with dissolved oxygen and superficial pH peaks (algal blooms). From March 22 to April 22 there was a general decrease in the average POC and PN concentrations down to 2.59% and 0.47%, respectively, in the uppermost trap, and down to 2.09% and 0.27%, respectively, in the lowermost trap, with a marked flattening of POC differences between the two traps. In the same period the minimum POC and PN concentrations were recorded: 2.21% and 0.32%, respectively, for the uppermost trap, and 1.63% and 0.20%, respectively, for the lowermost trap. These values are very close to the average POC and PN concentrations (1.49% and 0.19%, respectively) found in bottom sediments during the same period. This is attributable to the two major storms of late March and early April which resuspended greater amounts of residual organic particles from the sea floor. The highest average concentrations in both uppermost (POC=6.48%; PN=1.40 %) and lowermost traps (POC=3.93%; PN= 0.48%) were found in springtime, when also POC peaks in both traps (7.29%; 6.52 %) and the PN peak in the topmost trap (1.40 %) were recorded. Figures 3c-d illustrate organic carbon contents together with gross sedimentation rate data.

The short sampling intervals of the lowermost trap made it possible to obtain a good exponential correlation between total flux and POC concentrations (Figure 5); POC trends are inversely proportional to the gross sedimentation rate.

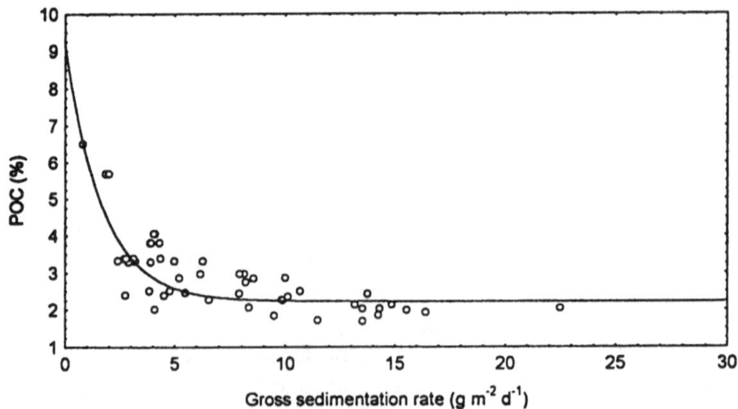

Figure 5. Exponential regression of POC and gross sedimentation rate for the trap at 21 m. The equation is: %POC=2.23+exp(1.94-0.59 gross sed. rate); the correlation coefficient is r=0.74.

3.4.3 Phosphorus

With respect to total P concentrations (Table I), the uppermost trap (1194 mg kg^{-1}) does not differ significantly from the lowermost trap (1237 mg kg^{-1}), whereas the sea floor has a considerably lower total P value (858 mg kg^{-1}). The topmost trap recorded the highest average P concentration in springtime (1445 mg kg^{-1}), when the Po river discharge increased; however, the average P concentration recorded by the lowermost trap was only 950 mg kg^{-1} in that season. The maximum average P concentration for the lowermost trap (1371 mg kg^{-1}) was found in winter, when the uppermost trap exhibited its minimum P value (969 mg kg^{-1}).

With respect to inorganic P, the easily exchangeable and Fe-bound P fractions presented increasing concentrations from the seafloor to the topmost trap, whereas the Ca-bound fraction showed the opposite trend (Figure 6).

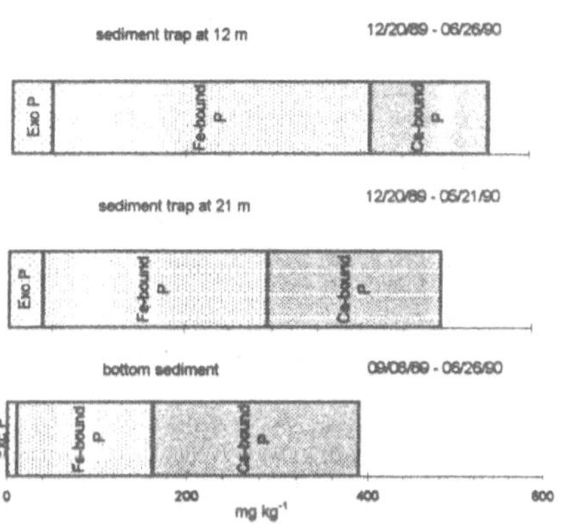

Figure 6. Average concentrations of the inorganic P fractions (mg kg^{-1}) in the traps at 12 m and 21 m, as well as in bottom sediments.

3.4.4 Metals

A good, direct correlation was found between the gross sedimentation rate and the Mn concentration recorded by the lowermost trap in winter (Fig. 7). In the investigated environment, where fine sediments are mainly anoxic, the Mn in deeper sediments becomes mobilized towards the seafloor and forms an enriched layer a few millimeters thick at the sediment-water interface (Barbanti *et al*, 1995). This leads us to suggest that high Mn concentrations are direct signs of resuspension of the first millimeters of bottom sediments, in a period with low meteomarine hydrodynamics and low water regime of the Po river (Figure 3e). Indeed, when meteomarine dynamics were more intense and a thicker sediment layer was resuspended accordingly, no correlation whatsoever was found between particle flux and Mn concentration, such as in the sample collected during the storm of late March. Our suggestion is supported also by the Mn contents of both topmost trap samples (501 mg kg^{-1}) and bottom sediment samples (613 mg kg^{-1}) which are lower than the Mn content recorded in the lowermost trap samples (766 mg kg^{-1}). Bottom sediment samples had a thickness exceeding 1 cm and, therefore, the Mn enrichment in their first millimetres could not be appreciated. The average total concentrations of Al and Fe are greater in the sea floor than in the particles collected by both traps (Table I), which underlines that particle samples are characterized by abiotic components.

3.5. ESTIMATION OF THE DIFFERENT COMPONENTS OF THE GROSS SEDIMENTATION RATE

The various components of the gross sedimentation rate were determined in the lowermost trap, a task made easier by the high sampling resolution of this trap. The upper trap did not give such accurate determinations because of its limited number of samples (see section 2.3.). There are three such components: the first involves the primary production in the water column, the second relates to the resuspension of bottom particulate matter, and the third is linked to the lateral transport (river input).

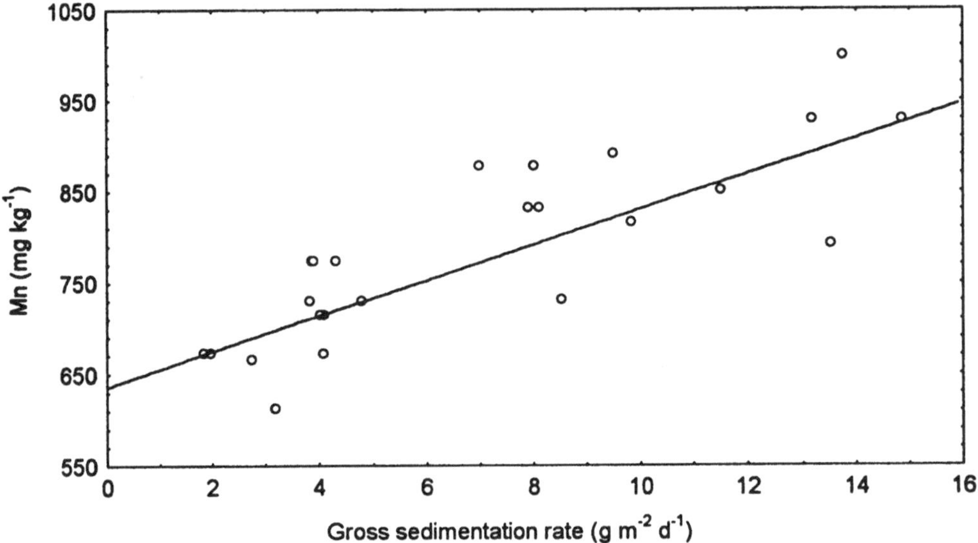

Figure 7. Linear regression of Mn and gross sedimentation rate in the trap at 21 m. Regression equation: Mn=636+19.5 gross sed. rate; the correlation coefficient is r=0.79.

3.5.1 Flux of primary production

The primary production flux was calculated with the label approach (Gasith, 1975):

$$F_p = F_t\,(C_r\text{-}C) / (C_p\text{-}C_r) \quad (1)$$

where F_p (g m^{-2} day^{-1}) is the sedimentation rate of the autochthonous fraction, F_t (g m^{-2} day^{-1}) is the measured gross sedimentation rate and C (% dry weight) is the concentration of the label substance with subscritps p and r referring to primary and resuspended materials, respectively.

POC is the label substance which was selected to identify primary production. The C_p and C_r end members were selected within the trap samples for a homogeneous data distribution (Blomqvist and Larsson, 1994). As the primary production end member we selected the POC content of 5.7% derived from a sample taken in early February (Figure 3d) during a superficial algal bloom with quiet meteomarine conditions and low Po river discharges. The sample with the lowest POC content (1.63%), which was collected during the late March storm, was chosen as opposite end member (Figure 3d).

3.5.2 - Flux of resuspended particles

The resuspended particle flux was calculated for the period from December '89 to March '90 (see section 3.4.4) again with the label approach:

$$F_r = F_t\,(C_r\text{-}C) / (C_p\text{-}C_r) \quad (2)$$

where F_r (g m^{-2} day^{-1}) is the sedimentation rate of the resuspended fraction.

Mn is the label substance used in this case. The minimum Mn concentration (613 mg kg^{-1}) was taken to represent the flux of non-resuspended material, while the maximum Mn concentration (1000 mg kg^{-1}) was taken to represent the resuspended particle flux. As for POC, the C_p and C_r values for Mn concentrations were selected within the trap samples.

3.5.3 - Flux of lateral transport

Once primary production fluxes (1) and resuspension fluxes (2) were determined, we calculated the lateral transport fluxes of sediments:

$$F_l = F_t - (F_r + F_p) \qquad (3)$$

4. Discussion

The lateral input mainly from the Po river and primary production play a major role in the uppermost trap with respect to the lowermost trap (as it can be seen from the hydrologic characteristics as well as the POC and PN percentages, Figure 3) while resuspension has a negligible effect on it (as it can be seen from the lack of direct correlation between gross sedimentation rate and seafloor dynamics and Fig.4) and becomes significant in the composition of samples only with stronger sea dynamics (lowest POC and PN values). Based on physical parameters (wave height, TSM, current speed) and compositional characteristics (POC and Mn), the resuspension events are more important in the lowermost trap, as evidenced by the good correlation between gross sedimentation rate and seafloor marine dynamics. This correlation together with the low Po river discharges substantiate the assumption that, in the lowermost trap, the riverine input of organic matter is negligible during the study period.

It is clear from the above that the particle fluxes recorded by the two traps have completely different implications. The uppermost trap, which was just above the pycnocline, measured fluxes of materials which would deposit onto the study area only to a small extent for two reasons: a) the strong influence of the lateral transport component predominantly consisting of fine and ultrafine riverine particulate matter (as confirmed by low flux values in conjunction with high turbidity); and b) the presence of temperature and salinity gradients (Håkanson *et al*, 1989). On the contrary, the lowermost trap measured fluxes which more accurately represent the depositing particles, be they primary settling or resuspended particles.

Therefore the differentiation of the gross sedimentation rate for the lowermost trap makes it possible to evaluate the relative influence of primary productivity, river input, and resuspension on bottom sedimentation. Figure 8 shows the flux components determined with the label approach together with the gross sedimentation rates. Although the label approach provides a simple way to evaluate the contribution of different source materials to sedimentation (Gasith, 1975; Blomqvist and Larsson, 1994; Heiskanen and Leppanen, 1995), this method is not easily applicable to the environment investigated, because lateral inputs may interfere with calculations (Håkanson *et al*, 1989). To obviate this problem, two label substances of different nature (POC and Mn) were used successfully since the two calculations validated each other. Only for the samples taken as end members (for 100% of primary production and resuspended particle fluxes) was the summation of fluxes greater than the gross sedimentation rate by less than 20%. The calculated resuspended particle flux is in line with the correlation between meteomarine dynamics and gross sedimentation rate. In fact, this flux shows that resuspension events strongly influence the seafloor even with moderate meteomarine events, such as in the period from December to late March. In this period the average resuspension flux value was 3.6 g m^{-2} d^{-1}. The Po river then had a low water regime (Figure 3e) and the lateral transport flux was 2.8 g m^{-2} d^{-1} on average with larger values occurring at the extremes of this time period (Figure8). On average the combination of lateral transport and resuspended particle fluxes accounts for approximately 80% of the gross sedimentation, a very large proportion in the light of the relatively calm meteomarine dynamics and the low Po river discharge; this demonstrates the limited primary production (1.3 g m^{-2} d^{-1}) during winter. Figure 9 shows primary production fluxes expressed in g m^{-2} d^{-1}, together with the gross sedimentation rate and the primary settling organic carbon (mg C m^{-2} d^{-1}) from December to early May. In contrast with the findings for Mn, the storms of late March and early April do not affect the calculations for the POC label substance. Figure 9 indicates that primary production is characterized by short-term algal blooms. These episodes take place mainly in the first meters of the water column (as evidenced by dissolved oxygen and pH values) and are facilitated by nutrient-rich riverine inputs. Fairly long-lasting, high POC fluxes caused by primary production are recorded from late January to mid-February and from late April onwards. Minimum POC primary fluxes occur during the colder season (late December-late January) and from late March to late April, the period with the two major storms.

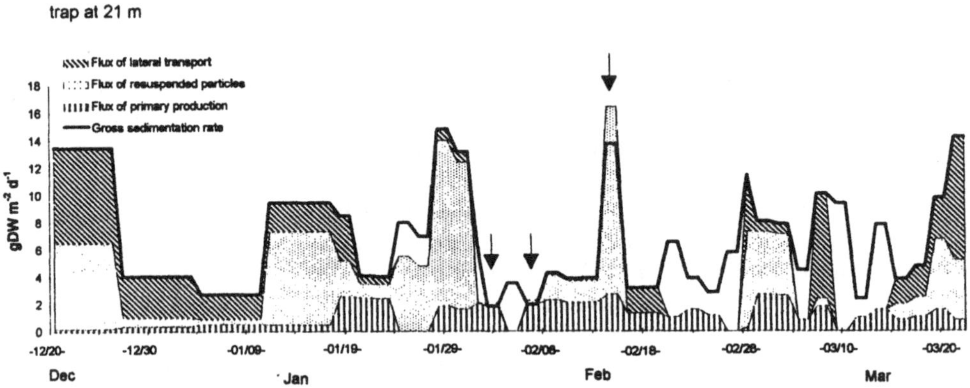

Figure 8. Gross sedimentation rate divided into its three components. Blank areas indicate that no data were available. The arrows indicate end members.

Al and Fe concentrations correlated directly to the fluxes of resuspended particles and lateral transport calculated as a percentage over the gross sedimentation rate for the lowermost trap (Al=3.67+0.035 (%Fr + %Fl) - r=0.79; Fe=1.76+0.20 (%Fr + %Fl) - r=0.66). Since Al and Fe can be considered as the main markers of the inorganic matrix of particles from both resuspension and riverine inputs (Fr and Fl), the above correlation further confirms the validity of the approach followed in determining the various gross sedimentation rate components.

Throughout the study period the phosphorus concentration in the samples of both traps was greater than the P concentration of bottom samples. The average P concentration was 29% greater in the lowermost trap from December to late March, while in the spring the trend was reversed so that the P concentration was 52% lower in the lowermost trap. The winter difference is likely to depend on the reduced primary production, whereas in spring the large river input together with the magnitude of primary production may play a greater role at the 12 m level.

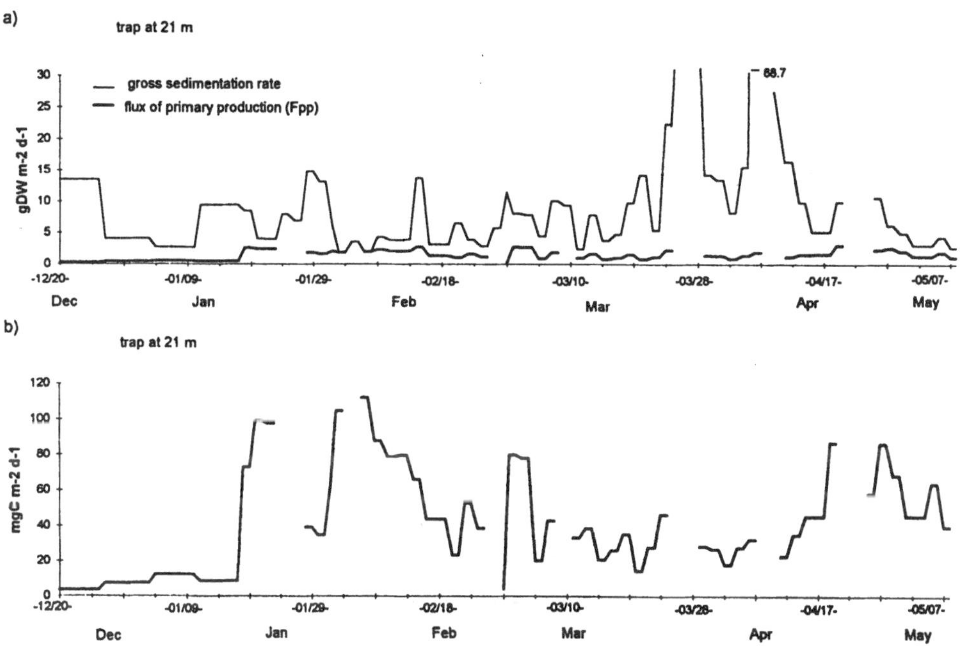

Figure 9. a) sedimentation rate of primary production particles and gross sedimentation rate in g of dry weight; b) primary production carbon flux calculated from the primary production flux.

A good direct correlation between the bioavailable inorganic P fractions (exchangeable P and Fe-bound P) and the primary production flux calculated as a percentage over the gross sedimentation rate was found for the lowermost trap (Figure 10). This is probably due to the sinking of dead organic matter which can aggregate the colloidal particles that are present in large quantities at the upper levels (Olsen *et al*, 1982). This explains also why the uppermost trap collected more of these fractions compared to the lowermost trap and the bottom sediments. Even if some exchangeable P can derive from fresh phytoplankton (Ruttenberg, 1990), the above bioaggregation hypothesis is additionally substantiated by the Fe-bound P fraction findings. On the contrary, the Ca-bound P concentration in bottom sediments is greater than that found in the two sediment traps (Figure 6), because Ca-bound phosphorus is contributed to from the river and the authigenic apatite in the sea bed. Given its origin, we found in the lowermost trap that the Ca-bound P concentration has not individual, direct correlation with neither resuspension with lateral transport. At the same time, the direct correlation found here between this Ca-bound P concentration and the resuspended particle and lateral transport fluxes calculated as a percentage over the gross sedimentation rate confirms that Ca-bound P is influenced by the two sources mentioned earlier (Figure 10).

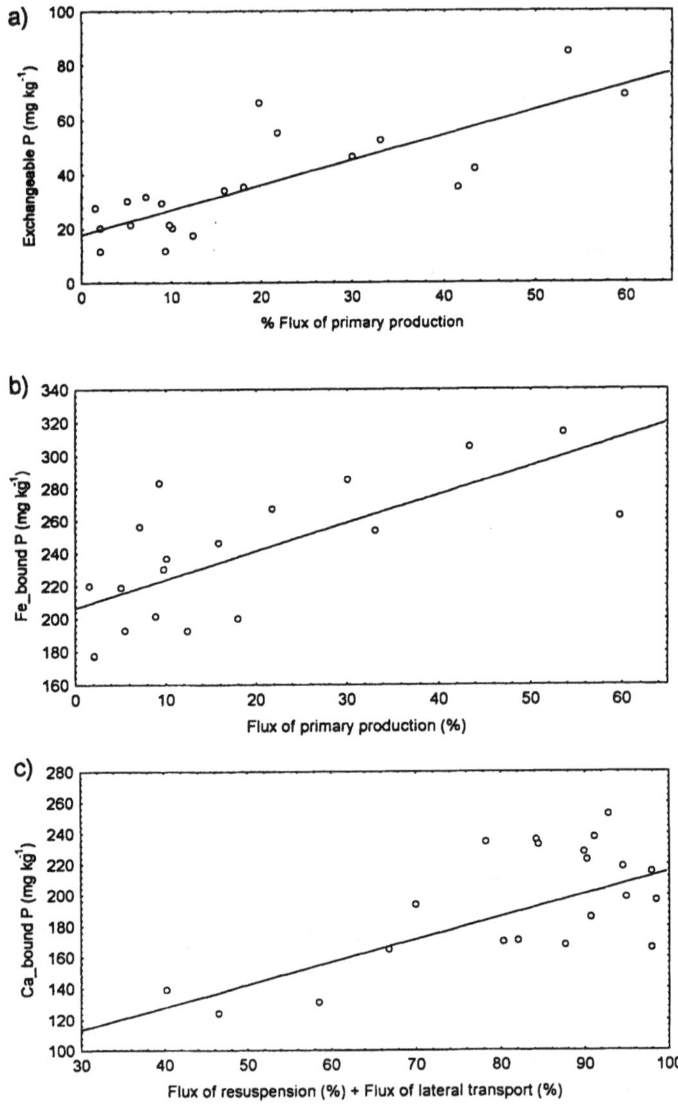

Figure 10. Linear regression of exchangeable P, Fe-bound P and Ca-bound P with the primary production flux (Fpp) in the trap at 21 m. Regression equations: Exc-P=17.86+0.91 %Fp, (r=0.80); Fe-bound P=206+1.74 %Fp, (r=0.77); Ca-bound P=224-1.64 (%Fr + %Fl), (r=0.75).

5. Conclusions

We differentiated and quantified the vertical fluxes of settling particles in an area with a strong riverine nutrient input which is the main cause of recurring eutrophication in the Northern Adriatic Sea. To achieve this important objective, we used an approach based on the identification of label substances and end members with a seasonal character. This was facilitated by the use of instruments which monitored environmental conditions continuously (salinity, temperature, dissolved oxygen, pH, river discharge, wave height, current speed and direction, and TSM).

This integrated study of suspended particles and bottom sediments lead to the identification of two strata in the water column. The stratum above the pycnocline (0-15 metres) is richer in riverine particles, belongs to a photic zone which is active even in winter, as witnessed by various algal blooms, and occasionally includes particles resuspended by strong storm events. The deeper stratum is influenced by sediment resuspensions caused by even moderate storms, limited direct supplies of materials from the Po river, and sinking organic matter which already partly mineralized.

It was found also that the uppermost sediment trap had the highest bioavailable inorganic phosphorus fractions which then decreased with depth, with the lowest values occurring in the seabed sediments. However, the values of such fractions increased in the lowermost strata when primary production fluxes increased. This is probably due to bioaggregation processes which facilitate the deposition of bioavailable inorganic P fractions. The impoverishment in bioavailable P fractions in bottom sediments is attributable to frequent resuspension events: they can easily mobilize a large portion of these fractions which then re-enter the trophic chain.

Acknowledgments

We thank Prof. Dave Cacchione for the oceanographic data measured by GEOPROBE on the benthic boundary layer; Dr. Stefano Miserocchi and Dr. Andrea Barbanti for their assistance during field work and their help in laboratory analyses; Prof. Claudio Fuschini for the English translation. This program has been supported by the Istituto di Geologia Marina, C.N.R. (Bologna, Italy) and U.S.Geologial Service. (Menlo Park, California). This paper is contribution N° 1069 of Istituto per la Geologia Marina - CNR (Bologna, Italy).

References

Anderson, R. J.: 1977, *Limnol. Oceanogr.* **22**, 423-433.

Artegiani, A., Bregant, D., Paschini, E., Pinardi, N., Raichic, F. and Russo A.: *in press*.

Barbanti, A., Bergamini, M.C., Frascari, F., Miserocchi, S. and Rosso, G.: 1994, *Jour. Environ. Qual.* **23**, 1093-1102.

Barbanti, A., Bergamini, M.C., Frascari, F., Miserocchi, S., Ratta, M. and Rosso, G.: 1995, *Mar. Freshwater Res.* **46**, 55-67.

Blomqvist, S. and Larsson, U.: 1994, *Limnol. Oceanogr.* **39**, 880-896.

Boldrin, A., Bortoluzzi, G., Frascari, F., Guerzoni, S. and Rabitti S.: 1988, *Mar. Geol.* **79**, 159-170.

Cacchione, D. A. and Drake, D. E.: 1979, *Mar. Geol.* **30**, 299-312.

Dal Cin, R.: 1983, *Boll. Soc. Geol.* **102**, 9-56.

Drake, D. E., Cacchione, D. A., Frascari, F., Bortoluzzi, G., Tate, G., Miserocchi, S., Thede, J. and Viall, R.: 1992, *Rapp. Comm. Int. Mer Medit.* **33**, 128.

Faganeli, J., Gacic, M., Malej, A. and Smodlaka N.: 1989, *J. Plankton Res.* **11**, 1129-1141.

Franco, P., Jeftic, L., Malanotte Rizzoli, P., Michelato, A. and Orlic, M.: 1982, *Oceanol. Acta* **5**, 379-389.

Frascari, F., Frignani, M., Guerzoni, S. and Ravaioli, M.: 1988 *Annals of the New York Accademy of Sciences* **534**, 1000-1020.

Froelich, P. N.: 1980, *Limnol.. Oceanogr.* **25**, 242-248.

Gasith, A.: 1975, *Verh. Int. Ver. Limnol.* **19**, 116-122.

Golterman, H. L. and Booman, A.: 1988, *Verh. Int. Ver. Limnol.* **23**, 904-909.

Guerzoni, S., Frignani, M., Giordani, P. and Frascari, F.: 1984, *Environ. Geol.* **6**, 111-119.

Håkanson, L., Floderus, S. and Wallin, M.: 1989, *Hydrobiologia* **176/177**, 481-490.
Hedges, J. I. and Stern, J. H.: 1984, *Limnol. Oceanogr.* **29**, 657-663.
Heiskanen, A.S. and Leppänen, J.M.: 1995, *Hydrobiologia* **316**, 211-224.
Nelson, B.W.: 1970, *S.E.M.P. Spec. Publ.* **15**, 152-184.
Olsen, C. R., Cutshall, N. H. and Larsen, I. L.: 1982, *Mar. Chem.* **11**, 501-533.
Orlich, M., Gacic, M. and La Violette, P.E.: 1992, *Oceanologica Acta* **15**, 109-124.
Pigorini, B.: 1968, *Mar. Geol.* **6**, 187-229
Puškaric, S., Fowler, S.W. and Miquel J.C: 1992, *Est. Coast. Shelf Sci.* **35**, 267-287.
Regione Emilia Romagna: 1989, *Eutrofizzazione delle acque costiere dell'Emilia Romagna, Rapporto Annuale*, p.166.
Regione Emilia Romagna: 1990, *Eutrofizzazione delle acque costiere dell'Emilia Romagna, Rapporto Annuale*, p.222.
Ruttenberg, K. C.: 1990, *Ph. D. diss. Yale University* (Diss. Abstr. 90 - 34244).
Strickland, J.D.H. and Parsons, T.R.: 1972, *A practical Handbook of seawater anakysis, Fish. Res. Board Canada, Ottawa*, p. 310.
U.S. G.O.F.S., 1989 *Sediment trap technology and sampling* **10**, 94.

SEDIMENT RECORDS OF FALLOUT RADIONUCLIDES AND THEIR APPLICATION TO STUDIES OF SEDIMENT-WATER INTERACTIONS

P.G.APPLEBY

Department of Mathematical Sciences, University of Liverpool, P.O.Box 147, Liverpool L69 3BX, UK.

Abstract. The fallout radionuclides ^{210}Pb and ^{137}Cs are widely used to date environmental records contained in lake sediments. Since the radionuclide records are themselves the outcome of the transformation of atmospheric fallout by mediating transport processes from the catchment, through the water column and post-depositional migration via pore waters, reliable models of these processes are crucial to accurate dating. The large quantities of data on ^{210}Pb and ^{137}Cs in lake sediments accumulated through their widespread dating applications may be used to study transport models. Their advantages as tracers of transport processes include widespread dispersal through the environment, relatively simple and well known input functions, and ease of measurement.

One of the principle factors controlling the transport of any species through the water column is its distribution between aqueous and particulate phases. The relatively solubility of ^{137}Cs in the water column is demonstrated by the reduced ^{137}Cs/^{210}Pb inventory ratios in sediments compared to values expected from direct fallout. Using sediment records from a wide range of Cumbrian lakes, calculations based on simple models indicate that the particulate fraction of weapons fallout ^{137}Cs in the water column ranged from 5-22%, and was proportional to the square root of the sedimentation rate (determined by ^{210}Pb). The K_D value for weapons ^{137}Cs in the water column is estimated to be in the range 1-2x10^5 L kg^{-1}. This is comparable with K_D values for Chernobyl ^{137}Cs in these lakes (Smith *et al.* in press) obtained from direct measurements in the water column.

Key Words: sedimentation rates, lead-210, cesium-137, particulate phase

1. Introduction

Many lake sediment cores contain high quality environmental records that clearly and faithfully record known deposition events (c.f. Appleby *et al.* 1990). The quantitative validity of these records is demonstrated by the high level of consistency between dates calculated from ^{210}Pb and those determined using independent chronostratigraphic markers such as fallout ^{137}Cs and ^{241}Am (Appleby *et al.* 1991).

One of the increasingly important objectives of palaeolimnology is the use of sediment records for making quantitative reconstruction of past changes in the level of environmental pollution. This can only be achieved by using models for predicting the fate of pollutants deposited in lakes and on their catchments. Essential to these models is reliable validation on timescales appropriate to the parameters being studied. The objective of this paper is to examine these models using our large data base on ^{210}Pb and ^{137}Cs in sediments, and to use show how the results can be used to help establish transport parameters within the catchment/lake system.

Water, Air and Soil Pollution **99**: 573-586, 1997.

2. Palaeolimnological transport models

Concern about the fate of radiocaesium in aquatic systems following the 1986 Chernobyl accident stimulated a great deal of work on the development of transport models, exemplified by the VAMP project (Håkanson *et al.* 1996). The timescale of much of that work is in the first instance necessarily related to the residence time of pollutants in the water column, normally measured in months to years. Since sections of sediment cores almost invariably span at least 2-3 years, *palaeolimnological models* for predicting sediment records necessarily operate on much longer timescales, typically years to decades, and are generally much simpler and less detailed by comparison. Although the precise outcome of a particular fallout event will depend on a large number of subtle and complex processes, since even well-instrumented sites have relatively little detailed information about the lake or its catchment on these timescales, realistic models are necessarily simple, and limited in their objectives.

The basic elements of the palaeolimnological model are illustrated in a very simplified way in e.g. Appleby & Smith (1993). A part of the atmospheric flux deposited on the catchment will be transported into the lake via runoff or erosion. Fallout entering the lake directly via atmospheric deposition or indirectly via transport from the catchment will eventually be lost from the lake via the outflow, or transported to the bed of the lake and incorporated in the sediment record. Achievment of the objectives of the model demands appropriate representation and parameterisation of:

- the catchment/lake transport rate,
- the residence time of the pollutant in the water column,
- the water column/bottom sediment transfer fraction,
- post-depositional transport processes horizontally over the bed of the lake and vertically in the sediment column.

In most cases the only data available to the model will be parameters such as lake area, volume and water residence time, catchment area, rainfall, and basic water chemistry.

2.1. Model Validation using Fallout Radionuclides as Tracers

Models can be used with confidence only if they have been adequately tested in a range of different contexts. The fallout radionuclides ^{210}Pb and ^{137}Cs are in many ways ideally suited to this purpose.

- They have well defined but contrasting input functions to environmental systems.
- They have contrasting geochemical behaviour, ^{210}Pb being strongly particle bound and ^{137}Cs relatively more soluble.
- They are easily and unambiguously determined, with little or no problem of confusion with non-atmospherically derived sources.
- There is a large existing data-base on ^{210}Pb and ^{137}Cs in sediment records from numerous palaeolimnological studies

All models are based on a set of operating assumptions and of equal importance is the development of criteria for assessing their validity at a given site. One of the crucial questions in any palaeolimnological investigation is the quality of the sediment record, which can be degraded by processes such as sediment erosion or slumping, or even the act of coring. Radionuclides play an increasingly important role in assessing the extent to which sediment records may have been influenced by extraneous processes not included within a particular model under consideration.

2.2. Atmospheric fluxes

Use of fallout radionuclides for validating palaeolimnological models demands accurate information on the atmospheric inputs to the system, as well as the outputs to the sediment record.

Atmospheric ^{210}Pb derives from the decay of gaseous ^{222}Rn introduced into the atmosphere by diffusion from soils, and is normally assumed to have a constant flux at any given locality. Inputs of ^{137}Cs have been due mainly to fallout from the atmospheric testing of nuclear weapons, and more recently in some parts of the world, from the 1986 Chernobyl reactor fire. Following the initial thermonuclear tests, global fallout of weapons ^{137}Cs rose rapidly after 1954 to reach a peak in 1963, after which it declined rapidly following implementation of 1963 test ban. 96% of the total weapons fallout inventory had been deposited by 1970. Measured on the timescales of ^{210}Pb and weapons ^{137}Cs, Chernobyl fallout can be regarded as virtually instantaneous.

Although different locations have similar fallout histories, the amount of fallout may vary from place to place by an order of magnitude, depending on factors such as rainfall and geographical location. Variations in the ^{210}Pb flux are illustrated by data summarised in Appleby & Oldfield (1992). The influence of rainfall on ^{137}Cs fallout is shown by the results presented in Cawse & Horrill (1986) and Cawse *et al.* (1988).

2.3 Validation against long times-series sediment records

Some of the earliest studies of ^{210}Pb and ^{137}Cs in lake sediments were carried out by Pennington *et al.* (1973 & 1976) on cores from lakes in the Windermere catchment in Cumbria (UK), including Blelham Tarn, Esthwaite Water, Elterwater and Windermere itself. Further analyses carried out during the 1980s and 1990s by the Liverpool Environmental Radioactivty Research Centre have resulted in a record of measurements from these lakes that now spans a period of about 25 years. Data from these (and other similar sites) offer a means for testing models on long-time scales.

3. Transport equations

The supply of fallout products to the sediments is immediately determined by processes in the water column. Letting

$\Phi(t)$ = atmospheric flux (per unit area)
$\Psi_C(t)$ = transport rate from catchment
$Q(t)$ = inventory stored in water column

$\Psi_O(t)$ = loss via outflow
$\Phi_S(t)$ = mean flux per unit area to sediments

the mass balance equation for the radionuclide (or pollutant) in the water column may be written

$$\dot{Q} = A\Phi + \Psi_C - A\Phi_S - \Psi_o - \lambda Q,$$

where λ is the radioactive decay constant and A is the lake area. The objective of paleolimnological models is to use this equation to determine the flux $\Phi_S(t)$ to the sediments. Solution depends on the way in which we represent the catchment inputs Ψ_C, the outflow loss Ψ_O and the relation between Φ_S and the water column inventory Q.

3.1. Transport from the Catchment

For the most part, radionuclides falling directly onto the catchment are retained in the surface soils and only slowly released into the lake on time scales of hundreds to thousand of years. Rates of transport from the catchment can be represented in two ways (Appleby & Smith, 1993):

- in terms of a *transport coefficient*

$$\kappa_C = \Psi_C / Q_C$$

relating the transport rate Ψ_C to the catchment inventory Q_C;

- in terms of a *catchment/lake transfer function*

$$\mathcal{T}\Phi = \Psi_C / A$$

expressing indirect inputs via the catchment as an augmentation of the direct atmospheric flux onto the lake surface. The total input to the lake can then be written

$$A(1+\mathcal{T})\Phi.$$

The reciprocal

$$T_C = 1/\kappa_C$$

of the transport coefficient is a measure of the time-scale of residence in the catchment, though this will not be constant since the value of κ_C may be expected to decline for older deposits which are less available for transport. This time dependence is represented more precisely using the concept of a *unit response function* $h(t)$, defined as the decay corrected transport rate due to a unit amount of fallout on the catchment. The transport rate due to an arbitrary fallout history can then be written

$$\Psi_C(t) = A_C \int_0^\infty \Phi(t-\tau) h(\tau) e^{-\lambda\tau} d\tau,$$

where A_C is the catchment area. The catchment/lake transfer function is

$$\{\mathcal{T}\Phi\}(t) = \alpha \int_0^\infty \Phi(t-\tau) h(\tau) e^{-\lambda\tau}\, d\tau,$$

where $\alpha = A_C / A$ is the catchment/lake area ratio.

Precise determination of the response function is impractical. All that is really possible is its representation by an appropriate function, typically exponential, involving a small number of parameters to be estimated from experimental or theoretical considerations. From a palaeolimnological point of view the essential parameters are:

I. the ultimate fraction

$$\eta = \int_0^\infty h(\tau)\,d\tau$$

of the original fallout that is delivered to the lake, and

II. the timescale over which this occurs.

Steady state transport of radionuclides such as ^{210}Pb with a constant fallout are characterised by just a single parameter. The transport rate in this case can be written

$$\Psi_C(t) = A_C \eta_s \Phi,$$

where

$$\eta_s = \int_0^\infty h(\tau)\,e^{-\lambda\tau}\,d\tau$$

is the decay-weighted area under the unit response function. There is thus a constant transfer function

$$\mathcal{T}\Phi = \alpha\eta_s\Phi$$

and a constant transport coefficient

$$\kappa_C = \frac{\lambda\eta_s}{1-\eta_s}.$$

The parameter η_s represents the fraction on the annual fallout onto the catchment that is delivered to the lake. The reciprocal of κ_C in this case defines a mean (^{210}Pb) residence time in the catchment.

For radionuclides such as ^{137}Cs with a variable flux, it is common to assume an exponential response function such as

$$h(t) = \eta\,k\,e^{-kt},$$

or a combination of such functions to represent processes on different time-scales (c.f. Helton *et al.* 1985). The parameter k characterises the time-scale of transfer. Its reciprocal is the time taken to transport 63% of the removable fraction (η).

3.2 Transport through the water column

Transport through the water column is crucially dependent on the extent to which the species is bound to settling particulates. Partitioning between the particulate and dissolved phase is commonly characterised by the *distribution coefficient*

$$K_D = C_S / C_W,$$

where C_W denotes the dissolved concentration per unit volume in the lake waters and C_S denotes the concentration per unit mass of the suspended solids, or by a partition fraction

$$f_D = sC_S/C = \frac{1}{1+(sK_D)^{-1}}$$

characterising the fraction of species on particulates, where s is the suspended sediment concentration.

Assuming that concentrations at the outflow are close to the mean value for the lake, losses from the lake via the outflow may be written

$$\Psi_O = Q/T_W,$$

where T_W is the lake water rsidence time. Assuming that losses to the bottom sediments are controlled largely by the process of sedimentation, the flux to the bottom sediments may be written

$$A\Phi_S = f_D Q/T_S,$$

where T_S is a typical settling time for suspended particulates To a first approximation

$$T_S = d/v$$

where d is the mean water depth and v is a mean settling velocity (c.f. Santschi & Honeyman, 1989). Although more precise descriptions may be obtained using e.g. two box models (Imboden & Schwarzenbach, 1985) to decribe the lake during stratified periods, apart from the fact that detailed information on stratification is just not available at many sites used for palaeolimnological studies this approach may be less important for modelling long-term sediment records.

Substituting the above expressions for inputs and outputs into the water column balance equation gives

$$\dot{C} + \kappa C = \frac{1}{d}(1+\mathcal{T})\Phi,$$

where

$$\kappa = \frac{f_D}{T_S} + \frac{1}{T_W} + \lambda = \frac{1}{T_L} + \lambda,$$

and T_L is a residence time of the pollutant in the water column, defined by

$$\frac{1}{T_L} = \frac{f_D}{T_S} + \frac{1}{T_W}.$$

We may assume that $T_L \approx 1/\kappa$, except when the radionuclide half-life is short compared to the residence time T_L.

4. Water column/bottom sediment transfer fraction

Precise time-dependent solutions of the water column balance equation are irrelevant to palaeolimnogical studies. Except on time-scales comparable to the residence time T_L the flux to the sediments can be estimated from the residence times alone. At each instant losses to the bottom sediments and via the outflow are in the ratio

$$\frac{A\Phi_S}{\Psi_O} = \frac{f_D T_W}{T_S}.$$

Hence the fraction of any input ultimately transferred to the sediments is

$$\mathcal{F} = \frac{f_D T_L}{T_S}.$$

On this quasi-equilibrium time-scale the flux to the sediments can be written

$$\Phi_S = \mathcal{F}(1+\mathcal{T})\Phi.$$

Using the approximate formula

$$T_W = \frac{d}{R(1+\alpha)}$$

for the water residence time, where R is the mean annual rainfall and α the catchment/lake area ratio (perhaps modified by a runoff factor), we obtain the approximate equation

$$\mathcal{F} = \frac{f_D}{f_D + \frac{R}{v}(1+\alpha)}.$$

for the transfer fraction. Assuming a typical rainfall of 1 m y^{-1}, since v is ~ 1 m d^{-1}, R/v will be ~0.003. Fig.1 plots $\mathcal{F}$ against f_D for values of the catchment/lake area ratio α ranging from 0 to 50. This shows that for relatively insoluble species (f_D>80%) the bulk of the input (>85%) is always transported to the bottom sediments. Furthermore, for a given value of α the transfer fraction is relatively insensistive to f_D. For relatively soluble species (f_D<10%), the bulk of the input will be incorporated in the sediment record when the catchment/lake area ratio is relatively small (α<5), but for larger values of α may drop to as little as 40%. In these cases the transfer fraction $\mathcal{F}$ is very sensistive to f_D.

4.1 Application to ^{210}Pb and weapons fallout ^{137}Cs

Assuming a constant atmospheric flux P, ^{210}Pb inputs to the lake from the catchment will be (Appleby & Smith, 1993)

$$\Psi_C = A_C \eta_{Pb} P = A\alpha\eta_{Pb} P.$$

The mean ^{210}Pb concentration in the water column will have the equilibrium value

$$C_{equ} = \frac{1}{d} T_L(1+\alpha\eta_{Pb})P.$$

The mean ^{210}Pb flux to the sediments, given by

$$P_S = \mathcal{F}_{Pb}(1+\alpha\eta_{Pb})P,$$

can be estimated from the measured sediment inventory I_S using the formula

$$P_S = \lambda I_S.$$

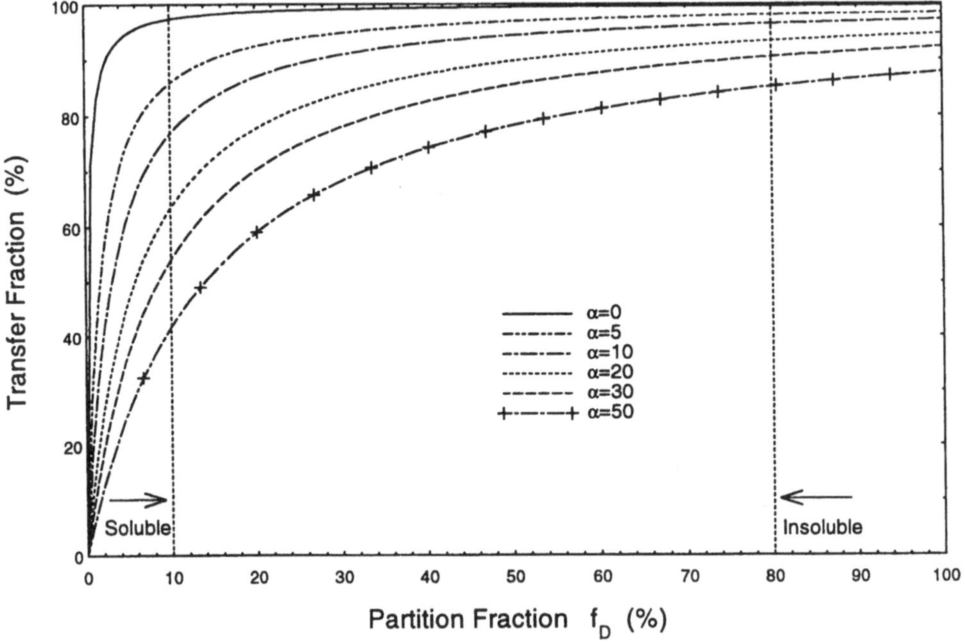

Fig.1 Water column/sediment transfer fraction v partition fraction for different catchment/lake area ratios (assuming c.1000 mm y^{-1} annual runoff)

From todays perspective we may suppose that the total weapons ^{137}Cs fallout F per unit area on any catchment was delivered in a relatively short period of time around 1963. The cumulative transport from the catchment to the lake during the intervening t years and corrected for decay will be $A_C \eta_{Cs} F$, where

$$\eta_{Cs}(t) = \int_0^t h(\tau)\,d\tau,$$

and $h(t)$ is the catchment unit response function for ^{137}Cs. The weapons ^{137}Cs inventory of sediment cores after t years (corrected for decay) will thus be

$$F_S(t) = \mathcal{F}_{Cs}(1 + \alpha\eta_{Cs})F.$$

Since inventories at particular sites may be influenced by post-depositional processes such as sediment focussing, transport processes in the catchment and through the water column can be investigated by considering the *relative Cs/Pb transport factor*

$$\mathcal{R} = \frac{F_S/F}{P_S/P}.$$

From the above results this factor can be written

$$\mathcal{R}(t)=\frac{\mathcal{F}_{Cs}(1+\alpha\eta_{Cs})}{\mathcal{F}_{Pb}(1+\alpha\eta_{Pb})}.$$

Its value will depend on the relative catchment transport parameters, and relative solubility of ^{137}Cs and ^{210}Pb in the water column

4.2 Catchment/lake transport parameters

The enhanced ^{210}Pb inventories observed in many lake sediment cores (see e.g. Fig.2 below) have two likely causes:

- transport from the catchment,
- sediment focussing.

Since ^{210}Pb is strongly particle reactive it is generally thought that transport rates are relatively low, except where the catchment is ice covered or largely devoid of soil. In these exceptional circumstances almost the entire catchment load may be shed annually into the lake during the spring thaw (Appleby *et al.* 1995). Otherwise, ^{210}Pb will be fixed in the surface soils and transported mainly by soil erosion, though even then much will be deposited in flood plains and other up-catchment storage sites. Results from a number of studies summarised in Table 1 give ^{210}Pb removal rates from catchments of c.1-2% of the annual fallout and mean residence times in the catchment of the order of thousands of years.

TABLE 1

^{210}Pb catchment/lake transport parameters

Catchment	*Response coefficient* η_s	*Transport coefficient* κ_c (y^{-1})	*Residence time* T_c (y)	*Reference*
Susquehanna	0.016	0.00052	1930	Lewis (1977)
Mississipi	0.012	0.00039	2580	Scott et al. (1985)
Alpine Rhône	0.023	0.00072	1400	Dominik et al. (1987)

NB The values for Mississipi have been calculated from the data on ^{210}Pb transport rates given in Scott *et al.* (1985).

Determination of reliable weapons ^{137}Cs transport parameters is even more difficult in view of the need to monitor inflows over a decade or more. Limited results from a small number of studies comparing weapons and Chernobyl ^{137}Cs removal rates (c.f. Bonnett & Appleby 1994) suggest that the fraction of catchment ^{137}Cs transported to the lake is comparable to that of ^{210}Pb and that the 90% removal time is not more than a few decades, though higher removal rates may occur in catchment dominated by peaty soils (Hilton *et al.*, 1993).

5. Application to Cumbrian lakes

The above equations indicate that fallout radionuclide inventories in sediments could be related to the catchment/lake area ratio, and that weapons ^{137}Cs inventories could vary with time. To investigate these possibilities, fig.2 plots (a) the ^{210}Pb flux and (b) the weapons ^{137}Cs/^{210}Pb inventory ratio versus catchment/lake area ratio for 9 lakes in Cumbria (England) ranging from major lakes such as Windermere to small upland tarns. Fig.3 plots (a) weapons ^{137}Cs/^{210}Pb and (b) ^{241}Am/^{210}Pb inventory ratios against time. The data were obtained from a total of more than 40 cores collected during 1974-95.

Although the ^{210}Pb flux (normalised against rainfall) is almost invariably equal to or greater than the atmospheric value (Smith *et al.* in press (a)), excluding those sites in west Cumbria that have obviously been influenced by inputs from the Sellafield plant the ^{137}Cs/^{210}Pb inventory ratio is consistently lower than in direct fallout. The inventory ratio is also noticeably less scattered, with a mean value of 1.02, or 46% of the fallout ratio. In contrast, the ^{241}Am/^{210}Pb inventory ratio is comparable to that expected from direct fallout following ingrowth from ^{241}Pu (Appleby *et al.* 1991). This radionuclide is strongly bound to particulates and is expected to behave in a manner comparable to that of ^{210}Pb. There is no apparent relationship to the catchment/lake area ratio or to the time, suggesting either that transport rates are quite varied, or less significant than expected.

From these results we draw the following conclusions:

1) cumulative inputs of weapons ^{137}Cs from the catchment at these sites have not changed significantly during the past 20 years,
2) total cumulative inputs of catchment ^{210}Pb and weapons ^{137}Cs, if significant, have been at a comparable level,
3) the reduced weapons ^{137}Cs /^{210}Pb ratio in sediments is largely due the higher solubility of weapons ^{137}Cs in the water column.

5.1 Estimation of the radionuclide partition fractions

From the above considerations we assume that for the Cumbrian lakes, catchment input terms have had a negligible effect on the relative Cs/Pb transport factor $\mathcal{R}$, and that this ratio can accordingly be approximated by the formula

$$\mathcal{R}(t) \approx \frac{\mathcal{F}_{Cs}}{\mathcal{F}_{Pb}} = \frac{1+\left(T_S/f_D T_W\right)_{Pb}}{1+\left(T_S/f_D T_W\right)_{Cs}}.$$

Given values of the water residence time T_W, the particle residence time T_S and the ^{210}Pb partition fraction $(f_D)_{Pb}$, this expression can be used to make estimates of the ^{137}Cs partition fraction $(f_D)_{Cs}$ in the water column from sediment records.

^{210}Pb K_D values determined from a number of studies of lake and fluvial waters (Talbot & Andren 1984; Dominik *et al.* 1987; Benoit & Hemond 1987 & 1990; Wang & Cornett 1993) show a strong dependence on the suspended sediment concentration s. A best fit to all their data gives the relation

$$K_D = 17.4\, s^{-0.48} \times 10^5 \text{ L kg}^{-1},$$

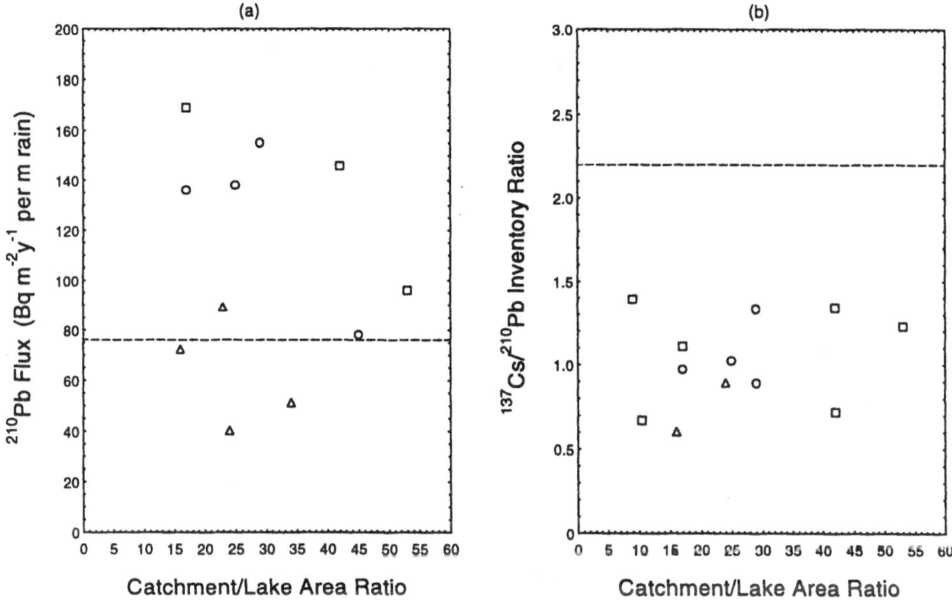

Fig.2 ^{210}Pb flux (a) and ^{137}Cs/^{210}Pb inventory ratio (b) versus catchment/lake area ratio for small tarns (Δ), medium tarns (□) and major lakes (O) in Cumbria, UK. Values determined by direct fallout are indicated by the dashed lines.

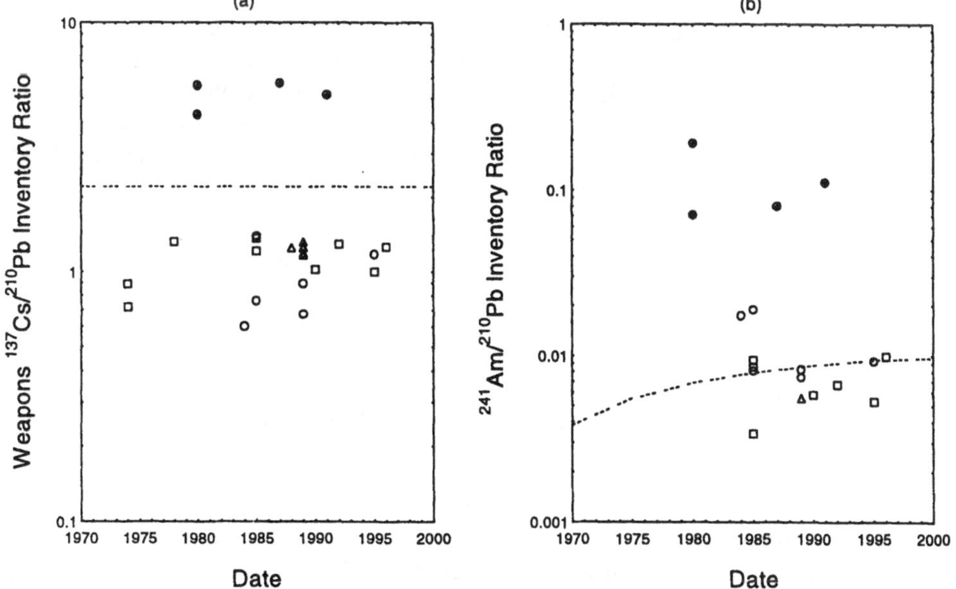

Fig.3 Weapons ^{137}Cs/^{210}Pb (a) and ^{241}Am/^{210}Pb (b) inventory ratios v time for Brotherwater (Δ), Windermere (□), and Eskdale/Wasdale (O) catchments. Values determined by direct fallout are indicated by the dashed lines. Also shown are data from coastal sites (⊕) influenced by local discharges.

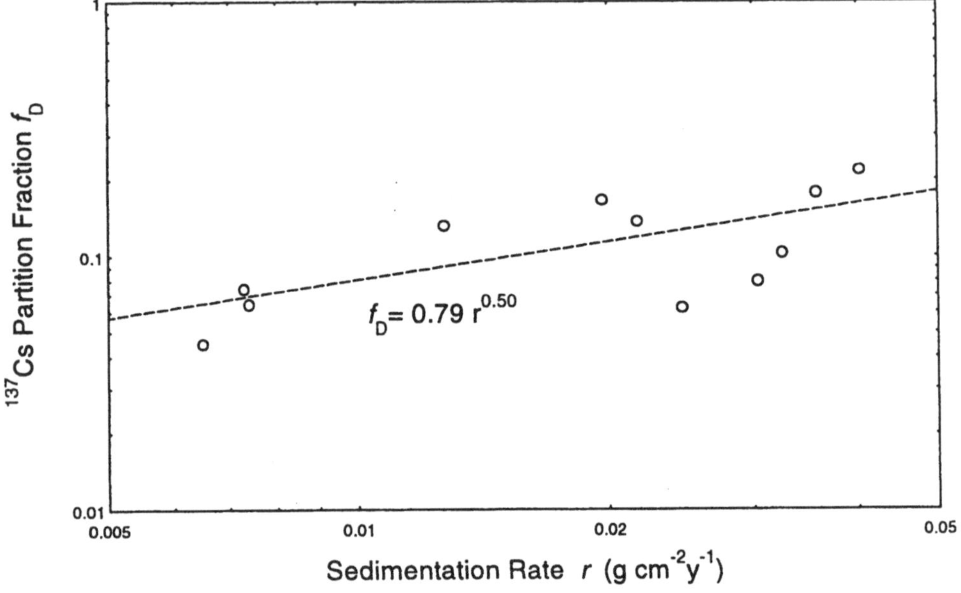

Fig.4 Mean long-term ^{137}Cs partition fractions for Cumbrian lakes versus sedimentation rate, calculated from the sediment records.

where s is measured in mg L^{-1}. The mean ^{210}Pb f_D value is calculated to be 0.75. This value was used to calculate ^{137}Cs f_D values. The results range from 0.05-0.22. They also suggest that the fraction on particulates increases with the accumulation rate, and hence presumably with the suspended sediment concentration. Fig.4 plots the ^{137}Cs f_D value against mean sedimentation rates r (g $cm^{-2}y^{-1}$) for each lake calculated from the ^{210}Pb chronology. A least squares fit to the data gives the relation

$$f_D = 0.79\, r^{0.5} .$$

Since for small values of sK_D the equation for f_D in terms K_D reduces to

$$f_D = sK_D,$$

assuming that the accumulation rate is proportional to the suspended sediment concentration these results further suggest that as for ^{210}Pb, the ^{137}Cs K_D value must be inversely proportional to $\sqrt{s}$. Assuming a typical settling velocity of c.0.8 m d^{-1}, we obtain the relationship

$$K_D = 1.4\, s^{-0.50} \times 10^5 \text{ L kg}^{-1}.$$

This result is comparable with directly measured K_D values for Chernobyl ^{137}Cs in a number of the same lakes (Smith 1994; Smith *et al.*, in press (b)). For particle concentrations in the range 0.5-2.2 mg L^{-1}, the direct measurements gave K_D values in the range 0.3-12 x 10^5 L kg^{-1}, with an average value of 2.7 x 10^5 L kg^{-1}.

6. Conclusions

1) Sediment records offer a useful independent means for validating catchment/lake transport models on long timescales.

2) ^{210}Pb fluxes to sediments suggest that lakes retain the bulk of the ^{210}Pb inputs, which are predominantly bound to the particulate fraction. Enhanced ^{210}Pb inventories may largely reflect sediment focussing.

3) Weapons ^{137}Cs inventories suggest that the bulk of this radionuclide was associated with the soluble phase, with a substantial fraction being lost via the outflow.

4) Weapons ^{137}Cs/^{210}Pb inventory ratios in sediments are relatively stable. Using simple models of transport through the water column they can be used to estimate time-average ^{137}Cs partition coefficients. The results obtained by this method are comparable to values from direct measurements in the water column.

References

Appleby,P.G., Jones,V.J. & Ellis-Evans,J.C.: 1995, *J. Paleolimnology*, **13**, 179-191.

Appleby,P.G. & Oldfield,F.: 1992. In: *Uranium series disequilibrium* (eds. M.Ivanovich & R.S.Harmon), OUP, pp.731-778.

Appleby,P.G., Richardson,N., Nolan,P.J. & Oldfield,F.: 1990, *Phil. Trans. R. Soc. Lond.* **B327**, 233-238.

Appleby,P.G., Richardson,N. & Nolan,P.J.: 1991, *Hydrobiologia*, 214:35-42.

Appleby,P.G. & Smith,J.T.: 1993, Proc. UNESCO Workshop on the "Hydrological Impact of Nuclear Power Plant Systems", UNESCO, Paris, pp.264-275.

Benoit,G. & Hemond,H.F.: 1987, *Geochim. Cosmo. Acta*, **51**, 1445-1456.

Benoit,G. & Hemond,H.F.: 1990, *Environ. Sci. Technol.*, **24**, 1224-1234.

Bonnett,P.J.P. & Appleby,P.G.: 1994, *Environmental Pollution*, **.83**, 327-334.

Cawse,P.A. & Horrill,A.D.: 1986, A survey of ^{137}Cs and Pu in British soils in 1977. AERE Harwell Report R-10155, HMSO, London, 55pp.

Cawse,P.A., Cambray,R.S., Baker,S.J. & Burton,P.J.: 1988, A survey of background levels of environmental activity in Wales, 1984-86 (Pre-Chernobyl).. AERE Harwell Report R-12535, HMSO, 21pp.

Dominik,J., Burrus,D. & Vernet,J.-P.: 1987, *Earth Plan. Sci. Lett.*, **84**, 165-180.

Håkanson,L., Brittain,J.E., Monte,L., Heling,R., Bergström,U. & Suolanen,V.: 1996, *J. Environ. Radioact.*, **33**, 255-308.

Helton,J.C., Muller,A.B. & Bayer,A.: 1985, Health Physics, **48**, 757-771.

Hilton,J, Livens,F.R., Spezzano,P. & Leonard,D.R.P.: 1993, *Sci. Total Environ.*, **129**, 253-226.

Imboden,D.M. & Schwarzenbach,R.P.: 1985, In: *Chemical Processes in Lakes* (ed. W.Stumm), Wiley Interscience, pp.1-30.

Lewis,D.M.: 1977, *Geochim. Cosmo. Acta*, **41**, 1557-1564.

Pennington,W, Cambray,R.S. & Fisher,E.M.: 1973, *Nature*, **242**, 324-326.

Pennington,W, Cambray,R.S., Eakins,J.D. & Harkness,D.D.: 1976, *Freshwater Biology*, **6**, 317-331.

Santschi,P.H. & Honeyman,B.D.: 1989, *Radiat. Phys. Chem.* **34**, 213-240

Scott,M.R., Rotter,R.J. & Salter,P.F.: 1985, *Earth Plan. Sci. Lett.*, **75**, 321-326.

Smith,J.T.: 1994, Mathematical modelling of ^{137}Cs and ^{210}Pb transport in lakes, their sediments and the surrounding catchment. PhD thesis, University of Liverpool.

Smith,J.T., Appleby,P.G, Hilton,J. & Richardson,N.: In press (a), *J Environ. Radioact.*

Smith,J.T., Leonard,D.R.P., Hilton,J. & Appleby,P.G.: In press (b), *Health Physics.*

Talbot,R.W. & Andren,A.W.: 1984, *Geochim. Cosmo. Acta*, **48**, 2053-2063.

Wang,K. & Cornett,R.J., 1993.: *J.Paleolimnology*, **9**, 179-188.

A SPACIO-TEMPORAL PATTERN OF POLLEN SEDIMENTATION IN A DIMICTIC LAKE WITH LAMINATED SEDIMENTS

TOMASZ MIESZCZANKIN

Nicholas Copernicus University, Department of Hydrobiology, Gagarina 9, PL-87-100 Toruń, Poland

Abstract. Resuspension in the main basin of Lake Gościąż results approximately a doubling of the total sedimentation rate in the deepest region. The increase pollen flux in bottom traps during overturn was more four times higher than in the upper traps. The most intensive pollen sedimentation was recorded before and just after freeze-up. This could only have been the result of previous sediment resuspension. Mechanisms of sedimentation in the studied lake showed that resuspension and resedimentation can change "the environmental record" of events even in a lake with laminated sediments. Higher sedimentation of pollen during autumn circulation confirmed that in the sediment layers (varves or laminae) particles from previous seasons also could have been found. In the bay pollen sedimentation was lower and depended on plant flowering, there was no resuspension.

key words: pollen flux, resuspension, resedimentation, laminated sediments

1. Introduction

Sedimentation in lakes responds to both external and internal forces. It is closely related to the parameters controlling the water dynamics, which influence wind/wave action. Sedimentation also is related to the morphology of the lake basin. In some lakes (e.g. Mirror Lake) the intensity of sedimentation is determined by resuspension and resedimentation processes. Davis *et al.* (1984) stressed that these processes can play a fundamental role in the sediments formation. The sediments, especially the varved once, contain a specific, long chronological record of environmental changes. The laminated gyttja of Lake Gościąż comprises more than 13,000 couplets consisting of a light, mostly calcite layer, and a dark layer rich in organic detritus (Ralska - Jasiewiczowa *et al.*, 1992). Due to their exceptionally high time resolution, annually laminated sediments have became of major interest lately (e.g. Lotter, 1991; Nuhfer *et al.*, 1993).

Determination of the composition and proportions of pollen grains in lake sediments is a priority for paleoecological interpretations. By the time, however, these fossils are used as indicators of vegetational and climatic parameters, it is essential to understand the mechanisms influencing sediment formation on the lake bottom. As was observed by Mieszczankin (*in preparation*) resuspension was the main process governing sediment accumulation in Lake Gościąż. The present study aimed to discuss the influence of sediment redeposition processes on pollen flux and the importance of observed phenomena for paleoecological record in the lake sediments. For this purpose sediment traps were used as a unique tool to investigate particles flux (Rosa *et al.*, 1991). Sedimentation studies (e.g. Likens and Davis, 1975; Davis and Ford, 1982) demonstrate that the mechanisms of pollen accumulation should be considered for paleoecological interpretation. It is to be hoped that present - day observation will be helpful for a more thorough understanding of the factors affecting pollen deposition.

Water, Air and Soil Pollution **99**: 587-592, 1997.

2. Study area

Lake Gościąż (lat. 52° 30', long. 10° 20', alt. 64.4 m a.s.l.) is small (area 45.5 ha), but rather deep (depth 24 m). Its direct surroundings and catchment are overgrown by fresh conifer forest. The bay (area *ca* 5 ha and max. depth 1.9 m), somewhat isolated from the main basin, is situated in the northern part of the lake (Figure 1).

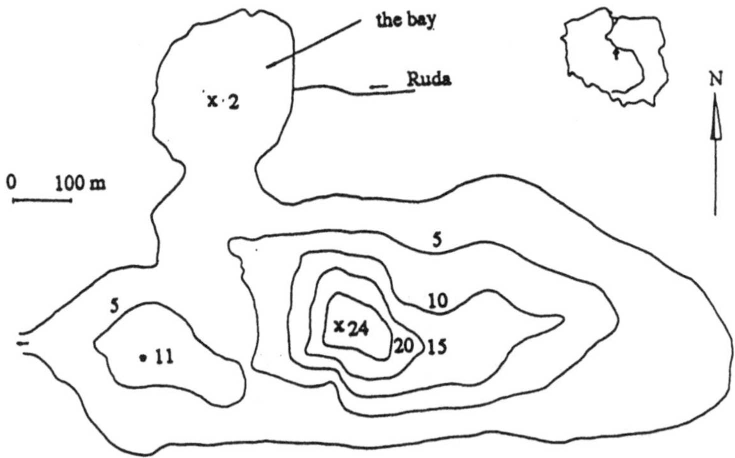

Based on morphological parameters Mieszczankin (*op. cit.*) divided the main basin of Lake Gościąż into two regions with differing water dynamics: a shallow, "dynamic" region (down to *ca* 6 m depth) classified as the resuspension zone and a deep, "static" region (below 6 m depth), classified as the accumulation zone with oxygen deficits during summer stratification. During traps experiments the summer stratification included periods: 03.06. - 08.09.1993 and 11.06. - 19.10.1994.

The bay has its individual features:

- it is shallow, but nearest neighborhood of forest reduces the wind effect on water mass dynamics,
- biofilm develops on the sediment surface (Mieszczankin, *op. cit.*); it is well - known that microbial communities of variable composition and productivity, and its extracellular polymeric substances (EPS) increase the sediments stability (Krumbein *et al.*, 1995).

Mentioned above factors control sediments dynamics in the bay and as a consequence protect against sediment resuspension (Mieszczankin, *op. cit.*).

3. Materials and Methods

Sedimenting matter was collected using simple, cylindrical traps (∅ 105 mm and height 315 mm). In the central part of the main basin traps were installed at the maximum depth station at two levels: 1.5 m (lower traps) and 10 m (upper traps) above the bottom. From the difference of hypolimnetic trap flux and epilimnetic trap flux, the amount of resuspended material can be calculated (Bloesch, 1995). Traps were also deployed in the central part of the bay (in the year 1994 - 1995) 0.5 m above the bottom. Traps in Lake Gościąż were exposed from 1 to 10 weeks.

Subsamples were prepared from material collected from each sample. The palinological analyses were carried out according to the acetolysis method (Berglund and Ralska - Jasiewiczowa, 1986). Results of pollen sedimentation are presented as a number of pollen grains per cm^2 per day (pollen grains $cm^{-2} \cdot d^{-1}$).

4. Results

Annual pollen deposition in the central part of the lake was: *ca* 19,500 pollen grains $cm^{-2} \cdot y^{-1}$ in the upper traps and *ca* 38,500 pollen grains $cm^{-2} \cdot y^{-1}$ in the lower ones. Mean value of one day deposition was respectively: 50 pollen grains cm^{-2} and 105 pollen grains cm^{-2}. Maximum deposition was recorded in the lower traps (Figure 2) during circulation periods (mean 185 pollen grains $cm^{-2} \cdot d^{-1}$), whereas minimum - during summer stagnation (mean 25 pollen grains $cm^{-2} \cdot d^{-1}$) in the upper traps.

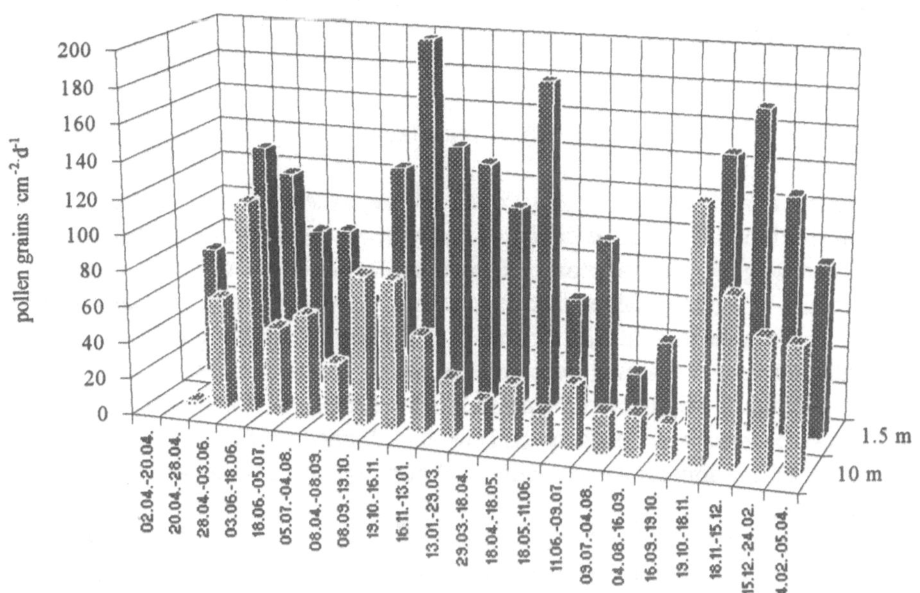

Changes in pollen flux in the bay is presented in Figure 3. Annual pollen deposition was *ca* 11,500 pollen grains $cm^{-2} \cdot y^{-1}$. Maximum pollen flux was observed in the summer:

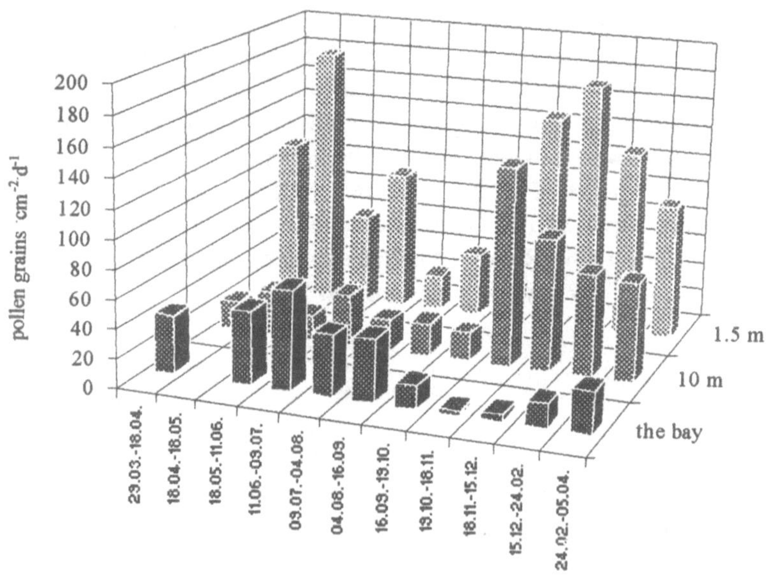

69 pollen grains $cm^{-2} \cdot d^{-1}$ (11.06.-9.07.1994), minimum: in the late autumn and winter - 4 pollen grains $cm^{-2} \cdot d^{-1}$ (19.10.-15.12.1994).

5. Discussion

The number and kinds of pollen grains are dependent on surrounding terrestrial vegetation. Such particles can be displaced easily by air currents and spread over a long distance. The range of that transport is determined by power and direction of winds, and by surrounding topography. Many elements hovering in the air reach the water surface of the lakes, which is the natural trap for them. As a result of gravitation all particles settle in the water column. A seasonal deposition of pollen grains in the sediments allows for the reconstruction of the history of vegetation. However, some events occurring in lakes, such as resuspension and focusing (Likens and Davis, 1975) can change both the conditions of lake sediments accumulation and also deposition some of their components i.e. pollen grains.

It was assumed that in Lake Gościąż primary sedimentation in the pelagic zone of the main basin as well as in the bay was similar. On the basis of the morphology of lake basin (Figure 1) it was expected also, that resuspension occurred mainly in the shallow part of the main basin (down to the depth of 6 m), transport and resedimentation predominated in the accumulation zone. Certainly, above - mentioned factors distort the simple relationship between flowering periods of the plants and pollen composition in the lake sediments. In the main basin of Lake Gościąż, pollen flux in the lower traps (mean value 105 pollen grains $cm^{-2} \cdot d^{-1}$) was much higher than in the upper ones (50 pollen grains $cm^{-2} \cdot d^{-1}$). During circulation, differences between traps were higher, than in the summer stagnation periods (Figure 2), because of intensive sediment resuspension. Davis (1973) recorded, that the increase in pollen flux followed resuspension of material deposited in the littoral zone and redeposition in the profundal zone. The same author (1968) noted, that pollen grains are deposited an average of two to four times before being buried deeply enough to escape further disturbance.

The recording of most intensive pollen flux not in the period of plant flowering, but at the time just before freeze - up and even under ice cover (Figure 2 and 3) is an another argument that resuspension occurred in Lake Gościąż. The observed phenomenon showed that the ice cover affected "calm" pollen sedimentation (resedimentation) following a period of intensive mixing of water and sediments resuspension.

The dynamics of sedimentation processes in the bay were distinctly different than in the main basin, because the bay is much less exposed to the external forces. The pattern of seasonal changes of pollen flux in the bay (Figure 3) is the evidence that resuspension did not occur there. Pollen flux in the summer periods was highest (69 pollen grains $cm^{-2} \cdot d^{-1}$) and correlated with the flowering season. Lowest pollen flux was recorded during circulation (4 pollen grains $cm^{-2} \cdot d^{-1}$), whereas in the main basin this flux was highest.

Frequent and irregular resuspension in shallow lakes brings about many problems for paleoecologists in particular (Anderson and Odgaard, 1994). The "calm" regions of the bay of Lake Gościąż are a much better place for paleoecological explorations, because of the lack of resuspension.

Resuspension and the annual stirring may involve the uppermost layer of sediment containing pollen grains deposited in one or few years. Thus processes responsible for translocation of lake sediments can affect their homogenization and formation of the sediment layer representing a mean value of even several years' deposition. Resuspension and mixing cause the paleolimnological record in the lake sediments to be less distinct even in deep lakes with annual laminations. According to the present study the statement that paleolimnological studies involving varves allow the highest possible temporal resolution, i.e. single years or even seasons (Lotter and Sturm, 1994), does not have to be true. Further information regarding, how resuspension can influence the record of environmental history and the temporal resolution of laminated sediments, will be published independently.

References

Anderson, N.J. and Odgaard, B.V.: 1994, *Hydrobiologia* **275/276**, 411-422.

Berglund, B.E. and Ralska - Jasiewiczowa, M.: 1986, *Pollen analysis and pollen diagrams*. In Berglund, B.E. (ed.) *Handbook of holocene palaeoecology and palaeohydrology*, John Wiley and Sons, New York, 455-484.

Bloesch, J.: 1995, *Mar. Freshwater Res.* **46**, 295-304.

Nuhfer, E.B. Anderson, J.P. and Dean, W.E.: 1993, *Geol. Soc. of America Spec. Paper* **276**, 75-96 pp.

Davis, M.B.: 1968, *Science* **162**, 796-799.

Davis, M.B.: 1973, *Limnol. Oceanogr.* **18**, 44 - 52.

Davis, M.B. and Ford, J.: 1982, *Limnol. Oceanogr.* **27**,137-150.

Davis, M.B. Moeller, R.E. and Ford, J.: 1984. *Sediment focusing and pollen influx*, In Haworth, E.Y. and Lund, J.W.G. (ed.) *Lake sediments and Environmental History*, Leicester University Press, 261-293.

Krumbein, W.E. Paterson, D.M. and Stal, L.J.: 1994. *Biostabilization of sediments*. Verlag Rösemeier Bad Zwischenahn, p. 529.

Likens, G.E. and Davis, M.B.: 1975, *Verh. Internat. Verein. Limnol.* **19**, 982-993.

Lotter, A.F.: 1991, *Hydrobiologia* **214**, 53-57.

Lotter, A.F. and Sturm, M.,: 1994, *Journal of Paleolimnology* **11**, 311-312.

Mieszczankin, T. in preparation. *Processes of sedimentation in Lake Gościąż - On genesis of the modern lake sediments and role in the functioning of the lake ecosystem*, PhD thesis, Dept. of Hydrobiology, N. Copernicus University Toruń, Poland.

Ralska - Jasiewiczowa, M. van Geel, B. Goslar, T. and Kuc, T.: 1992, *Sveriges Geologiska Underskning, Ser. Ca* **81**, 257-268.

Rosa, F. Bloesch, J. and Rathke, D.E.: 1991, *Sampling the settling and suspended particulate matter (SPM)*, In Mudroch, A and MacKnight, S.D. (ed.) *Handbook of techniques for aquatic sediment sampling*, CRC Press: Boca Raton, 97-130.

PALAEOENVIRONMENTAL RECONSTRUCTION OF LAGO DI ALBANO (CENTRAL ITALY) DURING THE LATE PLEISTOCENE USING FOSSIL OSTRACOD ASSEMBLAGES

CLAUDIO A. BELIS

CNR-Istituto Italiano di Idrobiologia, 28048 Pallanza Verbania, Italy

Abstract. A fine resolution study on fossil ostracods in late glacial sediments from Lago di Albano was performed within the framework of the European project PALICLAS. Four cores from two coring sites were analysed. At Site 1, located at 70 m depth, cores PALB94 1E and PALB94 1C were collected, while cores PALB94 6A and PALB94 6B were collected at Site 6 at 30 m depth. The autoecology of the ostracod species, cluster analysis, PCA and CA ordination numerical analyses were performed in order to identify changes in environmental conditions within the sedimentary record. Three assemblages were recognized: (a) dominated by *C. neglecta* representing sublittoral environments, (b) dominated by *C. neglecta* accompanied by *Potamocypris* spp. related to spring influx and (c) dominated by *Cyclocypris* sp. associated with littoral and relatively warmer conditions.

A reconstruction of lake water level, oxygen concentration near the bottom and lake productivity , based on ostracod autoecology and supported by fossil pigment data, was accomplished. Sharp fluctuations in water level and productivity seem to indicate climatic oscillations between ca 30 to ca 17 kyr BP.

Key words: ostracods, autoecology, crater lake, palaeolimnology, Central Italy, numerical analyses

1. Introduction

Predicting future global climatic changes is one of the major research priorities of this decade. Intensive studies on physical, chemical and biological processes are being carried out in order to predict, by modelling, the rates and direction of biotic change that may occur in the near future (Delcourt & Delcourt, 1991).

This contribution is part of the EU project PALICLAS (Palaeoenvironmental Analysis of Italian Crater Lake Sediments) whose main objectives are to reconstruct the palaeoclimate during the last 30 kyr, to recognise the responses of natural systems to climate changes and to disentangle natural from human effects during the Holocene (Oldfield, 1996).

Many organisms leave morphological remains in sediments which are useful for palaeoecological analyses because of the strong relationship existing between biota and environmental conditions (Charles & Smol, 1994). In general, the overall composition of biotic communities is a good reflection of the overall environment. Among the biological records, ostracods have been employed extensively in palaeoecological research to provide information from both their autoecology and their shell chemistry (Palacios-Fest, 1994; Holmes, 1996).

Numerical analyses are commonly performed with the aim of extracting relationships hidden by the complexity of the data. Classification is aimed at defining groups of similar elements; in ordination, the object is simply to arrange the elements in two or three dimensional plots in which geometric distances represent their dissimilarity (Birks, 1987).

The most up-to-date knowledge in ostracod ecology and simple numerical analyses

Water, Air and Soil Pollution **99**: 593-600, 1997.

will be combined in this study in order to provide a fine resolution reconstruction of some lacustrine palaeoenvironmental parameters in the sediments from the small (6 km^2 surface) and deep (176 m depth) crater Lago di Albano (Central Italy).

2. Materials and methods

Four cores (ca 9 to ca 14 m long) were taken from two sites of Lago di Albano along a bathimetric transect by the staff of the ETH-Zentrum (Zürich). The coring device was a piston Kullenberg system (Kelts *et al.*, 1986). Cores PALB94 1E (1E) and PALB94 1C (1C) were collected at Site 1, located at 70 m depth; cores PALB94 6B (6B) and PALB94 6A (6A) were collected at the shallower Site 6, located at 30 m depth (Fig. 5). Cores were cut longitudinally and sectioned into 2-cm slices. Only selected slices, on the basis of lithological changes, were analysed. Continous sub-sampling carried out only in core 1C (1264-1329 cm depth) enabled us to perform a ca 37 yr time-resolution study.

For ostracod analysis, 10-20 g of wet sediment was sieved through a 200 µm mesh screen and the valves were removed, identified and enumerated under a binocular microscope (magnification 20x - 100x). A detailed taxonomic description of the species and their autoecology has been presented elsewhere (Belis *et al.*, 1997).

Depth index was computed as the ratio between the absolute abundaces of profundal species (*Candona neglecta* and *Cytherissa lacustris*) and littoral species *(Cyclocypris* sp., *Ilyocypris bradyi, Cypria ophtalmica* and *Herpetocypris reptans*). Primary Productivity was estimated using a regression equation based on fossil pigments (Lami *et al.*,1997; this issue). Qualitative oxygen concentration in bottom waters was inferred from the presence of *C. lacustris* and *Potamocypris* spp. in the ostracod assemblages.

The chronolgy of Site 6 was established by ^{14}C calibrated ages on pollen grains and varve counting. Extrapolated dates, based on ^{14}C cal. age at 1040 cm (1E) and a dated tephra at the bottom, were used for the chronology of Site 1 (Chondrogianni *et al.*, 1997). For a more detailed description of pigment, geochemistry and lithological analyses see Lami *et al.* (1997; this issue) and Chondrogianni *et al.*, (1997). Numerical analyses were carried out following Davis (1986) and Jongman *et al.* (1987).

3. Results and discussion

OSTRACOD ASSEMBLAGES

A summary description of ostracods assemblages and other parameters along the cores of sites 1 and 6 is presented in tables 1 and 2 respectively. The ostracod sequence of cores from Site 1 are confined to their lower portions, no ostracods were found above 1100 cm in core 1E (1264 cm in core 1C). Worth to mentioning is the fact that cores 1E and 1C differ in length (Figure 1). The latter is shorter but presents higher accumulation rates in the Pleistocene-Holocene boundary; for that reason a difference of 1.6 m is observed in the core correlation (Chondrogianni *et al.*, 1997)

Only a preliminary study was carried out in core 6A, whereas core 6B was studied thoroughly (Figure 2).

Table 1. Summary biostratigraphy of Site 1

Zone	1C (cm)	1E (cm)	Ostracods	Other parameters
11	1271-1264	1107-1100	very few ostracods; disappearance	Higher pigments conc.
10	1279-1271	1115-1107	*Cyclocypris* sp peak., *C. neglecta*, *I. gibba*, *C. ophtalmica*, *C. candida*, *Leptocythere* sp.	Peak of pigment conc. Non planktonic diatoms
9	1304-1279	1140-1115	*C. neglecta*, *C. lacustris*, *Potamocypris* spp., *I. gibba*, *Cyclocypris* sp.	Higher chlorophyll derivative conc.
8	1314-1304	1150-1140	*C. neglecta* peak, , *C. candida*, *Cyclocypris* sp., *Potamocypris* spp, *I gibba*, *C. lacustris*, *C. ophtalmica*	Increase in chlorophyll derivative conc.
7	1326-1314	1162-1150	*C. neglecta*, *I. gibba*, *C. lacustris*, *Cyclocypris* sp., *C. ophtalmica*, *C. candida*, *Potamocypris* spp.,	
6	-1326	1210-1162	*C .neglecta*, *I. gibba*, *Cyclocypris* sp., *C. lacustris*	
5		1220-1210	*C. neglecta*, *I. gibba*	Peak of pigment conc.
4		1256-1220	*C. neglecta*, *Cyclocypris* sp. *C. lacustris*, *L. inopinata*, *I. gibba*	
3		1307-1256	very few valves	Peak of pigments, low carbonate conc.
2		1332-1307	*Cyclocypris* sp.	Planktonic diatoms (here on)
1		1342-1332	*I .gibba*	Benthic diatoms

Table 2. Summary biostratigrapy of Site 6

Zone	6B (cm)	6A (cm)	Ostracods	Other parameters
9	10-0	10-0	*C. neglecta*, *Cyclocypris* sp. *F. cf fabaeformis*, *L. inopinata*, *C. opthalmica*	high carbonate and pigment conc.
8	90-10	170-10	*Cyclocypris* sp.	Medium to high carbonate and pigment conc.
7	193-90	240-170	not found	low carbonate and pigment conc.
6	327-193	370-240	*Cyclocypris* sp., *C. ovum*	high carbonate and pigment conc.
5	454-327	505-370	not found	low carbonate and pigment conc.
4	540-454	600-505	*C. neglecta*, *H. reptans*, *Cyclocypris* sp., *Potamocypris* spp., *C. opthalmica*	high carbonate and pigment conc.
3	631-540	660-600	not found	low carbonate and pigment conc.
2	726-631	770-660	*Cyclocypris* sp.	low carbonate and high pigment conc.
1B	765-726		few valves	low carbonate and pigment conc.
1A	820-765		*I. gibba*, *C. neglecta*	high carbonate and pigment conc.

CLUSTER ANALYSIS

In the cluster analysis the relative abundance, as percentages, of all the sections bearing ostracods from cores 1E, 1C, 6B and 6A (116) have been included (Figure 3). The difference between samples (sections) was computed as euclidean distances and Ward's method was used to link them (van Tongeren, 1987).

Seven main clusters were identified at a distance (cut off) of 200 units. The principal division of the diagram into two large sets of clusters coincide with the dominance of *C. neglecta* and is probably related to variations in the lake water level.Within the group characterised by the dominance of *C. neglecta* two clusters can be observed. The first of these (on the right of the diagram) is characterised by a low

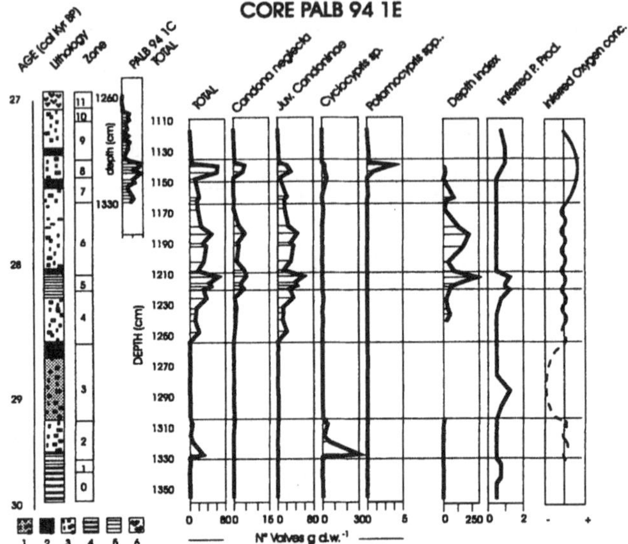

Fig. 1.Total and specific ostracod abundances in core 1E. Biostratigraphic zones are indicated with dashed lines. A total ostracods abundance profile from core 1C has been included. Depth index, inferred primary productivity and inferred oxygen concentration are also shown. Chronology and lihtology are presented on the left side of the diagram
(1) light grey spotted muds; (2) dark grey spotted muds; (3) yellowish grey spotted silts rich in calcite; (4) distinct lamination of whitish calcite and olive grey detrital silts; (5) fine laminated yellowish grey muds, rich in calcite; (6) light grey muds with moss remains.

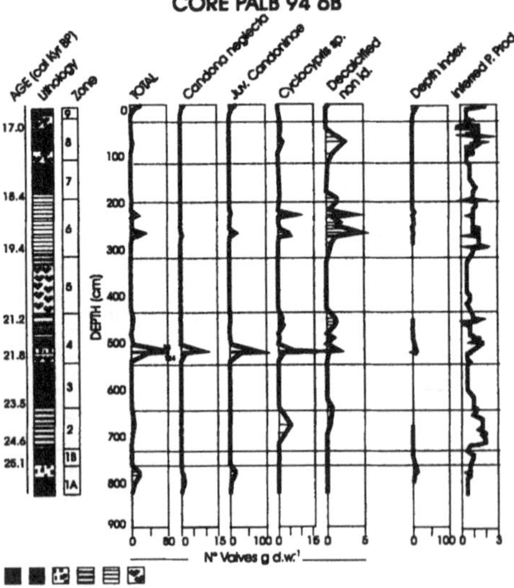

Fig.2. Total and specific ostracod abundances in core 6B (only selected species). Biostratigraphic zones are indicated with dashed lines. Depth index and inferred primary productivity are also included. Chronology and lithology are presented on the left side of the diagram.
(1) light brown rich in organic matter; (2) dark grey muds; (3) yellowish grey silts rich in calcite; (4) dark brown laminated muds; (5) fine laminated yellowish grey muds, rich in calcite; (6) massive olive grey with moss remains.

abundance of ostracods and a low number of species. Mainly sections from intermediate zones of core 1E are present in this cluster. Higher values of valve abundance and species richness are recorded among the sections in the second cluster; they belong mainly to core 1C. In this cluster species of the genus *Potamocypris* (*P. zschokkei, P. pallida* and *P. fallax*) are well represented.
In the second main group of clusters (5) in which *C. neglecta* is not dominant, sections of core 6B are more abundant. The cluster on the extreme left of the dendrogram comprises sections from zones 2B and 2E (the letter indicates the core) in which only *Cyclocypris* sp. was found. The neighbouring clusters include the sections of the zones 6B, 8B, 4B and

10C in which *Cyclocypris* sp. is dominant and *C. neglecta* is also present. Finally, there is a little cluster (two sections at the bottom of core 1E), in which only *I. gibba* was found.

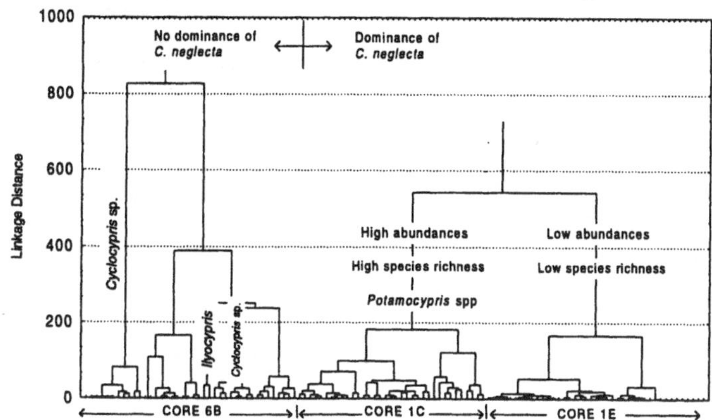

Fig 3. Tree diagram of the cluster analysis of 116 sections from cores 1E, 1C, 6A and 6B. The three major groups repoted on *x*-axis largely correspond to the indicated cores (see text).

In general, sections belonging to the same site tend to lie together, highlighting the identity of the basins of the lake. However, in many cases sections of different sites are close together, probably stressing stratigraphic or ecological similarities.

The ostracod sequences of the replicate cores 1E and 1C overlap only in zone 8 and part of zones 7 and 9. This feature is reflected in the dendrogram by the close linkage between sections of zones 8E and 8C. Nevertheless, the sections of zones 7E, 7C and 9E, 9C are more distant. The last is probably due to discrepancies in the of peaks of *Potamocypris* spp. and *C. lacustris*. These inconsistencies could be accounted for by the different scale of resolution performed in cores 1E and 1C.

PRINCIPAL COMPONENTS ANALYSIS

The analysis was run using relative abundances. In the biplot presented in figure 4 the samples are distributed along an arch. At the end of the left branch are located the sections with dominance of *Cyclocypris* sp. (zones 2E, 2B, 4C, 8B and 6B). At the center there are samples with a strong dominance of adults of *C. neglecta* and juveniles of Candoninae (zones 4E, 5E, 6E, and 9B). In the right branch of the arch there are scattered sections rich in species with dominance of *C. neglecta,* and also a strong presence of *Potamocypris* species (zones 7E, 9C, 7C, 8C and 4B).

Only the more representative species have been plotted as vectors in order to identify by visual inspection their influence on the distribution of the core samples in the correlation biplot. The first and second axes accounted for 23% of the total variability. These apparently low values are common in analyses based on data with many zeroes due to the presence of rare species (Stevenson *et al.*, 1991). A Correspondence Analysis (not shown) was also performed on the same data. The first and second axes accounted for 32% of the total variance. The patterns observed in the joint plot are analogue to those already described in the PCA. The information provided by the cluster analysis and the PCA highlighted the importance of the dominance of *C. neglecta* and juveniles of Candoninae as the main segregating factor. However, the dominance of *Cyclocypris* sp., the abundance of *Potamocypris* spp. and the richness of species are factors that contribute substantially to the variability.

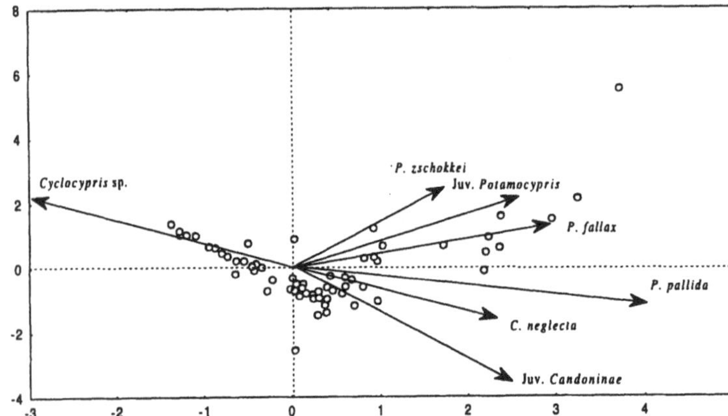

Fig. 4. Correlation biplot of the Principal Components Analysis of sections from cores 1E, 1C, 6A and 6B. Circles represent sections. The arrows represent species vectors. The length and orientation of vectors reflect the variability of the species and the angle between vectors is associated with the correlation between species.

The biplot diagram highlighted three well-defined species assemblages from the biostratigraphic point of view: a community dominated by *C. neglecta*, one dominated by *Cyclocypris* sp. and one dominated by *C. neglecta* with the presence of well-structured population of *Potamocypris*. Each of them is related to well-defined environmental conditions: sublittoral, littoral and littoral-sublittoral with input of cold springs, respectively (Meisch, 1984; Nüchterlein, 1969; Belis *et al.*, 1997). The succession of different lake environmental conditions was reconstructed through the identification of these assemblages along the sequences.

4. Environmental reconstruction

Potamocypris zschokkei and *C. lacustris* were considered as indicative of cold conditions. The last-named was also used as an indicator of deep, well-oxygenated waters (Geiger, 1990). *C. neglecta* is characteristic of sublittoral environments whereas assemblages dominated by *Cyclocypris* sp. were considered representative of littoral zones. *Potamocypris zschokkei, P. pallida and P. fallax* are indicators of the presence of springs or water flowing from springs (Meisch, 1984). In particular *P. pallida* is always present in waters poor in electrolytes.

Although the reconstruction presented in this paper was built up mainly on the basis of fossil ostracod data, integration with other biological proxy records was useful to improve the result. In fact, algal and bacterial pigment concentrations were considered to reconstruct paleoproductivity as well as redox conditions (Lami *et al.*, 1997; this issue). In addition, geochemical ($CaCO_3$, organic matter, etc.) and lithological features were contemplated in the elaboration of the inferred parameters (Chondrogianni *et al* 1997; Lami *et al.*, 1996; Lami *et al.*, 1997; this volume).

A summary scenario of the evolution of lake environment from its origin to 17 kyr BP is presented below:

Site 1 spans a time interval between ca 30 and ca 27 kyr BP. The lake water level fluctuated from very low at the beginning of the sequence to a higher level from zone 3 onwards and then back to low at the top of the ostracod sequence. Deep waters were generally oxygenated. Oxygen concentrations are high in zones 7, 8 and 9. Three main peaks of productivity accompanied by seasonal anoxic conditions are observed in zones 3, 5 and 8-9-10.

At **Site 6** the lake water level was shallower than in Site 1, fluctuating from very shallow (swampy) to deeper (littoral). Phases of high and low productivity alternated. Four main productive periods, with lower oxygen content, are observed in zones 2, 4, 6 and 8. Excepting the uppermost 10 cm, this sequence represents sediments deposited between ca 25 and ca 17 kyr BP.

The correlation of the sequences of Site 1 and Site 6 , on the basis of fossil pelagic diatoms, demonstrated that 10 cm in core 6B can be correlated to 720 cm in core 1E, with the bottom of the first corresponding to ca 1050 cm in the second. Hence the whole sequence of ca 8 m of core 6B represent an expanded record of the section 720-1050 cm in core 1E (Ryves *et al.*, 1996).

Assuming this correlation a schematic reconstruction of major changes in the water level of Lago di Albano during the latest Würm is presented in figure 5. A low level persisted in the time window represented by the ostracod sequence of Site 1 (30 to 27 kyr BP). A rapid rise of around 40 m in the level occurred after that period leading to the beginning of the deposition of lacustrine littoral and sublittoral sediments in Site 6 during the time interval between 25 and 17 kyr BP. At the same time a deep lake environment with increased input of suspended inorganic particles (turbidity), reduced supply of carbonates, temporarily stratified water column and seasonal oxygen stress became established in Site 1 causing the disappearance of ostracods from the sedimentary record.

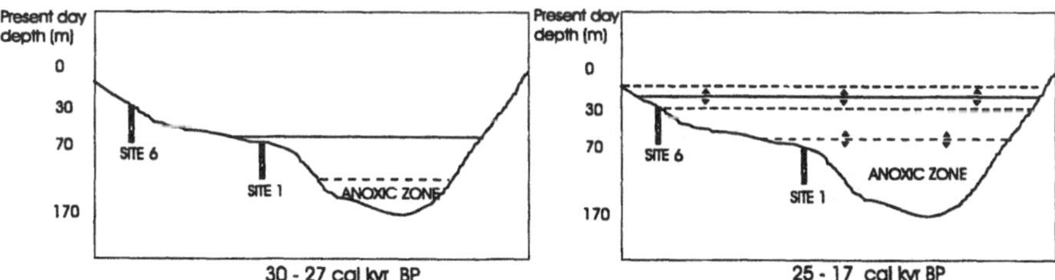

Fig. 5. Hypothetical scenario of water level evolution at Lago di Albano during the late Würm.

A recent study on fossil ostracods from the nearby Valle di Castiglione (Gliozzi & Mazzini, 1997) suppport the scenario presented in figure 5. In this sequence, the interval 33-26 kyr BP, that present an ostracod community dominated by *C. neglecta*, forms part of an arid phase and is followed by a cold semiarid oscillation (26-17 kyr BP), slightly more humid.

5. Concluding remarks

Ordination analyses highlighted assemblages whose ecological meaning is well defined. As a whole, the numerical analyses contributed significantly to describing and finding patterns within the dataset. There is good agreement between replicate cores 1E and 1C despite minor discrepancies. The ca 37 yr temporal resolution achieved in core 1C made it possible to recognise high frequency fluctuations of environmental conditions. The proposed model of the filling of the lake provides a satisfactory explanation of the complex pattern of sequences and assemblages of the lake sediments in sites 1 and 6. Considering the importance of the integration of biological and chemical features of ostracod valves, a study of trace elements and stable isotopes of fossil ostracods in this lake is one of the prioritary objectives for future research.

Acknowledgements

The author is grateful to P. Guilizzoni, A. Lami and A. Marchetto for the critic review of the manuscript and helpful comments. Thanks to all colleagues participating in the PALICLAS for the access to their data. Two anonymous reviewers improved the first version of the MS. This study was partially supported under the EU Environmental Programme contract n° EV5V-CT93-0267.

References

Belis, C.A., Geiger, W., Lami, A., Guilizzoni, P., Ariztegui, D. & Chondrogianni, C.: 1997, *J. Paleolim.* (submitted)

Birks, H.J.B.: 1987, *Chemometrics and Intelligent Laboratory Systems* **2**: 15.

Charles, D.F. & Smol, J.P.: 1994, in: Baker L. (ed) *Environmental Chemistry of Lakes and Reservoirs.* American Chemical Society. pp 3-31.

Chondrogianni, C., Ariztegui, D., Niessen, F., Ohelndrof, C., Lister, G.: 1996, *Mem. Ist. ital. idrobiol.* **55** (in press).

Davis, J.C.: 1986, Statistics and data analysis in Geology. John Wiley & Sons, New York.

Delcourt, H.R. & Delcourt, P.A.: 1991, Quaternary Ecology, A paleoecological perspective. Chapman & Hall, London. 239 pp.

Geiger, W.: 1990, *Bull. Inst. Geol. Bassin d'Aquitaine* **47**: 167.

Gliozzi, E. & Mazzini, I.: 1997, "Palaeoenvironmental analysis of the 250,000-year sediment core of Valle di Castiglione (Latium, Italy) using ostracods". *Proceedings of the 3rd European Ostracodologists Meeting* Paris-Bierville July 8-12. (in press)

Holmes, J.A.:1996, *J. Paleolimnol.* **15**:223-235.

Jongman, R.H.G., ter Braak, C.J.F. & van Tongeren O.F.R.: 1987, Data analysis in community and landscape ecology. Wagenigen, The Netherlands.

Kelts, K., Brieguel, U., Ghilardi, K. & Hsü, K.:1986, *Schweiz. Z. Hydrol.* **48**:104-115.

Lami, A., Guilizzoni, P., Bettinetti, R., Belis, C. A., Manca, M., Comoli, P., Marchetto, A., Ariztegui, D. & Chondrogianni, C.: 1996, *Il Quaternario* (in press)

Lami, A. , Guilizzoni, P., Battarbee, R.W., Belis, C.A., Bettinetti, R., Manca, M., Comoli, P., Marchetto, A., Ryves, D.B., Jones, V.J. and Nocentini, A.M.: 1997, (this issue).

Meisch, C.: 1984, "Revision of the recent western Europe species of the genus *Potamocypris* (Crustacea, Ostracoda) Part I". Travaux Scientifiques du Museé d'Histoire Naturelle de Luxembourg 3. Luxemburg. 55 pp.

Nüchterlein, H.: 1969, *Int. Revue ges. Hydrobiol.* **54**: 223-287.

Oldfield, F. (ed): 1996, PALICLAS *Final Report.* Liverpool. 702 pp.

Palacios-Fest, M.R.:1994, *Geoarcheology: An international Journal*, **9**: 1-29.

Ryves, D.B., Jones, V.B., Guilizzoni, P., Lami, A., Marchetto, A., Bettinetti, R. & Devoy, E.: 1996, *Mem. Ist. ital Idrobiol.* **55** (in press)

Stevenson, A.C., Juggins, S., Birks, H.J.B., Anderson, D.S., Anderson, N.J., Battarbee, R.W., Berge, F., Davis, R.B., Flower, R.J., Haworth, E.Y., Jones, V.J., Kingston, J.C., Keiser, A.M., Line, J.M., Munro, M.A.R., & Renberg, I.:1991, *The Surface Waters Acidification Project Palaeolimnology Programme: Modern Diatom/Lake-Water Chemistry Data-Set.* Ensis London 86 pp.

van Tongeren, O.F.R.: 1987, in: Jongman, R.H.G, ter Braak, C.J.F. and van Tongeren, O.F.R. (eds.) *Data analysis in community and landscape ecology.* Wagenigen, The Netherlands.

A LATE GLACIAL AND HOLOCENE RECORD OF BIOLOGICAL AND ENVIRONMENTAL CHANGES FROM THE CRATER LAKE ALBANO, CENTRAL ITALY: AN INTERDISCIPLINARY EUROPEAN PROJECT (PALICLAS)

A. LAMI[1], P. GUILIZZONI[1], D.B. RYVES[2], V.J. JONES[2], A. MARCHETTO[1], R.W. BATTARBEE[2], C.A. BELIS[1], R. BETTINETTI[1], M. MANCA[1], P. COMOLI[1], A. NOCENTINI[1] AND L. LANGONE[3].

[1]CNR Istituto Italiano di Idrobiologia, Verbania Pallanza, Italy
[2]Environmental Change Research Centre, University College London, UK
[3] CNR Istituto per la Geologia Marina, Bologna, Italy

Abstract. This paper reports the results of biological analyses (pigments, diatoms, chrysophyte cysts, cladocerans, chironomids and ostracods) of a ca. 14 m-long sediment core recovered from Lake Albano (Central Italy) in the course of the EU-funded project PALICLAS (PALaeoenvironmental analysis of Italian Crater Lake and Adriatic Sediments).

A reconstruction of the environmental evolution and ecosystem response of Lake Albano during the last ca. 30 kyr was possible. Additional information on lake level oscillation is obtained from benthic and planktonic palaeocommunities. Several oscillations in the productivity and the level of the lake were detected in the oldest sediment layers (from ca. 30 kyr BP to ca. 17 kyr BP), followed by a long (ca. 5 kyr BP) period of low productivity in which cold, holomictic conditions prevailed. A period of high biological activity and, probably, meromictic conditions during the early-mid Holocene was detected. A clear impact of human activities in the catchment was found at ca. 4 kyr BP in the form of increased erosion, associated with a decline in the abundance of biological remains. Further signs of human impact on the lake ecosystem are recorded during the Roman period. Although large-scale environmental changes (e.g. regional climate changes) caused many of the observed biological changes, human activities were important during the mid-late Holocene.

Key words: pigments, diatoms, cladocera, ostracods, chironomids, chrysophytes, sediments, Late Pleistocene, Holocene, crater lake, Italy

1. Introduction

Lake Albano is a small (6 km^2) and deep (175 m) crater lake in Central Italy located at 293 m a.s.l. (Margaritora, 1992). It was studied in the framework of an EU funded project (PALICLAS) which considered the vegetational sequence through pollen analysis, palaeoenvironmental reconstruction as revealed by studies of geochemistry, tephra, and magnetic properties, and biological remains: algal and bacterial pigments, diatoms, chrysophyte cysts, cladocerans, ostracods and chironomids (Guilizzoni and Oldfield, 1997).

Sedimentary pigments have proved valuable indicators of the past environment and can yield approximate indices for bioproductivity (Guilizzoni *et al.*, 1983). Pigments of plankton algal and microbial populations are partially sedimented and leave traces in the sediments (Züllig, 1982). A number of biological, physical and chemical factors influence their deposition and abundance (Leavitt, 1993), but they are useful indices of trophic conditions (e.g. Swain, 1985; Sanger, 1988), lake acidification (Guilizzoni *et al.*, 1992), and

Water, Air and Soil Pollution **99**: 601-613, 1997.

climate changes (Smol, 1988) and of permanent or transitory anoxic conditions at the sediment surface or at the sediment-water interface (Züllig, 1986).

The siliceous frustules of diatoms also are generally abundant and well preserved in lake sediments. Diatoms are sensitive ecological indicators and have been used widely to reconstruct changes in pH (Battarbee, 1984), salinity (Fritz, 1990) and nutrients (Hall and Smol, 1992; Bennion, 1994). Fossil cladocerans, ostracods and chironomids also have been used extensively for palaeoenvironmental reconstructions (Berglund, 1986; Lami *et al.*, in press).

This paper is concerned mainly with the palaeolimnology of the site as revealed by sedimentary pigments from algae and photosynthetic bacteria, fossil diatoms, ostracods, cladocerans and chironomids.

2. Methods

A 1387.5 cm sediment core (PALB 94-1E) was obtained from 70 m of water with a Kullemberg piston corer (Kelts *et al.*, 1986) in June 1994 by colleagues from the ETH-Zentrum, Zürich. Cores were cut longitudinally and sectioned into 2-cm slices. For this study, 146 samples were selected on the basis of core stratigraphy. An aliquot of wet sediment was weighed and freeze dried for the water content determination. Organic matter was estimated by loss on ignition (LOI). Organic carbon (C) and nitrogen (N) were measured by a CHN analyser (Carlo Erba NA1500).

Pigments were extracted and analyzed following Lami *et al.* (1994): chlorophylls and their derivatives (CD) and total carotenoids (TC) were measured spectrophotometrically and the single carotenoids by HPLC. CD are expressed as spectrophotometric units per gram organic matter (U g_{LOI}^{-1}) (Guilizzoni *et al.*, 1983). TC and single carotenoids are expressed as mg g_{LOI}^{-1} and nmoles g_{LOI}^{-1}, respectively (Züllig, 1982). Samples for the diatom, ostracod, cladocera and chironomid analyses were prepared and counted following Renberg (1990), Battarbee and Kneen (1982), Belis *et al.* (1997), Manca *et al.* (1996a), Brundin (1949) and Hofmann (1986). The data were expressed as cumulative percentages or in absolute counts per gram dry matter.

3. Results and discussion

3.1. CHRONOLOGY

The base of the sediment sequence has a 38 cm thick non-air fall tephra layer (Chondrogianni *et al.*, in press), dated at ca. 25 ± 5 kyr BP by the Ar^{40}/Ar^{39} method (Calanchi *et al.*, 1996). Based on correlation of paleomagnetic directional data and a geomagnetic master curve (Mackereth, 1971) from other dated cores (e.g. Windermere, U.K. and L. Bouchet, France), this tephra layer can be more precisely dated at ca. 30 kyr BP (Oldfield, 1996). Two air-fall tephras, one of Campanian origin (Avellino) at 354 cm, the second from Mt. Etna at 720 cm were dated (Narcisi, 1994) at 4.1 kyr BP and 17 kyr BP, respectively. The first 5 m of the core sequence, which comprise the Holocene period, were correlated with a dated core (collected at 120 m water depth) using magnetic susceptibility, pollen and diatoms profiles. This correlation allowed us to ascertain a hiatus in the sediment sequence between ca. 4.5 and 7 kyr BP (Oldfield, 1996). All the reported ages are calibrated (Stuiver and Reimer, 1993),

3.2. DRY MATTER AND ORGANIC CARBON AND NITROGEN

The dry matter content (d.m.) reaches its maxima at the core base and at 753 cm corresponding to the tephra layers and high erosional rates, respectively. High values of dry mass which reflect high mineral content, as well as the oscillations in the magnetic properties (Oldfield, 1996) are attained also during the Full Glacial Period (Fig. 1).

Organic C and N increase from very low concentrations (<5% d.m. and <0.5% d.m., respectively) in the Full Glacial period (with some exceptions) to about 12% and 1.1% d.m., respectively in the early-mid Holocene (Fig. 1). The Late Glacial rapid increase of these elements was interrupted during a period which can be probably associated with the Younger *Dryas*, and followed between 487 and 410 cm by a series of peaks.

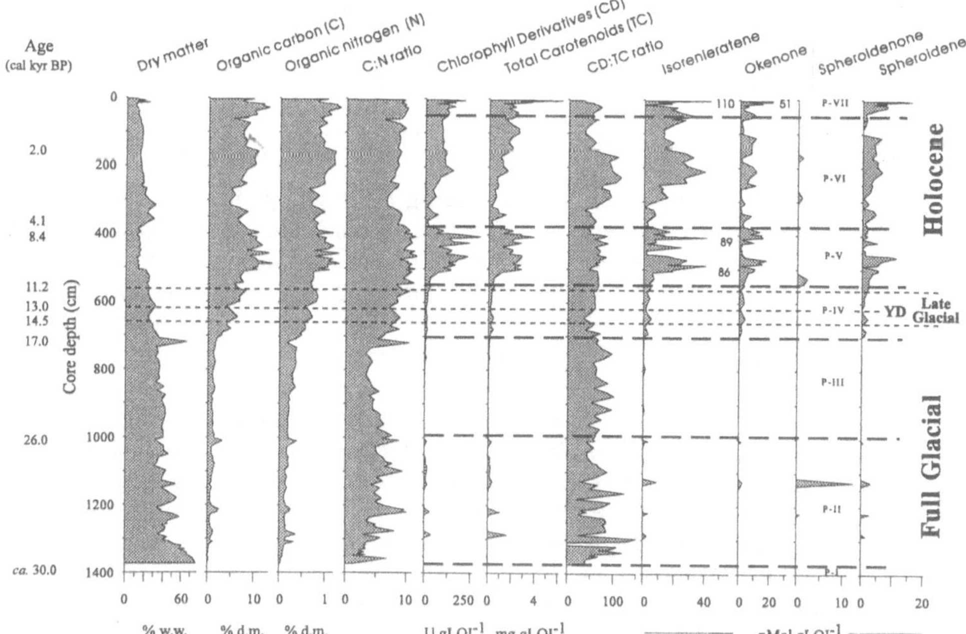

Fig. 1. Depth profiles of some geochemical records. The transitions Full Glacial/Late Glacial/Holocene are indicated by thin brocken lines. YD= Younger *Dryas*. Thick broken lines divide the sedimentary sequence in zones based on pigment distribution (P I-VII).

The C and N level fall at ca. 3.5-3.0 kyr, perhaps due to increased erosion (the percentage of dry matter increases), from human impacts on the catchment vegetation. Pollen, magnetic, mineralogical data (Oldfield, 1996) together with the pigment data (see below) substantiate this interpretation. High values are shown also during the past 2.0 kyr. A minimum at 65 cm during approximately the *Little Ice Age* is followed by a rapid increase of organic C and N which apparently reflects the onset of cultural eutrophication.

In general the C:N ratio is low (< 10) which indicates that the contribution of organic matter from the catchment was unimportant. The peaks during the Full Glacial Period and the increase during the Late Glacial/Holocene (Fig. 1) reflects the increased organic matter contribution due to changing vegetation (van der Kaars *et al.*, 1995) in a generally unproductive period (see below).

3.3. ALGAL AND BACTERIAL PIGMENTS

For most of the Full Glacial period, pigments (CD and TC) were found only in trace concentrations (zone P-I and P-III, Fig. 1). Five peaks of pigment concentrations were detected in zone P-II. This may reflect rapid climate change in the Full Glacial, which has been reported also by Alessio *et al.* (1986), Abbazzi *et al.* (1996) and Huntley *et al.* (1996) for other sites in Central and Southern Italy. Alternatively, these changes might be driven by geothermal activity in the crater lake basin (Martini *et al.*, 1994). Pigment concentrations increase during the Late Glacial (zone P-IV). Both CD and TC curves show a series of maxima and wide oscillations during the Early Holocene, reaching the levels of high productive lakes (e.g. Guilizzoni *et al.* 1983). Algal biomass is then markedly depressed at 377 cm (base of zone P-VI, ca. 3.5 kyr BP) due to turbid waters and high catchment erosion (Calanchi *et al.*, 1996). Considering all the other proxy-records investigated, this is interpreted as the first clear sign of human impact (Oldfield, 1996).

The CD:TC ratios are generally low during the high productivity phases and when the photosynthetic bacteria are also high. This indicates prevailing anoxic conditions. In addition, when both CD:TC ratio and the organic carbon are high, an allochthonous organic detritus input is indicated (e.g. between 300 cm and 190 cm).

If it is accepted that on a long-term basis, fossil pigment concentrations are generally proportional to the production (Leavitt, 1993), this can be modelled for the past ca. 30 kyr BP, using a regression equation established between cholorophyll derivatives and *in situ* rates of primary production as measured by the ^{14}C technique (Guilizzoni *et al.*, 1983). Six relatively high productivity peaks were found during the Full Glacial (cf. Fig. 3) associated with relatively high values of phototrophic bacterial pigments (Fig. 1) which indicate that anoxic conditions occurred temporarily. The values of isorenieratene, okenone and spheroidene rise dramatically at the Pleistocene/Holocene boundary, indicating strictly anaerobic conditions and the presence of sulphide.

According to Züllig (1985) the presence of spheroidene when spheroidenone is absent is indicative of meromictic conditions. Annual laminations and gelatinous, pigment-rich sediments also are represented in these zones of the core (zones P-IV, P-V). The remarkable presence of spheroidenone indicates that temporary oxic conditions were found possibly on the lake bottom.

Therefore, during the Holocene, three short periods of full lake circulation could have interrupted the chemical stratification at this location.

Higher concentrations of isorenieratene (green sulphur bacteria, Chlorobiaceae) over okenone (purple sulphur bacteria, Chromatiaceae) along most of the core indicate a more efficient penetration of light to the lake bottom (Takahashi and Ichimura, 1970). In fact, in many meromictic lakes two horizontally stratified populations of photosynthetic bacteria frequently occur: the upper composed of Chromatiaceae, and the lower by Chlorobiaceae (Brown *et al.*, 1984).

3.4. Chrysophyte cysts

Remains of these algae in Lake Albano are rare, especially when compared with the number of diatom frustules (Fig. 2). The stability and the low value of the Chrysophyte:diatom ratio (Ch:Di) throughout the sediment sequence is probably a consequence of lake conditions favouring a rich diatom phytoplankton community during most of the Late Pleistocene and Holocene despite major environmental changes in the lake system. By contrast, in a nearby shallow (5 m deep), crater lake (Lago di Monterosi), the chrysophyte flora was very rich and abundant during the glacial period (Leventhal, 1970) responding quickly to changes in lake habitat, which was probably low in level and ice-covered.

However, it should be noted that there is a single sample, at 794 cm depth (around 18 kyr B.P.), which shows a Ch:Di ratio distinctly higher than the rest of the core. This sample is characterized also by an anomalous diatom flora, with a high percentage of *Cyclotella pseudostelligera*, which is not present in any other sample. The occurrence of the two signals in the same sample is not however related to any other evidence from different records for an environmental change.

3.5. Diatoms

Lake Albano has been dominated by planktonic taxa throughout its history, except at the very base of the core (zone D-I, 1387-1322 cm) where epiphytic and non-open water forms (expecially *Achnanthes minutissima* and its varieties) briefly peak (Fig. 2). Between this level and 647 cm (zones D-II, D-IV), floral diversity notably increases, and the diatom assemblage is dominated by the centric *Cyclotella* sp. 1, a form which has similarities to *C. cyclopuncta*. Successional phases among *Cyclotella* sp.1., *Fragilaria brevistriata* and *Asterionella formosa* characterize zones D-II, D-III and D-IV. Above zone D-IV, *Cyclotella* sp. 1 essentially disappears, being replaced by significant but fluctuating proportions of other *Cyclotella* and *Stephanodiscus* taxa. *Fragilaria*

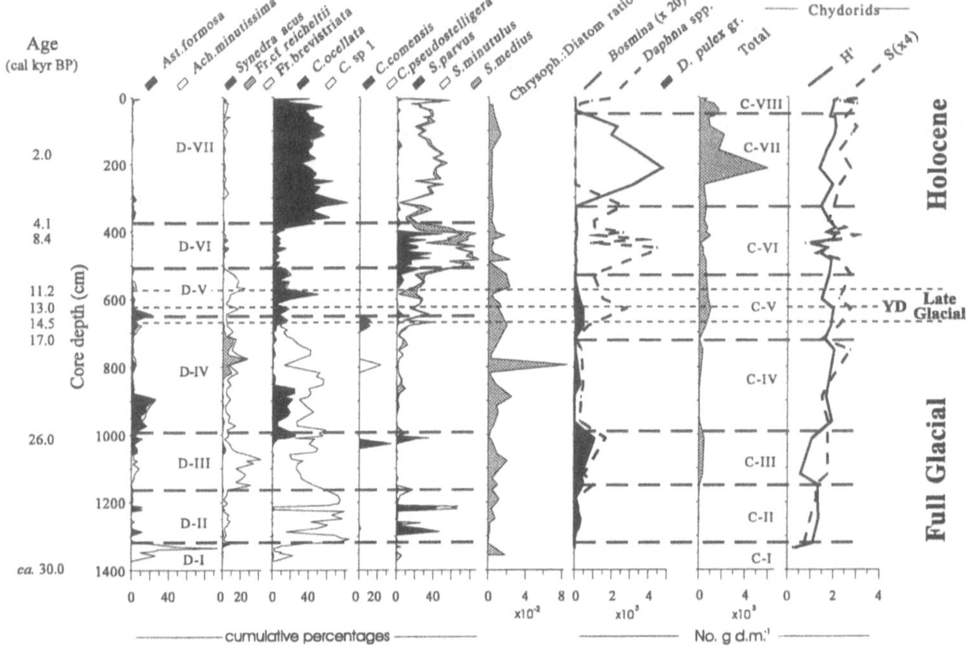

Fig. 2. Depth profiles of diatoms, chrysophycean cysts and cladocera (cumulative percentages). Shannon and Weaver diversity index (H') and number of species (S) of the chydorid community are also indicated. The transitions Full Glacial/Late Glacial/Holocene are indicated by thin brocken lines. YD= Younger *Dryas*. Thick broken lines divide the sedimentary sequence in zones based on diatoms and caldocera remains (D I-VII, C I-VIII).

and *Synedra* are relatively important from about 1147 to 1047 cm (zone D-III) and at 577-507 cm (zone D-V).

Between 507 cm and 377 cm (zone D-VI), the assemblage is characterized by the rise of several small *Stephanodiscus* taxa, typical of eutrophic lakes, suggesting that during the early Holocene the lake became much more productive, especially in zone VI, and has remained so to the uppermost levels. The rise in small *Stephanodiscus* taxa in zone VI, when catchment clearance was not apparent from the pollen diagram, may point to a very early human influence with little anthropogenic vegetational change, and may be linked to the climatic amelioration of the Early Holocene (The Hypsithermal). When pollen diagrams show evidence of forest clearance (397 cm, 310 cm) and during the Roman Period (190-170 cm) (Accorsi *et al.*, 1996), the *Stephanodiscus* taxa decline. The decline of *Fragilaria* and *Synedra* may in part reflect increasing turbidity of water, linked to algal blooms, as a result of an increase in the trophic level of the lake. In the uppermost zone (zone D-VII, 377-0 cm), the assemblage largely consists of *Cyclotella ocellata* and *Stephanodiscus* spp. Although small *Stephanodiscus* taxa are linked to highly eutrophic waters in contemporary systems, it should be noted that there are significant peaks of these forms in the lower 200 cm of the core (zone II). These may be related to short periods of warm conditions or periods of geothermal activity. These levels coincide with peaks in the preserved pigments and chironomids (Fig. 3), implying higher productivity at these periods.

A total phosphorus (TP) transfer function (Wunsam and Schmidt, 1994), has been applied to the Lake Albano diatom data (Fig. 3). Apart from taxonomic issues, there are some no-modern analogue problems, mainly for samples with significant proportions of *Cyclotella* sp.1 (most of the pre-Holocene). As there still remain questions of applicability of a transfer function developed for modern Alpine lakes to a crater lake sequence, the profile of reconstructed TP should be interpreted cautiously as a guide to trends in lake water phosphorus concentration. Despite these misgivings, the reconstruction (Fig. 3) suggests that TP levels were generally high throughout the Holocene, but especially in the early Holocene (487-427 cm), falling unevenly from this point until a more sustained high level from about 297 cm (somewhat after the pollen-inferred onset of anthropogenic catchment clearance at 380 cm) and remained high until the uppermost 20 cm.

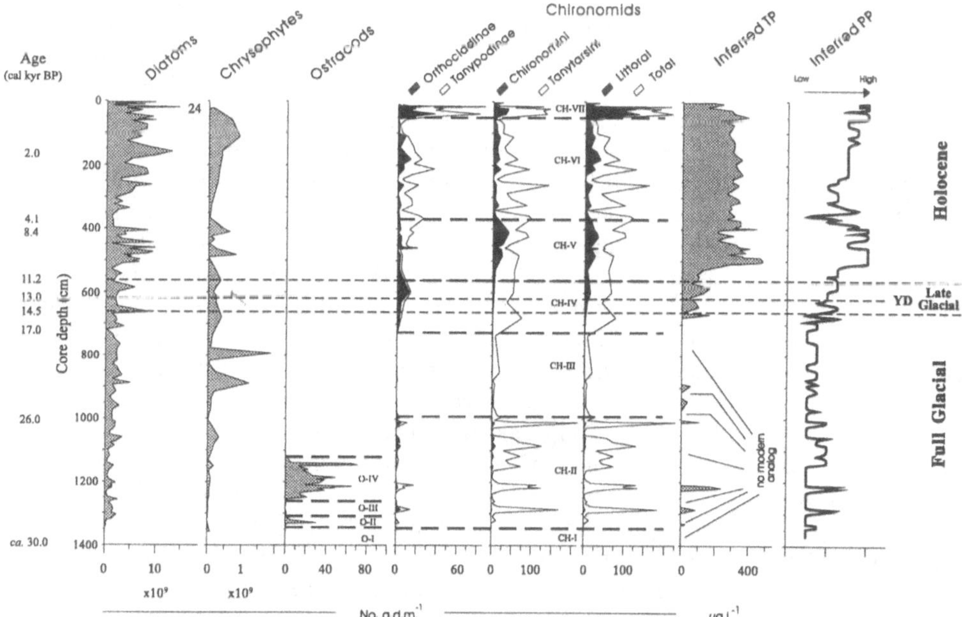

Fig. 3. Depth profiles of concentrations of diatom, chrysophyte cyst, ostracod and chironomid (cumulative) remains. Inferred total phosphorus (TP) concentration and primary productivity (PP) are also shown. The transitions Full Glacial/Late Glacial/Holocene are indicated by thin brocken lines. YD= Younger *Dryas*. Ostracod and chironomid biozones are indicated by thick broken lines (O I-IV; CH I-VII).

3.6. CLADOCERAN REMAINS

At the bottom of the core (1349-1320 cm, zone C-I), the very few remains found were only of chydorid species, with a Chydorus-Alona-Monospilus community (Manca and Comoli, 1996). Above this, the zone C-II (1320-1150 cm) is clearly defined by a steep increase in the number of chydorid taxa (S) and in the Shannon-Wiener diversity index (H'), indicating early stages of colonisation, mainly due to taxa associated with aquatic vegetation or to benthic diatoms. In this zone, the large zooplanktonic herbivore *Daphnia pulex* group is first recorded. At the base of zone C-III (1150-995 cm) a sharp decline in the diversity index is recorded, essentially due to the dominance of *Chydorus cf. sphaericus*, supporting the lowering of the water level suggested by the pigment, diatom and ostracod record. Then the abundance of *Daphnia* remains increases with a peak at ca. 1010 cm, corresponding to the *Stephanodiscus* peak and to the minimum of the Ch:Di ratio. The increase in *Daphnia* abundance might be indicative of an increase in productivity and TP (Jeppesen *et al.*, 1996).

Between 995 and 720 cm (zone C-IV) the number of remains is generally low, with a sharp decrease in *Daphnia* remains (mainly *D. longispina* group), and the appearance of the smaller-sized *Bosmina* (970-900 cm). At the same time, the diversity of the chydorid assemblage drops and the relative abundance of *Leydigia* spp. increases, indicating an increase in primary productivity and TP (Jeppesen *et al.*, 1996). Between 850 and 720 cm new chidorid species appear, both benthic and associated with macrophytes (Manca and Comoli, 1996).

After a minimum at 720 cm, there is an increase in the chydorid taxa (Zone C-V, 720-530 cm) with the appearance of *Pleuroxus striatus*, a benthic species that is found in modern L. Albano at about 50 m water depth (Margaritora, 1983). Among the Macrothricidae, a species of cold, relatively deep waters (*Ilyocryptus sordidus*) is found at 630 cm level, where the maximum development of *Monospilus dispar* is also recorded. These findings suggest a period with deep, low productive, transparent or acidic waters, with a cold, deep zone well established. Among zooplankton, there is an increase in concentration of *Daphnia* remains, particularly of the *D. longispina* group (Fig. 2), and a *Ceriodaphnia* phase. A decrease in H' and S is recorded at ca. 600 cm, and might be related to the Upper Phase of Younger *Dryas* (Fig. 2). The transition Late Glacial-Holocene is recorded by marked changes in the cladocera assemblage: an initial increase in H' and in S of the chydorid community, together with a decrease in *Daphnia* remains and an almost complete disappearance of *Daphnia pulex* group from 530 cm level onward.

Zone C-VI (530-330 cm) is characterised by the maximum development of the *Daphnia longispina* group, related to the maximum levels of CD (cf. Fig. 1) and the *Stephanodiscus* phase (Fig. 2). From 490 cm to 425-370 cm there is a marked decrease in *Daphnia* size, an increase in relative abundance of the small-sized *Bosmina* spp. and the appearnce of the large zooplanktonic predator *Leptodora kindtii* appears (Manca and Comoli, 1996). Altogether these changes indicate a period of increased productivity and TP. At ca. 430 cm a marked decrease in total remains, H' and S is recorded, with values that were found only during the Glacial Period (Fig. 2). A steep increase in the number of chydorid taxa (S) is recorded at ca. 410 cm level.

The transition to zone C-VII (330-50 cm) is marked by the shift from a *Daphnia*-dominated to a *Bosmina*-dominated community. *Bosmina* spp. increase exponentially from ca. 330 cm, probably in relation to forest clearance (Kerfoot, 1974) which marks the Bronze Age (Accorsi *et al.*, 1996). The maximum concentration of *Bosmina* remains is at ca. 210 cm, where also chydorids attain their maximum abundance (Fig. 2) and a peak in pollen concentration is recorded (Accorsi *et al.*, 1996). The presence of *Diaphanosoma* at 80 cm level (Manca and Comoli, 1996), is indicative of summer temperatures of ca. 20-22°C and of a well established summer thermocline (Herzig, 1994; Manca *et al.*, 1996b). An increase in the concentration of *Daphnia* remains characterize the most recent phase (C-VIII), where CD also increase (cf. Fig. 1). Chydorids are also relatively more represented, both in terms of total concentration and the number of taxa recorded (Fig. 2).

3.7. OSTRACODS

A high resolution study was carried out in the bottom 3 metres of the sequence, the only part of the core where ostracod remains were found (Fig. 3). Improved accuracy of the analysis was obtained by the use of a section from a replicate core (Belis, 1997) correlated by magnetic susceptibility (Oldfield, 1996).

At the very beginning of the sequence (above zone O-I, completely devoid of animal remains) the ostracod community is dominated by *Ilyocypris gibba* which is replaced by *Cyclocypris* spp. (*C. cf taubachensis, C. cf. impressopunctata* and *C. cf neumarkensis*) in the interval 1327-1307 cm (zone O-II). These pioneer assemblages coincide with the filling of the basin. After a zone very poor in ostracods (1307-1252 cm, zone O-III), the dominance of *Candona neglecta*, extends until 1127 cm. Two peaks of abundance were well correlated with the pigments peaks and could be interpreted as fluctuations in the lake productivity (Belis, 1997). At 1162-1127 cm (zone O-IV) the appearance of well preserved remains of *Potamocypris zschokkei*, *P. pallida* and *P. fallax* are interpreted as a signal of the input of cold and dilute spring water into the lake (Meisch, 1984). The bottom of zone O-IV coincides with a decrease in the relative importance of planktonic diatoms, which continues above the level at which ostracods disappear. The presence of *Cytherissa lacustris* in this zone indicates rather deep, well oxygenated and cold conditions. In the layer 1117-1107 cm the substitution of *C. neglecta* by *Cyclocypris sp.* suggests the return to an environment similar to the one existing at the beginning of the sequence, i.e. shallower. No ostracod have been found above 1107 cm.

3.8. CHIRONOMIDS

Low chironomid head capsule densities were observed at the beginning of the lake's life (Zone CH-I, Fig. 3). In zone CH-II (ca. 30-26 kyr BP), several peaks of chironomid remains were found, indicating a cold phase interrupted by temperate periods, which favoured the development of the chironomid community. The chironomids remains found in these periods are composed mainly of Tanytarsini of the genus *Micropsectra* and to a lesser extent of Chironominae *Microtendipes* and *Paracladopelma* group *nigritula* and of the Tanypodinae *Procladius*. These taxa suggest a generally low productivity of the lake (Saether, 1979; Wiederholm, 1981). Low temperature and low productivity also characterized the subsequent phases of the glacial period up to the beginning of the Late Glacial (960-735 cm of the core), in which the chironomid remains were scarce, belonging mostly to the genus *Micropsectra*.

Between the Late Glacial and the beginning of the Holocene, a large number of chironomid remains, mainly belonging to the genus *Micropsectra*, were observed (zone CH-IV and CH-V, 679-483 cm). The chironomid remains were also abundant up to the core depth of 359 cm (Zone CH-V). In this period of the Holocene important changes occurred in the qualitative composition of the chironomid remains, which consisted mainly of remains of the genera *Micropsectra* and *Procladius*, as well as some genera of Chironomiinae (*Dicrotendipes and Polypedilum*), of Tanytarsiinae (*Cladotanytarsus and Tanytarsus*) and of Orthocladiinae (*Cricotopus and Psectrocladius*) coming from the littoral zone (Frey, 1988), indicating lake level rising and greater development of the littoral vegetation (Warwich, 1980; Walker and Mathewes, 1989).

Important fluctuations in the chironomid remains were observed in zone CH-VI and CH-VII (last ca. 4 kyr BP; Fig. 3) when the anthropogenic influence overlapped the effects of the climate and other natural events. Two noticeable decreases in chironomid head remains occurred at a core depth of 341 and 253-243 cm, probably caused by a deterioration of the environmental quality, due to anthropogenic impact (deforestation and cultivation; Accorsi *et al.* 1996). The negative influence on the chironomid settlement was probably exerted by the onset of periodic anoxia, related to the increased load of organic material, and by the accumulation on the bottom of a thick layer of tephra of Campanian origin (Avellino). The low number of chironomid remains between 157 and 57 cm indicates a decrease in population density between 2 and 1 kyr BP (Fig. 3), which can be ascribed to climatic deterioration, in the form of colder and drier periods, which probably affected the Mediterranean region, as well as northern and central Europe (Brodin, 1986). A marked increase in the head capsule remains in the subsequent layers of the core, up to 14 cm, is followed by an important decrease in the top 12 cm of the core, i.e. the last 50-60 years (zone CH-VII), probably due to a slight increase in the trophic status of the lake. In this period, the qualitative structure of the chironomid fauna also changed: most of the remains belong to the Tanipodiinae (*Procladius*) as well as to some genera of Tanytarsiinae (*Cladotanytarsus, Tanytarsus*), Chironomiinae and Orthocladiinae, which derived from the littoral and sub-littoral zone, while the genus *Micropsectra*, characteristic of fairly unproductive, well-oxygenated environments, sharply dropped.

4. Conclusions

The Full Glacial Period (ca.30-14.5 kyr BP) is characterized by maximum catchment erosion and generally low concentrations of biological remains. This suggests that the lake remained largely oligotrophic, and that generally oxic conditions prevailed in the water column and at the mud surface for much of the time. However, in the lowest ca. 400 cm of the core, five periods of relatively high productivity with temporarily anoxic phases during the year, alternating with cold periods of lower water level show rapid cycle duration of ca.200-500 years. These rapid oscillations of pigments and other biological remains are in phase with the magnetic and mineralogical properties (Oldfield, 1996) and can be interpreted as evidence of alternations between periods of high particulate, minerogenic terrigeneous influx and periods of reduced flux and lake water fluctuations, which led to changes in the distribution and extent of different habitats (e.g. littoral *vs.* open water habitats) that affected the relative and absolute inputs of biogenic and non-biogenic sediments (Lami *et al.*, 1996). The lake productivity oscillations during the Glacial Period can be ascribed to rapid climate changes, as shown by the pollen concentration stratigraphy from Lake Albano (van der Kaars *et al.*, 1995), Lago di Monticchio (Huntley *et al.*, 1996), and from a study on mammal changes in southern Italy (Abbazzi *et al.*, 1996).

According to Huntley *et al.* (1996) and Oldfield (1996) these full glacial oscillations bring to mind similar oscillations from North Atlantic, Greenland and ice core studies (e.g. Dansgaard *et al.*, 1993). However, we cannot exclude at present the effect of geothermal activity on the observed oscillations or pulses of clastic input triggered by seismic activity, directly or indirectly affecting the limnetic ecosystem.

High productivity levels are detected during the Early Holocene (580-482 cm), indicated by peaks in pigments concentrations and specifically in those of sulphur reducing bacteria (e.g. isorenieratene). All the biological indications point to a maximum aquatic productivity and maximum, largely broad-leaved forest development at this time (van der Kaars *et al.*, 1995; Oldfield, 1996). Contemporary and similar important vegetational changes are reported from Valle di Castiglione and Lagaccione (central Italy, Magri and Follieri, 1989) and Lago Grande di Monticchio (Huntley *et al.*, 1996).

A clear discernible impact of human activity in the catchment in the form of a sudden increase in erosion, the end of dominance of reductive diagenesis, a sharp decline in pigment concentrations, a shift in diatom populations and increase of non-tree pollen (e.g. Gramineae), can be observed at 380 cm (ca.4.1 kyr BP; Oldfield, 1996). Other minor human impacts are shown during the Bronze Age and the Roman Period (Oldfield, 1996; Lami *et al.*, 1996).

The links between catchment process and lake productivity recorded during the Holocene are very interesting. They tend to support the view that eutrophication is not an inevitable response to all type of human impact on the catchment since, for instance, the sharp increase in particulate flux implied by the change at the ca. 380 cm did not lead to eutrophication (Oldfield, 1996).

Acknowledgments

We are grateful to D. Ariztegui, C. Chondrogianni and F. Niessen for collecting the cores and for many fruitful discussions and to R. Ferro (Centro Federale C.O.N.I., Castelgandolfo) and O. Catarci (Nemi) for their kind support during field work. L. Corbella helped in chironomid analyisis PALICLAS research was partially funded by EU 3rd Environment Programme, contract No. EV5V-CT93-0267.

References

Abbazzi, L., Delgado Huertas, R., Iacunin, P., Longinelli, A., Ficcarelli, G., Masini, F. and Torre, D.: 1996, In: Evans, S.P., Frisia, S., Borsato, A., Cita, M.B., Lanzinger, M. and Sala, B. (Eds), "*Late-Glacial and early Holocene climatic and environmental changes in Italy*", AIQUA-MTSN Conference, Trento (Italy), 7-9 Feb., 1996, 86-88.

Accorsi, C.A., Bandini Mazzanti, M., Forlani, L., Mercuri, A. M., Rivalenti, C. and Torri, P.: 1996, In: Oldfield, F. (Ed.), *Palaeoenvironmental Analysis of Italian Crater Lake Sediments (PALICLAS)*, EU, Technical final Report, contract N. EV5V-CT93-0267 (DGXII-FZPA), Bruxelles.

Alessio, A., Allegri, L., Bella, F., Calderoni, G., Cortesi, C., Dai Pra, G., De Rita, D., Esu, D., Follieri, M., Improta, S., Magri, D., Narcisi, B., Petrone, V. and Sadori, L.: 1986, *Geol. Romana* **25**, 287-308.

Battarbee, R.W.: 1984, *Phil. Trans. R. Soc. Lond. B.* **305**, 303-345.

Battarbee, R.W. and Kneen, M.J.: 1982, *Limnol. Oceanogr.* **27**, 184-188.

Belis, C.A.: 1997, *Wat. Air. Soil. Pollut.*, this volume.

Belis, C.A., Geiger, W., Lami, A., Guilizzoni, P., Ariztegui, D. and Chondrogianni, C., 1997: *J. Paleolimnol.*, pending revision.

Bennion, H.: 1994, *Hydrobiologia* **275/276**, 391-410.

Berglund, B.E., (Ed.):1986, *Handbook of Holocene palaeoecology and palaeohydrology*, Wiley and Sons, New York, p.869.

Brodin, Y.W.: 1986, *Int. Rev. ges. Hydrobiol.* **71**, 371-432.

Brown, S.R., McIntosh, H.J. and Smol, J.P.: 1984, *Verh. Internat. Verein. Limnol.* **22**, 1357-1360.

Brundin, L.: 1949, *Rep. Inst. Freshwat. Res.* **30,** 1-915.

Calanchi, N., Dinelli, E. and Lucchini, F.: 1996, In: Oldfield, F. (Ed.), *Palaeoenvironmental Analysis of Italian Crater Lake Sediments (PALICLAS)*, EU, Technical final Report, contract N. EV5V-CT93-0267 (DGXII-FZPA), Bruxelles.

Chondrogianni, C., Ariztegui, Niessen, F., Lami, A., Guilizzoni, P., Ohlendorf, C., Belis, C.A. and Lister, G.: in press, *Mem. Ist. ital. Idrobiol.* **55**.

Dansgaard, W., Johnsen, S.J., Clausen, H.B., Dahl-Jensen, D., Gundestrup, N.S., Hammer, C.U., Hvidberg, C.S., Steffensen, J.P., Sveinbjörnsdottir, A.E., Jouzel, J. and Bond, G.: 1993, *Nature* **364**, 218-220.

Frey, D.G.: 1988, *J. Paleolimnol.* **1**, 179-192.

Fritz, S.C.: 1990, *Limnol. Oceanogr.* **35**, 1771-1781.

Guilizzoni, P. and Oldfield, F.: 1997, *Mem. Ist. ital. Idrobiol.* **55,** in press

Guilizzoni, P., Bonomi, G., Galanti, G. and Ruggiu, D.: 1983, *Hydrobiologia* **103**, 103-106.

Guilizzoni, P., Lami, A. and Marchetto, A.: 1992, *Limnol. Oceanogr.* **37**, 1565-1569.

Hall, R.I. and Smol, J.P.: 1992, *Freshwat. Biol.* **27**, 417-434.

Herzig, A.: 1994. *Hydrobiologia* **275/276**, 81-96

Hofmann, W.: 1986, In: Berglund, B.E. (Ed.), *Handbook of Holocene palaeoecology and palaeohydrology*, Wiley, J. & Sons, 715-727.

Huntley, B., Allen, J. and Watts, B.: 1996, In: Evans, S.P., Frisia, S., Borsato, A., Cita, M.B., Lanzinger, M. and Sala, B. (Eds), "*Late-Glacial and early Holocene climatic and*

environmental changes in Italy", AIQUA-MTSN Conference, Trento (Italy), 7-9 Feb., 1996, 104-107.
Jeppesen, E., Madsen, E.A., Jensen, J.P. and Anderson, N.J.: 1996, *Freshwat. Biol.*, 115-127.
Kelts, K., Briegel, U., Ghilardi, K. and Hsu, K.: 1986, *Schweiz. Z. Hydrol.* **48**, 104-115.
Kerfoot, W.C.: 1974, *Ecology* **55**, 51-61.
Lami, A., Niessen, F., Guilizzoni, P., Masaferro, J. and Belis, C.A.: 1994, *J. Paleolimn.* **10,** 181-197.
Lami, A., Guilizzoni, P., Bettinetti, R., Belis, C.A., Manca, M., Comoli, P., Marchetto, A., Ariztegui, D. and Chondrogianni, C.: in press, *Il Quaternario.*
Leavitt, P.R.: 1993, *J. Paleolimnol.* **9**, 109-127.
Leventhal, E.A.: 1970, *Trans. Am. Phil. Soc.* **60**, 123-142.
Mackereth, F.J.H.: 1971, *Earth Planet Sci. Letts.* **12**, 332-338.
Magri, D. and Follieri, M.: 1989, *Mem. Soc. Geol. Ital.* **42**, 147-153.
Manca, M. and Comoli, P.: 1996, *Mem. Ist. ital. Idrobiol.* **54**, 61-67.
Manca, M., Comoli, P., Lami, A. and Guilizzoni, P.: 1996a, *S.It.E. Atti* **17**, 115-118.
Manca, M., Canale, C., Beltrami, M. and de Bernardi, R.: 1996b, In: Commissione Internazionale per la Protezione delle Acque Italo-Svizzere, *Ricerche sull'evoluzione del Lago Maggiore. Aspetti Limnologici. Campagna 1995 e Sintesi 1993-1994.* In press.
Margaritora, F: 1983, Consiglio Nazionale delle Ricerche AQ I/197: 169 pp.
Margaritora, F: 1992, *Mem. Ist. ital. Idrobiol.* **50**, 319-336.
Martini, M., Giannini, L. Prati, F., Tassi, F., Capaccioni, B. and Iozzelli, P.: 1994, *Geochem. J.* **28**, 173-184.
Meisch, C.: 1984, Trav. Sci.Mus.Hist. Nat. Luxemb. **3**, 1-55.
Narcisi, B.: 1994, *Il Quaternario* **7**, 545-554.
Oldfield, F.: 1996, *Palaeoenvironmental Analysis of Italian Crater Lake Sediments (PALICLAS)*, EU, Technical final Report, contract N. EV5V-CT93-0267 (DGXII-FZPA), Bruxelles.
Reavie, E.D., Hall. R.I. and Smol, J.P.: 1995, *J. Paleolimnol.* **14**, 49-67.
Renberg, I.: 1990, *J. Paleolimnology*, **12**, 513-522.
Saether, O.A: 1979, *Holarctic Ecol.* **2**, 65-74.
Sanger, J.E.: 1988, Palaeogeogr., Palaeoclimat., *Palaeoecol.* **62**, 343-359.
Smol, J.P.: 1988, *Paleogeogr. Paloeclimat. Paleoecol.* **62**, 287-297.
Stuiver, M. and Reimer, P.J.: 1993, *Radiocarbon* **35**, 215-230.
Swain, E.B.: 1985, *Freshwat. Biol.* **15**, 53-75.
Takahashi, M. and Ichimura, S.: 1970, *Limnol. Oceanogr.* 15, 929-944.
van der Kaars, S., Lowe, J.J., Accorsi, C.A., Bandini, Mazzanti, M., Mercuri, A. and Torri, P.: 1995, *INQUA, XIV International Congress, Berlin,* 282-283.
Walker, J.R. and Mathewes, R.W.: 1989, *J. Paleolimn.* **2**, 1-14.
Warwich, W.F.: 1980, *Can. Bull. Aquat. Sci.* **206**, 1-117.
Wiederholm, T.: 1981, *Schweiz.z.Hydrol.*, **43**, 140-150.
Wunsam, S. and Schmidt, R.: 1994, *Mem. Ist. ital.Idrobiol.* 53, 85-99.
Züllig, H.: 1982, *Schweiz. Z. Hydrol.* **44**, 1-98.
Züllig, H.: 1985, *Schweiz. Z. Hydrol.* **47**, 87-126.
Züllig, H.: 1986, *Hydrobiologia* **143**, 315-319.

BIVALVES AND HEAVY METALS IN POLLUTED SEDIMENTS: A CHEMOMETRIC APPROACH

G. ADAMI[1], F. ALEFFI[3], P. BARBIERI[1], A. FAVRETTO[2], S. PREDONZANI[3] and E. REISENHOFER[1]

[1] *Department of Chemical Sciences, University of Trieste, Via Giorgieri 1, 34127 Trieste, Italy,*
[2] *Department of Geography, University of Trieste, Via Tigor 32, 34127 Trieste, Italy,*
[3] *Laboratory of Marine Biology, Strada Costiera 336, 34136 Trieste, Italy.*

Abstract. Copper, lead, cadmium, zinc, chromium, manganese, iron and nickel were determined in near-shore sediments in the harbour of Trieste (Northern Adriatic), in an area highly exposed to urban and industrial wastes, where severe alteration of benthic population was observed. A typical bivalve of this area, *Corbula gibba*, was used as bioindicator of sea-bottom pollution. Multivariate statistical analysis of the chemical data interpret concentrations and distributions of heavy metals in these sediments, attesting anthropogenic source for Cu, Pb, Cd and Zn. Using labile fractions of heavy metals in sediments as 'predictors', and length or biomass of the bivalve as dependent variables, we obtained, by a multiple regression procedure, a predictive model showing the influence of metals on this benthic organism of this polluted area.

Key words: Near-shore sediments, heavy metals, bioindicators, bivalves, benthic populations, multivariate analysis, multiple regression analysis, predictive models.

1. Introduction

Near-shore sediments are found in a wide variety of environments (bays, lagoons, deltas and so on): the chemical, physical, and biological conditions of these environments are much more variable than those of the open sea. The typologies of these sediments can be very various, reflecting also local topography and structures. Near-shore sediments have found recent interest because they can be considered as repositories for many chemical species. As a rule, the accumulation rates of pollutants are much higher in coastal sites, nearby urban and industrial settlements, than in open-sea environments.

In previous works we have determined some heavy metals - as indicators of anthropogenic pollution - both in seawaters and sediments sampled in selected locations of the Gulf of Trieste (Favretto *et al.*,1988; Reisenhofer *et al.*, 1996; Favretto *et al.*, 1996). The results of these works allowed us to verify the trends of the heavy metal distribution in this area of the Northern Adriatic Sea along a time span of about 20 years (Costa *et al.*, 1982). We found that the present quality of the seawaters in the Gulf of Trieste is comparable to those of similar areas of the Mediterranean Sea (Martin *et al.*, 1994; Fagioli *et al.*, 1994).

In the present work, we focus our attention on a critical area, the Bay of Muggia. This area is semi-enclosed between an external three-dam system and the wharfs of an industrial zone, so the body of these waters receives urban sewages as well as

Water, Air and Soil Pollution **99**: 615-622, 1997.

industrial wastes. The renewal of the waters is scarcely given by an ascending, counter-clockwise current, that brings cleaner waters from the Istrian coast. We have taken samples here in 7 sites, following a transect going from the wharf in front of an industrial settlement to the external dam (Fig.1). In each sites, some heavy metals, of environmental relevance, were determined in sediment in order to evaluate their accumulation in a well-characterized environmental situation, *i.e.*, in an area densely inhabited, devoted to industrial and good-stocking activities, where the dilution of pollutants is scarce, due to the shallowness of the waters (10-20 m) and to the enclosure by the dam-system. In this area, profound alterations on the distributions of the benthic populations were observed (Orel, 1996). For a quantitative evaluation of the influence of heavy metals on the marine biota, we selected a marine organism typically found in environmentally unsteady areas, the bivalve *Corbula gibba*, which was present in all sampling sites. Main purpose of the present work is to verify the relevance of the determined metal pollutants in conditioning this particular ecosystem, to detect an anthropogenic origin for metal pollution by means of appropriate statistical methods, and to suggest a predictive model to assess a correlation between the labile fraction of metals and the benthic populations.

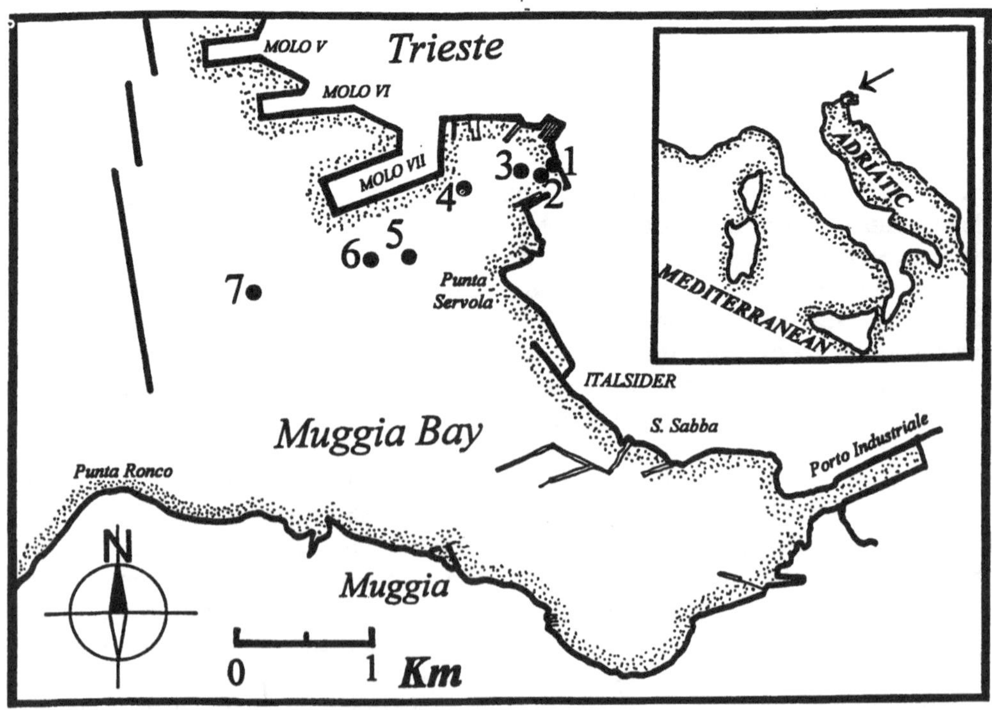

Fig.1. Sampling sites in *the* inner harbour of Trieste.

2. Material and methods

Sampling. Sediments were collected in the inner harbour of Trieste (Oct. 20th, 1995) with a Van Veen grab, which took samples over about 0.1 m^2, and penetrated up to a depth of about 15 cm, at 7 different sites, going from an internal wharf to the external three dam system (Fig.1). We followed the standard sample preparation procedures (Loring and Rantala, 1992). Coarse materials (> 2 mm) were removed, then the sediment was homogenized before subdivision into subsamples, and freeze-dried. The bivalve molluscs were collected with a grab on a bottom-surface of about 0.4 m^2. The individuals of *Corbula gibba* were counted. The mean values of the biomass (dry weight, in g) and shell length (cm) were determined on the specimens of each sampling site.

Chemical analysis of sediments. We have performed three decomposition procedures, in parallel:

- i) *Weak acid extraction.* This procedure is often used for chemical partition studies (Huerta-Diaz and Morse, 1990). 2.5 g of sediment were stirred in 20 ml of HCl 6N for 16 hours at room temperature.

- ii) *Strong acid digestion.* This procedure, using nitric acid, results in incomplete attack, because silicates and other refractory oxides are not dissolved (Krumgalz and Feinstein, 1989). 0.5 g of sediment were added to 3 ml of HNO_3 65%, 0.1 ml of $HClO_4$ 70% and 0.1 ml of H_2O_2 30%. The attack was performed by a Milestone MLS1200 programmed microwave System.

- iii) *Total decomposition.* In this case, 0.1 g of sediment were treated with 2 ml of HF 48% and 2 ml of HNO_3 65% in teflon decomposition vessels, by a Milestone microwave System, as in the former procedure.

The resulting solutions were filtered, except in the case of total decomposition, and then diluted to 50 ml with Milli-Q water (ρ = 18.2 MΩ cm). The metal contents were measured by atomic absorption spectrometry: Varian AA20 Spectrometer. The analytical precision (± RSD) was normally between 2 and 5%. The quantitative measurement was made using calibration curves obtained with three standard solutions of the element under analysis. All reagents were of ACS analytical grade.

Statistical treatment of data. We consider both 8 chemical variables (the metal concentrations), and 2 biological variables, *i.e.* biomass and length of molluscs, for a quantitative evaluation of the influence of heavy metals on the marine biota. The distance of the sampling sites from shore was also considered as a parameter. Consequently, the starting data matrix contained 11 variables on the whole. Basic statistics, principal component and multiple regression analyses were performed by a SPSS program ('SPSS Statistical Algorithms', 1985; Norusis, 1990).

3. Results and discussion

The environmental importance of the three metal content fractions is, obviously, different. In Fig.2, we plotted the trends of metal contents in the 7 sampling sites as detected in the sediments, by weak acid extraction (×), strong acid digestion procedure (Δ) or total decomposition of the sediment (□), respectively. Cu, Pb, Cd and Zn contents exhibit a sharp decrease going from the shore towards open-sea, whereas Cr, Mn, Fe and Ni have constant values along the transect. These different trends suggest an anthropogenic origin for the first group of metals. In fact, the higher metal contents were measured for the sampling points which are closer to the shore line, that can be likely considered as a source for these toxic species. Sampling points at increasing distance seem to show a dilution in contents of this group of metals. The concentration trends of the other metals suggest that their contents in these near-shore sediments are mainly determined by natural processes (wheathering of rocks, and so on), also in this densely inhabited and industrialized area.

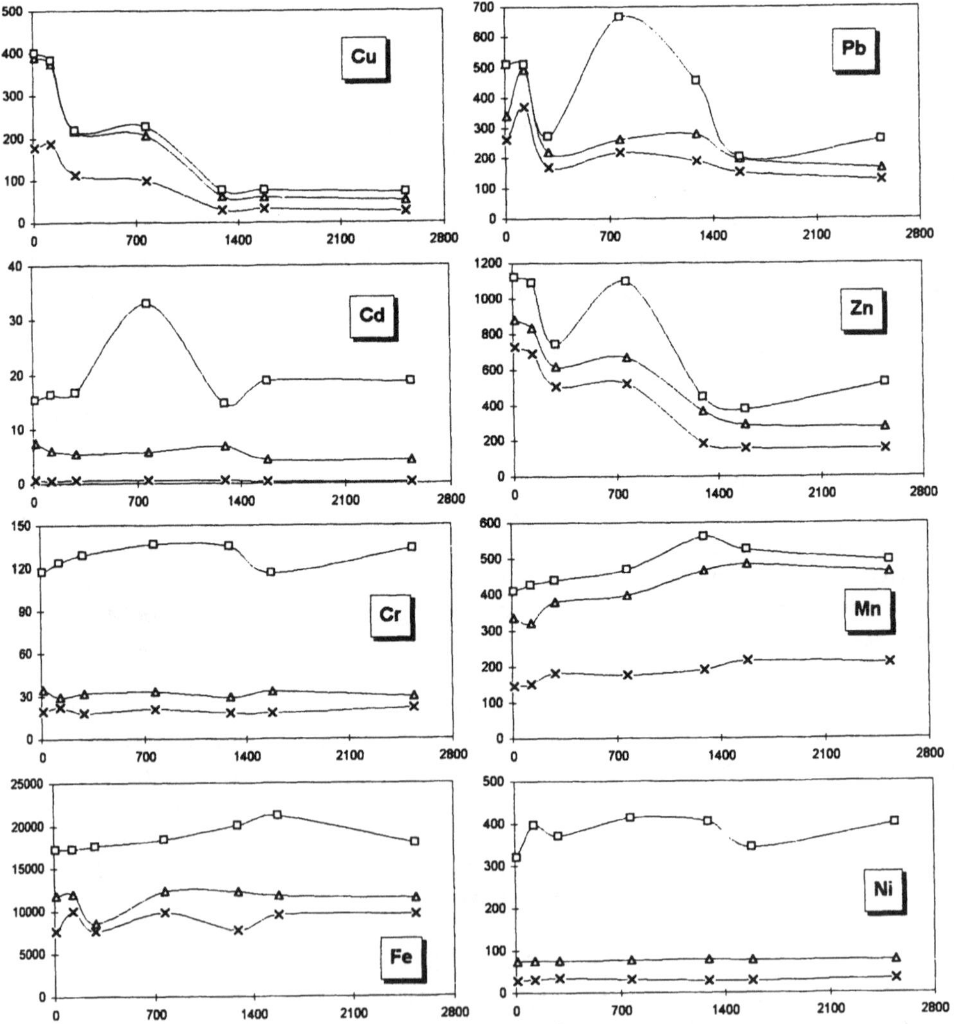

Fig.2. Concentrations (ppm) of Cu, Pb, Cd, Zn, Cr, Mn, Fe and Ni in the 7 sampled sediments, obtained by HCl extraction (×), by HNO_3 digestion (Δ) and by HF decomposition (□). The distance of sites from shore (m) is reported in abscissa.

The anthropogenic origin of the metals in sediments can also be tested, considering the basic statistics of the analytical data. In Table I, we report only the statistics relative to the labile metal fraction (obtained by HCl acid extraction), retained as the more bioavailable, and therefore more interesting for environmental considerations. The range of concentrations, with very high upper limits, and the coefficients of variation (standard deviation/mean ratios) confirm that the distributions of Cu, Pb, Cd and Zn in sediments have a higher variability, in comparison with the second group of metals (Cr, Mn, Fe and Ni), that have more homogeneous distributions.

TABLE I

Basic statistics of the chemical and biological variables.

Variable	Mean	SD	SD/Mean	Min	Max
Cu	94.2	68.5	0.73	26.2	185.8
Pb	213	82	0.39	127	370
Cd	0.50	0.17	0.34	0.21	0.73
Zn	420	252	0.60	154	729
Cr	19.6	1.8	0.09	17.8	21.9
Mn	181	27	0.15	146	216
Fe	8869	1109	0.13	7660	10000
Ni	31.6	2.3	0.07	29.3	35.0

In Table II, we report the correlation matrix relative to the labile metal fraction. We observe that Cu, Pb, Cd and Zn are the only metals that are inversely, and highly, correlated with biomass and length of the molluscs, and with the distance: *i.e.*, the contents of these metals in sediments *decrease* going from shore toward open-sea, whereas biomass and length of the molluscs *increase* in the same direction (*i.e.*, where the metal pollution is less severe). On the other hand, the generally low correlation coefficients between bioindicators and Cr, Fe and Ni confirm again that these metals have little influence on *Corbula*. The correlation matrices relative to the fractions obtained by stronger acid attacks (not here reported) confirm this pattern.

TABLE II

Correlation matrix for chemical parameters (labile fraction of metals), biomass and length of *Corbula gibba*, and distance from shore.

	Cu	Pb	Cd	Zn	Cr	Mn	Fe	Ni	Biom.	Leng.	Dist.
Cu	1.000										
Pb	0.850	1.000									
Cd	0.511	0.412	1.000								
Zn	0.983	0.792	0.573	1.000							
Cr	0.286	0.420	-0.431	0.261	1.000						
Mn	-0.944	-0.856	-0.688	-0.949	-0.249	1.000					
Fe	-0.067	0.178	-0.677	-0.105	0.698	0.216	1.000				
Ni	0.082	-0.137	-0.438	0.112	0.346	0.051	0.139	1.000			
Biom.	-0.824	-0.663	-0.784	-0.885	0.084	0.854	0.298	-0.009	1.000		
Leng.	-0.893	-0.671	-0.813	-0.928	0.082	0.913	0.387	0.122	0.937	1.000	
Dist.	-0.885	-0.728	-0.762	-0.903	0.121	0.880	0.335	0.025	0.968	0.955	1.000

On the basis of these observations, we retained as variables for a principal component analysis (PCA) only the metals that show a high correlation with the bioindicators, *i.e.*, Cu, Pb, Cd and Zn, and the distance along the considered transect. The eigenanalysis, performed on the labile fraction of the 4 selected heavy metals, allows us to obtain 2 principal components, that explain more than 90% of the variance. The PCA results are reported in Table III. Considering the Varimax rotated factor matrix, we observe that Cu, Pb and Zn are correlated with the first, more important factor (PC1, with 80.1% of explained variance), whereas cadmium is correlated with the secondary factor (PC2, with 13.9% of explained variance). All metals display factor loadings of opposite sign respect to the distance from the shore. The pattern of heavy metal pollution can be described with 2 factors.

TABLE III

Principal component analysis.

Final statistics

Variable	Communality	Factor	Eigenvalue	Pct of Var	Cum Pct
Cu	0.974	1	4.004	80.1	80.1
Pb	0.857	2	0.695	13.9	94.0
Cd	0.975				
Zn	0.945				
Distance	0.948				

Varimax Rotated Factor Matrix

	PC1	PC2
Cu	0.932	0.325
Pb	0.912	0.160
Cd	0.227	0.961
Zn	0.882	0.409
Distance	-0.721	-0.654

We would also verify the validity of a predictive model for assessing the inference of the labile fraction of metals in sediments on the bivalve populations. For this purpose, we searched an ordinary least squares (OLS) multivariate regression model (Draper and Smith, 1982), retaining a biological parameter, the length of *Corbula*, as dependent variable and chemical parameters, the labile fraction of toxic metals (Cu, Pb, Cd, Zn) in sediments, as 'predictors'. We examined values of cross-validated coefficient of determination ($\mathbf{R}^2_{cv}$) for models, considering all possible combinations of predictors. $\mathbf{R}^2_{cv}$ is a parameter adequate to test the predictive attitude of the models, and not only their fitting performance, and it is computed as:

$$R^2_{cv} = 1 - \frac{PRESS}{TSS} \tag{1}$$

In Equation 1, TSS is the total sum of squares and PRESS (predictive error sum of squares) is in turn computed as:

$$PRESS = \sum_i (y_i - \hat{y}_{i/i})^2 \tag{2}$$

In Equation 2, y_i is the i-me experimental value of the dependent variable, and $\hat{y}_{i/i}$ the predicted value for the i-me sample in a model which does not consider the predictors' values of that sample, in building model computations.

The selected results of OLS multivariate regression analysis are reported in Figure 3. In this Figure, $\mathbf{R^2}$ are the coefficients of determination, $\mathbf{R^2_{adj}}$ are the coefficients of determination adjusted for the degrees of freedom, $\mathbf{R^2_{cv}}$ have been defined above, **F** is the Fisher's ratio for the analysis of variance (testing the hypothesis that all the regression's coefficients **B** are zero), and s are the standard errors of the estimates. Moreover, **B** are the regression coefficients, **SE B** their standard errors, **Beta** the coefficients computed with standardized variables. The **tolerance** is the proportion of variance not explained by other variables.

	R^2	R^2_{adj}	R^2_{cv}	F	s
Cd-Zn	0.9791	0.9687	0.9050	93.88	0.3120
Cd-Cu-Pb	0.9910	0.9819	0.9034	109.62	0.2372

MODEL 1

	Cd	Zn	constant
B	-4.3980	-0.0048	12.1080
SE B	0.9252	0.0006	0.3979
Beta	-0.4188	-0.6883	
tolerance	0.6720	0.6720	

MODEL 2

	Cu	Pb	Cd	constant
B	-0.0229	0.0060	-4.9770	11.2443
SE B	0.0029	0.0022	0.6717	0.4090
Beta	-0.4739	0.2804	-0.4739	
tolerance	0.2455	0.2760	0.7368	

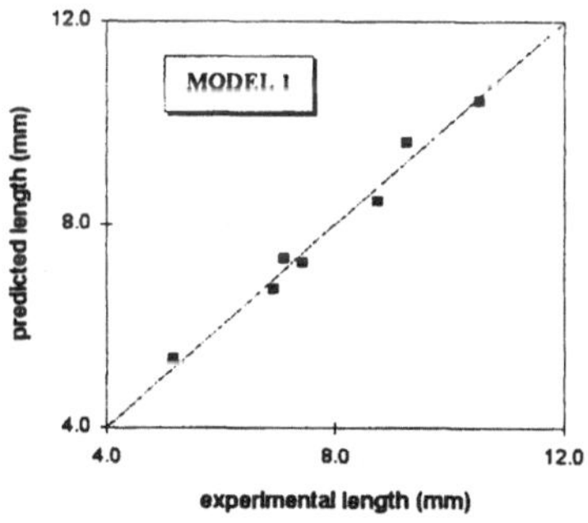

Fig. 3. Multiple regression analysis parameters for model 1 (Cd-Zn) and model 2 (Cd-Cu-Pb). Predicted vs. experimental lengths of *Corbula gibba* according to the Cd-Zn model.

The two models which achieve the best predictive performances according to this evaluation procedure, are the model 1 (Cd-Zn), and the model 2 (Cd-Cu-Pb). Since the above reported PCA has demonstrated that Cu, Pb and Zn are highly correlated and loaded on the same component (PC1), it is not surprising to find now in this multiple regression model either Zn or Cu and Pb, besides Cd, as 'predictors'. We consider the Cd-Zn model as the best choice because it realizes the best economy of description, allowing a simple linear relation between a biologic parameter (possible indicator of environmental stress), and chemical variables, to be considered, in this case, as anthropogenic pollutants[1]. In Fig.3 we plotted also the

[1] Cadmium and copper uptake by some test organisms, as mussels and clams, in estuarine sediments was object of very recent studies by Rule and Alden (1996), that confirm the environmental significance of these heavy metals in relation to the benthic populations.

predicted lengths of *Corbula gibba* vs. the experimental lengths: this plot shows the goodness of fit for the selected Cd-Zn model. This model results also in an elimination-selection stepwise method (using a probability of F-to-enter of 0.05, and a probability of F-to-remove of 0.10 as criteria to select the variables).

4. Conclusion

The chemometric approach to a set of analytical data allows us to evaluate an environmental situation, on condition that the analytical data, even if with limited numerosity, are adequately accurate, and that the sampling was opportunely planned, selecting significant parameters. The shore line can be considered as source for anthropogenic pollution by Cu, Pb, Zn and Cd. The multivariate statistical approach can also suggest a predictive model for attesting or anticipating the phenomena occurring in an ecosystem, for instance, in terms of influence of chemical species on marine biota. We have verified a model for the labile fraction of metals in near-shore sediments. The evidence of the strong linear relation between biometric parameters for sediment dwelling organisms and the heavy metal contents in this environmental compartment allow us to assess the relevance of anthropogenic Cu, Pb, Zn and Cd as a factor of pollution, conditioning the benthic life in the harbour of Trieste.

Acknowledgements

The Authors thank the MURST (*Italian Ministry for University and Scientific and Technological Research*) for its financial support (40%).

References

Costa, G., Honsell, G. and Reisenhofer, E.: 1982, *Defence Research Inf. Centre Trans.*, 6548.

Draper, N.R. and Smith, H.: 1981, *Applied Regression Analysis*, John Wiley & Sons, N.Y.

Fagioli, F., Locatelli, C. and Landi, S.: 1994, *Annali di Chimica* **84**, 129-140.

Favretto, L., Gabrielli-Favretto, L. and Reisenhofer, E.: 1988, *Z. Lebens. Unters. Forsch.* **187**, 8-14.

Favretto, L., Campisi, B., Reisenhofer, E. and Adami G.: 1996, *Anal. Chim. Acta*, in proofs.

Huerta-Diaz, M.A. and Morse, J.W.: 1990, *Marine Chemistry* **29**, 119-144.

Krumgalz, B.S. and Fainshtein, G.: 1989, *Anal. Chim. Acta* **218**, 335-340.

Loring, D.H. and Rantala, R.T.T.: 1992, *Manual for the geochemical analyses of marine sediments and suspended particulate matter*, Earth-Science Reviews, **32**, 235-283.

Martin, J., Wei Wen Huang and Yi Yong Yoon: 1994, *Marine Chemistry*, **46**, 371-386.

Norusis, M.J.: 1990, *The SPSS Guide to Data Analysis*, SPSS Inc., Chicago.

Orel, G.: 1996, personal communication.

Reisenhofer, E., Adami, G. and Favretto, A.: 1996, *Fresenius' J. Anal. Chem.* **354**, 729-734.

Rule, J.H. and Alden, R.W.: 1996, *Environmental Toxicology and Chemistry* **15**, 466-471.

'SPSS Statistical Algorithms': 1985, by SPSS Inc. Chicago.

COLONIZATION PATTERNS AND DENSITIES OF ZEBRA MUSSEL Dreissena IN MUDDY OFFSHORE SEDIMENTS OF WESTERN LAKE ERIE, CANADA.

John P. Coakley [1], Glenn R. Brown [2], Stefan E. Ioannou [2], and Murray N. Charlton [1]
[1] National Water Research Institute, 867 Lakeshore Road, Burlington, Ontario, Canada, L7R 4A6
[2] Department of Geology, University of Toronto, 22 Russell Street, Toronto, Ontario, Canada, M5S 3B1.

Abstract. Zebra mussels (Dreissena) have expanded rapidly throughout most of the Laurentian Great Lakes since their inadvertent release in 1986. These exotic molluscs now occur in great numbers on the bottom of western Lake Erie where they are found increasingly in deeper areas of the basin (average depth: 10 m), on soft, muddy substrates. This study is aimed at quantifying the density and the distribution patterns of mussel colonization in the basin as a first step in investigating the effect on sediment properties of such an abrupt change in benthic community structure. Underwater video imagery and diver-collected samples taken from representative offshore areas (seven sites) in western Lake Erie showed colonization levels of up to 20,000 live mussels per m^2 in soft sediments (adults with shells >10 mm comprised 47 %). Digital side-scan sonar records confirmed that colonization patterns were not random, but showed distinctive spatial signatures ranging from 30-m-long parallel stripes, to large ovate masses. Broad irregular mats were found in association with hard bottoms (bedrock, boulders, or wrecks and large debris). Mussel densities were averaged from the sites, assuming consistent relationships with substrate type and were combined with digitized percentage of areal coverage of major bottom types in western Lake Erie. This resulted in the first population figure of 10^{13} in the basin. This figure includes molluscs of all sizes > 0.84 mm.

KEY WORDS: Dreissena, zebra mussel, side-scan sonar, western Lake Erie,

1. Introduction

Since their inadvertent introduction into Lake St. Clair (Fig. 1) in 1986, the zebra mussel (Dreissena sp.) has expanded its range at an alarming rate throughout the Great Lakes and into adjacent watersheds.

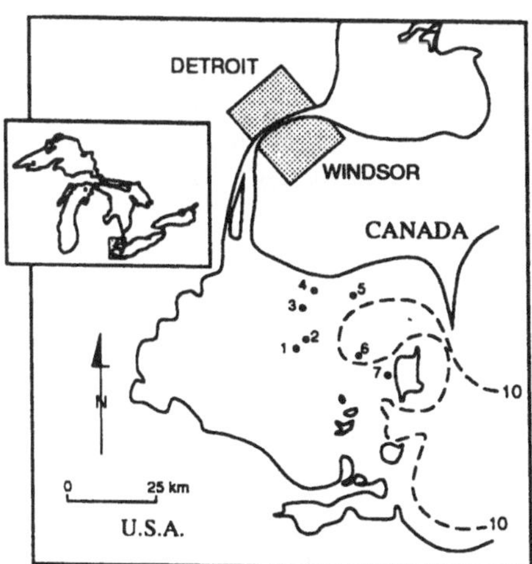

Fig. 1. Location map of western Lake Erie showing sites surveyed: (1) Bret; (2) Footballs; (3) Stripes; (4) Susan; (5) Barge; (6) Zebra; (7) Pelee. Position of 10 m depth contour is also shown.

In recent years, colonies of zebra mussels have established themselves outside the Great lakes watershed. Thriving populations now exist in the Mohawk / Hudson River systems of western New York, the Ohio River, and as far south as the Tennessee and Arkansas Rivers of the southern U.S.. They have been observed in the lower Mississippi River as far south as Baton Rouge (New York Sea Grant, 1993). In addition, the smaller Canadian lakes to the north of the Great Lakes and the Trent - Severn waterway now also are sites of growing zebra mussel infestation.

Initially, zebra mussels colonized hard substrates (bedrock and boulder deposits) and man-made coastal structures located in waters 2 to 12 m in depth. These are the common depths they inhabit in European ecosystems (Stanczykowska, 1964; Binelli *et al.*, 1997,

Water, Air and Soil Pollution **99**: 623-632, 1997.

this volume). The impact of zebra mussels is most pronounced in the lower Great Lakes, especially in western Lake Erie where, the combination of relatively warm, shallow water and eutrophic conditions provides ideal conditions for population growth. Mussel densities exceeding 342,000 mussels per m^2 have been recorded (Leach 1993; Fitzsimons *et al.*, 1995). In recent surveys, dense mats of mussels were observed in deeper waters on soft muddy sediments (Hunter and Bailey, 1992; Dermott and Munawar, 1993; Berkman *et al.*, 1995). The expansion of zebra mussels into such habitats that were once considered inhospitable is cause for concern, as such a move would open up for colonization the largest surface areas of the lakes (Table 1).

TABLE I

Proportions of various substrates and first-order estimate of live mussel population, western Lake Erie

Substrate type	Area (sq.km)	% of total	Representative site	Rep. density mussels x m-2	Total Population	Est. substrate Coverage (%)	Adjusted Population
Mud (with isolated druses)	2000	42.6	Midbasin 1	0	0.00E+00	95	0.00E+00
Mud (stripes & footballs)	580	12.3	BARGE	13000	7.54E+12	1	7.54E+10
Bedrock	200	4.3	(Fitzsimons et al.)	340000	6.80E+13	60	4.08E+13
Sand & gravel	930	19.8	PIPPE	20000	1.86E+13	5	9.30E+11
Mixed substrates (till)	990	21.1	ERIE	13000	1.29E+13	5	6.44E+11
TOTAL	4700	100.0					4.24E+13

Some researchers (MacIsaac *et al.*, 1992) have suggested that by their filtering activity, these mussels have permanently affected the nutrient dynamics and spawning success of higher, more commercial aquatic species such as walleye . Fitzsimons *et al.* (1995) disagree. Furthermore, material filtered from the water column that they do not ingest, is expelled as pseudofeces to the adjacent sediment surface. The effect of such transfers on sediment properties and contaminant cycling could be important, especially if zebra mussels are in fact invading new substrates and depths in Lake Erie.

This paper seeks to supplement the little that is known about how zebra mussels are spreading into areas of muddy sediment in western Lake using a combined geophysical and geologic approach. to map the distribution of population densities in the basin as a first step in identifying trends and patterns of their infestation.

2. Methodology

2.1 SIDE-SCAN SONAR SURVEYS

The survey began in the summer of 1994, using a digital side scan sonar system manufactured by Marine Sonics Technology of Virginia, U.S.A.. The side scan sonar transmits a 300 kHz acoustic pulse at right angles to the direction of the moving survey vessel; the returning echoes is collected and processed by computer to build an image of the bottom surface topography, including mussel beds (Carey, 1978). Sonar surveys focused on several areas in the western basin (Fig. 1), characterized by different substrates. In 1995 and 1996 we returned to the sites and remapped target areas. For the 1994 surveys, positioning was by GPS, with an accuracy of about 100 metres. In 1995 and 1996 we used differential GPS with an accuracy of about 1 metre. To coordinate the surface and diver surveys in the 1996 surveys, we placed distinct acoustic reflectors on the bottom at the study sites.

2.2. DIVER SURVEYS AND BOTTOM SAMPLING

Scuba divers, equiped with video and still-cameras, were used to "ground-truth" the side-scan data and to collect bottom samples. Two types of bottom samples were collected by divers. Quadrat samplers 0.0625 m^2 in area (25 cm x 25 cm x 4 cm) were used to collect surface samples of zebra mussels from areas deemed representative. The samplers were placed by the diver onto the sediment surface and pressed into the sediment. All the sediment and mussels within the quadrat were collected in a plastic bag and brought to the surface. The sample was washed on the boat through a 1 mm brass sieve and the mussels stored for later counting in the laboratory. The other samples consisted of short (10 - 15 cm long) plastic cores pushed by the diver into the sediment. Core samples were taken in pairs, one below a mussel mat, the other in bare sediment within 1 m of the first. These cores were kept cool until they were freeze-dried in preparation for later grain-size and chemical analysis.

2.3. ZEBRA MUSSEL COUNTING PROCEDURE

More than 100 selected video frames were chosen from the underwater video imagery collected during the surveys. For each frame, the number of living or dead mussels (disarticulated or bleached shells were counted as dead) were visually counted. In addition, the altitude of the camera and the surface area covered by the frame were calculated (Ioannou *et al.*, 1996). These numbers were then used to generate values representing mussels . m^{-2} for the different sites sampled. Regression of the video-based and the quadrat counts provided a conversion factor of approximately 10. This was used to gross-up the values to a 4 cm active mussel layer.

The technique was dependent on defining an accurate value for the median mussel size. This value was obtained from collected (quadrat) mussel samples in addition to an "on-screen" scaling method which made use of the video camera's protective frame as a size reference.

3. Results

3.1. VIDEO-CAMERA IMAGING OF ZEBRA MUSSEL DISTRIBUTIONS

Zebra mussel infestation of varying densities was observed on the underwater video footage taken at all the sites. Bottom coverage ranged from sparse (<1% of video frames) to dense (>90%). The densest coverage was noted in areas of rough bottom (bedrock, boulders, cobbles and wreck debris). At these sites we estimate that there were approximately 5000 mussels.m^{-2} at the surface, for an active total of 50 000.m^{-2} down to

4 cm. Less dense, patchy coverage was noted in areas of mud bottom (silty clay). Large expanses of bare sediment cover were noteed in the mussel distribution.

3.2 SIDE-SCAN IMAGING OF ZEBRA MUSSEL SITES

Over the past three years, a total of seven different sites have been investigated with the side-scan sonar (Fig. 1). Many of these sites have been revisited on an annual basis. The sonar records readily enabled distinction of the various major substrate types making up the sites: hard bedrock and boulder / cobble (very dark expanses, or stippled, irregular patterns, often with light-coloured shadow haloes indicating relief), granular sand (alternating dark and light linear bands suggesting dune and ripple forms), and soft mud (monotonous contrast). Representative images from the sites are presented in Figure 2.

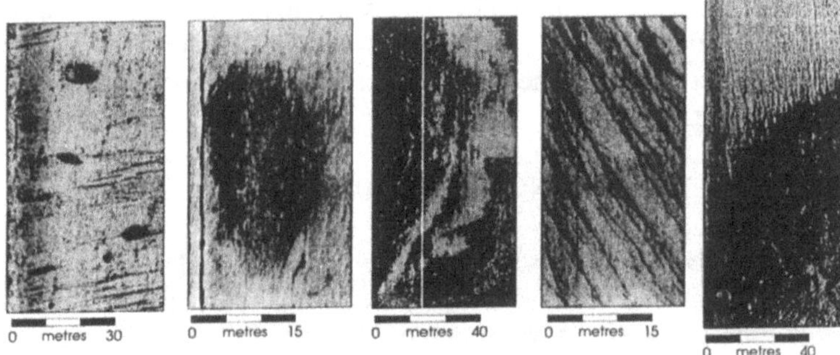

Fig. 2. Side scan records from five representative substrates in the western basin of Lake Erie. Left to right: A. Ovate masses ("footballs") with strong backscatter; B. Enlargement of ovate masses mussel mounds, C. Bedrock (mussel encrusted); D. Parallel linear bottom features with mussels in depressions ("stripes"); E. Transition from densely colonized bedrock (bottom) to stripes in adjacent soft sediments (top).

Because the estimated resolution of the technique was approximately 50 cm (depending on swath-width setting), it was impossible to image individual mussels. However, examination of surveys carried out over video-verified areas of muddy sediment allowed the identification of distinct density reflection patterns on the sonograph records that corresponded to mussel colonization. These patterns were classified into two major types: sub-parallel, linear "stripes" and rounded, ovate "footballs".

The former (Fig. 2D) is the name given to large areas of the sites whose side-scan record consisted of parallel, linear dark reflectors at a consistent angle to the boat track. They could be resolved out to a distance of up to half the swath width (up to 50 m). They apparently extended further, but could not be observed beyond this distance because of recording limitations. They are aligned consistently in a direction of N40 $^{\circ}$E to N85 $^{\circ}$E, averaging around N60 $^{\circ}$E) and were observed at most of the sites associated with a visible hard substrate or object (bedrock reef outcrop or a shipwreck) and extended to both sides. They were also found in association with the ovate patterns ("footballs") described below. The greatest elongation was consistently toward the NE, suggesting association with prevailing directional forcing processes, such as basin-wide circulation and sediment transport. Zebra mussels occupied the troughs of these features in dense colonies, although the divers noted often the presence of distinct demarcation borders or "fronts" of mussel coverage. A schematic of the perceived relationship between the "stripes" and the hard bottom areas is presented in Figure 3.

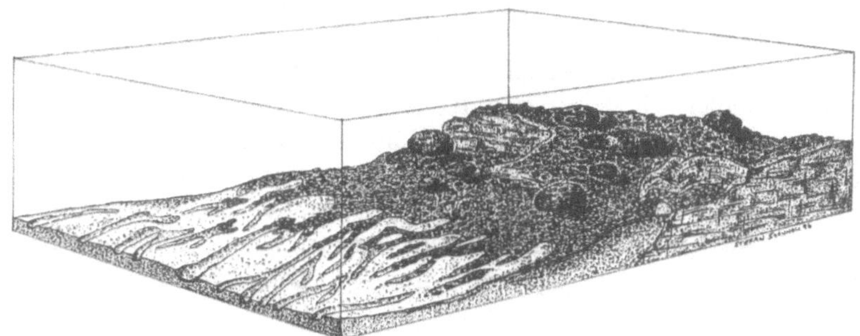

Fig. 3. Schematic of relationship between "stripes" pattern of zebra mussel colonization and nearby hard substrate. Note the expanding "apron" between the stripes and the bedrock nucleus.

The ovate "football" pattern (Fig. 2A, B) was observed less frequently and consisted of patches of dark reflectors (no shadow, indicating a lack of relief) with rounded margins. These patches were often ovate in shape and ranged in diameter from 10 to more than 50 m. Diligent search for "football" areas noted in previous years often failed to relocate them, suggesting that, unlike the "stripes", they appeared to be somewhat ephemeral and subject to alteration in form and location with time. These features, like the "stripes", were almost always associated with areas of hard bottom, but were formed in adjacent areas of muddy sediments. Diver descriptions, however, generally noted hard sediment close to the surface. Zebra mussel patterns were patchy and less dense than in the troughs of the stripes pattern.

Divers confirmed that the patterns were in fact due to zebra mussel colonies. The "stripe" patterns were caused by dense colonization of mussels within linear trough-like bottom features (Fig. 3). These troughs ranged from 0.25 to 1 metre in width and could be followed by the diver for more than 100 m. They were usually 5 - 15 cm deep, often with steep, at times undercut or excavated walls. The mussels occurred in concentrated mats along the bottom; disarticulated or empty shells could also be felt in the soft sediment to depths exceeding 10 cm.

The "football" patterns were observed by divers to correspond to areas where mussels occurred as partially buried, discontinuous patches. Like the "stripes" areas, they were consistently associated with areas where hard bottoms were nearby and within 20 cm of the surface. Also worthy of note was the fact that "footballs" observed and precisely located in previous years were absent or greatly modified in form in 1996.

3.3 ESTIMATION OF ZEBRA MUSSEL POPULATIONS

Measurements of zebra mussel colonization densities were taken using two techniques: Direct image analysis of the video frames, and by extrapolation of the spot quadrat samples. The video-based counting technique is detailed in Ioannou *et al.*, (1996). Individual videotape frames (105 in all), meeting criteria of high resolution and camera height above the bottom, were magnified and individual mussel and mussel shells (dead) were visually counted. The accuracy of the technique was dependent on defining an accurate value for the median mussel size. This value was obtained from size distribution measurements done on the quadrat samples (Fig. 4). Values ranged from 1000 to 2000 mussels / m^2 for soft sediment areas, but reached 5000 mussels / m^2 in rocky areas. These figures are most likely an underestimation, as they are based on only a 2-D picture of the mussel coverage, neglecting those below the surface layer.

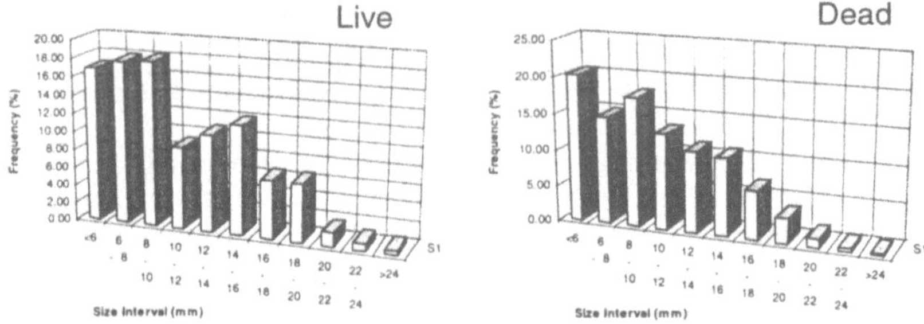

Fig. 4. Representative size frequency histogram of zebra mussel samples (both live and dead) based on a combined total of over 4000 shells from various locations.

Seven quadrat samples were counted manually and also provided size distributions for the zebra mussels sampled (Fig. 4). The size distributions were all bimodal, reflecting populations of different ages. Median size of the mussels sampled in the quadrats ranged between 10 and 14 mm, with maximum values up to 25 mm. Percentages of live mussels ranged from as low as 14% to as high as 82%. Using these values we obtained live mussel figures of as high as 20 000 mussels . m^{-2} (Table 1).

Both techniques offer advantages, predominantly that of speed and wide coverage for the video technique versus better resolution of the smaller (e.g. < 10 mm) individuals in the size distribution and better live / dead ratios for the quadrat sample technique. In the extrapolation of the counts for these representative areas to the lake as a whole, we decided to use the live counts from the quadrat measurements. A major factor was that these counts were more compatible with those on hard substrates (Fitzsimons et al. (1995).

A map of bottom substrates in the western basin (Fig. 5; Rasul *et al. 1996,* in prep.) was used to calculate the total area of the bottom covered by each of 4 types represented by the survey sites: mud, bedrock and hard ground, sand, and mixed sand and cobbles (Table 1). Combining these areas with the mussel per unit area counts described above, we arrived at a first-order approximation of 10^{13} for the entire western basin. It is difficult to assign an error margin for such an estimate (all of the components have considerable uncertainties), however, we would place the figure as within an order of magnitude of the true value.

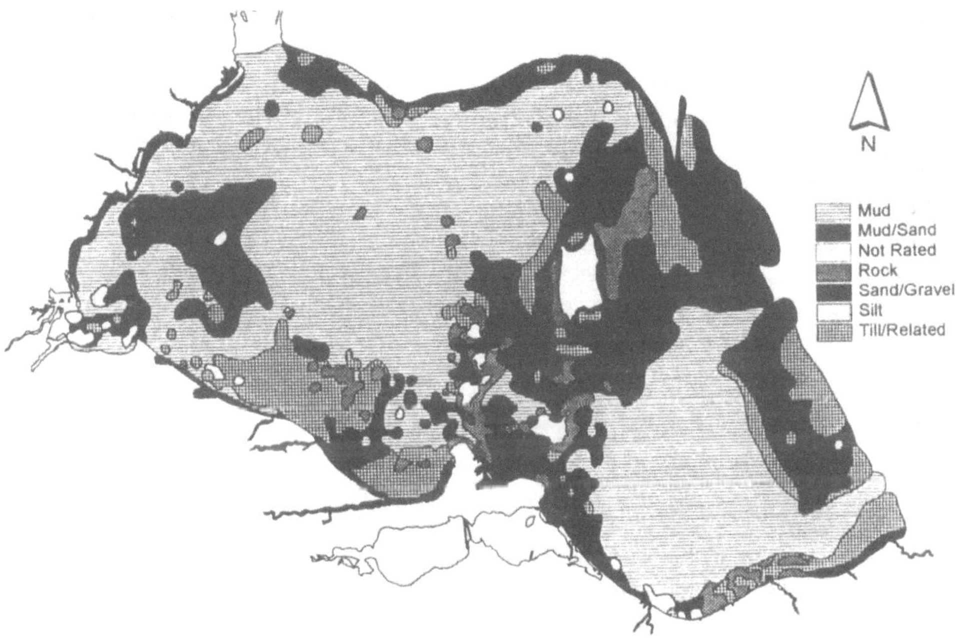

Fig. 5. Generalized substrate map of western Lake Erie.

4. Discussion

For the first time it was shown that regular patterns of colonization exist (ovate masses and linear features). This suggests that studies of mussel coverage in muddy sediments using random grab or dredge samples could be of doubtful validity unless a large number of samples were taken. Another factor not addressed here (and to our experience, of lesser importance) is the colonization of individual live or dead unionid clams in muddy offshore sediments.

4.1 HYPOTHESES FOR ORIGIN OF STRIPE AND FOOTBALL PATTERNS

It is noteworthy that with the exception of unionid clam colonization, zebra mussel colonies in muddy sediment substrates occur close to areas of bedrock outcrops. This coincidence leads to the hypothesis that the zebra mussels occupying soft substrates were physically derived from these hard-bottom nuclei.

STRIPES

The most likely origin of these features is through bottom scouring related to natural processes. Although deep-rooted ice keels can reach the bottom in the <10 depths of the basin, they tend to be much wider (several metres wide) than the stripes, and show no bifurcation. Sediments are also scoured by the activity of fishing boats, that commonly pull nets and anchors in the sediments. However, the resulting excavations would be more variable in direction and width, and show levees of ploughed material on the sides. In the absence of any direct studies of the stripe features, the most likely mode of formation for these linear troughs is by long-term exposure to leeside currents and eddies downstream from a bedrock exposure (reef). The steepness and undercut nature of the walls could be due to subsequent reworking by fish.

If this hypothesis is correct, it leads to the conclusion that the zebra mussels occupying soft substrates could have been physically exported from these hard-bottom nuclei by storm-induced current- or ice-scour processes. Dead mussel shells previously deposited in the troughs would then serve as attachment sites.

FOOTBALLS

The football patterns are interpreted as smaller, isolated hard-bottom areas that have a dense cover of mussels. They owe their rounded shape to the absence of relief and current scouring, which makes them vulnerable to periodic sedimentation. Their shape and precise location can thus change depending on whether they have been recently covered or exposed by erosional / depositional processes. Dead mussel shells at these sites could also serve as attachment sites for locally-derived veligers, causing the feature to grow outward..

SUMMARY MODEL OF ZEBRA MUSSEL COLONIZATION

Veligers carried along in the water column settle onto the bottom in a random manner. Those that settle onto hard substrates can attach their byssal threads firmly and thrive. This is the case in bedrock or cobble areas. If the veligers settle onto soft sediments, their only hope of surviving there is to attach themselves to hard objects, such as the shells of clams or zebra mussels, living or dead. In only a few cases were the zebra mussel clumps observed in soft sediments attached to clam shells; most were attached to disarticulated zebra mussel shells.

The association of the stripe pattern of zebra mussel colonization with current-mediated transport over densely colonized bedrock exposures suggests that the source of the shells used to initiate colonization were derived from the bedrock "reef" and deposited in the linear vortex-scours downstream. Veligers spawned from the reef would thus find settling sites nearby to form druse clumps. Patches of druses can join together over time to fill up the troughs, then infilled troughs would coalesce to form an apron adjacent to the bedrock nucleus, and thus the colonies would outward along the troughs (Figures 2E, 4). Enhanced current flow through the troughs would also prevent the colonies from being sedimented over.

4.3 ESTIMATION OF ZEBRA MUSSEL POPULATION OF WESTERN LAKE ERIE

One of the prime benefits of this study is that it provides an estimate of total zebra mussel populations in western Lake Erie that takes into account colonization in soft sediments. Estimation of impacts on phytoplankton stocks caused by zebra mussel filtration (MacIsaac et al. 1992) were based on on a figure for mean population density of zebra mussels and average clearance rates per mussel. They assumed densities averaging 270 000 individuals . m^{-2}, i.e. that measured for hard substrates (15% of bottom surface in western Lake Erie).

The new quantification technique applied here, i.e. direct counting from video footages calibrated by diver-collected quadrat samples, is still being tested and improved. Nevertheless, when combined with the updated and detailed map of the various substrates in the basin it is an improvement on earlier estimates of total colonization. Estimates of the statistical error is virtually impossible given the patchy distribution, the wide range in size of the individual mussels, and the difficulty in differentiating between live and dead individuals. These aspects will be improved as more areas are sampled. Our estimate of 10^{13} individuals for the entire basin provides an overall population density of approximately 10 000 mussels . m^{-2}. Such a reduction would make a considerable difference in the ecosystem impacts previously estimated.

ACKNOWLEDGEMENTS

We thank Najeeb Rasul for providing the substrate map for the western basin. Despina Brotea and Xiaowa Wang assisted in counting and measuring the shell samples and in other analytical tasks. Ship support was ably provided by the NWRI Technical Operations Unit; divers were Bruce Gray and Mark Dahl. The constructive critical review by the two anonymous referees aided the manuscript greatly.

References

Berkman, P.A.; Lewandowski, A.; van Bloem, S.; Garton, D.W.; and Kennedy, G. 1995. *Proc. 5th Internat. Zebra Mussel and Other Aquatic Nuisance Species Conf., Toronto, Can.*, 12.

Binelli, A.; Provini, A; and Galassi, S. 1997. *Water, Air, and Soil Poll.* (in press).

Dermott, R. and Munawar, M. 1993. *Can. Jour. Fish. Aquat. Sci.* **50,** 2298.

Carey, I. 1978. U.S. Naval Oceanographic Office TN 3432-01-78.

Fitzsimons, J.D.; Leach, J.H.; Nepszy, S.J.; and Cairns, V.W. 1995. *Can. Journal Fish. Aquat. Sci.* **52,** 578.

Hunter, R.D. and Bailey, J.F. 1992. *The Nautilus* ,**106,** 60.

Ioannou, S.E.; Coakley, J.P.; and Brown, G.R. 1995. *Zebra mussel quantification in the western basin of Lake Erie,* Environment Canada, NWRI Contribution 96-62, 11pp.

Leach, J.H., 1993. *Zebra Mussels: Biology, Impacts, and Control.*, Lewis Publishers Inc., Chelsea, Michigan, p. 381

MacIsaac, H.; Sproules, W.G.; Johannsson, O.E.; and Leach, J.H. 1992.. *Oecologia* **92**, 30.

New York Sea Grant. 1993. *North American range of the zebra mussel as of 15 December 1993. 2-sided map / information circular.*

Rasul, N.; Pippert, R.; and Coakley J.P. 1996 (in prep.). *NWRI Contribution Series.*

Stanczykowska, A. 1964.. *Ekol. Pol.* **12**, 653.

TROPHIC MODIFICATIONS IN LAKE COMO (N. ITALY) CAUSED BY THE ZEBRA MUSSEL (*DREISSENA POLYMORPHA*)

Binelli A., A. Provini, S. Galassi
Department of Biology - University of Milan - Via Celoria 26 - 20133 Milan - Italy

Abstract. *A large scale study on the western basin of Lake Como (N. Italy) was started in 1995 to examine the effects of the zebra mussel colonization which began in early '90. Our results have been related to '91-92 data (pre-Dreissena period), before the maximum colonization of zebra mussel. In spring and summer of the post-Dreissena period total phosphorus, P-PO$_4$,, nitrate and chlorophyll values decreased, while ammonium and transparency increased at every sampling station.*
Zebra mussel does not modify the trophic state of this sub-basin but it plays an important role in nutrient cycling. The entire population can filter epilimnetic waters 2.1 times per year and can produce 2.9 x 10^4 t/y of pseudofaeces which are transferred to sediments.

Key words: zebra mussel, *Dreissena polymorpha*, phosphorus, trophic state, recovery, Lake Como.

1. Introduction

Lake Como, like the majority of Italian sub-alpine lakes, has experienced eutrophication problems since the beginning of the 1960's. As a result of measures taken for the control of trophic conditions (contruction of waste water treatment plants and reduction of phosphorus content in domestic detergents), the water quality of the lake has improved and the mean phosphorus content in the lake during circulation period in 1992 has decreased to 41 µg/L (Chiaudani & Premazzi, 1993). During the early 1990's Lake Como was colonized by the zebra mussel (*Dreissena polymorpha*), but the highest density was reached probably in '94-95. Lake Como was suitable for colonization because of the presence of appropriate rocky substrate, productive plankton community and moderate temperatures.

Zebra mussel is widespread in lakes and, where it is present, is the most abundant mussel species and certainly the most important one as regards to biomass. The role played by *Dreissena polymorpha* in aquatic ecosystems has long been considered important. In recent years, special attention has been paid to the effects of its filtration activity. This organism consumes seston from the water, while its pseudofaeces and faeces drop to the lake bottom. The amount of faeces which deliver P and N to the sediments is also enormous (Stanczykowska & Lewandowsky, 1993). Recently, the effects of zebra mussel colonization were studied in large American waterbodies and significant changes in water quality were noticed after its establishment in the Great Lakes (Fahnenstiel *et al.*, 1995). These investigations documented dramatic increases in water clarity, drops in chlorophyll and phytoplankton abundances and changes in benthic invertebrate communities (Nalepa & Fahnenstiel, 1995).

Although zebra mussel was accidentally introduced in Lake Garda since the end of '60s (Giusti & Oppi, 1972) and more recently it was found in many northern Italian basins, the effects of the establishment of large populations of this mussel have never been investigated. This preliminary study represents the first attempt to investigate the possible influence of zebra mussel on trophic state modification in a deep basin. Trends in epilimnetic water quality parameters were examined during 1995-96 in the western basin of Lake Como and compared with 1991-92 data, at the beginning of colonization.

Water, Air and Soil Pollution **99**: 633-640, 1997.

1.1 Study area:

Lake Como (Fig. 1) is the deepest Italian lake and it can be divided in three different basins (northern, western and eastern). Table I shows its main morphometric and hydrological characteristics. Lake Como is oligomictic and the winter overturn generally does not go deeper than 150 m. The climate of the area is considered to be intermediate between continental and north-mediterranean.

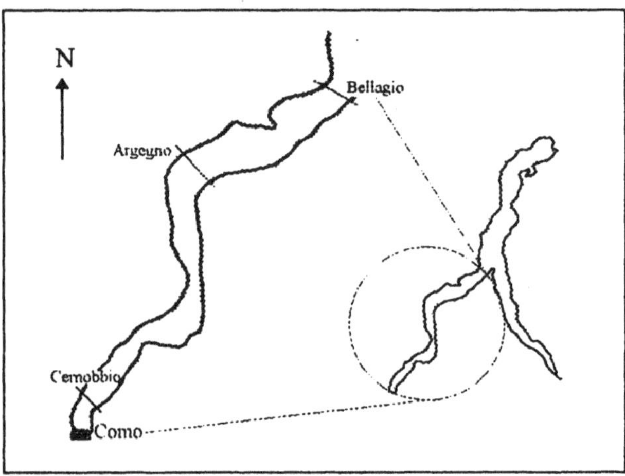

Fig. 1 Study area and sampling transects.

Phosphorus load is 304 t/y, 180 t/y of which are introduced by point sources (domestic discharges and industry), while the remaining 123 t/y come from various sources (agriculture, animal rearing and natural loads) (Chiaudani & Premazzi, 1993).

Table I Main characteristics of Lake Como.

Catchment area (km^2):	4,522
Lake area (km^2):	145
Mean depth (m):	161
Maximum depth (m):	410
Volume (x 10^6 m^3):	23,372
Theoretical renewal time (y):	4.5
Average residence time (y):	8.3-30 (w. basin)
	5.6-8.7 (n. basin)
	3.6-6.4 (e. basin)

In the western sub-basin, which has an area of 46.7 km^2, waste water treatment plants currently serve about 184,000 equivalent inhabitants. This represents only 36% of the polluting load (Chiaudani & Premazzi, 1993). This basin has an average residence time longer than the other two (Table I) because of its particular hydrological characteristics (lack of outlet and presence of an underwater ridge near Bellagio) that prevent a full water exchange with the rest of the lake. According to OECD classification Lake Como is mesotrophic.

2. Sampling and methods

Monthly surveys were carried out in the western basin since June '95, except during winter time, at three different crosswise transects: Cernobbio, near the most populated area of the lake, Argegno, at the maximum depth of the lake and Bellagio, the closing section of the basin (Fig. 1). Last samples were taken in September 1996. Each transect was sampled by a Van Dorn bottle at three stations, one in the middle of the lake and two near the shore. Total Phosphorus (TP), Total Nitrogen (TN), chlorophyll, transparency, temperature, P-PO_4, nitrate and ammonium were measured at depths of 1, 2.5, 5, 10, 20 m. These last three parameters were measured on 0.45 µm filtered water. Occasionally we also sampled at depths of 30 and 40 m. Two of the middle lake stations coincide with those selected by Chiaudani and Premazzi (1993) in 1991-92; the third one (Cernobbio) is about 1 km to the North of that selected in 1991-92 to minimize the influences of municipal discharges. The analytical protocol also was similar, except for chlorophyll determination. The 1991-92 samples were extracted by methanol (cold extraction), while we used hot extraction with acetone (24 h) after homogenization of the glass-fibre filter by a grinding mill. However comparable results are reported for the two methods (Nusch, 1980). Concentrations were reported as weighted means for the upper 10 m of the water column and averaged over each transect.

3. Results

Ortho-phosphate (P-PO_4) concentration, which amounted to 25-30 µg/l during winter overturn, sharply decreased between March and April and it remained below the detection limit (4 µg/l) till Autumn. TP concentrations for the survey period are given in Fig. 2. As expected, a decrease in TP levels was observed during the spring and summer months ; values in 1995 seemed to be slightly lower than those in 1996, but this was due probably to climatic conditions. Moreover, July 1996 was unusually cold and rainy, causing an extra nutrient load from the catchment through runoff.

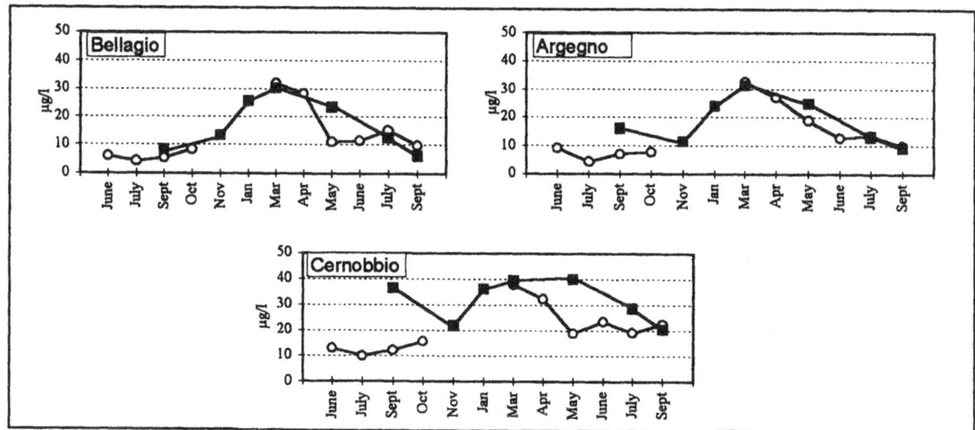

Fig. 2 Mean Total Phosphorus concentrations at Bellagio, Argegno and Cernobbio in '91-92 (filled squares) and in '95-96 (empty circles).

Although TP trend was similar in the three transects, Cernobbio concentrations were always higher than those of the other two stations; this agrees with the fact that the southern part of the basin has the greatest anthropogenic impact.

Differences among stations were even more relevant in the case of chlorophyll concentration (Fig. 3). An algal bloom was evident in April at Bellagio and Argegno. whereas at Cernobbio the chlorophyll concentration increased sharply in April and then more slowly till July.

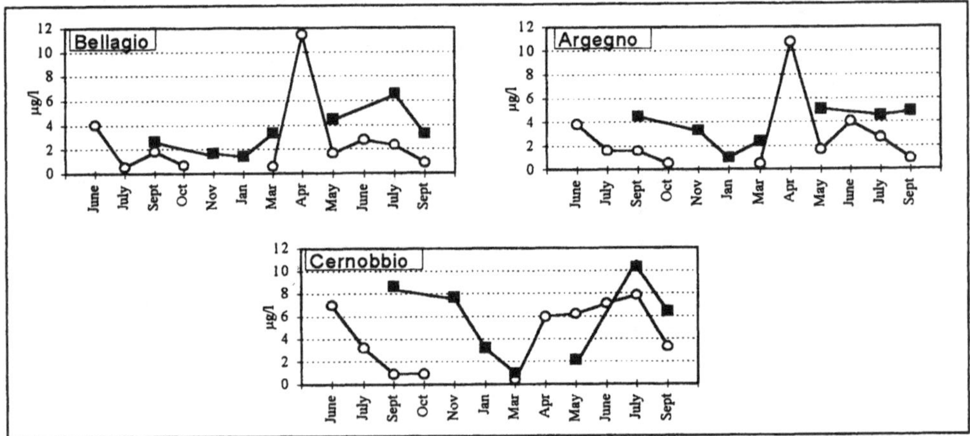

Fig. 3 Mean chlorophyll concentrations in '91-92 (filled squares) and '95-96 (empty circles) at the three sampling stations.

Transparency, measured by Secchi disk, had an opposite trend compared to chlorophyll, indicating that algae were the main component of the suspended matter. Figure 4 shows the transparency values at Bellagio, that were almost the same as those of Argegno, with a maximum of 17 m in March and a minimum of 2.8 m in April. At Cernobbio the maximum was only 13 m, and the minimum (3.2 m) was reached in June-July.

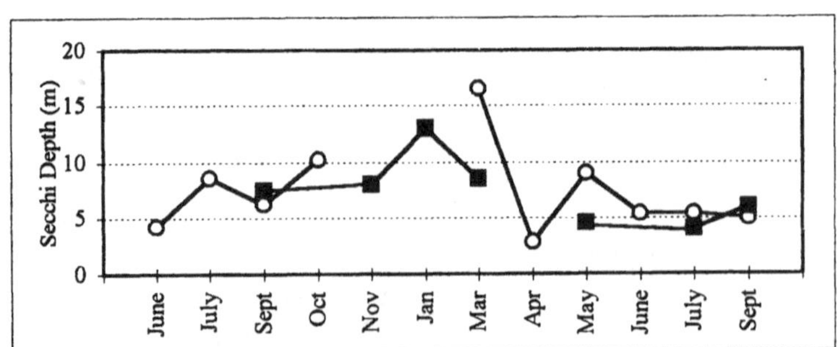

Fig. 4 Monthly transparency at Bellagio central station in '91-92 (filled squares) and in '95-96 (empty circles).

Regarding N compounds (ammonia and nitrate), Cernobbio differed from the other two sampling sites, that were very similar. Ammonium and nitrate trends for one of these two stations are shown in Figure 5. Nitrates were always the most important inorganic nitrogen compounds, since TN ranged between 400-600 µg/l at Bellagio and Argegno and 650-950 μg/l at Cernobbio from April to June. $N\text{-}NH_3$ presented a concentration peak in May '96, but a high level was maintained until the end of June at Cernobbio.

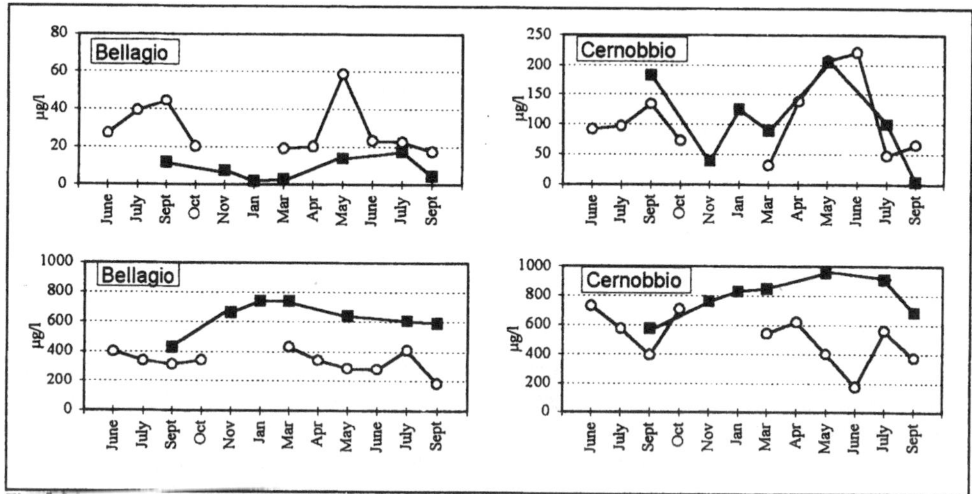

Fig. 5 Mean ammonium (above) and nitrate (below) concentrations in 1991-92 (filled squares) and 1995-96 (empty circles) at Bellagio and Cernobbio.

Nitrate concentration decreased during spring and summer in the lake; the increased concentration found in July '96 can be ascribed to meteorological conditions and, probably, to nitrification processes at Cernobbio.

4. Discussion

After the establishment of a large population of zebra mussel, waters of the western basin of Lake Como remained mesotrophic, as defined by their nutrient chemistry, because the concentration of trophic state parameters (TP, $P\text{-}PO_4$ and chlorophyll) were similar to those recorded in 1991-92 during winter circulation (Fig. 2 and 3).

In the spring of 1996 a dramatic decrease in TP was noticed (Fig. 2) at every station from March to May with a reduction of 66%, 42% and 50% of winter TP concentration going from north to south, respectively. Phosphorus uptake by algae is to be expected in spring, but a comparison with the 1991-92 data shows that the 1995-96 decrease was much higher. In our opinion, this phenomenon is to be ascribed to the filtering effect of *Dreissena*, since no further action for eutrophication control was undertaken in recent years. It is known from the literature that the filtration efficiency of *Dreissena* is temperature dependent with optimum filtration rate between 10 °C and 20 °C (Fanslow *et al.*, 1995). The threshold temperature value was reached in the epilimnetic waters in April, when the maximum reduction of phosphorus was observed. The hypothesis that zebra mussel plays an important role in phosphorus removal is confirmed by the fact that this phenomenon starts from the shores, where mussels grow. Figure 6 shows the TP pattern of the Argegno transect in May '96: it is evident that TP concentrations are low near the shores at all depths and only in the upper 2.5 m in open waters.

TP disappearance rates were quantified from March to May in '92 and '96 in the upper 10 m of water. During '96 and '92 the decrease was 12.1 kg/d and 4.6 kg/d at

Bellagio, 8 kg/d and 4.34 kg/d at Argegno, respectively. Instead, at Cernobbio there was a reduction of 10.9 kg/d in '96, while in '92 a low increase (0.52 kg/d) was detected.

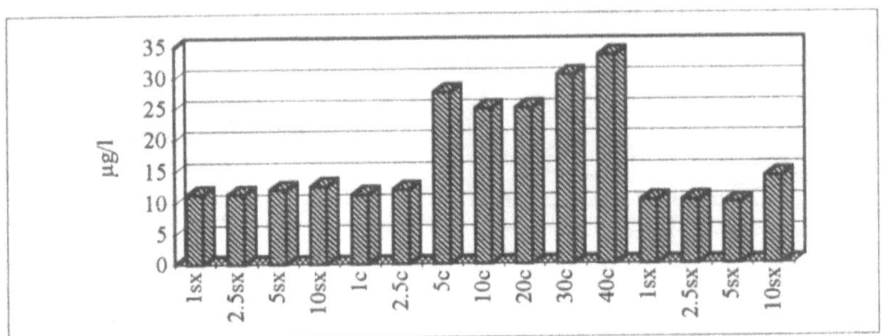

Fig. 6 TP concentrations in May '96 at Argegno transect. sx=left; c=center; dx=right.

Furthermore, the maximum TP reduction occured from May to July in 1992 and from April to May in 1996 (Fig. 2), indicating once again an active role of *Dreissena* in particulate phosphorus removal from the water column.

Assuming that all TP disappearance was due to *Dreissena* filtration, the maximum apparent phosphorus removal rate (k) can be calculated from the equation of Coughlan (1969):

$$k = (\ln C_1 - \ln C_2) / t$$

where C_1 and C_2 are the starting and final concentrations and t is the time period (days). Our value of k ($1.58 \times 10^{-2}\ d^{-1}$) agrees with the rate estimated on the basis of data presented by Holland *et al.* (1995) for the Hatchery Bay, western Lake Erie ($k = 1.64 \times 10^{-2}\ d^{-1}$).

Regarding differences in TP levels in June and July '95-96, at least two hypotheses are possible. The first one is the influence of climatic conditions; from June '96 numerous strong storm events might have caused disturbances to phosphorus dynamics. The second is linked with a different composition of age classes of zebra mussel in 1995. If this year were effectively the period of maximum population density, it should also correspond to the maximum occurrence of juvenile mussels, that show higher pumping rates (Bunt *et al.*, 1993). Several studies indicate that a large proportion of the nitrogen and phosphorus processed by *Dreissena* is excreted as ammonium and phosphate, detectable as soluble reactive phosphorus (Holland *et al.*, 1995). In Lake Como during spring and summer '95-96, very low $P\text{-}PO_4$ concentrations were measured. This is most likely due to algal growth, but changes in ammonium levels can be attributed to *Dreissena* activity (Fig. 5). Concentration peaks observed in spring are consistent with literature data, showing seasonal changes in nitrogen excretion by *Dreissena* with highest rates occurring in spring and early summer, closely related to water temperature (Quigley *et al.*, 1993).

Differences between pre and post-*Dreissena* invasion can not be appreciated at Cernobbio, but it is to be remembered that the location of this station is different compared to Como. In fact, occasional samplings carried out at Como in '96 showed ammonium concentrations even higher than those reported in '92.

Ammonium and nitrate trends in 1996 were almost opposite; moreover present $N\text{-}NO_3$ levels were always lower than those of '92. The spring decrease followed the P pattern at all stations and it was 10-15 times higher. This agrees with N/P ratios in algae and in *Dreissena* soft tissues where N and P are about 12% and 0.9% of dry weight, respectively (Stanczykowska & Lewandowsky, 1993).

Chlorophyll concentrations in 1995-96 were lower than those in 1991-92; differences observed in July '95-96 were ascribed to storm events. For this parameter the comparison between years is not easy because of the number of variables involved. In '92 the maximum chlorophyll peak was probably lost due to low sampling frequency. It is also possible that *Dreissena* filtration and nutrient excretion could modify phytoplankton composition (Tilman *et al.*, 1982).

Contradictory results are reported in several studies and more research is needed to evaluate the role of zebra mussel in changes of phytoplankton composition. In a short-term experiment, biovolumes and abundance of cyanophytes were unchanged over the experimental period, despite declines in other phytoplankton groups (Heath *et al.*, 1995; Lavrentyev *et al.*, 1995). Instead Nicholls and Hopkins (1993) reported that zebra mussel affected all phytoplankton groups equally. Our study of phytoplankton dynamics is still in progress and we cannot state if *Dreissena* affects the relative abundance of cyanophytes. In the literature, the values of filtration rate of zebra mussel vary considerably (2 to 287 ml/mussel/h); filtration rate and pseudofaeces production depend on suspended matter content, temperature and shell length (Noordhuis *et al.*, 1992). In Lake Como, suspended matter and water temperature are not limiting for zebra mussel filtration from April to November. Mean theoretical filtration rate (FR, ml/mussel/h) and pseudofaeces production (PSP, mg/mussel/d) were calculated from the shell length (L) according to the equations of Noordhuis *et al.* (1992):

$$FR = 15.43 / (0.293 + 52.38 \times e^{-0.367\,L}) \qquad PSP = 34.87 / (1 + 34.87 \times 0.83^{L})$$

Density population of zebra mussel was measured in the winter of '95-96, indicating that there were about 6000 mussels/m^2 with a mean shell length of 15 mm (M. Mariani, personal communication). Several observations showed that this density is maintained down to 15 m at least. A mean filtration rate of 0.73 l/mussel/d was calculated. This value corresponds roughly to 1.9×10^9 m^3/y for the entire population, considering only the 8 months of *Dreissena* maximum activity and a colonized surface of 1.75 km^2 calculated from the bathymetric map. The possible daily or seasonal rhytms of zebra mussel filtration were not taken into account (Noordhuis *et al.*, 1992). Thus, about 23% of the total volume of the western basin is filtered yearly and the epilimnion (8.9×10^7 m^3) is flushed 2.1 times per year. As a consequence of *Dreissena* activity, 11.14 mg/mussel/d of pseudofaeces should be produced with a total transport of 2.9×10^4 t/y to the sediments. Since the pseudofaeces have a higher sedimentation rate than algae, phosphorus and nitrogen, linked to this material, which are several times higher than that accumulated in the soft tissues (Stanczykowska & Lewandowsky, 1993), are transferred faster to the sediments, changing the availability of nutrients in the water column.

5. Conclusions

Seasonal changes in nutrient cycling were observed after *Dreissena* colonization of Lake Como. Similar findings were reported for other aquatic environments, even of larger size, but were never found for very deep lakes. In Lake Como the effects of

Dreissena activity were noticed only in the epilimnetic layer of water during spring-summer. The trophic state of the lake was not affected, since phosphorus concentration during winter circulation was the same as before *Dreissena* colonization. In any event, through its filtering activity, the zebra mussel transfers production and resources from the epilimnion to the benthic region.

The time scale of this study did not allow the observation of all the modifications that the invasion of an empty ecological niche by *Dreissena* can imply. However, besides indirect effects due to a different nutrient availability, we expect that this primary consumer could affect the energy flux through the trophic web. Biomass quantification and age classes study of zebra mussel in Lake Como will provide a more realistic evaluation of its pumping efficiency and will lead us to understand whether if *Dreissena* population has reached its carrying capacity.

The ecological role of this filter-feeder that occupies the middle of the food web is difficult to assess. Phytoplankton characterization over several years, also taking into account the seasonal variations, is needed to verify the filtration selectivity of *Dreissena*. Being a benthic organism, this mussel also can modify carbon and nutrient decomposition rates at the bottom of the lake through pseudofaeces and excretion. This fact should lead to a change in benthic community composition. Therefore, *Dreissena* may change the lake response to perturbation from both ends of the food web.

This survey will be carried on over a longer time span to confirm the results of the present study. Since now all the Italian subalpine great lakes are colonized by zebra mussel, it is important to monitor its impact on the water quality.

6. References

Bunt, C.M. MacIsaac, H.J. and Sprules, W.G.: 1993, *Can. J. Fish. Aquat. Sci.* **50**, 1017-1022.

Chiaudani, G. and Premazzi, G.: 1993, *Il lago di Como, EUR 15267 IT.* p. 237.

Coughlan, J.: 1969, *Mar. Biol.* **2**: 356-358.

Fahnenstiel, G. Lang, G.A. Nalepa, T. and Johengen T.H.: 1995, *J. Great Lakes Res.* **21** (4), 435-448.

Fanslow, D. Nalepa, T.P. and Lang, G.A.: 1995, *J. Great Lakes Res.* **21** (4), 489-500.

Giusti, F. and Oppi, E.: 1972, *Mem. Mus. Civ. St. Nat. Verona.* **20**, 45-49.

Heath, R.T. Fahnenstiel, G.L. Gardner, W.S. Cavaletto, J.F. and Hwang, S.J.: 1995, *J. Great Lakes Res.* **21**: 501-516.

Holland, R.E. Johengen, T.H. and Beeton, A.M.: 1995, *Can. J. Fish Aquat. Sci.* **52**, 1202-1209.

Lavrentyev, P.J. Gardner, W.S. Cavaletto, J.F. and Beaver, J.R: 1995, *J. Great Lakes Res.* **21**, 545-557.

Nalepa, T.F. and Fahnenstiel G.L.: 1995, *J. Great Lakes Res.* **21** (4), 411-416.

Nicholls, K.H. and Hopkins, G.J.: 1993, *J. Great Lakes Res.* **19**, 637-647.

Noordhuis, R. Reeders, H.H. and Bij de Vaate, A.: 1992, *The Zebra Mussel Dreissena p. Neumann/Jenner eds. Gustav Fisher Verlag.* 101-114.

Nusch, E.A.: 1980, *Arch. Beth. Ergebn. Limnol.* 14-36.

Quigley, M.A. Gardner, W.S. and Gordon, W.M.: 1993, *Zebra Mussel, Nalepa & Schloesser eds., Lewis Publishers.* 295-306.

Stanczykowska, A. and Lewandowsky, K.: 1993, *Hydrobiologia.* **251**, 73-79.

Tilman, D.S. Kilham, S.S. and Kilham P.: 1982, *Ann. Rev. Ecol. Syst.* **13**, 349-372.

ASSESSING BENTHIC IMPACTS OF ORGANIC ENRICHMENT FROM MARINE AQUACULTURE

B.T. Hargrave[1], G.A. Phillips[1] L.I. Doucette[1], M.J. White[1], T.G. Milligan[1], D.J. Wildish[2] and R.E. Cranston[3]

[1]*Marine Environmental Sciences Division, Maritimes Region, Fisheries and Oceans Canada, Bedford Institute of Oceanography, P.O. Box 1006, Dartmouth, Nova Scotia B2Y 4A2, Canada.* [2]*Marine Environmental Sciences Division, Maritimes Region, Fisheries and Oceans Canada, St. Andrews Biological Station, St. Andrews, New Brunswick E0G 2X0, Canada.* [3]*Natural Resources Canada, Bedford Institute of Oceanography, P.O. Box 1006, Dartmouth, Nova Scotia B2Y 4A2, Canada.*

ABSTRACT. Benthic observations were carried out at 22 stations in the Western Isles region of the Bay of Fundy on the east coast of Canada to evaluate impacts at salmon aquaculture sites. Eleven sites were located under salmon net-pens and 11 sites (reference or control locations) were at distances >50 m from net-pens. Total $S^{=}$ and redox potential (Eh) in surface sediment and benthic O_2 uptake and CO_2 release were sensitive indicators of benthic organic enrichment. High variability between replicate measurements of sediment gas exchange could reflect spatial patchiness in sedimentation of fecal waste and food pellets under fish pens. Biomass of deposit feeders was significantly increased at cage sites but total macrofauna biomass was similar at cage and reference locations. Surface sediment water content, modal grain size, pore water salinity and sulfate, and total biomass of macrofauna were the least sensitive indicators of enrichment.

Key Words: fish farms, fecal waste, sediment, geochemistry, metabolism, macrofauna

1. Introduction

The rapid development of salmonid aquaculture in the Western Isles region of the Bay of Fundy has resulted in several studies directed toward assessment of environmental impacts of this new industry (Hargrave et al. 1993; Strain et al. 1994; Trites and Petrie 1995; Wildish et al. 1993). Studies here and elsewhere have shown that benthic responses to organic enrichment from finfish mariculture occur as a result of enhanced sedimentation of particulate organic waste products (Hargrave 1994a). The degree of impact depends on many factors such as water current speed, sediment composition, seasonal storm-related resuspension, and the presence of benthic fauna (Findlay and Watling 1994; Gowen et al. 1994; Silvert 1994). However, it is not presently possible to predict the impacts of increased organic matter deposition at sites occupied or potentially available for aquaculture development. There is often no basis for comparing present sedimentary conditions with those in the past to assess the impacts of finfish aquaculture.

Measures of benthic enrichment would be useful for assessing long-term changes in ecological conditions within areas used for the expanding aquaculture industry. This paper compares geotechnical, geochemical, and biological variables in surface sediments under salmon aquaculture sites with similar measurements at nearby reference locations. Modal inorganic grain size, water content, salinity, redox potentials, ammonium, sulfides,

Water, Air and Soil Pollution **99**: 641-650, 1997.

sulfate, organic carbon and nitrogen, benthic fluxes of dissolved O_2 and CO_2, and macrofauna biomass were measured at all sites. The data were used to assess the sensitivity of different variables to detect benthic impacts of net-pen aquaculture.

2. Materials and Methods

2.1. STATION LOCATIONS

Samples were collected close to the time of low tide between June 3 and August 15, 1994, at 22 stations in the Fundy Isles region, New Brunswick at the southwest end of the Bay of Fundy on the Atlantic coast of Canada (Hargrave et al. 1995). Water depths (from chart datum) varied from 7 to 20 m. 11 stations were located under salmon pens and 11 other sites >50 m from pens served as reference sites. Licenced capacity (New Brunswick Department of Fisheries and Aquaculture) for various farms varied from 40 x 10^3 to 320 x 10^3 tons in 1991, but no data was available for actual fish biomass held at each site during our study. A previous study (Hargrave et al. 1993) showed that benthic effects of salmon pens in this area were not detected by measurements of sediment/water gas and nutrient exchange at distances >50 m. Fish pens at one location (Frye Island) were moved ~500 m north of their original position in July-August 1992. Thus, this sampling location ('old farm' site), had not received direct input of particulate wastes from fish pens for 2 years prior to our study. All other farm sites had been occupied for at least three years.

2.2. SAMPLE COLLECTION

Divers used acrylic plastic core liners at each station to collect three short cores (15 cm long, 5.7 cm dia, 2 cm water-filled head space) for benthic gas exchange measurements and three long cores (50 cm long, 6.5 cm dia) for sediment geochemical and geotechnical profiles. Macrofauna biomass determinations were made using five replicate Hunter 0.1 m^2 grab sampler (Hunter and Simpson 1976) at each station.

2.3. ANALYTICAL METHODS

Analytical methods are described in detail in Hargrave et al. (1993). Cores were kept upright to maintain an undisturbed sediment-water interface, and transported to the laboratory in a chilled cooler within a few hours after collection. Benthic metabolism was measured as the average of oxygen (O_2) uptake and carbon dioxide (CO_2) release in three short sediment cores. Cores were incubated with filtered seawater over the sediment for measures of total gas exchange for the first incubation. Seawater was then replaced with seawater containing 0.1% w/v mercuric chloride ($HgCl_2$) as a metabolic poison. Methods of O_2 and CO_2 analysis and calculations of gas exchange rates are described in Hargrave and Connolly (1978) and Hargrave and Phillips (1981). Fluxes before and after addition of $HgCl_2$ were reported as mg O_2 and CO_2 m^{-2} h^{-1}.

Long cores were stored (2 to 4°C) prior to subsampling (within 48 h of collection). Holes (1.9 cm dia.) drilled in corers at 2 cm intervals were covered with duct tape to prevent water and sediment loss. Orion® ion-specific electrodes were used for duplicate measurements on two cores from each site for oxidation reduction (redox)(Eh) potentials (corrected to a standard reference calomel electrode), ammonium (NH_4^+) and total sulfide

($S^{=}$). Sediment was withdrawn for determination of vertical profiles of sediment water content, inorganic grain size, organic carbon (C_{org}) and nitrogen (N) (Perkin Elmer 240B elemental analyzer), and pore water analysis of salinity, sulfate ($SO_4^{=}$), and NH_4^+ (Solarzano 1969) as described in Hargrave et al. (1995).

Sediment from grab samples was sieved (0.5 mm mesh) using cold running seawater. Macrofauna were removed, placed in 5% buffered formalin solution, and grouped into trophic types (deposit feeders, suspension feeders, predators, omnivores, and herbivores) (Wildish and Peer 1983). Molluscs were enumerated and dissected to remove tissue from shells for weight determination. Wet weights of specimens were determined by blotting individuals or tissues on absorbent paper (15 sec.) and rapid weighing (15 to 30 sec.) to reduce weight loss through desiccation. Biomass was reported as g m^{-2} wet weight.

For stations where it was possible to calculate significant ($p < 0.05$) increases in NH_4^+ with sediment depth, concentrations in pore water were used to derive gradients and carbon burial rates (CBR) as described by Cranston (1994). Derivation of the Benthic Enrichment Index (BEI), combining data for water content, Eh and C_{org} is described in Hargrave (1994b).

3. Results and Discussion

3.1. COMPARISON OF VARIABLES

Ranking to evaluate the sensitivity of different variables to detect benthic enrichment was made by performing a one-way ANOVA to determine if variables at cage sites were significantly different from those at reference sites. $S^{=}$, total CO_2 release , total O_2 uptake, chemical O_2 uptake and chemical CO_2 release were the variables with the greatest difference ($p < 0.00002$) between cage and reference sites (Table I).

3.2. WATER CONTENT

Surface sediment water content varied widely (19.1 to 72.7%) with no consistent pattern between cage and reference stations. There was no significant difference ($p=0.45$) between mean values for cage and reference locations (Table I). The range of values (51 to 74%) encompasses those measured previously in the study area (Hargrave et al. 1993). As a rule, water content decreased with sediment depth, reflecting compaction.

3.3. SALINITY

Averages for pore water salinity showed that 16 cores had mean salinity values below the range expected for bottom water in the region (30 to 31.6‰) (Trites and Petrie 1995) but no core had a mean salinity that exceeded the expected range. Residual deionized water in the filter disc used during extraction to remove particulate matter from pore water could have diluted these samples. The downcore variation in salinity should be limited to the bottom water salinity range within the precision of the method (±0.3‰). The standard deviation for two cores collected at Frye Island (0.4‰) was close to the expected variation due to analytical precision. All other cores had higher standard deviations (0.7 to 5.7‰). There was no significant difference ($p=0.644$) between average values for cage and reference sites.

TABLE I

Ranked ANOVA probability values (p) for different benthic variables in 0 to 2 cm surface sediment, benthic fluxes, benthic enrichment index and measures of benthic macrofauna biomass under salmon cages and at reference locations (>50 m from pens) at sampling sites in the Western Isles Region of the Bay of Fundy. The full data set is presented in Hargrave et al. (1995). $p < 0.05$ indicates significant differences between cage and reference sites.

Variable	p
$S^{=}$	**0.00001**
Total CO_2 Release	**0.00001**
Total O_2 Uptake	**0.00001**
Chemical O_2 Uptake	**0.00001**
Chemical CO_2 Release	**0.00002**
BEI (Benthic Enrichment Index)	**0.0003**
Deposit Feeders as Percentage of Total	**0.0090**
Macrofauna Biomass (Deposit Feeders)	**0.0052**
Eh	**0.0470**
N	**0.0188**
NH_4^+ Concentration	**0.0165**
CBR (Carbon Burial Rate)	**0.0129**
NH_4^+ Gradient	**0.0125**
C_{org}	**0.0110**
Percent Water	0.450
Macrofauna Biomass (Total)	0.499
Modal Grain Size	0.532
Pore Water Salinity	0.644
Pore Water $SO_4^{=}$	0.716
Macrofauna Biomass (Suspension Feeders)	0.800

3.4. INORGANIC GRAIN SIZE

Sediment at most sites consisted of silty mud. Inorganic grain size distributions, typical of other Atlantic coastal embayments, were characterized by two components - a flat portion (slope ~0) in the smallest particle size classes and a well-defined silt or sand peak at the coarse end. Additional coarse peaks occurred in some cores. The flat, fine-grain portion of the distribution represents material deposited due to flocculation while peaks are the result of material deposited as single grains (Kranck et al. 1996a,1996b).

The relationship between the floc and single grain components of a size distribution is an indicator of the depositional conditions under which the sediment was formed. The two extremes of floc and single grain settling were evident in samples from low and high energy regions with modal size ranging from 10 to 500 μm. In low-energy areas most of the material was deposited as flocs whereas in high energy regions sediment consisted almost exclusively of well-sorted coarse sands with only minor amounts of floc deposited sediment trapped between grains. Small variations in size distributions were observed downcore at most stations and both the modal size and the relationship between the single grain and floc settled portions of the curve remained constant, indicating that minimal historic changes have occurred in depositional conditions to the depth of core penetration.

3.5. REDOX POTENTIALS

With the exception of two sites, all reference stations had surface sediment Eh potentials $> +100$ mV (Fig. 1A). Values were also usually positive in subsurface sediments at these locations. In contrast, potentials at all but three cage sites were $< +100$ mV. The mean Eh potential of $+117$ mV at the Frye Island 'old farm' site is similar to values at this location between 1989 and 1990 (Wildish et al. 1990). Removal of the pens in 1992 has thus not resulted in an increased in Eh potentials at the former farm location.

3.6. TOTAL SULFIDES

Total $S^=$ concentrations in surface sediments at all cage sites were all >180 μM while values at all but one reference locations were <200 μM (Fig. 1A). $S^=$ concentrations increased between 4 and 18 cm in subsurface layers in most cores. There was considerable variation between replicate cores, but most profiles were similar in shape. Concentrations >2000 μM indicate that high rates of organic matter loading at these cage sites have created anoxic sediments with Eh potentials <0 mV (Fig. 1A). Maximum $S^=$ levels (6 to 7 mM) were observed in subsurface (2 to 8 cm) sediment layers at these cages sites. Surface sediment $S^=$ concentrations at Frye Island have fallen from high levels (>100 mM) in 1990-1991 (Hargrave et al. 1993) to <1 mM in 1994.

3.7. DISSOLVED SULFATE AND AMMONIUM GRADIENTS AND CARBON BURIAL RATES

Normalizing pore water $SO_4^=$ and NH_4^+ values to a constant salinity did not remove excessive variation in these variables. Within the variation of $SO_4^=$ determinations (± 1 mM) and considering the problems due to sample extraction indicated by salinity measurements, $SO_4^=$ concentrations in pore water in most cores did not decrease with depth as would be expected due to $SO4^=$ reduction. In contrast, while dissolved $NH4^+$ concentrations were as variable as those for $SO4^=$, perhaps due to sampling artifacts mentioned above, significant gradients of increasing $NH4^+$ with sediment depth (correlation coefficients >0.95) were measured in 15 of the 24 cores.

The analysis showed that the most impacted cage site had a CBR values of 0.22 g C m^{-2} d^{-1}. Lower CBR values (~ 0.1 g C m^{-2} d^{-1}) were calculated for other cage sites. A relative 95% confidence interval for these estimates (due to variance in sample handling, storage, and analysis) is $\pm$ 25%. In the present study, values of CBR were based on the analysis of $NH4^+$ gradients in a single core from each site. Heterogeneity in sedimentary properties under salmon net-pens would result in a higher confidence interval for CBR at cage sites.

3.8. AMMONIUM

NH_4^+ concentrations using an Orion® electrode were generally >200 μM at cage sites and <200 μM at reference sites. At most sites, concentrations increased with sediment depth, but there were some locations where a reverse trend occurred with maximum

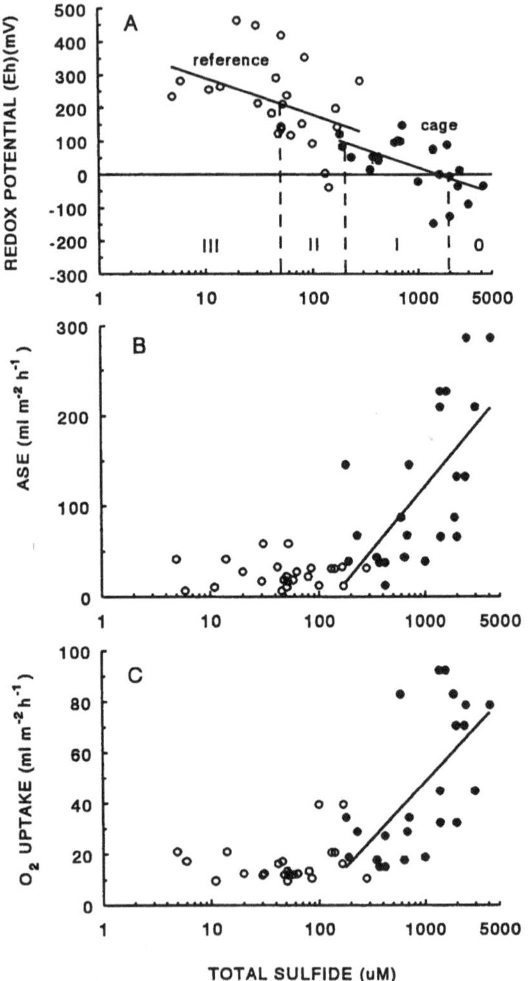

Fig. 1. Correlations between total $S^{=}$ (S) and Eh in replicate (n=2) samples of surface (0-2 cm) sediments and sediment/water fluxes of dissolved CO_2 (CO) and O_2 (O) under salmon cages (solid points) and at reference sites >50 m away (open circles) in the Western Isles Region of the Bay of Fundy. Lines represent least squares regressions: (A) open circles, Eh=261.2-0.61 S, r^2=0.10, p=0.132; solid points, Eh=74.0-0.039 S, r^2=0.25, p=0.013; (B) CO=52.6+0.057 S, r^2=0.45, p=0.0003; (C) O=25.7+0.018 S, r^2=0.38, p=0.0014.

values in surface sediment layers. There was a positive linear correlation (r^2=0.80, n=22) between in NH_4^+ concentrations determined by the electrode and direct chemical measurements of dissolved NH_4^+. Both techniques indicated that NH_4^+ concentrations were highest at the same cage site.

3.9. ORGANIC CARBON AND NITROGEN

The average value for C_{org} at cage sites was 40% higher than the value for reference sites, however, the range of values under cages (0.8 to 3.8%) was similar to that at reference sites (0.4 to 3.2%). A similar increment occurred for N between cage and reference sites, and the ranges of values for these two groups of stations overlapped (0.08 to 0.6). The increment in mean values for both measures at cage sites over those at reference sites (+40%) shows that organic matter accumulated in surface sediments under cages. This reflects organic matter remaining from direct sedimentation of particulate matter as waste food and feces and enhanced sedimentation of fine particulate matter from the water column through flocculation processes. Fish food pellets formed a visible upper layer on the surface sediment at two cage sites where values of C_{org} (22.7 to 27.3%) and N (3.02 to 3.83%) (C:N atomic ratios 8.3 to 9.2) were similar to those in dry food pellets fed to salmon (C_{org}=50.4%; N=8.5%, C:N atomic ratio of 6.92). Decomposition of organic matter in pellets plus mixing with surface sediments would decrease the percentage organic content and increase the C:N ratio. Inorganic grain size distributions indicated that organic matter accumulates with fine sediments at sites where hydrodynamic conditions cause sediments to be deposited.

3.10. BENTHIC ENRICHMENT INDEX

The average BEI values for both cage and reference sites was positive, but negative values appeared at four cage sites. Only one reference site had with a negative BEI value. Negative values of BEI are associated with areas receiving organic carbon sedimentation at rates >1 g C m^{-2} d^{-1} (Hargrave 1994b). BEI values calculated with data from under fish pens at Frye Island in 1990-1991 varied from -1750 to -2800. Positive values (84 and 268) were calculated from data collected in this study during 1994. Values > +1000 at the edge of the fish pen array and a proximate reference site in the earlier data set were comparable with the mean value (+660) in the present study.

3.11. BENTHIC GAS EXCHANGE

Mean values for total sediment O_2 uptake and CO_2 release at cage sites were 175% and 355% higher than values at reference sites (Figs. 1B and 1C). The ranges of values for O_2 uptake at cages sites (15 to 92 ml m^{-2} d^{-1}), and at reference sites (10 to 40 ml m^{-2} d^{-1}) are similar to those measured under and adjacent to pens at the Frye Island location in 1990-1991 (Hargrave et al. 1993). Between-site variation (coefficients of variation = σ/mean) (cv) for both O_2 and CO_2 flux was greater at cage sites (0.6 to 0.7) than at reference sites (0.5 to 0.6). Heterogeneity in benthic metabolism could reflect spatial patchiness of organic matter sedimentation of waste food pellets and feces under fish pens (Gowen et al. 1994; Hargrave 1994b; Silvert 1994).

Measures of gas exchanges in the presence of $HgCl_2$ (representing chemical fluxes in the absence of biological activity) also were higher at cage than at reference sites but the relative magnitude of enhancement was greater for chemical CO_2 release (+199%) than for chemical O_2 demand (COD) (+64%). The cv value for COD measurements (0.33) was lower than that observed for total gas exchange, and values were similar at both cage and reference sites. COD is due in part to accumulated reduced inorganic end products such as $S^{=}$ produced during anaerobic metabolism. Anaerobic respiration would increase dissolved CO_2 in sediment pore water above concentrations in overlying seawater.

Positive concentration gradients across the sediment-water interface would increase diffusion and result in higher CO_2 release rates.

Benthic COD can be calculated as a percentage of total O_2 uptake to examine changes in the proportion of uptake sensitive to poisoning with $HgCl_2$. In an earlier study at Site 6 during 1990-1991, when total $S^=$ under fish pens reached maximum levels of 500 mM, COD accounted for a large proportion (40 to 100%) of total O_2 uptake (Hargrave et al. 1993). In the present study, COD represented a smaller proportion of total O_2 demand at both cage (12 to 25%) and reference (19%) sites. Lower COD values as a proportion of total O_2 uptake at the Frye Island site are consistent with the reduced concentrations of $S^=$ in sediments at this location following movement of fish pens in 1992 as discussed above.

3.12. MACROFAUNA BIOMASS

The range of wet tissue biomass in benthic macrofauna observed in our study (6 to 2600 g m^{-2}) spans observations at other locations in the Bay of Fundy (Hargrave et al. 1993; Wildish and Peer 1983; Wildish et al. 1986). Total biomass was approximately 63% higher at reference sites (mean of 473 g wet weight m^{-2}) than at cage sites (mean of 290 g m^{-2}), but the difference was not significant (p=0.499) due to variance in the data. In contrast, the average biomass of deposit feeders at cage sites (22 g m^{-2}) was significantly higher (p=0.0052) than the value for reference sites (13 g m^{-2}), indicating that high rates of organic matter sedimentation under salmon pens enhanced food supply for this trophic group. The biomass of suspension feeders was not significantly different at cage and reference sites (p=0.80). A list of macrofauna species at all sites is presented in Hargrave et al. (1995).

3.13. RELATIVE SENSITIVITY OF VARIABLES FOR DETECTING ORGANIC ENRICHMENT

$S^=$, total and chemical CO_2 release and O_2 uptake and BEI values derived from measurements of Eh and C_{org} were the variables with maximum differences ($p<0.0001$) in ANOVA comparisons between cage and reference sites (Table I). The relationship between $S^=$ and Eh (Fig. 1A) can be used to quantify the concept of benthic organic enrichment zones based on differences in benthic microbial (Poole et al. 1978) and macrofauna (Pearson and Rosenberg 1978) communities. The zones correspond to regions 0 (anoxic, grossly polluted, $S^=>2000$ μm), I (hypoxic, polluted, $S^=>200$ μm), II (oxic, transitory, $S^=>50$ μm), and III (normal, $S^=<50$ μm) illustrated in Fig. 1A. Significant correlations between $S^=$ and CO_2 release and O_2 uptake only existed at cage sites with anoxic and hypoxic sediments (zones 0 and I)(Figs. 1B and C).

Mean values for some other variables (NH_4^+ measured by electrode in pore water, CBR, NH_4^+ gradients) at cage sites were >3 times values at reference sites with significant differences ($p<0.05$) between cage and reference locations (Table I). These variables would also be sensitive indicators for detecting organic enrichment. Sediment water content, pore water salinity and $SO_4^=$ were the least sensitive indicators of differences between cage and reference sites ($p>0.05$). These variables would not be useful for monitoring benthic enrichment effects.

Biomass of deposit feeders was the only measure of macrofauna biomass sensitive to organic enrichment. The observation is consistent with the increase in numerical abundance of deposit feeders along an organic enrichment gradient (Pearson and

Rosenberg 1978). This trophic group was also predominantly associated with fine-grained sediments in our study. However, Snelgrove and Butman (1994) reviewed studies that show that sedimentary grain size is not the only or even primary determinant of infaunal species distribution. Grain size co-varies with sedimentary organic matter content, pore water composition, bacterial abundance and species composition, all of which are controlled by the near-bed flow regime.

Pearson et al. (1983) proposed that the bivalve *Nuculana tennisulcata* could be used as an indicator species for organic matter loading beneath aquaculture sites since this deposit-feeding bivalve responds to increased sedimentation of organic material. *N. tennisulcata* was abundant at all cage sites but only occurred at four reference sites in our study.

The polychaete *Capitella* sp. is also often considered an indicator species of organic enrichment in marine sediments. The absence of *Capitella* sp. at all cage sites sampled in 1994 contrasts past studies in the Western Isles region of the Bay of Fundy and other coastal areas (Dubilier 1988; Hargrave et al. 1993; Grant et al. 1995). Cuomo (1985) showed that $S^{=}$ concentrations between 0.1 mM and 1 mM elicited optimal settlement, metamorphosis, and survival of *Capitella* sp.. Dubilier (1988) noted that no acute toxic effects, such as arrest of ciliary movement, occurred when *Capitella* was exposed to sulphide concentrations up to 2 mM. However, $S^{=}$ concentrations >2 mM adversely affected larvae and prevented settlement. $S^{=}$ tolerance by benthic fauna varies widely and may be correlated with $S^{=}$ levels in sediments (Bagarino 1992).

$S^{=}$ accumulation reaches a seasonal maximum in August in sediments under fish pens in the Western Isles region (Hargrave et al. 1993). Wildish et al. (1990) suggested that increased $S^{=}$ levels in sediments could lead to lower rates of anaerobic metabolism through poisoning of $SO_4^{=}$ reducing bacteria. This would explain the absence of *Capitella* sp. and other macrofauna at cage sites with high levels of organic enrichment as indicated by low Eh potentials and high $S^{=}$ levels.

4. Conclusion

Total $S^{=}$, benthic O_2 and CO_2 exchange and Eh potential were the most sensitive variables for detecting differences between sediments under salmon aquaculture cage sites and adjacent reference locations in the Western Isles Region of the Bay of Fundy. Total macrofauna and deposit feeder biomass and derived variables (BEI, CBR) based on organic carbon accumulation, NH_4^+ gradients and Eh potentials, were also sensitive indicators of organic enrichment. Measurements of modal grain size, sediment water content, pore water salinity and $SO_4^{=}$ were not significantly different between sites. Patchiness in sedimentation of uneaten food pellets and fish fecal waste products probably accounts for the high variability observed in some sediment geochemical properties under aquaculture cages.

Acknowledgements

We thank J. Hunt, diver and Captain of the *Sound Lady*, and W. Miner, Captain of the CSS *Pandalus III* for assistance with sample collection. K. Saunders and A. Prior carried out the particle size analyses and P. MacPherson, Department of Oceanography at Dalhousie University, assisted with measurements of sediment organic carbon and nitrogen. We thank K. Mann and J. Prena for comments on the manuscript.

References

Bagarino, T.: 1992, *Aquatic Toxicol.* **24**, 21-62.

Cranston, R.E.: 1994, *Can. Tech. Rep. Fish. Aquat. Sci.* **1949**, pp. 93-120.

Cuomo, M.C.: 1985, *Biogeochem.* **1**, 169-181.

Dubilier, N.: 1988, *Biol. Bull.* **174**, 30-38.

Findlay, R.H., Watling, L.: 1994, *Can. Tech. Rep. Fish. Aquat. Sci.* **1949**, p. 47-77.

Gowen, R.J., Smyth, D., Silvert, W.: 1994, *Can. Tech. Rep. Fish. Aquat. Sci.* **1949**, pp. 19-30.

Grant, J., Hatcher A., Scott, D.B., Pocklington, P., Schafer, C.T.,Winters, G.V.: 1995, *Est.* **18**, 124-144.

Hargrave, B.T. [ed.]: 1994a, *Can. Tech. Rep. Fish. Aquat. Sci.* **1949**, 125 pages.

Hargrave, B.T.: 1994b, *Can. Tech. Rep. Fish. Aquat. Sci.* **1949**, pp. 79-91.

Hargrave, B.T., Phillips, G.A., Doucette, L.I., White, M.J., Milligan, T.G.,Wildish, D.J., Cranston, R.E.: 1995, *Can. Tech. Rep. Fish. Aquat. Sci.* **2062,** 159 pages.

Hargrave, B.T., Connolly, G.F.: 1978, *Limnol. Oceanogr.* **23**, 1005-1010.

Hargrave, B.T., Phillips, G.A.: 1981, *Est. Coastal Shelf Sci.* **12**, 725-737.

Hargrave, B.T., Duplisea, D.E., Pfeiffer, E., Wildish, D.J.: 1993, *Mar. Ecol. Prog. Ser.* **90**, 249-257.

Hunter, B., Simpson, A.E.: 1976, *J. Mar. Biol. Ass. U.K.* **56**, 951-957.

Kranck, K., Smith, P.C., Milligan, T.G.: 1996a, *Sedimentol.* **43**, 589-596.

Kranck, K., Smith, P.C., Milligan, T.G.: 1996b, *Sedimentol.* **43**, 597-606.

Pearson, T.H., Rosenberg, R.: 1978, *Oceanogr. mar. Biol. Ann. Rev.* **16**, 229-311.

Pearson, T.H., Gray, J.S., Johannessen, P.J.: 1983, *Mar. Ecol. Prog. Ser.* **12**, 237.

Silvert, W.: 1994, *Can. Tech. Rep. Fish. Aquat. Sci.* **1949**: pp. 1-18.

Snelgrove, P.V.R., Butman, C.A.: 1994, *Mar. Biol. Ann. Rev.* **32**, 111-177.

Solarzano, L.: 1969, *Limnol. Oceanogr.* **14**, 799.

Strain, P., Yeats, P., Wildish, D.: 1994, *Mar. Poll. Bull.* **30**, 253-261.

Trites, R.W., Petrie, L.: 1995, *Can. Tech. Rep. Hydrogr. Ocean Sci.* **163**: 53 pages.

Wildish, D.J., Peer, D.: 1983, *Can. J. Fish. Aquat. Sci.* **40**, Suppl., 309-321.

Wildish, D.J., Peer, D.L., Greenberg, D.A.: 1986, *Can. J. Fish. Aquat. Sci.* **43**, 2410-2417.

Wildish, D.J., Martin, J.L., Wilson, A. J., Ringuette, M.: 1990, Can. Tech. Rep. Fish. Aquat. Sci. **1760**, 13 pages.

Wildish, D.J., Keizer, P.D., Wilson, A.J., Martin, J.L.: 1993, *Can. J. Fish. Aquat. Sci.* **50**, 303-311.

ACCUMULATION OF INORGANIC AND ORGANIC POLLUTANTS BY BIOFILMS IN THE AQUATIC ENVIRONMENT

MARCELL SCHORER and MICHAEL EISELE

Department of Hydrology, University of Trier, 54286 Trier, Germany

Abstract. In a partly urbanized catchment to the south of Trier, Germany, short term variations in river sediment compounds as well as the bioaccumulation of pollutants on surface associated microbial coatings (biofilms) were investigated weekly during a period of six months. Concentrations of selected heavy metals (Cu, Zn, Pb), polycyclic aromatic hydrocarbons (PAH), polychlorinated biphenyls (PCB) and for microbial characterisation protein, carbohydrate and uronic acid were analyzed. Sorption processes on biofilms were determined by temporal variations in pollutants and microbial parameters and through the comparison of sorbed substances in biofilms and sediments. The results show, that sorption events on biofilms play an important and dynamic role in spring and summer for transport and accumulation of the investigated pollutants in the aquatic environment. The amount of pollutants sorbed on sediment particles is not only dependent on the particulate bound or solved pollutants in the river water, but is strongly controlled by the changing conditions of the biofilms.

Key Words: River sediments, biofilms, organic micropollutants, heavy metals, temporal variations, biosorption, accumulation processes

1. Introduction

The adsorption on particle surfaces plays a decisive role in the transport of hydrophobic pollutants in river water (Karickhoff, 1981; Means *et al.*, 1980). Therefore, suspended particles and sediments are not only important sinks, but also a potential source of pollutants. Investigations carried out in 1992 (Schorer *et al.*, 1994; Symader *et al.*, 1994) have shown that in river sediments, temporal variations of heavy metals and organic micropollutants in the sediment surface layer, occurred during the year. These variations were not controlled by random processes. The observed phenomena can be explained in part by the different sources which supplied these pollutants at different times as well as mixing and dilution processes.

In the aquatic environment, the surfaces of mineral particles are, for the most part, covered with organic coatings which can drastically change the sorption behaviour of the particles. These organic coatings are made from humic substances and biofilms (Flemming *et al.*, 1996; Characklis & Marshall, 1990). A considerable amount of data concerning adsorption on humic substances is available from the literature (e.g. Frimmel & Christman, 1988). In contrast to this inanimate matrix, biofilms can actively influence the sorption, desorption and decomposition of pollutants. It must be expected, that a part of the accumulation of pollutants in river sediments is due to sorption on microbial coatings (Baughman & Paris, 1981; Urey *et al.*, 1976; Tsezos & Bell, 1989). The importance of sorption and desorption processes on biofilms with respect to temporal variations in pollutant contents in river sediments is complex and not understood (Evans *et al.*,

Water, Air and Soil Pollution **99**: 651-659, 1997.

1990; Flemming *et al.*, 1996). Nevertheless, sorption data on the accumulation of pollutants by biofilms are available only from lab studies with defined conditions and as a result, without the competition of a natural ecosystem. Unfortunately, investigations on natural river ecosystems are missing. This process could be an important key to the understanding of loadings, transport and sink of pollutants in the aquatic system. The sorption of pollutants on biofilms plays a very dynamic role, because they are not chemically inert. Changing conditions in the environment can induce changes both in the microbiota and their physiology. The sloughing off of biofilms and the breakage of binding sites in the extracellular matrix can result in remobilization and in a new source of pollutants (Flemming *et al.*, 1990, Tsezos & Bell, 1989).

In our study, the importance and influence of the sorption properties of biofilms contributing to the accumulation of pollutants in river sediments were investigated in more detail. Short temporal variations in the sediment compounds as well as the bioaccumulation of pollutants on surface associated biofilms were determined. At two fixed sampling stations, concentrations of different inorganic and organic micropollutants in river sediments and in biological growth were analyzed weekly during a period of six months.

2. Study area

The study area represents the drainage basin of the Olewiger Bach with an area of 39 km^2. It is located to the south of Trier in the western part of Germany. Devonian shales of the Hunsrück mountains with quartz and diabas veins form the bedrock geology. In the northern part of the drainage basin, pleistocean terraces of the river Mosel lie on the geological underground (Wagner, 1983). Land use is predominantly agricultural with grassland and arable farming. Settlements take up 10 % of the whole area. Several roads, small industries and waste water from solitary farms, with no connection to a sewage plant, influence the quality of the river water.

3. Methodology

3.1 SAMPLING PROCEDURE

Sampling biofilms in their natural aquatic habitat from surfaces like mineral particles, stones or plants is impossible, because the biological coatings cannot be separated from the surface of the substrate without changing or disturbing their original sorbing properties (MacNicol & Beckett, 1989). Therefore, the biofilm samples were obtained from a chemically inert artificial silicate material, which is similar to natural surfaces because of its roughness and its chemical properties. The silicate substrates were packed into nets and were exposed in the river water near the sediment surface. Because the lag time for developing biofilms on surfaces takes several days to weeks, the silicate substrates were left in river water for five weeks to allow sufficient time for biofilms to develop and pollutants to accumulate. To observe short term variations, five week old silicate substrates with grown biofilms were taken out and examined weekly over a period of six months. Sediments were sampled weekly at the same site.

3.2 ANALYTICAL PROCEDURE

Material, which was lying loosely on the silicate substrates was carefully removed with deionized water. Then the adhering biofilm was dried. These samples were analyzed together with the substrate, since separation of the biofilm material is impossible. However, this procedure was not a hindrance, as the silicate substrate turned out to be chemically inert for our analyzing programme. The biofilm mass was calculated as the difference between the whole sample material and the cleaned substrate after analysis.

The biofilm and sediment samples were analyzed for selected heavy metals (Pb, Cu, Zn, Fe, Mn), nutrients (Ca, Mg, K), polychlorinated biphenyls (PCB), and polycyclic aromatic hydrocarbons (PAH). In addition proteins, carbohydrates, uronic acids, and organic matter (loss on ignition) were determined. After decomposition under pressure with concentrated nitric acid, heavy metals and nutrients were analyzed with an atomic absorption spectrometer. For the analysis of PAHs and PCBs the samples were spiked with internal standards and solvent extracted with acetone/hexane (1:1) in a Soxhlet system for eight hours. After rotary evaporation to near dryness, the solvent extracts were purified by column chromatography. Identification and quantification were achieved by gas chromatography/mass spectrometry operating in the selected-ion-monitoring mode. The protein, carbohydrate and uronic acid content were analyzed photometrically (Lowry *et al.*, 1951; Bailey, 1958; Blumenkrantz & Asboe-Hansen, 1973).

4. Results and Discussion

The concentrations of carbohydrates and organic matter (OM) in the biofilm samples are described in Fig. 1. The OM content of the sediment samples is shown for comparison. The content is about three times higher in the biofilms than in the sediments. This indicates, that compared to the sediment samples, a higher accumulation of biological substances like biofilms was achieved in the sampled material. The OM content of the biofilm samples increases strongly in May and for the rest of the sampling period, the concentrations remain at a lower level and only vary slightly. The carbohydrate contents, however, show greater variations during the sampling period and the temporal development between OM and carbohydrate content is quite different. The variations in carbohydrates, proteins and uronic acids (both not shown) indicate the continuous changing conditions of nutrient supplies and population development.

At the beginning of May lower hydraulic forces, lower turbidity with little suspended solid contents caused by lower discharge (Fig. 2), a good nutrient supply and an increased irradiation result in fast biological growth. At this stage nutrients are fixed in the biomass. This limited supply results in a decrease in the amount of biological growth in the middle of June. The changes in the contents of carbohydrates, proteins and uronic acids exhibit a change in the biomass in the sampling period. A further important factor is that over the year the discharge controls the levels of concentrations. Depending on the height and temporal length of the discharge peak, the discharge causes a drop in OM content in the sediments.

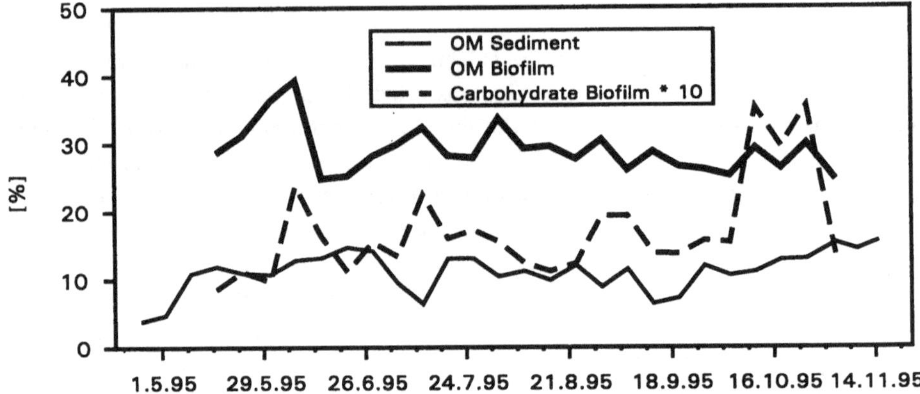

Fig. 1. Weekly organic matter (OM) and carbohydrate content in biofilms and weekly organic matter (OM) content in sediments (dry weight).

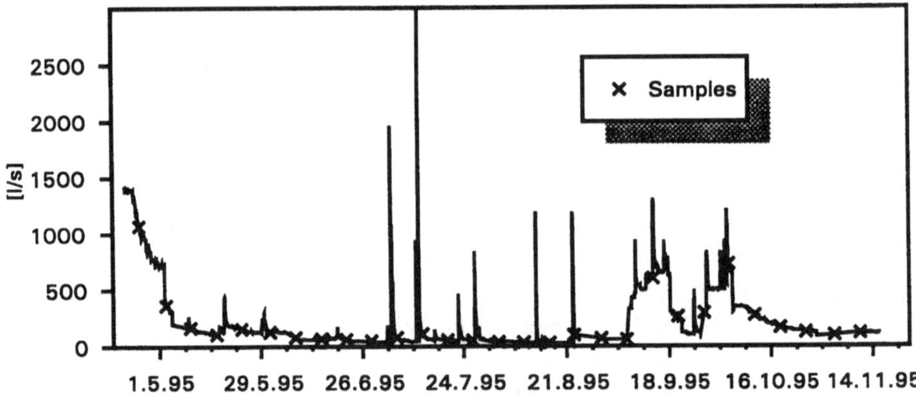

Fig. 2. Discharge Olewiger Bach 1995.

Fig. 3 shows the content of copper, lead and zinc in the sediments and biofilms. The heavy metals in sediments have a similar temporal structure to the OM in sediments. At the beginning of the sampling period in 1995 the pollutant contents are dominated by hydrodynamic conditions. At the end of a flood in April (Fig. 2) the sediments were replaced by mineral particles with very low contents of pollutants and OM. The low discharge in May leads to a strong increase in concentrations in the river sediments. For example, zinc increases from 100 mg/kg up to 270 mg/kg within two weeks. However, the short, but high discharges in July (2000 and more than 3500 l/s) partly eroded the sediment, which was rich in OM and pollutants. The content of heavy metals decreases. Both transport forces and time are insufficient to wash out the whole sediment from the river bed. Afterwards, the variations in heavy metal content can be more or less related to increases in the discharge. It is remarkable, that after the extreme flood waves in July the heavy metal content increases more than the OM content. It is the quality, rather than the quantity of OM that seems to be responsible for the sorption of heavy metals. The reason could be that a change in the composition of the biofilm within the sediment occurred. This is indicated by the change in the protein/carbohydrate-ratio.

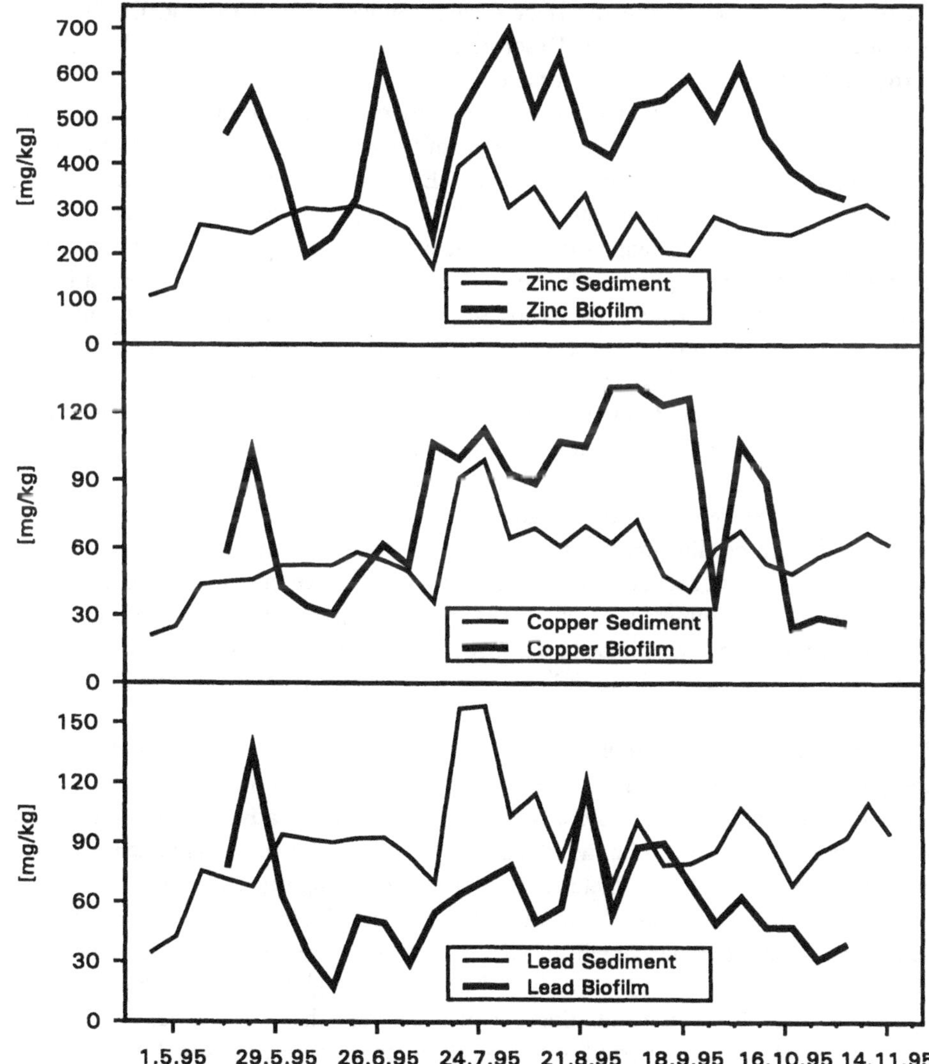

Fig. 3. Temporal variations in zinc, copper and lead in biofilms and sediments.

In general, concentrations of heavy metals in the biofilms show high temporal variations. In the middle of June a strong decrease of zinc, lead and copper follows a maximum content in May. The amount of OM and, for example, of carbohydrate shows quite a different pattern. With the sorption, the amount of biological material is not the decisive factor. During the course of summer, the contents generally increase but are interrupted at specific times by decreases, which are marked by the different heavy metals. On July 3rd, a flood wave (2000 l/s) causes a short term decrease in heavy metals. In spite of the next extreme flood (more than 3500 l/s) one week later, lead and copper increase, but zinc decreases. Ensuing peaks of individual heavy metals (e.g. Pb 23.8.:119 mg/kg) indicate inputs of polluted material into the river system. Sudden decreases (e.g. Cu 29.9.: 36 mg/kg) are difficult to understand. Desorption processes or a change in the sorption properties of the material may be the reason.

The high variations show, that sorption is not only dependent on the amount of sorbed substances available, but also on the conditions of the biofilms. Carboxyl- and hydroxyl-groups of the extracellular polymer substances (EPS) have a very high capacity for the sorption of heavy metals. A small amount of EPS could theoretically bind a large amount of a given metal (Geesey & Jang, 1989), but the affinities of EPS vary depending on the specific metal and micro-organisms involved (Harvey & Luoma, 1985). The variations in EPS composition, which were indicated by the biological parameters, seem to influence the sorption and are responsible for a part of the variance. A fast decrease after maximum concentrations shows, that desorption processes may also play an important role.

Compared to sediment contents, zinc and copper are strongly enriched in biofilms. Lead however, is not found in high quantities in biofilms. This must be due to the different mechanisms of enrichment and the sorption properties of the biofilm. In summer, copper content shows lower variations than lead and zinc (with the exception of 29.9.). More stable fixing by the biofilm matrix is the reason for lower desorption. When low water-levels occur, the biofilm seems to have an important influence on the level of concentrations of pollutants in sediments. At this time sorption of heavy metals by biofilms causes an increase in pollutant levels in sediments. If this sediment is resuspended or if the biofilm is disturbed by external conditions, the sediment can be a potential source for pollutants.

The sorption of organic hydrophobic compounds onto biofilms is a well known phenomenon (Decho, 1990). However, little research has been done on the binding mechanisms and sites of sorption. Fig. 4 shows the contents of two representative PAHs and one PCB in biofilms and sediments. While the PCBs of the sediments have the same temporal patterns as the OM and the heavy metals, the PAHs show a slightly different behaviour. In comparison to the heavy metals, the same mixing and dilution processes of different loaded materials and the enrichment of OM, influence the temporal course of PCB and PAH. The temporal pattern of the PAHs is difficult to recognize, because there are three exceptions of high PAH concentrations, which suppress the other variances: the first exception takes place over a period of two weeks after the high floods in July when concentrations of PAHs increase abruptly; the second and third divergences begin in October and November respectively with two weaker concentration peaks. Waste water effluents must be responsible for these peaks.

The PAH-and PCB-content in biofilms show high variations, with a general increase from May to September, and a sudden decrease in October. The variations between the individual PAHs are similar, but differences can be seen in the detailed structures. Although similar processes control the sorption, they occur to a different extent depending on the physico-chemical properties. The variations in PCB-content are lower and until August are similar to the PCB-content of the sediments. The temporal behaviour of organic micropollutants in biofilms is distinguished from that of heavy metals.

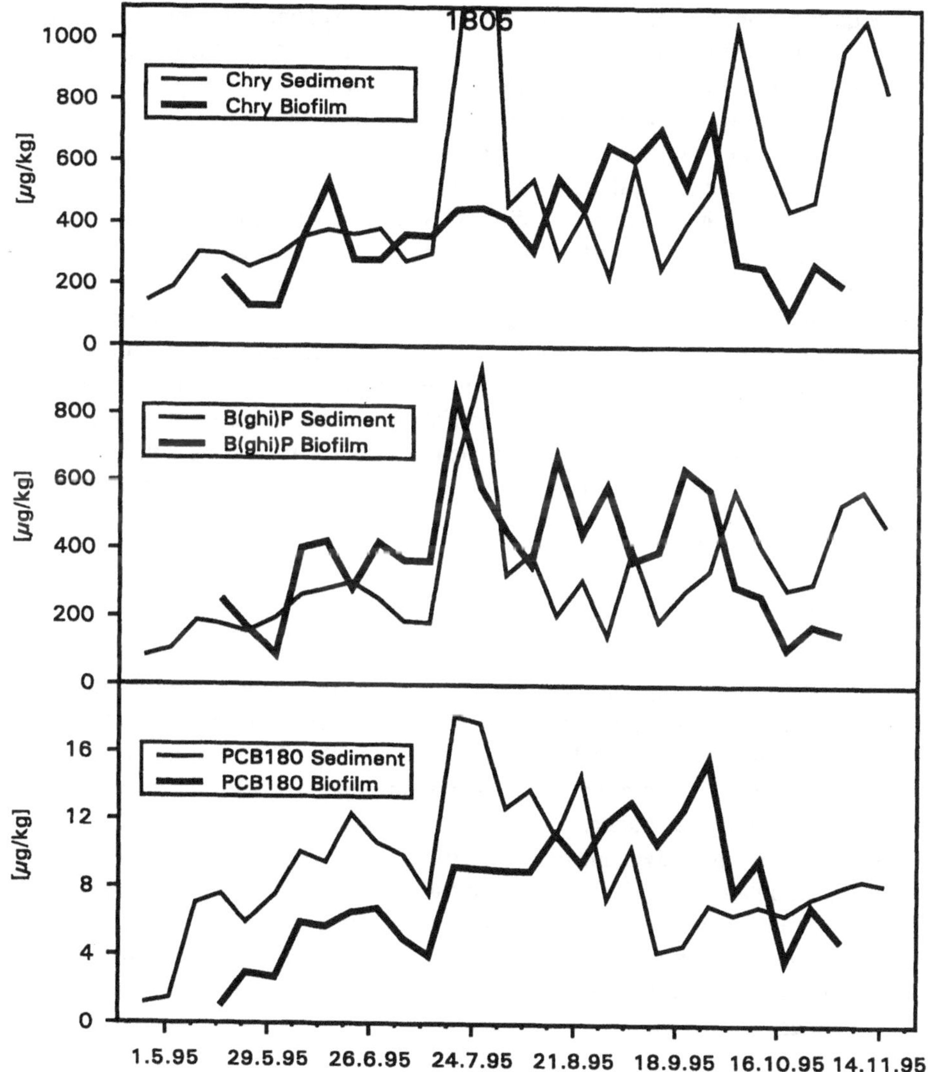

Fig. 4. Temporal variations in selected PAH and PCB in biofilms and sediments.

The sorption of organic micropollutants by biofilms occurs by interactions with particles or the solution by partition equilibriums (Baughman & Paris, 1981). In comparison to the PAH patterns in the sediment, the PAHs with higher molecular weight and higher partition coefficient (e.g. Benzo(ghi)perylen) are more enriched in biofilms than the PAHs with lower molecular weight (e.g. Chrysen). More rigid binding and less microbial decomposition may be the reasons (Mackay *et al.*, 1992). In August and September, the high values of PAHs and PCBs in biofilms exhibit, that at this time the conditions for sorption were optimal. The high concentrations correspond with high content of proteins and uronic acids. An increased biological activity results in an increased sorption of organic micropollutants. The concentrations of heavy metals in the biofilm are also highest at this time. The decrease of all pollutants in October cannot be explained by a simple desorption process. A possible explanation is the change in the biofilm matrix which due to seasonal changes, is now made up of a large quantity of dead material.

The differences between biofilm and sediment content are remarkable. In contrast to the sediments, the biofilm samples represent a period of time up to a maximum of five weeks. At different times of exposure, the growth of and sorption by the biofilms are influenced by meteorological and hydrological conditions. At the beginning of the exposure of the porous silicate material to river water a starting phase of several days is required for first pioneers of micro-organisms to adhere and grow. At this time an enrichment of pollutants is negligible, as there is no significant biomass. On the other hand episodic flood waves, depending on their intensities, can erode parts of the mature and thicker biofilms. The thickness and shaping of the biological growth also play an important role. Further important factors include the time of pollutant enrichment in the water phase, the supply of pollutants and the physico-chemical conditions (such as the pH-value) within the water phase (Flemming *et al.*, 1996). The longer and more intensive contact the biofilm makes with the pollutants, the more the pollutants can be enriched.

5. Conclusions

In the complex system of sorption in aquatic systems, biosorption is a highly dynamic and variable factor in pollutant transport and accumulation. It is controlled partly by the availability of pollutants and partly by the biological activity of the biofilm. The dynamic nature of biofilms gives them a distinctive role in sorption processes, because immobilization in biomass can be reversible.

Sediments have a highly dynamic temporal structure throughout the year. The temporal variations of pollutants in the biofilm are mostly hidden in the sediment samples by mixing and dilution processes and by effluents from point and nonpoint sources. If the sediments are undisturbed for several weeks, the importance of bioaccumulation is not negligible. During sediment deposition, the influence of biosorption is an important factor in the accumulation of pollutants in river sediments, especially during low water periods in summer.

There are clear differences between the investigated pollutants in the biofilms. Variable stability of binding on charged functional groups in the biofilm matrix and the transport of the pollutants onto various particles are responsible for the sorption of heavy metals. Physico-chemical properties play a central role for organic micropollutants. It can be shown that with decreasing water solubility there is an increased possibility that organic micropollutants can be bound. The more persistant substances such as certain PCBs are accumulated to a greater extent than others. In comparison to laboratory studies, with fixed conditions that do not include the competition of natural ecosystems for sorption, our results show, that laboratory results can be transferred only to a limited degree to the aquatic environment.

The high content of heavy metals and hydrophobic organic micropollutants show that biofilms represent an important and dynamic sink in the aquatic environment for sorbed substances. As biofilms are not inert chemical structures, the biological binding sites for pollutants will sometimes be degraded. If the physico-chemical conditions change, for example by resuspension through flood waves or by microbial degradation processes of the biological binding sites, the pollutants can be desorbed and remobilized in a more soluble form and can assume a potentially higher bioavailability. The highly dynamic biofilm will then change from a sink to a source.

Acknowledgments

The authors would like to thank the Deutsche Forschungsgemeinschaft for their financial support.

References

Bailey, R.W.: 1958, *J. Biochem.* **68**, 669.
Baughman, G. L., and Paris, D. F.: 1981, *Critical Reviews in Microbiology* **8**, 205.
Blumenkrantz, N., and Asboe-Hansen, G.: 1973, *Analyt. Biochem.* **54**, 484.
Characklis, W. G., and Marshall, K. C.: 1990, *Biofilms*. John Wiley. New York.
Decho, A. W.; 1990, *Oceanogr. Mar. Biol. Ann. Rev.* **28**, 73.
Evans, K. M., Gill, R. A., and Robotham, P. W. J.: 1990, *Wat., Air, and Soil Pollut.* **51**, 13.
Flemming, C. A., Ferris, F. G., Beveridge, T. J., and Bailey, G. W.: 1990, *Appl. Environ. Microbiol.* **56**, 3191.
Flemming, H.-C., Schmitt, J, and Marshall, K.: 1996, in Calmano, W., and Förstner, U. (eds.). *Sediments and Toxic Substances*. Springer, Berlin.
Frimmel, F. H., and Christman, R. F.: 1988, *Humic substances and their role in the environment. Dahlem workshop reports*. John Wiley & Sons, Chichester.
Geesey, G. G., and Jang, L.: 1989, in Beveridge, T. J., and Doyle, R. J. (eds.), *Metal ions and bacteria*. John Wiley, New York.
Harvey, R. W., and Luoma, S. N.: 1985, *Mar. Ecol.* **22**, 281.
Karickhoff, S. W.: 1981, *Chemosphere* **10**, 833.
Lowry, O.H., Rosebrough, N. J., Farr, A. L., and Randall, R. J.: 1951, *J. Biol. Chem.* **193**, 265.
Mackay, D., Shiu, W. Y., and Ma, K. C.: 1992, *Illustrated handbook of physical-chemical properties and environmental fates for organic chemicals*. Lewis Publishers, Chelsea.
MacNicol, R. D., and Beckett, P. H. T.: 1989, *Wat. Res.* **23**, 199.
Means, J. C., Wood, S. G., Hassett, J. J., and Banwart, W. L.: 1980, *Environ. Sci. Technol.* **14**, 191.
Schorer, M., Bierl, R., and Symader, W.: 1994, *Vom Wasser* **83**, 117.
Symader, W., Schorer, M., and Bierl, R.: 1994, *IAHS*, Publ. no. **224**, 491.
Tsezos, K., and Bell, J. P.: 1989, *Wat. Res.* **22**, 561.
Urey, J. C., Kricher, J. C., and Boylan, J.M.: 1976, *Bull. Envir. Contam. Toxicol.* **16**, 81.
Wagner, W.: 1983, *Mitteilungen der deutschen Bodenkundlichen Gesellschaft* **37**, 90.

Use of a 6-steps microcosm for studying a wastewater discharge in a freshwater ecosystem: a multidisciplinary study

MONTUELLE B.[1], LATOUR X.[1], VOLAT B.[1], and LAFONT M.[2]
CEMAGREF, Laboratoires (1) EcoDynamique des Sédiments and (2) Diagnose des Systèmes Aquatiques, 3 Quai Chauveau, 69336, Lyon, Cédex 09, France.

Abstract. An experimental microcosm has been designed for simulating and studying impacts of a wastewater treatment plant (WTP) discharge on a freshwater/sediment ecosystem. The study was focused on the changes in biodiversity of benthic populations, especially bacteria and oligochaetes. Effluents were discharged in the Saône river, near Lyon (France) from a small treatment plant which treated domestic raw water by an activated sludge process. Freshwater and sediments were sampled in the Saône river upstream of the discharge point and placed in microcosms. Following the WTP discharge, physicochemical parameters of the overlying water column and sediments exhibited only a slight change, as compared to a reference.

Characterization of the sediment bacterial populations was conducted with the Biolog and API systems. Strain identification and interpretation of data was difficult using thesetwo systems. Bacterial taxa in the sediments increased slightly below the WTP discharge. Gram negative strains dominated in the effluents, but G^+ and G^- bacteria were balanced in the sediments. *Pseudomonas sp.* and *Bacillus sp.*, were the dominant strains. Invertebrate populations indicated an effect of the WTP discharge, with increasing of pollution resistant strains (Tubificidae) and disappearance of pollution intolerant strains such as *Limnodrilus udekemianus* and *Quistadrilus multicoetosus*.

Taken as a whole, biological parameters indicated an environmental changes despite only slight changes in the physicochemistry of water. This experimental microcosm has proven to be a useful tool for studying impact of wastewater discharge on benthic populations.

Keywords: Microcosm, bacteria, oligochaete, wastewater treatment

1. Introduction

Many experimental devices have been developed for studying and assessing the functioning of aquatic ecosystems at different scales: batch systems with pure or mixed culture (bacteria, e.g.: Percherancier *et al.*, 1996), microcosms more or less complex (multiphasic, parameters regulation; e.g.: Pipke *et al.*, 1992; Benoit *et al.*, 1993; Montuelle *et al.*, 1996), mesocosms or experimental enclosures (for example: Maughan & Oviatt, 1993) and in different environments (fresh or marine water, soils) and with different trophic levels (e.g.: Kroer *et al.*, 1994). All these experimental levels give useful and complementary information, from cellular to biocenotic levels, which are necessary for linking laboratory results to field studies (Elliott *et al.*, 1986) and for ecological risk assessment (Shaw & Kennedy, 1996).

Many studies describe field ecotoxicological studies which focus on the effects of toxic compounds (e.g. Rosso *et al.*, 1994). There are also numerous studies of freshwater environments polluted by wastewater (Cemagref, 1994; Garric *et al.*, 1996; US EPA, 1991; Grothe *et al.*, 1996) describing relationships among effluent toxicity, ambient toxicity and receiving ecosystems impacts. In 1992, a research program was initiated by French public and private organizations, Research Ministry and Environment Ministry, to assess the impact of urban WTP on receiving water, in connection with biological and chemical quality of their effluents. Within this program, we have studied a common

Water, Air and Soil Pollution **99**: 661-669, 1997.

situation in French rivers, receiving treated effluents with a (in theory) slight residual polluting load.

A microcosm approach, focusing on sediments, was developed to assess:

1- the suitability of the microcosm for studying the effects of WTP discharge,
2- the changes through time of sediment quality and its overlying water,
3- the changes of the sediment biocenosis: bacterial communities and invertebrates populations.

2. Materials and Methods

2.1. EXPERIMENTAL MICROCOSM

The microcosm is composed of 6 glass dual-walls reactors, with temperature regulation (Lauda UKT 600), flows control (ISMATEC MV pumps) and oxygen probes (WTW) monitored by a microcomputer (InfoPhysic software, specially designed for this experimental system).

Each reactor has a volume of 5.7L and the 6 reactors could be supplied with water in series (these sets of experiments) or in parallel. Microcosms are biphasic with sediments and an overlying water column. Sediments with their biocenosis (microbial and invertebrate populations) were sampled with a dredge in the Saône river, well homogenized before being distributed among the 6 reactors: 2L in each (1,7kg dry weight) forming a layer of 5 cm depth with an overlying water capacity of 3.7L. Freshwater was also collected in the Saône river from the same sampling area, filtered throught a 80µm mesh Fitsch sieve and stored at 4 °C. WTP effluent samples were collected from a WTP discharge (activated sludge process), prefiltered through a 80µm mesh Fitsch sieve and stored at 4°C during the experiments (stock solutions). The freshwater and effluent were subsampled every 3 days and added to the microcosm. Microcosm flow rates were 835mL/h for freshwater and 417mL/h for the effluent, i.e. a ratio freshwater/effluent of 1/3, except for the reference (without effluent). Total flow rate was 1.2L/d with a hydraulic residence time HRT (HRT= Volume/ flow rate) of 3 days. The effluent enters reactor 1 and the flow is gravitational from reference to 5 (Figure 1): this succession of reactors resulted in several HRT's of 3, 6, 9, 12 and 15 days respectively for reactors 1, 2, 3, 4 and 5, which correspond to different Biodegradation Times (BdT).

Microcosm temperature was kept constant at 20°C (summer field temperature) and reactors were kept in the dark to avoid algal development during the 30 days of the experiment. A preliminary period of 9 days with freshwater alone was allowed for equilibration of the sediment-water system. The water column was stirred and saturated with dissolved oxygen by bubbling with air.

2.2 EXPERIMENTS AND ANALYTICAL PROCEDURES .

Water and sediment were characterized after 30 days of effluent input. Ammonia (NH_4^+), nitrites (NO_2^-), and nitrates (NO_3^-) were analyzed according to the following standards (AFNOR NFT 90-015, 90-013 and 90-012). Ntotal was measured on a Technicon Autoanalyzer; dissolved organic carbon (DOC) (<0.2µm) and Total organic Carbon (TOC) were measured with a Dohrman DC 80 Carbon Analyser. pH was measured with a

WTW 96 pHmeter; dissolved O_2 was measured with a WTW Oxymeter; Eh and rH were measured with WTW Redoxmeter. Sediment granulometry was determined using 4 vibrating wet Fitsch sieves: 2000, 500, 200 and 50μm mesh sizes.

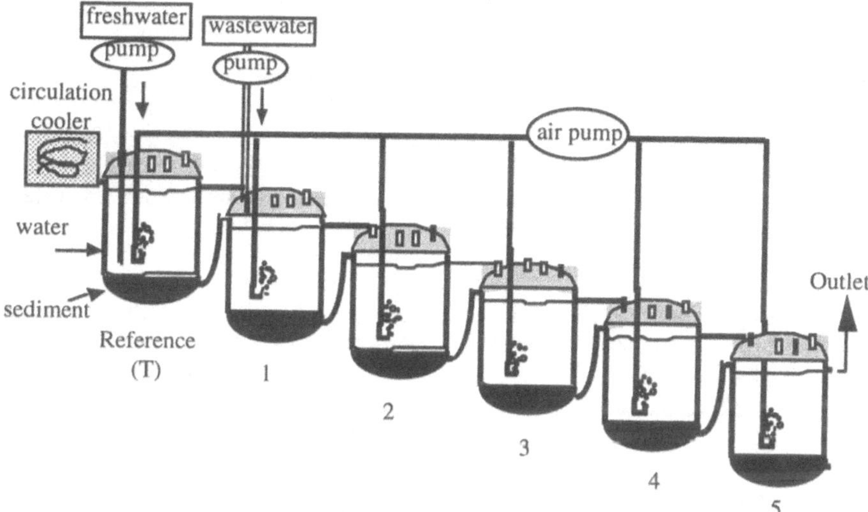

Fig. 1: Diagram showing the microcosm devices.

Extracellular enzyme activity (aminopeptidase, Montuelle & Volat, 1993) and INT reductase activity (Broberg, 1985) are used as characterizations of global bacterial activity. They were measured in the water column and in the sediment. Bacterial enumerations were made by direct microscopic counts on 0.2 μm Nuclepore black filters. Samples were stained with 4,6 di-amidino-2-phenylindol (DAPI) (Rebillard & Torre, 1993). Culture and isolation of bacterial strains was conducted on a non selective PGY growth medium and were carried out on 3 reactors (T, 1 and 5) and on the effluent. Colonies were examined visually, chosen for their unusual pigmentation or peculiar colony morphology. The first 40 individual colonies obtained were repeatedly transferred on growth medium. After obtaining pure cultures, systematic determinations were carried out following classical schemes of bacteriology. Strains were tested with two commercial kits: API System and Biolog system, according to the specifications of their respective manufacturer, and completed with some biochemical assays (Gram, spores, oxydase and catalase activity, respiratory type). Each strain was characterized in duplicate. Bianchi and Bianchi, (1982) have shown that, for one sampling of one type, between 20 and 30 strains from the sample described by approximately 30 characters yield a statistically valid measure of bacterial community diversity.

The Biolog Microplates test the ability of microorganisms to oxidize a selected panel of 95 different carbon sources. In a second step, we used API strips for identification. It is a standardised method combining enzymatic tests and assimilation tests (21 tests), that we have completed with 5 other biochemical and morphological tests.

Sediment samples for invertebrate identifications were collected from each reactor, preserved with formaldehyde (6%), washed over a 160μm sieve, and invertebrate specimens were sorted by hand under a binocular microscope. As only oligochaetes were found, we used specific sorting and species determination protocoles used by Rosso *et al.*

(1994). All the results presented here were obtained after 30 days of equilibrium in the microcosm.

3. Results

3.1. WATER AND SEDIMENT CHEMISTRY EVOLUTION

In the water column (Table 1), the effluent discharge has practically no effect: a slight decrease of the RedOx potential and a slight increase of the pH, from the reactor 1 (effluent input) were observed. Dissolved oxygen decreased slightly from 8.4 (reference reactor) to 8 in reactor 1, and then increased to the reference value.

TABLE I

Physico-chemical parametrers in the 5 reactors; Eh: mV; Dissolved Oxygen, DOC, TOC, N and P compounds: in mg/l. For sediments the two values shown are upper sediment layer/bottom sediment layer. Effluent input is in reactor 1.

Water column

	pH	Eh	D.O.	TOC	DOC	N tot.	NO3	NO2	NH4	PO4
T	7.3	292	8.4	2.6	2.2	1.7	0.88	0.02	0.07	0.09
1	7.6	292	8	3.3	2.9	1.3	0.88	0.02	0.13	0.8
2	7.45	286	8.2	3.1	2.7	1.5	0.95	0.01	0.08	0.76
3	7.74	292	8.4	3.3	2.6	1.8	1.32	0.01	0.07	0.64.
4	7.8	260	8.4	3.2	2.7	1.8	1.44	0.01	0.17	0.64
5	7.75	256	8.4	3.2	2.7	2.1	1.75	0.01	0.1	0.58

sediment

	pH	Eh	H2O %	P %	total N %	Cmin %	Corg %	NH4
T	7.4/6.9	267/236	54/28	0.06/0.06	0.03/0.03	2.3/2.2	1.3/0.8	1.3/10
1	7.4/7.0	290/240	46/29	0.09/0.06	0.04/0.02	2.1/2.2	1.1/0.7	1.1/12
2	7.4/7.1	298/255	45/28	0.10/0.08	0.06/0.03	2.2/2.1	1/0.6	1.2/10
3	7.5/7.1	275/255	40/30	0.10/0.07	0.02/0.03	2.2/2.2	0.9/0.5	1.1/16
4	7.4/7.1	262/240	47/27	0.1/0.07	0.04/0.03	2.3/2.1	1/0.6	1/11
5	7.4/7.1	269/250	44/30	0.07/0.08	0.03/0.03	2.2/2	1/0.8	0.9/14

The increase in organic carbon (+ 32% in DOC and + 26% in TOC) is due to the input of effluent in reactor 1, then a small decrease appeared due to biodegradation processes. DOC remained above the reference level because the organic carbon of the treated effluent is only partially biodegradable (Percherancier *et al.*, 1996). A marked nitrification process is observed from reactor 1 (BdT: 3d, 0.88mg NO_3^-/L) to the last one (BdT: 15 d, 1.75 mg NO_3^-/L).

The physicochemical composition of the sediment remained stable for most chemical parameters. The effluent input had no marked effect on C or N after reactor 2. Conversely phosphorus concentrations in the sediment increased after the input of effluent (+ 50 to 100% depending on the reactors) and stayed at high concentrations.

The sediments were bioturbated (visual observations, worm abundance) during the 30 days of equilibrium and this resulted in the formation of 2 distinct layers: the upper layer (2cm depth) was mainly made of particles <50µm (40-70%) and 50-200µm (35-55%). The bottom layer (3 cm), was made of coarser particles: <50µm (10%) and 50-200µm (80%). Other classes of size of particles (200-500 and 500-2000µm) were not significantly abundant.

3.2. CHARACTERIZATION OF BACTERIAL ASSEMBLAGES

Reactors T, 1 and 5 and effluent were chosen for a detailed characterization of bacterial assemblages: about 35 strains each.

TABLE II:
Identification tests of sediments strains using API kits and complementary biochemical assays, and BIOLOG kits. "Non interpretable" means : for API that the strains had no regrowth after storage (strains used for API identification have been kept in glycerol) and for BIOLOG that all the wells of the microplates (but the reference well) were positive.

		API				BIO LOG			
		T	1	5	E	T	1	5	E
G⁻	*Pseudomonadas sp.*	2	4	3	12				1
	Agrobacterium sp.			1	1				
	Flavobacterium sp.	1							
	Serratia sp.	2		1	1				
	Aeromonas sp.	3	1	3	1				
	Micrococccus sp.	1		2			1		
G+	*Bacillus sp.*	4	6	7					
	Sporolactobacillus sp.		1						
	Sporosarcina sp.			1					
	Lactobacillus sp.		1						
	Kurthia sp.		1						
	Arthrobacter sp						1	3	
	Rhodococcus sp.					2	3		
	Corynebacterium sp.	2		1	1	1	1	1	1
	Total identified strains	15	14	19	16	4	6	8	3
	Unidentified strains	6	9	10	7	4	7	5	4
	not interpretable	11	8	7	9	24	18	23	25
	Total isolated strains	32	31	36	32	32	31	36	32

The Biolog System constitutes a metabolic fingerprint of the capacities of the inoculated microorganisms. Results showed 66 % of the microplates were not identifiable and the remaining 34% have a metabolic fingerprint which did fit well with the Biolog database (567 sp G^- and 256 sp G^+). This system was first created for clinical use and human health applications, with some environmental use (Klinger *et al.*, 1992; Amy *et al.*, 1992). Data were then analysed and identified with the help of the Bergey's Manual of Systematic Bacteriology (1984) (TableII). To simplify Table II, all the identified strains were grouped together at the genus level.

In the water column, peptidase activity increased slightly (+10%) with effluent input and then decreased to the initial value (reactor 5). In the sediment no significant changes were observed. Bacterial counts evolved in the same way as peptidase activity: maximum in the overlying water and in the sediment at reactor 1 (Table III).

TABLE III

Changes in some bacterial parameters in sediments and in water column

	WATER			SEDIMENT		
	T	1	5	T	1	5
Aminopeptidase	255	271 (µM/L.h)	254	9.8	9.2 (µM/g.h)	10.4
Bacterial numeration	1.8	4.5 (10^6 bact/mL)	2.5	1.5	4 (10^{10}bact/g)	2.8

3.3 INVERTEBRATES

Thirteen oligochaete taxa were collected in the 6 reactors (*Limnodrilus hoffmeisteri, L. claparedeanus, L. udekemianus, Potamothrix moldaviensis, P. vejdovskyi, P. bavaricus, Aulodrilus limnobius, Spirosperma velutinus, Branchiura sowerbyi, Tubifex ignotus, Quistadrilus multisetosus, Haber* sp., and immatures of Tubificidae probably belonging to *Tubifex tubifex*). These taxa are quite common in the River Saône(Lafont,

TABLE IV

Changes of oligochaete communities in the sediments of the 6 reactors; Tub. w. h.s.: Tubificidae with hair setae; *P. moldaviensis* : *Potamothrix moldaviensis.*

	T	1	2	3	4	5
Total numbers	908	833	883	700	975	975
Number of species	7	6	5	4	7	5
Tub. w. h. s. (%)	15	12	5	9	11	10
Limnodrilus spp. (%)	72	61	88	60	59	78
P. moldaviensis (%)	13	27	7	31	30	12

1989). However, only *Limnodrilus* spp. and *P. moldaviensis* are abundant and predominate. If the number of species and the worm abundances per reactor are relatively high and the abundances remain relatively constant (Table IV), no invertebrates other than worms were found.

The following differences were considered significant: at least 2 species or 200 specimens or 10% of total number. The general structure of the oligochaete communities were very similar at the reference (T) and the last reactor (5), with the exception of the number of species. On the contrary, differences are marked between control reactor T and reactors 2 and 3, reactors 1 and 4 exhibiting intermediate situations. The most altered biological situation occured in reactor 3 (smallest abundance and species richness of the data set). The results suggest that the effect of the effluent input is maximum at reactors 2 and 3, with a tendency towards recovery to the control situation at the last reactor. It is interesting to note too that an opposite change exists between the abundances of the two predominant taxa, *Limnodrilus* spp. and *P. moldaviensis* (Table IV).

4. Discussion

4.1 WATER AND SEDIMENT PHYSICOCHEMISTRY

There have been some criticisms of microcosms due to large variability between replicate microcosms (Kroer & Coffin, 1992). In our case, though there is no replication, the evolution of the quality of the sediment and of the overlying water showed that none of the of the 6 reactors in series had an "individual" evolution and that the whole microcosm behaved like a continuum, with a credible evolution of the main physicochemical parameters in comparison to the natural environment: organic carbon degradation, nitrogen evolution with a nitrification process. These transformations, mainly under microbial control, are similar to field processes (e.g. for nitrification process: Montuelle *et al.*, 1996; Brion, 1996).

The effluent has a low nitrogen content (decrease in reactor 1) and the increase along the 5 reactors was probably due to a transfer from the sediment to the water. The formation of 2 layers in the sediment reflected the intense activity of worm populations leading to realistic exchange processes between water and sediment: for instance, NH_4^+ was higher in the bottom layer, as observed in the field (Piersotte, 1991).

4.2 BIOLOGICAL PARAMETERS

The comparison of bacterial activity and number between microcosm and field experiments sometimes leads to contradictory results: Kroer & Coffin (1992) found that there was no substantial differences in term of bacterial density between microcosm and field, but the rates of microbial processes were 25 to 40% lower in microcosms. Conversely, Kroer *et al.* (1994) obtained higher heterotrophic activity in microcosms and the complexity of the microcosms seemed to be an important parameter for the control of such variations.

In our set of experiments, the bacterial number (in water and sediment) and the activity of peptidase were in the same range as our field data and followed the same trend as in field sudies (Cemagref, 1994; Bérard, 1995). The diversity of the bacterial strains found in our microcosm is quite similar to that identified in field studies (Peduzzi *et al.*, 1992). So, as far as these general bacterial parameters are concerned, no obvious differences appeared between microcosm and field data. A notable evolution in bacterial characteristics (activities and assemblages) needs a more drastic change of environmental conditions than macrofauna (Burton, 1991): for example higher polluting discharge, presence of toxicants. However, special attention must be given to some conflicting results obtained with API and Biolog. The results obtained on the Biolog system are opposite to the results of Klinger *et al.* (1992) who succeeded in identifying 93% of water isolates.

However, they worked on a water recycling system and not in organically polluted freshwater and sediment. Perhaps our specific environment is the cause of our disappoiting results (10 to 20% of identified strains). Discrepancies appeared between the API System and the Biolog System for strain identification as noted by Amy *et al.* (1992). Some identifications obtained with Biolog were not credible (for example *Kingella kingae*, *Rhizobium meliloti* or *Xanthomonas sp.* are not found in water-sediment environments and so they are not indicated in Table II). Taken as a whole, we have identified less strains with the Biolog than with the API System.

The species richness and the abundances of oligochaetes were relatively high in the reactor sediments, a result which was not *a priori* expected. The oligochaete communities collected from the reactors are very similar to those that can occur in river sediments polluted by toxic substances (Rosso, 1995).The reactor system used here seems to be convenient for the study of oligochaetes but not of other invertebrates. It is not possible to explain here the negative relationship between *Limnodrilus* spp and *P. moldaviensis* abundances. However, this relationship suggests that, besides a pollution effect, competition for food (including bacteria) might occur, a fact already well known in confined experimental systems.

It must be noted that the biological effects of the effluent input are not marked in reactor 1 but rather slightly shifted back in reactors 2 and 3. Similar observations have been made from field studies with pollution effects expressed not exactly at the effluent discharge point, but downstream of the effluent discharge point (Lafont *et al.*, 1996). A negative confining effect (spatial effect due to a small quantity of sediment) is also possible: Spencer and Warren (1996) have shown, in fact, that habitat size could influence the diversity and the density of a trophic network. But they worked on 163d experiments and compared small habitats (250 and 1000mL). In our set of experiments, time was 30d and habitat volumes were greater (2L sediment and 3.7L overlying water). However if a confining effect appeared in our set of experiments, it was likely weak because it did not hide the toxic effect of the effluent on the invertebrates. Such a toxic effect was not shown for the bacterial populations or for the water quality (resulting from biological processes, e.g. nitrification, organic matter oxidation).

4.3 MICROCOSM FUNCTIONING

Exact simulations in microcosms of the natural conditions are unlikely. Microcosms cannot encompass all the complexity that exists in the field. The effects of nutrient deficiency, temperature fluctuations, trophic level interactions, biogeochemical cycling and other environmental parameters on microbial degradation cannot be totally assessed in microcosm studies but microcosm studies are often a necessity (Carson *et al.*, 1990). It was not attempted in this set of experiments to mimic exactly a field situation by manipulating experimental conditions. The aim rather was to consider chemical and biological evolutions in a biphasic system of sediment and overlying water under conditions of slight pollution. The input of energy represented by the effluent in this study did not radically disrupted the physico-chemical equilibrium nor the benthic biocenosis. But progressive and realistic changes appeared in water and sediment quality and in the microbial and macrofaunal populations.

As a conclusion, these results suggest that our microcosm may represents a realistic and satisfying simplification of field condition such as dilution rates, biodegradation time, dissolved O_2, and effluent composition. The data constitute a network of convergent facts and it seems that the microcosm is well adapted for studies on WTP (or toxics) discharge impact, at micro- and macrofaunal levels. The experimental data describing the evolution of invertebrate populations or some bacterial-mediated processes (e.g. organic matter biodegradation, nitrification) showed that one can simulate the effect of "distance downstream from a discharge point" by the control of the hydraulic residence time (HRT) in our reactors. However, coupling experimental data obtained in experimental systems with data on natural environments is necessary to estimate the relevance of such studies.

References

AFNOR, 1979, Recueil de normes françaises Eaux; Méthodes d'essais, 342p.

Amy P.S., Haldeman D.L., Ringelberg D., Hall D.H. & Russell C.:1992, *Applied Environemental Microbiology*, **58**, 10,3367-3373.

Benoit D.A., Phipps G.L. & Ankley G.T.:1993, *Water Research*, **27**, 9, 1403-1412.

Bérard A., Volat B. & Montuelle B., 1995, Bacterial activity and its trophic role in a eutrophic pond, Arch. Hydrobiol., 134, 4, 499-513.Bianchi M. A.G. & Bianchi A.J.M.:1982, *Microbial Ecology*, **8**, 61-69.

Bergey's manual of Systematic Bacteriology,:1984, Vol. 1, 2, 3 & 4, Krieg.N.R. & Holt J.G., Williams & Wilkins, Baltimore, London, 2619p.

Brion N, 1996, Colloque "Seine Aval", 8-9/02/96, Paris.

Broberg A.: 1985, *Hydrobiologia*, **120**, 181-187.

Burton G.A. Jr.:1991, *Environmental Toxicology and Chemistry*, 10, 1585-1627.

Carson D.B., Saeger V.W. & Gledhill W.E.:1990, *Aquatic Toxicology and risk assessment: thirteen volume*, ASTM STP 1096,Landis W.G. & Van der Schalie W. Eds, pp48-59.

Elliott E.T., Hunt H.W., Walter D.E. & Moore J.C.:1986, Proceedings of the IVthISME, (Eds F. Megusar & M. Gantar), pp. 472-480. Ljubljana.

CEMAGREF: 1994, Report DRAEI 93196, Min. Environ.- Cemagref, 139p.

Grothe D.R., Dickson K.L. & Reed-Judkins D.K.: 1996, SETAC Press, 350p.

Garric J., Vollat B., Nguyen D.K., Bray M., Migeon B. & Kosmala A.: 1996, *Water Science and Technology*, **33**, 6, 83-91.

Klinger J.M., Stowe R.P., Obenhuber D.C., Groves T.O., Mishra S.K. & Pierson D.L.: 1992, *Applied Environemental Microbiology*, **58**, 6, 2089-2092.

Kroer N., Coffin R.B. (1992) *Microbial Ecology*, **23**, 143-157.

Kroer N., Coffin R.B. & Jorgensen N.O.G.: 1994, *Environmental and Toxicological Chemistry*, **13**, 2, 247-257

Lafont M.: 1989, Thèse de Doctorat d'Etat ès Sciences, Université LyonI, 311p + annexes.

Lafont M., Camus J.C. & Rosso A.:1996, *Hydrobiologia* , **334**, 147-155.

Maugham J.T., Oviatt C.A.:1993, *Water Environment Research*, **65**, 7, 879-889.

Montuelle B., Volat B.; 1993, *Revue des Sciences de l'Eau*, **6**, 251-268.

Montuelle B., Volat B., Torio-Fernandez M., Navarro E.: 1996, *Water Research*, **30**, 5, 1057-1064.

Peduzzi R., Demarta A. & Tonolla M.: 1992, *Aquatic Sciences*, **54**, 3/4, 331-337.

Percherancier H., Volat B. & Montuelle B.: 1996, *Water Science and Technology*, **33**, 6, 221-229.

Piersotte N., 1991, Report Univ. Namur (B) - Cemagref, 89 p.

Pipke R., Wagner-Döbler I., Timmis K.H. & Dwyer D.F.: 1992, *Applied and Environmental Microbiology*, **58**, 4, 1259-1265.

Rebillard J. & Torre M.: 1993, *Revue des Sciences de l'Eau*, **6**, 153-174.

Rosso A., Lafont M & Exinger A.: 1994, *Water Science and Technology*, **29**, 3, 241-248.

Rosso A.: 1995, Thèse Université Lyon I, 176p + annexes.

Shaw J.L. & Kennedy J.H.:1996, *Environmental Toxicology and Chemistry*, **15**, 5, 605-607.

Spencer M. & Warren P.: 1996, *Oikos*, **75**, 419-430.

US Environmental Protection Agency: 1991, EPA Report 600/4/-90-027, 4th Ed.

ARYLSULFATASE AND ALKALINE PHOSPHATASE (APASE) ACTIVITY IN SEDIMENTS OF LAKE KINNERET, ISRAEL

O. HADAS and R. PINKAS

Israel Oceanographic & Limnological Research, Yigal Allon Kinneret Limnological Laboratory P.O.Box 345, Tiberias, Israel 14102

Abstract. Arylsulfatase and APase activities were monitored in the upper sediment layer, of Lake Kinneret, Israel, a warm, freshwater, monomictic lake characterized by a heavy spring bloom of the dinoflagellate *Peridinium gatunense*. Activity of both enzymes varied with depth and season. Highest activity was measured in July and high activities were monitored during the stratified period. Low values were observed in winter, when oxic conditions prevail in the water column and no organic sedimentation occurred. The values for APase ranged from 14-438 nmol PNP $g^{-1}h^{-1}$ and for arylsulfatase from 103 to 843 nmol PNP $g^{-1}h^{-1}$. Highest APase activity was recorded at 29°C, and most of it took place on mud particles and not in the interstitial waters. There were differences in enzyme activity at different stations in the lake, corresponding to differences in nutrient pore water concentrations. Enzymatic activity in Lake Kinneret sediments was related to lake trophic status, water levels, and climate conditions, all of which have an impact on the amount of organic matter reaching the sediments.

Keywords: Alkaline phosphatase, arylsulfatase, sediments.

1. Introduction

Hydrolytic enzymes have a major role in the decomposition and biochemical transformation of organic matter in aquatic environments. Microheterotrophs, particularly heterotrophic bacteria, can transform inorganic and both dissolved (DOM) and particulate (POM) organic matter. The oligomeric end products become available substrates for uptake by microbial cells (Chrost 1991; Gude 1978; Hoppe *et al.* 1988; 1993; Munster 1991; Overbeck 1991).

In thermally stratified lakes the highest ectoenzyme activities are observed in the epilimnion where highest primary production of phytoplankton occurs (Chrost *et al.* 1989). In the sediments ectoenzyme activity has been correlated with sedimentation rates of particulate organic matter (Meyer-Reil 1987) and showed the highest values in the 0-5 cm surface layer (King 1986). The regeneration of nutrients in the sediments is the result of the enzymatic hydrolysis of inorganic and organic matter by bacteria, determining the concentrations of nutrients in the pore water and their further utilization in organic carbon oxidation, and/or their release through the sediment water interface. The upper layer (0-2.5 cm depth) of Lake Kinneret sediments is characterized by high sulfate reduction rates (Hadas and Pinkas 1995a, b). The activity of sulfohydrolases might provide a mechanism for the maintenance of sulfate reducing bacteria when inputs of sulfate are low (Cooper 1972; Jarvis and Lang 1987; King and Klug 1980; Landers and Mitchell 1988).

Water, Air and Soil Pollution **99**: 671-679, 1997.

The concentration of PO_4-P in Lake Kinneret water is low (1-10 $\mu g\ l^{-1}$). The release of phosphorus through the sediment water interface contributes to the available phosphorus in the lake, especially in winter after overturn. Although release of phosphorus in Lake Kinneret is related to redox potentials (Serruya *et al.* 1974, 1980; Staudinger *et al.* 1990) the contribution of phosphatases to the phosphorus pool cannot be excluded. Regeneration of phosphorus through APase activity increases productivity at times when allochthonous inputs of P are minimal (Cotner and Wetzel 1991). In this study we present the activities of arylsulfatase and APase in sediments of Lake Kinneret in different seasons, and discuss their role in the mineralization process of organic matter and regeneration of nutrients.

2. MATERIALS AND METHODS

Lake Kinneret, Israel, a freshwater monomictic lake, is situated at -210 m below sea level, with mean and maximum depths of 24 and 42 m, respectively. The lake is stratified from May to December. Mixing occurs between December and January, resulting in a homogenous, oxygenated water column. Epilimnion temperatures range from a low of 15° during mixing to a high of 29°C during stratification. The *in situ* temperature of the hypolimnion at Station A is approximately 14-16°C. Phytoplankton biomass in the lake is dominated by the dinoflagellate *Peridinium gatunense* which blooms every spring, reaching a biomass of 250 g wet weight m^{-2} and lakewide annual production of 150,000 tons (wet weight). The total biomass during the second half of the year is about 7 times lower than during the *Peridinium* bloom. The upper sediment layer of Lake Kinneret is composed of black, jelly-like low density material 2-5 cm thick. The sediments are mainly clay or silty clay with 40-50% $CaCO_3$.

Sediment cores (40-50 cm in length) were taken monthly at Station A (Fig. 1), the central station of the lake, at maximum depth of 42 m, using perspex sediment gravity sampler (Tessenow *et al.* 1977). The cores were transferred to the laboratory within 30 minutes, sliced at 0.3 cm intervals, and processed for APase and arylsulfatase activities, SO_4^{-2}, PO_4-P concentrations, pH, and sulfate reduction.

In addition APase activity was measured at Station G, influenced by the Jordan River (22 m) and a shallow station S (10 m), with oxygenated water during the whole year.

APase activity in the particle fraction and in the pore water (free enzyme) was assayed by measuring the hydrolysis of p-nitrophenol phosphate (PNPP) as described by Wynne (1981). The reaction mixture for sediments contained: 1 g sediment, 2.6 ml 0.05 M Tris-buffer pH 9.0, 0.03 ml 0.1 M $MgCl_2$, and 0.1 ml 10mM PNPP. The samples were incubated for 1 h at 37°C. Reaction was terminated by addition of 0.3 ml of 1M NaOH. In order to get a clear solution the reaction mixture was filtered through a 0.45 μm filter, and read at 410 nm using a Contron Spectrophotometer.

The reaction mixture for pore water contained: 0.7 ml of tris-buffer, 1.3 ml of pore water and 1 ml of 1 mM PNPP. Incubation time was 48-96 h at 37°C. Results were read as above.

The effect of temperature on the activity of APase in the sediments was studied in May, when replicate cores were incubated for 24 h at 5°C, 17°C and 29°C. The assay was made as described above.

Arylsulfatase activity was assayed by measuring the hydrolysis of p-nitrophenolsulfate (PNPS), using the methodology described by Tabatabai and Bremner (1970) and King and Klug (1980). For each sample a control, without the addition of PNPS, was run in order to correct for unspecific hydrolysis or hydrolysis originating from substrates in the sediments. The reaction mixture used was: 2 g sediment, 10 ml of 0.5 M Na-acetate, flushing for 20 min under nitrogen, 0.6 ml of toluene, 1 ml of 15mM PNPS. Samples and controls were incubated for 30 min at 36°C. The reaction was terminated by addition of 10 ml of 0.5 M NaOH. Samples were filtered (0.45 μm) and measured at 410 nm.

To prevent possible oxidation of sulfide to sulfate, sediment samples for pore water sulfate were processed immediately after slicing. The respective slices were extracted on GF/C filter under vacuum into a test tube containing 0.2 g of zinc-acetate and left overnight at 4°C. The solution was then filtered through GF/C filters to remove precipitates of zinc, and sulfate was determined by the turbidometric method according to Standard Methods (APHA, 1989).

Pore water of the respective slice was extracted into a test tube and assayed for PO_4-P according to Standard Methods (APHA, 1989).

3. RESULTS

3.1. PO_4-P

The dissolved PO_4-P in the epilimnion of the lake is low (1-10 μg P l^{-1}), with highest values after overturn. The PO_4-P in Lake Kinneret pore water was in the range of 12-800 μg P l^{-1}. At Station A, the deepest station of the lake (40 m), concentrations were at the higher end of the range and at Stations S (10 m) and G (22 m) at the lower level.

In winter (February) during mixing, phosphate was evenly distributed throughout the 2.4 cm of the sediments at Station S (16-50 μg P l^{-1}). In July at the beginning of stable stratification, PO_4-P concentrations increased with depth from 12 to 128 μg P l^{-1} at 0.3 and 2.4 cm depth, respectively (Tables 1 & 2). At Station G, which is mostly influenced by the Jordan River, there was no prominent difference in dissolved phosphorus in the upper and deeper layers of the sediment. The highest value was ≈80 μg P l^{-1}. PO_4-P in the sediments at Station A ranged between 50-800 μg P l^{-1} with increased dissolved phosphorus concentrations in the deeper layers, below 1.5 cm. Apparently lake water level has an impact on dissolved PO_4-P concentrations in the sediment pore water. When lake water was relatively high, the concentration of PO_4-P was 400-800 μg P l^{-1} in 1989 (Fig. 2) as compared to a year with a low water level (1991), when 220-450 μg P l^{-1} was monitored (Table 1).

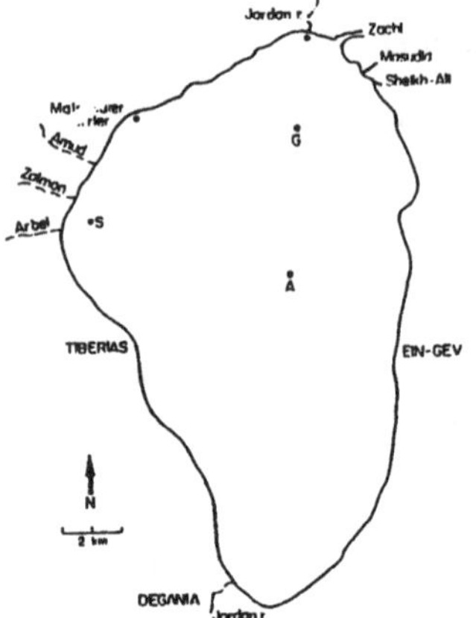

Fig 1. Stations A , G , S in Lake Kinneret.

TABLE 1

APase activity at stations A ,G ,S in sediments of Lake Kinneret , February 1991

	A		G		S	
Depth cm	PO_4-P µg l^{-1}	APase nmol g^{-1} h^{-1}	PO_4-P µg l^{-1}	APase nmol g^{-1} h^{-1}	PO_4-P µg l^{-1}	APase nmol g^{-1} h^{-1}
0.3	256	112	27	201	33	362
0.6	217	90	59	232	16	218
0.9	116	94	48	310	29	290
1.2	247	41	61	348	54	256
1.5	227	18	71	223	27	336
1.8	385	41	38	263	33	214
2.1	385	66	42	439	46	168
2.4	456	74	68	813	32	75

3.2. APASE

APase activity was correlated usually with PO_4-P concentrations, although not always. At Station A, high APase activity was measured at 0-0.9 cm depth during the stratified period, reaching values of 438 nmol PNP g^{-1}(wet weight)h^{-1} in July (Fig. 3). Minimal values (14-18 nmol PNP $g^{-1}h^{-1}$) were observed during the mixing period (January-February).

In winter at low water levels at Station A, higher APase activity was measured than when lake water was high and allochthonous P is brought to the lake by the Jordan River (Table 1; Fig. 3). At Stations G and S, higher APase activity was monitored than at Station A and APase activity correlated well with the lower concentrations of dissolved phosphorus in the pore water. In July 1991, high APase activity was found with low PO_4-P concentrations in the 0-0.6 cm depth in the sediments at Stations A and S. Station G showed a high and evenly distributed activity of the enzyme throughout the 2.4 cm depth.

3.3. APASE ACTIVITY IN INTERSTITIAL WATERS

Compared to APase activity bound to particles, the free enzyme activity was 20-200 times lower. Lowest enzyme activity in the interstitial water was observed during the winter mixing period. During stratification and anaerobiosis, an increase in free APase activity was recorded reaching a peak in December just before overturn (data not shown).

3.4. THE EFFECT OF TEMPERATURE ON APASE ACTIVITY

In the temperature experiment sediment ambient temperature was 17°C, but the highest enzymatic activity was measured at 29°C in the upper layers. The results of the APase activity in the interstitial water extracted from these sediments showed highest activity in the upper layers at 5°C and 17°C. Surprisingly sediments incubated at 5°C showed high activity throughout the whole 2.4 cm depth (Fig. 4).

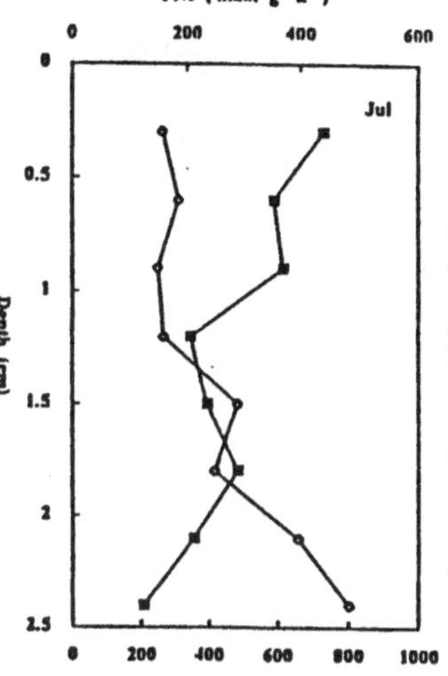

Fig 2. APase and PO_4-P concentrations in Lake Kinneret sediments at high water levels in 1989.

◊ —— ◊ PO_4-P $\mu g\ l^{-1}$

■ —— ■ Apase activity nmol PNP $g^{-1}\ h^{-1}$

TABLE 2

APase activity at stations A ,G ,S in sediments of Lake Kinneret , July 1991

Depth	A		G		S	
cm	PO_4-P $\mu g\ l^{-1}$	APase nmol $g^{-1}\ h^{-1}$	PO_4-P $\mu g\ l^{-1}$	APase nmol $g^{-1}\ h^{-1}$	PO_4-P $\mu g\ l^{-1}$	APase nmol $g^{-1}\ h^{-1}$
0.3	52	215	39	245	12	171
0.6	64	119	57	138	52	153
0.9	106	69	74	197	52	112
1.2	247	39	60	247	128	105
1.5	324	28	79	251	88	80
1.8	391	78	63	282	84	72
2.1	274	105	33	311	104	66
2.4	440	36	22	256	128	70

3.5. SULFATE

The distribution of SO_4^{-2} in the sediments at Station A ranged between 0-500 μM, depending on season and depth in the sediment. Usually higher concentrations of SO_4^{-2} were measured in the upper layers of the sediments (0-0.6 cm) in winter (Fig. 5) reaching values of about 500 μM, similar to values in the overlying water. In deeper layers or before overturn, lower concentrations ($\approx$50 μM) were observed (Fig. 5).

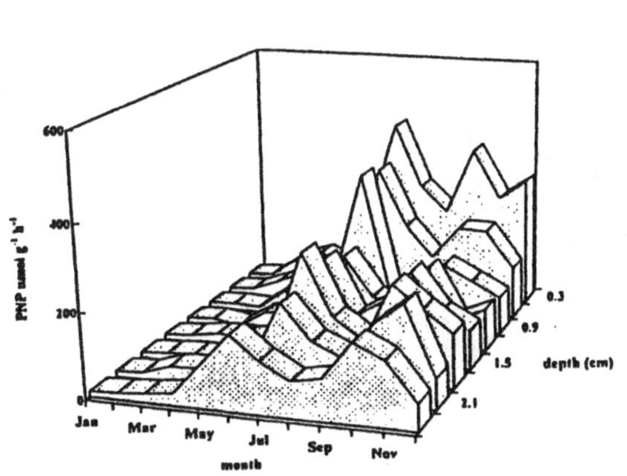

Fig 3. APase activity in the sediments at Station A .

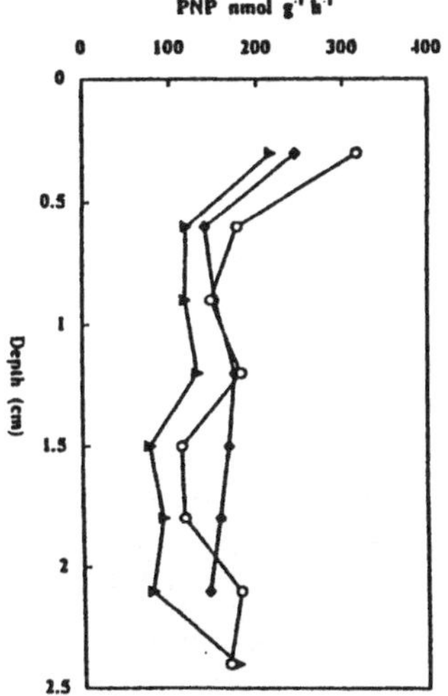

Fig 4. APase activity at different temperatures in the sediments at Station A. ◆ —— ◆ 5°C ► —— ► 17°C O —— O 29° C

3.6. ARYLSULFATASE ACTIVITY

The activity of arylsulfatase changed with depth and season (Fig. 5 & 6). During the stratified period, intensive sulfate reduction resulted in a decrease in SO_4^{-2} concentration and high arylsulfatase activity. The low SO_4^{-2} concentrations in the sediments just before overturn (December) were accompanied by high values of arylsulfatase activity (Fig. 5 & 6).

At the decline of the *Peridinium* bloom, high arylsulfatase activity was induced, providing the sediments with additional sulfate for the sulfate reduction process, which was at its maximum rate during this period. After overturn during the mixing period, high SO_4^{-2} concentrations were observed in the upper sediments accompanied by low arylsulfatase activity (Fig. 5).

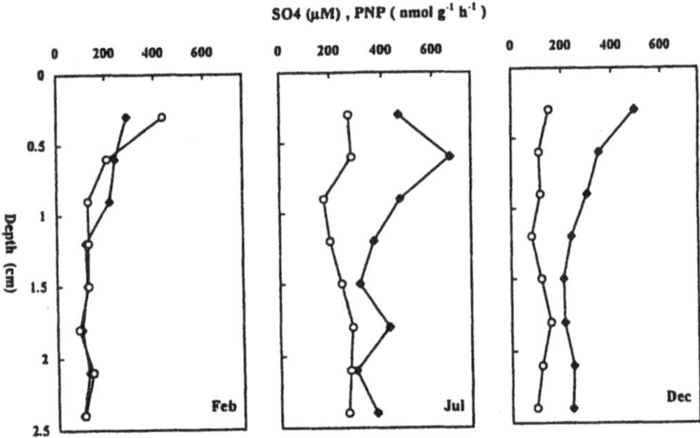

Fig 5. Arylsulfatase and SO_4^{-2} concentrations at different depths and seasons in the sediments at Station A.

■ —— ■ Arylsulfatase activity nmol PNP g^{-1} h^{-1}

O —— O SO_4^{-2} µM

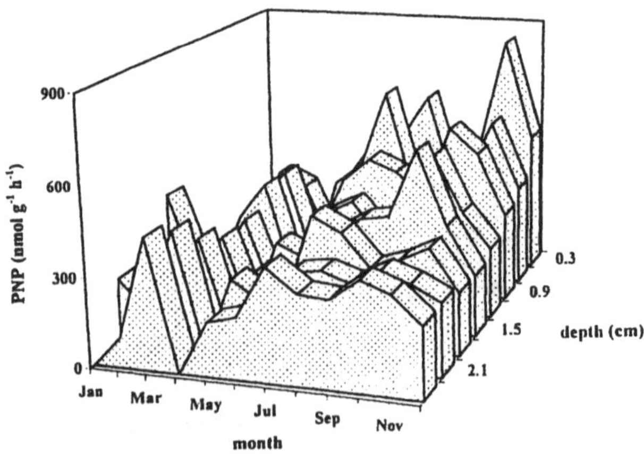

Fig 6. Arylsulfatase activity in the sediments at Station A.

4. DISCUSSION

Phosphorus and sulfate ions are important factors in the distribution of microbial activities in the sediments of Lake Kinneret. The sediments of Lake Kinneret are exposed to fluctuating environmental conditions such as major organic matter sedimentation in May-July, oxic-anoxic overlying water, highly variable redox potential and variable temperature at shallow stations, all of which influence the enzymatic activity of the microbes living in this habitat. APase and arylsulfatase activity varied with depth in the sediment and season. The lowest activities were observed in winter during mixis; the highest rates occurred at the beginning of summer during the breakdown and degradation of the *Peridinium* bloom. High activities were recorded during the whole stratified period, usually, but not always in the upper layers of the sediment. In winter after lake overturn, as a result of vertical convection currents generated by the winter storms, turbulence and bioturbation, a certain mixing of the upper sediment layers occurs and lake water penetrates into the sediment, causing an increase in sulfate concentration. The winter is characterized by very low organic sedimentation, so in January-February and during the bloom phase, the increase in sulfate with lack of available organic matter resulted in low arysulfatase activity in the sediments at Station A (Figs. 5 & 6). The mechanism of bioturbation and enzymatic activity related to biogenic structures may play a role in the littoral zone of Lake Kinneret as was reported for the Norwegian-Greenland Sea (Koster *et al.* 1991). Phosphorus concentrations in Lake Kinneret are redox related and may change between years. In winter when redox is high, and sulfate is available (no H_2S), phosphorus is bound in iron compounds (Serruya, 1978; Staudinger *et al.* 1991). During low redox (stratification) or in deeper layers of the sediment, higherPO_4-P concentrations were found (Tablc 1 & 2). Among the three stations, Station A had the highest PO_4-P concentrations. The binding of iron with sulfides over longer periods (Station A) allowed a more efficient release of phosphate. The deep sediments of the central area of the lake contribute a larger amount of PO_4-P to the overlying lake water than do the shallow areas, at least during the anoxic period (Serruya *et al.* 1974). At all three stations APase activity showed an inverse correlation with PO_4-P concentrations.

Peridinium gatunense accumulates phosphorus in the form of polyphosphate bodies (Elgavish *et al.* 1980a, b), which at the breakdown of the bloom are subject to intense hydrolytic activity. In the water column, APase activity rises dramatically only towards the end of the bloom, even though Pi concentrations in the lake are always low (Wynne, 1977; 1981). At the beginning of summer, the high rates of organic sedimentation are followed by enhanced microbial activity of heterotrophic bacteria (Cavari and Hadas, 1979), providing the ecosystem with high amounts of dissolved organic matter (DOM), inducing the synthesis of ectoenzymes. The ability of bacteria to utilize a wide variety of DOM and DOP (dissolved organic phosphorus compounds) is reflected in the amount of enzyme produced (Cotner and Wetzel, 1991). That is exactly the case in Lake Kinneret sediments for APase and arylsulfatase activities in summer and autumn (Figs. 3, 5 & 6; Tables 1 & 2). The results are similar to those observed in Kiel Bight (Germany), where the input of organic matter into the sediments following the breakdown of phytoplankton caused an immediate response in enzymatic activity (Meyer-Reil, 1987; 1991). In autumn when thermal stability prevails, no rainfall and low sedimentation rates were observed (Serruya *et al.* 1974; Zohary *et al.* submitted). Nevertheless, intense sulfate reduction processes continue, causing a decrease in sulfate in the sediments, which in turn results in high arylsulfatase activities (Fig. 6).

Lake water levels had an impact on APase activity and PO_4-P concentrations in the pore water. At low lake water level (e.g. 1991), intensive sulfate reduction occurred in the hypolimnion, less organic matter reached the sediments, less SO_4^{-2} diffused and lower rates of sulfate reduction were recorded in the sediments. The result was lower concentrations of H_2S and less PO_4-P liberated into the pore water. At relatively high water levels (1989) more organic matter reached the sediments and high sulfate reduction and APase activities were found (Hadas and Pinkas, 1995b). High APase activity was accompanied by a higher release of H_2S, with consequent precipitation of FeS, resulting in higher release of phosphorus in pore water (800 $\mu g\ l^{-1}$) (Fig. 2). In December before overturn there was a drop in arylsulfatase and APase activities, probably resulting from less organic matter being available (Hadas and Pinkas, 1992). Although the ambient temperature of Lake Kinneret sediments is 15-17°C, the highest APase activity occurred at 29°C. This result is consistent with findings that the optimal temperature for hydrolytic enzymes in natural sediments was between 30°-40°C, even for sediments from the cold Norwegian-Greenland Sea (Meyer-Reil, 1991). Most of the activity of APase in Lake Kinneret sediments was related to the particle fraction. Bacteria probably settle on the surface of particles in the sediments and only a minor portion is present in the water phase (Chrost *et al.* 1989; Chrost and Overbeck 1987; Meyer-Reil 1991; Rego *et al.* 1985). Close proximity of the hydrolysis process and uptake system is beneficial to the survival of microorganisms when substrates may be limited (Chrost 1986; Gude 1978). Nevertheless, APase activity measured in Lake Kinneret sediments was higher than reported for stream waters and sandy sediments in Germany (Marxen and Witzel 1991). Activities found for Australian billabongs (Boon, 1991) were close to APase activities in Lake Kinneret in winter. In summer much higher activities of APase were observed in Lake Kinneret sediments.

In conclusion, the high APase and arylsulfatase activities recorded in Lake Kinneret sediments emphasize their significance in the mineralization process of organic matter, sulfate reduction and phosphorus flux. APase, besides providing the ecosystem with phosphorus, also can supply the bacterial demand for carbon (Chrost 1990; Chrost and Overbeck 1987; Martinez *et al.* 1996). The activities of both enzymes in Lake Kinneret sediments were sensitive to trophic status, lake water levels, climate conditions and time of thermal stabilization, because all these factors can influence the amount of organic matter reaching the sediments.

References

APHA: 1989, *Standard Methods for Examination of Water and Waste Water.* 17th Ed. American Public Health Association, Washington, D.C., USA.
Boon, P.I.: 1991, *Microbial Enzymes in Aquatic Environments,* Chrost, R.J. (ed.), Springer Verlag, p. 286.
Cavari, B.Z. and Hadas, O.: 1989, *Freshwat. Biol.* **9**, 329-338.
Chrost, R.J.: 1986, *Perspective in Microbial Ecology,* Megusar & Gantar, Ljubljana, p. 360.
Chrost, R.J.: 1990, *Aquatic Microbial Ecology,* Overbeck, J. and Chrost, R.J. (eds.), Springer Verlag, p. 47-78.
Chrost, R.J. 1991. *Microbial Enzymes in Aquatic Environments,* Chrost, R.J. (ed.), Springer Verlag, p. 29-59.
Chrost, R.J. and Overbeck, J.: 1987, *Microb. Ecol.* **13**, 229-248.
Chrost, R.J., Munster, U., Rai, H., Albrecht, D., Witzel, P.K. and Overbeck, J.: 1989, *J. Plank. Res.* **11**, 223-242.
Cooper, J.B.: 1972, *Soil Biol. Biochem.* **4**, 333-337.
Cotner, J.B. and Wetzel, R.G.: 1991, *Microbial Enzymes in Aquatic Environments,* Chrost, R.J. (ed.), Springer Verlag. p. 183-205.
Elgavish, A., Elgavish, G.A., Halman, M., Berman, T. and Shomer, I.: 1980a, *FEBS Letters,* **117**: 137-142.
Elgavish, A., Elgavish, G.A., Halman, M. and Berman, T.: 1980b, *J. Phycol,* **16**, 626-633.
Gude, H.: 1978, *Arch. Hydrobiol.* **55**, 157-185.
Hadas, O. and Pinkas, R.: 1992, *Hydrobiol.,* **235**, 295-301.
Hadas, O. and Pinkas, R.: 1995a, *Freshwat. Biol.* **33**, 63-72.
Hadas, O. and Pinkas, R.: 1995b, *Microb. Ecol.* **30**, 55-66.
Hoppe, H.-G., Ducklow, H. and Karrasch, B.: 1993, *Mar. Ecol. Prog. Ser.* **93**, 277-283.
Hoppe, H.-G., Kim, S.J. and Gocke, K.: 1988, *Appl. Environ. Microbiol.* **54**, 784-790.
Jarvis, B.W. and Lang, G.E.: 1987, *Soil Biol. Biochem.* **19**, 107-109.
King, G.M.: 1986, *Appl. Environ. Microbiol.* **51**, 373-380.
King, G.M. and Klug, M.J.: 1980, *Appl. Environ. Microbiol.* **39**, 950-956.
Koster, M., Jensen, P. and Meyer-Reil, L.A.: 1991, *Microbial Enzymes in Aquatic Environments,* Chrost, H.J. (ed.), Springer Verlag, p. 298-310.
Landers, D.H. and Mitchell, M.J.: 1988, *Hydrobiol.* **160**, 85-95.
Martinez, J., Smith, D.C., Steward, G.F. and Azam, F.: 1996, *Aquat. Microbial. Ecol.* **10**, 223-230.
Marxsen, J. and Witzel, K.P.: 1991, *Microbial Enzymes in Aquatic Environments,* Chrost, H.J., (ed.), Springer Verlag, p. 270-297.
Meyer-Reil, L.A.: 1987, *Appl. Environ. Microbiol.* **53**, 1748-1755.
Meyer-Reil, L.A.: 1991, *Microbial Enzymes in Aquatic Environments,* Chrost, H.J. (ed.), Springer Verlag, p. 84-95.
Munster, U.: 1991, *Microbial Enzymes in Aquatic Environments,* Chrost, H.J. (ed.), Springer Verlag, p. 96-122.
Overbeck, J.: 1991, *Microbial Enzymes in Aquatic Environments,* Chrost, H.J. (ed.)., Springer Verlag. p. 1-5.
Rego, J.V., Billen, G., Fontigny, A. and Sommville, M.: 1985, *Mar. Ecol. Prog. Ser.* **21**, 245-249.
Serruya, C.: 1978. *Lake Kinneret Monographiae.* Dr. Junk Publishers, The Hague. 502 p.
Serruya, C., Edelstein, M., Pollingher, U. and Serruya, S.: 1974, *Limnol. Oceanogr.* **19**, 489-508.
Staudinger, B., Peiffer, S., Avnimelech, Y. and Berman, T.: 1990, *Hydrobiologia,* **207**, 167-178.
Tabatabai, N.A. and Bremner, J.A.: 1970, *Soil Sci. Soc. Amer. Proc.* **34**, 225-229.
Tessenow, U., Frevert, T., Hofgastner, W. and Moser, A.: 1977, *Arch. Hydrobiol., suppl.* **48**, 438-452.
Wynne, D.: 1977, *Physiol. Plant.* **40**, 219-224.
Wynne, D.: 1981, *Hydrobiol.* **83**, 93-99.
Zohary, T., Pollingher, U., Hadas, O. and Hambright, K.D. (submitted to Limnol. Oceanogr.).

SEDIMENT FEATURES, PRIMARY PRODUCERS AND FOOD WEB STRUCTURE IN TWO SHALLOW TEMPORARY LAKES (MONEGROS, SPAIN)

Alcorlo, P.; Díaz, P.; Lacalle, J.; Baltanás, A.; Florín, M; Guerrero, M.C., and Montes, C.
Department of Ecology, Universidad Autónoma de Madrid, E-28049 Madrid, Spain.

Abstract. The aim of this study is to describe general features of sediment, primary producers and both benthic and planktonic consumers in two shallow saline lakes (*Salada* de La Muerte and *Salada* de Piñol) in order to detect main factors influencing food web structure.

The lakes are located in Los Monegros district, in the central area of the Ebro River catchment, NE Spain. Both lakes are temporary with salinity well above 30 g L^{-1}. Although they are situated close to each other (distance: 300 m), their communities of primary producers differ dramatically. One lake (La Muerte) is dominated by microbial mats and seems to function through the benthic pathway. The other (Piñol) has some macrophytes but phytoplankton is the main source of primary production. Two cycles (1994/95 and 1995/1996), quite different in their hydrological characteristics, have been studied.

PCA demonstrated the major influence of hydrologic features (e.g. water level) over biotic and abiotic parameters. The presence of microbial mats in La Muerte played a key role in stabilizing the sediments. A comparison of food web structure and dynamics in both lakes has been performed and the influence of sediment features is discussed.

Keywords: sediments, shallow lakes, microbial mats, salinity, primary production, halophilous.

1. Introduction

Scheffer *et al.* (1993) proposed that turbidity of shallow lakes may not be a smooth function of their nutrient status. Instead, two alternative equilibria dominated by submerged macrophytes (clear) and phytoplankton (turbid) determine water-sediment interactions in such systems. Field evidence supporting that hypothesis generally comes from studies on permanent, moderately shallow lakes. However, very little is known about water-sediment interactions in temporary lakes, systems which are common in Mediterranean and semiarid areas. More than 50% of natural lentic habitats in Spain are inland shallow-water environments other than floodplains (Casado *et al.*, 1992). Many of these habitats are temporary and saline. Hydrological disturbances are known to alter population dynamics in aquatic ecosystems (Grimm and Fisher, 1989; Fisher and Grimm, 1991; Grimm and Fisher, 1992) and to control the number of trophic levels that occur in them (Pimm and Lawton, 1977; Pimm, 1982). On a broad scale, salinity is known to be negatively correlated to species richness (Williams *et al.*, 1990). However, primary producers can be remarkably diverse in saline lakes (Javor, 1989).

A main question to be addressed in such extreme environments is the role played by sediments. How do sediments affect basic levels of food webs? Florín *et al.* (1994) assess that water and organic matter content in the sediment control the dominance of major functional types of primary producers in saline temporary lakes; and that dominance is also affected by patterns of inorganic carbon availability resulting from hydrochemical interactions between water column and sediment subsystems (Florín and Montes, in press).

Water, Air and Soil Pollution **99**: 681-688, 1997.

The main aim of this study is to address the influence of sediment-water interactions in food web structure in two temporary shallow saline lakes (*saladas*) which differ greatly in dominant primary producers. To achieve that goal we have focused on two questions: (1) which environmental factors are related to the presence of the major functional groups in both primary producers and consumers (zooplankton and zoobenthos) and (2) what are the differences between those systems and similar ones elsewhere in the world.

2. Study area

The lakes studied are the Salada de Piñol (41°26'36"N 0°15'42"W, 11.9 ha surface, 0.4 m maximum depth) and the Salada de La Muerte (41°26'05"N 0°16'04"W, 11.5 ha surface, 0.4 m maximum depth), located just 300 m apart from each other, in Los Monegros district, in the central area of the Ebro River catchment (NE quarter of the Iberian Peninsula). Those *saladas* lie on an almost horizontal platform that extends between 300 and 360 m above sea level. It is a part of an evaporite basin with limestone materials at the top, underlain by marl and gypsum (Quirantes Puertas, 1965). The climate is Mediterranean semiarid (average annual values of rainfall and potential evapotranspiration are 383 and 778 mm, respectively). Dry conditions are caused by a rain shadow from the mountain ranges forming its boundaries. A strong persistent wind blowing from the NE (*cierzo*) increases evaporation (Pueyo, 1978).

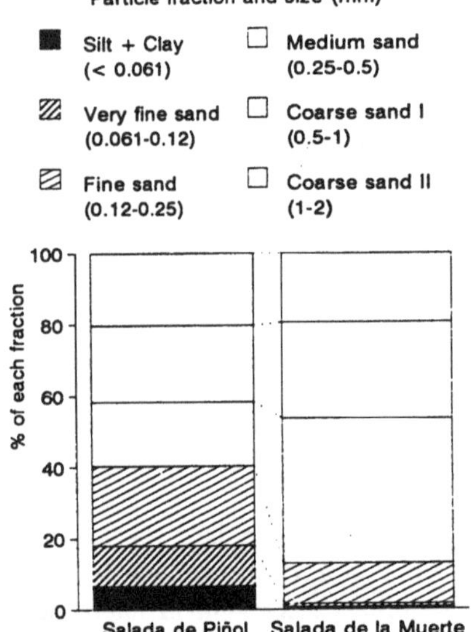

Fig. 1. Size distribution of sediment -particles (top 10 cm) in the studied *saladas* (modified from

Water persistence is favored by impeded drainage, due to the low permeability of lake sediments and because the water table is very close to the ground surface. Main water sources are runoff, drainage of surrounding croplands, and direct rainfall. The onset and span of flooding is highly unpredictable because of the broad variability of the Mediterranean climate, here enhanced by strong wind action (Pueyo & Inglés, 1984).

Water salinity is high (Piñol: 14.8-149.4 g L^{-1}; La Muerte: 50.5-87.3 g L^{-1}), with an average ionic formula Na-(Mg)-Cl-SO_4 and Na-Mg-Cl-(SO_4), respectively (unpublished data). Salt balance has three main controls: groundwater as the source, flooding and desiccation as the geochemical control, and wind action as the cause of losses.

The sediments of both *saladas* may be distinguished by looking at their granulometry (Figure 1), that serves to clasify La Muerte among the systems having the sediment type that allows the development of more complex and compact microbial mats (Guerrero *et al.*, 1994).

3. Methods

The *saladas* were visited monthly from November 1994 till May 1996, but the basins were filled only during two short periods: Nov.'94-Feb.'95 and Dec.'95-March '96. Both cycles differed dramatically in the amount of rainfall, the former was much drier than the latter (Note: although official rainfall data from the National Meteorological Institute were not available while writing this paper, differences between hydrologic cycles were large enough at a regional, and even at a national scale, to be considered a main source of variability). Water and organic matter content in sediment (Duarte *et al.,* 1988) and total phosphorus concentration (Andersen, 1976) were measured from three replicate samples of sediment taken from the top 5 cm with a Plexiglas corer (inner diameter Ø=5cm) along a shore-center transect. Other physico-chemical variables were estimated from water samples taken at the center and the in-shore areas. Biomass of primary producers was estimated as chlorophyll *a* concentration in both planktonic and microbenthic communities (APHA, 1989). Available information on macrophyte biomass has not been considered here due to the low abundance and patchiness of such producers and sampling problems related to turbidity. Primary productivity of microbenthos, plankton and submerged macrophytes was estimated on each sampling date measuring ^{14}C fixation rate for 2 hours (Vollenweider, 1974). Abundance of zooplankton and zoobenthos was estimated from quantitative samples (10 sites per lake, 3 replicates each) of top 1-2 cm of sediments and the water column above, taken with a 5 cm Ø Plexiglas core. Samples were filtered *in situ* with a plankton net (60 µm mesh size), fixed with 4% neutralized formalin and preserved afterwards in ethanol (70%) (Downing & Rigler, 1984).

Variables measured *in situ* included depth, water temperature, electric conductivity at 25° (WTW conductivity meter calibrated against a KCl standard), pH (Crison 506 pH-meter, with a Metrohn electrode-ORP), total alkalinity and phenolphthalein alkalinity (Aminot & Chaussepied, 1993). Total phosphorus (total-P), total nitrogen (total-N) and total dissolved solids (TDS) were measured in the lab according to standard techniques (APHA, 1989).

Principal Component Analysis (PCA) was performed as an heuristic tool to explore temporal changes in the main features of the lakes. Data input was a matrix including abiotic and biotic variables from water and sediments.

4. Results

Temporal changes of selected environmental descriptors are displayed in Figure 2. Flooding period is similar in both basins spanning between early autumn and late winter- early spring. Salinity (TDS) is high and inversely related to water level, increasing at the end of all flooding periods in an exponential way. Water content of sediments increases in both *saladas* at the beginning of flooding period and at the middle of the dry period, with values slightly less variable in La Muerte (Figure 2a). In this *salada*, where microbenthos is the dominant primary producer, organic matter content (OM_s) is lower (average 10.8±1.4 %) than in Piñol (13.2±1.4 %).

PCA results are displayed in Figure 3. In order to better see the pattern, samples from each lake are displayed in different plots although both refer to the same analysis using the whole data set. The first and second axes of PCA account for 51% of the variance of the data. The first axis (26%) is related to changes in water level thus reflecting temporal evolution of the system through the wet season. The second axis (24%) summarizes changes

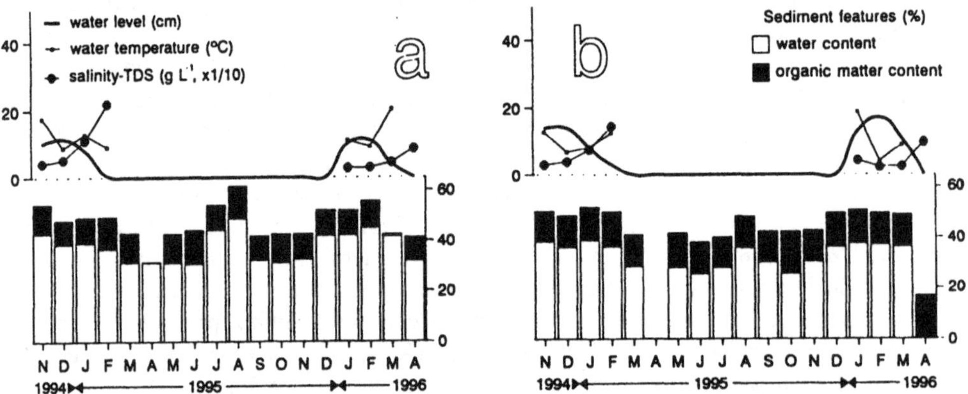

Figure 2. Evolution of main physico-chemical variables in the studied *saladas* in the two consecutive flooding periods that were recorded (1994/95 and 1995/96): a) Salada de La Muerte; b) Salada de Piñol.

in biological activity, mainly the dominance of either phytoplankton or phytobenthos.

Differences between lakes are patent. Piñol is more variable, less buffered in limnological features, than La Muerte especially in the driest period (1994-95). Temporal changes are positively correlated with the first axis as flooding periods progress. Hence, both lakes evolve towards higher salinities and increasing total-N concentration.

In La Muerte, total-nitrogen is significantly correlated with both phytoplanktonic primary production and microbenthic chlorophyll **a** ($r=0.94$, $p<0.05$).

In Piñol almost all significant correlations are related to nutrients: chlorophyll **a** of phytobenthos to total alkalinity ($r= 0.91$, $p<0.05$); water level to total nitrogen ($r=0.93$, $p<0.01$) and temperature to chlorophyll **a** of phytobenthos ($r= 0.82$, $p<0.01$). Nutrients (total-N and total-P) show similar temporal dynamics ($r=0.94$, $p<0.01$) (Figure 4).

Temporal changes in primary production are shown in Figure 4. Water level and nutrient concentrations are also included in order to stress relationships among them.

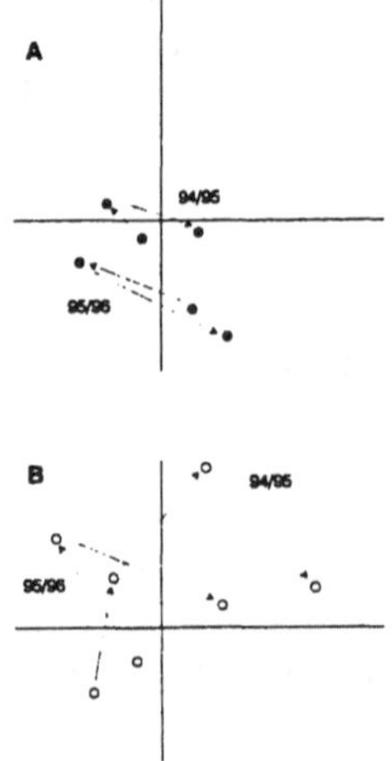

Figure 3. Principal Component Analysis plot: (a) Salada de La Muerte; (b) Salada de Piñol (see text)

In Piñol, the dominant primary producers are planktonic species like *Navicula* sp, *Thalassiosira weiisflugii* sp., *Cymbella* sp. and *Oocystis borjeii* with a secondary contribution from the submerged macrophyte *Ruppia drepanensis* and a minimal presence of the liverwort *Riella helicophylla*. In La Muerte, primary production is mainly due to well developed microbial mats, dominated by *Microcoleus chtonoplastes*, that cover almost the entire bottom of the salada , whereas *R. drepanensis* is scarce. Phytoplankton includes as common representatives *Entomoneis alata, Thalassiosira weissflugi, Aphanotece salina, Chlamydomonas* sp., *Oocystis* spp., *Microcystis pupuerea* and *Pseudoanabaena* sp. Most of those taxa are halophilous (Javor, 1989; Symoens, 1988).

Animal communities (plankton and benthos) are species-poor and very similar in both lakes, although abundances are much higher in Piñol (Figure 5 for the period 94/95). Most of the species are halophilous: *Fabrea salina* (Ciliata); *Hexarthra fennica* (Rotifera); *Arctodiaptomus salinus* (Copepoda: Calanoida), *Cletocamptus retrogressus* (Copepoda: Harpacticoida) (Reddy, 1994); *Branchinectella media* (Anostraca); *Prionocypris aragonica* (Ostracoda), an endemism; and *Ephydra* sp. (Diptera). Tubularia and Nematoda species were also present but were not identified to the species level. Immediately after flooding (early autumn of 1994), there is a burst of life forms but with a low abundance. From that moment onwards diversity decreases and some taxa become dominant.

5. Discussion and Conclusions

Although temporal dynamics in water level, salinity and some sediment features are similar in both *saladas*, their biota and food-webs differ dramatically. Granulometric composition of sediments and the consistency of sediment water content seem to be responsible for such a pattern. Biotic (trophic) and abiotic features of both water and sediments show a broader variability in Salada de Piñol (phytoplankton-dominated) than in Salada de La Muerte (microphytobenthos-dominated). Microbial mats are of paramount importance in La Muerte. While the basin is dry they prevent sediment removal by wind action and reduce water loss from sediments. During the flood period they keep the water clear by avoiding sediment resuspension induced by water turbulence. On the contrary, water is commonly turbid in Piñol because it has no mats.

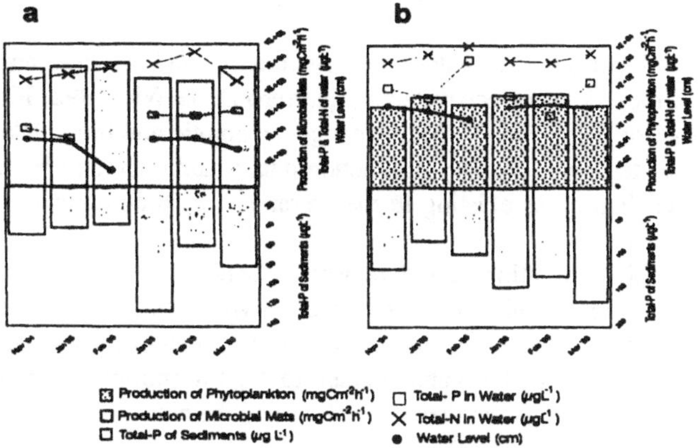

Figure 4. Temporal dynamics of primary producers in relation to water level and nutrients during flood periods. a) La Muerte, b) Piñol.

It is surprising that animal communities do not reflect major differences between both systems, e.g. Piñol is not dominated by planktonic feeders but by benthic consumers. The high organic matter content in the sediments suggests that that is the main food source

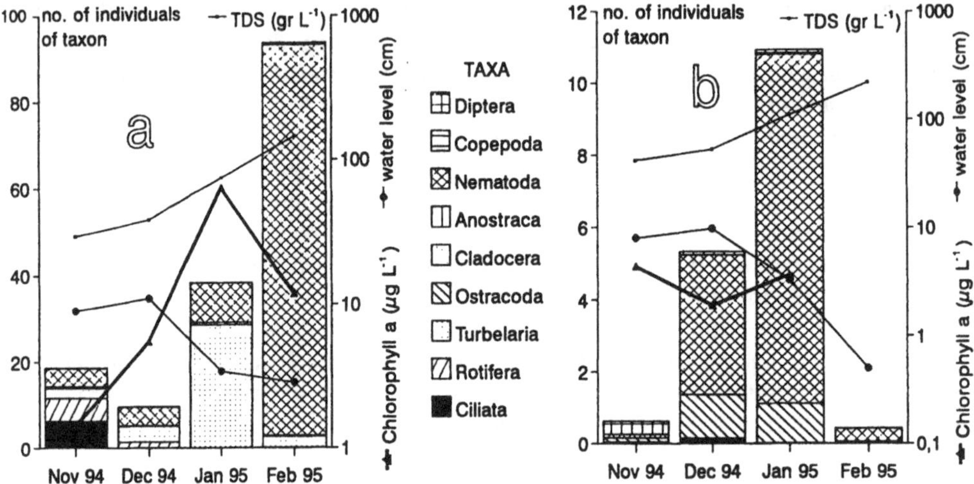

Figure 5. Animal taxa abundance in 1994/95: a) Piñol and b) La Muerte. Zooplankton abundance in numbers per volume (cm^3) of water column; zoobenthos abundance in numbers per volume (cm^3) of sediment.

controlling food-web composition. Obviously, well-structured mats as found in La Muerte allow fewer grazing strategies, but this phenomenon may be also linked to two relevant processes: (1) the higher temporal instability (higher recycling of both organic matter and nutrients) in Piñol; (2) despite benthic primary production is higher in La Muerte, overall resources of benthic organic matter are both more abundant and more available in Piñol; and (3) eolic deposition (dry season) together with turbidity due to sediment resuspension (flood season) prevent the development of microbial mats in Piñol. Benthic communities are in a continuous recolonization process abruptly interrupted by light limitation, which explains the high values of chlorophyll and organic matter found in this *salada* and the development of benthic trophic pathways.

The role played by water level fluctuation and its inter-annual rhythm, a key disturbance in these kind of ecosystems (Guerrero & de Wit, 1992; Florín *et al.*, 1994), is shown by axis I of PCA (Fig. 3). Main changes in abiotic and biotic lake features are linked to changes in water level. Primary production data (planktonic and benthic) in both *saladas* are much lower than those existing in the literature for some other saline lakes throughout the world (Table 1).

The high correlation found between Total-N and primary producers activity is similar to observations made in other aquatic environments of arid regions (Reuter *et al* 1993). The sparse and scarce vegetation of their catchments allows only a small amount of allochthonous nutrients and organic matter to be transported from terrestrial to aquatic environments. Correlations between Total-N and primary production (measured as ^{14}C

uptake and chlorophyll **a** concentration ($r=0.94$, $p<0.05$), although far from conclusive, suggest a relevant dependence of *salada* ecosystems on nitrogen.

There is a relative association between well-structured microphytobenthos communities and oligothrophic conditions. Alternative sources of N are available for these communities. Nitrogen production by bacterial degradation activities and not biological degradation processes in the deeper layers of microbial mats can be more important than nitrogen removed from sediments, water column or atmospheric uptake (Bebout *et al.*, 1993). On the contrary, phytoplankton depend mainly on dissolved N and N-fixation

Table I.

Comparison of Primary Production rates due to benthic and planktonic communities in several saline lakes (modified from Javor 1989).

Primary producer	Productivity	Saline Lakes	Reference
PHYTOPLANKTON	802.296 ($mgCm^{-2}d^{-1}$)	La Muerte, Spain	unpublished data
	1012.608 ($mgCm^{-2}d^{-1}$)	Piñol, Spain	unpublished data
	184.0 ($mgCm^{-2}d^{-1}$)	Pink Lake, Australia	Hammer, 1981 in: Javor, 1989
	1188.0 ($mgCm^{-2}d^{-1}$)	Little Manitou Lake, Canada	Haynes and Hammer, 1978 in: Javor, 1989
	2040.0-7440 ($mgCm^{-2}d^{-1}$)	Mono Lake, U.S.A.	Mason,1967; Winkler,1977 in: Javor, 1989
MICROBIAL MATS	226.035 ($gCm^{-2}d^{-1}$)	La Muerte, Spain	unpublished data
	5.0-12.0 ($gCm^{-2}d^{-1}$)	Solar Lake , Israel	Krumbein *et al*, 1977 in: Javor, 1989

processes in the water column. (note the significant correlation between total-N and primary production of phytoplankton ($r=0.94$, $p<0.05$) and total-N to chlorophyll **a** of phytobenthos ($r=0.94$, $p<0.05$)).

We conclude that each lake here studied has a unique trophic pattern, where sediment resuspension together with the occurrence of different communities of primary producers and water level fluctuation play an important role in the organization and functioning of the system. In these extremely shallow temporary lakes, water level fluctuations are a main driving force. But there are also strong interrelations between water and sediments affecting the cycling of nutrients, salinity evolution, food supply and sediment bank (storage of animal eggs, seeds of macrophytes, resistant animal stages and other physico-chemical parameters). All that biological information accumulated in the sediments plays a key role at the beginning of the flooding period and determines the subsequent evolution throughout the year.

Acknowledgements

Special thanks to S. Álvarez for the revision of the phytoplankton material. The study was funded by the Spanish Interministerial Commission of Science and Technology (ref. AMB94-0827). M. Florín was supported by EU project EV5V-CT94-0559 and by the Spanish Ministry of Education and Culture (grant PF96-0050048879). P. Díaz was supported by a scholarship of the Instituto de Cooperación Iberoamericana (ICI).

References

Aminot, A. & Chaussepied, M : 1993, *Manuel des analyses chimiques en milieu marin*. CNEX, Paris.
Andersen, J.M.: 1976, *Water Research* **10**, 329-331.
APHA (ed): 1989, *Standard Methods for the Examination of Water and Wastewater*. Washington.
Beboult, B.M.; Fitzpatrick, M.W & Paerl H.W.: 1993, *Appl. Envir. Microbiol.* **59**, 1495-1503.
Casado, S. Florín, M. Mollá, S. and Montes, C.: 1992, *Managing Mediterranean Wetlands and their Birds*, IWRB Special Publications No. 20. IWRB & Istituto Nazionale di Biologia della Selvaggina, 56-58.
Downing, J.A.& Rigler, F.H.: 1984, *A manual on Methods for the assessment of Secondary Productivity in Fresh Waters*. Balckwell Scientific Publications, Oxford, 106-110.
Duarte, C.M.; Bird, D.F. & Kalff, J.: 1988, *Verh. Internat.Verein.Limnol.*, **23** (1), 271-281.
Fisher, S.G. and Grimm, N.B.: 1991, *Comparative analysis of ecosystems: patterns, mechanisms, and theories*.Springer-Verlag, New York, 196-221.
Florín, M., Priebe, Ch. and Besteiro, A.G.: 1994, *Verh. Int. Verein. Limnol.* **25**, 1342-1344.
Florín, M. and Montes, C.: (in press), *Proc. Int. Assoc. Limnol.* **26**
Grimm, N.B. and Fisher, S.G.: 1989, *J.N. Am. Benthol. Soc.* **8**, 293-307.
Grimm, N.B. and Fisher, S.G.: 1992, *Global climate change and freshwater ecosystems*, Springer-Verlag, New York, 211-233.
Guerrero, M.C. and de Wit, R.: 1992, *Limnetica* **8**, 197-204.
Guerrero, M.C. Tadeo, A.-B. and de Wit, R.: 1994, *Microbial Mats*, NATO ASI Series, Vol. G **35**, 85-90.
Javor, B: 1989, *Hypersaline environments. Microbiology and biogeochemistry*, Springer-Verlag, London, 328.
Pimm, S.L.: 1982, *Food webs*. Chapman & Hall, London.
Pimm, S.L. and Lawton, J.H.: 1977, *Nature* **268**, 329-331.
Pueyo, J.J., 1978, *Revista Investigaciones Geológicas*, **34**: 195-207.
Pueyo, J.J. & Inglés, M., 1984, *Geochemistry and mineral formation in the earth surface*. Rodriguez Clemente R. & Tardy, Y. Eds.
Quirantes Puertas, J., 1965, *Estudios Geológicos*, vol XXVII (4): 355-362.
Reddy, R.: 1994, *Copepoda:Calamoida: Diaptomidae*, SBP Academic Pub.
Reuter, J.E., Rhodes, C.L., Lebo, M.E., Kotzman, M. And Goldman, C.R.: 1993, *Hydrobiologia*, **267**, 179-189.
Scheffer, M. Hosper, S.H. Meijer, M.-L. Moss, B. and Jeppesen, E.: 1993, *TREE* **8**(8), 275-278
Symoens, J.J. (ed).:1988, *Vegetation of Inland Waters*, Kluwer Ac.Pub.
Vymazal, J.1995. *Algae and Element Cycling in Wetlands*, Lewis Pub.
Williams, W.D.; Boulton, A.J. & Taaffe.: 1990, Hydrobiologia **197**: 257-266.

INFLUENCE OF PARTICLE SIZE DISTRIBUTION AND CONTENT OF ORGANIC MATTER ON THE TOXICITY OF COPPER IN SEDIMENT BIOASSAYS USING *CAENORHABDITIS ELEGANS* (NEMATODA)

S. HÖSS[1,2], M. HAITZER[1,2], W. TRAUNSPURGER[2], H. GRATZER[3], W. AHLF[3] and C. STEINBERG[1]

[1] Zoologisches Institut der LMU, Abt. Limnologie, Karlstr.23-25, 80333 München, FRG, [2] Institut für Gewässerökologie und Binnenfischerei, Müggelseedamm 310, 12587 Berlin; FRG, [3] TU Hamburg-Harburg, AB Umweltschutztechnik, Eissendorferstr. 40, 21071 Hamburg, FRG

Abstract. The influence of particle size distribution and organic matter on the toxicity of copper was investigated using the nematode *Caenorhabditis elegans* as testorganism. Sediments taken at various depths from three lakes of different trophic status and artificial sediments were spiked with sublethal concentrations of $CuSO_4$. After an exposure of 72 h to spiked sediment or liquid medium, body length of the nematodes was determined. Both artificial and natural sediments reduced the effect of copper, with natural sediments being more effective. In natural sediments worms grew normally at concentrations of copper up to 63.5 mg/L, whereas in artificial sediments body length was reduced at concentrations of 11.3 mg Cu/L or higher. Body length was positively correlated with content of fine particles and organic matter, indicating that particle size distribution and organic matter are determinant factors for the ecotoxicology of sediments.

Keywords: Sediment, particle size distribution, organic matter, *Caenorhabditis elegans*, Cu, toxicity

1. Introduction

Heavy metals discharged into natural waters partly become associated with the sediment, as a result of physical, chemical and biological processes (Tessier and Campbell, 1987). Consequently, sediments accumulate these contaminants to concentrations several orders of magnitude higher than those in the water column (Lee and Jones, 1984). The partitioning dynamics of metals between sediments, interstitial and overlying water are particularly complex, which makes predictions of partitioning difficult (Burton, 1991). Factors such as redox gradients, pH, temperature, adsorption, complexation, precipitation and particle size influence the fate of metals in sediments (Förstner, 1990). Attempts to fully normalize sediment sorption coefficients by using single sediment characteristics have not been successful (Podoll and Mabey, 1984), indicating the complexity of sediments. The partitioning dynamics also determine the availability of metals to aquatic organisms, since only particular species of metals are available to the biota.

Several studies (e.g. Allen et al., 1980, Cairns et al., 1984) have shown that free ions (e.g. Cu^{2+}) generally are the most toxic forms of heavy metals. In laboratory studies (Green et al., 1993, Donkin and Dusenbery, 1993), the toxicity of copper to benthic organisms decreased when sediment was added. In both studies copper was more toxic in treatments with only water than in treatments with sediment. While pelagic organisms are affected primarily by the free soluble forms of metals, benthic organisms, such as nematodes or oligochaetes, also ingest heavy metals associated with small sediment particles (Callahan et al., 1991, Donkin and Dusenbery, 1993). This additional route of uptake can increase

Water, Air and Soil Pollution **99**: 689-695, 1997.

the dose of the contaminants, because adsorbed contaminants can become available under altered physicochemical conditions inside the digestive system (Tessier and Campbell, 1987). Another pathway for the uptake of contaminants by benthic organisms is direct contact with solid particles (Knezovich et al., 1987). In order to consider all routes of uptake of heavy metals into aquatic organisms, benthic organisms should not be neglected in the assessment of pollutants in aquatic systems.

In this study we examined the influence of particle size distribution and organic matter on the toxic effect of copper on nematodes. A bioassay with the nematode *Caenorhabditis elegans* was performed by using sediments taken at various depths in three lakes of different trophic status, and three artificial sediments, in order to obtain a gradient in the sediment properties investigated. In total, twelve sediments differing in particle size distribution and content of organic matter were used in the bioassay. *C. elegans* was used as testorganism, because this species has been shown to be well suitable for testing in either liquid medium (e.g. van Kessel et al., 1989 , Williams and Dusenbery, 1990, Traunspurger et al., 1997), soil (Donkin and Dusenbery, 1993, 1994) and sediment (Traunspurger et al., 1997). The sublethal endpoint of change in body length was used as an indication of toxicity.

The aims of this study were to investigate (1) the toxicity of Cu to nematodes in experimental systems containing sediments or liquid medium and (2) the influence of sediment particle size distribution and content of organic matter on the toxicity of copper.

2. Materials and methods

2.1 SEDIMENT SAMPLES

Twelve sediments, nine natural and three artificial, were used in the toxicity test. The natural sediments represent gradients for the physico-chemical attributes of interest, particle size distribution and content of organic matter. The sediments were taken from three unpolluted lakes of different trophic status in Germany (South Bavaria), the meso-eutrophic Lake Klostersee, the oligo-mesotrophic Lake Brunnsee, and the mesotrophic Lake Starnberg. The lakes are situated in nature reserves and have only groundwater input. Sediment samples (three replicates from each site) were collected with a gravity corer. In order to obtain defined amounts of sediment from all sites, the top 2 cm of each core were used. The three replicate samples were mixed and then used for analysis and toxicity testing. Samples were taken from depths of 1, 5, 10, and 15 meter at Lake Klostersee (K1, K5, K10 and K15) and Lake Brunnsee (B1, B5, B10 and B15). At Lake Starnberg (STA) samples were taken only from 2 meter water depth. Sediments for the toxicity test were stored at 4 °C and used within one week. Sediments were not treated in any way prior to the test. Samples destined for analysis of organic matter were collected separately, stored at 4 °C and analyzed on the same day. Quartz sand (W4; Quarzwerke-Millisil, Frechen, Germany), quartz gravel (Masa Natursteinhandel GmbH, Hamburg, Germany), and bird sand (Pitti Heimtierprodukte, Willich, Germany) were used as artificial sediments.

TABLE 1

Characteristics of the different sediment types: particle size distribution (intervals in µm), median particle size, organic matter (% of dry weight), and pH, LM=liquid medium, QS=quartz sand, QG=quartz gravel, BS=bird sand, K=Lake Klostersee, B=Lake Brunnsee (figure represent depth in m), STA=Lake Starnberg at 2 m

	particle size distribution: content in %								median particle size	organic matter	pH
	>1046	1046-506	506-263	263-127	127-66.1	66.1-32.0	32.0-15.5	<15.5	µm	% of dw	
LM											7.5
QS	0	0	4	18	29	19	10	20	65	<0.1	8.2
QG	98.0	2.0	0	0	0	0	0	0	>1000	0.4	8.1
BS	67.9	30.4	1.7	0	0	0	0	0	>1000	0.2	8.3
K1	0.1	17.6	29.5	24.1	12.7	10.1	4.0	1,9	245	25.2	7.7
K5	0.3	3.1	10.2	34.2	25.8	16.5	7.1	2.8	147	40.4	8.2
K10	0	4.8	19.8	25.5	18.7	18.8	8.7	3.7	127	26.2	8.1
K15	0	3.3	9.6	12.8	22.9	28.7	13.7	9.0	39.9	19.6	8.0
B1	0.1	7.0	16.1	21.9	20.5	18.2	9.3	6.9	110	71.6	8.0
B5	0.2	2.5	7.6	16.7	27.3	24.9	13.5	7.3	76.5	43.6	8.0
B10	0.2	0.2	1.7	4.8	22.4	40.2	18.1	12.4	46.0	64.8	8.1
B15	0	0.3	1.4	5.2	18.0	30.5	27.9	16.6	63.6	61.0	8.2
STA	0	10.3	28.1	27.5	14.9	9.3	4.9	5.0	197	2.4	7.9

2.2 SEDIMENT ANALYSIS

The properties of the sediments are summarized in Table 1. Particle size distribution was analyzed with a laser analyzer (Malvern SB.09). Organic matter was determined as loss on ignition. Dry weight and loss on ignition were analyzed according to German standard methods (Deutsche Einheitsverfahren 1996).

2.3. PREPARATION OF SPIKED SEDIMENT

Nominal concentrations of copper (0, 2.0, 11.3 and 63.5mg/L) in the test vessels were calculated for the total volume (1 ml) of the test matrix (g wet weight = ml sediment). The test matrix was composed of 0.5 ml sediment, 0.25 ml of a solution of $CuSO_4 \cdot 5\,H_2O$ in K-Medium and 0.25 ml of a suspension of *Escherichia coli* in K-medium (density of bacteria in the final test matrix: 2×10^9 cells/ml). In liquid controls, the 0.5 ml of sediment was replaced by 0.5 ml of K-medium. K-medium was prepared according to Williams and

Dusenbery, (1990) (3.1g/L NaCl, 2.4 g/L KCl). The spiked sediments were thoroughly mixed and incubated in a shaker for 24 h at 20 °C prior to the start of the test.

2.4 TEST ORGANISMS AND TOXICITY TESTS

Toxicity tests were carried out using the nematode *Caenorhabditis elegans* according to Traunspurger et al. (1997). Briefly, at the start of a test ten juvenile worms of the first stage (mean body length ± standard deviation: 270±16 µm) were transferred to each test vessel and incubated at 20 °C. After 72 h the test was stopped by heat-killing of the worms and the body lengths of all worms were determined. Five laboratory replicates per treatment were used. For statistical comparison of the body lengths in the different treatments, one-way ANOVAs (with posthoc Bonferroni t-tests) were run for each level of contaminant (Traunspurger et al., 1997). One-way ANOVA and Pearson linear correlation analysis (Zöfel, 1988) were carried out using SPSS® microcomputer software.

3. Results

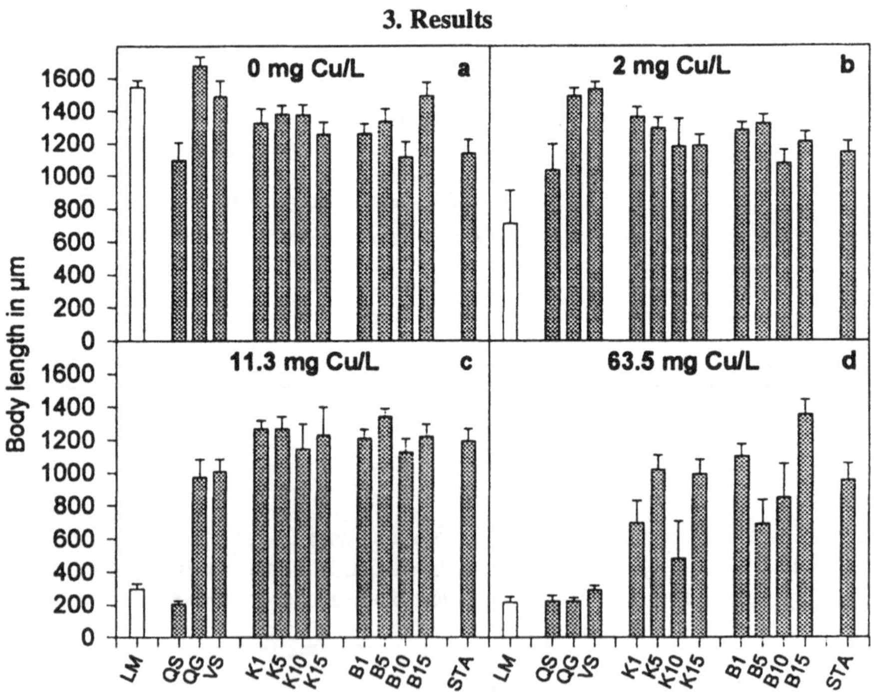

Fig 1: Body length (in µm) of *C. elegans* after 72 h exposure to liquid medium (white bars) and different types of sediments (gray bars) spiked with different concentrations of $CuSO_4$: (a) 0 mg Cu/L, (b) 2 mg Cu/L, (c) 11.3 mg Cu/L, (d) 63.5 mg Cu/L; LM=liquid medium, QS=quartz sand, QK=quartz gravel, BS=bird sand, K=Lake Klostersee sediment, B=Lake Brunnsee sediment (numbers represent the depth in m), STA=Lake Starnberg sediment at 2m; error bars represent standard deviation.

With no copper added, the presence of sediment in the test resulted in reduced growth of the worms, with the exception of quartz gravel (Fig. 1a; one-way ANOVA: $p<0.05$ for bird sand and B15, $p<0.01$ for other sediments). Large variation was observed for artificial sediments. While in quartz gravel and bird sand the worms were longer or only slightly shorter than in liquid medium, a distinct decrease of growth was apparent in quartz sand. Growth of the test organisms was quite similar within natural sediments, with the exception of B10, B15 and STA, being significantly different (one-way ANOVA: $p<0.01$) from the other natural sediments (Fig. 1a). At 2 mg Cu/L the nematodes showed better growth in all sediments, particularly in quartz gravel and bird sand, than in liquid medium (Fig. 1b; one-way ANOVA: $p<0.01$). At 11.3 mg Cu/L the worms in liquid medium and quartz sand showed no or only marginal growth. Also the worms in quartz gravel and bird sand were smaller than the worms in the natural sediments (one-way ANOVA: $p<0.01$; Fig. 1c). At the highest concentration of copper (63.5 mg/L) a distinct difference in body length between artificial and natural sediments occurred. At this concentration also the most pronounced differences in toxicity among the natural sediments were observed. The minimum values occurred at K10, the maximum values at B15 (Fig. 1d).

TABLE II

Pearson correlation coefficients: Body length of *C. elegans* at different copper concentrations (0, 2, 11.3 and 63.5 mg/L) vs. proportion particles in various particle size classes (intervals in μm) and content of organic matter; only gray fields show significant values ($p<0.05$).

Particle size	Body length			
(in μm)	0 mg Cu/L	2 mg Cu/L	11.3 mg Cu/L	63.5 mg Cu/L
<15.5	-0.51	-0.77	-0.49	0.25
15.5-32	-0,22	-0.57	0.15	0.65
32-66.1	-0.49	-0.71	0.14	0.55
66.1-127	-0.67	-0.76	0.08	0.37
127-263	-046	-0.37	0.07	0.23
263-506	-0.33	-0.13	-0.34	-0.20
506-1046	0.19	0.61	-0.12	-0.27
>1046	0.73	0.74	-0.17	-0.60
org. matter	-0.14	-0.24	0.48	0.71

4. Discussion

Our results show that sediments alter the conditions for *C. elegans* in bioassays, in both, presence and absence of a contaminant.

In the controls (0 mg Cu/L) the worms were longer in the treatments where no fine particles were present (liquid medium, quartz gravel, bird sand) than in the other sediments, with the exception of B15. This can be explained by the fact that *C. elegans* ingests unselectively bacteria and other fine particles (Donkin and Dusenbery, 1993, own observations). These fine particles reduce the density of bacteria per volume ingested and thus make food less available. In coarse sediments nematodes will only ingest bacteria, because the particles are too large to be taken up.

We found that all sediments reduced the toxicity of copper to *C. elegans*. Similar results were obtained by Green et al. (1993) for cadmium using the harpacticoid copepod *Amphiascus tenuiremis*. They calculated LC_{50} values in aqueous phase, pore water and whole sediment of 240, 608, and 860 μg/L, respectively. Also in a study by Donkin and Dusenbery (1993) LC_{50} values for copper were elevated by a factor of at least one order of magnitude in different soils as compared to liquid medium, using the nematode, *C. elegans*.

We were interested in the factors responsible for reducing the toxicity of heavy metals in sediments. In this study, we focused on the impact of two sediment properties, particle size distribution and organic matter, on the outcome of the toxicity test. The results show that in the natural sediments the toxicity of copper was reduced more effectively than in the artificial sediments, which is probably due to the higher content of organic matter in the natural sediments. Organic matter is an important factor for controlling the bioavailability of copper (Luoma and Davis, 1983, Malueg et al., 1986). Freely dissolved ions can be adsorbed to organic particles, particle surfaces coated by organic substances (Tessier and Campbell., 1987, Patrick et al., 1977), or to dissolved organic carbon in the interstitial water (Alberts et al., 1984). Bioassays in aquatic media have shown that the toxicity of copper was reduced when dissolved organic carbon was present (e.g. Winner, 1984, Meador, 1991).

In the present study the content of organic matter in the sediments was positively correlated with the body length of *C. elegans* at the highest concentration of copper (63.5 mg/L) (for correlation coefficients see Tab. 2). This is in agreement with spiking experiments where the toxicity of copper to *Daphnia magna* and *Chironomus tentans* was lower in sediment with a higher content of organic matter (Cairns et al., 1984). Also Donkin and Dusenbery (1993), using copper-spiked soils and *C. elegans*, found the lowest toxicity in the soil with the highest content of organic matter.

Interestingly, the correlation between body length and content of fine particles in this study was negative at low concentrations of copper (0 and 2 mg Cu/L; see Tab. 2), while it was positive in sediments with a high concentration of copper (63.5 mg Cu/L; see Tab. 2). We hypothesize that particle size distribution influenced the growth of the nematodes in different ways, depending on the concentration of copper. At 2 mg Cu/L all sediments may have been able to bind enough copper so that no toxic effects occurred. At this concentration the food may have been the limiting factor for growth. At 63.5 mg Cu/L the different binding capacities of the various sediments for copper became obvious. At this concentration copper was probably not completely bound by all sediments, so that the effect of the remaining free copper may have limited the growth of *C. elegans*. Fine sediment has been shown to bind metals more effectively than coarse sediment (Campbell et al., 1988). Heavy metals have been found to be mainly present in the fine particle fraction, essentially consisting of silt and clay minerals, as well as insoluble organic complexes and ferric and manganic oxyhydroxides (Patrick et al., 1977, Förstner, 1990). Therefore we assume that in sediments with finer particles copper was less available to the nematodes. Cairns et al. (1984) compared two types of sediments and found a lower toxicity in the sediment with higher clay content which was explained by the higher binding capacity of the fine sediment. In other studies no correlation was found between particle size distribution and toxicity, which was probably due to the small number of investigated sediments or soils (Wentsel et al., 1977, Donkin and Dusenbery, 1994).

5. Summary

(1) Sediments reduced the toxic effect of copper on the nematode *Caenorhabditis elegans*. (2) This reducing effect on toxicity was stronger in natural than in artificial sediments. (3) Particle size distribution and organic matter are determining factors for the outcome of toxicity tests with sediment.

Acknowledgments

We want to thank Christian Baumer and Patrick Neustedt for the laboratory work and Willem Goedkoop, Sebastian Diehl and Matthias Höss for critically reviewing the manuscript. This study was supported by DFG (Deutsche Forschungsgemeinschaft, Schwerpunkt ROSIG) and UBA (Umweltbundesamt).

References

Alberts, J.J., Giesy, J.P. and Evans, D.W.: 1984, *Environ. Geol. Water Sci.* **6**, 91.
Allen, H.E., Hall, R.H. and Brisbin, T.D.: 1980, *Environ. Sci. Technol.* **14**, 441.
Burton, Jr., G.A.: 1991, Environ. Toxicol. Chem. **10**, 1585.
Cairns, M.A., Nebeker, A.V., Gakstatter, J.N. and Griffis, W.L.: 1984, *Environ. Toxicol. Chem.* **3**, 435.
Callahan, C.A., Menzie, C.A., Burmaster, D.E., Wilborn, D.C., and Ernst, T.: 1991, *Environ, Contam. Toxicol.* **10**, 817.
Campbell, J.A., Towner, J.V., and Vallurupalli, R.: 1988, in J.J. Lichtenberg, J.A. Winter, C.I. Weber,, and L. Fradkin (eds.), *Chemical Characterisation of Sludges, Sediments, Dredge Spoils, and Drilling Muds, ASTM STP 976*, American Society for Testing and Materials, Philadelphia, pp. 93-101.
Deutsche Einheitsverfahren zur Wasser, Abwasser- und Schlammuntersuchung: 1996.
Donkin, S.G. and Dusenbery, D.B.: 1993, *Arch. Environ, Contam. Toxicol.* **25**, 145.
Donkin, S.G. and Dusenbery, D.B.: 1994, *Water, Air and Soil Pollut.* **78**, 359.
Förstner, U.: 1990, in R. Baudo, J. Giesy and H. Muntau (eds.), *Chemistry and Toxicity of In-Place Pollutants*. Lewis publisher, Boca Raton, FL, pp. 61-106.
Green, A.S., Chandler, G.T. and Blood, E.R.: 1993, *Environ. Toxicol. Chem.* **12**, 1497.
Knezovich, J.P., Harrison, F.L. and Wilhelm, R.G.: 1987, *Water, Air and Soil Pollut.* **32**, 233.
Lee, G.F. and Jones, R.A.: 1984, in K.L. Dickson, A.W. Maki and W.A. Brungs (eds.), *Fate and Effects of Sediment-Bound Chemicals in Aquatic Systems*, Pergamon Press, Elmsford, NY, pp. 1-34.
Luoma, S.N. and Davis, J.A.: 1983, *Mar. Chem.* **12**, 159.
Malueg, K.W., Schuytema, G.S. and Krawczyk, D.F.: 1986, *Environ. Toxicol. Chem.* **5**, 245.
Meador, J.P.: 1991, *Aquat. Toxicol.* **19**, 13.
Patrick, Jr., W.H., Gambrell, R.P. and Khalid, R.A.: 1977, *J. Environ. Sci. Health*, **A12** (9), 475.
Podoll, R.T. and Mabey, W.R.: 1984, in K.L. Dickson, A.W. Maki and W.A. Brungs (eds.), *Fate and Effects of Sediment-Bound Chemicals in Aquatic Systems*, Pergamon Press, Elmsford, NY, pp. 83-98.
Tessier, A. & Campbell, P.G.C.: 1987, *Hydrobiologia* **149**, 43.
Traunspurger, W., Haitzer, M., Höss, S., Beier, S. and Ahlf, W. and Steinberg, C.: 1997,*Environ. Toxicol. Chem.* **16** (2).
van Kessel, W.H.M., Brocades Zaalberg, R.W., and Seinen, W.: 1989, *Ecotoxicol. Environ. Safety* **18**, 181.
Wentsel, R., McIntosh, A. and Atchison, G.: 1977, *Hydrobiologia* **56** (2), 153.
Williams, P.L. and Dusenbery: 1990, *Environ. Toxicol. Chem.* **9**, 1285.
Winner, R.W.: 1984, *Aquat. Toxicol.* **5**, 267.
Zöfel, P.: 1988, Statistik in der Praxis (ed.), Gustav Fischer Verlag, Stuttgart, Germany.

HEAVY METAL DISTRIBUTION IN SEDIMENT CORES FROM WESTERN ROSS SEA (ANTARCTICA)

M. RAVANELLI[1], O. TUBERTINI[1], S VALCHER[1], W. MARTINOTTI[2]
[1] *Centro di Radiochimica Ambientale, Università di Bologna, Italy*
[2] *ENEL - CRAM S.p.A, Milano, Italy*

Abstract: Seven sediments cores collected in the Western Ross Sea (Antarctica) were investigated for heavy metals (Cr, Ni, Pb, Cd, Co, Zn Cu and Fe, Mn), several major elements (Ca, Ba and Al), total and organic carbon, and isotopic composition in order to investigate vertical variability. Samples were analyzed by ET-AAS and ICP-AES for metals and by CF-IRMS for carbon content. Generally, concentrations at the background levels were higher with respect to recent ones excluding in this area any kind of anthropogenic contamination. Cluster and Factor Analysis were applied. Best correlations were obtained for the elements showing geochemical affinities.

1. Introduction

The Antarctic marine ecosystem is characterized by absence of river inputs, low humidities and cold temperature. Precipitation falls mainly as snow and only some coastal areas receive rain in summer. Because of the remoteness from human impacted areas, Antarctica can be studied in order to establish a "reference state" for monitoring global change with respect to worldwide anthropogenic contamination. The Antarctic Treaty Protocol aims at differentiating between natural changes and changes caused by human activities in or outside of Antarctica. Basic environmental research can be useful to collect information about background environmental quality criteria and its dependence on abiotic and biotic processes and, further, its temporal variation over a defined temporal scale (at least 100 years). The major concern has been to obtain quantitative information on biogeochemical cycles of carbon and silica (DeMaster *et al.*, 1991) in order to estimate fluxes and primary production budgets. Less attention has been dedicated so far to investigate metal distribution and variability.

Heavy metals such as Pb, Cd and Zn are good markers of contamination from human activity and can be detected by different techniques. Recent studies have reported data on heavy metal concentrations in different components of the Antarctic environment such as snow-ice (Wolff, 1992), soil (Claridge *et al.*, 1995) and sediments (Hieke Merlin, 1989; Cosma *et al.*, 1991; Caramella Crespi *et al.*, 1993; Cosma *et al.*, 1994; Abollino *et al.*, 1996). Sediment sampling has been limited generally to 0 - 10 cm depth and no studies of heavy metal contamination have been reported.

Marine sediments are the ultimate sink for most heavy metals released into the aquatic environment and can record information on long and medium-term metal fluxes. Hence, vertical distribution profies of heavy metals in sediments cores reflect the geochemical history of a given region, including changes due to anthropogenic impact (Szefer and Skwarzec, 1988) assuming minimal post-depositional movement of metals. Generally, heavy metals become associated with organic matter through adsorption by suspended particles sinking to the sea bed and biological uptake.

To support the correlation between bioavailable metals and organic matter, total and organic carbon content in sediments are also determined. Furthermore, stable isotopes of carbon can be very useful for tracing and quantifying the amounts of anthropogenic and natural terrigenous organic matter in the marine environment, providing the determination

Water, Air and Soil Pollution **99**: 697-704, 1997.

of sources (Gearing, 1988). In fact organic matter has a $^{13}C/^{12}C$ ratio characteristic of its origin and little changes in this value in terms of $\delta^{13}C$ notation are significant of a variation of sediment origin. Since sediments provide information on the integrated result of organic matter deposition in the course of the time, the stable isotope ratio represents the best way for determining the origin of sedimentary material. The objective of this investigation was to determine concentration and distribution of bioavalaible fraction of some metals in a series of sediment cores from the Ross Sea in order to establish a "reference state" and to monitor environmental changes.

2. Study area

The Ross Sea continental shelf is unusually wide and deep and consists of a series of highs and troughs trending north-easterly in the west and more northerly in the east. Detailed morphology of the area and deposition settings have been widely studied by Anderson *et al.* (1984) and Ledford-Hoffman *et al.* (1986). Sediment transport in the Ross Sea is primarily as large particles sinking vertically and most of the material moving vertically is biogenic in origin (Dunbar and Leventer, 1994).

Recent Antarctic continental shelf sediments can be divided in two major classes according to their origin: reworked lithogenic glacial sediments and biogenic sediments. The biogenic component, which is generally more abundant than the glacial one, is mainly composed of biogenic silica. The biogenic silica is a consequence of the local high productivity by diatoms population, whereas carbonated residues are rare and due to preservation of benthic fauna skeletons (Brambati *et al.*, 1989). A classification based on multiparameters Q-model Factor Analysis of the composition and behavior of Ross Sea sediments has been proposed by Frignani *et al.* (1992). According to these authors, sediments studied in this work are more than 80 % characterized by biogenic and coarse sediments. Lithology and biogenic silica content of these sediments have been previously reported by Labbrozzi (1991).

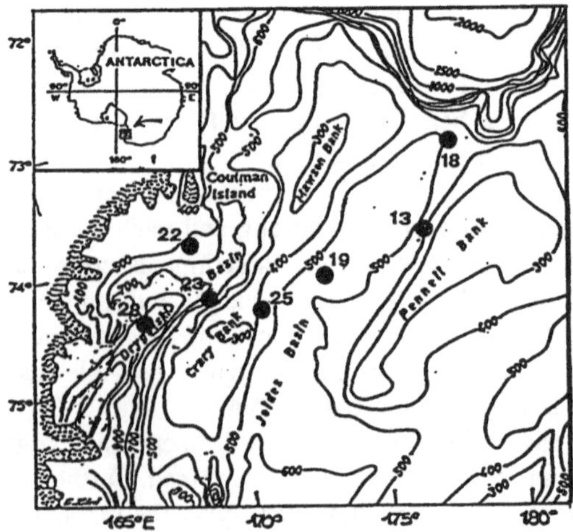

Fig. 1 Locations of sediment sampling stations in Western Ross Sea

3. Materials and methods

Sediment samples were collected in the North-West Ross Sea during the Italian Oceanographic Expedition in 1990-91. Locations of the seven cores analyzed are shown in Figure 1. Samples, collected by box corer, were sub-sampled with polyethylene tubes (diameter (ϕ) 54 mm; l 400 mm). Sub-cores were stored at 4°C and then sectioned in different thickness from 0.5 to 1 cm. Sediment samples were dried at 105°C for 12 hours and finely ground. Core locations and water depths are listed in Table I.

TABLE I
Sample locations, water depth and length of sediment cores

Station	Latitude S	Longitude E	Water depth (m)	Core length (cm)
13 BC	74° 3.72'	176° 48.78'	403	32
18 BC	73° 25.93'	177° 43.57'	449	30
19 BC	74° 25. 94'	173° 47. 99'	556	26
22 BC	73° 59.70'	169° 1.72'	528	32
23 BC	74° 25.69'	169° 35.38'	678	24
25 BC	74° 38. 61'	171° 32. 29'	462	27
28 BC	74° 38.91'	167° 14.70'	981	30

3.1 METAL EXTRACTION

The leachable metal fraction is usually defined as that including exchangeable, carbonate-bond, iron and manganese oxide-bound and organically-bound fractions. The leachable fraction has often been extracted by dilute acid (Malo, 1977; Salomons and Forstner, 1980; Hamilton-Taylor *et al.*, 1984; Qucirazza and Guzzi, 1987; Angelidis and Aloupi, 1995). The procedure adopted in this work involved a weak cold leaching. An aliquot of 500 mg of dry sediment was treated with an excess of 160 ml of 0.3 N HCl in conditioned polyethylene vessels. After stirring with a horizontal shaker for two hours at room temperature, the suspension was filtered on Nuclepore 0.4 μm polycarbonate filters. Filters were previously treated with 50 ml of 6 M HCl, 100 ml of ultrapure water, 10 ml of 1 M CH_3COONH_4 and 100 ml of ultrapure-Q water in sequence. Extracts were analyzed by a ET-AAS (Varian AA 1475) for Pb, Cd, Co, Ni and Cr and by a (ICP-AES Varian Liberty 200) for Zn, Fe, Mn, Cu, Ca, Ba and Al. Concentrations are expressed as μmol/g dry sediment.

The quality of the instrumental results was checked by the analysis of a reference material (BCSS-1, marine sediment), which was totally digested by aqua regia/HF/$HClO_4$ solution. Precision of the analytical procedure was tested by analyzing two samples in triplicate from the top and the bottom of each core. Uncertainty is expressed as a percentage calculated on the basis of the maximum standard deviation. The following coefficients of variation (%) were obtained: 9.2 (Pb), 5.3 (Cd), 5.1(Co), 12.4 (Ni), 2.6 (Cr) for ET-AAS and 13.5 (Zn), 7.6 (Fe), 8.4 (Mn), , 4.8 (Cu), 5.5 (Ca), 3.6 (Ba), and 6.3 (Al) for ICP-AES. Detection limits (μmol/g) calculated as 3 times the σ value calculated from 10 blank determinations were 6.7×10^{-3} (Pb), 7.1×10^{-4} (Cd), 5.1×10^{-3} (Co), 1.7×10^{-2} (Ni), 1.7×10^{-2} (Cr), 9.2×10^{-2} (Zn), 3.6×10^{-2} (Fe), 5.5×10^{-2} (Mn), 1.4×10^{-1} (Cu), 1.5×10^{-1} (Ca), 2.2×10^{-3} (Ba) and 1.1×10^{-1} (Al).

3.2 DETERMINATION OF TOTAL AND ORGANIC CARBON AND $\delta^{13}C$

Total and organic carbon content and $\delta^{13}C$ were determined by a fully automated CF-IRMS (ANCA-MS, Europa Scientific, UK). Sample preparation and instrumental optimization were performed according to Nieuwenhuize *et al.* (1994) and Gioacchini *et al.* (1995), respectively. Uncertainty was 0.01% for both organic and total carbon and 0.01 ‰ for $\delta^{13}C$ determinations.

4. Results and Discussion

In table II minimum, maximum and average values of metal concentrations together with standard deviation as a percentage, as well as total and organic carbon content and isotopic composition are reported for each core. Table II also reports the mean values averaged over all the sediments.

Results show that the variability within the vertical profile is high: relative standard deviations can span from several units to more than 100 %. In some cases Cd is below the detection limits: in core 13 and 18 detectable concentrations were measured only in the deepest sections. The same trend is observed for Pb in core 13. For core 23, Cd is undetectable with the only exception being the surficial (0-1 cm) section. This is due to its high tendency to be remobilized and released from sediments through oxidation of organic matter that is incompletely regenerated in the water column (Niagru and Sprague, 1987).

Average Cr concentrations span from 0.026 to 0.11 μmol/g and the average total content is about 0.05 μmol/g. Core 13 shows the highest Cr content, followed by cores 18 and 23 respectively. Cores 19, 22, 25 and 28 have homogeneous Cr levels. Ni, Pb and Co concentrations are comparable in all the cores whereas Zn content in core 25 is higher than in the other cores. Also Ca and Fe contents are similar except in cores 13 and 18 which are one order of magnitude higher. This points out that geochemical composition of these cores is different from the others. In fact both cores 13 and 18 lying on the slope of Joides Basins and will be subjected to the same processes of deposition, erosion and resuspension.

Total carbon content is usually low (no more than 1.1 %) and organic fraction prevails as reported by Frignani *et al.* (1992). Organic $\delta^{13}C$ values (- 24.73 ‰ - -27.57 ‰), characteristic of phytoplankton, confirm the biogenic origin of Antarctic sediments. Cores 13 and 18 show a different feature if compared to the other ones, showing higher total $\delta^{13}C$ suggesting an important input of terrigenous materials in recent times.

Metal profiles do not show any particular trend and at depth below 13-15 cm concentrations are on average higher than in the upper layers. Only Cr and Ni show similar profiles within each core. The higher concentration values in the uppermost levels can be related to early diagenetic processes and to sediment mixing. The lower concentrations can be caused by a modification of bottom current transport which might have implemented the erosion. Only in the case of cores 13 and 18 can recent variations in metal content be related to a different supply of terrigenous material as confirmed by higher values of total $\delta^{13}C$. For other cores, the explanation for metal variability is not so clear, depending on the mutual interactions of mixing, erosion and resuspension processes.

TABLE II

Maxima, minima, average and relative standard deviation (%) of parameters measured for the seven cores analyzed and mean values. Metal concentrations are expressed in μmol/g; total and organic carbon in % and $\delta^{13}C$ in ‰, (PDB).

13	Cr	Ni	Pb	Cd	Co	Zn	Fe	Mn	Cu	Ca	Ba	Al	C tot	δ13C	C org	δ13C
max	0.37	0.480	0.025	0.007	0.053	1.234	107.32	1.449	0.242	624	0.586	132	0.92	-6.55	0.36	-25.14
min	0.02	0.015	0.000	5.1E-3	0.003	0.087	14.20	0.183	0.044	211	0.329	75	0.41	-16.10	0.11	-29.76
average (n = 36)	0.11	0.153	0.013	0.001	0.029	0.337	48.64	0.938	0.127	357	0.455	114	0.66	-13.21	0.26	-26.84
s.d. %	72.1	71.6	74.2	152.7	48.4	67.4	49.6	33.6	37.3	21.2	12.7	12.9	22.8	-15.8	30.6	-4.2
18	**Cr**	**Ni**	**Pb**	**Cd**	**Co**	**Zn**	**Fe**	**Mn**	**Cu**	**Ca**	**Ba**	**Al**	**C tot**	**δ13C**	**C org**	**δ13C**
max	0.10	0.20	0.08	0.02	0.170	5.04	128.29	1.71	0.30	424	0.903	130	1.26	-4.49	0.32	-25.08
min	0.05	0.01	0.01	5.1E-3	0.002	0.29	36.60	0.57	0.06	145	0.352	104	0.46	-19.21	0.02	-29.08
average (n = 43)	0.07	0.11	0.04	0.00	0.037	0.703	81.38	1.14	0.18	209	0.446	121	0.68	-10.45	0.18	-26.97
s.d. %	23.5	56.7	63.0	233.4	126.6	159.9	34.6	33.0	35.0	33.3	27.5	6.2	21.3	-43.4	39.1	-4.3
19	**Cr**	**Ni**	**Pb**	**Cd**	**Co**	**Zn**	**Fe**	**Mn**	**Cu**	**Ca**	**Ba**	**Al**	**C tot**	**δ13C**	**C org**	**δ13C**
max	0.040	1.828	0.017	0.003	0.023	0.418	36.88	0.53	0.20	52.6	1.988	75	1.54	-26.83	0.98	-26.84
min	0.013	0.764	0.011	5.1E-3	0.010	0.168	14.63	0.37	0.16	41.6	1.311	65	0.12	-29.51	0.62	-28.15
average (n = 17)	0.026	1.222	0.014	0.002	0.014	0.308	19.27	0.44	0.17	47.7	1.634	69	1.12	-28.21	0.77	-27.57
s.d. %	24.6	24.7	11.2	64.6	22.9	24.0	34.6	11.4	9.5	5.4	11.9	4.3	26.7	-2.7	14.5	-1.8
22	**Cr**	**Ni**	**Pb**	**Cd**	**Co**	**Zn**	**Fe**	**Mn**	**Cu**	**Ca**	**Ba**	**Al**	**C tot**	**δ13C**	**C org**	**δ13C**
max	0.035	0.109	0.025	0.007	0.025	0.384	44.73	1.09	0.09	60.6	1.490	159	0.71	-20.41	0.60	-22.62
min	0.019	0.073	0.015	5.1E-3	0.017	0.273	26.08	0.68	0.05	46.6	0.482	105	0.19	-26.44	0.18	-33.80
average (n = 23)	0.026	0.084	0.017	0.002	0.021	0.335	34.42	0.92	0.07	52.7	0.960	124	0.53	-25.04	0.43	-25.48
s.d. %	16.4	11.0	16.6	85.3	10.8	11.1	15.4	13.3	17.1	6.4	36.9	11.5	31.9	-5.2	30.2	-9.3
23	**Cr**	**Ni**	**Pb**	**Cd**	**Co**	**Zn**	**Fe**	**Mn**	**Cu**	**Ca**	**Ba**	**Al**	**C tot**	**δ13C**	**C org**	**δ13C**
max	0.118	0.079	0.002	5.1E-3	0.026	0.272	76.79	1.353	0.112	52.5	0.430	88	0.35	-21.70	0.28	-22.94
min	0.018	0.030	0.002	5.1E-3	0.004	0.116	41.24	0.084	0.039	39.0	0.308	59	0.17	-26.56	0.13	-27.26
average (n = 31)	0.059	0.054	0.002	5.1E-3	0.015	0.189	57.67	1.193	0.055	45.8	0.362	73	0.27	-24.41	0.19	-25.22
s.d. %	52.4	28.2	0.0	0.0	36.9	25.0	17.1	26.0	33.3	8.0	9.7	13.0	21.6	-4.4	15.7	-4.0
25	**Cr**	**Ni**	**Pb**	**Cd**	**Co**	**Zn**	**Fe**	**Mn**	**Cu**	**Ca**	**Ba**	**Al**	**C tot**	**δ13C**	**C org**	**δ13C**
max	0.030	0.104	0.015	0.002	0.011	20.90	25.10	0.45	0.13	45.5	1.500	71	1.09	-19.17	0.84	-25.25
min	0.019	0.058	0.008	5.1E-3	0.006	9.42	11.00	0.23	0.04	34.5	0.573	54	0.83	-27.72	0.49	-28.65
average (n = 19)	0.025	0.083	0.010	0.001	0.008	14.96	13.96	0.30	0.10	39.6	1.161	64	0.93	-25.39	0.69	-26.93
s.d. %	13.6	17.5	18.0	82.7	19.2	20.9	30.7	15.5	24.0	7.2	19.8	6.8	8.4	-8.2	15.2	-3.0
28	**Cr**	**Ni**	**Pb**	**Cd**	**Co**	**Zn**	**Fe**	**Mn**	**Cu**	**Ca**	**Ba**	**Al**	**C tot**	**δ13C**	**C org**	**δ13C**
max	0.042	0.136	0.025	0.005	0.029	0.565	34.88	0.556	0.118	67.8	1.566	130	1.21	-22.84	0.91	-23.61
min	0.023	0.071	0.015	0.001	0.019	0.325	24.56	0.462	0.081	58.7	0.911	109	1.02	-25.86	0.16	-27.81
average (n = 25)	0.031	0.103	0.018	0.003	0.024	0.410	28.70	0.515	0.103	62.2	1.276	122	1.12	-24.32	0.79	-24.73
s.d. %	17.9	16.8	18.9	58.3	13.6	16.0	11.3	5.0	8.2	3.6	13.7	4.3	4.3	-3.6	17.6	-3.3
Mean	0.049	0.258	0.016	0.0014	0.021	2.46	40.6	0.779	0.115	116.4	0.899	98.0	0.760	-21.58	0.471	-26.25

To further support the absence of sediment contamination, metal data were assembled in surficial (0-1 cm section) and natural level (background) (Table III). Background values correspond to sections below the base of excess ^{210}Pb profiles (Cantelli *et al.*, 1995). In fact, since the activity of ^{210}Pb into the sediment can be related only to recent sedimentation and mixing processes, the depth of sediment that do not show any excess ^{210}Pb can be defined as background level. On the basis of accumulation rates of 0.5-1.5 10^{-2} cm/y calculated from ^{14}C for Ross Sea (De Master,1990; in press), only the upper one-two centimeters should be indicative of anthropogenic inputs.

For Pb, Cr, Ni, Cd, Co, and Zn, surficial values are the same or lower than deep ones with the exception of core 23 where Pb, Cu and Zn show an opposite trend. The prevailing tendency is also confirmed by the pattern of the major elements (Ca, Ba, Fe, Mn, Al). These results support the hypothesis that there is no contamination of Antarctic marine sediments due to long range transport of pollutants from human activities.

TABLE III

Surficial and background metal content. Concentrations are expressed in μmol/g.

		Cr	Ni	Pb	Cd	Co	Zn	Fe	Mn	Cu	Ca	Ba	Al
13	Surficial	0.031	0.055	0.000	5.1E-3	0.018	0.228	14.2	1.36	0.063	239	0.586	110
	Backgr. (n = 29)	0.160	0.165	0.011	0.001	0.027	0.245	51.5	0.68	0.122	506	0.476	123
18	Surficial	0.038	0.095	0.014	5.1E-3	0.029	0.105	32.0	2.38	0.069	575	0.706	112
	Backgr. (n = 36)	0.071	0.287	0.045	4.4E-05	0.039	0.460	74.0	4.43	0.203	330	0.758	123
19	Surficial	0.022	0.764	0.015	5.1E-3	0.023	0.203	27.2	0.50	0.164	47.7	1.988	73.5
	Backgr. (n = 5)	0.030	1.388	0.013	0.002	0.011	0.363	15.0	0.39	0.167	47.4	1.529	67.2
22	Surficial	0.038	0.094	0.017	0.0007	0.023	0.376	36.5	1.11	0.097	54.7	1.50	120
	Backgr. (n = 18)	0.031	0.096	0.021	0.0012	0.021	0.336	35.0	0.93	0.070	52.8	0.99	124
23	Surficial	0.016	0.013	0.020	5.1E-3	0.007	0.287	34.0	0.97	0.196	50.0	0.601	78.0
	Backgr. (n = 20)	0.058	0.053	0.006	0.0002	0.016	0.188	56.6	1.17	0.055	45.6	0.365	73.2
25	Surficial	0.016	0.065	0.013	5.1E-3	0.015	16.9	25.6	0.92	0.104	41.3	1.433	70.3
	Backgr. (n = 13)	4.23	15.7	7.29	0.349	1.552	15.3	12.3	0.30	0.095	39.6	1.104	62.8
28	Surficial	0.020	0.104	0.017	0.0004	0.015	0.325	32.0	0.55	0.106	62.2	1.340	123
	Backgr. (n = 12)	0.032	0.104	0.019	0.0033	0.024	0.429	27.8	0.50	0.105	59.6	1.211	122

4.1 CLUSTER AND FACTOR ANALYSIS

To carry out a quantitative investigation on biogeochemical variations of metals in Antarctic sediments results were analyzed statistically by means of cluster and factor analysis. Factor and cluster analysis were applied to metal and carbon data sets. Data exceeding mean values ± 3σ were replaced by values interpolated from the near figure pair. The main results of those analyses are shown in Table IV. The variability of metal content in sediments is explained by at least four factors, corresponding to a total variance between 80 and 85%.

Four cores show a good correspondence between Fe and Mn distributions: in particular this is evident for core 13 (euclidean distance (e.d.) = 1.5 and Fe/Mn ratio equal to 56.8) and for cores 19, 23 and 25 (e.d. = 1.9, 2.2 and 2.5 respectively and Fe/Mn ratios equal to 43.6, 42.8 and 46.4 respectively). This is in agreement with chemical and geochemical affinity. Core 22 shows a good correlation between organic and total carbon content, whereas the correspondence between organic carbon content and $\delta^{13}C$ is not significant.

In Core 18 and, less clearly, also in others, a significative correlation is observed between the Mn and Fe content (with a major evidence for Pb and Co); in fact two distinct linear trends are observed for deeper sections and for more recent ones (Fig.2). This can be related to a temporary definite variation of fluxes due to a change of relative importance of sources or to the current action. Analogy in distribution of Fe and Mn for cores 13,19,23 and 25 was not confirmed to further clusters but this fact requires further investigation.

Table IV
Cluster and Factor analysis results. Clusters in brackets have a low significance.

Core	Cluster	Euclidean Distance	Number of Factors and total variance	Max. Eigenvalue and variance
13	Fe-Mn	1.5	4 80%	4.9 41%
	Corg-δ^{13}Corg	4		
18	Ca-Ba	1.6	4 81%	5.4 41%
	Ni-Mn	2.7		
	(Fe-Mn)	(5.2)		
	Corg-δ^{13}Corg	4.5		
19	Fe-Mn-Co	1.9	4 85%	6.3 45%
	Ni-Zn-Cd-Cr	2.3-2.4-2.6		
22	Ca-Al-Fe	2.3	3 81%	6.0 43%
	(Fe-Mn)	(5.5)		
	Corg-δ^{13}Corg	1.1		
23	Zn-Cu	1.9	3 83%	6.4 53%
	Cr-Ni	2.0		
	Fe-Mn	2.2		
	Ctot- δ^{13}Ctot	5.4		
25	Fe-Co-Mn	2.3 -2.5	4 83%	4.8 37%
	Cd-Zn	2.9		
	Corg-δ^{13}Corg	3.5		
28	Cd-Co	3.3	4 81%	3.9 30%
	Zn-Ca	3.5		
	(Fe-Al-Mn)	(4.2 -4.3)		
	Ctot-δ^{13}Ctot	3.6		

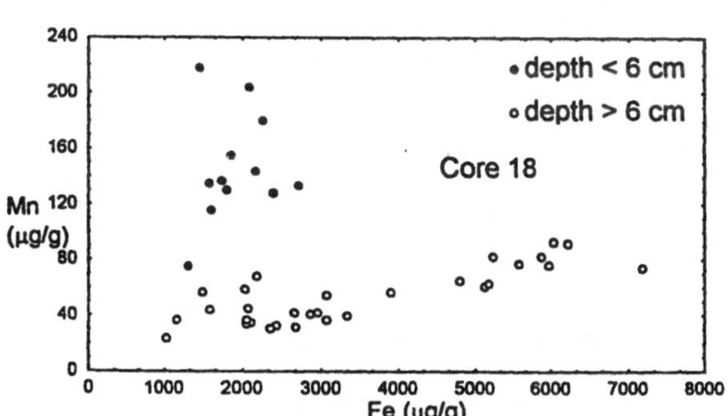

Fig. 2 Concentration of Mn as a function of Fe content in core 18. Full dots represent more recent data.

5. Conclusions

This study outlines the variability in bioavailable metal content in Ross Sea sediments. Moreover it points out the difficulty in establishing a reference state. Nevertheless, detailed information obtained on Antarctic sediments may be used as a baseline to monitor possible future environmental change as well as provide a preliminary data set to study biogeochemical processes.

According to the results of the distribution of several metals (Cr, Ni, Pb, Cd, Co, Zn, Cu, Fe, Mn, Ca, Ba and Al) as well as total, organic carbon content and carbon isotope distribution in seven Antarctic sediment cores, we can draw the following conclusions:

1. Most of the considered cores show metal profiles increasing with depth. Generally concentrations begin to increase at about 13 - 15 cm depth.
2. Carbon content is typically low and organic carbon is dominant. Organic $\delta^{13}C$ confirm the biogenic origin of sediments and total $\delta^{13}C$ show that only two cores (13 and 18) have been affected by an important supply of terrigenous material, probably due to glacial transport.
3. Results of inter-metal correlations, obtained by Cluster analysis, are location dependent. No significant correlations among anthropogenic metals were found.
4. The comparison between surficial and background levels shows that no contamination of any kind is present.

Acknowledgments

Authors acknowledge Mr. G. Righetti (CRA Montecatini, Ravenna) for instrumental ICP-AES support and Dr. A.M. Gioacchini (Dip. di Scienze Farmaceutiche, Università di Bologna) for CF-IRMS measurements. A special thanks to Dr. M. Frignani, Dr. L. Tositti and Dr. L. Cantelli. Financial support for this research was provided by the Italian "Progetto Nazionale di Ricerche in Antartide".

References

Abollino, O., Aceto, M., Sacchero, G., Sarzanini, C., Mentasti, E.: 1996, *Ann. Chim.*, **86**, 229-243.
Anderson, J.B. Brake, C.F. Myers, N.C.: 1984, *Mar. Geol.*, **57**, 295-333.
Angelidis, M.O. Aloupi, M.: 1995, Mar. Poll. Bull., **31**, 273-276.
Brambati, A., Fanzutti, G.P., Finocchiaro, F., Simeoni, U.:1989, *Boll. Ocean. Teor. Appl.* **VII** (1-2), 159-188.
Cantelli, L., Ravanelli, M., Tubertini, O., Valcher, S.: 1995, *Proceedings of 4°Convegno Nazionale Contaminazione Ambientale del Programma Nazionale di Ricerche in Antartide* - Venezia, 6-7 Dicembre 1995
Caramella Crespi, V., Genova, N., Tositti, L., Tubertini, O., Bettoli, G., Oddone, M., Meloni, S., Berzero, A.: 1993, J. *Radioanal. Nucl. Ch. Ar.* **168** (1), 107-114.
Claridge, G.G.C. Campbell, I.B. Powell, H.K.J. Amin, Z.H. Balks, M.R.: 1995, *Antarctic Science*, **7** (1), 9-14.
Cosma, B., Frache, R., Mazzuccotelli, A., Soggia, F.: 1991, *Ann. Chim.*, **81**, 371-382.
Cosma, B. Soggia, F., Abelmoschi, M.L., Frache, R.: 1994, *Int. J. Environ. Anal. Chem.* **55**, 121-128.
DeMaster, D.J., Harden, S., Pope, R.: 1990, *Antarctic Journal*, **XXV** (5), 104-105.
DeMaster, D.J, Nelson, T. M., Harden, S. L., Nittrouer, C.A.: 1991, *Mar. Chem*, **35**, 489-502.
DeMaster, D.J., Ragueneau, O., Nittrouer, C.A.: *J. Geoph. Res.* (in press)
Dunbar, R.B. Leventer A.R.: 1994, *EOS. Trans. Amer. Geophys.* Un., **75** (3), 139.
Frignani, M., Labbrozzi, L., Langone, L., Ravaioli, M.,: Proceedings of 10° Congresso A.I.O.L. 4-6 Novembre 1996, Alassio (Italy), 699-707.
Gearing, J.N.: 1988, *Lecture Notes on Coastal and Estuarine Studies*, **22**, 69-101
Gioacchini, A.M., Roda, A., Parenti M., Cipolla, A., Baraldini, M.: 1995, *Rap. Commun. Mass Sp.*, **9**, 1106-1108.
Hamilton-Taylor, J. Willis, M Reynolds C.S.: 1984, *Limnol. Oceanogr.*, **29** (4), 695-710.
Hieke Merlin, O., Longo Salvador, G., Menegazzo Vitturi, L., Pistolato, M., Rampazzo, G.: 1989, *Boll. Ocean. Teor. Appl.*, **VII** (1-2), 97-108.
Ledford-Hoffman, P.A. DeMaster, D.J. Nittrouer, C.A.: 1986, *Geochim. Cosmochim. Acta*, **50**, 2099-2110.
Labbrozzi, L.: 1991, in *"Sedimentazione attuale e recente in Antartide, con particolare riguardo alla componente biogena:parte Nord-Occidentale del Mare di Ross"* Thesis, University of Bologna
Malo B.A.: 1977, *Environ. Sci. Techn.*, **11**, 277-282.
Nieuwenhuize, J., Maas, Y.E.M., Middelburg, J.J.: 1994, *Mar. Chem.*, **45**, 217-244.
Nriagu, J.O., Sprague, J.B.: 1987, *Cadmium in the aquatic environment*, J.Wiley & Sons Ed., pp.13
Queirazza, G. Guzzi, L.: 1987, *Sci. Tot. Environ.*, **64**, 191-209.
Salomons, W. Förstner U.: 1980, *Environ. Technol. Lett.*, **1**, 506-518.
Szefer, P., Skwaezec, P.: 1988, *Mar. Chem.*, **23**, 109-129
Wolff, E.: 1992, *Mar. Poll. Bull.*, **25** (9-12), 274-280

SCAVENGING PROCESSES AND EXPORT FLUXES CLOSE TO A RETREATING SEASONAL ICE MARGIN (ROSS SEA, ANTARCTICA)

L. LANGONE[1], M. FRIGNANI[1], J.K. COCHRAN[2], M. RAVAIOLI[1]

[1] *Istituto di Geologia Marina CNR., via Gobetti 101, 40127 Bologna, Italy.* [2] *Marine Sciences Research Center, SUNY, Stony Brook, NY, 11794-5000, USA.*

Abstract. The distribution of dissolved and particulate ^{234}Th in the upper 200 m of the water column was obtained for three stations in the Ross Sea off Victoria Land and Terra Nova Bay. At site 24a, close to the retreating ice margin, all the sampled depths showed deficiencies in ^{234}Th relative to the equilibrium with ^{238}U. These are related to uptake of ^{234}Th onto sinking particles. Residence times of ^{234}Th in solution and of particulate ^{234}Th were 130-247 days and 8.1-6.6 days, respectively. A high particle flux (1.23-5.03 g m^{-2} d^{-1}) was calculated at this station. At the other two sites (11c and 15c), ^{234}Th depth profiles are irregular, probably due to the release of dissolved ^{234}Th by decomposing particles at certain depths, or to the contribution from lateral advection. Bulk mass fluxes measured by floating traps at stations 11c and 15c are very low (66-138 mg m^{-2} d^{-1}). Also fluxes of organic carbon and nitrogen, and biogenic silica are reported for these two sites. The calculated fluxes are discussed with respect to methodologies and to the dynamics of the ice margin retreat.

Keywords: ^{234}Th, particle flux, Ross Sea, Antarctica, organic C, N, Si, scavenging rates

1. Introduction

The Southern Ocean in general, and the Ross Sea in particular, appear to be areas where great amounts of biogenic silica are accumulating in modern sediments (DeMaster, 1981; Ledford-Hoffman *et al.*, 1986). According to DeMaster (1981), over one half of the silica (about 2.5 10^{14} g SiO_2 yr^{-1}) supplied annually to the ocean by rivers and hydrothermal emanations is removed in the Southern Ocean. Furthermore, Nelson and Smith (1986) estimated that 50% or more of the surface-produced silica in the western Ross Sea is deposited on the bottom at depths greater than 500-600 m. The rates of silica accumulation in the western Ross Sea may be among the highest in the ocean (up to 7 mol Si m^{-2} yr^{-1}; Ledford-Hoffman *et al.*, 1986). Previous biological and geochemical studies, including time-series data from sediment traps (DeMaster *et al.*, 1992), have shown that organic C fluxes through the mid-water column are <5% of the export from the photic zone. The fractionation between biogenic silica and organic matter is evident from the fluxes as well as from the consistent increase in the biogenic silica to organic carbon ratio with depth. DeMaster *et al.* (1992) found that in the Ross Sea, at one site located close to Ross Island, 12% of the biogenic silica produced in surface waters accumulates in the sediments below, which is considerably greater than the percentage for the world ocean. In contrast, at the same site, only ca. 0.5% of the surface organic carbon production in the Ross Sea accumulates in the seabed, which is comparable to the world ocean value. This implies a decoupling of the cycles of siliceous and organic matter in the Southern Ocean, arising from marked differences in the rates of key processes: inorganic dissolution reactions for biogenic silica and oxidative destruction for carbon compounds.

The Ross Sea is characterized by a deep, broad shelf, with basins which act as sediment traps, and by the recurrent development of a large coastal polynya close to the Ross Ice Shelf and a smaller one at Terra Nova Bay. The advance and retreat of sea ice

Water, Air and Soil Pollution **99**: 705-715, 1997.

are thought to control the initiation and termination of phytoplankton blooms and to play a key role in maintaining high productivity in polar regions (e.g., Smith and Nelson, 1985). Although the Ross Sea supports unusually high levels of new production (Smith and Dunbar, submitted; Nelson *et al.*, 1996), recent results show that the annual production is low, and intense blooms are episodic and limited in time. They occur more likely at the beginning of the austral summer, in connection with the seasonal retreat of the ice margin. This situation gives rise to a stratification of the water column corresponding with the availability of light and nutrients (Smith and Nelson, 1985). Within this period, marginal ice zones (MIZ) are subject to continuous modifications in position, length and shape.

Due to these characteristics of the system, it is important, in particular, to determine present fluxes of carbon and biogenic materials, and to understand the processes controlling their magnitude and variability. Despite the number of studies focused on the cycling of biogenic silica and organic carbon in the water column and upper sediments in the Ross Sea, many open questions remain. The spatial and temporal variability of vertical flux rates of biogenic silica and organic carbon must be investigated to understand the quantitative relationships involved in export from the photic layer and sediment accumulation. The particle-reactive short lived thorium isotope ^{234}Th (half life = 24.1 days), produced in seawater from decay of dissolved ^{238}U, has proven to be a useful tracer for particle transport and cycling (Bruland and Coale, 1986; Buesseler *et al.*, 1992; Cochran *et al.*, 1995). Because ^{234}Th is scavenged onto particles while its parent radionuclide remains in solution, a disequilibrium is formed which is a sensitive indicator of the thorium scavenging and particle fluxes, processes which change on a time scale from days to a few months (McKee *et al.*, 1984; Cochran *et al.*, 1995). In this paper we use ^{234}Th distribution in the upper 200 m of the water column, together with data collected by floating traps at the same sampling depths, to obtain information on scavenging rates and particle fluxes in an area of the Ross Sea studied during the seasonal sea ice retreat.

2. Methods

Sampling was carried out in December 1994, during the cruise ROSSMIZE of the Italian Progetto Nazionale Ricerche in Antartide, on board of the R/V Italica. The sampling scheme was designed to perform oceanographic and biologic research following the retreat of the seasonal sea ice. Therefore, most samples were collected along a south to north transect. Our experiments, however, were carried out on the return leg and the three stations (Figure 1) were occupied in the following order: 24a (December 07, 1994), 15c (December 12, 1994) and 11c (December 13, 1994).

CTD deployments at each station provided data on the depth distribution of temperature, salinity, chlorophyll, and light transmission (Artegiani *et al.*, 1996). Suspended particulate matter (SPM) concentrations were obtained by filtration of 1 liter of water through pre-weighed Nucleopore 0.4 μm filters (P. Povero, unpublished data). Filters were rinsed with 100 ml of deionized water to remove salts before being stored.

Large volumes of water (600-800 l) were filtered using *in situ* pumps at four depths (50, 100, 150, and 200 m) in the upper water column. ^{234}Th was retained by a series of cartridge filters: the first for particles and the other two, MnO_2 impregnated, for the dissolved form (Livingston and Cochran, 1987; Buesseler *et al.*, 1992).

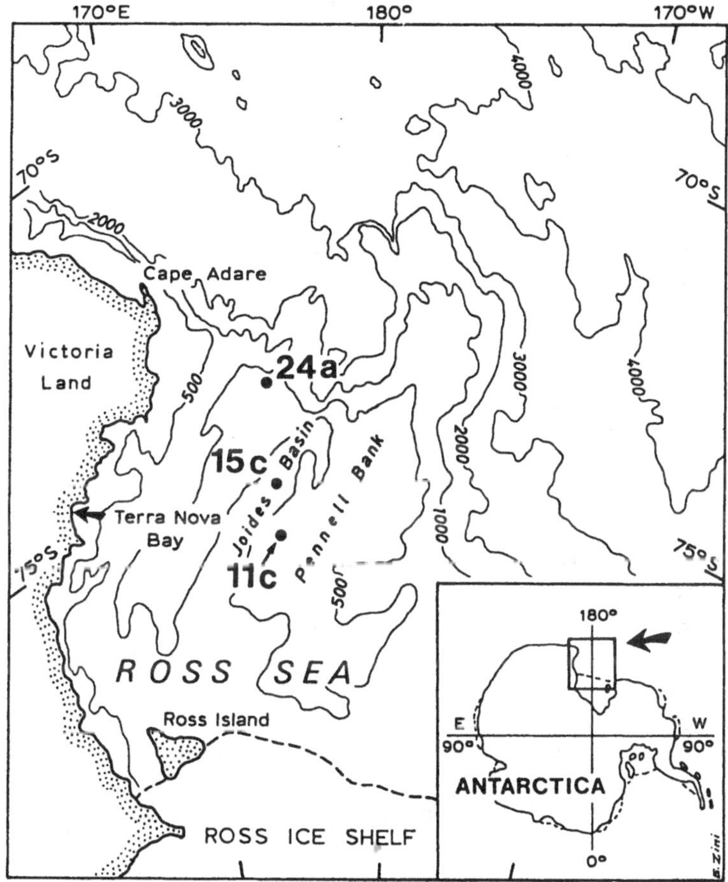

Fig. 1 Study area and sample locations.

We used special polypropylene cartridges with porosity of 0.5 μm (prefilters) and 1.0 μm (MnO_2 filters). The cartridges were shipped to the MSRC-SUNY at Stony Brook at the end of the cruise. In the lab they were ashed at 500°C for 4-8 hours. The ash was sealed into polyvinyl chloride jars and ^{234}Th was measured by gamma spectrometry using a high purity, planar-intrinsic germanium detector according to the method described by Buesseler *et al.* (1992). Errors associated with the activities are one sigma. The average precision is 7.13% and ranges between 5.70 and 11.6%.

During the same cruise floating traps were deployed, at depths of 50 m, 100 m, 150 m, and 200 m to measure export fluxes from the photic zone. The traps had a conic shape, with a collecting area of 0.5 m^2. They were covered with a baffle with square cells of 2.5x2.5 cm section and an aspect ratio of 2.5. These samplings were almost synchronous with those of thorium, and lasted 13 hours. Due to the shortness of the experiments the collection cup was not poisoned. Station 24a was not sampled due to the interference of the pack ice. After retrieval, the water left in the cone was slowly drained off to minimize the loss of the particulate material caught by the trap but not yet fallen into the cup. The suspension from the collection cup was first passed through a nylon net of 1 mm to remove swimmers (pteropods), and then split into several fractions for analysis. Swimmers were found only at station 15c, at 50 m depth. Pre-weighed 0.45 μm MF-Millipore filters were used to measure particle density and calculate the bulk mass flux. Biogenic silica was determined according to Brzezinski and Nelson (1989). A fraction was filtered on a glass fiber filter to obtain material for POC and PON determinations, which were carried out using a CHN Analyzer.

3. Results and Discussion

In the Ross Sea, the seasonal ice retreat starts from the polynya close to the Ross Ice Shelf and continues northward. This process occurs between mid November and mid December. Every location in the Ross Sea successively experiences different situations, from pack ice to MIZ, then to open sea conditions. This pattern is approximately represented by stations 24a, 15c and 11c.

3.1. HYDROGRAPHY

Figure 2 shows the depth profiles of temperature, salinity, fluorometry and light transmission, which is inversely correlated with turbidity, in the upper 300 m of the water column. The profiles of station 24a show the development of a strong pycnocline at about 80 m depth. Major variations in temperature, turbidity, and chlorophyll a occur at the same depth. Turbidity is nearly constant through the upper 80 m, whereas fluorescence shows a maximum between 35 and 80 m. At station 15c, the maximum turbidity occurs within the upper 50 m, a fluorescence maximum occurs at about 70 m depth and a thermocline lies between 50 and 70 m depth. At site 11c, both turbidity and fluorescence peak within the top 50 m, above the pycnocline. Turbidity in the upper levels decreases from station 11c to station 24a.

These hydrological characteristics trace different water masses, which can be classified following Jacobs *et al.* (1985). At station 24a, close to the retreating pack ice and to the shelf edge of the Ross sea, three water masses are clearly visible. At the surface, cold and fresher water indicates that the ice is melting. Below 80 m, the warm Circumpolar Deep Water (CDW) is intruding onto the shelf. At the bottom, the Low Salinity Bottom Water (LSBW) is probably flushing out from the shelf. At station 15c, in the Joides Basin, below the surface water, the Low Salinity Shelf Water (LSSW) overlies the High Salinity Shelf Water (HSSW). At station 11c, on the western flank of the Pennell Bank, the structure of the water column is determined by surface water, modified CDW, LSSW and HSSW.

Another trend is shown by chlorophyll a distributions. At station 24a the peak is narrow, whereas it enlarges going to station 15c and 11c: this pattern suggests that the depth penetration of the biomass is a function of the time elapsed from the release of material from the ice melting and/or the development of a phytoplankton bloom following the retreat of the ice margin. During the cruise, a time-series of primary production measurements was obtained by Saggiomo *et al.*, (submitted). At station 11 some 20 days before our sampling, high levels of primary production accompanied a maximum

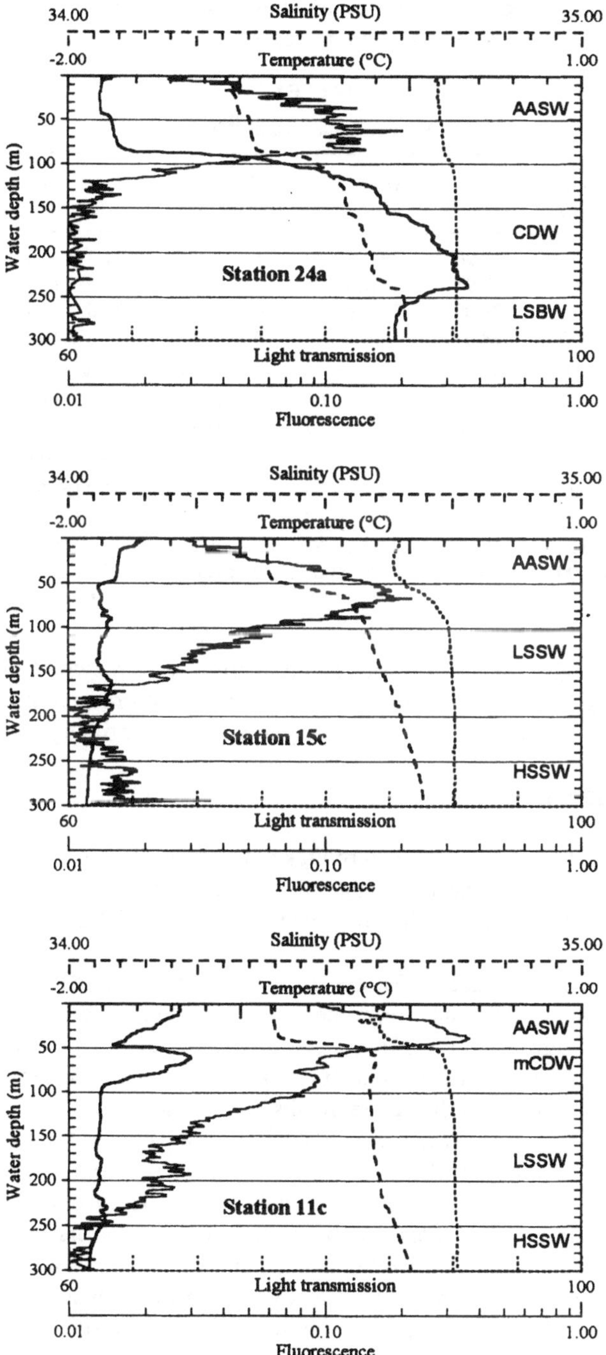

Fig. 2 Hydrographic structures at the sampling stations and depth distributions of fluorescence and turbidity.

presence of zooplankton (krill), sea birds and Minkey whales, pointing to the late stage of the algal bloom. After that, the production decreased.

TABLE I

Water column data of ^{234}Th

Station	Depth (m)	^{238}U (Bq m^{-3})	Diss. ^{234}Th (Bq m^{-3})	Part. ^{234}Th (Bq m^{-3})	Total ^{234}Th (Bq m^{-3})	$^{234}Th/^{238}U$
24a	50	37.90	29.83±2.00	1.500±0.50	31.33±2.06	0.83±0.06
	100	37.95	34.67±2.33	0.033±0.50	34.67±2.38	0.91±0.06
	150	38.00	36.00±2.33	0.667±0.50	36.67±2.38	0.97±0.06
	200	38.07	33.50±2.17	0.333±0.33	33.83±2.19	0.89±0.06
15c	50	37.90	37.33±2.50	1.667±0.83	39.00±2.63	1.03±0.07
	100	38.02	33.17±3.00	3.667±0.67	36.83±3.07	0.97±0.08
	150	38.08	48.17±3.17	1.167±0.83	49.33±3.28	1.30±0.09
	200	38.12	34.67±2.17	0.667±0.50	35.33±2.22	0.93±0.06
11c	50	38.02	35.83±4.17	10.50±0.67	46.33±4.22	1.22±0.11
	100	38.03	41.15±2.67	3.667±0.33	45.17±2.69	1.19±0.07
	150	38.03	28.33±1.83	0.667±0.33	29.00±1.86	0.76±0.05
	200	38.05	47.33±3.67	0.667±0.33	48.00±3.68	1.26±0.10

At station 15, the highest values of primary production were recorded at the time of our water sampling (early bloom stage), whereas station 24a, corresponding to the pack ice covered area, was characterized by significantly low levels of primary production.

3.2. $^{234}Th/^{238}U$ DISEQUILIBRIUM

Dissolved, particulate and total ^{234}Th activities are summarized in Table 1, and Figure 3 shows the depth profiles at the three stations. ^{238}U concentrations were calculated as a function of salinity (S) using the relation (Coale and Bruland, 1985): ^{238}U (Bq m^{-3}) = 1.18 S (PSU).

Activities of particulate ^{234}Th are 0.1-4.8%, 1.9-9.9%, and 1.4-22% of the total activities at stations 24a, 15c and 11c, respectively. The highest particulate ^{234}Th activity is found at 50 m depth of station 11c. At stations 15c and 24a, particulate ^{234}Th shows lower concentrations, with a maximum at 100 m at 15c, and at 50 m depth at 24a. Profiles of dissolved ^{234}Th are less regular: three depths at station 11c (50, 100, and 200 m) and one at station 15c (150 m) show higher activities of dissolved and total ^{234}Th with respect to the parent ^{238}U. It is likely that no or very little deficit is present at these sites. Indeed, the mean $^{234}Th/^{238}U$ activity ratio is 1.08±0.2, indistinguishable from equilibrium, and the observed fluctuations in the profiles may be due to analytical errors larger than indicated by counting statistics alone. On the other hand, station 24a shows a consistent depletion of ^{234}Th in the upper 100 m. In a different way, an extra amount of dissolved ^{234}Th at certain depths could be ascribed to the sinking and decomposition of thorium-bearing particulate matter. This may be the case at station 11c, where the extra thorium at 200 m depth can be supplied by the water column immediately above, causing the relatively strong deficit at 150 m depth. More difficult to explain is the feature at 150 m depth of station 15c, which would require a lateral advection of dissolved ^{234}Th that is not supported by the structure of the water column.

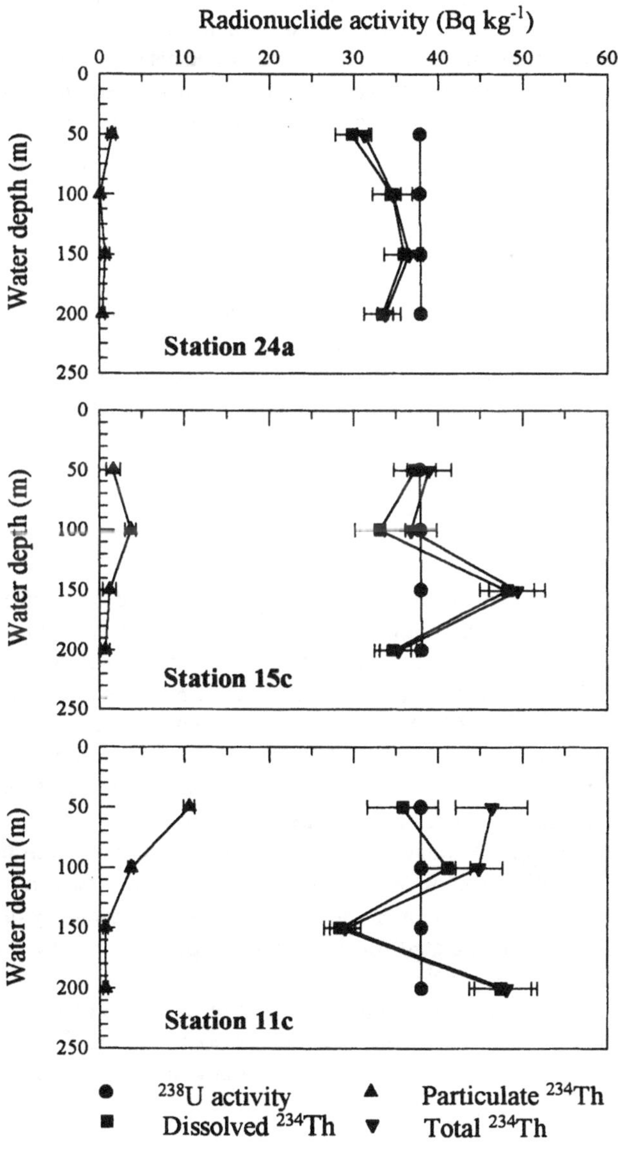

Fig. 3 Activity-depth profiles of ^{238}U, dissolved, particulate, and total ^{234}Th at the three stations.

If the ^{234}Th profile at station 24a is due to uptake to ^{234}Th onto particles that sink out of the surface water, the ^{234}Th/^{238}U scavenging model of Coale and Bruland (1985, 1987) can be used to calculate the residence times of dissolved and particulate thorium and the flux of particles out of the euphotic zone. The model is based on the assumptions that advection and diffusion of ^{234}Th are negligible with respect to the vertical transport via sinking particles and that the activity of ^{234}Th is not changing significantly during the course of our observations (Coale and Bruland, 1987). Due to ship-time constrains, it was not possible to obtain time-series data and it was necessary to assume steady state conditions. This assumption may not be strictly valid because the opening of the polynya was in progress, causing changes in particle production and sinking-driven processes. Mass balances for ^{234}Th in the dissolved and particulate forms can be written as:

$$J_{Th} = \lambda\,(A_U - A^d_{Th}) \qquad (1)$$

and

$$P_{Th} = J_{Th} - \lambda\, A^p_{Th} \qquad (2)$$

where A_U , A^d_{Th} and A^p_{Th} (Bq m^{-3}) are the integrated activities of ^{238}U, dissolved and particulate ^{234}Th above a certain depth, and λ (d^{-1}) is the ^{234}Th decay constant. Therefore, J_{Th} (Bq m^{-2} d^{-1}) represents the flux of dissolved ^{234}Th to particles and P_{Th} (Bq m^{-2} d^{-1}) is the flux of sinking or grazed particles. Residence times can be calculated as follows:

$$\tau_d = A^d_{Th} / J_{Th} \qquad (3)$$

and

$$\tau_P = A^P_{Th} / P_{Th} \qquad (4)$$

where τ_d (d) and τ_P (d) are the residence times of dissolved and particulate thorium, respectively. As a first approximation, concentrations of dissolved and particulate ^{234}Th from 0 to 50 m were assumed to be similar to those measured at 50 m depth. Table 2 shows the results of model calculations for station 24a. Residence times of ^{234}Th due to scavenging and sinking were 130-247 days and 8.1-6.6 days, respectively.

Particle concentrations at station 24a range between 0.2 mg l^{-1} at the top of the water column, to 0.1 mg l^{-1} at 200 m depth. These values were used to estimate particle fluxes assuming that the ^{234}Th residence time and bulk particle residence time are the same. This leads to an upper estimate of particle fluxes because thorium may be removed more efficiently than bulk particles. We use the relation: $F = \Sigma P / \tau_P$, where F (g m^{-2} d^{-1}) is the downward flux of particulate matter and ΣP (g m^{-2}) is the inventory of particles in the water column. The fluxes, calculated for the various depths, are reported in Table 2, and range between 1.24 and 5.03 g m^{-2} d^{-1}.

These values are very high but comparable to short term sedimentation pulses measured using floating traps by DeMaster *et al.* (1992) in the southwestern Ross Sea (1.8 g m^{-2} d^{-1}). At the same site, using moored sediment traps, Langone *et al.* (1996) measured time-integrated fluxes up to 0.88 g m^{-2} d^{-1}.

TABLE II

Model calculations for station 24a. J_{Th} values are fluxes of thorium from solution onto particles, and P_{Th} values are the fluxes of particulate thorium due to particle sinking. The flux value at a certain depth was calculated taking into account the integrate deficiencies in the water column above

Depth (m)	A_U (Bq m^{-2})	A^d_{Th} (Bq m^{-2})	A^p_{Th} (Bq m^{-2})	A^t_{Th}* (Bq m^{-2})	J_{Th} (Bq m^{-2} d^{-1})	P_{Th} (Bq m^{-2} d^{-1})	τ_d (d)	τ_P (d)	mass flux (g m^{-2} d^{-1})
50	1895	1496±141	75±35	1571±146	11.5±4	9.3±4	130±47	8.1±5	1.24±0.8
100	3791	3108±209	113±50	3221±215	19.6±6	16.4±6	190±60	6.9±3	2.81±1.2
150	5690	4875±266	131±31	5006±273	23.4±8	19.7±8	247±86	6.6±3	4.13±1.9
200	7592	6612±310	156±68	6768±317	28.2±9	23.7±9	234±75	6.6±4	5.03±3.0

* corresponds to total Th activity = $A^d_{Th} + A^p_{Th}$

TABLE III

Floating trap data

Station	Depth (m)	OC (wt%)	N (wt%)	C/N	mass flux (mg m^{-2} d^{-1})	OC flux (mg m^{-2} d^{-1})	N flux (mg m^{-2} d^{-1})
15c	50	11.0	1.51	7.3	106.1	11.7	1.6
	100	7.08	0.47	15.1	66.2	4.69	0.31
	150	6.30	0.32	20.0	109.3	6.89	0.34
	200	3.60	0.46	7.9	138.4	4.98	0.63
11c	50	11.9	1.34	8.9	87.9	10.4	1.17
	100	4.99	0.55	9.1	122.5	6.11	0.67
	150	4.75	0.45	10.5	104.4	4.96	0.47
	200	6.07	0.35	17.4	118.5	7.19	0.41

3.3. FLOATING TRAPS

Table 3 shows the results obtained from samples collected by floating traps at stations 15c and 11c. Traps were deployed for short times (12-13 hours) and, therefore, samples represent an instantaneous situation. Overall mass fluxes are similar at the two sites: 66-138 and 88-122 mg m^{-2} d^{-1}, respectively. These values are low by an order of magnitude when compared with those calculated for site 24a. The lowest values at each station have been recorded just below the fluorescence peak. Concentrations and fluxes of organic carbon, nitrogen and biogenic silica show fluctuations. In most cases at the top of the water column concentrations of biogenic components are high. Then, the composition of sinking material varies roughly regularly with depth, showing a decrease in carbon and nitrogen contents. The maximum change occurs in both cases between 50 and 100 m depth. The highest organic carbon fluxes (from 10 to 12 mg C m^{-2} d^{-1}) are low in comparison with those (from 25 to 93 mg C m^{-2} d^{-1}) measured during January 1990 at two stations in the southern part of the Ross Sea (Smith and Dunbar, 1995). C/N ratios increase with depth, from ca. 7 to 20, pointing out the progressive decomposition of the organic matter. The C/N ratio in Antarctic diatoms is ca. 7 (Brzezinski, 1985). The highest C/N ratios could be ascribed also to the contribution of old organic matter derived from melting ice.

Particle fluxes both from ^{234}Th and sediment traps are generally consistent with prior work, although problems arise in the present study when a comparison is made between calculated particle fluxes from ^{234}Th at station 24a and those measured at stations 15c and 11c by traps. The discrepancy may be related to a variable and low trap efficiency, as discussed by Buesseler (1991). Many factors, including hydrodynamics, can affect material collection and bias the results of floating traps. On the other hand, low levels of turbidity, chlorophyll a and primary production at station 24a do not support the hypothesis of a high vertical mass flux, and it is possible that the ^{234}Th deficiency at this station is caused more by lateral removal than particle sinking.

4. Conclusions

This paper provides the first insight into the evolution of ^{234}Th removal and particle flux during the opening of the polynya in the Ross Sea. The distribution of dissolved and particulate ^{234}Th in the upper 200 m of the water column was obtained for three stations in the Ross Sea off Victoria Land and Terra Nova Bay. At two of these stations measurements of fluxes of particulate matter were obtained using sediment traps. In summary the results indicate that:

1) One site (24a) showed deficiencies of ^{234}Th due to the uptake onto particles and particle removal. This permitted the calculation of thorium fluxes and residence times, assuming steady state conditions. Residence times of ^{234}Th due to scavenging and sinking were 130-247 days and 8.1-6.6 days, respectively;

2) Upper estimates of sinking fluxes of particles can be obtained from particulate ^{234}Th residence times and particle concentrations. The values are high, ranging from 1.23 to 5.03 g m^{-2} d^{-1};

3) The other two stations (11c and 15c) do not show significant depletion of ^{234}Th relative to ^{238}U;

4) The lack of removal of ^{234}Th at stations 11c and 15c is consistent with the low particle fluxes measured by sediment traps at these stations. The trap fluxes are low when compared with those calculated by ^{234}Th deficiencies for station 24a. The impossibility of obtaining trap-based fluxes at station 24a, makes it difficult to understand if the trap underestimated the fluxes or if fluxes based on ^{234}Th deficiencies were overestimated.

5) Fluxes of organic carbon and nitrogen, and biogenic silica measured by sediment traps at stations 11c and 15c show decreasing trends and C/N modifications with depth, emphasizing the different preservation efficiencies of the biogenic particles in the water column.

Acknowledgments

Research was carried out in the framework of Project "Ecology and Biogeochemistry of the Southern Ocean" (subproject Biogenic Sedimentation) of the Italian "Programma Nazionale di Ricerche in Antartide". We are indebted to O. Radakovitch for the critical review of the manuscript. We thank E. Lipparini, who helped in sampling operations. D. Hirschberg and L. Labbrozzi carried out part of the analytical work. A. Russo provided CTD data and P. Povero SPM concentrations. G. Zini made the drawings. This is contribution n. 1062 of the Istituto di Geologia Marina, CNR, Bologna, and n. 1059 of the MSRC-SUNY, Stony Brook.

References

Artegiani, A., Budillon, G., Moretti, M., Paschini, E., Russo, A., Spezie, G.: 1996, *International Workshop "Ross Sea Ecology"*, Taormina, 14-16 May 1996, 149-151.

Bruland, K.W. and Coale, K.H.: 1986, In: *Dynamic Processes in the Chemistry of the Upper Ocean* (J.P. Burton, P.G. Brewer and R. Chesselet, eds.) Plenum Publ. Corp., 159-172.

Brzezinski, M.A.: 1985, *Mar. Biol.*, **12**, 347-357.

Brzezinski, M.A. and Nelson, D.M.: 1989, *Deep Sea Res.*, **36**, 1009-1030.

Buesseler, K.O.: 1991. *Nature*, **353**, 420-423.

Buesseler, K.O., Cochran, J.K., Bacon, M.P., Livingston, H.D., Casso, S.A., Hirschberg, D., Hartman, M.C. and Fleer, A.P.: 1992, *Deep-Sea Res.*, **39**, 1103-1114.

Coale, K.H. and Bruland, K.W.: 1985, *Limnology and Oceanography*, **30**, 22-33.

Coale, K.H. and Bruland, K.W.: 1987, *Limnology and Oceanography*, **32**, 189-200.

Cochran, J.K., Barnes C., Achman D. and Hirschberg D.J.: 1995, *J. Geophys. Res.*, **100**, 4399-4410.

DeMaster, D.J.: 1981, *Geochim. Cosmochim. Acta*, **45**, 1715-1732.

DeMaster, D..J., Dunbar, R.B., Gordon, L.I., Leventer, A.R., Morrison, J.M., Nelson, D.M., Nittrouer, C.A. and Smith, W.O. Jr.: 1992, *Oceanography*, **5**, 146-153.

Jacobs, S.S., Fairbanks, R.G. and Horibe, Y.: 1985, In Jacobs S.S. (ed), *Oceanology of the Antarctic Continental Shelf*: Antarctic Research Series, **43**, American Geophysical Union, 59-85.

Langone, L., Dunbar, R.B., Labbrozzi, L., Ravaioli, M., Frignani, M.: 1996, *International Workshop "Ross Sea Ecology"*, Taormina, 14-16 May 1996, 98-100.

Ledford-Hoffman, P.A., DeMaster, D.J. and Nittrouer, C.A.: 1986, *Geochim. Cosmochim. Acta*, **50**, 2099-2110.

Livingston, H.D. and Cochran, J.K.: 1987 - *J. Radioanalyt. Nucl. Chem.*, **115**, 299-308.

McKee, B.A., DeMaster, D.J. and Nittrouer, C.A.: 1984, *Earth Planet. Sci. Lett.*, **68**, 431-442.

Nelson, D.M. and Smith, W.O.: 1986, *Deep Sea Res.*, **33**, 1389-1412.

Nelson, D.M., DeMaster, D.J., Dunbar, R.B. and Smith, W.O., Jr.: 1996. *Journ. Geophysic. Res.*, **101**, 18519-18532.

Saggiomo, V., Carrada, G.C., Mangoni, O., Ribera d'Alcalà, M. and Russo, A.: Submitted to *Journ. Mar. Systems.*

Smith, W.O., Jr. and Dunbar, R.B.: Submitted *to Journ. Mar. Systems.*

Smith, W.O., Jr. and Nelson, D.M.: 1985, *Science*, **227**, 163-166.

THE DISTRIBUTION OF PCB's AND CHLORINATED PESTICIDES IN TWO CONNECTED HIMALAYAN LAKES

S. GALASSI[1], S. VALSECCHI[2] AND G. A. TARTARI[3]

[1] University of Milan, Department of Biology, Via Celoria 26, 20133 Milano, Italy

[2] Istituto di Ricerca sulle Acque, C.N.R., Via Mornera 25, 20047 Brugherio (MI), Italy

[3] Istituto Italiano di Idrobiologia, C.N.R., L.go Tonolli 50/52, 28048 Pallanza (VB), Italy

Abstract. PCBs and organochlorine pesticides were determined in water, sediment and zooplankton of two Himalayan lakes, located at different altitudes and connected to each other in such a way that Superior Lake acts as a sedimentation basin for Inferior Lake. Surficial sediments of both lakes show PCB contamination comparable to lakes of industrialised areas. Biota appear to be the main machanism responsible for micropollutant burial in the sediments of Inferior Lake, whereas inorganic particles are more relevant in Superior Lake. Physical and chemical properties of individual chemicals, particularly Henry's law constant and K_{ow} values, seem to regulate distribution in different environmental compartments.

Keywords: chlorinated pesticides, PCB, sediment, Himalayan lakes.

1. Introduction

Many studies have been conducted in recent years with the aim of evaluating micropollutant levels in remote areas (Barrie *et al.*,1992; Muir *et al.*, 1992; Bidleman *et al.*, 1993). These studies are appealing for two main reasons: 1) remote environments seem to be more vulnerable to this kind of pollution because climatic conditions enhance residence times in abiotic compartments and 2) trophic webs in cold environments are very simple, characterised by lipid-rich top predators in which lipophilic compounds are readily accumulated.

Atmospheric transport and deposition is the most likely route of entry of organochlorine compounds (OCs) into remote environments (Cotham and Bidleman, 1991). There is evidence in the literature indicating long-range transport as a source of pollution for pristine areas far from the production and use of synthetic chemicals (Golderg , 1975).

According to the "cold condensation" theory, compounds with higher vapor pressure condense at colder latitudes and altitudes (Wania and Mackay, 1993). However, once trapped in snow and ice during the cold season, OCs can evaporate from the snowpack when the temperature rises, constituting a "local" source of pollution (de Voogt and Jansson, 1993). The role of this "secondary" pollution source in remote areas is still unclear. Seasonal variations have not been investigated extensively in remote regions due to difficulties in undertaking expeditions in harsh conditions.

In this respect, the availability of a permanent laboratory at 5000 m a.s.l. (Pyramid) in the Khumbu Valley represents a unique opportunity to undertake environmental studies in an area which can be considered as the "third pole" (Bahadur, 1993). In the framework of the Ev-K^2-CNR project, scientific expeditions were organised yearly beginning in 1992 and environmental samples have been collected from several high altitude lakes (Gosso *et al.*, 1993; Manca *et al.*, 1994, 1995; Guilizzoni *et al.*, in press; Tartari *et al.*, in press) for biological, chemical and limnological characterisation.

In the present study, sediments, water and zooplankton were collected in two Himalayan lakes. Sample preparation for micropollutant analyses was carried out at the Pyramid laboratory in order to avoid contamination and to reduce the amount of material to be transported. Results of PCB and chlorinated pesticide determinations are presented with the aim of contributing to the understanding of the role of "local" OC cycling in the more general movement of these compounds from polluted to remote areas.

2. Materials and Methods

2.1 STUDY AREA

Superior Lake and Inferior Lake are clear-water high altitude water bodies located in the Khumbu Valley, Sagarmatha National Park, Nepal, at 5,050 m a.s.l. (Figure 1). Superior Lake is located in a basin on the eastern side of a ridge and has an outflowing stream that connects it with Inferior Lake, at 150 m lower altitude. Their main characteristics are reported in Table 1 (De Vito, 1994). The sedimentation rate for Superior Lake (150 mg $cm^{-2}y^{-1}$) is about 10 times higher than for Inferior Lake (15 mg cm^{-2} y^{-1}); the rate of deposition of organic matter was 8 and 2 mg $cm^{-2}y^{-1}$ in Superior and Inferior Lakes, respectively (Guilizzoni *et al.* in press).

Figure 1. Map of lakes in the Himalayan region (▲ Ev-K²-CNR laboratory)

TABLE I
Geographic and geomorphological characteristics of the two Himalayan lakes

	Superior Lake	Inferior Lake
Altitude (m. a.s.l.)	5,213	5,067
Longitude	86° 48' 42''	86° 48' 58''
Latitude	27° 57' 55''	27° 57' 45''
Width (m)	67	91
Length (m)	112	246
Area (m^2)	5,669	16,718
Maximum depth (m)	8.2	14.8
Watershed (m^2)	794,474	372,565

2.2 SAMPLING

In 1992, one sediment core was collected from both lakes and sectioned into 1.4-cm-thick slices for Inferior Lake and into 1-cm slices for Superior Lake. On October 6th 1994, a 5 l water sample was collected along with a vertical profile in the middle of Inferior Lake and preconcentrated by Empore™ Extraction Disk C_{18} (STEPBIO Bologna, Italy), according to Hagen *et al.* (1990), after a pre-filtration on Nucleopore filters of polycarbonate (0.40 μm). On the same day, zooplankton were collected from Inferior Lake with a net (126 μm) and filtered through a Whatman GF/C filter (1.2 μm) obtaining a fresh biomass of about 2 g. Sediment samples, Empore™ disk and filters were transported frozen to Italy for freeze drying and solvent extraction.

2.3 ANALYTICAL PROCEDURE

Frozen sediment and zooplankton samples were lyophilised under vacuum at 20 °C for 24 hours. A 1 g sample of freeze dried sediment for each slice was extracted for 8 hr in a Soxhlet apparatus with 100 ml of pesticide-grade n-hexane. The extracts were concentrated to 1 ml under reduced pressure using a rotary evaporator. The same procedure was applied for the freeze dried zooplankton. The zooplankton lipid content was determined by weighing the residue of zooplankton extract dried under nitrogen flux. The lipid fraction was dissolved in 1 ml of n-hexane and this solution was digested by adding 2 ml of concentrated H_2SO_4 at room temperature. The Empore™ disk was extracted according to Hagen *et al.* (1990) and the solvent was concentrated to 1 ml under reduced pressure using a rotary evaporator.

Sediment, zooplankton and water extracts were passed through a Florisil column (4 x 0.7 cm). In the case of sediment extracts, a layer of Cu powder (0.1 g) was put on top of the Florisil column to retain the sulphur compounds. Cu powder was previously treated with HCl (18%) and washed with acetone and n-hexane. The Florisil column was eluted with 25 ml of n-hexane and the eluate was concentrated to 1 ml under reduced pressure using a rotary evaporator.

Cleaned-up extracts were injected into a gas chromatograph (Carlo Erba 8000 Mega Series) equipped with a ^{63}Ni electron capture detector (ECD). The injection was performed in the on-column mode. Instrumental parameters and operational conditions were as follows: a fused silica capillary column (50 m x 0.25 mm i.d.), CP-Sil-8 CB, film thickness 1.9 - 2.0 μm, was used with a temperature programme from 60°C to 180°C at 15°C min^{-1} followed by a run from 180°C to 270°C at 1.5°C min^{-1} . The carrier gas was helium at 1 ml min^{-1}, and nitrogen was used as auxiliary gas for the detector at 30 ml min^{-1}. The detector temperature was fixed at 270°C.

Pure reference standards (SIC, Roma) were used for the chlorinated pesticides quantification. Aroclor 1260 (Altech) was used as a reference standard for PCBs quantification. The 16 PCB congeners in the reference mixture were identified and measured against pure reference PCB standards (BCR, Brussels): 101, 110, 151, 149, 153, 141, 138, 187, 183, 185, 174, 177, 180, 170, 201, 196. Detection limits for chlorinated pesticides and individual PCB congeners were: 0.1 ng l^{-1} in water, 0.4 ng g^{-1} dry wt in sediment, and 4 ng g^{-1} dry wt in zooplankton. Blanks were analysed according to the same procedure as the samples and no contamination was revealed.

3. Results

Total PCB and chlorinated pesticide concentrations in sediments for the two Himalayan lakes are given in Table II. PCB profiles in sediment layers are shown in Figure 2a. Sediment ^{210}Pb dating was performed by Guilizzoni *et al.* (in press) who analysed two sediment cores taken simultaneously to our cores. All OCs but γ-HCH are more concentrated in the first sediment layer in agreement with an higher pollution load in recent years than in the past and with a higher concentration of organic carbon than in deeper sediments (Guilizzoni *et al.* in press).

Concentrations measured in water and zooplankton from Inferior Lake are shown in Table III. Only nine of the 16 PCB congeners were detected in the zooplankton and even less in water although it was concentrated 10,000 times. α-HCH was undetectable either in water or in zooplankton and sediments samples.

The PCB profile (Figure 2a) of the four layers of sediment of Superior Lake, covering a period of about ten years, from 1979 to 1992 was similar to that of commercial mixtures of Aroclor 1254 and Aroclor 1260. Lower chlorinated PCBs of Aroclor 1242 and 1248 mixtures were not present in our Himalayan samples. Recent sediments of Superior Lake (1986-1992) are richer in the more chlorinated congeners. Since persistence increases with chlorination we expected to find more chlorinated congeners in oldest sediments. Probably, this contradictory result might be abscribed to changes in uses. Inferior Lake sediments showed a different pattern (Figure 2b) with an enrichment of less chlorinated congeners in both layers (1992-1972, 1972-1952). As these two layers are very similar to each other, although they represent the historical records of very different periods of PCB use and diffusion in the world, differences between the two lakes are more likely to be explained by phenomena occurring on a local scale than from long-range transport.

Zooplankton of Inferior Lake have a PCB congener distribution very similar to that of its sediment (Figure 2c) . Apart from congener 180 and 187 which have seven chlorine atoms, only penta and hexachlorobiphenyls are accumulated in zooplankton at detectable levels.

TABLE II
PCB congeners and chlorinated pesticides in lake sediment layers

	Superior Lake ng g^{-1} (dry wt)				Inferior Lake ng g^{-1} (dry wt)	
Period	1992-1989	1989-1986	1986-1983	1983-1979	1992-1972	1972-1952
Layer	0 - 1 cm	1 - 2 cm	2 - 3 cm	3 - 4 cm	0 - 1.4 cm	1.4 - 2.8 cm
total PCBs	430.9	70.6	21.2	21.2	204.6	132.9
pp'DDT	5.84	0.80	0.44	1.46	3.06	1.54
pp'DDE	*1.62	*0.51	<0.40	<0.40	*10.48	*8.18
pp'DDD	0.63	0.59	<0.40	<0.40	<0.40	<0.40
γ - HCH	2.13	1.49	1.17	1.30	1.76	2.89
HCB	1.10	0.81	0.63	0.52	n.d.	0.88

n.d. = not determined; * overestimated due to the presence of interfering compounds

TABLE III
PCB congeners and chlorinated pesticides in Inferior Lake water and zooplankton

	Water	Zooplankton			log BAF
	ng l^{-1}	ng g^{-1} (dry wt)	ng g^{-1} (wet wt)	ng g^{-1} (lipid)	
PCB 101	0.27	28.78	2.98	110.78	4.04
PCB 110	<0.1	33.04	3.42	127.15	
PCB 151	<0.1	10.13	1.05	38.99	
PCB 149	<0.1	28.43	2.95	109.41	
PCB 153	0.34	32.88	3.41	126.56	4.00
PCB 141	<0.1	7.59	0.79	29.21	
PCB 138	<0.1	16.31	1.69	62.78	
PCB 187	0.51	11.68	1.21	44.95	3.37
PCB 180	<0.1	11.46	1.19	44.09	
pp'DDT	<0.1	57.66	5.98	221.94	
pp'DDE	<0.1	31.60	3.27	121.61	
pp'DDD	<0.1	7.46	0.77	28.70	
γ - HCH	0.28	16.52	1.71	63.59	3.79
HCB	n.d.	4.41	0.46	16.98	

PCB 183, PCB 185, PCB 174, PCB 177, PCB 170, PCB 201, PCB 196 were under the detection limit
n.d. = not determined; BAF = bioaccumulation factor in zooplankton (wet wt)

4. Discussion

Total PCB concentration in the superficial sediments of the two Himalayan lakes is higher than those determined in Lake Como (161 ng g^{-1}dry wt) and Lake Maggiore (64 ng g^{-1}dry wt) sediments (Northern Italy), for the same deposition period (Provini *et al.*,1995). Since the organic matter content of the most recent layer of Superior Lake sediment is the lowest

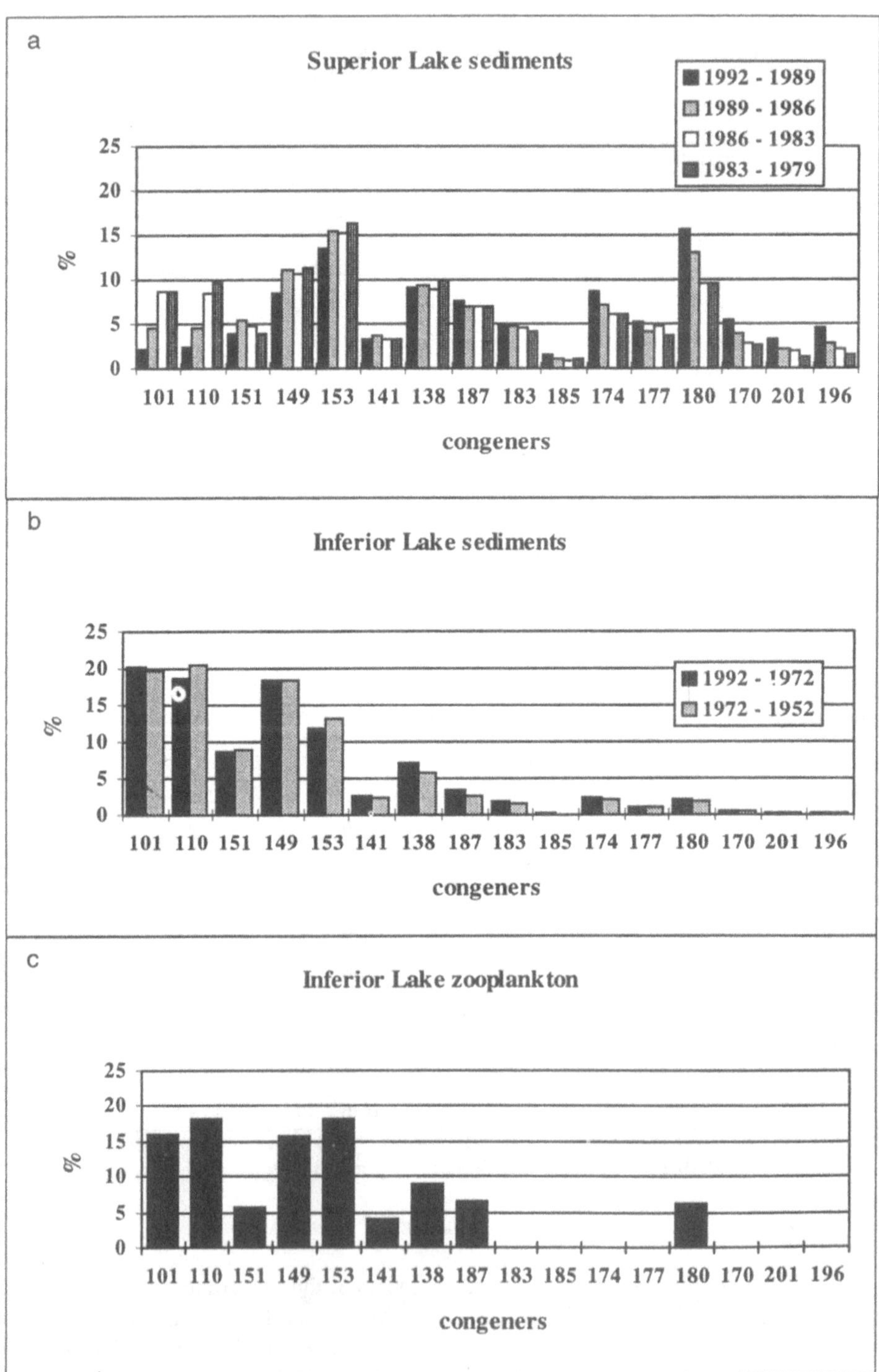

Figure 2. Distribution of PCB congeners in lake sediments and zooplankton

(53 mg g^{-1}dry wt) of the above mentioned lakes its higher contamination should be due to the "cold condensation" phenomenon.

Interesting findings can be drawn comparing the results obtained for the two Himalayan lakes. Comparing the PCB congener distribution (Figure 2a and 2b), it is evident that Superior Lake sediments reflect the composition of commercial PCB mixtures of Aroclor 1254 and 1260, while Inferior Lake sediments are enriched in the lower chlorinated and more volatile congeners occurring in these mixtures. As long range contamination sources must be the same for the two aquatic environments, only local recycling could explain the observed differences (de Voogt and Jansson,1993).

Since air-water and particle-water exchanges should be correlated with Henry's law constant (H_c) and octanol/water partitioning coefficient (K_{ow}), respectively, the ratio between these two constants, that is the ratio between vapour pressure and solubility in n-octanol, should regulate cycling and distribution of semivolatile pollutants from particulate matter or sediment to air and vice-versa through the water interface.

TABLE IV
Octanol/water partition coefficient (K_{ow}) and Henry's law constant (H_c) of some PCB congeners, pp'DDT and γ-HCH

	K_{ow} (10^6)	H_c Pa m^3 mol^{-1} (15-20°C)	H_c/K_{ow} (10^{-5})
PCB 101	2.39 [a]	12.9 [d]	0.4
PCB 110	3.02 [a]	6.6 [d]	0.2
PCB 153	8.32 [a]	6.6 [d]	0.08
PCB 180	22.9 [a]	11.1 [d]	0.05
PCB 187	14.8 [a]	15.5 [d]	0.10
PCB 201	41.7 [a]	23.5 [d]	0.05
pp'DDT	1.55 [b]	1.31 [e]	0.08
γ - HCH	0.011 [c]	0.22 [e]	2.0

[a] Hawker & Connell, 1988; [b] Chiou et al., 1977; [c] Ernst, 1977; [d] Baker & Eisenreich, 1990; [e] Atkins & Eggleton, 1971

Looking at H_c/K_{ow} ratios (Table IV) they decrease from γ-HCH to PCB 201, (an octachlorobiphenyl); differences between γ-HCH and PCB 101 and 110 are one order of magnitude or more greater between γ-HCH and more chlorinated congeners. It is reasonable to suppose that γ-HCH has a greater affinity for the atmospheric compartment and a net flux from Superior Lake particles to air occurs during the warm season. Congeners 101 and 110 are released from suspended particles and superficial sediment into water but do not escape into the atmosphere (in terms of net flux). Dissolved and particle-bound micropollutants are transported from Superior into Inferior Lake since the first is a tributary of the latter. Compounds having very low H_c/K_{ow} ratios tend to be retained by particles. As the warm season is very short at this altitude, probably there is not enough time to reach equilibrium conditions for desorption of extremely lipophilic compounds.

OCs accumulated in the snowpack in the drainage basin of the Superior Lake appear to be transported by particles, as is indicated by the high sedimentation rate (150 mg $cm^{-1}y^{-1}$). Superior Lake acts as a settling pond for Inferior Lake, where the sedimentation rate is ten times lower. Therefore, this latter lake is probably much more influenced by local re-volatilisation and deposition phenomena than by pollutant transported from long distances and trapped in snow and ice.

Probably most of the PCB burden of surficial sediments is released into water when the organic matter decomposes. The remaining PCBs are diluted in the large amount of inorganic material in the case of Superior Lake and the PCB profile is not modified by the biota. Conversely, settling biota are more important in determining sediment composition in Inferior Lake, where organic deposition prevails. In fact the sedimentation rate is very low in Inferior Lake and the organic matter concentration is higher (133 mg g^{-1} dry wt) than in Superior Lake. Furthermore, the PCB composition of zooplankton collected in Inferior Lake is very similar to that of its sediment (Figure 2c).

Skoglund *et al.* (1996), studied PCB bioaccumulation in phytoplankton. Their results show that this component of aquatic ecosystems reaches equilibrium concentrations in a short time and that concentrations in phytoplankton were higher than those predictable on the basis of partitioning in its lipid fractions for congeners with log K_{ow} lower than 4.5. Looking at the bioacculation factors in zooplankton (Table III), log BAF of congener 187 is lower than congeners 153 and 101 in spite of its highest lipophilicity (log K_{ow} 7.36). As a matter of fact, phyto and zooplankton should vehiculate OCs from water to sediment discriminating among different compounds as a function of their K_{ow}.

5. Conclusions

Since no limnological and chemical data on Himalayan lake sediments were published after the Yale North India Expedition (Hutchinson, 1937; Hutchinson *et al.*, 1943), the results presented in this work represent the first contribution to the knowledge of the contamination state of the "third pole" of the world.

Although the results of the present study do not to allow a complete understanding of the factors regulating the circulation of semi-volatile micropollutants in this remote environment, some hypotheses are proposed to stimulate further investigations. OC loads calculated from sediment concentrations can be compared for different environments or used for budget purposes provided individual chemicals are considered instead of classes, (e.g. PCBs and DDTs), encompassing a number of compounds with a wide range of physico-chemical properties.

Since lake sediments show a different trapping capacity as a function of their organic carbon content and composition, OC concentration in the sediment is not likely to reflect the bioavailable concentration in the water column. Therefore, the risk to aquatic life is better evaluated by biomonitoring. As habitat and energetic flow dramatically change during the year in cold climate environments, attention should be paid to the role of the biotic component in OC cycling and its seasonal variability.

Acknowledgements

We express our gratitude to Dr. G. Tartari, project leader of Ev-K^2-CNR scientific and technological research at high altitude, environmental science, who invited us to undertake this study and co-ordinated the scientific expeditions and to Dr P. Guilizzoni who provided us with sedimentation rate of the sediments from the two lakes.

References

Atkins D. H. F. and Eggleton A. E. J.: 1971, *Studies of atmospheric washout and deposition of γ-BHC, dieldrin and pp'DDT using radiolabelled pesticides*. Internat. Atomic Agency Symposium. Vienna, SH-1429/32, 521-533.

Bahadur J.: 1993, *Snow and Glacier Hydrology*. Young G. J. (ed.), IAHS **218**, 181-190.

Baker J. E. and Eisenreich S. J.: 1990, *Environ. Sci. Technol.* **24**, 342-352.

Barrie L. A., Gregor D., Hargrave B., Lake R., Muir D., Shearer R., Tracey B. and Bidleman T.: 1992, *Sci. Total Environ.* **122**, 1-74.

Bidleman T., Walla M. D., Roura R., Carr E. and Schmidt S.: 1993, *Mar. Pollut. Bull.* **26**, 258-262.

Chiou C. T., Freed V. H., Schmedding D. W. and Kohnet R. L.: 1977, *Environ. Sci. Technol.* **11**, 475-478.

Cotham W. E. and Bidleman T. F.: 1991, *Chemosphere* **22**, 165-188.

De Vito C.: 1994, *Realizzazione del catasto dei laghi himalaiani della valle del Khumbu (Nepal orientale)*, Degree Thesis, University of Torino, Torino, p. 177.

de Voogt P. and Jansson B.: 1993, *Rev. Environ. Contamin. Toxicol.* **132**, 1-27.

Ernst W.: 1977, *Chemosphere* **7**, 731-740.

Golderg E. D.: 1975, *Proc. Roy. Soc. Lond.* **B189**, 277-289.

Gosso E., Tartari G., Valsecchi S. Ramponi S. and Baudo R.: 1993, *Verh. Internat. Verein. Limnol.* **25**, 800-803.

Guilizzoni P., Lami A., Smith J. D., Belis C., Bianchi M., Bettinetti R., Marchetto A. and Muntau H.: *Top of the World Environmental Research*, in press.

Hagen, D. F., Markell, C. G., Schmitt, G. A. and Blevins, D. D.: 1990, *Anal. Chim. Acta* **236**, 157-164.

Hawker D. W. and Connell D. W.: 1988, *Environ. Sci. Technol.* **22**, 382-387.

Hutchinson G. E.: 1937, *Int. Rev. ges. Hydrobiol.*, **35**, 134-177.

Hutchinson G. E., Wollack A. and Setlow J. K.: 1943, *Am. Jour. Sci.*, **241**, 533-542.

Manca M., Cammarano P. and Spagnuolo T.: 1994, *Hydrobiologia*, **287**, 225-231.

Manca M., Nocentini A. M., Ruggiu D., Panzani P., Bonardi M., Cammarano P. and Spagnuolo T.: 1995, *Mem. Ist. ital. Idrobiol.*, **53**, 17-25.

Muir D. C. G., Wagemann R., Hargrave B. T., Thomas D. J., Peakall D. B. and Norstrom R. J.: 1992, *Sci. Total Environ.* **122**, 75-134.

Provini A., Galassi S., Guzzella L. and Valli G.: 1995, *Mar. Freshwater Res.* **46**, 129-136.

Skoglund R. S., Stange K and Swackhamer D. L.: 1996, *Environ. Sci. Technol.* **30** (7), 2113-2120.

Tartari G., Tartari G. A., Valsecchi S. and Camusso M.: *Verh. Internat. Verein. Limnol.*, in press.

Wania F. and Mackay D.: 1993, *Ambio* **22**, 10-18.

The Interactions Between Sediments and Water

Baveno, Italy

22-25 September 1996

List Of Participants

1. Session Chairs

The following people kindly chaired sessions:

1A	B. DAVISON	Experimental Devices for Sediment Investigations
1B	M. CAMUSSO	Remote Areas
1C	M. STURM	Sediment Dating and Chemical Characterization
1D	A. N. NOZHEVNIKOVA	Microbial Processes I
2A	T. N. ANGELIDIS	Methodological Problems
2B	L. HÅKANSON	Evaluation/Remediation of Contaminated Sediments
2C	P. GUILIZZONI	Long-term Records from Sediment Cores
2D	A. PROVINI	Hydrodynamics and Sediment Transport
3A	M. PETTINE	Transition Zones I
3B	J. S. CHEN	Modelling Toxicant-Sediment Interactions
3C	R. D. EVANS	Nutrient Partitioning and Lake Modelling
3D	E. L. PETTICREW	Hydrodynamics and Sediment Transport
4A	B. T. HART	Release of Heavy Metals from Polluted Sediments
4B	M. BATTEGAZZORE	Microbial Processes II
4C	P. JONSSON	Ecosystem and Field Studies
4D	D. WALLING	Resuspension and Deposition of Sediments
5A	J. SKEI	Interactions Between Sediment and Benthic Organisms I
5B	J. BLOESCH	Particle and Sediment Dynamics in the Marine Environment
5C	J. DROPPO	Sediment-Water Interactions
5D	W. LICK	Suspended Matter and Sedimentation
6A	B. T. HARGRAVE	Interactions Between Sediment and Benthic Organisms II
6B	S. GALASSI	Monitoring Micropollutants by Sediment Analysis
6C	A. TURNER	Transition Zones II
6D	B. G. KRISHNAPPAN	Particle Size and Aggregation in Deposition Phenomena

2. Participants

Abdullah Mohamed
University Of Oslo
Oslo, Norway
Abdullah.Mohamed@Bio.Uio.No

Aigars Juris
Stockholm University
Stockholm, Sweden
Rolf.Hallberg@Bgkem.Su.Se

Airoldi Laura
University Of Pisa
56126 Pisa, Italy
Lairoldi@Discat.Unipi.It

Alcorlo Paloma
Univ. Autonoma
28049 Madrid, Spain
Aperez@Ccuam3.Sdi.Uam.Es

Andersen H.E..
Nat. Environ. Res. Inst.
Silkeborg, Denmark
Fvhea@Dmu.Dk

Angelidis Thomas
Aristotle University
54006 Thessaloniki, Greece
Fax +30 31 99 76 96

Appleby P.G.
University Of Liverpool
L69 3BX Liverpool, UK
Appleby@Liverpool.Ac.Uk

Araujo Fatima
Dep. Quimica Itn
2695 Sacavem, Portugal
Faraujo@Itu1.Itu.Pt

Atkinson Joseph
S.U.N.Y. At Buffalo
Buffalo New York, USA
Atkinson@Acsu.Buffalo.Edu

Aziz Nadim
Clemson University
Clemson, SC 29634, USA
Aziz@Clemson.Edu

Bai Zhan-Guo
Chinese Academy Of Agricultural Sciences
Beijing, P.R. Of China
Fax 00 86 010 6217 65 3822

Baldwin D. S.
Freshwater Research Ctr.
Albury NSW 2640, Australia
Dbalwin@Mdfrc.Canberra.Edu.Au

Barbieri Alberto
Lab. Studi Ambientali
6900 Lugano, Switzerland
Barbieri@Stse.Trevano.Ch

Bartoli Marco
Dip. Scienze Amb.
43100 Parma
Bartoli@Eagle.Bio.Unipr.It

Battegazzore Maurizio
Amm. Comun. Di Alessandria,
15100 Alessandria, Italy
Mbeco@Pop.Tor.It

Belis Claudio
Ist. Ital. Idrobiologia
28048 Pallanza, Italy
Fax +39 323-55 65 13

Berge John Arthur
Norw. Inst. For Water Res.
0411 Oslo, Norway
John.Berge@Niva.No

Bergtold Matthias
Zoologisches Inst. Der Lmu
80333 Munchen, Germany
Traunsp@Zi.Biologie.Uni-Muenchen.De

Bertuzzi Alessandra
Lab. Di Biologia Marina
34010 Trieste, Italy
Fax +39 40-22 44 37

Bettiol Cinzia
Univ. Ca Foscari
30123 Venezia, Italy
Bettiol@Unive.It

Bettoli M. Giovanna
Univ. Di Bologna
40126 Bologna, Italy
Tube@Ciam01.Ciam.Unibo.It

Bierl Reinhard
University Of Trier
D-54286 Trier, Germany
Bierl@Uni-Trier.De

Binelli Andrea
Universita' Di Milano
20133 Milano, Italy
Fax. +39 2 26604361

Bloesch Jurg
EAWAG
Ch-8600 Dubendorf, Switz.
Bloesch@Eawag.Ch

Boero V.
Universita' Di Torino
10126 Torino, Italy
Boero@Valnet.It

Bott Markus
EAWAG/ETH
Ch-6047 Kastanienbaum, Switz.
Markus.Bott@Eawag.Ch

Bradley Stephen
Westlakes Research Inst.
Cumbria, England
Steveb@Westlakes.Ac.Uk

Breen Peter
Monash University
Caulfield E. 3145, Australia
Peter.Breen@Sci.Monash.Edu.Au

Broberg Anders
Uppsala Univ.
S-752 36 Uppsala, Sweden
Anders.Broberg@Limno.Uu.Se

Buis Kerst
Eawag
Ch 6047 Kastanienbaum
Kerst.Buis@Eawag.Ch

Calmano Wolfgang
Tu Hamburg-Harburg
21071 Hamburg, Germany

Camusso Marina
Cnr-Irsa
Milano, Italy
Camusso@Irsa1.Irsa.Rm.Cnr.It

Capart Herve
Nat. Taiwan Univ.
Taipei 10764, Taiwan
Fax +2 363 9258

Capodaglio Gabriele
Univ. Di Venezia
30123 Venezia, Italy
Capoda@Unive.It

Cappenberg Thomas E.
Netherlands Inst. Of Ecology
4401 Ea Yerseke, Netherlands
Cappenberg@Cemo.Rijoo.Lenaw.Nl

Casper Peter
Inst. Of Freshwater Ecol And Inland Fisheries.
16775 Neuglobsow Germany
Pc@Igb-Berlin.De

Castaldelli Giuseppe
Univ. Di Ferrara
44100 Ferrara, Italy
Col@Ifeuniv.Unife.It

Cermelj Branko
Morska Postaja Piran
6330 Piran, Slovenija
Branko.Cermelj@Uni-Lj.Si

Chen Jingsheng
Peking Univ.
Beijing, China
Jsc@Urbanms.Urban.Pku.Edu.Cn

Christensen Kasper K.
Odense University
Dk-5230 Odense M., Denmark
Kasper@Dou.Dk

Clifton Julian
Westlakes Research Inst.
Cumbria Ca24 3ln, England
Julian@Westlakes.Ac.Uk

Coakley John P.
National Water Res. Inst.
L7R 4A6 Burlington, Canada
John.Coakley@Cciw.Ca

Contado Catia
Universita' Di Ferrara
44100 Ferrara, Italy
Kat@Dns.Unife.It

Crivelari Orlando Richard
Centro De En. Nuclear Na
13400-970 Piracicaba, Brazil
Richard.Orlando@Iaag.Geo.Uni-Muenchen.De

Davison Bill
Lancaster University
Lancaster, England
W.Davison@Lancaster.Ac.Uk

De Nadai Fernandes Elisabete
Centro De En. Nuclear Na
13400-970 Piracicaba, Brazil
Eadnfern@Pira.Cena.Usp.Br

Delle Donne Arturo
Univ. Parma
43100 Parma, Italy
Arturo@Eagle.Bio.Unipr.It.

De Pinto Joseph
University Of Buffalo
14260-4400 Buffalo, N.Y. USA
Depinto@Eng.Buffalo.Edu

Diaz Paula Andrea
Univ. Autonoma
Madrid, Spain
Paula@Ccuam3.Sdi.Uam.Es

Dolenec Tadej
University Of Ljubljana
61000 Ljubljana, Slovenia
Fax +61 1258 114

Dominik J
Universite De Geneve
Ch-1290 Versoix, Suisse
Dominik@Sc2a.Unige.Ch

Donnelly Terrence
Csiro
Camberra Act 2601, Australia
Terry@Cbr.Dwr.Csiro.Au

Droppo Ian G.
Nat. Water Research Inst.
Burlington, ON, L7R 4A6, Canada
Ian.Droppo@Cciw.Ca

Duiker J.M.C.
Utrecht University
3508 TC Utrecht, Netherlands
Fax +30 25 40 604

Eadie Brian
NOAA
Ann Arbor Mi. 48105, 2205 USA
Eadie@Glerl.Noaa.Gov

Eckert Werner
IOLR
14-102 Tiberias, Israel
Werner@Inter.Net.Il

Eisenreich Steven J.
Rutgers University
New Brunswick, Nj 08903, USA
Eisenreich@Aesop.Rutgers.Edu

Evans Hayla E.
Roda Environmental Research
Box 447 Lakefield, K0L2H0, Can.
Hevans@Knet.Flemingc.On.Ca

Evans R. Douglas
Trent University
Peterborough, K9J 7B8, Can.
Devans@Trentu.Ca

Faganeli Jadran
Marine Biological Station
66330 Piran, Slovenia
Fax +66 746 367

Fattore Elena.
Ist. Di Ricerche Farmacol.
20157 Milano, Italy
Fax +39 2 39 00 19 16

Feibicke Michael
Technical Univ. Berlin
14195 Berlin, Germany
Fax 4930 8 2309397

Flindt Mogers
University Of Copenhagen
Dk-3400 Hillerod, Denmark
Flabms@Inet.Uni-C.Dk

Floderus Soren
Inst.Of Earth Science
S-752 36 Uppsala, Sweden
Soren@Bahnhof.Se

Forstner Ulrich
T U Hamburg Harburg
D-21071 Hamburg, Germany

Frascari Franca
Ist. Geologia Marina Cnr
40129 Bologna, Italy
Frascari@Boigm2.Igm.Bo.Cnr.It

Frignani Mauro
Ist. Geologia Marina, Cnr
40129 Bologna, Italy
Frignani@Boigm2.Igm.Bo.Cnr.It

Galassi Silvana.
Universita' Di Milano
20133 Milano, Italy
Alfpro@Imiucca.Csi.Unimi.It

Gambi M. Cristina
Stazione Zoologica Di Napoli
80077 Ischia Napoli, Italy
Gambimc@Alpha.Szn.It

Genon Giuseppe
Politecnico Di Torino
10100 Torino, Italy
Fax.+39 11 5644699

Giani Michele
Ist. Cent.Ricerca Appl.Icata
30015 Chioggia, Italy
Icramve@Ipolumidx.Umipd.It

Giordani Gianmarco
Dip. Scienze Ambientali
43100 Parma, Italy
Ggiord@Eagle.Bio.Unipr.It

Giordani Paolo.
Ist. Geologia Marina, Cnr
40129 Bologna, Italy
Fax +39 51 639 89 40

Goedkoop Willem
Uppsala University
S-752-36 Uppsala, Sweden
Willem.Goedkoop@Limno.Uu.Se

Gonsiorczyk Thomas
Inst. Of Freshwater Ecology 16775
Neuglobsow, Germany
Pc@ Igb-Berlin.De

Gratzer Hermann
Tu Hamburg-Harburg
21073 Hamburg
Gratzer@Tu-Harburg.D400.De

Gries Thomas
University Of Constance
78467 Konstanz, Germany
Thomas.Gries@Uni-Iconstanz.De

Guarino Sandro
I.S.B.L. Cnr
71010 Lesina, Foggia, Italy
Fax +39 881-91 352

Gude Hans
Institut For Lake Research
88085 Langenargen, Germany
07543 8030@+-Online.De

Guerrero M. Carmen
Univ. Autonoma
Madrid, Spain
Fax +1 397 80 01

Guilizzoni Piero
Cnr, Ist. Ital.Idrobiologia
28048 Verbania, Italy
Guilizzo@Iii.To.Cnr.It

Gulmini Monica
C\O Dip. Chimica Analitica
10125 Torino, Italy

Gunnarsson Jonas
Goteborg Universitet
45034 Fiskebackskil, Sweden
J.Gunnarsson@Kmf.Gu.Se

Guo Jin Yi
State Key Lab.
Cas Guiyang Guizhou 550002,
P.R. Of China
Fax +851 584 1609

Guzzella Licia
Irsa Cnr
Milano, Italy
Guzzella@Irsa1.Irsa.Rm.Cnr.It

Hadas Ora
Iolr Kinneret Limn. Lab.
14102 Tiberias, Israel
Kinneret@Shani.Net

Håkanson Lars
Uppsala University
752 36 Uppsala, Sweden
Lars.Hakanson@Natgeo.Uu.Se

Hall Gwendy
Geological Survey Canada
Ottawa K1A OE8 ON, Canada
Hall@Gsc.Nrcan.Ca
Hanaoka Ken'ichi
National Fisheries Univ.
Shimonoseki 759-65,Japan
Hanaoka@Fish-U.Ac.Jp

Hargrave Barry
Dept. Of Fish. And Oceans
Dartmouth, NS B2Y 4A2 Can.
Barry.Hargrave@Maritimes.Dfo.Ca

Hart Barry T.
Monash University
3245 Melbourne, Australia
Barry.T.Hart@Sci.Monash.Edu.Au

He Qingping
University Of Exeter
EX4 4RJ Exeter, Devon, England
Qphe@Exe.Ac.Uk

Hollerez Karin
Goteborg University
Yso 34 Fiskebackskil, Sweden
K.Hollerez@Kmt.Gu.Sc

Holmer Marianne
Odense University
Dk 5230, Odense M. Denmark
Holmer@Biology.Ou.Dk

Hoss Sebastian
Zoologisches Inst. Der Lmu
80333 Munchen, Germany
Hoss@Zi.Biologie.Uni-Muenchen.De

Hoyal David C.J.D.
S.U.N.Y. At Buffalo
Buffalo, NY 14260, USA
Hoyal@Mailhub.Acsu.Buffalo.Edu

Huet Marie Chantal
OECD
75775 Paris, France
Marie-Chantal.Huet@Oecd.Org

Hylland Ketil
Norw. Inst. For Wat. Res.
N-0411 Oslo, Norway
Ketil.Hylland@Niva.No

Jansson Mats
Umea University
S-901 87 Umea, Sweden
Mats.Jansson@Geography.Umu.Se

Jepsen Rich
Univ. Of California
Santa Barbara CA 93106 USA
Raj@Engineering.Ucsb.Edu

Jonsson Anders
Umea University
S-901 87 Umea, Sweden
Jonars95@Student.Umu.Se

Jonsson Jan
Uppsala University
S-752 36 Uppsala, Swed.
Jan.Jonsson@Sediment.Uu.Se

Jonsson Per
Uppsala University
S-752 36 Uppsala, Sweden
Per.Jonsson@Geo.Uu.Se

Jouanneau Jean Marie
Ura Cnrs Univ. Bordeaux I
33405 Talence Cedex France
Jouanneau@Geoceam.
U- Bordeaux.fr

Kamp-Nielsen L.
University Of Copenhagen
3400 Hillerod, Denmark
Fax +48241476

Kellner Silke
Ufz-Centre For Environ. Res.
39104 Magdeburg, Germany
Kellner@Water.Gm.Ufz.De

Kern Ulrich
University Of Stuttgart
70550 Stuttgart, Germany
Kern@Iws.Uni-Stuttgart.De

Kersten Michael
Institut Fur Ostseeforschung
18119 Rostock, Germany
Kersten@Io-Warnemuende.De

Kleeberg Andreas
Btu Cottbus
15526 Bad Saarow, Germany
Fax +33 631 5200

Konitzer Kerstin
Uppsala University
S 752 36 Uppsala, Sweden
Kerstin.Konitzer@Sediment.Uu.Se

Kozerski Hans-Peter
Inst. Freshwater Ecology
12562 Berlin, Germany
Fax 0049 30 648 407 10

Kraaij Rik
Utrecht University
3508 Td Utrecht, Netherlands
R.Kraaij@Ritox.Dgk.Ruu.Nl

Krishnappan B.G.
National Water Research Inst.
Burlington, On L7r 4a6 Can.
Krish.Krishnappan@Cciw.Ca

Kristensen Erik
Odense University
5230 Odense M., Denmark
Ebk@Biology.Ou.Dk

Kronvang Brian
National Env. Res. Inst.
8600 Silkeborg, Denmark
Fvbk@Dmu.Dk

Kumar Ashok
Banaras Hindu Univ.
221 005 Varanasi, India
Fax +542 311 997

Kusuda Tetsuya
Kyushu University
812-81 Fukuoka, Japan
Kusuda@Civil.Kyushu-U.Ac.Jp

Lami Andrea
CNR Ist. It. Idrobiologia
28048 Verbania, Novara, Italy
Fax +39 323 556513

Langner Constanze
Ufz-Centre For Environ. Res.
39104 Magdeburg, Germany
Langner@Water.Gm.Ufz.De

Langone Leonardo
Ist. Geologia Marina Cnr
40129 Bologna, Italy
Langone@Bolgm2.Bo.Cnr.It

Larsson Helen
Stockholm University
106 91 Stockholm, Sweden
Rolf.Hallberg@Bgkem.Su.Se

Leeming Rhys
Csiro Divison Oceanog.
Hobart 7001 Tasmania, Aust.
Leeming@Ml.Csiro.Au

Lehtoranta Jouni
Finnish Environ. Inst.
00251 Helsinki, Finland
Jouni.Lehtoranta@Vyh.Fi

Lick Wilbert
University Of California
Santa Barbara, CA 93106 USA
Willy@Ferkel.Ucsb.Edu

Lindstrom Martin
Uppsala University
752 36 Uppsala Sweden
Martin.Lindstrom@Natgeog.Uu.Se

Lojen Sonja
J.Stefan Institute
61000 Ljubljana, Slovenia
Sonja.Lojen@Ijs.Si

Lopez Pilar
Univ. Barcelona
08028 Barcelona, Spain
Fax +3 411 1438

Lotter Andre F.
EAWAG
Ch-8600 Dubendorf Switz.
Lotter@Eawag.Ch

Lund-Hansen Lars C.
Aarhus Univ.
520 8000 Arhus C, Denmark
Geollclh@Aau.Dk

Lyons Michael
Westlakes Research Inst.
Moor Row CA24 3JZ Cumbria, UK
Mickl@Westlakes.Ac.Uk

Madruga Maria Jose
Dga / Dpsr
2685 Sacavem, Portugal
Fax + 351 1 994 1995

Maggi Chiara
Univ. La Sapienza
00185 Roma, Italy
Fax +39 6 49913723

Mahlich Birgit
Brandenburgische
Technische Univ. Cottbus
03013 Cottbus, Germany
Mahlich@Umwdt.Tu-Cottbus.De

Mariani Mauro
Acquario Civico
Stazione Idrobiologica
20121 Milano, Italy

Martincic Barbara
Lab. Biologia Marina
34010 Trieste, Italy
Labbioma@Univ.Trieste.It

Martinotti Walter
Enel Spa Cram
20134 Milano, Italy
Fax +39 2-722 43 915

Mattice Jack
E. P. R. I.
Palo Alto, California Usa
Fax +1 415 855 1069

Mattiuzzo Laura
C\O Politecnico
10129 Torino, Italy

Mauri Marina
Dip. Biologia Animale
41100 Modena, Italy
Fax +39 59 581 069

Meade Robert H.
U.S. Geological Survey
Denver, Colorado 80225, USA
Fax +30 32365034

Meili Markus
Uppsala Univ.
752 36 Uppsala, Sweden
Markusmeili@Limno.Uu.Se

Mezzanotte Valeria
Univ. Milano
20126 Milano, Italy
Fax +39 2 644 74 200

Michael Eisele
Rosenstr. 20
54295 Trier, Germany

Mieszczankin Tomasz
N. Copernicus Univ.
87-100 Torun, Poland
Mieszcz@Biol.Uni.Torun.Pl

Migon Christophe
Univ. De Corse
06230 Villefranche Sur Mer
France
Migon@Ccrv.Obs-Vlfr.F.

Milligan Tim
Bedford Inst. Oceanog.
Dartmouth, NS B2Y 4A2 Canada
Tim.Milligan@Scotia.Dfo.Ca

Mitchell A.
Charles Sturt University
Wagga Wagga
NSW, 2651, Australia

Montuelle Bernard
Cemagref
693336 Lyon Cedex 09 France
Bernard.Montuelle@Cemagref.Fr

Moreira Isabel
Puc-Rio
Cep 22453-900 Gavea
Rio De Janeiro, Brazil

Mosello Rosario
C.N.R. Ist. Ital. Idrobiol.
28048 Verbania, Italy
Fax +39 323 556513

Muntau Herbert
J.R.C. Environ. Inst.
21020 Ispra, Italy
Fax +39 332 785212

Negro Giovanni
Ass. Amb. Reg. Piemonte
10123 Torino, Italy
Fax. +39 11 4324632

Ni Jinyen
Peking University
Beijing 100871, China

Nishri Aminadav
I.O.L.R.
14-102 Tiberias, Israel
Kinneret@Shani.Net

Nozhevnikova Alla
Russian Acad. Of Science
7 K 2 Moscow, Russia
Allan@Imbran.Msk.Su

Ogrinc Nives
J.Stefan Institute
61000 Ljubljana, Slovenia
Nives.Ogrinc@Ijs.Si

Omlin Martin
EAWAG
8600 Dubendorf, Switzerland
Omlin@Eawag.Ch

Ostacoli Giorgio
Univ. Di Torino
10125 Torino, Italy
Fax +39 11-670 76 15

Ostrovsky Ilia
IOLR Yigal Allon
Tiberias 14102, Israel
Kinneret@Shani.Net

Packman Aaron
Caltech 138-78
Pasadena, CA 91125 USA
Apack@Cco.Caltech.Edu

Paulsen Susan
Caltech 138-78
Pasadena, CA 91125 USA
Spaulsen@Cco.Caltech.Edu

Pellegrini David
I.C.R.A.M.
00197 Roma, Italy
Mc6460@Mclink.It

Persson Johan
Uppsala University
752 36 Uppsala, Sweden
Johan.Persson@Kultur.Lul.Se

Peters Gregory
Univ. Of Sydney
2006 Sydney, Australia
Slice@Chem.Eng.Usyd.Edu.Au

Petersen Wilhelm
Gkss Res. Centre
21502 Geesthacht, Germany
Wilhelm.Petersen.@Gkss.De

Petronio Bianca Maria
Univ. La Sapienza
00185 Roma, Italy
Fax +39 6-499 13 723

Pettersson Kurt
Uppsala University
761 73 Norrtalje, Sweden
Kurt.Pettersson@Limno.Uu.Se

Petticrew Ellen
Univ. Northern B.C.
Prince George BC V2N 4Z9, Can.
Ellen@Unbc.Edu

Pieters Henk
Inst. For Fisheries Research
P.O. Box 68 Ymuiden 1970 Ab
The Netherlands
H.Pieters@Rivo.Dlo.Nl

Pirrone Nicola
Univ. Of Michigan
Ann Arbor MI 48109-2029, USA
Npirrone@N.Imap.Itd.Umich.Edu

Predonzani Sergio
Lab. Biologia Marina
34010 Trieste, Italy
Delnegro@Univ.Ts.It

Provini Alfredo
Universita' Di Milano
20133 Milano, Italy
Alfpro@Imiucca.Csi.Unimi.It

Pusceddu Antonio
Ist. Scienze Amb. Marine
16038 S. Margherita Ligure
Genova, Italy
Ligursea@Promix.Shiny.It

Puschel Roland
Tu Hamburg-Harburg
21071 Hamburg, Germany
Puschel@Tu-Harburg.D400.De

Ravanelli Marzia
Univ. Di Bologna
40126 Bologna, Italy
Tube@Ciam01.Ciam.Unibo.It

Rawling Mark Carl
Univ. Plymouth
Plymouth Devon PL4 8AA, UK
C.Rawling@Plymouth.Ac.Uk

Regnell Olof
University Of Lund
223 62 Lund, Sweden
Olof.Regnell@Ecotox.Lu.Se

Reisenhofer E.
Univ. Studi Trieste
34127 Trieste, Italy
Reisen@Univ.Trieste.It

Reynoldson Trefor B.
CCIW
Burlington, ON L7R4A6 Can.
Trefor.Reynoldson@Cciw.Ca

Richards Carl
Univ. Of Minnesota
Duluth MN 55811 USA
Crichard@Sage.Nrri.Umn.Edu

Rosenberg Rutger
Goteborg University
450 34 Fiskebackskil, Sweden
R.Rosenberg@Kmf.Gu.Se

Rukavina Norm
National Water Res. Inst.
Burlington, ON L7R 4A6 Can.
Norm.Rukavina@Cciw.Ca

Rydin Emil
Uppsala University
752 36 Uppsala, Sweden
Emil.Rydin@Limno.Uu.Se

Rzepecki Marek
Inst. Of Ecology
05-092 Lomianki, Poland
Ekolog@Warman.Com.Pl

Sara Gianluca
Ist. Zoologia
90123 Palermo, Italy
Upamblgs@Mbox.Vol.It

Scarlatos Panagiotis
Florida Atlantic Univ.
33431 Boca Raton, Florida, USA
Pscarlat@Oe.Fau.Edu

Schaanning Morten
Niva
0411 Oslo, Norway
Morten.Schaanning@Niva.No

Schorer Marcell
Univ. Of Trier
54286 Trier,Germany
Schorer@Uni-Trier.De
Seritti Alfredo
C.N.R. Ist. Di Biofisica
56127 Pisa, Italy
Seritti@Ib.Pi.Cnr.It

Skei Jens
Niva
0411 Oslo, Norway
Jens.Skei@Niva.No

Sijim Dide T.K.M.
Utrecht University
3508 Td Utrecht, Netherlands
D.Sijm@Ritox.Dgt.Ruu.Nl

Skold Mattias
Provincial Government
Of Goteborg And Bohus
S-40340 Goteborg, Sweden
Mask@O.Lst.Se

Spagnoli Federico
Ist. Geologia Marina Cnr
40129 Bologna, Italy
Federico@Boigm2.Igm.Bo.Cnr.It

Spence K.J.
Univ. Of Sheffield
Sheffield SH1 3JD, UK
Fax +14 2728910

Stheinman Boris
Ygal Alon Kinneret Lab.
Tiberias 14-102, Israel
Kinneret@Shani.Net

Stone Peter
Univ. Of Exeter
Exeter Devon EX4 4RJ, UK
Pmstone@Exeter.Ac.Uk

Sturm Michael
EAWAG-ETH
8600 Dubendorf Switz.
Sturm@Eawag.Ch

Svensson Jonas M.
Lund Universitet
22362, Lund, Sweden
Jonas.Svensson@Limnol.Lu.Se

Symader Wolfhard
University Of Trier
54286 201-2232 Trier, Germany

Tartari Gianni
Cnr-Irsa
Milano, Italy
Tartari@Irsa1.Irsa.Rm.Cnr.It

Teixeira Elba
Fundacao Estadual De
Protecao Ambiental
Cep 90245-000, Brazil
Fax +51 342 02 24

Tessier Andre
Univ. Du Quebec
Sainte-Foy, PQ G1V 4C7 Canada
Atessier@Inrs-Eau-Uquebec.Ca

Thierfelder Tomas
Uppsala University
752 36 Uppsala, Sweden
Tomas.Thierfelder@Natgeog.Uu.Se

Thomas Christine
Simon Fraser Univ.
Burnaby BC V5A 1S6 Canada
Christit@Sfu.Ca

Thoolen Pauline
P.O. Box 177
2600 MH Delft, The Netherlands
Fax +31 15 261 96 74

Tomaszek Janusz
Rzeszow Univ. Of Technology
35-959 Rzeszow, Poland
Tomaszek@Ewa.Prz.Rzeszow.Pl

Tornblom Erik
Uppsala Univ.
752 36 Uppsala, Sweden
Fax +48 24 1476

Traunspurger Walter
Zool. Inst. Der Univ. Munchen
80333 Munchen, Germany
Traunsp@Zi.Biologie.Uni-Muenchen.De

Tsai Cheng-Han
National Taiwan Ocean Univ.
Keelung 202, Taiwan
Chtsai@Ntougg.Ntou.Edu.Tw

Tubertini Ottavio
Univ. Di Bologna
40126 Bologna, Italy
Tube@Ciam01.Ciam.Unibo.It

Turner Andrew
Univ. Of Plymouth
Plymouth PL4 8AA Devon, UK
Aturner@Plymouth.Ac.Uk

Valeur Jens R.
Danish Hydraulic Inst.
2970 Horsholm, Denmark
Fax +76 2567

Viaroli Pierluigi
Univ. Di Parma
43100 Parma, Italy
Pier@Eagle.Bio.Unipr.It

Wai Onix W.H.
Hong Kong Polytech. Univ.
Hung Hom Kowloon,
Hong Kong
Ceonyx@Polyu.Edu.Hk

Walling Desmond
Univ. Of Exeter
EX4 4RJ Exeter, UK
Fax +1392 263 342

Wan Guojiang
State Key Lab. Environ.
Guizhou 550002, P.R. Of China
Fax + 851 584 1609

Westrich Bernhard
University Of Stuttgart
70550 Stuttgart, Germany
Westrich@Iws.Uni-Stuttgart.De

Weyhenmeyer Gesa
Uppsala University
752 36 Uppsala, Sweden
Fax + 18 182 737
Gesa.Weyhenmeyer@Natgeog.Uu.Se

Williams Ron
Clemson University
Clemson SC, 29634 USA
Wronald@Clemsa.Edu

Winkels Herman
Rijkswaterstaat
8200 Aa Helystad Netherlands
Fax +31 320298339

Wisniewski Ryszard
N. Copernicus Univ.
87-100 Torun Poland
Wisniew@Biol.Uni.Torun.Pl

Wortelboer F.G.
Nat. Inst. For Public Health And
Environ. Protection
3720 Ba Bilthoven, Netherlands
Rich.Wortelboer@Rivm.Nl

Wu Fengchang
Chinese Academy Of Sciences
Guiyang 550002, P.R. China
Fax 0086 0 851 5841 609
Wuest Alfred
Eawag
Ch-8600 Dubendorf
Switzerland
Wuest@Eawag.Ch

Yen Chin-Lien
National Taiwan Univ.
10617 Taipei, Taiwan
Clyen@Ccms.Ntu.Edu.Tw

Young Der-Liang
National Taiwan Univ.
10617 Taipei, Taiwan
Dlyoung@Hy.Ntu.Edu.Tw

Zelano Vincenzo
Univ. Di Torino
10125 Torino, Italy
Fax +39 11-670 76 15

SUBJECT INDEX

AUTHOR INDEX

SPRINGER NATURE

GPSR Compliance

The European Union's (EU) General Product Safety Regulation (GPSR) is a set of rules that requires consumer products to be safe and our obligations to ensure this.

If you have any concerns about our products, you can contact us on ProductSafety@springernature.com

In case Publisher is established outside the EU, the EU authorized representative is:

Springer Nature Customer Service Center GmbH
Europaplatz 3
69115 Heidelberg, Germany

Zeitfracht Medien GmbH
Ferdinand-Jühlke-Straße 7
99095 Erfurt, Deutschland
produktsicherheit@kolibri360.de